Mechanisms in
Fibre Carcinogenesis

NATO ASI Series

Advanced Science Institutes Series

A series presenting the results of activities sponsored by the NATO Science Committee, which aims at the dissemination of advanced scientific and technological knowledge, with a view to strengthening links between scientific communities.

The series is published by an international board of publishers in conjunction with the NATO Scientific Affairs Division

A	Life Sciences	Plenum Publishing Corporation
B	Physics	New York and London
C	Mathematical and Physical Sciences	Kluwer Academic Publishers
D	Behavioral and Social Sciences	Dordrecht, Boston, and London
E	Applied Sciences	
F	Computer and Systems Sciences	Springer-Verlag
G	Ecological Sciences	Berlin, Heidelberg, New York, London,
H	Cell Biology	Paris, Tokyo, Hong Kong, and Barcelona
I	Global Environmental Change	

Recent Volumes in this Series

Series A: Life Sciences

Mechanisms in Fibre Carcinogenesis

Edited by

Robert C. Brown and John A. Hoskins

Medical Research Council Toxicology Unit
Carshalton, Surrey, United Kingdom

and

Neil F. Johnson

Inhalation Toxicology Research Institute
Albuquerque, New Mexico

.

Plenum Press
New York and London
Published in cooperation with NATO Scientific Affairs Division

Proceedings of a NATO Advanced Research Workshop
on Mechanisms in Fibre Carcinogenesis,
held October 22–25, 1990,
in Albuquerque, New Mexico

Library of Congress Cataloging-in-Publication Data

NATO Advanced Research Workshop on Mechanisms in Fibre Carcinogenesis
  (1990 : Albuquerque, N.M.
    Mechanisms in fibre carcinogenesis / edited by Robert C. Brown and
John A. Hoskins and Neil F. Johnson.
        p.   cm. -- (NATO ASI series. Series A, Life sciences ; v.
223)
    "Proceedings of a NATO Advanced Research Workshop on Mechanisms in
Fibre Carcinogenesis, held October 22-25, 1990, in Albuquerque, New
Mexico"--T.p. verso.
    "Published in cooperation with NATO Scientific Affairs Division."
    Includes bibliographical references and index.
    ISBN 0-306-44091-1 (hardbound)
    1. Lungs--Cancer--Etiology--Congresses. 2. Mesothelioma-
-Etiology--Congresses. 3. Inorganic fibers--Carcinogenicity-
-Congresses. 4. Asbestosis--Congresses.   I. Brown, R. C. (Robert
Charles), 1945-   . II. Hoskins, John A. III. Johnson, Neil F.
IV. North Atlantic Treaty Organization. Scientific Affairs
Division. V. Title. VI. Series.
    [DNLM: 1. Carcinogens--toxicity--congresses. 2. Minerals-
-toxicity--congresses.   QZ 202 N2785m 1990]
  RC280.L8N374 1990
  616.99'4071--dc20
  DNLM/DLC
  for Library of Congress   91-39485
                                                                 CIP

ISBN-13: 978-1-4684-1365-6   e-ISBN-13: 978-1-4684-1363-2
DOI: 10.1007/978-1-4684-1363-2

© 1991 Plenum Press, New York
Softcover reprint of the hardcover 1st edition   1991
A Division of Plenum Publishing Corporation
233 Spring Street, New York, N.Y. 10013

INTRODUCTION

The Editors are sorry that the production of this volume was delayed by the ill-health of one of them and we hope that this does not detract from the value of the contents. For once this delay is not the fault of any of the authors only the editors are to blame.

Many of the workers in the field of fibre toxicology became convinced by the middle 1980's that the worst of the furore over asbestos was over although we were left with an intriguing problem - how does asbestos cause disease? It was expected that the future impact of fibres on human health would be very small since asbestos exposure would be controlled and there was little chance that man-made fibres would prove hazardous. These man-made fibres are much thicker than asbestos and, in most cases, they are less durable in the body. Both of these properties are believed to make them much less likely to cause disease. However many of us had fallen into the habit of calling these materials "asbestos substitutes" and thus they have acquired a little of the notoriety attached to the natural fibrous minerals. Very few of these man-made fibres are actually used as replacements for asbestos. Asbestos was not suitable for the uses to which the insulation wools are usually put and the ceramic fibres are replacements for fire brick not asbestos which is destroyed at the temperatures at which these materials are used. In most cases the man-made fibres are amorphous or vitreous and thus they do not cleave into finer fragments, so that fibres do not become respirable. Even where they are crystalline the crystals are small (that is the fibres are polycrystalline) and the crystal axes are not along the axis of the fibres, thus cleavage destroys the fibrous nature of the product.

Such epidemiology as there is does not suggest that these mam-made fibres are dangerous. There has only ever been a small excess of lung cancer in persons exposed to one type of insulation wool, and this at a period when the modern techniques of production were not in use; even this excess could be due to other causes. However society, that is both the industries making and distributing man-made mineral fibres and those responsible for the regulation of toxic materials, have learned from asbestos. The industry has been willing to spend time and money on the safety assessment of their products. This has been done and often to higher standards than the requirements of the legislators - a situation almost unique in modern toxicology. Taking advice mainly from clinicians and epidemiologists the industry spent most of it resources on studies of morbidity and mortality in the manufacturing industries. Epidemiological studies can only reveal small effects when a large number of humans have been exposed. It is unlikely that such human studies will reveal an effect in widely spread and sometimes poorly documented users. Therefore more recently the industry has anticipated the requirements for pre-marketing toxicology

v

and have moved away from this concentration on the health effects of materials already finding their way into commerce. In one sense we are all following the example set by the career of Chris Wagner - our "guest of honour" at the meeting on which this volume is based. Having discovered the fact that crocidolite (blue) asbestos caused human mesothelioma he then pioneered the development of predictive animal testing, using both the injection and inhalation routes of exposure.

As an example of the present day use of such animal testing we are pleased to include in this volume a partial report of one study financed by the industry through TIMA (the Thermal Insulation Manufacturer's Association) (Hesterberg *et al.* page 531). In this paper and other reports on the same group of studies the industry may have inadvertently obtained results causing probably **too much** concern over their product safety. The fibres chosen for study, while in some sense "typical of those found in the workplace", are exactly those which other experiments would predict to be the most active. The methodology used was exquisitely designed to maximise the impact of these fibres on experimental animals - and that is what happened. This type of study and the activities of a Californian pressure group - "Victims of fibreglass" have caused a great deal of pressure on both sides of the Atlantic to control "potentially dangerous" fibrous materials. One of the most active **scientists** attempting to devise control procedures is Professor Pott (see his report on page 547). Professor Pott's suggestions have caused some controversy since he would base control on the activity of fibre in rodent intracavity injection experiments. One of the editors (RCB) is on record as supporting this type of experiment; without Chris Wagner's use of the intrapleural route we would not have known the hazardous nature of blue asbestos. However, as even Professor Pott recognises the intracavity routes (intraperitoneal in his case) can give false positive results. Contrarily inhalation studies (at least those carried out in the past) can give false negative results.

What does the future hold? Will glass fibre products be banned from buildings - and if so what is to replace them as insulation? Are the pottery and metallurgical industries to revert to the use of fire brick rather than ceramic fibre to line their furnaces and kilns? If so will world industry move *en masse* to the third world where even asbestos is uncontrolled. Clearly the job of the experimental toxicologist is going to be to work with the fibre industry helping to create safe, or at least safer, materials. Those of us working with fibres are at the moment in the vanguard of this movement as far as the materials industries are concerned but we are still some way behind our colleagues in the drug and agrochemical industries. It seems that with fibres there are two choices - make fibres too large to be respirable or make them so soluble that they are not durable in the mammalian body. The former solution is difficult - some would say impossible - since control of size distribution cannot be achieved with 100% guarantee of success - reduced *in vivo* durability is the key to safe and effective insulation materials.

The meeting in Albuquerque ended with agreement that there should be a further meeting on fibre durability (this will take place at IARC, Lyon, France 7-9/9/'92). We wish them every success and we hope that in the future our thermal environment will be better controlled, fuel consumption will be reduced and the high temperature industries have access to refractory materials all relying on products of demonstrated utility **and safety**. It remains likely that even relatively short lived fibres will be the subject of some public and governmental concern. To eliminate all these worries we must discover the mechanisms operating during fibre induced pathogenesis. Most of the papers presented here attempt to throw some light on the possible mechanisms operating during the initiation and development of fibre associated disease. Thus we include reports on the changes in the target cells during the development of disease or cellular changes directly resulting from fibre/cell interaction. The chemical and physical properties of the fibres themselves influence both the nature of these interactions and their biological consequences and several papers address this subject or even describe new fibres which may be of use in experimental toxicology, if not in commerce.

It is with some regret that we leave this subject with nearly as many questions open as when we conceived the idea of organising a workshop. We hope that this document will provide a background for those entering the field or needing to assess data on fibre toxicology. Finally we would like to thank NATO for their support and the staff at Plenum Press for their help in publishing the manuscript.

CONTENTS

THE CARCINOGENICITY OF MINERAL FIBRES

PHYSICAL AND CHEMICAL PROPERTIES

1. Fibre Size and Chemistry

2. Minerology

MECHANISMS OF PATHOGENESIS

1. Genotoxic Effects

2. Effects on Gene Expression

3. Fibres and Free Radicals

4. Other Effects and Promotion

HUMAN RISK ASESSMENT

THE CARCINOGENICITY OF MINERAL FIBRES

1. Human studies

GAPS IN KNOWLEDGE OF FIBRE CARCINOGENESIS:
AN EPIDEMIOLOGIST'S VIEW

Douglas Liddell

Department of Epidemiology and Biostatistics
McGill University
1020 Pine Avenue West, Montreal
Quebec, Canada,   H3A 1A2

Despite intensive research over several decades, there remain many gaps in knowledge of fibre carcinogenesis. This paper proposes nine questions which need to be answered to fill the more important gaps, and attempts to provide the epidemiologic background, question by question.

Question 1. Can fibres cause latent disease which becomes apparent only after they have been eliminated from the lung?

Since the nineteen-sixties it has been suspected that crocidolite is more toxic than chrysotile. It now seems likely that there is a 'fibre gradient', particularly marked for mesothelioma, but still quite steep for lung cancer, although modified by certain industrial processes. The approach has to be through proportional mortality, despite well-known drawbacks arising from differences in age distributions, duration and intensity of exposure, length of follow-up, etc.

Nevertheless, the proportional mortality from mesothelioma in 33 male cohorts varied very much more between than within fibre types, to an extent that simply could not be due to chance. The between-fibre differences are so enormous that they quite transcend the effects of variations in methodology. The results presented in Table 1 have been summarized from Liddell (1989), but exclude entries for which total deaths were less than 500. The extended table recorded a total of 483 deaths from mesothelioma among 17,832 deaths from all causes, or 2.7%. A proportional mortality rate of 2.7% would have yielded 137 mesotheliomas (where just 11 were observed) among the 5062 deaths in the chrysotile cohorts, and only 18 (43 observed) in the 653 deaths in the

TABLE 1. Proportional Mortality from Mesothelioma in the Major Cohorts of Male Asbestos-exposed Workers

Fibre type (and process)	No. of Cohorts	Total deaths	Deaths from mesothelioma	Proportional mortality (%)
Chrysotile	5	5,062	11	0.2
Chrysotile with small proportion of amphibole - manufacture	6	3,925	25	0.6
- textiles	2	1,622	26	1.6
Amosite	3	871	25	2.9[*]
Mixed - shipyard work	2	1,077	31	2.9
Crocidolite	3	653	43	6.6[*]
Mixed - insulation work	5	3,055	235	7.7
Chrysotile with large proportion of amphibole	2	926	77	8.3[*]

[*] Marked heterogeneity

crocidolite cohorts. In none of the 11 chrysotile cohorts in the first two lines of Table 1 was the proportional mortality rate as high as the lowest rate for an amosite cohort.

Hughes (1991) has summarised lung cancer risks, and Table 2 is a further condensation, but with the addition of what purport to be confidence intervals around the SMRs. Where the SMRs for several cohorts were heterogeneous, the quoted confidence interval is too narrow. Even so, it is obvious that, except for textiles, the chrysotile cohorts had much the lowest lung cancer SMRs; those for insulators were particularly high (as were their PMRs for mesothelioma). As with the results in Table 1, the difference between the lines of Table 2 could not be due to chance nor can they be fully explained by methodologic variations.

It is, however, clear that, for lung cancer, the 'fibre gradient' is modified when chrysotile was used for textiles; also in insulation workers. Further, some mixtures of fibres appeared, in these rather crude comparisons, more toxic than would have been expected as a combination of the effects of 'pure' fibres.

There are major differences between the types of asbestos fibre: in chemistry, in dimensions, and in durability in tissue. This last is shown most clearly by Sébastien and co-workers (1989), who found about three times as many tremolite fibres as chrysotile in lung tissue from miners and millers at Thetford Mines, and almost the same number of tremolite fibres as chrysotile in the lungs of former textile workers at Charleston, South Carolina, who had worked fibre only from Thetford. This fibre was known to have been slightly contaminated with perhaps 1-2% of tremolite. Therefore, the retained proportion of inhaled tremolite fibres was at least an order of magnitude greater than that of chrysotile. Amosite and crocidolite are also known to have high durability (in man as well as *in vivo*).

Wagner (1986) has written that he and his colleagues were endeavouring to assess the significance of the selective retention of fibres. He was convinced that the diseases associated with exposure to mineral fibres are due to the fibres that are retained in the lungs, so that the analysis of the lung mineral content is essential when searching for a potential occupational hazard. He also expressed the view that chrysotile (if uncontaminated by amphiboles) is probably a material causing little disease, but pointed out that this was not accepted by all workers, although the number of converts was increasing. He claimed that his group had provided the major evidence for the innocence of chrysotile to humans, although their animal experiments had suggested that chrysotile is both fibrogenetic and carcinogenetic. His conclusion was: "It is now established that the inhalation of all mineral fibres of a specific diameter and length size range may be associated with the development of diffuse pleural and peritoneal mesotheliomas. Pulmonary fibrosis is associated with the inhalation of mineral fibres which can be coarser than those involved in the development of mesotheliomas. In these situations the importance of the selective retention of fibres must be stressed. It is the fibre that is retained in the lung tissue that is responsible for the disease." Most epidemiologists agree with the last sentence (added emphasis), but when we cite evidence from lung tissue analysis, we are frequently asked how we know that the disease was not caused by fibres before they had been eliminated from the lung.

Question 2. Why does a given exposure to chrysotile when used in textile manufacture cause 35 times more lung cancer (but no more mesothelioma) than in other uses?

In cohorts of workers exposed to chrysotile (only) in mines and mills, in friction products factories and in asbestos-cement, lung cancer risks have been universally low, except where exposures have been exceptionally high. However, in textile factories, lung cancer risks have been very high, even when not accompanied by mesothelioma risks that indicated substantial presence of amphiboles.

In the Charleston textile plant, using fibre from Thetford Mines, the lung cancer risk was much greater than in other chrysotile exposures. Sébastien and co-workers (1989) related fibre counts in lung tissue (72 subjects at Charleston from the cohort studied by McDonald *et al.*, 1983; 89 at Thetford Mines from the Quebec cohort of McDonald *et al.*, 1980) to duration and intensity [million particles per cubic foot, or mpcf] of exposure, accumulated exposure [(mpcf) x years], time since cessation, etc. The lengths of the retained fibres were no greater at Charleston than at Thetford, and on average (by a heuristic calculation, because the published data do not allow a precise computation), the numbers of retained fibres were roughly 2 - 3 per thousand inhaled, in both locations. No precision could be claimed for these estimates, but there is no evidence that at Charleston, relative to Thetford Mines, the inhaled fibre *per se* was longer or more durable, and strong evidence that the airborne concentrations were an order of magnitude less severe. Thus, there is no explanation from these factors for the very much higher risk of lung cancer, per unit asbestos exposure, at Charleston than at Thetford Mines. Liddell and Hanley (1985) found that the point estimates of the slopes of the exposure-response lines were in the ratio 35:1.

One explanation proffered is that aerosols of mineral oil, used to suppress dust and to make fibres more malleable, may have acted as co-carcinogens, perhaps interacting with the residues of smoking rather than with the asbestos. Such aerosols were used in all the asbestos textile factories which have been subject to epidemiologic investigation, but this remains an area of controversy.

TABLE 2. Lung Cancer risks in 29 Cohorts of Asbestos Workers

Fibre type (and process)		No. of cohorts	Deaths from lung cancer	SMR
Chrysotile[a]	- manufacture, cement	4	62	1.02 (0.78-1.31)[b]
	- mining/milling, gas mask filters, friction			
	products	5	406	1.23 (1.12-1.36)
	- textiles	2	94	1.69 (1.37-2.07)[c]
Mixed	- cement	7	351	1.56 (1.40-1.73)[c]
	- textiles	3	342	1.81 (1.62 2.01)[c]
	- manufacture	1	77	2.71 (2.14-3.39)
	- insulation	1	397	4.24 (3.83-4.67)
Amphibole	- gas mask assembly	3	33	2.09 (1.44-2.93)
	- mining/milling	1	91	2.64 (2.12-3.24)
	- insulation	2	84	4.80 (3.83-5.94)

[a] Exclusively or very predominantly chrysolite

[b] Figures in brackets are 95% confidence intervals

[c] Marked heterogeneity of SMRs; see text

Question 3. Why is there such a wide variation, 15 to at least 50 years, in latency, which is not inversely related to dose?

There is a common belief that time to tumour on latent period is dependent, in man as well as in animals, on the intensity of exposure, lower levels requiring a longer latent period. Indeed, it has sometime been stated that the relationship:

$$(time\ to\ tumour)\ varies\ as\ (dose)^{1/3}$$

had been authenticated. In fact, this belief arises from a theoretical exercise in which the relationship was taken as axiomatic and six other more-or-less plausible assumptions were made. The 'results' gained acceptance, but there has never been epidemiologic confirmation.

Liddell (1980) studied the intervals from first exposure to death in the 244 cases of lung cancer among 4463 deaths before 1976 in a cohort of 10,939 men, born 1891 through 1920, who had worked for at least a month in the Quebec chrysotile mines and mills. The interval averaged 39.6 years (s.d. 10.9); not unexpectedly, the latent period was related to age at first employment and to age at death. However, the interval did not appear to be associated in any important way with other measures of exposure or with smoking habit. The average latent periods in a 5 x 4 cross-classification by cumulative asbestos exposure and smoking habit varied from 34.0 to 46.0 years, but with no discernible pattern; ignoring smoking, the latent period was shortest (37.7 years) for the lowest exposure class and longest (43.2 years) for the highest exposures.

6

The intervals ranged from 6 to 61 years. Over five classes of latent period [*i.e.* up to 20 years (16 men); 20-29 years (26 men); 30-39 years (69 men); 40-49 years (91 men); 50 years and more (42 men)], distributions by duration of employment, by intensity of asbestos exposure, and by smoking habits, were all remarkably similar.

It seems clear that the variation in latency was not inversely related to dose. It is perhaps a manifestation of variation in susceptibility; see Question 8.

Question 4. As epidemiology is unlikely to be capable of detecting cancer risks from exposure to all but a few naturally-occurring fibres, MMMF, and organic fibres, what guidance is possible from experimental studies?

Gilson (1982) pointed out that most occupational diseases, including asbestos-induced lung cancer and mesothelioma, were first detected by astute observation by clinicians or pathologists, and it has usually taken many years before their severity and extent have been revealed by systematic surveys. Further, the ability of any epidemiologic study to detect a hazard will be slight unless the hazard leads to a high risk of disease or there is a high (absolute) incidence of the disease - or both. Examples of truly successful detection arising from high risk are the studies of mesothelioma in workers who assembled crocidolite filters for gas masks (McDonald & McDonald, 1978; Jones *et al.*, 1980), although the numbers of cases were not great. The very large numbers of deaths from cancer (lung: 250; abdominal: 328; larynx: 21) ensured that the cohort of 10,939 Quebec chrysotile production workers is particularly informative (McDonald *et al.*, 1980). High risks combined with fairly large numbers of deaths were important features of the Mount Sinai study of 17,800 insulation workers (Selikoff *et al.*, 1979).

Those of the Quebec cohort who had had 20 years or more employment had all suffered very considerable asbestos exposure: the men were placed in four more-or-less equal classes by the intensity of their exposure, roughly estimated as:- Low (average 15 f/ml); Medium (32 f/ml); High (67 f/ml); Very High (160 f/ml). This meant that, in the most heavily exposed groups, risks of certain cancers were clearly demonstrable. However, "if the only subjects studied had been the 1904 with at least 20 years' employment in the lower dust concentrations, averaging 6.6 million particles per cubic foot (or about 20 f/ml), excess mortality would not have been considered statistically significant, except for pneumoconiosis" (McDonald *et al.*, 1980). Moreover, the lung cancer SMR for the low dust exposure group (1.21) was higher than that (1.08) for the medium exposure group; only the greatly enhanced SMRs for those with High and Very High exposure allow the conclusion that there was a response to exposure. Nevertheless, the lung cancer SMR for all 1904 men was 1.15, in close conformity with the relative risk of 1.16 from the line that had been fitted to the SMRs for ten sub-cohorts defined by exposure to age 45 years.

By their nature, epidemiological surveys, however large, cannot achieve greater precision than the one described here. Another important issue is, clearly, whether the linear exposure-response model, which has commonly been adopted, is indeed appropriate. Some progress has been made (Vacek & McDonald, 1990), and further analysis is in hand, but the answers may well be indeterminable from epidemiologic data.

Although, fibre for fibre, mineral wool could well be at least as carcinogenic as chrysotile (Meek, 1991), the low concentrations of airborne fibres, except in the earliest days of the industry, mean that even the very large cohort studies are unlikely ever to produce convincing evidence - one way or the other - of the risks.

Question 5. How do fibre carcinogens migrate from lung to pleura and perhaps to peritoneum? Why do they affect the larynx and gastro-intestinal tract so seldom (if at all)?

It is well known that pleural *mesothelioma*, although otherwise a very rare condition, is a sadly frequent sequela to exposure to crocidolite; the risk is less, but not much less, than that of lung cancer, and SMRs have been immeasurably high. As can be seen from Table 1, exposure to amosite also presents an important risk, while with chrysotile the risk is comparatively low. Peritoneal mesothelioma is virtually unknown in chrysotile-exposed workers, but there are many authenticated cases - but fewer than pleural tumours - among those exposed to amphiboles.

Findings from the six case-referent studies of *laryngeal cancer* were mutually inconsistent. Those from the cohort studies were not demonstrably inconsistent, but certainly did not indicate a major excess of laryngeal cancer mortality; indeed, in most cohorts laryngeal cancer was not studied, and there were no excesses in several others. Estimates of the relative risk of laryngeal cancer obtained from the various surveys were so divergent that no merged value could be obtained - except by violation of long-standing principles; meta-analysis, although advocated by Smith and colleagues (1990), is not applicable. Exposure-response relationships were either unobtainable or equivocal. Almost all cases had been smokers, and there is no evidence as to whether any cases who were non-smokers were also non-drinkers. No experimental evidence has been proffered. By long-established criteria for causation (Hill, 1965), the link between exposure to asbestos and laryngeal cancer has not been established. But even if such excesses as were observed were to be considered asbestos-induced, they remain small in relative terms and minuscule (compared with other asbestos diseases) in absolute terms.

There is no reliable evidence that directly-ingested asbestos can cause *gastro-intestinal cancer*. Nevertheless, it is difficult to accept that this condition can be excluded as an outcome of exposure to asbestos. Indeed, some explanation is needed for the findings in relation to inhaled asbestos, such as in Quebec production workers (Liddell *et al.*, 1984). However, this is not to imply that abdominal cancer can be caused by low or even moderate exposures. It is not clear whether there is any threshold, but positive relationships have been demonstrated only where exposure has been severe.

Question 6. How do smoking and asbestos exposure interact in the development of lung cancer?

Saracci (1977) reviewed the relative roles of asbestos and smoking in the causation of lung cancer. He examined three models: (a) multiplicative risks; (b) additive risks; and (c) that exposure to asbestos did not cause lung cancer except in smokers. To general acclaim he found that the multiplicative model provided the most plausible explanation. The next results to be published (Hammond *et al.*, 1979) were closely in accord with this hypothesis, but they were too good to be true, as can be seen from careful reading of the evidence. Nevertheless, most workers in the field were willing to wield Occam's Razor and accept the multiplicative model.

Later findings have been much less supportive. Berry and colleagues (1985) calculated what they called the "relative asbestos effect" (RAE) *i.e.* the ratio of the asbestos risk in non-smokers to that in smokers; if the interaction were truly multiplicative,

RAE would, of course, be unity. In the 1979 study (by Hammond *et al.*) it was 1.1, but the 95% confidence limits were very wide: 0.3 - 2.6. This illustrates the inherent unreliability of estimates of RAE, arising from the fact that one of the terms in the ratio is inevitably very small, and so has a major destabilizing effect. In the six studies reviewed by Berry and colleagues (1985) this term (usually the number of lung cancer deaths in non-smokers) was 0, 1, 4, 4, 5 and 6. The point estimate of RAE from all six studies combined was 1.8; this is within the 95% confidence limits for all six separate values of RAE. Thus, some credence is lent to the view that the true RAE is somewhere near this level, despite caveats and the wide confidence interval (1.1 - 2.8) around even the combined estimate.

It would appear that the multiplicative model fits the available data less well than was previously believed, and that the relative risk of lung cancer due to exposure to asbestos is rather greater for non-smokers than for smokers. Nevertheless, it remains evident that smokers have a much higher absolute risk of lung cancer than have non-smokers.

Question 7. Can exposure to asbestos induce lung cancer in the absence of asbestosis?

It has been known since the turn of the century that asbestosis (interstitial pulmonary fibrosis) was one life-threatening outcome of severe exposure to asbestos. By at latest 1934, it was suspected that lung cancer was another. It has been pointed out recently (Browne, 1986) that, in 1938, E.R.A. Merewether raised the question whether these two conditions are independent hazards or if lung cancer, when asbestos-induced, always follows asbestosis.

Hughes (1991) discusses the findings of seven studies all of which support the hypothesis that the risk of lung cancer is higher in subjects with histologic or radiologic evidence of asbestosis than in asbestos-exposed subjects in whom such signs are absent. It suffices in the present context to present only the most convincing evidence (Hughes & Weill, 1988). In a cohort of 642 asbestos cement workers, lung cancer mortality was related to earlier radiographic findings in the ILO (1980) classification. For most subjects (420), no radiologic abnormalities were recorded; these men suffered 10 deaths from lung cancer, a tiny excess (SMR = 1.05). For 62 men, pleural changes were read in the X-ray, but no parenchymal opacities; the SMR was 1.33, but based on only two deaths. Small opacities were observed in 160 subjects: the profusion was 0/1 in 83, and the SMR was 4/2.3 = 1.78; for the other 77, the profusion was 1/0 or greater and the SMR 4.33 (9/2.1).

What we may perhaps call the Hughes' hypothesis can, I consider, be accepted, but there is no *proof* that asbestosis is always a precursor of asbestos-induced lung cancer. If it were, lung cancer could not arise as a result of non-occupational exposure to any form of asbestos, because the levels of exposure are so low that asbestosis is an unthinkable outcome.

Question 8. Why is it that such a small proportion of grossly insulted lungs are affected by malignancy?

Some findings from the study of the cohort of Quebec chrysotile miners and millers have already been presented (see Question 3). It was stated that, among those who had been employed for at least 20 years in the High and Very High classes of intensity of exposure, risks of certain cancers had been demonstrated clearly.

Occupational exposure to 1 fibre/millilitre for a year implies the inhalation of:-

2×10^4 f/minute, assuming minute volume of 20 litres/min;
120×10^4 f/hour;
6×10^6 f/day, assuming exposure for 5 hours/day;
3×10^7 f/week, assuming a 5-day week;
144×10^7 f/year, assuming a 48-week working year.

Thus for each fibre/millilitre, a worker must have inhaled 3×10^{10} fibres over 20 years of employment. In the Very High class, net service and intensity averaged more than 30 years and over 150 f/ml, respectively (McDonald *et al.*, 1980), so that, on average, each of the 636 men in this category had inhaled 7×10^{12} (regulatory) fibres - and this is a gross understatement because all the assumptions above are deliberately conservative.

Of course many fibres either do not penetrate the lung or are 'eliminated' from it. Some idea of what proportion is retained *post mortem* can be obtained from the results of Sébastien and colleagues (1989); a heuristic calculation suggests it is around 0.3%. For the particularly severely exposed group just discussed, this would give the order of magnitude of retained fibres in the lung as 2×10^{10}, or 70 f/μg. (This is an order of magnitude higher than the geometric mean number of fibres per microgram in the 89 lung tissue samples from Thetford Mines studied by Sébastien and colleagues (1989) and two orders of magnitude higher than in the 72 samples from Charleston.)

How could any lung stand up to such insult? Yet most of the 636 men in the long service, Very High intensity, class did survive at least until 1976. There had, of course, been much excess mortality: 98 more deaths (all causes) than expected, but 20 of the excess deaths were due to pneumoconiosis, 20 to lung cancer, and 7 to other respiratory conditions, excluding tuberculosis, while 12 were caused by cancer of oesophagus or stomach, and 18 by heart disease and cerebrovascular disease. Much of the excess could be attributed to asbestos exposure, but only about one-third of it was due to cancer.

What protected the survivors?

Question 9. What are the determinants of disease in the few cases of cancer associated with comparatively mild exposures?

Berry (1977) carried out an unmatched analysis of dust exposure in 215 lung cancer cases and 1075 referents in the Quebec cohort (Liddell *et al.*, 1977). He found that the Likelihood Ratio statistic of 38.76 with 9 degrees of freedom (d.f.) could be subdivided into 38.00 (1 d.f.) representing a linear trend and 0.76 (8 d.f.) for deviations about the trend - and commented that this was almost too good to be true. When there were 245 cases and 735 referents, he found (Berry, 1980) the corresponding statistics as 19.61 (1 d.f.) and 3.43 (8 d.f.) - still good, but not so astounding. Liddell and Hanley (1985) fitted two-parameter linear relationships between lung cancer SMRs and cumulative exposure in eight cohorts of asbestos-exposed workers; the sum of the goodness of fit statistics for the deviations from the lines was 17.94 with 30 d.f., giving no indication of inadequate fit (although there was one doubtful fit).

If the linear exposure-response model is appropriate, there is some risk of lung cancer even when exposure has been so mild that the risk of significant asbestosis is minimal. It is also possible that fairly mild exposure to amphiboles can be associated with mesothelioma and excesses, even if small, of lung cancer. It is also clear that any excess of lung cancer due to exposure to mineral wools will be quite small.

ENVOY

The epidemiology of fibre carcinogenesis is highly developed. The fibres that have been investigated include:- asbestos [chrysotile, amosite, crocidolite, mixtures]; other naturally occurring fibres [erionite, attapulgite]; man-made mineral fibres [mineral wool, glass wool, glass fibres]. The best studies have involved the evaluation of the exposure of each subject, usually in terms of temporal factors and environmental measurements, but also through lung burdens and sputum analysis. Cancer risks have been assessed in relation to earlier signs (radiologic and functional) and symptoms as well as exposure to fibre and smoking. Mortality, especially due to cancer, has been investigated in well over 50 cohorts or workers exposed to mineral fibres, there have been case/referent studies of mesothelioma and laryngeal cancer, and ecological and other approaches have been followed.

It must therefore be clear that gaps in epidemiologic knowledge are not due to lack of effort. Nor are they due to lack of skill. Indeed, the most important methodologic advances of the last four decades have been made by epidemiologists working in the field of fibre carcinogenesis: the subject-years method, the case/referent-within-a-cohort (ineptly named "nested case-control") approach, and rigorous methods for their analysis, have all become staples in many branches of epidemiology.

However, as mentioned in relation to Question 4, what statisticians call the "power" of a study depends on its size (in practice the number of deaths expected) and/or the magnitude of risk. [Note that power (related to sensitivity) is the ability to avoid false negative findings; in the present context it is more important than "significance" (akin to specificity), the avoidance of false positives.] It was unavoidable that many studies were of low power; most of the true positive findings arose because exposures were at intensities that are simply inconceivable today. Thus even the enormous Quebec cohort (4463 deaths, 869 due to cancer) has so far provided useful information on carcinogenesis only in relation to extremely high exposures. It is, however, being extended: the current vital status of the 5387 men known to be alive in 1976 has been determined, and death certificates are being collected; findings will be available in about a year's time. As later exposures were very much less severe than earlier, and as the number of deaths will be close to 7,000 [any excess over numbers expected on the basis of mortality in Quebec will be quite small], much more light will be cast on the subject, so important from the point of view of public health, of the ill-effects - if any - of comparatively mild exposure to chrysotile, by far the most common mineral fibre.

Theory is tested by experiment but will not be modified unless the results are convincing; if there is doubt, the experiment will be replicated. When experimental findings are tested by epidemiology, its results also need to be convincing. Replication, however, is usually impossible: any other study group will have been selected into its environment differently; the agent may be nominally the same but is unlikely to be identical; exposures will usually be quite different; and it is extremely difficult to create a study design that is better than similar. Thus, it is seldom that study results can even be combined satisfactorily. Indeed, a group from the International Commission on Occupational

Health failed in six years to find one body of studies with findings sufficiently consistent that they could be used even for illustrating pooling processes.

Perhaps epidemiology has already contributed almost as much as it can. Nevertheless, questions remain, and experimentation may provide potential answers. But, just as theory tumbles before experimental findings, even the best experimental results must be subject to confirmation in man - through epidemiology or otherwise.

We must not return to the era in which observational findings were denied because they were in conflict with laboratory results.

REFERENCES

Berry,G. (1977) Discussion on the paper by Professors Liddell and McDonald and Dr Thomas. *J. Royal Statist. Soc. A*. **140**:485-486.

Berry,G. (1980 Dose-response in case-control studies. *J. Epidemiol. Community Hlth*. **34**:217-222.

Berry,G., Newhouse,M.L. and Antonis,P. (1985) Combined effect of asbestos and smoking on mortality from lung cancer and mesothelioma in factory workers. *Br. J. Indust. Med*. **42**:12-18.

Browne,K. (1986) Is asbestos or asbestosis the cause of the increased risk of lung cancer in asbestos workers? *Br. J. Indust. Med*. **43**:145-149.

Gilson,J.C. (1982) Introduction: aims of conference. In: "Biological Effects of Man-Made Mineral Fibres". Proceedings of a WHO/IARC Conference, WHO Regional Office for Europe, Copenhagen.

Hammond,E.C., Selikoff,I.J. and Seidman,H. (1979) Asbestos exposure, cigarette smoking and death rates. *Ann. NY Acad. Sci*. **330**:473-490.

Hill,A.B. (1965) The environment and disease: association or causation. *Proc. Royal Soc. Med*. **58**:295-300.

Hughes,J.M. (1991) Epidemiology of lung cancer in relation to asbestos exposure. In: "Mineral Fibres and Health", Liddell,D. and Miller,K. (eds.) CRC Press, Boca Raton, FA, pp. 135-145.

ILO (1980) Guide lines for the use of the ILO international classification of radiographs of pneumoconioses, revised edition, 1980. *Occupational Safety and Health Series*, No. 22 (rev. 80), International Labour Office, Geneva.

Jones,J.S.P., Pooley,F.D., Sawle,G.W., Madeley,R.J., Smith,P.G., Berry,G., Wikgnall,B.K. and Aggarwal,A. (1980) The consequences of exposure to asbestos dust in a wartime gas-mask factory. In: "Biological Effects of Mineral Fibres", Wagner,J.C. (ed.) (IARC Scientific Publications No. 30) International Agency for Research on Cancer, Lyon. pp. 637-653.

Liddell,D. (1989) Epidemiological observations on mesothelioma and their implications for non-occupational exposure to asbestos. In: "Symposium on Health Aspects of Exposure to Asbestos in Buildings", Spengler,J.D., Ozkaynak,H., McCarthy,J.F. and Lee,H. (eds.) December 14-16, 1988, Harvard University Energy and Environmental Policy Centre, Cambridge, MA.

Liddell,F.D.K. (1980) Latent periods in lung cancer mortality in relation to asbestos dose and smoking. In: "Biological Effects of Mineral Fibres" , Wagner,J.C. (ed.) (IARC Scientific Publications 30). International Agency for Research on Cancer, Lyon. pp. 661-665.

Liddell,F.D.K. and Hanley,J.A. (1985) Relations between asbestos exposure and lung cancer SMRs in occupational cohort studies,*Br. J. Indust. Med.* **42**:389-396.

Liddell,F.D.K., McDonald,J.C. and Thomas, D.C. (1977) Methods of cohort analysis: appraisal by application to asbestos mining. *J. Royal Statist. Soc. A* **140**:469-491.

Liddell,F.D.K., Thomas,D.C., Gibbs,G.W. and McDonald,J.C. (1984) Fibre exposure and mortality from pneumoconiosis, respiratory and abdominal malignancies in chrysotile production in Quebec, 1926-75, *Ann. Acad. Med. Singapore* **13**, No. 2 (Suppl.):340-344.

McDonald,A.D., Fry,J.S., Woolley,A.J. and McDonald,J.C. (1983) Dust exposure and mortality in an American Chrysotile textile plant. *Br. J. Indust. Med.* **40**:361-367.

McDonald,A.D. and McDonald,J.C. (1978) Mesothelioma after crocidolite exposure during gas mask manufacture. *Environ. Res.* **17**:340-346.

McDonald,J.C., Liddell,F.D.K., Gibbs,G.W., Eyssen,G.E. and McDonald,A.D. (1980) Dust exposure and mortality in chrysotile mining, 1910-75. *Br. J. Indust. Med.* **37**:11-24.

Meek,M.E. (1991) Lung cancer and mesothelioma related to MMMF: the epidemiological evidence. In: "Mineral Fibres and Health", Liddell,D. and Miller,K. (eds.), CRC Press, Boca Raton, FA, pp. 175-186.

Saracci,R. (1977) Asbestos and lung cancer: an analysis of the epidemiological evidence on the asbestos-smoking interaction, *Int. J. Cancer* **20**:323-331.

Sébastien,P., McDonald,J.C., McDonald,A.D., Case,B. and Harley,R. (1989) Respiratory cancer in chrysotile textile and mining industries: exposure inferences from lung analysis. *Br. J. Indust. Med.* **46**:180-187.

Selikoff,I.J., Hammond,E.C. and Seidman,H. (1979) Mortality experience of insulation workers in the United States and Canada, 1943-1976. *Ann. NY Acad. Sci.* **330**:91-116.

Smith,A.H., Handley,M.A. and Wood,R. (1990) Epidemiological evidence indicates asbestos causes laryngeal cancer. *J. Occup. Med.* **32**:499-507.

Vacek,P. and McDonald,J.C. (1990) Effect of intensity in asbestos cohort exposure-response analyses. In: "Occupational Epidemiology", Sakkrai,H., Okazaki,I. and Omae,K., (eds.), Elsevier, Amsterdam, pp. 189-193.

Wagner,J.C. (1986) Mesothelioma and Mineral Fibres. In: "Accomplishments in Cancer Research 1985, Fortner,J.G. and Rhoads,J.E. (eds.) Lippincott, Philadelphia, pp. 60-72.

ADDENDUM

During the Workshop of which this volume is the proceedings, a better appreciation developed of why it must appear to experimental scientists that the epidemiology of fibre carcinogenesis is largely *ad hoc*: the observation of mortality, mainly not due to malignancies, in relation to imperfect measures of exposure, in human subjects who have been selected into their (often changing) environments in quite tenuous ways. Appearances are, however, misleading. Good epidemiology requires careful research design, just as important for surveys as in the laboratory; after all, latency necessitates observation over several decades, which implies major design problems. Also required is a statistical methodology specific to epidemiology; this is now well-developed (although not all epidemiology uses the best statistical methods).

It should not be surprising that the experimental approach to carcinogenesis also requires a separate statistical methodology. Such a methodology does not seem to be in general use, but Julian Peto in discussion during the workshop called attention to the text

TABLE A.1. Probabilities (%) of Tumour-free Experiments

Background rate (b)	n = 30	n = 40	n = 50	n = 60
0.5%	86.0	81.8	77.8	74.0
1%	74.0	66.9	60.5	54.7
2%	54.5	44.6	36.4	29.8
3%	40.1	29.6	21.8	16.1

by Gart *et al.* (1986). What is clear is that the elementary methods of "classical" statistics are not always applicable - and, as in many fields, they are only too easily misused.

However, they can illuminate the earlier discussion of "negative" and "positive" experiments. The probability of a tumour-free experiment is:

$$(1 - b)^n \qquad [1]$$

where b is the background risk of tumour and n is the number of animals in the experiment, quantified in Table A.1.

Even with low background rates, the probabilities that tumours will occur in control animals are not negligible; for $b = 0.5\%$, at least one animal will develop a tumour in over a quarter of control groups of 60 animals.

But, given control animals free of tumours, what constitutes a "positive" experiment? Take the numbers of exposed and control animals as both n, with t tumours among exposed animals (none in controls). The recognised statistical test of the hypothesis that the tumour risk is the same in experimental and control groups (Fisher, 1935) yields the probability:

$$P = \{(n)! \ (2n - t)!\}/\{(n - t)! \ (2n)!\} \qquad [2]$$

that a finding as divergent as t:0 would arise. For t = 1, the value of P is 1/2; for t = 2, 3, ..., it is slightly less than 1/4, 1/8, ..., respectively.
But these are one-sided P-values, and statistical theory demands (whatever some pundits may say) that P-values be two-sided; they are given in Table A.2 for n = 35; for greater n and t > 1, the P-values are closer to 1/2, 1/4, 1/8, etc.

The convention of a two-sided P-value of 5% would require 6 tumours in 35 exposed animals to yield a "significant" experiment. This is, however, yet another situa-

TABLE A.2. Two-sided P-Values[*] when 35 Controls are Tumour-free and there are t Tumours in 35 Exposed Animals

t = 1	t = 2	t = 3	t = 4	t = 5	t = 6
100%	49%	24%	11%	5.4%	2.6%

[*] Twice the probability given by expression [2]

tion in which "P < 0.05" appears quite unsuitable. It is not for an epidemiologist even to suggest what P-value should be used to define a positive experiment - but it is clear that a single tumour in exposed animals is insufficient to be "positive".

ADDITIONAL REFERENCE

Gart,J.J., Krewski,D., Lee,P.N., Tarone,R.E. and Wahrendorf,J. (1986). *Statistical Methods in Cancer Research III: The Design and Analysis of Long-term Animal Experiments.* (IARC Scientific Publication No. 79). International Agency for Research on Cancer, Lyon.

HEALTH EFFECTS OF INSULATION WOOLS (ROCK/SLAG WOOL AND GLASS WOOL) AND FUTURE RESEARCH NEEDS

David Douglas

Consultant in Occupational and Environmental Medicine
GPO Box 653
Sydney, Australia

ABSTRACT

This paper first summarises the data from the US and European epidemiological studies upon which many claims about the carcinogenicity of insulation wools (rock wool, slag wool, and glass wool) have been made. Emphasis has been given to the conclusions of the researchers rather than the conclusions of reviewers who have pooled the data from both continents. The results of the most recently published epidemiological study of crocidolite miners and millers (the Wittenoom cohort) have then been summarised and compared with the man-made mineral fibre studies.

In doing this, it is demonstrated that some of the statements in the past about the comparative toxicity of insulation wools and asbestos have been nonsensical. It is also demonstrated that the US and European historical cohort studies contain too many unknowns to sort out the precise reasons for the elevated lung cancer SMRs in those cohorts.

Reasons are given for the need for further research. It is recommended that future research be based on a large (minimum 5000 persons) prospective cohort for which essential personal, employment, radiographic, environmental and smoking data are available. The cohort must include people engaged in both the manufacture and use of insulation wools. Analyses of incidence and mortality data, using logistic regression and person-years techniques, would be carried out at five-yearly intervals. It is further recommended that case-referent and other studies using state of the art methods be nested within the main cohort.

INTRODUCTION

At the 1982 WHO/IARC Conference on the "Biological effects of man-made mineral fibres" the Conference Chairman stated in his introduction (Gilson, 1984):

"In the past, most occupational diseases were first spotted in individuals by clinicians or pathologists. Only later, and sometimes after many years, have their severity and extent been revealed by systematic surveys. In the case of MMMF the pattern has been quite different. There have been only isolated clinical reports of possible chronic ill health, and these reports have not increased in number or certainty with time. However, by the early 1970s experimental studies in animals, primarily aimed at investigating the mechanisms of action of asbestos, showed that mesothelial tumours of the pleura, similar to those seen in man following exposure to asbestos, could be induced in animals by intrapleural implantation of a variety of fibres, including some MMMF."

The extensive worldwide investigations throughout the 1980's have not revealed the existence of clinical disease attributable to MMMF, other than skin, eye, and respiratory irritation due to coarse (diameter >4 μm) fibres and related dusts. The above statement, therefore, is still valid. But serious questions about the safety of insulation wools (glasswool rockwool and slagwool), and other man-made mineral fibres, have been raised following the release of the findings of the two large scale epidemiological studies of insulation wool production workers in the United States (Enterline *et al.*, 1987) and Europe (Simonato *et al.*, 1987). Both studies were reviewed at the WHO/IARC Conferences held in Copenhagen in 1982 and 1986.

At the 1982 Conference, McDonald stated that there was a "disturbing consistency" in the finding of an increase in lung cancer mortality 20 or more years after first exposure, but that there was no clear relationship with intensity or duration (McDonald, 1984). He was drawn to the conclusion that: "Bearing in mind what is known about response to asbestos exposure in man, the present findings with man-made fibres point to a similar order of effect, a conclusion also compatible with the experimental data."

In summarising the 1986 Conference, Sir Richard Doll (1987) concluded "that, taking into account also the results of animal experiments, the experience of the asbestos industry and the experience of the glass filament sector of the MMMF industry, MMMF are not more carcinogenic than asbestos fibres and exposure to current mean levels in the manufacturing industry of 0.2 f/ml or less is unlikely to produce a measurable risk after another 20 yr have passed."

These conclusions by McDonald and Doll, together with the classification of insulation wools by the International Agency for Research on Cancer (IARC, 1988) as category 2B (possibly carcinogenic to humans), have had the effect of linking insulation wools very closely with asbestos in the minds of many people. Particularly in Australia, many workers now regard insulation wools as barely indistinguishable from asbestos in their health effects.

Because of these perceptions, and because of the uncertainties associated with many of the findings of the two major epidemiological studies, it is important to continue with research into the health effects of insulation wools. It is particularly important to carry out research into the health status of people manufacturing and using insulation wools under modern conditions, with quantitative estimates of their occupational exposures, with accurate smoking histories, and incorporating state of the art research methods.

But before making recommendations about future research, it is important to review briefly the US and European studies, as well as reviewing some of the most recent epidemiological studies of asbestos workers.

U.S. MORTALITY STUDY

The results of the mortality study of 16,661 male workers who had been employed in the US insulation wool production industry for one year or more during the years 1945 to 1963 were reported by Enterline *et al.* (1987). Exposure estimates suggested that cumulative exposures to respirable glass fibres ranged from 0.04 to 3.1 f/ml years, and to respirable mineral fibres ranged from 1.3 to 6.8 f/ml years. No quantitative estimates of other occupational exposures were provided.

There was a 98% follow-up of the cohort, and death certificates were obtained for 97% of the 4986 deaths recorded between 1946 and 1982. The standardised mortality ratio (SMR) was 102 for all causes of death using US National death rates to calculate expected deaths. The most important SMRs were for those first exposed 20 or more years before death, and when expected deaths were based on local instead of national death rates. There were statistically significant raised SMRs for all cancers of 108 and for lung cancer of 113. The lung cancer SMRs when analysed by fibre type were not statistically significant but were elevated, being 111 for glass wool and 131 for mineral wool.

There was no correlation between lung cancer SMRs and duration of exposure, except for an inverse relationship for duration of mineral wool exposure with ten year increases in duration of exposure, the SMRs were 145, 142, 109 and 114. Non-statistically significant raised lung cancer SMRs were found in those ever exposed to fine diameter glass fibres, with an SMR of 198 for those first exposed 30 or more years before death.

There was no relationship between estimated cumulative doses of glass fibres and lung cancer SMRs; but an inverse relationship existed between dose and lung cancer SMRs for mineral wool workers - with four increasing cumulative doses ranging from <0.4 to >8.0 f/ml years, the SMRs went 185, 164, 119 and 104.

No smoking histories were available for the study cohort, so in order to examine the possible confounding effect of smoking, a case-referent study nested within the cohort was used. The cases of lung cancer and nonmalignant respiratory disease were compared with the referents which were a 4% random sample of other cohort members stratified by plant and year of birth. Although the referents were a small sample and the overall response rate to the telephone request for smoking information was 72%, there was reasonable evidence to support the conclusions that: smoking habits of the cohort were similar to the US population; early hires smoked less than later hires; and smoking was a confounding factor in the raised lung cancer SMRs in mineral wool workers.

There were no trends in SMRs that suggested a relationship between insulation wool exposure and non-malignant respiratory disease. There was no evidence of an excess of deaths due to mesothelioma nor any evidence of deaths due to pneumoconioses. SMRs for stomach cancer, cirrhosis of liver, and accidents were not elevated.

There were several features which were not consistent with a causal relationship between MMMF exposure and raised lung cancer SMRs: there was a lack of relationship with duration of exposure or cumulative dose of fibres; there were inconsistencies in the relationship with time since first exposure and death from lung cancer; and in mineral wool workers, the most recent hires experienced the greatest lung cancer excess.

The authors concluded that: "On balance, our findings argue for some kind of a relationship between work in the man-made mineral fibre industry and health. However, they throw little light on the question of whether this relationship is due to the fibres themselves or is related to other contaminants that might be present in the environment of man-made mineral fibre workers."

EUROPEAN MORTALITY STUDY

The results of a mortality and incidence study of 24,609 workers employed between 1933 and 1961 in the production of insulation wools at 13 plants in 7 European countries was reported by Simonato *et al.* (1987).

There were 117 deaths from all causes, with an SMR of 74, in the 2642 office workers. Seven deaths, SMR 73, were due to lung cancer.

There were 2719 deaths from all causes in the 21,967 plant workers resulting in statistically significant raised SMRs for all causes (111), all malignancies (111), lung cancer (125), accidents, poisoning and violence (153), and suicide (130). There was no evidence of increased mortality or incidence rates due to mesothelioma, non-malignant respiratory disease, or pneumoconioses.

When those employed for less than one year in the insulation wool industry were analysed separately, it was clear that raised SMRs for all causes, non-malignant respiratory disease, digestive system diseases, and violent deaths were due to the short duration workers.

As with the US study, the most important results were for lung cancer occurring 20 or more years since first exposure and compared with expected deaths calculated from locally adjusted death rates. There were 189 deaths from lung cancer in the total cohort with a nonstatistically significant raised SMR of 110. For those in which the lung cancer occurred 30 or more years since first exposure, the SMR of 152 was statistically significant.

In those employed in rock/slag wool production, the lung cancer SMR was 124 for all workers (81 deaths) and 185 for those first employed 30 or more years before death (12 deaths). Neither result was statistically significant. The corresponding non-statistically significant SMRs for glass wool workers were 103 (93 deaths) and 138 (17 deaths). In neither glass wool nor rock/slag wool workers was there any relationship between duration of exposure and SMR.

The strongest relationship demonstrated was that between the technological phase in which people were employed in rock/slag wool production and the SMRs for lung cancer. Ten lung cancer deaths, with a statistically significant SMR of 257, occurred in the "early technological phase", 6 of which (SMR 295) were first employed 30 or more years before death. Nonstatistically significant SMRs of 141 (14 deaths) and 111 (57 deaths) were observed in the "intermediate" and "late" technological phases. Similar patterns were not observed in the glass wool production workers.

The authors reported that standardised incidence ratios (SIRs), where available, followed patterns similar to the SMRs; but the ratios were generally lower and not statistically significant.

Whereas exposures to respirable glass wool fibres throughout the study period were probably comparable to those currently experienced (0.1 f/ml or less), respirable rock/slag wool fibre exposures have been estimated by Dodgson *et al.* (1987) to be

1-2 f/ml (up to 10 f/ml in the dustiest plants) during the early technological phase. Since the average employment duration of the cohort was 2.7 years, the average cumulative dose of rock/slag wool during the "early" technological phase could have ranged from <3.0 to >30 f/ml years. However, it was highly likely that the cumulative exposures were at the lower limit of the range.

There were no data on the smoking practices of the cohort, and, in the absence of information to the contrary, it was assumed that the smoking habits in the cohort were similar to those in the reference populations, particularly where local rates were used.

As with the US study, there was no evidence of a dose-response relationship when duration of exposure was used as an estimate of dose. Such a relationship was not demonstrated either in the total cohort or when the cohort was separated into the different technological phases.

The data were analysed for confounding due to asbestos, bitumen, and formaldehyde and the conclusion reached that confounding due to these agents did not occur. But the absence of accurate and detailed occupational hygiene data from the early and intermediate technological phases made the validity of such analyses doubtful. Other possible confounders such as arsenic and polycyclic aromatic hydrocarbons could not be allowed for.

The authors concluded that: "it is plausible that fibre exposure in the early phase of rock/slag wool production - alone or in combination with other factors - may have contributed to the observed lung cancer excess."

The European study was a collaborative effort involving seven different countries. Since the combined study indicated an elevated SMR for lung cancer in rock/slag wool production workers, the separate results from the major rock/slag wool producing countries have also been reviewed.

Westerholme and Bolander (1986) studied the 524 deaths occurring in 3600 production workers and reported that the Swedish cohort alone contributed most to the raised SMRs described in the total European cohort. The highest lung cancer SMRs were seen in those with 30 or more years latency and who were employed in the "early" technological phase - 3 deaths with SMR 227 for rock/slag, and 5 deaths with SMR 205 for glass wool. There were 9 deaths (SMR 150) due to stomach cancer in rock/slag wool workers.

The authors concluded that the elevated SMRs were highly susceptible to confounding factors, both inside and outside the factory, and stated: "It is a reasonable hypothesis that the work conditions, including exposure to MMMF during the early production phase have contributed to the risk of lung cancer in the Swedish cohort. This study does not, however, in our opinion, provide sufficient evidence to draw the scientific conclusion that this is the true explanation for the findings."

Claude and Frentzel-Beyme (1986) studied 2092 rock wool workers, none of whom had been employed in the "early" technological phase. The SMR for all causes (110) and lung cancer (120) were raised but the increase was not statistically significant. The lung cancer excess was not related to time since first exposure nor to duration of employment, but was higher in those employed prior to 1953 (SMR 138) compared with those employed after 1953 (SMR 100). However, statistically significant excesses were found for mental disorders and other conditions including alcoholism.

The authors concluded that: "the evidence from this study indicates some adverse effects from being ever employed in rock wool production, but it appears difficult to relate these causally to exposure to MMMF."

RECENT ASBESTOS EPIDEMIOLOGY

Because of the importance of the conclusions of McDonald and Doll who both made comparisons between asbestos epidemiology and the US and European MMMF studies, it is important to review the recently published results of the studies of workers employed in the mining and milling of crocidolite asbestos.

In 1988, Armstrong *et al.* (1988) reported the results of a mortality analysis of all known deaths to the end of 1980 occurring in a cohort of the 6505 men and 411 women ever employed at the Wittenoom crocidolite asbestos mine in Western Australia from 1943 to 1966. The vital status of 73% of the men was known at the end of 1980.

Fifty per cent of the cohort worked at Wittenoom for less than 100 days, but the exposures to respirable fibres were intense. Hence, estimated cumulative exposures were: 56% of cohort - up to 10 f/ml years; 29% of cohort - 10-100 f/ml years; 5% of cohort - >100 f/ml years; and 10% of cohort - unknown.

Expected deaths were calculated using age, period, cause, and sex specific death rates for Western Australia. SMR1 was calculated with censoring at 31 December 1980, and SMR2 was calculated with censoring at the date last known to be alive:

	SMR1	OBS	SMR2
all causes	96	820	153
asbestosis	1510	34	2550
lung cancer	160	91	264
peptic ulcer	158	12	248
alcoholism	316	20	487
cirrhosis	253	25	394
other accidents	162	53	236

All the SMRs, with the exception of SMR1 for all causes, were statistically significant. There were 32 deaths from mesothelioma.

The authors described evidence for a dose-response for lung cancer, mesothelioma, and asbestosis, with increasing ratios for years since first exposure and intensity of exposure.

In 1989, de Klerk *et al.* (1989) carried out a case-control study on the same population in which each case of lung cancer, mesothelioma, and stomach cancer was matched with up to 20 controls from within the total cohort. As with the earlier study, there were no data on smoking.

After logistic regression analysis, the authors concluded, *inter alia*, that: there was no evidence of an increase in deaths from lung cancer for employment periods of under two years; there was little evidence that year of first employment or worksite had any effect on death from lung cancer; there was clear evidence that duration of exposure had a much stronger effect than intensity; and there was some evidence that the lung cancer effect was greatest 10 to 20 years after first employment.

In another publication, de Klerk *et al.* (1989), made predictions about the future incidence of crocidolite related diseases. In relation to lung cancer he stated: "it must be remembered that all such cases are being considered, most of which will be caused by tobacco smoke. In a follow-up to the end of 1980, the SMR for lung cancer was either 160 or 264 depending on which of the two censoring schemes was used. Taking the SMR of 160 to be better implies that about 40% of the cases of lung cancer could be attribulable to crocidolite exposure. However, in view of the high prevalence of smoking in this cohort, even this is likely to be an overestimate".

Baker (1985), in an unpublished thesis "Lung cancer incidence amongst previous employees of an asbestos mine in relationship to crocidolite exposure and tobacco smoking" has obtained detailed smoking histories of 1746 out of 2390 workers from the original cohort who were traced as being in Western Australia.

His principal conclusions were: "analyses indicated strong associations between lung cancer incidence, intensity of asbestos exposure, net duration of employment, and usual daily consumption of cigarettes; and analyses strongly suggested that lung cancer incidence rate ratios associated with crocidolite exposures rose to a peak less than 25 years from first exposure and then fell."

CONCLUSIONS FROM THE EPIDEMIOLOGY

Comparisons between the MMMF cohorts and the crocidolite cohort have been regarded as valid given that any long term health risks associated with MMMF were thought to be related to the respirable fibres, and all authors have concentrated on this in their analyses. Conclusions from the epidemiology are:

(a)   Apart from lung cancer, none of the other "asbestos- related diseases", *viz* mesothelioma and pneumoconiosis (asbestosis) were observed as raised SMRs or raised SIRs in the MMMF cohorts. On the other hand, the crocidolite cohort not only had raised SMRs for lung cancer, but a massive increase in mortality due to mesothelioma and asbestosis.

(b)   The SMRs for lung cancer in all the cohorts were similar - 160 to 264 for crocidolite, 125 to 257 in the European study, and 131 to 158 in the US cohort. The authors of the MMMF studies claimed that smoking differences between their cohorts and the reference populations would not account for the excess lung cancers. But the authors of the crocidolite study suggested that their lung cancer SMRs could be inflated by the high smoking patterns in their cohort, with a corresponding smaller percentage of the excess being due to the asbestos exposure.

(c)   The estimates of cumulative doses of respirable fibres were approximations only, with the true doses likely to be lower than stated in the MMMF cohorts, and up to an order of magnitude higher than stated in the crocidolite cohort (Rogers, 1990).

(d)   If the practice of comparing the SMRs and estimated fibre doses in one study with those in another is valid, then it can be deduced from points (b) and (c) that MMMF have been demonstrated to be many times more potent in causing lung cancer than crocidolite. Such a proposition cannot be supported when made with the knowledge of the relative toxicity of MMMF and crocidolite in animal studies, the information on differences in fibre diameters, and the absence of any evidence of the other relevant diseases, mesothelioma and pneumoconiosis.

(e)   If it were valid to regard the exposure to respirable fibres as the cause of, or a major contributor to, the raised lung cancer mortality, then it must be equally valid to postulate that the MMMF were operating in biological systems in a totally different way from asbestos. That is, they were increasing the lung cancer risk without increasing the risk of fibrosis or mesothelioma, at doses many times less than crocidolite, and in a random non-dose-related fashion.

(f)   A more rational explanation for the raised SMRs in the MMMF cohorts would be that the smoking habits, the poor socioeconomic status of the workers in the 1940s and 50s, and the generally poor and polluted working environments combined to cause the excess lung cancer rather than the fibres *per se*.

Because of the perceptions and assumptions which now exist about the safety of insulation wools, it is important to continue with research. But the complexity of the possible confounders outlined in point (f) above make it very unlikely that more useful information can be obtained from the exisiting historical MMMF cohorts in the US and Europe.

It is recommended therefore, that future insulation wool research be focussed on a long term prospective study. This could contain "nested" case-referent and other studies designed to test a range of hypotheses derived from other epidemiological studies, clinical medicine, and experimental toxicology.

PROPOSALS FOR FUTURE RESEARCH

1. Main Study

A prospective study based on a cohort of no less than 5000 people occupationally exposed to rock/slag wool and glass wool should be established. Essential data on each person in the cohort must include:

detailed employment histories including asbestos and other dust exposures;
chest radiograph - ideally pre-employment and five-yearly thereafter;
lifetime smoking history;
alcohol history; and environmental exposures including respirable fibre and dust levels.

Because of the improvements in occupational hygiene during the past decade, knowledge of occupational exposures should be adequate to identify occupational groups for which the necessary data are available for the past ten years. Accordingly, it may be possible for the prospective cohort to commence with all those employed on 1 January 1980. But the practicability of this would also depend on the availability of chest radiographs in the potential cohort.

The cohort should include both manufacturers and users of insulation wools, with emphasis on the user industries and those exposed to rock/slag wool. Good baseline data must be got on all members of the cohort, and the data updated as changes in employment, exposure, or any other vital status occur.

All people who leave employment in the insulation wool industry should be traced and included in future analyses, so that the eventual outcomes of all those who ever enter the cohort are followed.

The incidence of the nominated disease outcomes, as well as mortality, must be recorded. Nominated disease outcomes would include lung cancer, mesothelioma, non-malignant respiratory disease, and pneumoconioses.

An analysis of incidence and mortality data, using logistic regression and person-years techniques, should be carried out in 1995 and at five-yearly intervals thereafter.

Analyses must allow for the confounding effects of asbestos, smoking, and all occupational exposures.

A cohort of 5000 would be the minimum number to enable detection after ten years of statistically significant relative risks (RR) for the following mortality outcomes: all causes, RR of 1.2; coronary heart disease, RR of 1.3; lung cancer, RR of 2.0; blood neoplasms, RR of 2.3; and bladder cancer, RR of 6.0.

2. Subsidiary Studies

The development of this prospective cohort would offer an excellent opportunity for developing other "nested" studies:

(a)   Case-referent studies are the most obvious, where those with the nominated disease (or cause of death) are compared with matched referents in the cohort who do not have the disease. The principles of the case-referent study could be applied to other variables, such as never smokers compared with ever smokers, or one exposure level compared with another.

(b)   Chest radiographs should be read periodically by a panel of experienced readers using the ILO classification. The results would be analysed against the principal variables including smoking and fibre exposures.

(c)   Morphometry (Barry and Crapo, 1985) is a developing science which has particular application to the study of diffuse and focal lesions in the lung caused by a wide range of agents or fibres. It applies the principles of a group of geometrical and stereological formulas and statistical methods to solve problems of microscopic anatomy. It uses the information derived from two-dimensional sections of lung tissue to yield information about the tissue in three dimensions. In the present context it offers a new method for studying exactly what changes, if any, have occurred in the lungs of people exposed to insulation wools.

The objectives of such research would be the morphometric examination of a small number (initially) of lungs of deceased members of the cohort. The findings would be related to any pulmonary disease and to the known history and environmental exposures. Over time, a precise picture of what happens to the lungs of workers breathing various levels of insulation wools would be established.

(d)   Bronchoalveolar lavage (BAL) is a method for sampling inflammatory cells of the human respiratory tract. Data from studies of asbestos exposed individuals (Robinson *et al.*, 1986) suggest that there may be a place for BAL in the early detection of asbestosis. That is, before the inflammatory response to asbestos has resulted in clinically and radiologically detectable fibrosis. While the relationship between BAL results and pulmonary fibrosis has not yet been validated, the results have been sufficiently important to warrant the use of such an investigatory method in a cohort of insulation wool workers. The procedure is invasive and its use as a research tool would require particular sensitivity, but along with morphometry, it could contribute to the understanding of the pulmonary effects of insulation wools well in advance of any findings from radiology, disease incidence or mortality.

(e)   Irritation of the skin, eyes and upper respiratory tract could be studied in more detail on a subset of the prospective cohort. Whereas most of the concern about insulation wools has focused on the long term effects and the respirable fibres, irritation due to the coarse fibres, the unfiberised particles, and the related dusts is of most concern to many users of the products. There have been very few studies of the incidence and severity of the irritation aspects and a better understanding of this problem and its prevention would contribute to improved user acceptance of insulation wools.

CONCLUSIONS

I suggest that a key to obtaining meaningful data on the health effects of insulation wools is the establishment of a prospective cohort of no less than 5000 workers in the manufacturing and user sectors of the industry. With such a cohort as a research base, periodic analyses of disease incidence and mortality could be carried out, and concurrently, the detailed data base of information on each member of the cohort could support subsidiary research utilising advanced methods derived from the disciplines of epidemiology, clinical medicine and animal toxicology.

REFERENCES

Armstrong,B.K., de Klerk,N.H., Musk,A.W. and Hobbs,M.S.T. (1988) Mortality in miners and millers of crocidolite in Western Australia. *Br. J. Indust. Med.* **45**:5-13.

Baker,J.E. (1985) Lung cancer incidence amongst previous employees of an asbestos mine in relationship to crocidolite exposure and tobacco smoking. Thesis presented for the degree of Doctor of Philosophy of the University of Western Australia. Perth.

Barry,B.E. and Crapo,J.D. (1985) Application of morphometric methods to study diffuse and focal injury in the lungs caused by toxic agents. *CRC Crit. Rev. Toxicol.* **14**:1-32.

Claude,J. and Frentzel-Beyme,R. (1986) Mortality of wokers in a German rock-wool factory - a second look with extended follow-up. *Scand. J. Work Environ. Hlth* **12**:53-60.

de Klerk,N.H., Armstrong,B.K., Musk,A.W. and Hobbs,M.S.T. (1989) Cancer mortality in relation to measures of occupational exposure to crocidolite at Wittenoom Gorge in Western Australia. *Br. J. Indust. Med.* **46**:529-536.

de Klerk,N.H., Armstrong,B.K., Musk,A.W. and Hobbs,M.S.T. (1989) Predictions of future cases of asbestos-related disease among former miners and millers of crocidolite in Western Australia. *Med. J. Aust.* **151**:616-620.

Dodgson,J., Cherrie,J. and Groat,S. (1987) Estimates of past exposure to respirable man-made mineral fibres in the European insulation wool industry. *Ann. occup. Hyg.* **31**:567-582.

Doll,R. (1987) Symposium on MMMF, Copenhagen, October 1986; overview and conclusions. *Ann. Occup. Hyg.* **31**:805-819.

Enterline,P.E., Marsh,G.M., Henderson,V. and Callahan,C. (1987) Mortality update of a cohort of US man-made mineral fibre workers. *Ann. occup. Hyg.* **31**:625-656.

Gilson,J.C. (1984) Introduction: Aims of the Conference. Biological effects of man-made mineral fibres, Proceedings of a WHO/IARC Conference, Copenhagen, 20-22 April, 1982.

IARC (1988) IARC Monographs on the Evaluation of Carcinogenic Risks to Humans, Vol **43**, Man-made mineral fibres and radon. Lyon, pp. 39-171.

McDonald,J.C. (1984) Peer review: mortality of workers exposed to MMMF - current evidence and future research. In, "Biological effects of man-made mineral fibres", Proceedings of a WHO/IARC Conference, Copenhagen, 20-22 April, 1982.

Robinson,B.W.S., Rose,A.H., James,A., Whitaker,D. and Musk,A.W. (1986) Alveolitis of pulmonary asbestosis. Bronchoalveolar lavage studies in crocidolite and chrysotile exposed individuals. *Chest* **90**:396-402.

Rogers,A. (1990) Cancer mortality and exposure to crocidolite. *Br. J. Indust. Med.* **47**:286.

Simonato,L., Fletcher,A.C., Cherrie,J.W., Anderson,A., Bertazzi,P., Charnay,N., Claude,J., Dodgson,J., Esteve,J., Frentzel-Beyme,R., Gardner,M.J., Jensen,O., Olsen,J., Tappo,L., Winkelmann,R., Wetserholm,P., Winter,P.D., Zocchetti,C. and Saracci,R. (1987) The International Agency for Research on Cancer historical cohort study of MMMF production workers in seven European countries: extension of the follow-up. *Ann. Occup. Hyg.* **31**:603-623.

Westerholm,P., Bolander,A.-M. (1986) Mortality and cancer incidence in the man-made mineral fiber industry in Sweden. *Scand. J. Work Environ. Hlth* **12**:78-84.

NON-ASBESTOS FIBRE BURDEN IN INDIVIDUALS EXPOSED TO ASBESTOS

R.F. Dodson, M.G. Williams, Jr., C.J. Corn, A.Brollo[1]
and C. Bianchi[1]

Department of Cell Biology and Environmental Sciences
The University of Texas Health Center at Tyler
P.O.Box 2003, Tyler, Texas, U.S.A.

[1]Laboratory of Pathological Anatomy
Hospital of Monfalcone
34074 Monfalcone, Italy

ABSTRACT

The characteristics of asbestos, which cause it to be classified as a carcinogen, have been the subject of numerous investigations. While various mechanisms might contribute to its overall cancer producing potential, the physical characteristics of asbestos are widely recognized as major factors. The relationship of fibrous form to tumour production is often referred to as conforming to the "Stanton Hypothesis." The model used by Stanton incorporated non-asbestos fibres which were of a similar dimensions to asbestos fibers. Appreciable attention has been given to assessing lung tissue asbestos burdens in a number of cohorts; however, limited information exists on the amount of "non-asbestos" fibres in the lung. Since the "Stanton Hypothesis" includes fibres of a given dimension and not mineral type, it is important to gain more information on the presence and characteristics of non-asbestos fibres and their possible contribution to the development of disease.

The present study evaluates tissue from eight former shipyard workers who were exposed to asbestos but would also have been expected to have been exposed to other fibrous dusts. A comparison is made between asbestos and non-asbestos fibre burden in lung parenchyma, lymph nodes, and pleural plaques. The analytical techniques used incorporated both light microscopy and transmission electron microscopic assessments. The data indicate that non-asbestos fibres are found in all three sites and that a small percentage of these fibres are within the dimensions conforming to the "Stanton Hypothesis" for being potentially carcinogenic.

INTRODUCTION

The relationship between asbestos exposure and an increased risk of developing cancer has been established (Becklake, 1976). The mechanisms which contribute to this carcinogenic potential may include chemical composition, surface charge, and radical formation; however, a link has been established with regard to physical dimensions. Stanton and Wrench (1972), Wagner and co-workers, (1973) and Pott and co-workers (1987), have shown that fibre length is an important parameter for not only asbestos as well as similarly sized non-asbestos fibres. The data from these studies suggested that fibres longer than 8 μm in length (the "Stanton Hypothesis") constitute a more tumourigenic fraction of respirable dust. While these studies have been carried out in animal models, recent reviews pose continuing concern regarding non-asbestos fibre exposure in man (Doll, 1987; Dunnigan, 1989).

An increasing base of information is available on the tissue burden of asbestos in occupationally exposed individuals (Whitwell *et al.*, 1977; Dodson *et al.*, 1985) and some information exists on concentrations in lungs from the general population (Dodson *et al.*, 1984; Churg & Warnock, 1980). Many of these work histories suggested the potential existed for exposure to various non-asbestos fibres, but few studies have reported their numbers, dimensions, and types in tissue (Dodson *et al.*, 1985; Churg, 1983). Since most of these fibres are thin and uncoated, their detection requires transmission electron microscopy for quantitation. The present study will thus characterize the non-asbestos fibre burden in lung, thoracic lymph nodes, and pleural plaques from eight individuals with known asbestos exposure. The characteristics of these non-asbestos fibres will be compared with known data from the asbestos burden at these sites (Dodson *et al.*, 1990). These comparisons include numbers and types as well as the percentage which conforms to the "Stanton" dimensions.

METHODS

Autopsy tissue from eight former shipyard workers was obtained from the General Hospital of Monfalcone, Italy. The characteristics of the group are shown in Table 1. The tissue collected on each subject included lung parenchyma, tracheal lymph node, and pleural plaque. To avoid contamination, pre-filtered (0.2 μm pore size) solutions were used and both nodes and plaques were prewashed and debrided before sampling. Multiple sites from each lung were pooled to assure adequate sampling. All tissues were digested following the Williams' procedure referred to below (Williams *et al.*, 1982). For each tissue type, individual samples were split into two portions, one pool was used to determine dry weight and the other pool was used for the digest. The tissue was digested in 9.2% pre-filtered Wright's Laundry Bleach, (Wright, Inc., Fort Worth, TX). Aliquots of each digest were filtered using 0.2 μm polycarbonate (Nuclepore) filters for uncoated fibre evaluations and 0.22 μm mixed cellulose ester filters (Millipore) for ferruginous body counts.

TABLE 2.[1] Former Shipyard Workers [**]

Case	Sex	Age	Smoke	Occupation	Disease	Plaque
1	M	58	15-20 Cig/ day	Shipyard Plumber and Painter 30yr Miner	Asbestosis;Small Cell Ca[*] of rt lung	Bilat.
2	M	61	20Cig/ day	Shipyard Welder 37yr	Asbestosis;Adeno Ca of the rt lung; Arteriosclerosis	Bilat.
3	M	63	40Cig/ day	Shipyard Laborer 6 yr Navy; Oil Mill; Crane Oper.	Larygeal Ca; Arteriosclerosis	Bilat.
4	M	70	60Cig/ day	Shipyard work-er & Draftsman 46 yr	Lung Fibrosis; Emphysema; Arteriosclerosis	Bilat.
5	M	72	Ex after 1973	Shipyard Tracer 27yr	Arteriosclerosis; Meningioma	Bilat.
6	M	76	No	Shipyard welder 45yr	Asbestosis; Arteriosclerosis	Left
7	M	78	No	Shipwright 25yr Chemical Ind.	Asbestosis; Emphysema; Arteriosclerosis	Bilat.
8	F	82	No	Shipyard Cleaner for some months. Husband ship-yard welder	Arterioscleosis	Bilat.

[*] Ca: carcinoma

[**] Data reprinted with permission (Dodson *et al.*, 1990).

Filter residues were treated sequentially with 4% potassium permanganate, 8% oxalic acid, a second bleach exposure, and a second 8% oxalic acid. Each treatment was followed by a water rinse. A final alcohol treatment was done and the filters were air dried. Both filter backgrounds and reagent blanks were determined.

Ferruginous bodies were assessed by light microscopy. For this evaluation, one fourth of the mixed cellulose ester filter was mounted on a glass slide, cleared (made

TABLE 2.[1] **Former Shipyard Workers**

Case No.	Tissue[2] Location	FB[3]/g	CHRY[3]/g	AMPH[3]/g	ASB[3]/g	AlSi[4]/g	MixSi[4]	Si[4]/g	Ti[4]	Fe[4]/g	Others[5]/g	All non-ASB/g
1	PL	ND[3]	9,900	1,500	11,400	630	420	ND	ND	630	210	1,890
	LU	853.000	2,000	190,000	192,000	2,000	8,100	2,000	ND	ND	2,000	14,100
	NO	68.600	ND	1,700,000	1,700,000	33,000	ND	ND	16,000	ND	16,000	65,000
2	PL	ND	5,500	1,300	6,800	1,1001320	160	ND	160	ND	1,740	
	LU	7.080	3,800	4,600	8,400	150	ND	ND	ND	310	ND	460
	NO	3.420	15,000	62,000	77,000	51,600	6,700	3,300	6,700	1,700	ND	70,000
3	PL	ND	16,000	2,900	18,900	1,300	830	420	ND	ND	420	2,970
	LU	1.300	500	330	830	ND	170	ND	ND	80	ND	250
	NO	11.100	ND	2,900	2,900	223,300	ND	ND	17,400	ND	ND	240,700
4	PL	ND	17,000	1,600	18,600	390	ND	780	ND	1,200	ND	2,370
	LU	0.269	140	340	480	200	ND	ND	ND	200	410	810
	NO	17.200	13,000	9,000	22,000	111,300	5,400	5,400	32,300	1,800	3,600	159,800
5	PL	ND	4,000	1,400	5,400	400	400	200	ND	200	200	1,400
	LU	3.030	5,500	720	6,220	580	430	ND	290	1,000	580	2,880
	NO	4.370	12,000	5,200	17,200	19,500	5,100	3,100	6,200	2,100	1,000	37,000
6	PL	ND	5,200	630	5,830	2,100	470	160	320	790	ND	3,840
	LU	9.340	2,400	90	3,330	170	80	ND	ND	340	ND	590
	NO	ND	18,000	7,700	25,700	9,900	4,400	ND	51,600	ND	2,200	68,100
7	PL	ND	3,900	160	4,060	160	320	ND	ND	160	320	960
	LU	22.300	2,800	1,100	3,900	170	ND	340	80	510	ND	1,100
	NO	10.200	6,600	33,000	39,600	62,100	8,900	ND	48,800	ND	4,400	124,200
8	PL	ND	21,000	ND	21,000	3,800	420	ND	840	840	840	6,740
	LU	1.120	ND	2,400	2,400	ND	ND	ND	ND	ND	1,600	1,600
	NO	0.610	5,500	4,100	9,600	2,800	ND	ND	1,400	ND	ND	4,200

Footnotes to TABLE 2.

1. Data x 1000. All concentrations are /g dry weight
2. PL = Plaque; LU = Lung; NO = Node
3. Asbestos fiber abbreviations: FB = Ferruginous Bodies; CHRY = Chrysotile; AMPH = Amphibole; ASB = Total Asbestos; ND = None detected. Data reprinted with permission (Dodson *et al.*, 1990)
4. Non-asbestos abbreviations: AlSi = Aluminium Silicate; MixSi = Silicates with multiple cations; Si = Silica; Ti = Titanium oxides; Fe = Iron oxides
5. Others = Non-asbestos fibers containing Al, AlFe, Ca, MgSi, NaSi, CaS, CaSi, and CaSn.

transparent) using acetone vapour and then scanned by light microscopy at 200x using the Zeiss Photomicroscope III.

Fibre analysis was done by transmission electron microscopy. A strip was cut from the filter, carbon coated, mounted on 100/200 mesh folding grids and the filter matrix dissolved by the use of chloroform, producing a carbon extraction replica containing the entrapped fibres and other particulates. Evaluations were made at 16,000x to 100,000x in a JEOL 100CX Temscan equipped with a Tracor Northern TN)2000 energy dispersive X-ray analyzer. All fibres were counted which conformed to a 3:1 aspect ratio and had parallel sides for the majority of their length. The fibres were evaluated for elemental composition as well as by selected area diffraction and both length and width measurements were obtained. The latter two measurements were used to differentiate the "longer" versus "shorter" populations within the sites.

RESULTS

Uncoated asbestos fibres, in subjects in this study, have been shown to relocate from the lung to extrapulmonary sites (lymph node and pleural plaque) (Dodson *et al.*, 1990). Likewise, uncoated non-asbestos fibres were found not only in the lung tissue from these asbestos exposed individuals but also in each of the extrapulmonary samples. These non-asbestos fibres were characterised by elemental composition as shown on Table 2.

The distribution of each type of fibre among the three sites (lung parenchyma, node, and pleural plaque) was determined by calculating the percentage of positive samples for the eight subjects, (Table 3). All the sites were positive for uncoated asbestos fibres while the percent of positive sites per each type of non-asbestos fibre varied. The most common type of non-asbestos fibre was aluminium silicate and this was found in 91.7% of all sites sampled whereas silica fibres were observed in only 41.7% of the sampled areas. The uncoated asbestos fibres in the lung exceeded the number of non-asbestos fibres in seven of the eight patients (Table 2) but this was true for only three of the eight lymph node sites. However, the non-asbestos fibre burden was greater in the nodes than from the lung tissue in all patients.

The typical non-asbestos fibre was quite short when compared with those of chrysotile or amphiboles (Table 4). This was further emphasised by the rarity of fibres longer than 5 μm. Since width is a major determinant of detectability by light microscopy, the uncoated non-asbestos fibres were compared with the asbestos fibres using a counting scheme which included only those at or above the light microscopy limit of detectability (Table 5). As is obvious, most of these non-asbestos fibres would not have

TABLE 3. Percentage of sites positive for each uncoated fiber type

Fiber Type	Plaque %	Lung %	Node %	All Sites %
Amphiboles:	87.5	100.0	100.0	95.8
Chrysotile:	100.0	87.5	75.0	87.5
Aluminum Silicate:	100.0	75.0	100.0	91.7
Mixed Silicates:	87.5	50.0	62.5	66.7
Silica:	62.5	25.0	37.5	41.7
Titanium Oxides:	25.0	25.0	100.0	50.0
Iron Oxides:	87.5	75.0	37.5	66.7
All Other:	62.5	50.0	62.5	58.3

TABLE 4. Uncoated Chrysotile Fibers[***]

	Av. Length[*]	Av. Width	% > 5 μm[**]	% > 10 μm[**]
Plaque:	1.34 ± 1.50	.09 ± .15	3.1	0.0
Lung:	2.87 ± 1.07	.07 ± .06	14.0	4.0
Node:	1.27 ± 1.07	.08 ± .06	0.0	0.0

Uncoated Amphibole Fibers[***]

Plaque:	1.98 ± 3.13	.15 ± .07	10.0	8.0
Lung:	5.82 ± 6.71	.19 ± .21	41.0	20.0
Node:	2.22 ± 3.36	.21 ± .12	6.0	2.5

Uncoated Non-asbestos Fibers

Plaque:	1.01 + 1.32	.22 ± .54	1.0	1.0
Lung:	1.35 + 1.43	.19 ± .10	1.6	0.0
Node:	1.11 + 1.06	.19 ± .12	1.1	0.2

[*] All dimensions in μm

[**] Length

[***] Data reprinted with permission (Dodson et al., 1990)

TABLE 5. Percentage total uncoated asbestos and non-asbestos fibers counted by light microscopy

(Fibers longer than 5 micrometers)

Resolution:	0.2 μm	0.25 μm
Amphiboles:[*]	10%	5%
Chrysotile:[*]	0%	0%
Non-asbestos:	1.4%	1.2%

[*] Data reprinted by permission (Dodson *et al.*, in press)

TABLE 6. Percentage uncoated asbestos and non-asbestos fibers conforming to the Stanton Hypothesis

(Length eight micrometers or greater)

Location	Chrysotile[*]	Amphiboles[*]	Non-asbestos
Plaque:	0.3%	10%	1.0%
Lung:	6%	23%	0.8%
Node:	0%	3%	0.5%

[*] Data reprinted with permission (Dodson *et al.*, in press)

been included in such a count by light microscopy. In fact, the ability of an analyst to detect such non-asbestos fibres (<0.2 μm diameter and longer than 5 μm) by optical microscopy would be only slightly better than for chrysotile and not nearly as efficient for the amphiboles. The latter point must be appreciated with the recognition that only 10% or less of the amphiboles would be seen under the conditions used on Table 5 as compared with 1.4% non-asbestos and 0% chrysotile asbestos.

In assessing the non-asbestos burden by another parameter, those conforming to the "Stanton hypothesis" are show on Table 6. The percentage of such fibres is less than those reported for amphiboles in all sites and approximates the percentage for chrysotile in the plaques and node. More chrysotile fibres in the lung tissue conform to these dimensions than non-asbestos fibres. Even with the abundance of non-asbestos fibres in the lung, none were found as cores of ferruginous bodies. A reasonable explanation of this is that the selection process for ferruginous body formation involves fibres 8 μm and longer. Few non-asbestos fibres were sufficiently long to meet these physical requirements for stimulating the tissue reaction necessary for the formation of ferruginous coatings.

DISCUSSION

Ferruginous bodies have been reported in extra pulmonary sites including stomach, liver, kidney, lymph nodes, bile duct tumour, and pleural plaques (Churg & Warnock 1981). The presence of asbestos fibres in lung parenchyma, lymph node and pleural plaques of the subjects in our study (Dodson et al., 1990) is consistent with this concept for translocation.

Animal experimentation has confirmed that a number of fibre types are potentially carcinogenic (Pott et al., 1987). While other factors may be important contributors to this tumor inducing potential, those defined by the "Stanton hypothesis" are recognised as having a set of common features. These include the physical attributes of being long, thin, and durable.

An appreciable base of information exists on the tissue burden of asbestos fibres in humans but little comparable information exists for the non-asbestos fibres in lung tissue (Dodson et al., 1985; Churg, 1983), even though many such potentially respirable dusts exist. Even less information exists relating the extrapulmonary relocation of such inhaled fibres to their physical dimensions. Non-asbestos fibres were found to be deposited in the lung in appreciable numbers and to be relocated into the same extrapulmonary sites (lymph node and pleural plaques) as were the asbestos fibres. The non-asbestos fibres in all three sites (parenchyma, nodes, and pleural plaques) were typically short fibres which were more dimensionally irregular and more like chrysotile than amphiboles. The overwhelming majority of the fibres were therefore below the 8 μm length. Thus, most were below the length that Stanton attributed to being more potentially carcinogenic fibres. Some scientists have proposed that only longer asbestos fibres pose a risk to man. However, it is rather difficult to accept such a rigid position since exposure to long fibres often includes exposure to higher levels of shorter fibres. Recent studies by Churg (1983) have challenged the concept of attributing low risk to short fibre exposure.

The issue of length, as related to fibre toxicity, is only one of several factors which must be considered as contributors to the development of disease. These include surface charges (Jolicoeur & Poisson, 1987) and chemical changes (Hahon et al., 1986). Some investigations suggest the latter process may result in the eventual dissolution of chrysotile in the tissue (Becklake, 1982). These factors as well as radical formation and other, yet undefined, mechanisms constitute the total toxicity which results in asbestos induced diseases and must be of concern when exposure occurs to non-asbestos fibres.

The present study confirms non-asbestos fibres are found in both lung and extrapulmonary sites. The physical characteristics of the fibres are more like those associated with the chrysotile burden rather than the amphiboles in tissue. However, "longer" fibres of the non-asbestos burden do pose a concern as potential carcinogens and the risk of the total non-asbestos burden for contributing to the development of disease in man remains an open question.

REFERENCES

Becklake,M.R. (1976) Asbestos-related diseases of the lung and other organs: their epidemiology and implications for clinical practice. *Am. Rev. Respir. Dis.* **114**:187-227.

Becklake,M.R. (1982) Asbestos-related diseases of the lungs and pleura. *Am. Rev. Respir. Dis.* **126**:187-194.

Churg,A. (1983) Non-asbestos pulmonary mineral fibers in the general population. *Environ. Res.* **31**:189-200.

Churg,A., Warnock,M.L. (1980) Asbestos fibers in the general population. *Am. Rev. Respir. Dis.* **122**:669-678.

Churg,A., Warnock,M.L. (1981) Asbestos and other ferruginous bodies. *Am. J. Pathol.* **102**:447-456.

Dodson,R.F., Greenberg,S.D., Williams,M.G., Corn,C.J., O'Sullivan,M,F. and Hurst,G.A. (1984) Asbestos content in lungs of occupationally and nonoccupationally exposed individuals. *J. A. M. A.* **252**:68-71.

Dodson,R.F., Williams,M.G., O'Sullivan,M.F., Corn,C.J., Greenberg,S.D. and Hurst,G.A. (1985) A comparison of the ferruginous body and uncoated fiber content in the lungs of former asbestos workers. *Am. Rev. Respir. Dis.* **132**:143-147.

Dodson,R.F., Williams,M.G., Corn,C.J., Brollo,A. and Bianchi,C. (1990) Asbestos content of lung tissue, lymph nodes and pleural plaques from former shipyard workers. *Am. Rev. Respir. Dis.* (in press).

Doll,R. (1987) Symposium on MMMF, Copenhagen, October 1986: Overview and conclusions. *Ann. Occup. Hyg.* **31**:805-819.

Dunnigan,J. (1989) Comparing biological effects of mineral fibres. *Br. J. Indust. Med.* **46**:681-682.

Hahon,N., Vallyathan,V., Booth,J.A., Sepulveda,M.J. (1986) *In vitro* biologic responses to native and surface-modified asbestos. *Environ. Res.* **39**:345-355.

Jolicoeur,C. and Poisson,D. (1987) Surface physico-chemical studies of chrysotile asbestos and related minerals. *Drug Chem. Toxicol.* **10**:1-47.

Pott,F., Ziem,U., Reiffer,F.J., Huth,F., Ernst,H. and Mohr,U. (1987) Carcinogenicity studies on fibres, metal compounds, and some other dusts in rats. *Exp. Pathol.* **32**:129-152.

Stanton,M.F. and Wrench,C. (1972) Mechanisms of mesothelioma induction with asbestos and fibrous glass. *J. Natl. Cancer Inst.* **48**:797-821.

Wagner,J.C., Berry,G., Timbrell,V. (1973) Mesothelioma in rats after inoculation with asbestos and other materials. *Br. J. Cancer* **28**:173-185.

Whitwell,F., Scott,J. and Grimshaw,M. (1977) Relationship between occupations and asbestos-fibre content of the lungs in patients with pleural mesothelioma, lung cancer, and other diseases. *Thorax* **32**:377-386.

Williams,M.G., Dodson,R.F., Corn,C. and Hurst,G.A. (1982) A procedure for the isolation of amosite asbestos and ferruginous bodies from lung tissue and sputum. *J. Toxicol. Environ. Hlth* **10**:627-638.

INFLUENCE OF LONG-LASTING ASBESTOS EXPOSURE ON IMMUNOLOGICAL STATUS OF ASBESTOS EXPOSED SHIPYARD WORKERS

I.Trosic, M.Saric, Z.Pisl and L.Stilinovic

University of Zagreb
Croatia
Yugoslavia

INTRODUCTION

Investigation into the fibrogenic and possibly carcinogenic effects of asbestos fibres have been centred largely on the structural and aerodynamic properties of the fibres (Timbrell, 1973). However, other factors, particularly the body's defences and altered immune competence, have also been linked with asbestos exposure. Epidemiological and immunological studies of asbestos workers have documented abnormalities in both humoral and cell-mediated immunity (Doll *et al.*, 1983). Immune response to asbestos has been suggested as an important component in the pathophysiology of asbestosis (Pernis & Vigliani, 1982), however, the effect of asbestos inhalation on local or systemic immunity is not well understood. The majority of workers developing asbestosis have been exposed to high dust concentrations for a prolonged periods of time. There are, however, many workers who remain healthy in spite of the significant exposure to asbestos (Turner-Warwick, 1973). The aim of this study was to compare the immunological alterations that occur in an asbestos-exposed diseased group and in an asbestos-exposed non-diseased group.

SUBJECTS AND METHODS

This retrospective study summarises immunological findings of 560 male shipyard workers with a mean age of 46.0 years who were occupationally exposed to asbestos dust on average for 22.5 years (Table 1). Standard laboratory methods for the identification of T-lymphocytes (Holm *et al.*, 1975) and the level of serum immunoglobulins (Radialimmunodifusion, Imunoloski zavod, Zagreb) were used. The total group was divided into two subgroups. Group GI consisted of 170 asbestos-exposed non-diseased subjects, and

TABLE 1. Groups of asbestos-exposed shipyard workers

Asbestos exposed workers	N	age $\bar{x}$	age min.	age max.	exposure $\bar{x}$	exposure min.	exposure max.
Total	560	46.3	22	65	22.5	1	40
GI non-diseased workers	170	45.1	27	62	20.6	1	36
GII diseased workers	390	46.8	22	65	23.6	2	40

group G II consisted of 390 asbestos-exposed diseased workers. The statistical significance was assessed by Mann-Whitney-Wilcoxon test (Willemsen, 1974). The level of significance was at $p < 0.05$.

RESULTS AND CONCLUSION

In the asbestos exposed group asbestosis was verified in 70% of subjects and not found in 30% of workers. Under the same of similar conditions of exposure only some of the workers had become ill. A significant reduction in the percentage and absolute numbers of T-lymphocytes was observed in the whole group, particularly in the asbestos-exposed diseased group (figure 1).

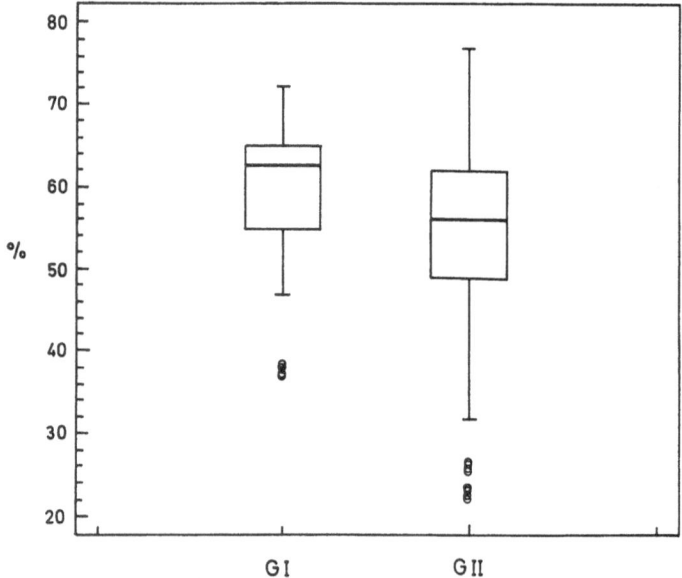

Figure 1. Interquartile range and median of T-lymphocyte percentage in the peripheral blood of the examined groups.

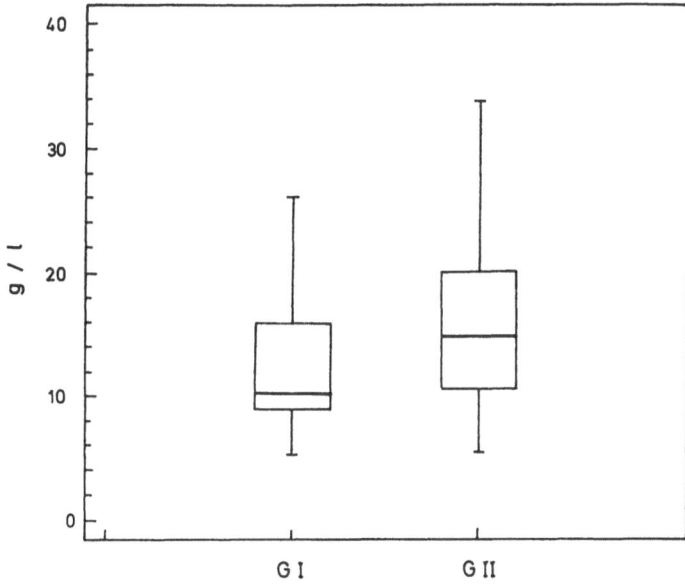

Figure 2. Interquartile range and median of serum I_gM levels in the examined groups

In 80% of the subjects a lowered skin reaction to PPD antigen was observed. A significant increase in the level of I_gM, I_gA and I_gE was seen in diseased subjects in relation to healthy asbestos exposed workers (figures 2,3,4).

Considering the results of this study, we conclude that in the pathophysiological development of asbestosis, an important role is played by the immunological status of the

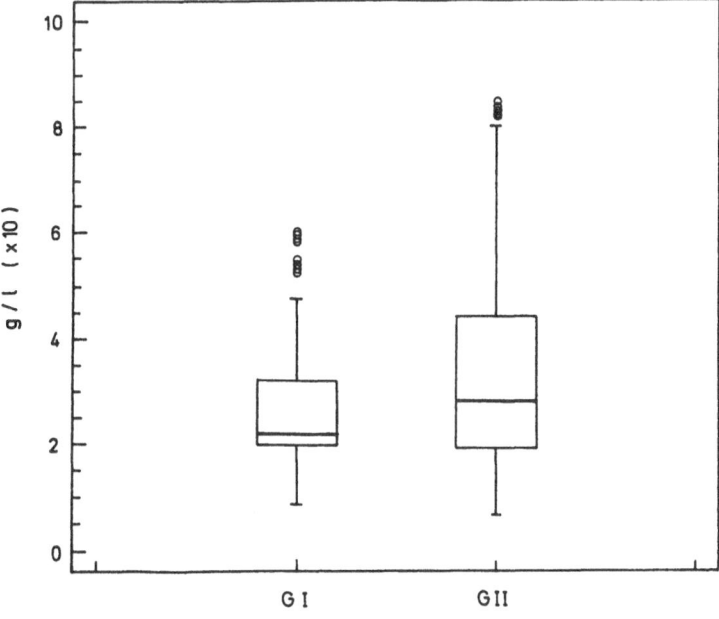

Figure 3. Interquartile range and median of serum I_gA levels in the examined group

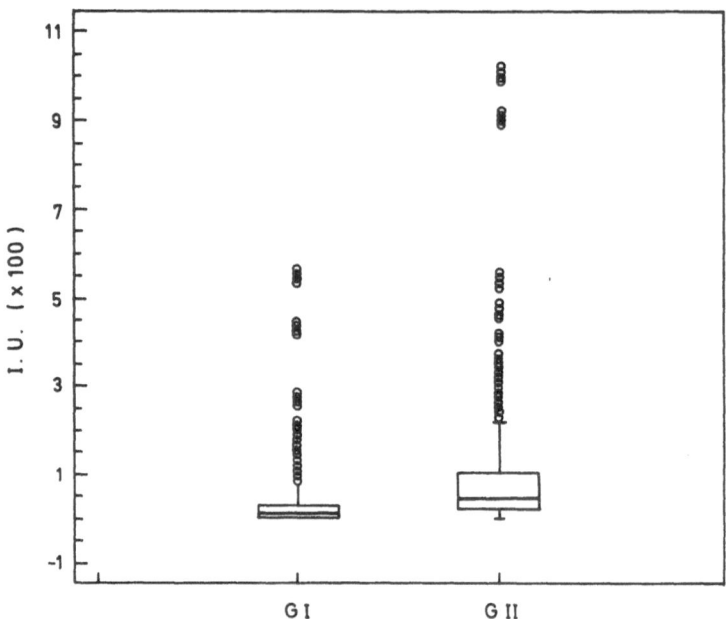

Figure 4. Interquartile range and median of serum I_gE levels in the examined groups

subjects. The results also indicate that regular examination of the immunological status of asbestos exposed workers could be of prognostic and diagnostic importance in the development of asbestosis. In the group of asbestos exposed workers a quantitative and qualitative reduction of T-lymphocytes was found, particularly in the group of workers in whom the disease was diagnosed. The immunological system with a tendency to an increased level of serum immunoglobulins appears to be involved in the development of asbestosis. The local effect of the respirable fraction of asbestos dust can thus result in a change in immunological parameters of peripheral circulation and systemic immunological reaction.

REFERENCES

Doll,N.J., Stankus,R.P. and Barkman,H.W. (1983) Immunopathogenesis of Asbestosis, Silicosis, and Coal Workers' Pneumoconiosis. *Clinics in Chest Medicine* **4**:3-14.

Holm,G., Pettersson,D., Mellstedt,H., Hedfors,E. and Bloth,B. (1975) Lymphocyte subpopulations in peripheral blood of healthy persons. Characterization by surface markers and lack of selection during purification. *Clin. Exp. Immunol.* **20**:443-457.

Pernis,B., Vigliani,E.C. (1982) The Role of Macrophages and Immunocytes in the Pathogenesis of Pulmonary Diseases Due to Mineral Dusts. *Am. J. Indust. Med.* **3**:133-137.

Timbrell,V. (1973) Physical factors as etiological mechanisms. Biological Effects of Asbestos - Proceedings of a Working Conference, held at the International Agency for Research on Cancer, Lyon, France, 2-6 Oct. 1972.

Turner-Warwick,M. (1973) Immunology and Asbestosis. *Proc. Roy. Soc. Med.* **66**:927-930.

Willemsen,E.W. (1974) Understanding Statistical Reasoning. Atkinson,R.C., Freedman,J. and Thompson.G.F., eds. Library of Congress Cataloging in Publication Data. San Francisco, (1974) pp. 168-196.

THE CARCINOGENICITY OF MINERAL FIBRES

2. Animal studies

MESOTHELIOMAS IN MAN AND EXPERIMENTAL ANIMALS

J.C. Wagner

59 Combe Valley Road
Weymouth, Dorset DT3 6N4
UK

In considering the use of experimental animals in the assessment of risk to man, it is paramount that the interest remains with the possible effects on man and that precise interpretation of the fate of the animal must be considered in this light. Therefore it is essential to use the epidemiological evidence that has been obtained from human exposure in known situations to calibrate the results obtained experimentally. When human evidence is lacking we are fortunate in having an end-point against which other fibers can be assessed. This material is a form of zeolite known as erionite. Where this material occurs in the form of fine fibers, that is, less than 0.25 microns in diameter and greater than 5.0 microns in length, it causes an extremely high rate of mesotheliomas in man, as reported by Baris *et al.*, (1978) and by Simonato and his colleagues (1989). In experimental studies with this fiber in rats and mice exposure intrapleurally results in 100% mesotheliomas (Wagner *et al.*, 1985; Maltoni *et al.*, 1982) and 100% by inhalation (Wagner *et al.*, 1985). When these results with erionite in man and experimental animals are compared with the other mineral fibers, the effects of the latter are far less severe.

ASBESTOS

The various forms of asbestos should be considered separately, especially the differing effects of exposure to the amphiboles, compared with chrysotile.

The results in man can be summarised as follows:-

(a)   **Asbestosis**
All forms of asbestos will produce asbestosis in man providing the exposure is sufficiently excessive.

(b)   Carcinoma of the lung

Carcinoma of the lung occurs in people who have developed asbestosis and are cigarette smokers. There is a low rate of carcinoma of the lung in workers with severe asbestosis who are non-smokers.

(c)   Mesotheliomas

Mesotheliomas occur in a small number of people exposed to excessive amounts of amphibole asbestos dust. In this situation the amphibole fibers, providing the size range is in the "final" Stanton criteria of fiber size, that is less than 0.25 microns in diameter and greater than 8 microns in length, will produce a few mesotheliomas of the pleural and possibly the peritoneal cavity. The low rate is demonstrated by considering the large populations at risk. For example, crocidolite is shown to be the most dangerous of the asbestos fibers, yet *in vitro* testing it is a low rated carcinogen, which is just as well if one considers the extremely large populations at risk in South Africa, Western Australia and Europe that have been exposed to this type of fiber. This can be appreciated if the papers by Sluis-Cremer (1991), Hobbs (1983), Anspach (1962) and Newhouse (1981) and their colleagues are studied.

Tremolite and amosite appear to be responsible for a much smaller number of mesotheliomas in man, and only when the dosage has been excessive.

In spite of the fact that chrysotile accounts for 95% of the total asbestos production, there is absolutely minimal evidence that chrysotile itself is the cause of malignant mesotheliomas in man. A few cases have been traced to "commercial chrysotile", but in these people the amphibole fiber tremolite has been found as the major fiber in the lung residue (McDonald & McDonald, 1980; Churg & Wright, 1989).

MINERAL FIBERS OTHER THAN ASBESTOS

It is here that our major interest lies, as we are looking for ways of predicting the possible hazard associated with exposure to these fibers, which can be grouped as follows:-

(a)   Natural mineral fibers

Those of possible economic value are few.
(i)   Wollastonite. There is no evidence of human disease.
(ii)   The absorbent clays. These are Palygorskite, Attapulgite and Sepiolite. No evidence of human disease.

(b)   Synthetic fibers

These are:
(i)   The man-made mineral fibers (vitreous) glass, slag wool, and rock wool. Glass is produced in three forms, filament, wool and microfibers.
(ii)   Refractory fibers which include those with ceramic, aluminium, or zirconium base.

There is little evidence that any of the other fibers have an effect on man. There is some evidence that, in the 1930's, exposure to a certain form of rock-wool was associated with subsequent development of carcinoma of the lung, but this occurred on a small scale and the findings are far from conclusive.

CONCLUSIONS FROM HUMAN EVIDENCE

It is on these asbestos substitutes that further evidence is required, in spite of the fact that the majority have not been shown to be dangerous, and there has been a reasonable epidemiological follow-up.

It is unlikely that fibrosis and subsequent carcinoma of the lung are going to be a danger in the future; providing the principals of good house-keeping are applied, and exposure to straight fibers is less than five fibers per cc. It is most important to discourage smoking.

This leaves us with the problem of diffuse malignant mesotheliomas. There is no evidence that any of the above mentioned fibers have caused these tumours in man. However, it can still be argued that in some cases the exposure has not been long enough, as the follow-up should be at least fifty years. Novel fibers will continue to be developed which will require testing.

TESTING FIBERS FOR POSSIBLE ASSOCIATION WITH MESOTHELIOMAS

(a)   Physical properties
If the fiber is greater than 0.5 microns in diameter and less than five microns in length, then such fibers are not going to produce mesotheliomas.

(b)   Experimental Evidence
(i)   Natural incidence of mesotheliomas. It must be remembered that there is a natural incidence of mesotheliomas of the pleura and the peritoneum in both man and experimental animals (Ilgren & Wagner, 1990).

(ii)   Inhalation Method. All the authorities have agreed when assessing the biological effects of dusts in experiments that the inhalation method is the only satisfactory method of study: Jotten & Klosterkotter from Germany, Vigliani from Italy, Policard from France, Haldane and Wright from Britain and Mavrogardata from South Africa. These experiments are complex to set up, extremely expensive to run, and long term, with no significant results for three years. The results, when available, will, however, provide evidence of the fibrogenicity and carcinogenicity of the fiber. This latter would include both carcinomas and possibly mesotheliomas. The low rate of mesotheliomata produced in the inhalation studies correlate with the human experience, providing the known background incidence is taken into consideration.

(iii)Intra-tracheal Method. This method was developed as a possible cheaper method of exposure than inhalation and a means of obtaining more rapid results. This is not a satisfactory method for exposure, particularly when studying fibrous dusts. The distribution in the lung varies greatly with the fiber blocking the air passages and causing obstructive non-specific disease.

(iv)   Implantation Method. When I first developed the intra-pleural method of exposing animals to asbestos fibers my hypothesis was that: **"Asbestos fibers, if present in the pleural cavity, will cause mesotheliomata"**.

It was never considered that this should be used as a method for calibrating the carcinogenic effect, for as a method it is unnatural, for it bypasses all the body's defence mechanisms. At all possible doses, there is an intense localised exposure which is far greater than would even happen in man.

The results from intraperitoneal studies have indicated that the peritoneal cavity of experimental rats is far more active and produces such a high rate of mesotheliomas, that the results become non-specific, and even particulate non-fibrous materials can produce mesotheliomas.

CONCLUSIONS

In my opinion the mineral fibers which should cause concern are those that (1) are retained in the lung parenchyma; (2) are durable; (3) are acicular in shape.

I subscribe to the view that (1) fibers over 3.0 microns in diameter are aerodynamically unacceptable in the acini; (2) fibers less than 5.0 microns in length are probably innocuous; (3) fibers of less than 0.25 microns in diameter are most clearly associated with the development of mesotheliomas. All these are factors which can be tested in the laboratory, without resort to biological assay.

When experiments with animals are necessary, the only satisfactory way to test new fibers is by inhalation, which is expensive and time-consuming. However, the other treatments may be of value when used in conjunction with *in vitro* techniques, to elucidate the basic mechanisms of the biological effects of exposure to mineral fibers.

REFERENCES

Anspach,M. (1962) Sind PleuraverkalKungen pathognonisch fur eine Asbestose? *Int. Arch: GewerbePath* 20:396.

Baris,Y., Sahin, A.,Ozesmi,M., Kerse,I., Ozen,E., Kolacan,B., Altinors,M. and Goktepeli,A. (1978) An outbreak of pleural mesotheliomas and chronic fibrosing pleurisy in the village of Karain/Urgup in Anatolia. *Thorax* 33:181-192.

Churg,A. and Wright,J.L. (1989) Fibre content of lung in amphibole and chrysotile - induced mesotheliomas; implications for environmental exposure. In: "Non-occupational exposure to mineral fibres", Bignon,J., Peto.J and Saracci,R., eds. IARC Publication No. 90, Lyon, pp. 314-318.

Hobbs,M.S.T., Woodward,S.D., Murphy,B., Musk.A.W. and Elder,J.E. (1983) The incidence of pneumoconiosis, mesothelioma and other respiratory cancer in men engaged in mining and milling crocidolite in Western Australia. In: "Biological effects of mineral fibres", Vol **2**. Wagner,J.C., ed. IARC Publication No. 30, Lyon, pp. 615-625.

Ilgren,E.B. and Wagner,J.C. (1991) Background incidence of mesotheliomas. Animal and human evidence. *J. Reg. Toxicol. Pharm.* **13**:133-149.

Maltoni,C., Minardi,F. and Morisi,L. (1982) Pleural mesotheliomata in Sprague-Dawley rats; first experimental evidence. *Environ. Res.* **29**:238-244.

McDonald,A. and McDonald,J.C. (1980) Malignant mesothelioma in North America. *Cancer* **46**:1650-1656.

Newhouse,M.L. (1981) Epidemiology of Asbestos-related tumours. *Sem. Oncol.* **8**:250-257.

Simonato,L., Baris,Y., Saracci,R., Skidmore,J. and Winkelmann,R. (1989) Relation of environmental exposure to erionite fibres to risk of respiratory cancer. In: "Non-occupational exposure to mineral fibres", Bignon,J., Peto,J. and Saracci,R. IARC Publication No. 90, Lyon, pp. 398-405.

Sluis-Cremer,G.K. (1991) The mortality experience of amphibole miners in South Africa. (Submitted for publication).

Stanton,M.F., Layard,M., Tegeris,A., Miller,E., May,M. and Kent,E.J. (1981) Relation of particle dimension to carcinogenicity in amphibole asbestos and other fibrous minerals. *J. Nat. Cancer Inst.* **67**:965-975.

Wagner,J.C., Skidmore,J.W., Hill,R.J. and Griffiths,D.M. (1985) Erionite exposure and mesotheliomas in rats. *Br. J. Cancer* **51**:727-750.

EXPERIMENTAL STUDIES ON MINERAL FIBRE CARCINOGENESIS: AN OVERVIEW

J.M.G. Davis

Institute of Occupational Medicine
Roxburgh Place
Edinburgh, Scotland

The major properties of mineral fibres that relate to carcinogenesis are the length and diameter of the fibres, a fact known for many years, and their chemical composition that enables them to remain in tissues for long periods without dissolving. This has been appreciated more recently. In stating that carcinogenicity depends on length, diameter and durability, however, it is important to examine the evidence on which these suggestions are based and consider if other factors may also be involved.

Considerable evidence has been accumulated from experimental studies that the pathogenicity of fibres varies with their length, the long fibres being the most dangerous. The first work on fibre dimension was included in the original programme of Gardner with unexpected results. Because it was known that quartz-related pathology increased with a reduction in particle size, Gardner's group included finely ground (short fibre) asbestos in their experiments. Not only did the short fibres fail to increase pulmonary fibrosis compared to long fibre material but reducing fibre length significantly reduced the level of fibrosis in the treated animals (Vorwald et al., 1951). Further substantiation of these findings are found in the work of King and co-workers (1946), when asbestos cut into different lengths on a special microtome was administered to rabbits by intratracheal injection. Far more pulmonary fibrosis was produced by long fibres (length approximately 15 μm) than short (length approximately 2.5 μm). Similar findings were subsequently reported by other groups of workers (Klosterkotter, 1968; Scymczykiewicz & Wiecek, 1960). That pulmonary fibrosis, resulting from glass fibre administration, also depended on fibre length was demonstrated by Wright and Kuschner (1977) who used intratracheal injection into rats of specially-sized glass fibres.

The importance of fibre length in relation to neoplasia was demonstrated in two laboratories by Stanton and co-workers (1972, 1977, 1978) and Pott and co-workers (1972, 1978). These workers demonstrated that following the intrapleural or intraperitoneal implantation of asbestos and other mineral fibres the development of mesotheliomas was most closely related to the number of fibres in any dust preparation that were >8 μm in length and <0.25 μm in diameter. The type of mineral appeared unimportant in these studies; fibre glass was almost as carcinogenic as asbestos and no minerals containing long thin fibres produced negative results. Some studies have, however, suggested that short fibre asbestos may be carcinogenic when injected (Kolev, 1982; Le Bouffant, 1985). These results are difficult to interpret since short fibres are usually only reported as a proportion of the total. With the ability of asbestos to split longitudinally as well as transversely it is quite possible to reduce the mean length of a sample greatly while actually increasing the number of long fibres/unit mass. Wagner and co-workers (1984) demonstrated that, following the injection of finely ground short crocidolite, the small proportion of long fibres present were selectively retained in the pleural granulomas produced by this dust.

Since injection techniques exaggerate tissue response to fibres by overcoming or bypassing clearance mechanisms it is important to determine whether or not fibrosis and neoplasia are related to fibre length when dust is inhaled. The difficulties in preparing suitable dusts in large quantities have inhibited this work since, as indicated, simple grinding of long fibre samples does not necessarily reduce the number of long fibres present. So far very few studies with satisfactory dusts have been possible. In 1986, Davis and co-workers reported long-term inhalation and injection studies in rats in which a sample of amosite with almost all fibres <5 μm in length was compared to a normal amosite dust preparation containing many long fibres. The short fibre material produced no fibrosis and only a single mesothelioma at the highest dose injected, while the long fibres were highly pathogenic. An attempt was made to duplicate this study using chrysotile (Davis *et al.*, 1988) but the intended short fibre preparation did contain a small proportion of long fibres and some fibrosis and pulmonary tumours were produced in rats although the response was much less than with long fibred chrysotile. When injected, short chrysotile produced fewer tumours at all doses than did the long. Perhaps the most dramatic demonstration of the importance of fibre length was reported by Wagner (1985, 1988). The mesothelioma incidence of almost 100% following the inhalation of erionite in rats was reduced to zero when short fibre preparations of the same material were used.

Combined data from two experimental inhalation studies published by Davis and co-workers (1978, 1986) may indicate that different fibre lengths are needed to produce disease in the pulmonary parenchyma than to cause mesothelioma. Of the amosite samples tested, both a very short sample and the medium length UICC Reference Sample produced minimal pulmonary fibrosis and almost no pulmonary tumours. In contrast a long fibre sample produced large amounts of fibrosis and significant numbers of pulmonary tumours. When the three dusts were injected at the same dose, both the long sample and the UICC material produced mesotheliomas in almost all animals while the short fibre material produced none (Davis, 1989). The implications of these results are that very short fibres produce no disease at any site, medium length fibres (8-10 μm) can cause mesotheliomas but long fibres (perhaps 15-25 μm) are needed to produce disease in the pulmonary parenchyma.

While experimental studies have clearly demonstrated the importance of fibre length in the production of both pulmonary fibrosis, pulmonary tumours and mesotheliomas the exact hazard for fibres of any dimension compared to any other could only be determined with fibre preparations where all fibres were of uniform length and these have not so far been available. The routine counting procedures used for asbestos monitoring in the workplace count only fibres >5 μm in length as seen by phase contrast optical microscopy (PCOM). The five micron cut-off was based on very little evidence at the time it was introduced but seems to have been a fortunate choice. Fibres below this length are certainly much less harmful than longer ones and they may be innocuous. This latter possibility is of major practical importance since the vast majority of airborne fibres are short, particularly in environmental situations. Elucidation of this issue should form one of the major research fields of the future. The techniques for biological testing already exist but suitable dust samples do not although they could, no doubt, be produced if sufficient research effort were applied.

While fibre dimensions are obviously of major importance, the importance of durability was highlighted by apparent differences between the pathogenicity of different asbestos types and particularly in relation to their ability to produce tumours in man and experimental animals. There is good evidence from human epidemiological studies that certainly mesothelioma production and probably the production of pulmonary carcinomas as well is more closely related to the amphibole fibres, crocidolite and amosite, than to chrysotile. Initially animal experimentation appeared to contradict these findings with chrysotile preparations being reported as equally hazardous as the amphiboles when administered by inhalation and injection (Wagner et al., 1973,1974,1980; Davis et al., 1978,1986,1988; Bolton et al., 1982).

The elucidation of this problem was difficult since human studies indicating greater hazard from amphibole asbestos were based on industrial situations where dust figures were either almost meaningless or non-existent. Since amphiboles are known to be much dustier materials than chrysotile it was presumed that the airborne concentrations of amphibole fibres had been much higher. In contrast, the dose of fibre administered in animal experiments was carefully monitored although in most experimental studies dose was compared by mass of dust. Experiments by Davis and co-workers (1978) compared crocidolite, amosite and chrysotile at equal fibre numbers (length >5 μm by PCOM) and found that chrysotile was the most pathogenic dust. However, as discussed previously, the UICC amphibole preparations used in this study are now known to be far too short to produce maximal damage to the pulmonary parenchyma. The UICC chrysotile preparation used was longer so that the fibre number comparisons were not really valid. Elucidation of the problem began with reports that following dust inhalation in experimental animals at equal mass dose much more amphibole than chrysotile was present in the lung tissue at the end of inhalation. Initially it was suggested (Timbrell, 1973) that the curly chrysotile fibres penetrated less readily to the smallest airways but later studies indicated that following the end of dusting, chrysotile could be removed from rat lung tissue over four times faster than amphibole (Davis, 1989). A similar situation exists in humans since it has been reported that most asbestos fibres found in human lungs at autopsy are amphibole even where chrysotile exposure was known to have predominated (Pooley, 1976; Rowlands et al., 1982; Gylseth et al., 1983).

From this combined data it appears most likely that while exaggerated experimental exposures can maintain chrysotile levels high enough and for long enough in short-lived rats to produce maximum carcinogenicity, in longer-lived humans exposed at much lower levels the rapid clearance of chrysotile significantly reduces the effects of this mineral type.

There are differences of opinion as to whether the rapid clearance of chrysotile from lung tissue depends on chemical dissolution or some other factor which stimulates the mechanical removal of fibres from the lung tissue. Certainly magnesium is rapidly leached from chrysotile in physiological solutions. Chemical dissolution certainly appears to occur with many types of man-made mineral fibre (MMMF) (Forster, 1984; Klingholz and Steinkopf, 1984; Leineweber, 1984; Morgan and Holmes, 1986) and appears likely to explain the lack of carcinogenicity found with these materials in animal inhalation studies (Wagner *et al.*, 1984; McConnel *et al.*, 1984; Smith *et al.*, 1987; Le Bouffant *et al.*, 1987). Only two MMMFs have been shown to produce tumours when inhaled by experimental animal. These dusts were ceramic fibre (Davis *et al.*, 1984;) and aramid fibre (Lee *et al.*, 1988). Both these fibres have chemical compositions that would be expected to make them extremely durable in lung tissue.

The low durability of MMMFs has, I believe, resulted in one of the apparent anomalies found in experimental studies of fibre carcinogenicity. While the non-durable fibres appear innocuous by inhalation, all of them if they have fibres of appropriate dimensions produce mesotheliomas when injected directly into the pleural or peritoneal cavities (Stanton *et al.*, 1977). When fibres are injected they tend to become localised in large dust-packed granulomas and it seems likely that in these situations dissolution is much slower than when fibres are deposited separately in the alveoli after inhalation. For this reason it is my opinion that injection studies should be used only as a preliminary screening device for pulmonary carcinogenicity and that only positive inhalation studies should be accepted as being an indication of potential hazard to humans.

While the dimension of mineral fibres is obviously of great importance in carcinogenicity and they must remain in lung tissue for a significant period of time, evidence is accumulating that other factors can affect or possibly increase the risk of tumour production. It has been known for many years that a synergism exists between asbestos exposure and smoking but in this instance two known carcinogens are acting in concert. It has now been demonstrated in experimental studies that the presence of particulate dusts along with asbestos fibres in lung tissue can increase tumour production even though the particles are not carcinogenic on their own. This effect has been demonstrated both with the innocuous titanium dioxide and with quartz (a toxic dust) (Davis *et al.*, 1990). With pulmonary carcinomas, it is likely that fibres exert their carcinogenic effect close to the original site of their deposition in lung tissue. For mesothelioma production, however, transport to the pleura following deposition must occur and it is likely that factors other than fibre geometry may influence the speed of this transport. Evidence for this suggestion comes from a number of sources. The fibrous mineral erionite has been shown to possess a potential to cause mesotheliomas both in man and experimental animals far in excess of the main asbestos types, yet its fibre geometry and size distribution in dust clouds is similar to amphibole asbestos (Baris *et al.*, 1979; Wagner *et al.*, 1985). There is, however, no evidence that erionite has a greater potential to produce pulmonary carcinomas than would be expected from its fibre geometry and this leads to the speculation that there may be something about the surface properties of erionite fibres that stimulate or facilitate their transport to the pleura. Evidence that dust penetration of the pleura can be varied has been produced by Robertson and co-workers (1984) and Davis and co-workers (1990). These experimental inhalation studies used mixtures of coal and quartz and asbestos and quartz and demonstrated that granulomas developed on the rat pleural surface that contain significant amounts of dust particles or fibres. Neither coal nor asbestos penetrates the pleura in this way when administered alone.

The finding that fibre carcinogenicity depends largely on fibre dimensions and durability does not explain the exact mechanisms which enables long fibres to transform cells and, so far, information on this subject is limited. Many chemical carcinogens are genotoxic, that is to say they react directly with DNA to cause gene damage and in some instances at least carcinogenic mutations. The potential to produce these changes is routinely examined by using bacterial populations (Ames test) but asbestos has proved completely negative using these techniques (Chamberlain and Tarmy, 1977). This is not surprising since bacteria are not phagocytic and asbestos fibres could not be expected to make contact with bacterial DNA in any way. Mammalian cells are different but the major types that phagocytose mineral fibres, macrophages and to a much lesser extent neutrophils, are considered to be "end cells" with no carcinogenic potential. Electron microscope studies (Suzuki, 1974) have shown that asbestos fibres can be found in alveolar epithelial cells following dust inhalation but this is a rare occurrence. Following intrapleural injection asbestos is frequently found within mesothelial cells (Davis, 1974) but whether fibres reach these cells to be engulfed after inhalation has not been demonstrated. Certainly from electron microscope studies on cells containing fibres it appears that these fibres do not penetrate the nuclear membrane in interphase and thus contact between fibres and DNA would be limited to the period of mitosis. Alterations that could be produced at this time-point have been demonstrated in some studies with chromosomal aberrations reported in fibroblast cultures (Hesterberg & Barrett, 1985; Mikalsen *et al.*, 1988) but it remains to be demonstrated that these changes can lead to tumour production *in vivo*.

As an alternative to direct contact between asbestos fibres and the genetic material of the cell, some workers have suggested that enhanced secretion of biologically-active materials by asbestos containing cells may be the main mechanism of asbestos carcinogenesis. Certainly a number of studies have demonstrated the increased production of a range of growth factors from both human and animal alveolar macrophages (Mossman & Begin, 1989). As an alternative many workers (e.g. Hansen & Mossman, 1987) have also demonstrated the increased release of active oxygen species (AOS) from macrophages containing asbestos fibres and these materials are known to be extremely toxic to tissues. A consideration of all these factors leads to two separate theories of asbestos carcinogenesis. On the one hand it is possible that asbestos fibres, at least during mitosis, can produce inheritable abnormalities in the cells genetic material that results in neoplasia. On the other hand, asbestos may never react directly with DNA but cause the long-term secretion of factors that cause tissue damage or stimulate cell division. These secretory products would create a tissue environment where spontaneous mutations leading to neoplasia would occur with much greater frequency than in normal tissue.

Determination of which of these mechanisms results in asbestos carcinogenesis has very great potential importance. If asbestos causes genetic modification of cells then, theoretically, a single fibre could cause a tumour and there would be no safe threshold for asbestos exposure. If, however, secretion products are important then the maintenance of elevated levels of these over long periods should require a significant area of tissue reaction that could only be caused by a significant dose of fibres. In support of this latter theory it is reported that it is very rare to find pulmonary carcinomas in asbestos workers unless asbestosis is also present (Browne, 1982; Kipen *et al.*, 1987). Similarly, Kuschner (1987) has suggested that pleural fibrosis may be a necessary precursor to mesothelioma development. To back up the human data, Davis and Cowie (1990) reported that in animal inhalation studies with a variety of fibre types, high levels of pulmonary fibrosis were always found with a high incidence of pulmonary tumours and *vice versa*. On an individual basis, animals with pulmonary tumours had roughly twice as much pulmonary fibrosis as those without. The elucidation of the exact mechanism by which mineral fibres are able to cause neoplasia certainly remains the major field for future research.

REFERENCES

Baris,Y.I., Artvinli,M. and Sahin,A.A. (1979) Environmental mesothelioma in Turkey. *Ann. NY Acad. Sci.* **330**:423-432.

Bolton,R.E., Davis,J.M.G., Donaldson,K. and Wright,A. (1982) Variations in the carcinogenicity of mineral fibres. *Ann. Occup. Hyg.* **26**:569-582.

Browne,K. (1986) Is asbestos or asbestosis the cause of increased risk of lung cancer in asbestos workers? *Br.J.Indust.Med.* **43**:145-149.

Chamberlain,M. and Tarmy,E.M. (1977) Asbestos and glass fibres in bacterial mutation tests. *Mutat. Res.* **43**:159-164.

Davis,J.M.G. (1974) An electron microscope study of the response of mesothelial cells to the intrapleural injection of asbestos dust. *Br. J. Exp. Path.* **55**:64-70.

Davis,J.M.G., Beckett,S.T., Bolton,R.E., Collings,P. and Middleton,A.P. (1978) Mass and number of fibres in the pathogenesis of asbestos-related lung disease in rats. *Br. J. Cancer.* **37**:673-688.

Davis,J.M.G., Addison,J., Bolton,R.E., Donaldson,K., Jones,A.D. and Wright,A. (1984) The pathogenic effects of fibrous ceramic aluminium silicate glass administered to rats by inhalation or peritoneal injection. In, "Biological Effects of Man-made Mineral Fibres". Report of a WHO/IARC meeting. WHO, Copenhagen, pp. 303-323.

Davis,J.M.G., Addison,J., Bolton,R.E., Donaldson,K., Jones,A.D. and Smith,T. (1986) The pathogenicity of long versus short fibre samples of amosite asbestos administered to rats by inhalation and intraperitoneal injection. *Br. J. Exp. Path.* **67**:415-430.

Davis,J.M.G., and Jones,A.D. (1988) Comparisons of the pathogenicity of long and short fibres of chrysotile asbestos in rats. *Br. J. Exp. Path.* **69**:717-739.

Davis,J.M.G. (1989) Mineral fibre carcinogenesis: experimental data relating to the importance of fibre type, size, deposition, dissolution and migration. In, "Non-occupational exposure to mine fibres", Bignon,J., Peto,J. and Saracci,R., eds. IARC Publication No. 90. International Agency for Research on Cancer, Lyon. pp. 33-34.

Davis,J.M.G. and Cowie,H. (1990) The relationship between fibrosis and cancer in experimental animals exposed to asbestos and other fibres. *Environ. Hlth Perspect.* **88**:305-309.

Davis,J.M.G., Jones,A.D. and Miller,B.G. (1990) Experimental studies in rats on the effects of asbestos inhalation coupled with the inhalation of titanium dioxide or quartz, (in press).

Forster,H. (1984) The behaviour of mineral fibres in physiological solutions. In, "Biological Effects of Man-made Mineral Fibres." Report of a WHO/IARC meeting. WHO, Copenhagen, pp. 27-60.

Gylseth,B., Mowe,G. and Wannag,A. (1983) Fibre type and concentration in the lungs of workers in an asbestos cement factory. *Br. J. Indust. Med.* **40**:375-379.

Hansen,K. and Mossman,B.T. (1987) Generation of superoxide from alveolar macrophages exposed to asbestiform and non-fibrous particles. *Cancer Res.* **47**:1681-1686.

Hesterberg,T.W. and Barrett,J.C. (1985) Induction by asbestos fibres of anaphase abnormalities: mechanism for aneuploidy induction and possibly carcinogenesis. *Carcinogenesis* **6**:473-475.

King,E.J., Glegg,J.W. and Rae,V.M. (1946) Effect of asbestos and asbestos and aluminium on the lungs of rabbits. *Thorax* **1**:188-196.

Kipen,H.M., Lillis,R., Suzuki,Y., Valciukas,J.A. and Selikoff,J. (1987) Pulmonary fibrosis in asbestos insulation workers with lung cancer: a radiological and histopathological evaluation. *Br. J. Indust. Med.* **44**:96-100.

Klingholz,R. and Steinkopf,B. (1984) The reactions of MMMF in a physiological model fluid and in water. In, "Biological Effects of Man-made Mineral Fibres." Report of a WHO/IARC meeting. Copenhagen pp. 60-87.

Klosterkotter,W. (1968) *Experimentelle Untersuchunger uber die Bedeutung der Faserlange für die Asbest-Fibrose sowie Untersuchunger uber die Beeinflussung der Fibrose durch Poly-vinylpyridin-n-oxid.* In, *"Biologische Wirkungen des Asbestos",* International Conference, Dresden. Deutches Zentralinstitut Fur Arbeitsmedizin, Berlin 1968, pp. 47-49.

Kolev,K. (1982) Experimentally-induced mesothelioma in white rats in response to intraperitoneal administration of amorphous crocidolite asbestos. *Environ. Res.* **29**:123-133.

Kuschner,M. (1987) The effects of MMMF on animal systems: some reflections on their pathogenesis. *Ann. Occup. Hyg.* **31**:791-797.

Le Bouffant,L., Daniel,H., Henin,J.P. and Martin,J.C. (1985) Carcinogenic potency of chrysotile fibres of the length < 5 μm.*Cahiers de Notes Documentaires* **118**:83-90.

Le Bouffant,L., Daniel,H., Henin,J.P., Martin,J.C., Normand,C., Tichoux,G. and Trolard,F. (1987) Experimental study on long-term effects of inhaled MMMF on the lungs of rats. *Ann. Occup. Hyg.* **31**:765-791.

Lee,K.P., Kelly,D.P., O'Neal,F.O., Slader,J.C. and Kennedy,G.L. (1988) Lung response to ultrafine Kevlar aramid synthetic fibrils following 2-year inhalation exposure in rats. *Fundamental Appl. Toxicol.* **11**:1-14.

Leineweber,J.P. (1984) Solubility of fibres *in vitro* and *in vivo.* In, "Biological Effects of Man-made Mineral Fibres." Report of a WHO/IARC meeting. WHO, Copenhagen, pp. 87-102.

McConnell,E.E., Wagner,J.C., Skidmore,J.W. and Moore,J.A. (1984) A comparative study of the fibrogenic and carcinogenic effects of UICC Canadian chrysotile B asbestos and glass microfibre (JM100). In, "Biological Effects of Man-made Mineral Fibres." Report of a WHO/IARC meeting. WHO, Copenhagen, pp. 234-251.

Mikalsen,S., Rivedal,E. and Sanner,T. (1988) Morphological transformation of Syrian Hamster embryo cells induced by mineral fibres and the alleged enhancement of benzo[a]pyrene. *Carcinogenesis* **9**:891-899.

Morgan,A. and Holmes,A. (1986) Solubility of asbestos and man-made mineral fibres *in vitro* and *in vivo.* Its significance in lung disease. *Environ. Res.* **39**:475-484.

Mossman,B.T., Begin,R.O. (1989) Effects of mineral dusts on cells. Report of a NATO-sponsored workshop. NATO AS1 Series H Vol **30.** Springer-verlag, Berlin, New York, London.

Pott,F. and Friedrichs,K.H. (1972) Tumours in rats after intraperitoneal injection of asbestos dusts. *Naturwissenschaften* **59**:318-332.

Pott,F. (1978) Some aspects on the doseometry of the carcinogenic potency of asbestos and other fibrous dusts. *Staub-Reinholt der Luft* **38**:486-490.

Pooley,F.D. (1976) An examination of the fibrous mineral content of asbestos lung tissue from the Canadian chrysotile mining industry. *Environ. Res.* **12**:281-298.

Robertson,A., Bolton,R.E., Miller,B.G., Chapman,J.S., Dodgson,J., Jones,A.D., Niven,K.J. and Davis,J.M.G. (1988) The effect of quartz content on the pathogenicity of coalmine dusts. In, "Inhaled particles Vi", Dodgson,J., McCallum,R.I., Bailey,M.R., Fisher,D.R., eds. Pergamon Press, Oxford, New York.

Rowlands,N., Gibbs,G.W. and McDonald,A.D. (1982) Asbestos fibres in the lungs of chrysotile miners and millers. *Ann. Occup. Hyg.* **26**:411-417.

Scymczykiewicz,K. and Wiecek,E. (1960) The effect of fibrous and amorphous asbestos on the collagen content in the lungs of guinea pigs. In, "Proc. 13th Int. Conf. Occupational Hygiene," New York, pp. 801 *et seq.*

Smith,D.M., Ortiz,L.W., Archuleta,R.F. and Johnson,N.F. (1987) Long-term health effects in hamsters and rats exposed chronically to man-made vitreous fibres. *Ann. Occup. Hyg.* **31**:731-755.

Stanton,M.F. and Wrench,C. (1972) Mechanisms of mesothelioma induction with asbestos and fibrous glass. *J. Natl Cancer Inst.* **48**:797-821.

Stanton,M.F., Layard,M., Tegeris,A., Miller,M. and Kent,E. (1977) Carcinogenicity of fibrous glass: pleural response in the rat in relation to fibre dimension. *J. Natl Cancer Inst.* **58**:587-603.

Stanton,M.F. and Layard,M. (1978) The carcinogenicity of fibrous minerals, National Bureau of Standards special publication 506, Proceedings of the workshop on asbestos, Gaithersburg, MD, pp. 143-151.

Suzuki,Y. (1974) Interaction of asbestos with alveolar cells. *Environ. Hlth Perspect.* **9**:241-252.

Timbrell,V. (1973) Physical factors as etiological mechanisms. In, "Biological Effects of Asbestos", Bogovski,P., ed. IARC Scientific publication no. 8. International Agency for Research on Cancer, Lyon. pp. 295-304.

Vorwald,A.J., Durkan,T.M. and Pratt,P.C. (1951) Experimental studies of asbestosis. *AMA Arch. Ind. Hyg. Occup. Med.* **3**:1-43.

Wagner,J.C., Berry,G. and Timbrell,V. (1973) Mesotheliomata in rats after inoculation with asbestos and other minerals. *Br. J. Cancer* **28**:173-185.

Wagner,J.C., Berry,G., Skidmore,J.W. and Timbrell,V. (1974) The effects of the inhalation of asbestos in rats. *Br. J. Cancer* **29**:252-269.

Wagner,J.C., Berry,G., Skidmore,J.W. and Pooley,F.D. (1980) The comparative effects of three chrysotiles by injection and inhalation in rats. In, "Biological Effects of Mineral Fibres", Wagner,J.C., ed. IARC Scientific Publication No. 30. International Agency for Research on Cancer, Lyon, pp. 363-372.

Wagner,J.C., Berry,G.B., Hill,R.J., Munday,D.E. and Skidmore,J.W. (1984) Animal experiments with MMM(V)F. Effects of inhalation and intraperitoneal inoculation in rats. In, "Biological Effects of Man-made Mineral Fibres." Report of A WHO/IARC meeting WHO, Copenhagen, pp. 209-234.

Wagner,J.C., Griffiths,D.M. and Hill,R.J. (1984) The effect of fibre size on the *in vivo* activity of UICC crocidolite. *Br. J. Cancer* **49**:455-458.

Wagner,J.C., Skidmore,J.W., Hill,R.J. and Griffiths,D.M. (1985) Erionite exposure and mesothelioma in rats. *Br. J. Cancer* **51**:727-730.

Wagner,J.C. (1988) Significance of the fibre size of erionite. Paper presented at VII International Pneumoconiosis Conference, Pittsburgh. (Conference report in press.)

Wright,G.W. and Kuschner,M. (1977) The influence of varying lengths of glass and asbestos fibres on tissue response in guinea pigs. In, "Inhaled particles IV." Walton,H., ed. Pergamon Press, Oxford. pp. 455-474.

THE RELEVANCE OF ANIMAL BIOASSAYS TO ASSESS HUMAN HEALTH HAZARDS TO INORGANIC FIBROUS MATERIALS

Neil F. Johnson

Molecular and Cellular Toxicology Group
Inhalation Toxicology Research Institute
P.O. Box 5890, Albuquerque, NM 87185, U.S.A.

INTRODUCTION

The consequences of inhaling asbestos and erionite have alerted the scientific community and the general public to the potential for adverse health effects from other inorganic fibrous materials such as ceramic fibers and whiskers. The health effects of asbestos and erionite were initially suspected from human evidence. However, the long latency between first exposure and development of diagnosable lung disease makes it inappropriate to wait for evidence of human disease before remedial action is taken. While it is possible in some cases to substitute fibrous materials with their non-fibrous counterparts, fibrous materials will remain an integral part of our environment. It is, therefore, prudent to test the toxic properties of newly exploited natural and man-made fibers.

A variety of animal assays has been used to assess the pathogenic properties of fibrous materials. The results of these assays, however, can vary, and they can also differ from the results of known human exposures. It is important to determine the strengths and weaknesses of these assays, because their results can be used to support environmental exposure regulations. The utility of the various animal bioassays for hazard identification can be assessed by comparing animal and human responses to materials that have been well characterized and for which good human exposure and epidemiological data are available. Erionite, crocidolite, and fine glass fiber are examples of such materials that can be used to make a comparison between the various animal bioassays and man.

The major consequences of asbestos inhalation are primary lung cancer, mesothelioma, and fibrosis of the parenchyma and pleura. Lung cancer and mesothelioma are biologically distinct tumors. The embryonic origin of the cells giving rise to the tumors are different (Thomas, 1987); the mesothelium arises from the mesoderm, while the lining of the respiratory tract is derived from the ectoderm. The *in vitro* cytotoxicity of chrysotile, amosite, and crocidolite toward human bronchial and mesothelial cells is very different, with the mesothelial cells being more sensitive (Gerwin *et al.*, 1990). The induction of mesotheliomas in asbestos- exposed humans is independent of exposure to cigarette smoke, while lung cancer in asbestos workers who smoke is greatly increased compared to nonsmoking asbestos workers. This suggests that fibers act as an initiator of carcinogenesis for mesothelial cells, but as either an initiator or promoter for respiratory epithelial cells. The occurrence of mesotheliomas does not appear to be related to fiber exposure level or dose; limited or minimal exposure to crocidolite can lead to the induction of this tumor. This is in contrast to the situation reported for lung cancer in asbestos workers, where lung cancer incidence increases with exposure concentration and dose (Dunnill, 1982). As mesotheliomas and lung tumors are two biologically distinct tumor types, it cannot be assumed that a fiber type shown to be carcinogenic for one tissue is carcinogenic for the other.

The major animal assays used to determine the pathogenicity of fibrous materials are intratracheal instillation, intracoelomic injection (intrapleural and intraperitoneal), and inhalation exposure. The intracoelomic and intratracheal assays involve the administration of single or multiple high nonphysiological doses of test material in suspension directly to the tissues. This method circumvents the natural defense mechanisms of the body and can result in marked inflammatory reactions not typically seen when the material is given physiologically, such as by inhalation. In addition, some fibrous material when in suspension can agglomerate or form gelatinous masses, which can cause obstructive lung disease (Fisher *et al.*, 1989). Intratracheal administration of materials has been shown to result in an uneven distribution of the material within the lung, with the majority of the dose being delivered to the conducting airways and not the peripheral lung (Pritchard *et al.*, 1985). The distribution of inhaled fibrous materials such as crocidolite is fairly even throughout the lung tissue (Morgan *et al.*, 1975). While the evenness of deposition of inhaled fibers is expected to be similar for man and the rodents typically used in animal inhalation studies, the degree of particle deposition, retention, and clearance is markedly different. The result of these latter factors is a certain mass of fibers retained in the lung that can exert its biological effect. The tissue burden allows some degree of comparison of tissue dose between species. The problems of high nonphysiological doses, particle clumping, and uneven distribution must be considered when evaluating the results of the injection and instillation assays, because they will affect the actual dose delivered to individual cells or particular areas of the lung or coelomic cavity.

CROCIDOLITE

Exposure of humans to crocidolite asbestos is associated with the induction of mesothelioma and lung cancer (Wignall & Fox, 1982; Jones *et al.*, 1980). Lung cancer mortality rates for crocidolite-exposed workers show considerable variation; typically they are about 6 to 10% (Enterline & Henderson, 1973; Hughes & Weill, 1986) with out adjustment for the influence of cigarette smoking. There is a long latency (10 to 30 years in humans) between first exposure and the occurrence of diagnosable lung cancer. After about 20 years from first exposure, the risk of lung cancer increases in proportion to the cumulative dose. The mortality rate for mesothelioma has been reported to be as high as 18% (McDonald & McDonald, 1978). The risk of mesothelioma appears to depend primarily upon the time elapsed since first exposure (Newhouse & Berry, 1976).

TABLE 1. Tumor incidence (number) of lung and mesothelium for rats exposed to crocidolite by inhalation. The number of animals in each group is shown in parenthesis; exposure is given as fiber hours per ml (fhr/ml).

	Exposure 10^6 fhr/ml	Mesothelioma Control	Mesothelioma Exposed	Lung cancer Control	Lung cancer Exposed
Wagner *et al.*, 1974	6.5	0(42)	2(26)	0(42)	9(26)
	12.9	0(42)	0(18)	0(42)	4(18)
Smith *et al.*, 1987	9.3	0(59)	1(57)	0(59)	2(57)
Wagner *et al.*, 1987	6.5	0(40)	0(40)	0(40)	1(40)
Wagner *et al.*, 1985	6.5	0(28)	0(28)	0(28)	1(28)
Muhle *et al.*, 1987	2.0	0(50)	0(50)	0(50)	1(50)

Historically, crocidolite exposure levels have been high; for example, concentrations of 2,700 million particles per cubic foot (mppcf) have been reported in the crocidolite mining and milling industries (Gibbs & Du Toit, 1973). Exposure concentrations of 8 to 400 fibers (f)/ml have been reported in naval shipyards during the maintenance and removal of sprayed crocidolite asbestos products (Harries, 1971). Bossard and co-workers (1980) reported lung burdens of 0.7 to 720 x 10^6 f/g (dried lung) in males with known occupational exposures to crocidolite.

In contrast, rodents have been experimentally exposed to atmospheres of up to 3000 f/ml of crocidolite for 24 months (Smith *et al.*, 1987) which resulted in mean lung burdens of 387 x 10^6 f/g (dried lung). Tumor incidences for the lung and mesothelium from several studies are shown in Table 1. The incidence varies between 1 out of 50 and 9 out of 26 exposed animals for the individual investigations. All six of the inhalation experiments resulted in lung tumors in the exposed animal groups; no lung tumors were seen in the control animal groups. The incidence of mesothelioma varied between zero and 2 out of 28 exposed animals; only two of the six assays resulted in mesotheliomas in exposed animals. Mesotheliomas were not seen in any of the control groups of animals.

The intracoelomic and intratracheal assays in animals give uniformly positive results for tumorgenicity with crocidolite (Wagner & Berry, 1969; Wagner *et al.*, 1973; Pott *et al.*, 1987; Smith *et al.*, 1987). The peritoneal cavity seems more sensitive to crocidolite than the pleural cavity. The amount of crocidolite required to induce a 50% mesothelioma rate is 2.5 mg for the peritoneal cavity (Bolton *et al.*, 1984) and approximately 23 mg for the pleural cavity (Stanton & Wrench, 1972).

ERIONITE

The inhalation of erionite (a fibrous zeolite) in man is associated with a greatly increased risk of lung cancer and mesothelioma (Baris *et al.*, 1987). The number of deaths from lung cancer following either erionite or crocidolite exposure is similar;

TABLE 2. The total number of deaths associated with human populations exposed to crocidolite or erionite. Data from Coffin *et al.*, 1989.

	Total Deaths (%)	
	Lung Cancer	Mesothelioma
Crocidolite	10.9	7.1
Erionite	11.9	40.0

however, the deaths from mesothelioma are much more numerous for erionite than crocidolite exposure (Table 2). Limited information is available concerning the environmental concentrations of erionite fibers. In the Turkish villages associated with the high incidence of mesothelioma, airborne fiber concentrations between 0.004 to 0.175 f/ml were reported. The bulk of these fibers was described as zeolites (Baris *et al.*, 1987).

The high rate of mesothelioma induction seen in humans exposed to erionite has been reproduced in animals. Wagner and co-workers (1985) induced mesotheliomas in 28 of 29 animals exposed to erionite fibers; however, the rats died too soon to show evidence of lung cancer induction. The exposure concentrations for this animal study were 542 f/ml, and the animals were exposed for 7 hr/day, 5 days-per-week, for 52 weeks.

Intraperitoneal and intrapleural assays give positive results with erionite. The mesothelium of the pleural cavity appears to be more sensitive than that of the peritoneal cavity. The erionite dose required to induce a 50% tumor rate in the pleural cavity is 0.1 mg in the rat (Hill *et al.*, 1990). The dose of erionite required to induce a 50 % tumor rate in the peritoneal cavity has not been reported. However, equal mass doses of crocidolite and erionite (0.5 mg) produce similar tumor rates of approximately 45% (Pott *et al.*, 1987). These data suggest that the pleural cavity is approximately five times more sensitive than the peritoneal cavity.

GLASS MICROFIBER

Human exposure to fine glass fiber is not associated with a significantly increased risk of lung cancer or mesothelioma (Marsh *et al.*, 1990). Animal inhalation studies with fine glass fiber have been uniformly negative (Wagner *et al.*, 1984; McConnell *et al.*, 1984; Smith *et al.*, 1987; Le Bouffant *et al.*, 1987; Muhle *et al.*, 1987; Gross 1976). The animal experiments are conducted using airborne concentrations many times those encountered in the work environment. In four factories manufacturing fine glass fibers, the average intensity of exposure to fibers less than 3 μm in diameter is between 0.027 and 0.292 f/ml (Marsh *et al.*, 1990). In contrast, the study of Smith and co-workers (1987) exposed animals to an atmosphere of 3000 f/ml (diameter less than 3 μm) for 6 hr-per-day, 5-days-a-week for 24 months. The lung burden of workers exposed to man-made mineral fibers has never been reported to exceed 0.2 f/μg dry lung; the mean diameter of the recovered fibers is 1.2 μm, and the mean length is 8.7 μm (McDonald *et al.*, 1990). In the inhalation study of Smith and co-workers (1987), the lung burden of rats exposed to glass microfiber (mean diameter 0.45 μm, mean length 7.5 μm) was 1.87 x 10^3 f/μg dry lung and 2.85 f/μg dry lung for a coarser glass fiber (mean diameter 1.4 μm, mean length 37 μm). The animal inhalation studies show that even the high exposure levels and high lung burdens are insufficient to induce either lung or mesothelial tumors.

The results of the inhalation studies with glass microfiber contrast with the results from intrapleural, intraperitoneal, and intratracheal assays. Intrapleural studies (Wagner *et al.*, 1973, 1984; Stanton *et al.*, 1977; Lafuma *et al.*, 1984; Davis, 1974 and Smith *et al.* 1980) gave positive results for mesothelioma induction. The intraperitoneal studies of Smith and co-workers (1987) and Pott and co-workers (1987) also gave positive results for mesothelioma induction. The results of intratracheal assays with glass microfiber are equivocal. Mohr and co-workers (1984) reported a high incidence of both lung and mesothelial tumors, while in later but similar study by the same group of workers (Pott *et al.*, 1987), demonstrated an increase in lung tumors but no increase in mesothelial tumors. The studies of other groups (Smith *et al.*, 1987; Feron *et al.*, 1985; Gross, 1976; Wright & Kuschner, 1976 and Pickrell *et al.*, 1983) were negative for lung and mesothelial tumors.

COMPARISON OF HUMAN AND ANIMAL INHALATION EXPOSURES

Animal inhalation studies with crocidolite have achieved similar or greater aerosol exposure concentration levels than those reported for humans. In addition, lung tissue burdens in animals and man of are the same order of magnitude. This shows that in spite of differences in clearance and deposition of respirable particles between man and animals, the mass dose to the lung tissue is similar. Lung tumor rates in man and animals are similar for crocidolite exposure, whereas the mesothelioma rate in animals is much lower than that found in exposed humans. The UICC crocidolite reference sample is a short-length material (mean length 2.2 μm), while the crocidolite fibers recovered from human lung tissue are considerably longer (mean length 5.8 μm) (Gibbs *et al.*, 1990). This marked difference in length may account for the disparity in biological activity. Fiber length has been shown to be one of the major determinants of the induction of mesotheliomas (Stanton *et al.*, 1977). The animal inhalation model has been criticized, because it is thought to be insensitive to crocidolite (commonly used positive control material), in terms of mesothelioma induction and lung cancer induction (Pott *et al.*, 1987). As noted above, the short length of the UICC reference sample is probably the major reason for the low mesothelioma rate. The insensitivity of the rodent inhalation model in relation to lung cancer is not substantiated by the evidence from many animal experiments.

Animal inhalation exposures to erionite and glass microfibers have used exposure concentrations many times those encountered by human populations. Erionite appears to be equally as potent in inducing lung cancers as crocidolite, but is a more potent carcinogen for mesothelial cells. A high rate of mesotheliomas has been induced in animals with erionite showing that this activity is also seen in rodents and that the rat mesothelium is sensitive to appropriate carcinogens. The lack of carcinogenicity of glass microfibers following inhalation in rodents is similar to that in humans, in spite of much higher fiber concentrations and lung dust burdens achieved in the animal experiments. This shows that the animal inhalation model does not readily produce false positive results.

TABLE 3. Results of human and animal exposures to glass microfibers. The tumor response is shown (+) following the various routes of administration in humans and animals.

Exposure Route	Tumor Response		
	Positive	Equivocal	Negative
Human Inhalation			+
Animal Inhalation			+
Animal Intratracheal		+	
Animal Intrapleural	+		
Animal Intraperitoneal	+		

Comparison of human and animal exposures to crocidolite, erionite, and glass microfiber shows that the effects of fiber inhalation in animals mimic those found in man. Animal exposures to erionite and crocidolite produce the same tumor types as found in humans. The tumor rates in man and animals for crocidolite show that fibers are not strong carcinogens. The relatively low tumor rates have implications for the design of future inhalation studies. The number of animals in each exposure group may need to be more than the 25 to 50 animals typically reported in the current open literature. Consideration has to be given to the numbers of animals in exposure groups that will allow a separation of the spontaneous tumor rate from a 4 to 10% tumor rate in exposed animals. In addition, a suitable positive control material must be used; UICC crocidolite would seem to be a poor choice because of its short length. Tremolite is a long-fibered amphibole asbestos that may be a suitable positive control material (Davis *et al.*, 1985).

Animal inhalation studies are relevant to identifying potential human health hazards. However, this type of experiment is costly and time-consuming, and it may not be possible to test all fibrous materials in this way. The injection and instillation bioassays may provide means to screen the most active materials for eventual inhalation studies. The relevance of the various bioassays for the identification of hazardous fibrous materials can be examined by comparing the results of the bioassays with crocidolite, erionite, and glass microfiber. Chrysotile asbestos has not been used as an example for this comparison because of the low relative risk of mesothelioma associated in humans exposed to this material (Churg, 1988).

COMPARISON OF ANIMAL BIOASSAYS

A concern with bioassays is whether they faithfully predict human experience and do not give either false positive or false negative results. Inhalation exposures have been shown to mimic human experience. The instillation and injection assays introduce material in a manner that circumvents some of the natural defense mechanisms of the lung, and they deposit extremely high non-physiological doses to the tissue. It may, therefore, be expected that these types of assays will give false positive results. Comparing the negative human and animal inhalation data for glass microfibers and fine glass fibers to the injection or instillation assays shows this to be true (Table 3).

The results of the instillation and injection assays show that they can give positive tumor results when there is clear evidence from human exposure that the material is not associated with increased risks for lung cancer and mesothelioma.

The intracoelomic assay has also been considered as a suitable test for the overall carcinogenic activity of fibrous materials (Pott *et al.*, 1987), *i.e.* the results of the assay can equally apply to the induction of lung cancer because of the induction of mesotheliomas. This assumption may not be valid, as the two tumor types are biologically distinct. The intracoelomic assay cannot be expected to reveal the likelihood of a fibrous material being a lung carcinogen.

Comparing erionite and crocidolite responses in the pleural and peritoneal cavities after injection shows that there are differences in responsiveness of the two mesothelial surfaces. Intrapleural injection of erionite shows it to be 230 times as effective as crocidolite in inducing mesotheliomas (Stanton & Wrench, 1972; Hill *et al.*, 1990), while injections of similar masses of erionite and crocidolite into the peritoneal cavity produce a similar number of tumors (Pott *et al.*, 1987). The mesothelium of the peritoneal cavity is much less sensitive to erionite than the mesothelium of the pleural cavity. The reverse is true with crystalline silica. Peritoneal injection of crystalline silica can induce a significant number (22%) of mesotheliomas (Pott *et al.*, 1987); injection of silica into the pleural cavity does not result in the formation of mesotheliomas (Wagner *et al.*, 1980). The pleural cavity behaves as expected from human exposures in which silica is not associated with mesothelioma induction. Of the two intracoelomic assays, the intrapleural route would seem the most appropriate for identifying potential hazardous materials.

EVIDENCE OF MECHANISM

Comparing the results with various fibrous materials and exploring the differences among the various bioassays may also provide information about mechanisms of fiber carcinogenesis. Inhalation experiments with erionite and glass microfibers provide evidence that parameters other than fiber dimensions are important in the toxicity of a material. The atmospheres of these materials contain a large proportion of fibers considered to be highly carcinogenic, yet the two materials have markedly different abilities to induce mesotheliomas. The high biological activity of erionite is also in contrast to that of crocidolite although both materials contain similar numbers of respirable fibers longer than 5 μm (Johnson & Wagner, 1989).

Inhalation of erionite may be a good model system for studying induction of mesothelioma. A striking feature of erionite-exposed rats is the development of pleural fibrosis (Johnson & Wagner, 1989). The fibrotic potential of erionite is also evident following intraperitoneal injection (Suzuki & Kohyama, 1988). In contrast, the inhalation of crocidolite results only in mild pleural fibrosis (Johnson & Wagner, 1989). The pleural effects of erionite may be exerted through a greater accumulation of fibers than those of crocidolite. This deposition may be enhanced by the fibrotic reaction associated with erionite exposure. The accumulation of fibers will, in part, be governed by lymphatic drainage of the material. Pleural fibrosis may impair lymphatic drainage, which, in the case of erionite, will result in a high focal concentration of pathogenic fibers. Pleural fibrosis induced by non-fibrous particles such as silica does not result in mesotheliomas (Heppleston, 1982). Interestingly, a recent inhalation study with ceramic fibers has also demonstrated in a high mesothelioma rate and marked pleural fibrosis (Manville, 1989).

SUMMARY

In summary, animal inhalation experiments are relevant to assessing human health risks from inorganic fibers. Appropriately conducted inhalation experiments should be used to identify hazardous fibrous materials, because the other animal bioassays can give false positive results. Injection and instillation bioassays can be used to screen and identify the most biologically active materials for subsequent inhalation experiments.

ACKNOWLEDGEMENTS

This research was supported by the Office of Health and Environmental Research, U. S. Department of Energy, under Contract DE-AC04-76EV01013.

REFERENCES

Baris, I., Simonato, L., Artvinli, M., Pooley, F., Sarraci, R., Skidmore, J., Wagner, J. (1987) Epidemiological and environmental evidence of the health effects of exposure to erionite fibres: A four-year study in the Cappadocian region of Turkey. *Int. J. Cancer.* **39**:10-17.

Bolton, R. E., Davis, J. M. G., Miller, B., Donaldson, K., Wright, A. (1984) The effect of dose of asbestos on mesothelioma production in the laboratory rat. In *VI International Pneumoconiosis Conference*. Bochum: Bergbau-Berufsgenossenschaft. pp. 1028-1035.

Bossard, E., Stolkin, I., Spycher, M. A., Ruttner, J. R. (1980) Quantification and particle size distribution of inhaled fibres in the lung. In, "Biological Effects of Mineral Fibres". IARC Publication No. 30, Vol. 1 pp. 35-41.

Churg, A. Chrysotile, tremolite, and malignant mesothelioma in man. (1988) *Chest* **93**:621-628.

Coffin, D. L., Palekar, L. D., Cook, P. M., Creason, J. P. (1989) Comparison of mesothelioma induction in rats by asbestos and nonasbestos mineral fibers: Possible correlation with human exposure data. In, "Biological Interaction of Inhaled Mineral Fibers and Cigarette Smoke", Battelle Press, pp. 347-354.

Davis, J. M. G. (1974) Histogenesis and fine structure of peritoneal tumors produced in animals by injection of asbestos. *J. Natl. Cancer Inst.* **52**:1823-1828.

Davis, J. M. G., Addison, J., Bolton, R. E., Donaldson, K., Jones, A. D., Miller, B. G. (1984) Inhalation studies on the effects of tremolite and brucite dust in rats. *Carcinogenesis* **6**:667-674.

Dunnill, M. S. (1982) "Pulmonary Pathology". Churchill Livingston, London.

Enterline, P. E., Henderson, V. (1973) Type of asbestos and respiratory cancer in the asbestos industry. *Arch. Environ. Hlth* **24**:312-317.

Feron, V. J., Scherrenberg, P. M., Immel, H. R., Spit, B. J. (1985) Pulmonary response of hamsters to fibrous glass: chronic effects of repeated intratracheal instillation with and without benzo[*a*]pyrene. *Carcinogenesis* **6**:1495-1499.

Fisher, G. L., McNeill, K. L., Singer, A. W., Smith, J. T. (1989) Pulmonary response to intratracheally instilled silicon nitride in rats. *Inhalation Toxicol.* **1**:227-241.

Gerwin, B. I., Betsholtz, C., Linnainmaa, K., Pelin, K., Reddel, R., Gabrielson, E., Seddon, M., Greenwald, R., Harris, C. C., Lechner, J. (1990) *In vitro* studies of human mesothelioma. In, "Biology, Toxicology, and Carcinogenesis of Respiratory Epithelium". Hemisphere Publishing Corp. pp 112-128.

Gibbs, A. R., Griffiths, D. M., Pooley, F. D., Jones, J. S. P. (1990) Comparison of fibre types and size distributions in lung tissues of paraoccupational and ocupational cases of malignant mesothelioma. *Br. J. Indust. Med.* **47**:621-626.

Gibbs, G. W., Du ToiT, R. S. (1973) Environmental data in mining. (1973) In, "Biological Effects of Asbestos". IARC Publication No. 8 pp. 311-320.

Gross, P. (1976) The effects of fibrous glass dust on the lungs of animals. *HEW Publication (NIOSH)* **76-151**:169-178.

Harries, P. G. (1971) A comparison of mass and fiber concentrations of asbestos dust in shipyard insulation processes. *Ann. Occup. Hyg.* **14**:235-240.

Heppleston, A. G. (1982) Silicotic fibrogenesis: A concept of pulmonary fibrosis. *Ann. Occup. Hyg.* **26**:446-449.

Hill, R. J., Edwards, R. E., Carthew, P. (1990) Early changes in the pleural mesothelium following intrapleural inoculation of the mineral fiber erionite and the subsequent development of mesotheliomas. *J. Exp. Pathol.* **71**:105-118.

Hughes, J. M., Weill, H. (1986) Asbestos exposure-quantitative assessment of risk. *Am. Rev. Respir. Dis.* **133**:5-13.

Johnson, N. F., Wagner, J. C. (1989) Effects of erionite inhalation on the lungs of rats. In, "Biological Interaction of Inhaled Mineral Fibers and Cigarette Smoke". BattellePress, pp. 325-345.

Jones, J. S. P., Smith, P. G., Pooley, F. D., Berry, G., Sawk, G. W. Aggarwal, A., Wignal, B. K., Madeley, R. J. (1980) The consequences of exposure to asbestos dust in wartime gasmask factory. In, "Biological Effects of Mineral Fiber". IARC Publication No. 30 pp. 637-653.

Lafuma, J., Morin, M., Poncy, J. L., Masse, R., Hirsch, A., Bignon, J., Monchaux. (1984) Mesothelioma induced by intrapleural injection of different types of fibres in rats; synergistic effect of other carcinogens. In, "Biological Effects of Man-Made Mineral Fibers". WHO, Regional Office Copenhagen Vol. 2: pp. 311-320.

Le Bouffant, L., Daniel, H., Henin, J. P., Martin, J. C., Normand, C., Tichoux, G., (1987)Trolard, F. Experimental study on long-term effects of inhaled MMMF on the lungs of rats. *Ann. Occup. Hyg.* **31**:765-790.

Manville Corporation. (1989) Submission to the Office of Toxic Substances, Environmental Protection Agency Document Control Number 8HEQ-0485-0553.

Marsh, G. M., Enterline, P. E., Stone, R. A., Henderson, V. L. (1990) Mortality among a cohort of US manmade mineral fiber workers: 1985 follow up. *J. Occup. Med.* **32**:594-604.

McConnell, E. E., Wagner, J. C.,Skidmore, J. W., Moore, J. A. (1984) A comparative study of the fibrogenic and carcinogenic effects of UICC chrysotile B asbestos and glass microfibre (JM 100). In, "Biological Effects of Man-Made Mineral Fibers". WHO Regional Office Copenhagen Vol. 2: pp 234-250.

McDonald, A. D., McDonald, J. C. (1978) Mesothelioma after crocidolite exposure during gasmask maufacture. *Environ. Res.* **17**:340-346.

McDonald, J. C., Case, B. W., Enterline, P. E., Henderson, V., McDonald, A. D., Plourde,

M., Sebastien, P. (1990) Lung dust analysis in the assessment of past exposure of man-made mineral fibre workers. *Ann. Occyp. Hyg.* (in press).

Mohr, U., Pott, F., Vonnahme, F. J. (1984) Morphological aspects of mesotheliomas after intratracheal instillations of fibrous dusts in Syrian golden hamsters. *Exp. Path.* **26**:179-183.

Morgan, A., Evans, J. C., Evans, R. J., Hounam, R. F., Holmes, A., Doyle, S. G. (1975) Studies on the deposition of inhaled fibrous material in the respiratory tract of the rat and its subsequent clearance using radioactive tracer techniques. *Environ. Res.* **10**:196-207.

Muhle, H., Pott, F., Bellman, B., Takenaka, S., Ziem, U. (1987) Inhalation and injection experiments in rats to test the carcinogenicity of MMMF. *Ann. Occup. Hyg.* **31**:755-764.

Newhouse, M. L., Berry, G. (1976) Predictions of mortality from mesothelial tumors on asbestos factory workers. *Br. J. Indust. Med* **33**:147-151.

Pickrell, J. A., Hill, J. O., Carpenter, R. L., Hahn, F. F., Rebar, A. H. (1983) *In vitro* and *in vivo* response after exposure to man-made mineral and asbestos insulation fibers. *Am. Indust. Hyg. Assoc. J.* **44**:557-561.

Pott, F., Ziem, U., Reiffer, F. J., Huth, F., Ernst., H., Mohr, U. (1987) Carcinogenicity studies on fibres, metal compounds, and some other dusts in rats. *Exp. Path.* **32**:129-152.

Pritchard J. N., Holmes, A., Evans, J. C., Evans, N. Evans, R. J., Morgan, A. (1985) The distribution of dust in the rat lung following administration by inhalation and by single intratracheal instillation. *Environ. Res.* **36**:268-297.

Smith, D. M., Ortiz, L. W., Archuleta, R. F., Johnson, N. F. (1987) Long-term health effects in hamsters and rats exposed chronically to man-made vitreous fibres. *Ann. Occup. Hyg.* **31**:731-754.

Smith, W. E., Hurbert, P. D. and Sobel, H. J. (1980) Dimension of fibers in relation to biological activity. In, "Biological Effects of Mineral Fibers". IARC Publication No 30 pp. 357-360.

Stanton, M. F., Layard, M., Tegeris, A., Miller, E., May, M., Kent, E. J. (1977) Carcinogenicity of fibrous glass: pleural response in relation to fibre dimension. *Natl. Cancer Inst.* **58**:587-603.

Stanton, M. F., Wrench, C. (1972) Mechanisms of mesothelioma induction with asbestos and fibrous glass. *J. Natl. Cancer Inst.* **48**:797-821.

Suzuki, Y., Kohyama, N. (1988) Carcinogenic and fibrogenic effects of erionite, mordenite, and synthetic zeolite 4A. In, "Occurence, Properties and Utilization of Natural Zeolites". Akademiai Kiado, Budapest, pp. 829-840.

Thomas, N. W. (1987) Embryology and structure of the mesothelium. In, "Pathology of the Mesothelium". Springer - Verlag, London, pp. 1-2.

Wagner, J. C., Berry, G. (1969) Mesotheliomas in rats following inoculation with asbestos. *Br. J. Cancer* **23**:567-581.

Wagner, J. C., Berry, G. B., Timbrell, V. (1973) Mesotheliomata in rats after inoculation with asbestos and other materials. *Br. J. Cancer* **28**:173-179.

Wagner, J. C., Berry, G., Skidmore, J. W., Timbrell, V. (1974) The effects of the inhalation of asbestos in rats. *Br. J. Cancer* **29**:252-269.

Wagner, J. C., Berry, G. B., Hill, R. J., Munday, D. E., Skidmore, J. W. n(1984) Animal experiments with MMM(V) - Effects of inhalation and intrapleural inoculation in rats. In, "Biological Effects of Man-Made Mineral Fibers". WHO Regional Office Copenhagen. Vol **2**:209-233.

Wagner, J. C., Skidmore, J. W., Hill, R. J., Griffiths, D. M. (1985) Erionite exposure and mesotheliomas in rats. *Br. J. Cancer* **51**:727-730.

Wagner, J. C., Giffiths, D. M., Munday, D. E. (1987) Experimental studies with palygorskite dusts. *Br. J. Indust. Med.* **44**:749-763.

Wagner, M. M. F., Wagner, J. C., Davies, R., Griffiths D. M. (1980) Silica-induced malignant histiocytic lymphoma: incidence linked with strain of rat and type of silica. *Br. J. Cancer* **41**:908-917.

Wignal, B. K., Fox, A. J. (1982) Mortality of female gasmask assemblers. *Br. J. Indust. Med.* **39**:34-48.

Wright, G. W., Kuschner, M. (1976) The effects of intratracheal instillation of glass fiber of varying sizes in guinea pigs. *USHEW Publication No. (NIOSH)* **76-151**:151-168.

RELATIVE INTRINSIC POTENCY OF ASBESTOS AND ERIONITE FIBERS: PROPOSED MECHANISM OF ACTION

David L. Coffin and Andrew J. Ghio[1]

Center for Environmental Medicine and Lung Biology
University of North Carolina
Chapel Hill, NC, U.S.A.

[1]Duke University School of Medicine
Durham, NC
U.S.A.

ABSTRACT

Data from human exposure to naturally occurring mineral fibers indicates extreme variability in the rate of mesothelioma between various asbestos varieties and erionite. Experiments on rats have confirmed these differences even where dose responses are based on the number, size and shape of the fibers. *In vitro* parameters of cytotoxicity and chromosomal aberrations follow the same trend. In such studies the zeolite fiber erionite, which possesses unique internal voids in the crystal lattice, ranks as several orders of magnitude more potent than any asbestos. The data suggests that an intrinsic physico-chemical phenomenon is superimposed on the known length/width factor which is responsible for the variability in mesothelioma induction.

Tumorigenesis by asbestos has been ascribed to induction by free radicals. It has been shown that free radical effects are associated with exposure to asbestos and that these can be quenched by free radical scavengers. A mechanism is proposed here based on the co-ordination of endogeous ferric iron with surface silanol groups and formation of hydroxy radicals via the Fenton reaction. It is postulated that the greater activity of erionite is due to a sheltering of the reaction from free radical scavengers within the internal voids of the mineral. A possibility also exists that prolongation of the half-life of the radicals would result from such sequestration. Additionally this activity of the radical effect may be enhanced by proximity to the chromatin material after penetration of the cellular nucleus by long thin fibers following pinocytosis by the mesothelial cells.

INTRODUCTION

Data is emerging which suggests that the amphiboles, particularly crocidolite, are more potent inducers of mesothelioma than chrysotile (U.K. Advisory Committee on Asbestos, 1983). A recent review by Hughs and Weil (1986) has ranked mesothelioma induction by asbestos in ascending order, as follows: chrysotile, amosite, mixed asbestos, crocidolite. Contrary views however have been expressed stating that the evidence for any differentiation on the basis of fiber type is lacking (Peto *et al.*, 1982). The fiber erionite, a zeolite mineral, has been recently shown to possess unquestionably higher potency for mesothelioma induction than any asbestos fiber (Artvinli & Baris, 1982) (Table 1).

Because of the great effort currently being expended to promote containment or removal of asbestos in public buildings, and the differences of opinion regarding the relative potencies of the various types, it has appeared useful to construct experiments to correlate such information under controlled laboratory conditions. Furthermore it was felt that demonstration of the relative potencies, particularly of erionite, in experiments in which the various geometric fiber types were quantified would serve as a useful data base for mechanism studies.

Experiments employing Fischer-344 rats treated either by intraperitoneal injection or intratracheal instillation with a variety of asbestos and non-asbestos mineral fibers were conducted (Coffin *et al.*, 1989a; 1991). In these experiments mesothelioma incidence was compared to the total number of mineral fibers in the ingested dose or to the number of retained fibers at various post-instillation periods. Tumor fiber potency was constructed as follows.

$$\frac{\text{Average percent mesothelioma}}{\text{Number of mineral fibers x } 10^{-6}} = \frac{\text{Fiber}}{\text{potency}}$$

The data summarised here show the relative fiber potencies for UICC Chrysotile, UICC Crocidolite asbestos and Rome Oregon erionite (Coffin *et al.*, 1991). These data indicate that there is a strong gradient in potency in ascending order from chrysotile to erionite (Table 2). Such striking differences in relative mesothelioma induction cannot be explained by such variables as the number of fibers, length and width ratios, splitting or durability. In fact, if such factors were operating, an opposite gradient of effect would obtain. It also seems unlikely that such large differences in tumorigenic potency could be the result of an increase in the rate of translocation for erionite for two reasons: (1) erionite and crocidolite appear essentially the same in average length and width and are equally straight and inflexible, (2) the difference in survival times for animals dying with tumors between the intratracheal and intrapleural tests were greater for erionite than the asbestos fibers - namely chrysotile +78, crocidolite -10 and erionite +295 days. These data do not indicate a more rapid rate of penetration from the airway to the pleura for erionite. It would appear logical therefore to ascribe the differences in mesothelioma induction to inherent fiber potency due to physico-chemical characteristics (Coffin *et al.*, 1991).

In vitro studies for cytotoxicity in V79 cells and cytomutagenic effects (Palekar *et al.*, 1988; 1989) showed that on the basis of the number of fibers in the system, the gradient of effect was similar to that in the animal experiments (Figure 1). Since erionite was 800 times more potent than chrysotile and 22 times more potent than crocidolite in

TABLE 1. Relative ranking of asbestos and erionite for mesothelioma induction

	A	B	C	D
Chrysotile	12.1	1.0	0.22	1.0
Amosite	22.2	1.83	2.2	10.0
Mixed fibers	65.9	5.44	4.7	21.4
Crocidolite	165.1	13.63	7.1	32.37
Erionite	540.0	44.60	40.1	181.8

A = % difference from lung cancer
B = same rates normalized for chrysotile
C = % mortality with mesothelioma
D = same rates normalized for chrysotile
Data for asbestos (Hughs & Weil, 1986)
Data for erionite (Artvinli & Baris, 1982)

the intratracheal studies, it would be germane to consider the unique physico-chemical properties of the zeolites of which erionite is a member. Zeolites are hydrated aluminium silicates in which the aluminium and silicon tetrahedra are bound anionicaly by their shared oxygen atoms in a three dimensional lattice. These structures are so joined or "stacked" as to contain internal interconnecting channels or "voids" which communicate to the external surface by minute apertures or "pores". In the case of erionite the effective critical diameters of these pores is 3.6 x 5.2 Å. Erionite has a very high capacity for adsorbing cations and cation complexes but the structure excludes all compounds whose molecular diameter is greater than that of the pore (Peters, 1988). The surface area of erionite is 203 m^2/g (Eberly, 1969). Contrasting with that for chrysotile and crocidolite respectively as 24 m^2/g and 8-10 m^2/g (Fournier et al., 1985). Most of the surface area of erionite is internal.

When aluminium is substituted for silicon in the tetrahedral structure of a polysilicate anion there is an effective net gain of one electron per substitution. The tetrahedra $Al(III)O_4$ have, in isolation, a charge of minus five compared to the silicon tetrahedra $(Si(IV)O_4)$ which have a charge of minus four leaving a net difference of minus one. In order to balance the negative charge electrostructurally cations are associated with the zeolites, most commonly Na^+, K^+, Mg^{2+} and Ca^{2+}. While many other cations are associated with zeolites, iron is naturally present only in trace amounts.

Because of its unique structure and large surface area zeolites have extremely high adsorbtive properties for cations, cation complexes, and short straight chain hydrocarbons. Consequently much use has been made of natural and synthetic zeolites in the chemical and petroleum industries as adsorbants and catalytic agents.

Mechanistic studies with Erionite

Comparative cytotoxicity studies have shown that the number of mineral fibers with L ≥ 8 μM and W ≤ 0.25 μm x 10^{-6} required to produce a LC_{50} in V79 cells were 348.80, 6.28 and 1.16 respectively for chrysotile, crocidolite and erionite (Palekar et al., 1988). Chromosomal effects in V79 cells were approximately three orders of magnitude greater for erionite (Palekar et al., 1987, 1989). In this same study mineral fibres were

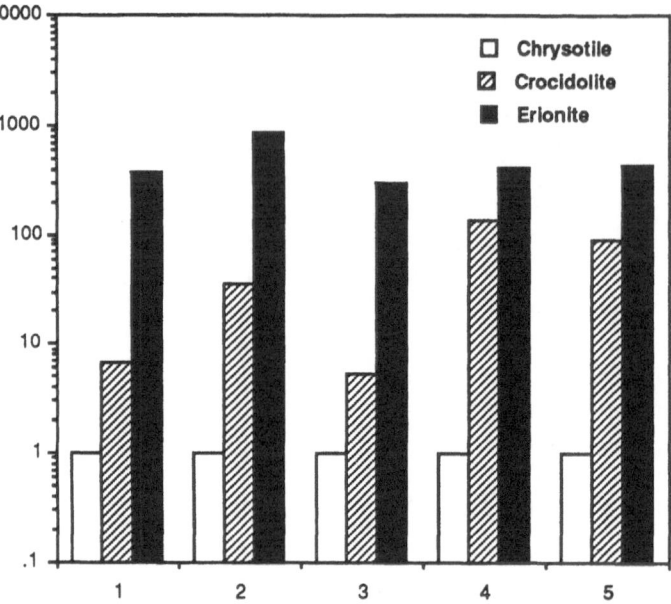

1   Intrapleural injection — percent death with tumor / number of mineral fibers Coffin, et. al. (1989, 1991)

2   Intratracheal instillations — percent death with tumor / average number mineral fibers retained one day to one year Coffin, et. al. (1991)

3   Relative toxicity to V79 cells Palekar, et. al. (1988)

4   Reduction in diploid cells / number mineral fibers Palekar, et. al. (1987)

5   Increase in aneuploid cells / number mineral fibers Palekar, et. al. (1987)

Figure 1. Comparison of relative potency of mineral fibres for *in vivo* and *in vitro* parameters. All potency calculations based on relative number of mineral fibres L $\geq$ 8 μm x W $\leq$ 0.25 μm. Plots normalized for chrysotile.

observed which had penetrated the cellular nucleus and were in contact with chromatin material in dividing cells. In order to determine the role of the void spaces experiments were conducted in which isopentane was encapsulated in the void spaces by forcing it through the apertures by high temperature and pressure. Encapsulation reduced the cytotoxic effect approximately 50%. The authors speculate that the reduction was due to blocking of the internal surface (Coffin *et al.*, 1989b).

These data appear congruent in both the *in vivo* tumorgenesis and the *in vitro* studies showing that erionite potency in mesothelioma inductions was two orders of magnitude or more than chrysotile. Furthermore the importance of the internal surface is suggested by reduction of the cytotoxicity when the internal voids were occluded by the presence of an inert substance, isopentane.

In view of the above data the following chemical mechanism is proposed to explain the unusual mesothelioma induction by erionite.

Possible mechanism of action

A number of authors have demonstrated that asbestos (and other particles) induce the formation of Active Oxygen Species in *in vitro* systems (Mossman *et al.*, 1990a). These reactions can be quenched by D.S.O. thereby blocking such reactions as lipid peroxidation. It is known that redox cycling of iron greatly promotes lipid peroxidation and that P450 or superoxide can reduce iron complexes (Gutteridge, 1987) and that iron may be released from heme protein by H_2O_2, (Gutteridge, 1982) or by superoxide organic radicals and by redox cycling chemicals (Aust, 1987). Aust has proposed a scheme in which ferrous iron is oxidized to the ferric form. Iron has been shown to complex with silicates and enhance their lipid peroxidation effect (Hochstein, 1964; Wills, 1965). Ghio and others (1990) have discussed the role of silicates in interaction with iron in lung disease via the Fenton reaction.

While the determinants of toxicity of asbestos and erionite remain unknown, injury resulting from exposure to them has been associated with oxidant production (Mossman *et al.*, 1986; 1990b). Silica and silicates can be involved in the formation of free radicals after heating to high temperature, grinding or by co-ordination of transition metals on their surfaces (Solomon, 1968; Iler, 1979). Only the last of these mechanisms can have significance in generating oxidants within a biological system. In an aqueous environment, oxides and oxide mineral are covered with surface hydroxyl groups and silica and all silicate surfaces have some concentration of silanol groups (-SiOH). The dissociation of these silanol groups, along with broken siloxane bonds and isomorphous substitution of Al^{3+} for Si^{4+}, contributes to a negative net charge on the surface of the particles which creates a capacity for adsorption of organic and inorganic cations (Grim, 1968). In addition, specific co-ordination of smaller molecules can account for adsorption to the particle surface. Co-ordination of these dusts depends on neither the ionic charge nor surface acidity but rather on the geometry of the co-ordination bonds formed. The resultant selectivity is the basis for silicate use in both water softeners and ion exchange chromatography (Grim, 1968). Ferric ion has a high affinity for oxygen donor ligands as evidenced by the large hydrolysis constant for $[Fe(H_2O)_6]^{3+}$ log $K_h = 2.56$ (Kragten, 1978). This propensity is the result of its high charge density (Crumbliss & Garrison, 1988). Ferric ions react with silanol groups of monomeric silicic acid and its amorphous polymers to produce a silicate-iron co-ordination complex (Hazel *et al.*, 1949; Dugger *et al.*, 1964; Olson & O'Melia, 1973).

$$Fe^{3+} + m\,(\text{-SiOH}) \longleftrightarrow Fe\,(OSi)_m^{3m+} + mH^+$$

Free energy release of silanol groups with ferric ion is greater than with any other metal ion in amorphous silica (Dugger *et al.*, 1964). Consequently, Fe(III) will displace all other metal cations from a silanol surface if no other restrictions or conditions other than thermodynamic drive apply. Dose-dependent co-ordination of iron has also been demonstrated for several crystalline silicates (Fordham, 1969; Herrara & Peech, 1970; Schindler & Stumm, 1987; Rees, 1982). Ferric ion is adsorbed onto a silicate surface even at a pH of 2.0, its isoelectric point where the silanol groups would not normally be ionized, suggesting that non-ionic forces assist in the formation of a silicate-iron complex (Dugger *et al.*, 1964).

Regarding the source of the complexed metal ion on the particle surface: unco-ordinated iron does not exist in a biological system because of its potential activity in reduction-oxidation reactions. The stability constants of ferric ion with silica and silicate

TABLE 2. Relative ranking of tumor/fiber ratios for mesothelioma induction in rats

	intrapleural inoculation		intratracheal instillation	
	ratio	normalized for chrysotile	ratio	normalized for chrysotile
chrysotile	0.076	1.0	0.044	1.0
crocidolite	5.2	68.4	1.58	35.9
erionite	28.8	378.9	38.0	863.6

data from Coffin *et al.*, 1991.

particles have been estimated to approach 10^{17} (Schindler & Stumm, 1987). Dusts are therefore unlikely to successfully compete with the transport and storage proteins of the body for iron. However simultaneous resolution of multiple equilibria to calculate products of silicate-iron complexes in the presence of body chelators requires consideration of ligand concentrations. Transferrin, lactoferrin and ferratin are in small concentration relative to the enormous burdens of silicate dusts which can accumulate in the lung (Rivers *et al.*, 1960). Under these conditions, silica and silicates should be able to compete successfully for body iron. Another potential source of iron which could be coordinated onto the crystal surface is the cellular pool associated with low molecular weight compounds (Jacobs, 1977). This is a poorly characterized yet recognized iron chelate whose critical stability constant with ferric iron is not quantified. It may be an important iron source available for co-ordination to these dusts. Following inhalation, coordination of iron from body sources onto the inert dusts over a prolonged period of time would account for both the low serum iron observed among workers and the high iron content of their lungs (Niculescu *et al.*, 1981; Guest, 1978). The ferruginous body, which is recognized as an index of exposure to many different silicates, provides a direct visual evidence that these dusts are successful in attaining and co-ordinating body iron.

Complexed iron, with at least one co-ordination site free, has a capacity to mediate electron exchange via the Fenton reaction when it is reduced from the ferric to the ferrous state (Cohen, 1985).

$$Fe^{3+} + O_2^- \longleftrightarrow Fe^{2+} + O_2$$
$$Fe^{2+} + H_2O_2 \longleftrightarrow Fe^{3+} + OH^{\cdot} + OH^-$$

The ability of silica and silicates to generate oxidants via the Fenton reaction has been established using silica, asbestos, clays and talcs suggesting a role for iron in oxidant generation after exposure to these dusts (Kennedy *et al.*, 1989). In addition to presenting incompletely co-ordinated iron on the surface, silicate crystals may catalyse electron transfer and the reduction-oxidation cycling of iron cation (Gerstl & Banin, 1980). Other requirements of a Fenton reaction include available hydrogen peroxide and reductants. Hydrogen peroxide can be produced by activated phagocytes or as a by-product of aerobic metabolism. In fact, production of hydrogen peroxide by alveolar macrophages is increased after exposure to asbestos when compared to controls (Rom *et al.*, 1987). An important reductant in biological systems is the superoxide anion O_2^-, which is formed by the respiratory burst of phagocytes and also by aerobic metabolism (Graf *et al.*, 1984). Another substance that can actively reduce iron is ascorbate which is present in the extra-cellular fluid of the lung (Slade *et al.*, 1985). The generation of hydroxyl radical can result in DNA strand breaks which is considered to be an initial step in genotoxicity and cancer (Floyd, 1981).

There appears to be no question that endogenous ferric iron plays a significant role in pulmonary biological alterations due to various fibers or non-fibrous silicates. Mesothelioma however appears to be induced (1) only by long thin fibers and (2) erionite, a long thin fiber with central voids. Erionite induces mesothelioma to an extent nearly three orders of magnitude greater than other long thin silicate fibers. We propose the following hypotheses:

1. Mineral fibers are transported via the lymphatics from the airway to the superficial plexi in the pleura. Pinocytosis by pleural cells places the fibers in proximity to the chromatin material. (Penetration of chrysotile fibers into pulmonary epithelial and endothelial cells *in situ* has been reported by Barry and others (1983), chromosomal alterations by Palekar and others (1987) and strand breaks by Libbus and others (1989).)

2. Tumoriginesis induction may be confined to long thin fibers either because of greater accumulation due to slower clearance or possibly because there is better opportunity for interaction with chromatin material by such fibers. This latter view may fit the apparent size cut off better than differences in clearance rates.

3. Superior mesothelioma induction by erionite is the result of: (a) increased surface: 210 m^2/g versus 10 for crocidolite for example, and (b) more particularly the fact that most of the erionite surface is contained in the internal void spaces which will freely admit iron ions or cations but blocks the large molecules of free radical scavengers, (c) the prolongation of free radical lifetimes by sequestration within the void space. This phenomenon has been shown by Pryor and others (1990) to occur in particles formed by the pyrolysis of perfluoro polymers.

CONCLUSIONS

Experimental data confirm the putative difference in human exposure to mineral fibers which rank chrysotile, crocidolite and erionite in ascending potency for mesothelioma induction. Intrapleural and intratracheal experiments have shown similar ranking on the basis of the number of fibers to tumor ratios. These relationships hold for cytotoxicity and cytogenicity studies. Erionite which is approximately 800 times more potent than chrysotile for mesothelioma induction possesses unique interior voids. When these interior spaces were filled by isopentane the cytotoxicity of the mineral was reduced by approximately 50%.

A mechanism of action is proposed based on co-ordination of endogenous ferric iron with surface silanol groups of the silicate and the subsequent formation of hydroxyl radicals via the Fenton reaction. Sheltering of this reaction from inactivation by radical scavengers within the interior spaces probably accounts for the unique potency of erionite. The half life of these radicals may also possibly be prolonged by such entrapment within a solid.

ACKNOWLEDGEMENTS

These studies were carried out under the aegis of the United States Environmental Protection Agency Health Effects Laboratory durable minerals program. The contents of this paper do not necessarily reflect the views and policies of the agency nor does mention of trade names or commercial products constitute endorsement or recommendation for use.

REFERENCES

Artvinli,M. and Baris,G.I. (1982) Environmental fiber-induced pleuropulmonary diseases in an Anatolian village: an epidemiologic study. *Arch. Environ. Hlth* **37**:177-181.

Aust,S.D. (1987) Source of iron for lipid peroxidation in biological systems. In "Proceedings of a Brook Lodge Symposium", ed. B.Halliwell. Augusta, Mich., U.S.A. April 27-29. Fed. Am. Socs. Exp. Biol., Bethesda MD, USA, pp. 27-33.

Barry,B.E., Wong,K.C., Brody,A.R. and Crapo,J.D. (1983) Reaction of rat lungs to inhaled chrysotile asbestos following acute and sub-chronic exposures. *Exp. Lung Res.* **5**:1-22.

Coffin,D.L., Cook,P.M. and Creason,J.P. (1991) Relative mesothelioma induction in rats by mineral fibers: correlation with physical and chemical fiber properties and epidemiology. *Inhalation Toxicol.* (in press).

Coffin,D.L., Palekar,L.D., Cook,P.M. and Creason,J.P. (1989a) Comparison of mesothelioma induction in rats by asbestos and nonasbestos mineral fibers: possible correlation with human exposure data. Proceedings of a workshop: "Biological Interaction of Inhaled Mineral Fibers and Cigarette Smoke." eds. A.P. Wehner & D. Felton. pp. 347-354.

Coffin,D.L., Peters,S.E., Palekar,L.D. and Stahel,E.P. (1989b) A study of the biological activity of erionite in relation to its chemical and structural characteristics. Proceedings of a workshop: "Biological Interaction of Inhaled Mineral Fibers and Cigarette Smoke." eds. A.P. Wehner & D. Felton, pp. 313-323.

Cohen,G. (1985) The Fenton rection. In: "CRC Handbook of Methods for Oxygen Radical Research." ed. R.A. Greenwald, CRC Press, Boca Raton, FA. pp. 55-64.

Crumbliss,A.L. and Garrison,J.M. (1988) Comparison of some aspects of the aqueous co-ordination chemistry of Al^{3+} and Fe^{3+} *Comments Inorganic Chemistry* **8**:1-26.

Dugger,D.L., Stanton,J.H., Irby,B.N., McConnell,B.L., Cummings,W.W. and Maatman, R.W. (1964) The exchange of twenty metal ions with the weakly acidic silanol group of silica gel. *J. Phys. Chem.* **68**:757-760.

Eberly,P.E. (1969) Adsorption of normal paraffins in erionite and 5A molecular seive. *Ind. Eng. Chem.* **8**:140-144.

Floyd,R.A. (1981) DNA-ferrous iron catalyzed hydroxyl free radical formation from hydrogen perioxide. *Biochem. Biophys. Res. Comm.* **99**:1209-1215.

Fordham,A.W. (1969) Sorption and precipitation of iron on kaolinite. Factors involved in sorption equilibria. *Aust. J. Soil Res.* **7**:185-197.

Fournier,J., Nangura,N.W.A.,Guignard,J. and Pezerat,H. (1985) Absorption properties for PAH for asbestos and iron oxides in relation with their activities in biological .pa medium. In, "*In vitro* effects of mineral dusts", eds. Bignon,J. & Beck,E.G. Third International Workshop, NATO ASI series **G3** Springer-verlag, Berlin, pp. 276-279.

Gerstl,Z. and Banin,A. (1980) Fe^{2+} - Fe^{3+} transformations in clay and resin ion-exchange systems. *Clays and Clay Minerals.* **28**:335-345.

Ghio,A.J., Kennedy,T.P., Schapéra,R.M., Crumbliss,A.L. and Hoidal,J.R. (1990) The lung diseases after silicate inhalation caused by oxidant generation. *Lancet* **336**:967-969.

Graf,E., Mahoney,J.R., Bryant,R.G. and Eaton,J.W. (1984) Iron-catalyzed hydroxyl radical formation. Stringent requirement for free iron coordination site. *J. Biol.Chem.* **259**:3620-3624.

Grim,R.E. (1968) "Clay Mineralogy." McGraw-Hill Book Company, New York.

Guest,L. (1978) The endogenous iron content, by Mossbauer spectroscopy, of human lungs - Lungs from various occupational groups. *Ann. Occup. Hyg.* **21**:151-157.

Gutteridge,J.M.C. (1987) Lipid peroxidation: some problems and concepts in oxygen radicals and tissue injury. Proceedings of a Brook Lodge Symposium, Augusta, Mich., U.S.A. April 27-29. Fed.Am. Soc.Exp. Biol., Bethesda MD,USA pp. 9-19.

Gutteridge,J.M.C. (1982) The role of superoxide and hydroxyl radicals in phospholipid peroxidation catalyzed by iron salts. *FEBS Letts* **150**:454-458.

Hazel,F., Schock,U. and Gordon,M. (1949) Interaction of ferric ions with silicic acid. *J. Am. Chem. Soc.* **71**:2256-2257.

Herrara,R. and Peech,M. (1970) Reaction of montmorillonite with iron (III). *Soil Sci. Soc. Amer. Proc.* **34**:740-745.

Hochstein,P., Nordenbrand,K. and Ernster,L. (1964) Evidence for the involvement of iron in the ADP.activated peroxidation of lipids in microsomes and membranes. *Biochem. Biophys. Res. Comm.* **14**:323-328.

Hughs,J.M. and Weil,H. (1986) Asbestos exposure - quantitative assessment of risk. *Amer. Rev. Resp. Dis.* **133**:5-13.

Iler,R.K. (1979) "The Chemistry of Silica." John Wiley and Sons, New York.

Jacobs,A. (1977) Low molecular weight intracellular iron transport compounds. *Blood* **50**:433-439.

Kennedy,T.P., Dodson,R., Rao,N.V., Ky,H., Hopkins,C., Baser,M., Tolley,E. and Hoidal,J.R. (1989) Dusts causing pneumoconiosis generate OH· and produce hemolysis by acting as Fenton catalysts. *Arch. Biochem. Biophys.* **269**: 359-364.

Kragten,J. (1978) "Atlas of Metal-Ligand Equilibria in Aqueous Solution." Halstead Press, New York.

Libbus,B.L, Illeny,S.A. and Craighead,J.E. (1989) Induction of DNA strand breaks in cultured rat embryo cells by crocidolite asbestos as assessed by nick translation. *Cancer Res.* **49**:5713-5718.

Mossman,B.T., Bignon,J., Corn,M., Seaton,A. and Gee,J.B.L. (1990a) Asbestos: scientific developments and implications for public policy. *Science* **247**:294-301.

Mossman,B.T., Marsh,J.P., Sesko,A., Hill,S., Shatos,M.A., Doherty,J., Petruska,J., Adler,K.B., Hemenway,D., Mickey,R., Vacek,P. and Kagan,E. (1990b) Inhibition of lung injury inflammation and interstitial pulmonary fibrosis by polyethylene glycol-conjugated catalase in a rapid inhalation model of asbestosis. *Am. Rev. Respir. Dis.* **141**: 1266-1271.

Mossman,B.T., Marsh,J.P., Hardwick,D., Gilbert,R., Hill,S., Sesko,A., Shatos,M., Doherty,J., Weller,A. and Bergeron,M. (1986) Approaches to prevention of asbestos-induced lung disease using polyethylene glycol (PEG) conjugated catalase. *J. Free Rad. Biol. Med.* **2**:335-338.

Niculescu,T., Dumitru,R. and Burnea,D. (1981) Changes of copper, iron and zinc in the *Environ. Res.* **25**:260-268.

Olson,L.L. and O'Melia,C.R. (1973) The interaction of Fe(III) with $Si(OH)_4$. *J. Inorg. Nucl. Chem.* **35**:1977-1985.

Palekar,L.D., Eyre,J.F., and Coffin,D.L. (1989) Chromosomal changes associated with tumorigenic mineral fibers. Proceedings of a Workshop: "Biological Interaction of Inhaled Mineral Fibers and Cigarette Smoke." eds. Wehner,A.P. & Felton,D. pp. 355-372.

Palekar,L.D., Most,B.M. and Coffin,D.L. (1988) Significance of mineral fibers in the determination of V79 cytotoxicity. *Environ. Res.* **46**:142-152.

Palekar,L.D., Eyre,J.F., Most,B.M. and Coffin,D.L. (1987) Metaphase and anaphase analysis of V79 cells exposed to erionite, UICC chrysotile and UICC crocidolite. *Carcinogenesis* **8**:553-560.

Peters,S.E. (1988) Surface modification of erionite. A tumorigenic fibrous zeolite. Master of Science Dissertation, N.C. state university, Raleigh, NC, USA.

Peto,J., Seidman,H. and Selikoff,U. (1982) Mesothelioma mortality in asbestos workers: implications for models of carcinogenesis and risk assessment. *Br. J. Cancer* **45**:124-135.

Pryor,W.A., Nuggenalli,S.K., Scherer,K.V. and Church,D.F. (1990) An electron spin resonance study of particles produced in the pyrolysis of perfluoropropane. *Chem. Rev. Toxicol.* **3**:2-7.

Rees,L.V.C. (1982) Mossbauer spectroscopic studies of ferrous ion exchange in zeolite A. Proceedings of a Workshop: "Metal Microstructures in Zeolites. Preparation - Properties - Applications", eds. P.A.Jacobs, N.I.Jaeger, P.Jiru and G.Schulz-Ekloff. Bremen, September 22-24 Elsevier Scientific Publishing Co., New York City. pp. 55-59.

Rivers,D., Wise,M.E., King,E.J. and Nagelschmidt,G. (1960) Dust content, radiology and pathology in simple pneumoconiosis of coal workers. *Br. J. Ind. Med.* **17**:87-108.

Rom,W.N., Bitterman,P.B., Rennard,S.I., Cantin,A. and Crystal,R.G. (1987) Characterization of the lower respiratory tract inflammation of nonsmoking individuals with interstitial lung disease associated with chronic inhalation of inorganic dusts. *Am. Rev. Respir. Dis.* **136**:1429-1434.

Schindler,P.W. and Stumm,W. (1987) The surface chemistry of oxides, hydroxides and oxide minerals. In: "Aquatic Surface Chemistry. Chemical Processes at the Particle-Water Interface", ed. W.Stumm. John Wiley and Sons, New York, pp.83-110.

Slade,R., Stead,A.G., Graham,J.A. and Hatch,G.E. (1985) Comparison of lung antioxidant levels in humans and laboratory animals. *Am. Rev. Respir. Dis.* **131**:742-744.

Solomon,D.H. (1968) Clay minerals as electron acceptors and/or electron donors in organic reactions. *Clays and Clay Minerals* **16**:31-39.

U.K. Advisory Committee on Asbestos. Vol 1: (1983) Final Report of the Advisory Committee Paragraph 176-2, pp. 74-78.

Wills,E.D. (1965) Mechanisms of lipid peroxide formation in tissue: role of metals and haematin proteins in catalysis of the oxidation of unsaturated fatty acids. *Biochem. Biophys. Acta* **98**:238-251.

HISTOPATHOLOGICAL ANALYSIS OF TUMOUR TYPES AFTER INTRAPERITONEAL INJECTION OF MINERAL FIBRES IN RATS

S. Rittinghausen, H. Ernst, H. Muhle, R. Fuhst, and U. Mohr

Fraunhofer Institute of Toxicology and Aerosol Research
Nikolai-Fuchs-Str. 1
D-3000 Hannover, F.R.G.

INTRODUCTION

Female rats were treated twice by an intraperitoneal injection of mineral fibres. The treatment led to malignant tumours in the peritoneal cavity. Most of these tumours were classified as malignant mesotheliomas. Some rats had undifferentiated sarcomas rather than mesotheliomas.

Three different types of malignant mesotheliomas can be distinguished: epithelioid, sarcomatoid and mixed. Cartilaginous or osseous differentiation occurring in mesotheliomas has only been described as a histological feature in a few cases of mesotheliomas of the pleura in man (Goldstein, 1979; McCaughey, 1958; Yousem and Hochholzer, 1987). Some of the affected persons were known to have been exposed to asbestos (Goldstein, 1979; Yousem and Hochholzer, 1987). In the following report histopathological features of malignant mesotheliomas occurring in the abdominal cavity after intraperitoneal injection of different types of mineral fibres in female Wistar rats are described. Additionally, the number of animals with mesotheliomas, according to the injected fibre type is presented.

MATERIAL AND METHODS

Different types of mineral fibres were injected twice intraperitoneally into female Han:WIST rats (Central Institute for Laboratory Animal Breeding, Hannover) with a time interval of one week between the injections. Fibres were suspended in 2 ml 0.9% sodium chloride solution (pH 6-8). For details see Pott et al., (1976). Types of mineral fibres and gravimetric doses are presented in Fig. 11.

The rats were 11-12 weeks old at the start of the study and weighed about 230 g. The number of animals was 50 in each group. They were kept in groups of a maximum of 8 in Macrolon^R cages (type IV) during the first year of the study. Afterwards, they were caged individually in Macrolon^R cages (type II) to avoid losses by cannibalism. Rats were maintained under standard laboratory conditions (22 ± 2 °C, relative humidity 55 ± 5%, air exchange rate 8 times/h) and were given a commercial cereal-based diet (RMH-GS, Hope Farms, Woerden, Netherlands) and water ad libitum.

The animals were killed when moribund or with signs indicating abdominal neoplasms. Surviving rats were killed 130 weeks after the treatment. A complete necropsy was performed on all animals. Tumours were fixed in 10% formalin and representative samples of the neoplasms were embedded in paraffin wax. Sections (4 μm) were stained routinely with hematoxylin and eosin.

RESULTS

The tumour rate and the survival time of the treated rats are presented in Table 1. Number of animals with mesotheliomas, according to the injected fibre type, are presented in Fig. 11. Malignant mesotheliomas were observed in 139 of 450 rats examined. In 23 rats (16.5%) mesotheliomas with cartilaginous or osseous differentiation occurred. No mesotheliomas were observed in control animals treated with pure sodium chloride solution or cement as well as in rats after intraperitoneal injection of wollastonite. The rate of mesotheliomas in rats treated with asbestos cement, crocidolite, UICC-amosite, and UICC-chrysotile-B ranged from 62 to 76%. After Calidria-chrysotile injection, frequency of mesotheliomas was 2% (1 mg) or 10% (3 mg).

Macroscopically, neoplasms occurred as multiple grayish-white nodules in the abdominal cavity. Infiltration of diaphragm, abdominal wall, liver, intestine, pancreas, omentum, mesentery, uterus, and ovary were frequently observed. Kidneys and spleen were seldom affected.

Thirtytwo point four percent of the tumours were classified as purely epithelioid mesotheliomas (Fig. 1-3, and 12). These had a predominantly papillary growth pattern (Fig. 1, 3) with solid or pseudoglandular foci. The sarcomatoid malignant mesotheliomas (24.5% of mesothelioma-bearing rats) consisted of numerous spindle cells with atypical nuclei, frequent mitotic figures, and mesothelial cells on the surface (Fig. 4, 5). Abundant collagenous fibres were found between the tumour cells, giving the neoplasms a fibrosarcoma-like appearance. In 26.6% of rats with mesotheliomas, the mixed type of the neoplasm occurred (Fig. 6, 7). Frequently, epithelioid parts were located at the surface of these tumours. Occasionally tubular or solid epithelial structures were observed deep within sarcomatoid areas. In both mixed and sarcomatoid mesotheliomas, foci of osseous or cartilaginous differentiation (Fig. 8-10) were seen (12.9% and 3.6% of animals with mesotheliomas, respectively).

Some rats had undifferentiated sarcomas in the peritoneal cavity rather than mesotheliomas. These tumours consisted of spindle cells devoid of any mesothelial foci. After wollastonite injection, frequency of undifferentiated sarcomas was 2%. The rate of sarcomas in rats treated with asbestos cement or UICC-Chrysotile-B was 3% and 4%, respectively.

Figure 1: Amosite-induced malignant mesothelioma (epithelioid type). Papillary growth in mesenteric location. The centres of the villous structures consist of stroma with a low cellular content.H.E.; x 128.

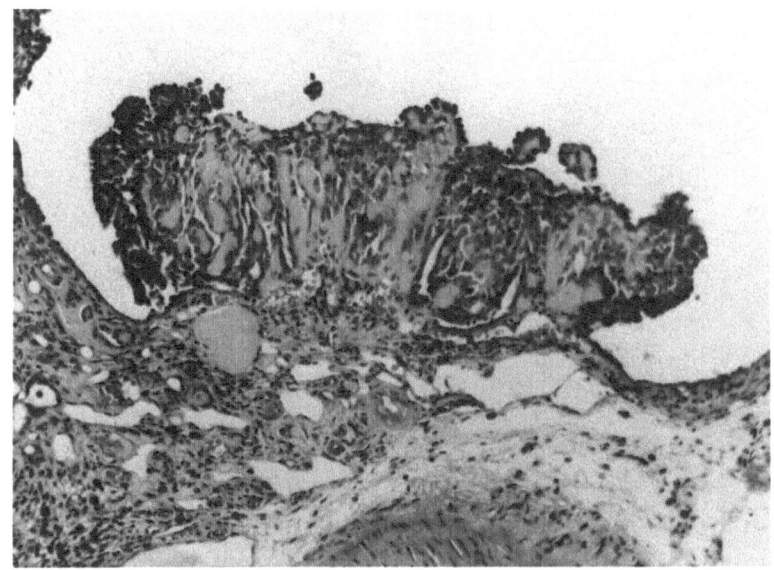

Figure 2: Plaque-like growth of a crocidolite-induced malignant mesothelioma (epithelioid type) on the pancreatic surface. H.E.; x 200.

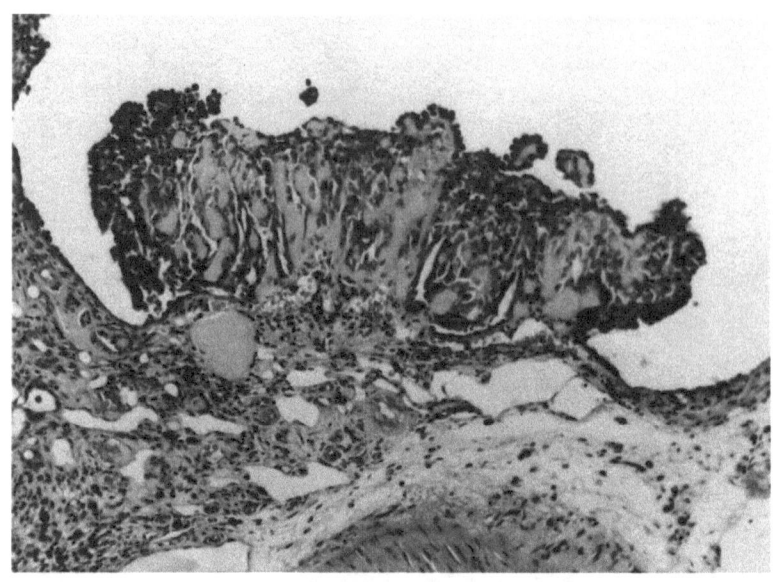

Figure 3: Amosite-induced papillary malignant mesothelioma (epithelioid type). Villous   tumour projections lined by neoplastic mesothelial cells. H.E.; x 240.

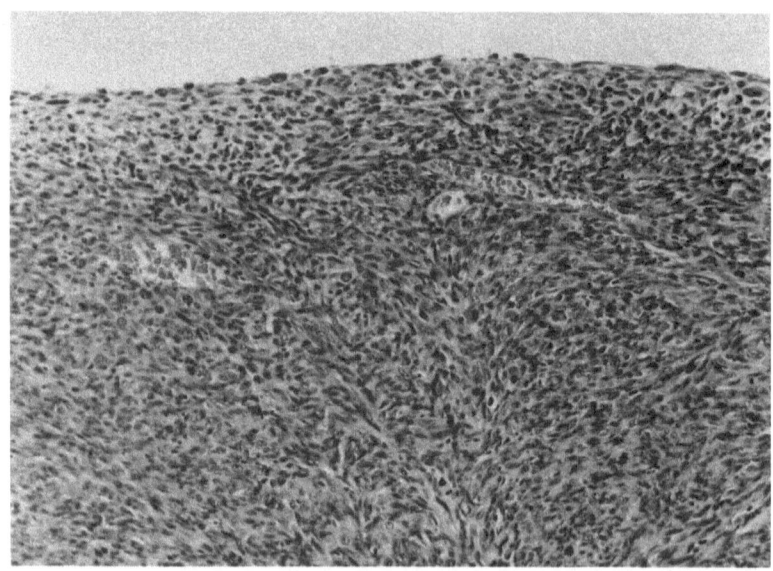

Figure 4: Malignant mesothelioma (sarcomatoid type) induced by asbestos cement. The   tumour which consists of interwoven bundles of oval to spindle-shaped cells has a surface lined by mesothelial cells. H.E.; x 200.

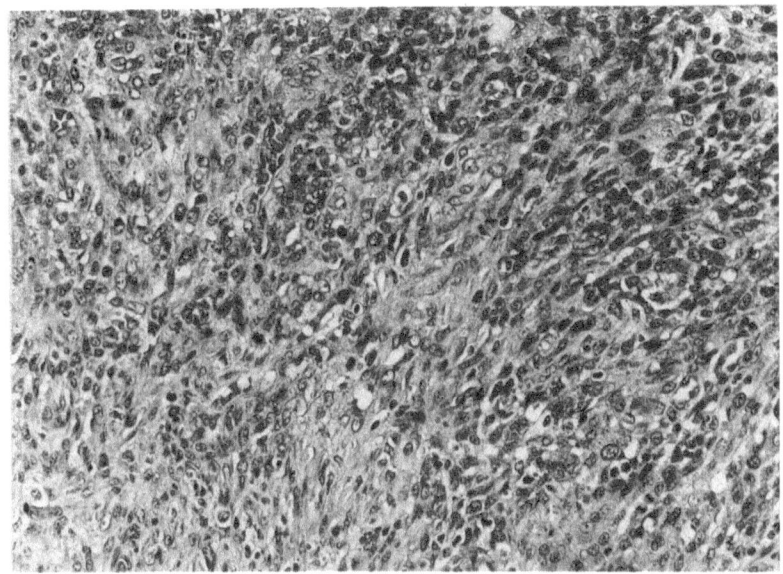

Figure 5: Amosite-induced malignant mesothelioma (sarcomatoid type). Note the high number of mitotic figures in spindle-cells of the neoplasm. H.E.; x 320.

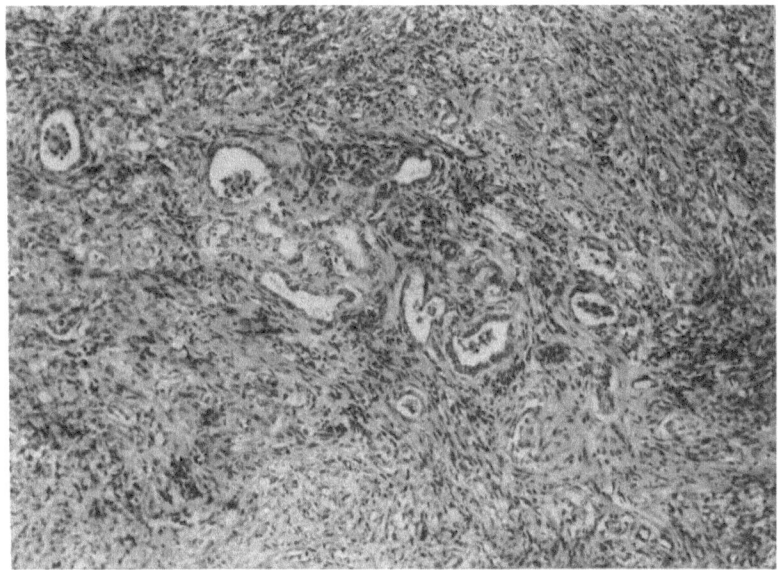

Figure 6: UICC-Chrysotile-B-induced malignant mesothelioma (mixed type) adjacent to the pancreas. Note pseudoglandular structures embedded in sarcomatous cells. H.E.; x 240.

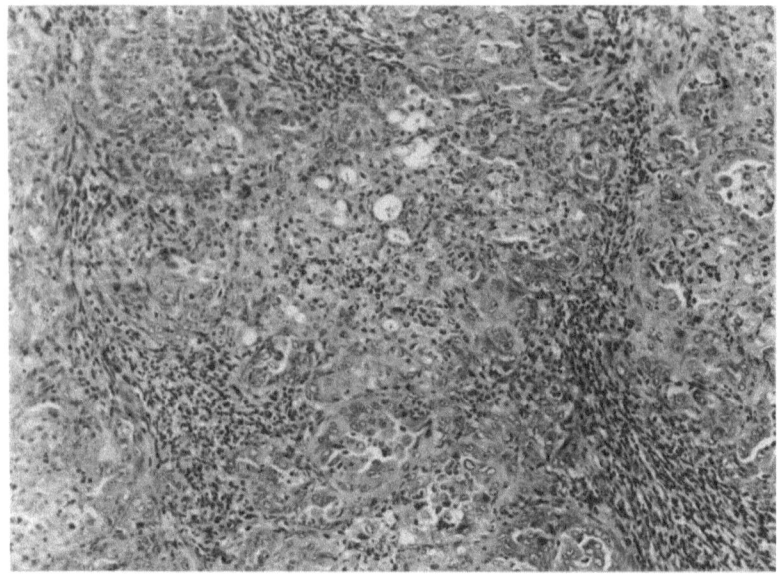

Figure 7: Malignant mesothelioma (mixed type) with carcinoma-like and sarcomatoid structures induced by UICC-chrysotile-B. H.E.; x 200.

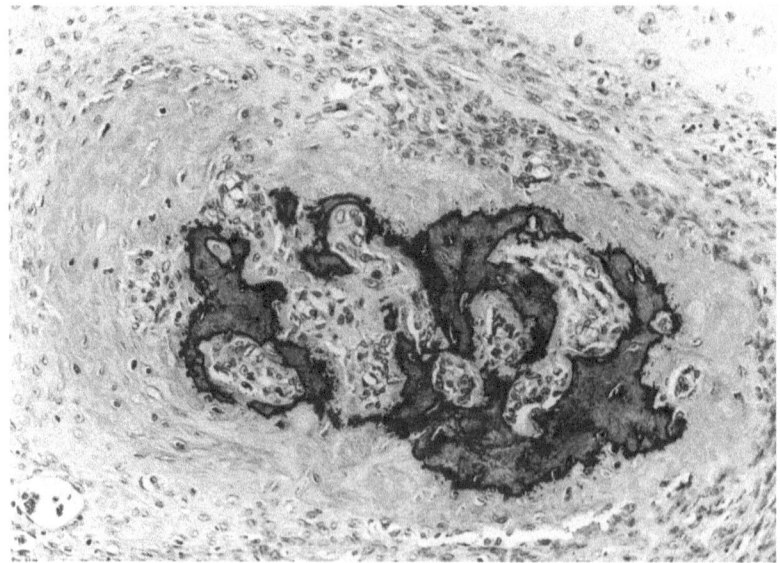

Figure 8: Focus of osseus differentiation in malignant mesothelioma (sarcomatoid type) induced by injection of asbestos cement. H.E.; x 200.

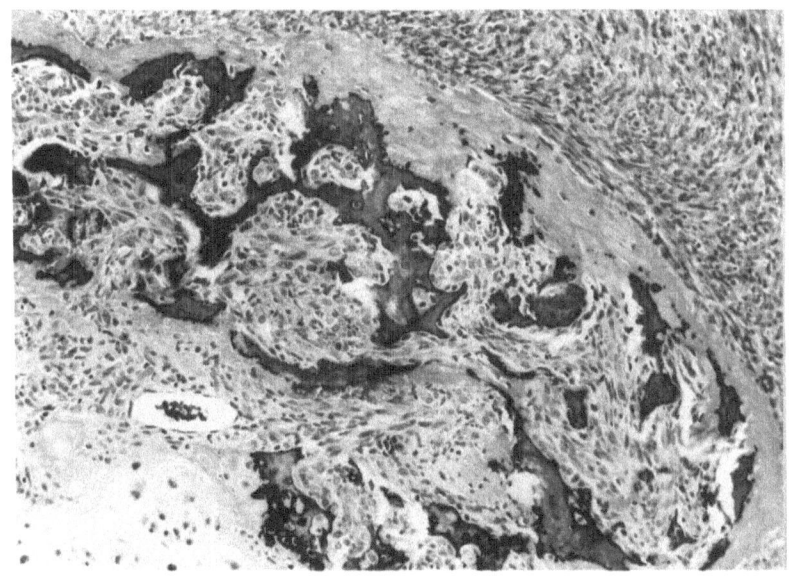

Figure 9: Sarcomatous portion of malignant mesothelioma (mixed type) with osseous and cartilaginous differentiation. H.E.; x 200.

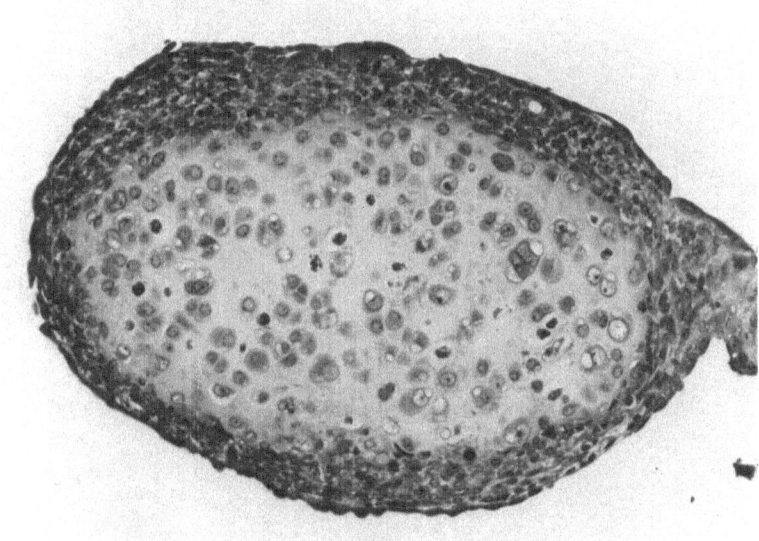

Figure 10: Villous proliferation of tumour cells with cartilaginous differentiation on the surface of a malignant mesothelioma (mixed type) induced by asbestos cement. H.E.; x 320

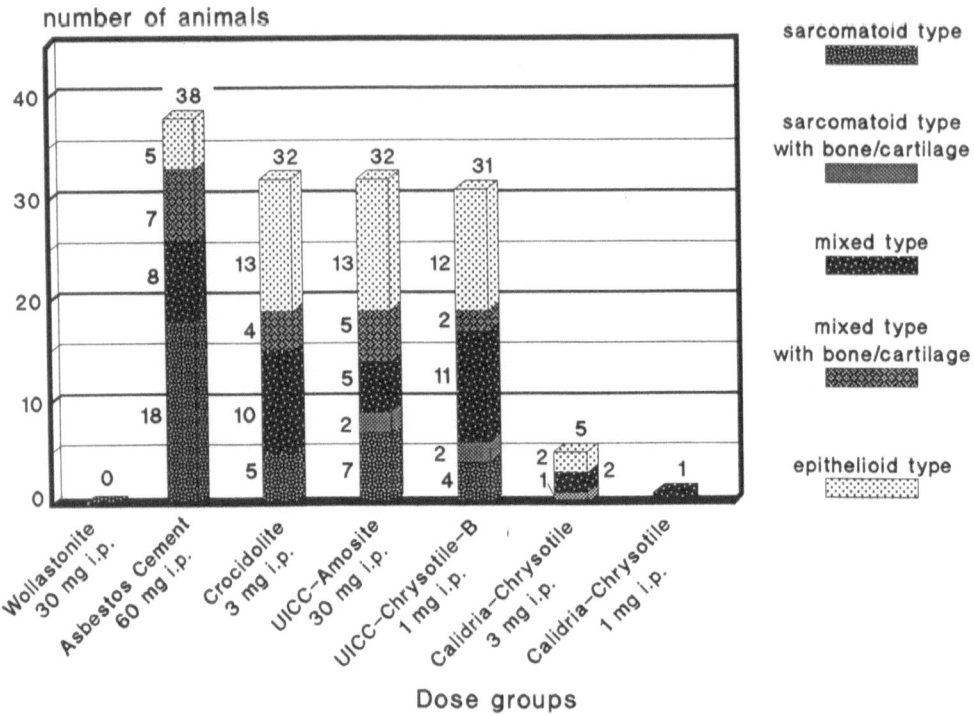

Figure 11: Number of animals with malignant mesotheliomas.

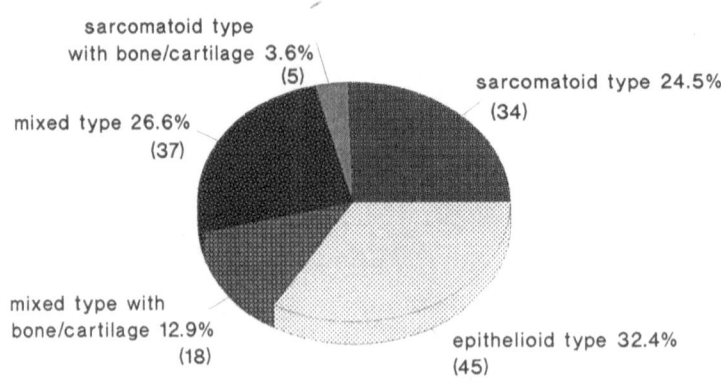

Figure 12: Types of malignant mesotheliomas.

CONCLUSION AND SUMMARY

Different histological types of malignant mesotheliomas and a few undifferentiated sarcomas were observed in Wistar rats after intraperitoneal injection of various types of asbestos fibres (asbestos cement, crocidolite, Calidria-chrysotile, UICC-amosite, UICC-chrysotile-B). Atypical peritoneal mesotheliomas containing bone and/or cartilage were found in 23 rat2 in the present study. The histological features of these mesotheliomas were different from most asbestos-induced mesotheliomas of rats described in the literature with respect to the frequency of osseous and cartilaginous differentiation in both sarcomatoid and mixed types. Mesotheliomas with calcification and bone formation have been described before by Davis (1974) after intraperitoneal injection of crocidolite asbestos. In one mesothelioma of a rat induced by inhalation of erionite large amounts of calcified spongy bone-like material were detected electron microscopically (Johnson et al., 1984). The capability of neoplastic mesothelial cells to form bone or cartilage was also confirmed by the study of Brown et al., (1985). Tumors arising in rats after subcutaneous injection of epithelial cell lines derived from a crocidolite induced mesothelioma revealed areas of cartilage as well as presence of bone formation.

The atypical mesotheliomas described in the literature and found in the present study resemble closely a rare type of pleural mesothelioma that occurs in man.

REFERENCES

Brown,D.G., Johnson,N.F., Wagner,M.M.F. (1985) Multipotential behavior of cloned rat mesothelioma cells with epithelial phenotype. *Br. J. Cancer* **51**:245-252.

Davis,J.M.G. (1974) Histogenesis and fine structure of peritoneal tumors produced in animals by injection of asbestos. *J. Natl. Cancer Inst.*. **52**:1823-1833.

Goldstein,B. (1979) Two malignant pleural mesotheliomas with unusual histology. *Thorax* **34**:375-379.

Johnson,N.F., Edwards,R.E., Munday,D.E., Rowe,N., Wagner,J.C. (1984) Pluripotential nature of mesotheliomata induced by inhalation of erionite in rats. *Br. J. Exp. Pathol.* **65**:377-388.

McCaughey,W.T.E. (1958) Primary tumors of the pleura. *J. Pathol. Bacteriol.* **56**:517-529.

Pott,F., Friedrichs,K.H. and Huth,F. (1976) Ergebnisse aus Tierversuchen zur kanzerogenen Wirkung faserförmiger Stäube und ihre Deutung im Hinblick auf die Tumorentstehung beim Menschen. *Zbl. Bakt. Hyg., Abt. I, Orig., Reihe B* **162**:467-505.

Yousem,S.A. and Hochholzer,L. (1987) Malignant mesotheliomas with osseous and cartilaginous differentiation, *Arch. Pathol. Lab. Med.* **111**:62-66.

RADIATION-INDUCED MESOTHELIOMAS IN RATS

F.F. Hahn, P.J. Haley, A.F. Hubbs, M.D. Hoover, and
D.L. Lundgren

Inhalation Toxicology Research Institute
P.O. Box 5890, Albuqurque
New Mexico, U.S.A.

Mesotheliomas have been reported in rats that inhaled plutonium, but have not been extensively studied (Dagle & Sanders, 1984). Rats injected intraperitoneally with 100 kBq of ^{239}PuO$_2$ developed mesotheliomas around focal, high concentrations of plutonium "hot spots" localized in the omentum (Sanders & Jackson, 1972). In this study, 27% of the rats developed peritoneal mesotheliomas (epithelial in morphology) and 38% developed sarcomas (sarcomatous mesotheliomas). The tumor incidence was dose dependent and a greater dose was required for the induction of epithelial mesotheliomas than for the sarcomas. When 3,4-benzo-[a]-pyrene or chrysotile asbestos fibers were injected with the ^{239}PuO$_2$, the resultant tumor incidence was additive (Sanders, 1973). The morphogenesis of the mesotheliomas from intraperitoneally injected ^{239}PuO$_2$ and chrysotile asbestos was similar.

Radiation-associated mesotheliomas have also been reported in human patients several years after therapeutic external radiation for Wilms' tumor, seminoma, thyroid ablation, teratocarcinoma, infiltrating breast carcinoma, or Hodgkin's disease (Peterson et al., 1984). Thorium dioxide, a radioactive, radiographic contrast media, has also been implicated in mesothelioma induction. Two cases have been described in which contamination of the peritoneal or pleural cavity was associated with mesothelioma appearing 25 and 36 years later.

To investigate a possible role for inhaled radionuclides in the induction of mesotheliomas, we reviewed four life-span studies conducted at ITRI in which rats had inhaled radioactive materials. A total of 3076 F344 rats were exposed by inhalation to aerosols of ^{239}PuO$_2$, mixed uranium-plutonium oxide, or ^{144}CeO$_2$. Results showed that a low incidence of pleural mesotheliomas was induced by either alpha- or beta-emitting radionuclides deposited and retained in the lung.

The first study reviewed involved 716 rats exposed once or repeatedly (7 times over one year) to monodisperse aerosols of $^{239}PuO_2$ (Lundgren *et al.*, 1990). The aerosols were about 1 μm AMAD with a geometric standard deviation (σ_g) of about 1.2. The initial lung burdens (ILB's) ranged from 0.11-1.5 kBq. Most of the plutonium particles were retained in the lung with a halftime of about 30 days (Fig. 1a). Nine to 11 percent of the initial lung burden was retained with a half-time of 790-1050 days.

The second study reviewed involved 461 rats exposed to aerosols of plutonium dioxide (69% $^{239/240}Pu$, 17% ^{238}Pu, 15% ^{241}Am), or uranium-plutonium oxides (43% $^{239/240}Pu$, 35% ^{238}Pu, 22% ^{241}Am) obtained from glove boxes in a nuclear reactor fuel fabrication facility (Mewhinney *et al.*, 1987). The aerosols were generated from dry powders and had an AMAD of about 3.5 μm (σ_g = 1.4) for PuO_2 and an AMAD of about 2.1 μm (σ_g 1.5) for the mixed uranium plutonium oxide. Most of the PuO_2 was retained with a half-time of about 41 days (Fig. 1b). Two percent of the ILB of PuO_2 was retained with an 810 day half-time. Most of the uranium-plutonium oxide material was retained with a half-time of about 15 days. Four percent of the ILB was retained with a halftime of 1700 days.

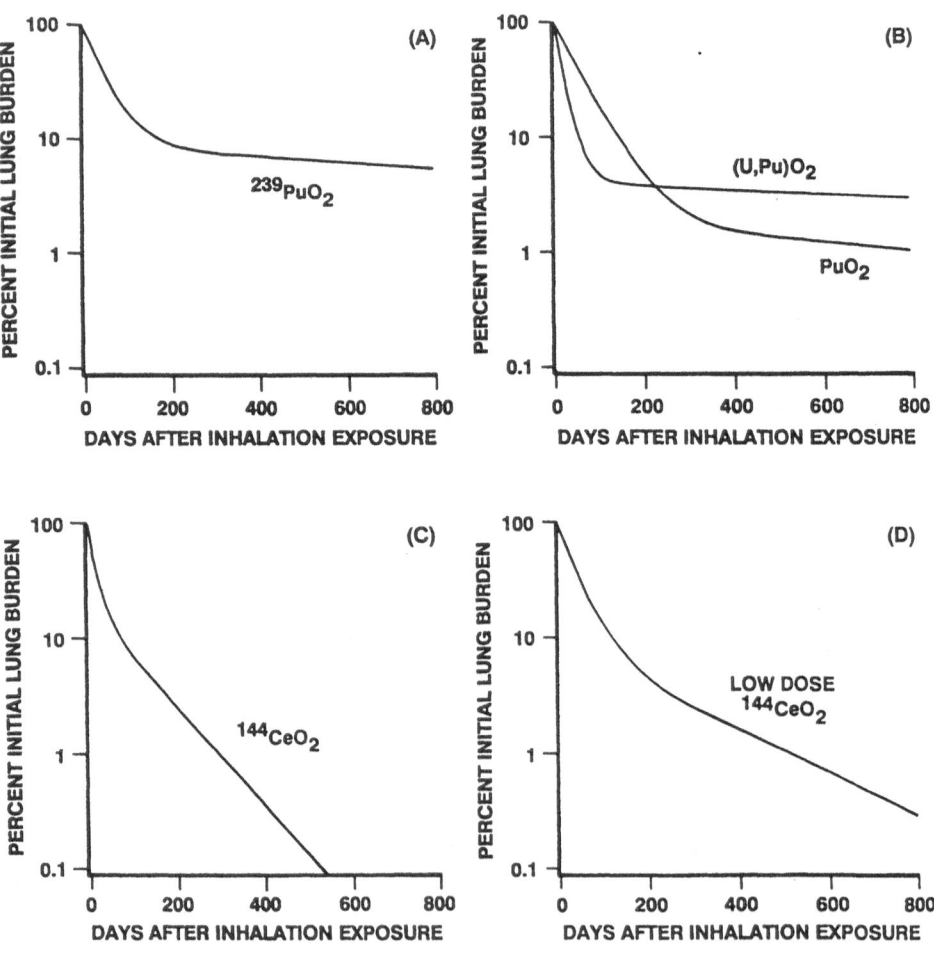

Figure   1. Retention patterns for radionuclides inhaled by F344 rats. a) $^{239}PuO_2$; b) PuO_2 or (U, Pu)O_2; c) $^{144}CeO_2$; d) $^{144}CeO_2$ low-dose exposure.

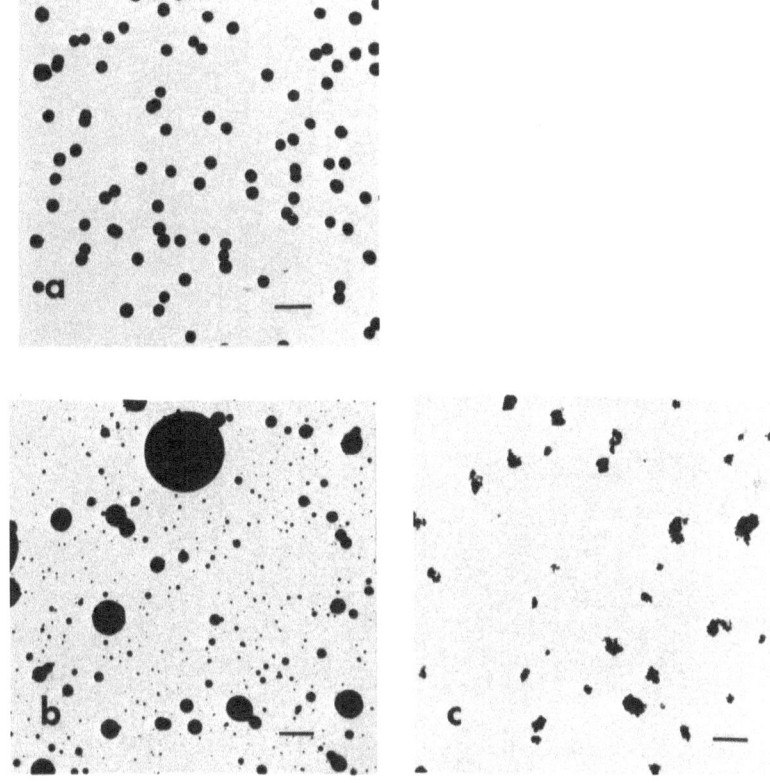

Figure 2. Exposure aerosol particles. a) Laboratory-produced monodisperse $^{239}PuO_2$ **particles - AMAD = 1.0 μm, σ_g 1.2. b) Laboratory-produced polydisperse** $^{144}CeO_2$ **particles - AMAD = 1.4 μm, σ_g 1.7. c) Mixed (U,Pu)O$_2$ particles from fuel fabrication facility - AMAD = 3.5 μm, σ_g 1.4. Bar = 1 μm.**

The third study reviewed involved 329 rats exposed once or repeatedly (7 times over one year) to aerosols of $^{144}CeO_2$ (Lundgren *et al.*, 1990). The aerosols had 0.9-1.4 μm AMAD (σ_g 1.4-2.0). The ILB's ranged from 13 to 40 kBq. For the rats exposed once, 82% of the cerium was retained in the lung with a half-time of 12 days and 18% with a half-time of 70 days (Fig. 1c). Repeated exposure prolonged the retention resulting in 48% being retained with a half-time of 15 days and 52% with a half-time of 100 days.

The fourth study reviewed involved 1570 rats exposed to relatively low doses of $^{144}CeO_2$ (Lundgren *et al.*, 1990). The aerosols were 0.9-1.4 μm (σ_g 1.4-2.0). The ILB's ranged from 13160 kBq. Most of the cerium was retained with a half-time of 29 days (Fig. 1d). Nine percent was retained with a half-time of 160 days.

A number of features were similar in all four studies. In all the exposures, aerosol particles were basically spherical in shape and respirable in size (Fig. 2). No fibers were present in the aerosols. The rats were exposed briefly (10-40 min) *per nasum*, to the aerosols. Approximately equal numbers of male and female F344 rats, raised in the Institute colony, were used in each study. The rats were observed twice daily until they died or were euthanized.

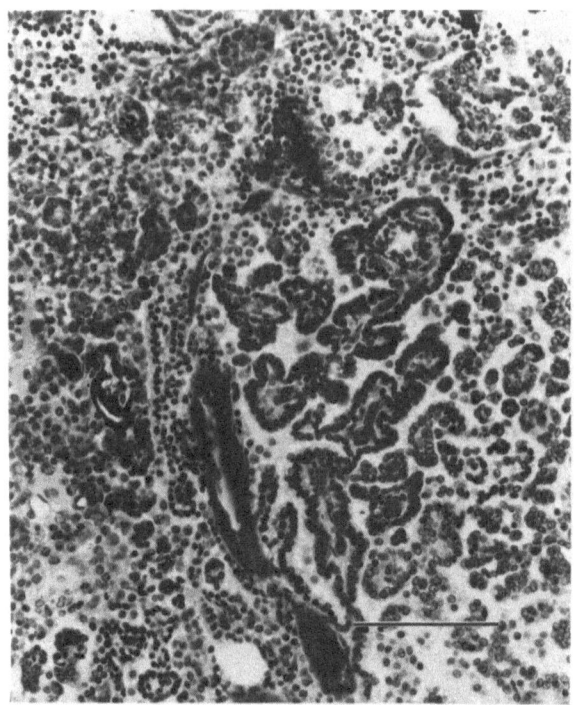

Figure 3. Mesothelioma, papillary epithelial pattern. Bar = 100 μm.

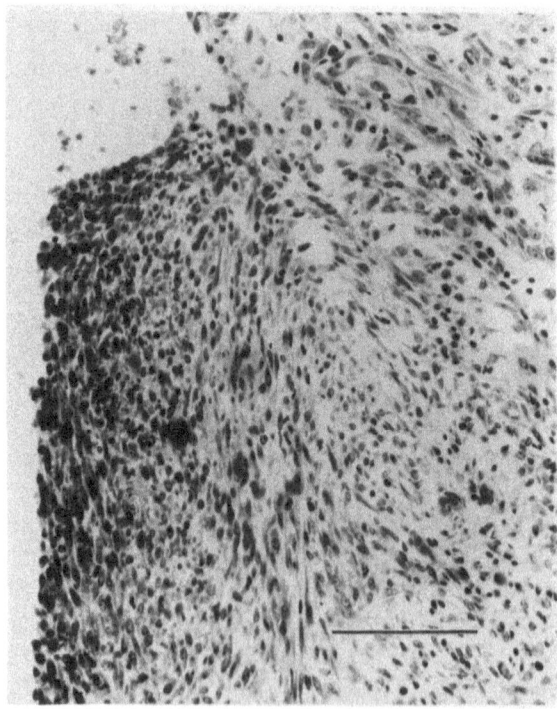

Figure 4. Mesothelioma, sarcomatous pattern. Bar = 100 μm.

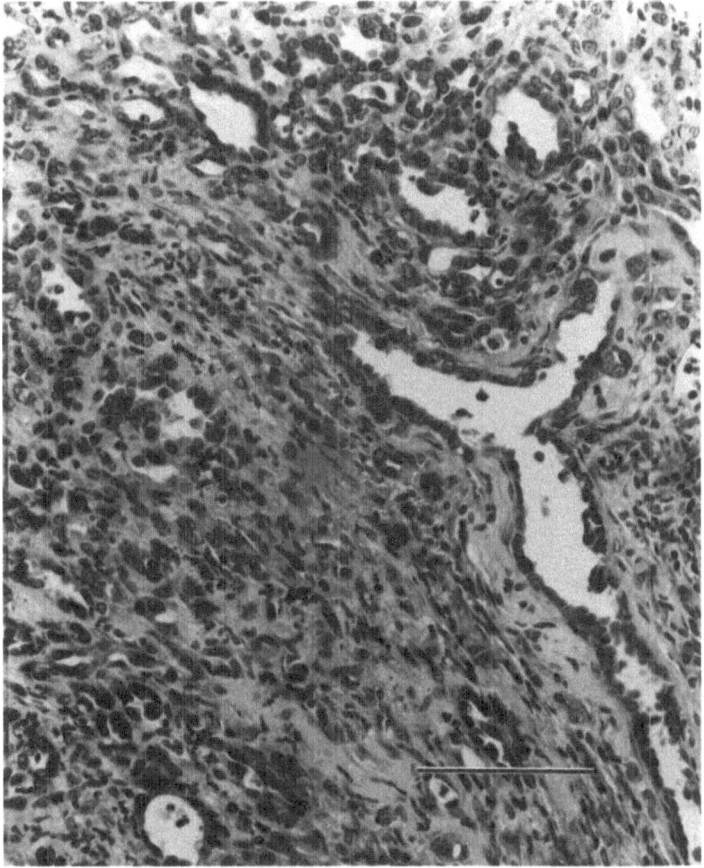

Figure 5. Mesothelioma, mixed pattern. Bar = 100 μm.

At death, each rat was necropsied and all major organ systems were examined. Routine paraffin-embedded, hematoxylin and eosin-stained sections were obtained from the lung and from lesions in the thoracic cavity. Sections of other organs were taken as required for diagnosis of lung lesions.

Over the range of doses used in the four studies, only in the highest dose groups of two of the studies ($^{239}PuO_2$ single exposure and $^{144}CeO_2$ repeated exposure) was the life span significantly reduced. These life spans were 86% and 80% of their respective control groups. This long survival indicates that competing risks were not an important factor in the tumor incidence analysis.

The morphologic patterns of the 28 pleural mesotheliomas found in the 3076 rats exposed to radionuclides are noted in Table 1. Epithelial mesotheliomas with a papillary pattern were the most frequently seen (23/28 cases) (Fig. 3). Tumor tissue was found spread over much of the mesothelial surface of the lung and caused thickening of the mediastinum. Most of these tumors were considered either fatal or contributory to death, based on the extensive growth of the tumors. Two of these tumors occurred in rats that also had squamous cell carcinomas of the lung. Four of the mesotheliomas occurred in lungs where focal fibrosis was prominent. This association of mesotheliomas in rats with other lesions occurred in rats receiving the highest radiation doses (> 17 Gy alpha or > 190 Gy beta).

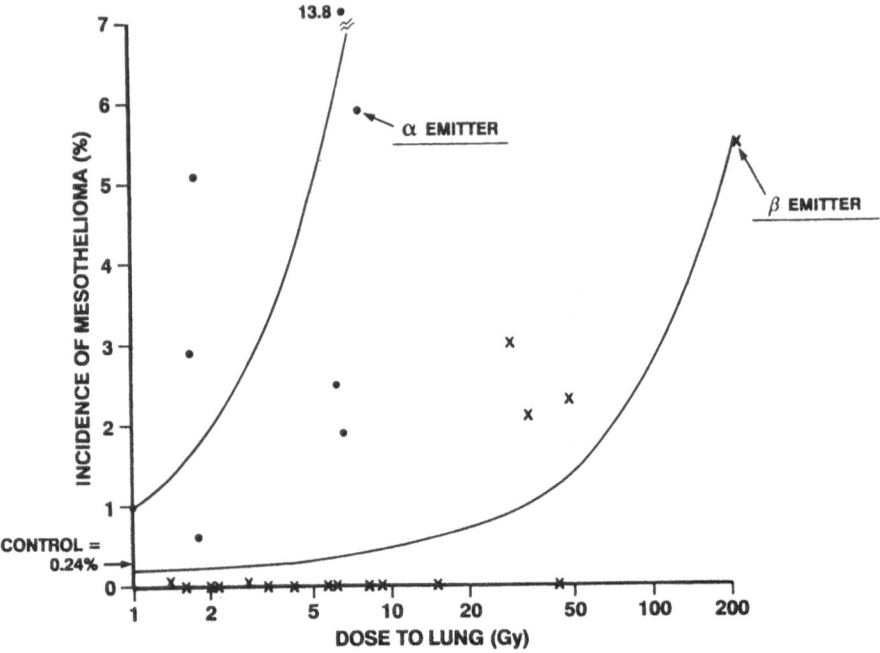

Figure 6. Dose response for mesothelioma after inhalation of alpha- or beta-emitting radionuclides. Mesothelioma incidence a linear function of dose according to the equation $Y = (0.95)X + 0.029$ for apha emitters (correlation coefficient = 0.72) and $Y = (0.025)X + 0.20$ for beta emitters (correlation coefficient = 0.87).

TABLE 1. Characteristics of Pleural Mesotheliomas in Rats with Lung Burdens of $^{144}CeO_2$, $^{239}PuO_2$, PuO_2 or $(U,Pu)O_2$

Morphologic Pattern	Total[a] Number	Number of Mesotheliomas				
		Fatal	Contributory to Death	Incidental to Death	With Focal Fibrosis	With Other Lung Tumors
Epithelial, Papillary, Focal	2	0	0	2	0	0
Epithelial, Papillary, Diffuse	21	13	7	1	4	2
Sarcomatous	2	1	1	0	1	2
Mixed	3	2	1	0	1	2

[a] In 3076 rats from four studies, including 716 rats exposed to $^{239}PuO_2$ once or repeatedly, 461 rats exposed to PuO_2 or $(U,Pu)O_2$, 329 rats exposed to $^{144}CeO_2$ once or repeatedly, and 1570 rats exposed to low doese of $^{144}CeO_2$.

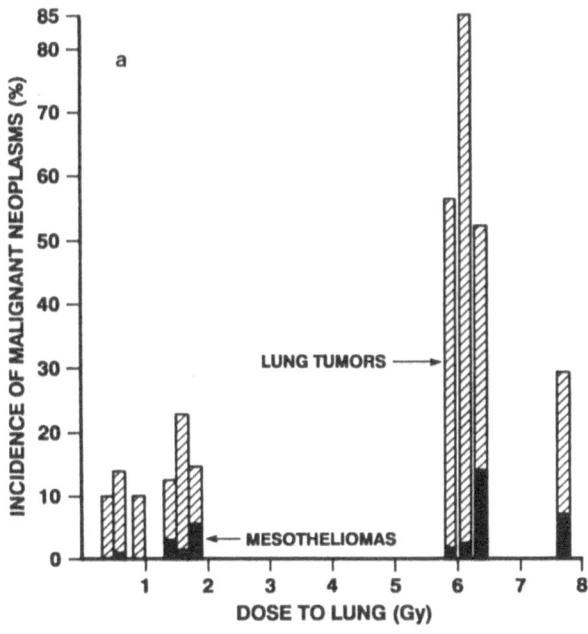

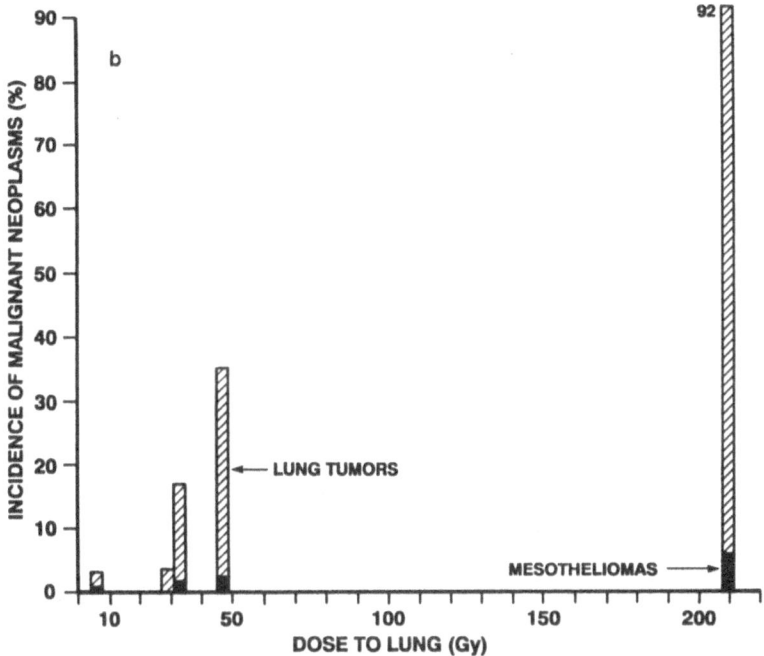

Figure 7. Incidence of malignant mesothelioma compared with incidence of other lung cancers in rats that inhaled: a) alpha-emitting radionuclides, or b) beta-emitting radionuclides. Control incidence = 4 mesotheliomas/1641 rats = 0.24%.

The sarcomatous mesotheliomas noted (2/28 cases) were characterized by marked proliferation of spindle-shaped cells with large accumulations of mature collagen (Fig. 4). The tumors appeared to arise in the hilus of the lung, invade the mediastinum and heart, and spread along the pleural surface of the lung. Other primary lung tumors were noted in the rats with sarcomatous mesotheliomas.

The three mixed mesotheliomas had both papillary epithelial elements and sarcomatous elements (Fig. 5). Two of the three rats with mixed mesotheliomas also had other primary lung tumors.

In none of the mesotheliomas was anaplasia marked or the mitotic rate more than minimally increased. Tumors were considered malignant if they invaded normal tissue, implanted along pleural surfaces, and were fatal or contributed to death. The incidence of malignant mesotheliomas is shown in Figure 6. The incidence for the combined control animals (4 Mesotheliomas/ 1641 control rats) was 0.24%. The incidence ranged from 0.8 to 13.8% for rats that inhaled alpha-emitting radionuclides and from 2 to 5% for rats that inhaled beta emitting radionuclides. The interrelationship of the incidence and dose to lung was tested with linear regression. For alpha emitting radionuclides, the relationship was described by $Y = (0.95)X + 0.029$ with a correlation coefficient of 0.72. For beta-emitting radionuclides, the relationship was $Y = (0.025)X + 0.20$ with a correlation coefficient of 0.87.

The incidence rate was only moderately correlated with increased radiation dose. The incidence of other primary lung tumors, however, was markedly increased by increased dose to lung and was much greater than the mesothelioma incidence (Fig. 7).

The chronic alpha irradiation was more effective than the chronic beta irradiation in causing mesotheliomas. The risk of a malignant mesothelioma for various groups of rats was determined using the formula:

$$Tumors/Gy = \frac{number\ of\ mesotheliomas}{average\ dose\ x\ number\ in\ group}$$

Such a calculation assumes a linear dose response. Figure 6 illustrates that a linear dose response is reasonable in this situation. Using this approach, the risk of mesothelioma for alpha-emitting radionuclides deposited in the lung was 90 tumors/ 104 Gy. For beta-emitting radionuclides, it was 6 tumors/ 104 Gy. The relative effectiveness ratio was:

$$\frac{Risk\ for\ \alpha\text{-}emitting\ radionuclides}{Risk\ for\ \beta\text{-}emitting\ radionuclides} = \frac{90\ tumors/10^4\ Gy}{6\ tumors/10^4\ Gy} = 15$$

The results of these four studies of rats with radionuclides deposited in the lung show that pleural mesotheliomas can be induced with either inhaled alpha- or beta-emitting radionuclides. Pulmonary fibrosis does not appear to be a prerequisite for these radiation-induced mesotheliomas as it appears to be for asbestos-induced mesotheliomas. The incidence of these tumors is dose dependent, but much less frequent than other radiation-induced lung tumors. Chronic alpha irradiation was more effective per unit dose in producing mesotheliomas than chronic beta irradiation of the lung by a factor of 15.

ACKNOWLEDGMENTS

Research supported by the Office of Health and Environmental Research, U. S. Department of Energy, under Contract No. DE-AC0476EV01013 and in facilities fully accredited by the American Association for Accreditation of Laboratory Animal Care.

REFERENCES

Dagle,G.E., Sanders,C.L. (1984) Radionuclide injury to the lung. *Environ. Hlth Perspect.* **55**:129-137.

Lundgren,D.L., Haley,P.J., Hahn,F.F., Griffith,W.C., Seiler,F.A., Diel,J.H. and McClellan,R.O. (1990) Pulmonary carcinogenicity of repeated inhalation exposure of rats to aerosols of ^{239}PuO$_2$. (in preparation).

Lundgren,D.L., Hahn,F.F., Griffith,W.C., Hubbs,A.F., Nikula,K.J., Newton,G.J., Cuddihy,R.C. and Boecker,B.B. (1990) Carcinogenic effects of relatively low doses of beta-irradiation to the lungs from inhaled ^{144}CeO$_2$ in rats. (in preparation).

Mewhinney,J.A., Eidson,A.F., Hahn,F.F., Scott,B.R., Seiler,F.A. and Boecker,B.B. (1987) Dose-response Study in F344 Rats Exposed to (U, Pu)O$_2$ or PuO$_2$. In: "Radiation Dose Estimates and Hazard Evaluation for Inhaled Airborne Radionuclides". Final Report to Office of Nuclear Regulatory Research, U. S. Nuclear Regulatory Commission, Washington, DC, NUREG/CR-4986, pp. 61-89.

Peterson,J.T., Greenberg,S.D. and Buffer,P.A. (1984) Non-asbestos-related malignant mesothelioma - A review. *Cancer* **54**:951-960.

Sanders,C.L., Jackson,T.A. (1972) Induction of mesotheliomas and sarcomas from "hot spots" of ^{239}PuO$_2$ activity. *Hlth Phys.* **22**:755-759.

Sanders,C.L. (1973) Cocarcinogenesis of ^{239}PuO$_2$ with chrysotile asbestos or benzo-[a]-pyrene in the rat abdominal cavity. In: "Radionuclide Carcinogenesis", Sanders,C.L., Busch,R.H. , Ballou,J.E. and Mahlum,D.D. (eds.), NTIS, Springfield, VA, CONF-720505, pp. 138-153.

CHEMICAL AND PHYSICAL PROPERTIES

1. Fibre Size and Chemistry

CARCINOGENIC EFFECT RELATED TO THE FIBER PHYSICS AND CHEMISTRY

K. R. Spurny

Fraunhofer-Institute für Umweltchemie und Ökotoxikolgie
D-5948 Schmallenberg-Grafsschaft
FRG

ABSTRACT

The results of biological and physico-chemical investigations relating to fiber toxicology during the last decade indicate that a "black box" model could be developed for the correlation of the physical and chemical properties of fibrous particles and their toxic and carcinogenic effects. Fiber size, fiber persistency, fiber charge and catalytic properties show correlations to their experimentally determined biological effects. Practical importance of such correlations is discussed and a series of standardized screening tests is proposed.

INTRODUCTION

Theoretical chemists and chemical toxicologists have tried for the last 10 years to develop cheaper procedures for testing the toxicological properties of new chemicals. The "QSAR-computation model" has been proposed in which the structural properties of molecules as well as some physico-chemical properties of different compounds are correlated to their known *in vitro* or *in vivo* toxicity. There is a real hope that models can be applied to determine the toxic and carcinogenic potency of newly produced chemicals in the near future.

The biological and physico-chemical investigations relating to fiber toxicology during the last decade indicate that in principle, analogical models could be also developed for the description of biological effects of inhaled mineral fibers.

Animal experiments have shown different carcinogenic potency of different types of natural and man-made mineral fibers.

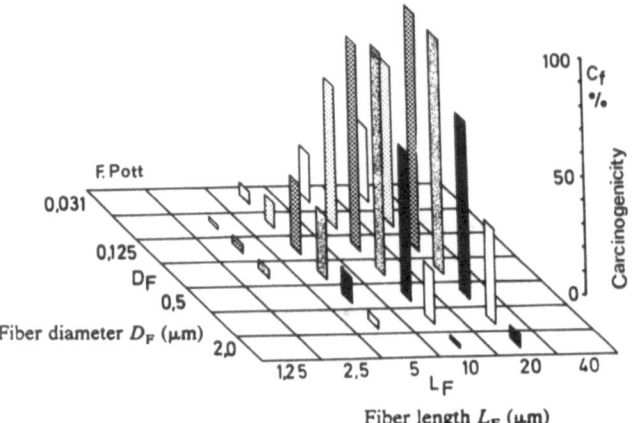

Figure 1. Chart depicting a hypothesis of carcinogenic potency of cylindrical fibers as a function of their lengths and diameters.

Following the hypotheses put forward by Stanton *et al.*, (1977) and Pott (1978) that respirable mineral fibers exert their effect largely by virtue of their dimensions, including width and length or surface. Further experimental investigation have shown that this "dimension-criteria" is a necessary factor but not the only factor. Other physico-chemical properties of single fibers deposited in the tissue may be responsible for their biological effects, these properties include, fiber persistency, fiber electric charge, catalytic activities by production of OH-radicals, etc (Walton *et al.*, 1987; Bignon *et al.*, 1989).

FIBER SIZE HYPOTHESIS

The results of experiments with asbestos and other mineral fibers carried out in the last 20 years have led to a general agreement that fibers exhibit their carcinogenic properties because of their elongated shape, if sufficiently long and thin. Following the experiments of Stanton as well as their own experimental results, Pott (1987) has

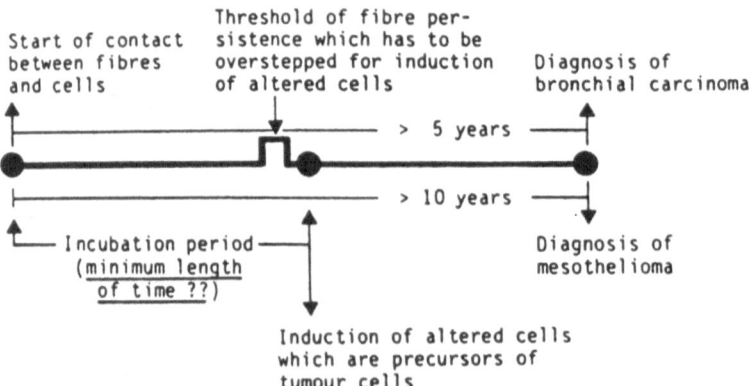

Figure 2. Diagram illustrating the question of the minimum length of the incubation period necessary for the induction of precursors of tumor cells.

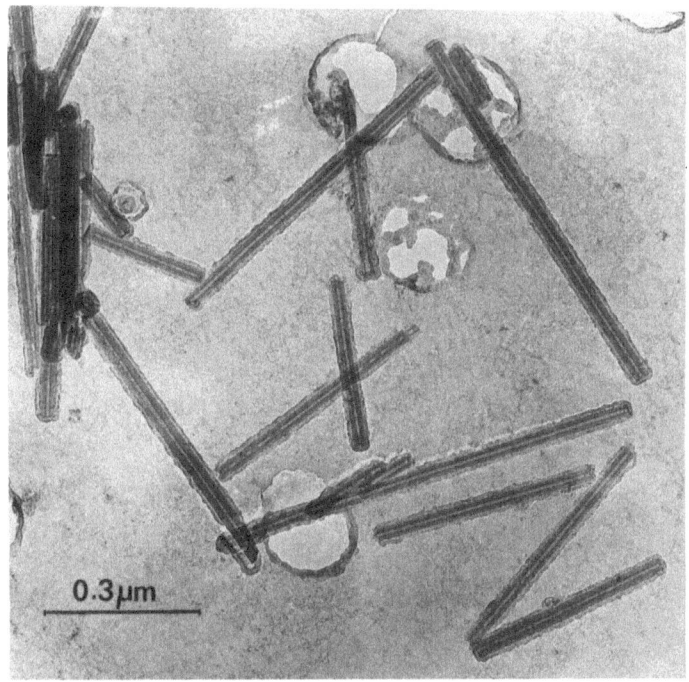

Figure 3. Transmission electron micrograph of chrysotile asbestos fibrils

proposed a semiquantitative hypothesis for the correlation between fiber size (Fiber length L_F and fiber diameter D_F), which is shown as a histogram in the Fig. 1. The carcinogenic potency C_f (%) is a function of L_F and D_F (μm). Following this hypothesis, we can see that also fibers shorter than 5 μm have some small carcinogenic potential.

BIOLOGICAL FIBER DEGRADATION

It was found that neither asbestos or manmade mineral fibers were resistant to biological fluids or inorganic solvents (Spurny *et al.*, 1983). The fibers undergo changes in their physical morphology and chemical composition. Biologically active fibers have to be chemically durable and to persist for a certain time in the tissue in order to induce a tumor. Figure 2 illustrates the question of the minimum time that fibers have to persist to alter cells (Pott 1987).

The size distribution as well as the physical and chemical persistence of new manmade or natural mineral fibers can be tested by methods, which have been already developed and described (Spurny *et al.*, 1983).

Figures 3 to 5 show some examples of the testing procedures and the results obtained.

The fiber size plays an important role not only in the evaluation of the possible carcinogenicity, but is also responsible for the deposition and retention of fibrous particles in the air ways (Fig. 6).

The degradation and solubility tests can be done by means of chemical procedures, such as dissolving fiber samples in strong or weak, organic and inorganic acids (Fig. 7).

The solubility, degradation or leaching of inorganic fibers can also be examined *in vivo* following either inhalation exposure or intratracheal instillation. Morphological and chemical changes are then measured as a function of exposure time (Fig. 8).

Another important parameter characterizing the fiber persistency in the tissue is the half-times of lung clearance. It can be estimated by animal instillation experiments. The faster the clearance, the smaller is the carcinogenic potency of the fiber type (Walton *et al.*, 1987, Bignon *et al.*, 1989).

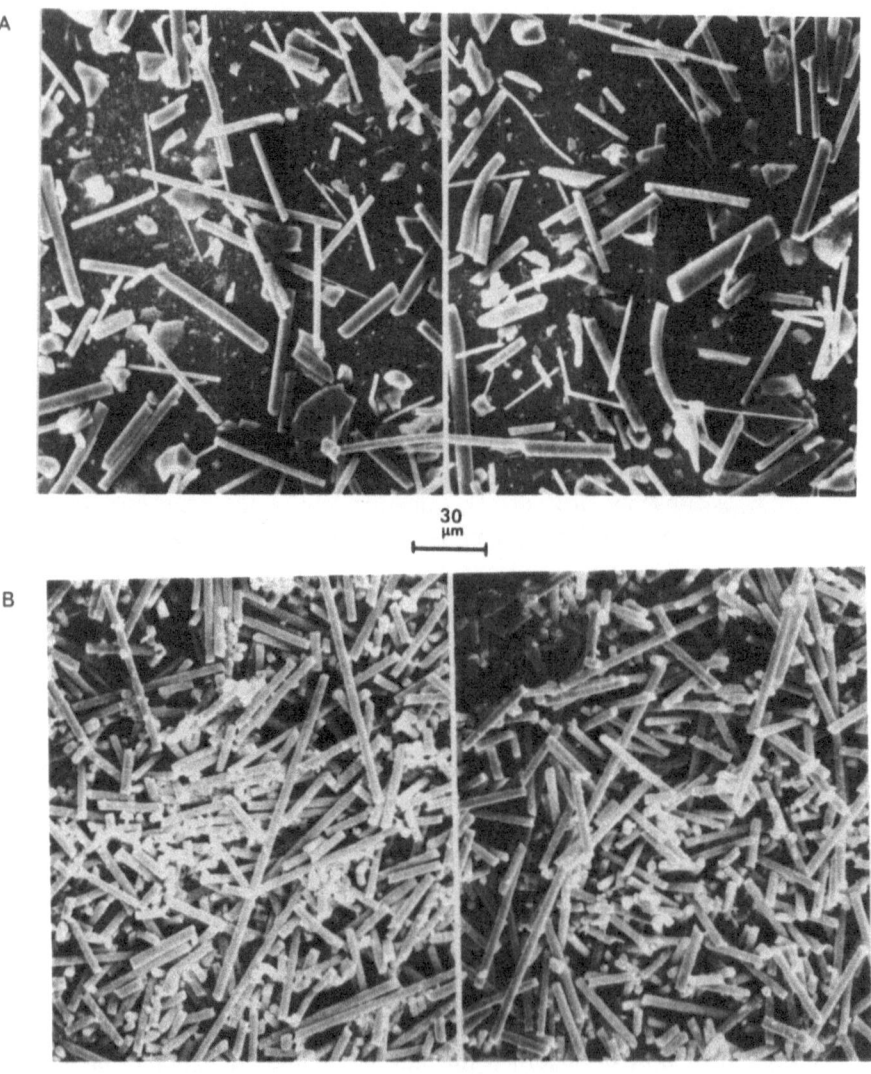

Figure 4. Scanning electron micrographs of man made mineral fibers: glass fibers (A) and mineral wool fibers (B)

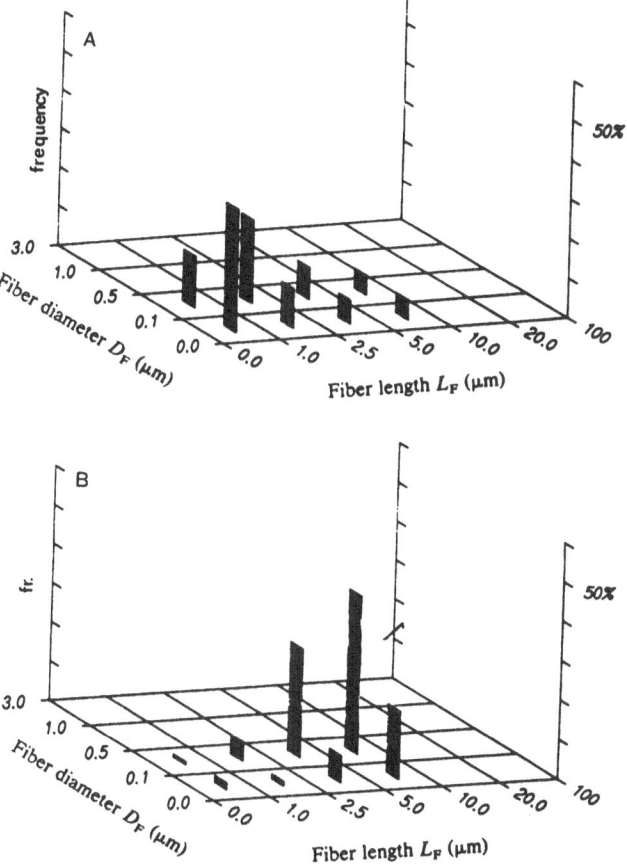

Figure 5. Diagrams illustrating the measured bivariate fiber size distributions: size distribution obtained by transmission electron microscopical evaluation of an ambient air chrysotile sample (A) as well as a transformation of this size distribution (B) done by multiplying the frequency values by carcinogenic factors (Fig. 1).

SURFACE FIBER PROPERTIES

The physical and chemical persistency is included in the second of "Pott's hypothesis", which states that the fiber pathogenicity depends also on the chemical persistence of mineral fibers deposited in the lung. Results of new physico-chemical and biological investigations support this hypothesis. The role of other physico-chemical and catalytic properties of fiber surface is not very well understood as yet, but seem to be of non-negligible importance.

Some recently published results indicate an important role for the electric charge of inhaled fibers (Light *et al.*, 1977) in relation to their toxicity.

The haemolytic activity of asbestos and other mineral fibers may be determined by the absolute magnitude of the surface electric charge that the fibers acquire in solution.

The surface electric charge is represented and measured by the zeta potential (mV) and depends on the fiber length (aspect ratio) (Light *et al.*, 1977).

The surface electric charge, which can be positive or negative, seems to play an important role in the adsorption of nucleic acids on inhaled fibers. It appears that DNA and RNA are adsorbed on fibers. The deletion or other changes at the DNA level by adsorption or catalysis may be involved in fiber carcinogenicity.

Mineral fibers, containing different trace metals, have been shown to work as biochemical catalysts and may generate the toxic hydroxyl radicals from a normal by-product of tissue metabolism, hydrogen peroxide. The more biological durable a fiber is, the higher is the production of hydroxyl radicals (Case *et al.*, 1986).

The interaction of chrysotile asbestos with alveolar macrophages *in vitro* has been shown to stimulate cellular superoxide anion production (Scheule *et al.*, 1989). Chrysotile asbestos fibers can catalyze the formation of hydrogen peroxide in water-air-rich medium. Its catalytic properties are dependent transitional metal ions, e.g. Fe (Korkina *et al.*, 1988).

The potential consequences of the all surface properties of inhaled fibrous particles in relation to their biological activity, toxicity and carcinogenicity are numerous, such as:

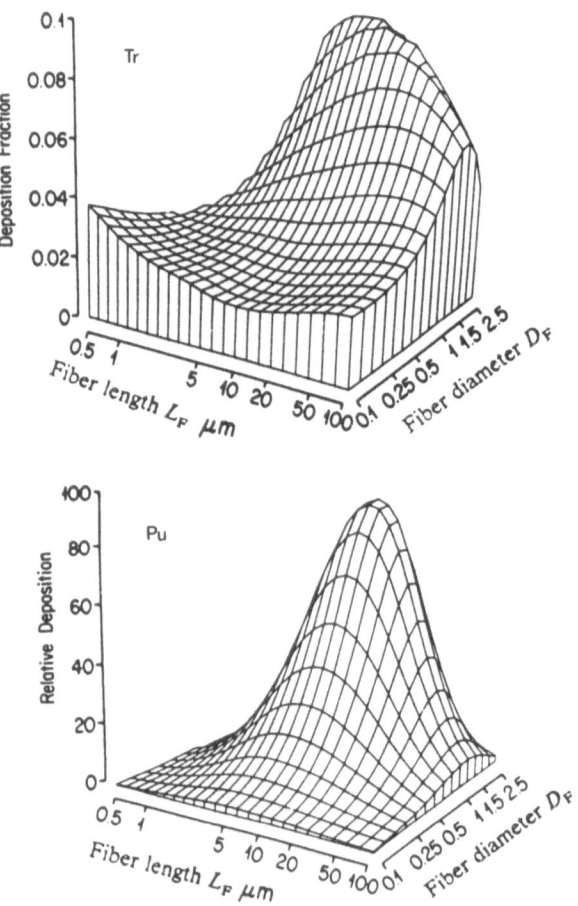

Figure 6. Pulmonary deposition of fibrous particles *via* rat trachea (Tr) and lung (Pu).

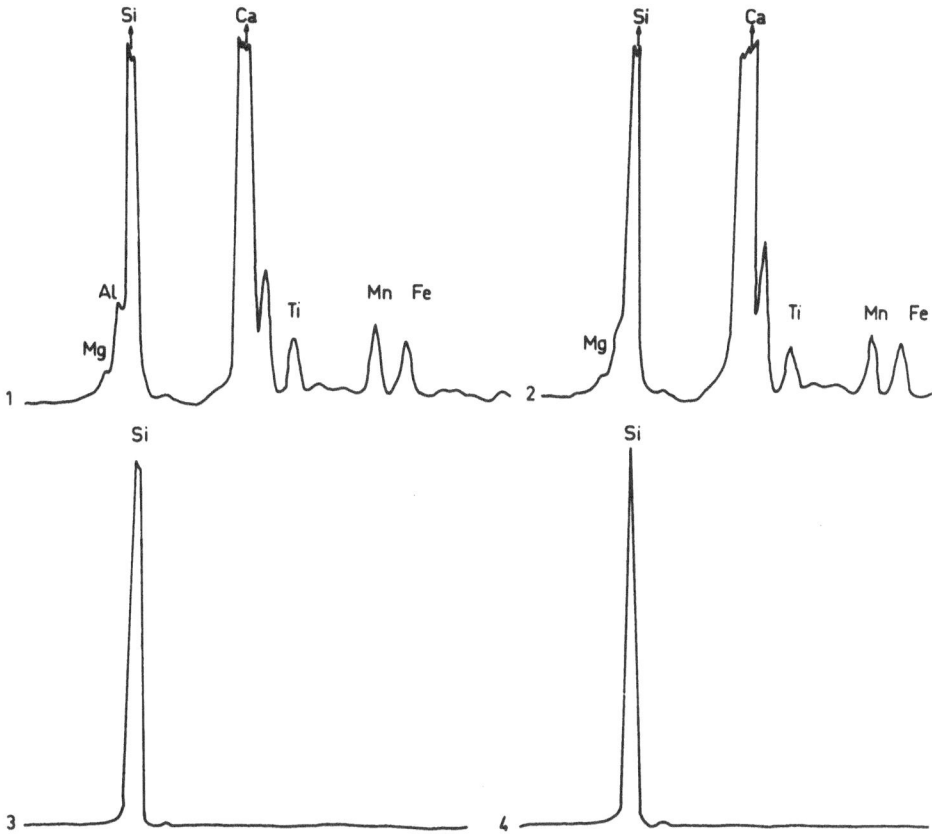

Figure 7. X-ray fluorescence (XRF) multielement analysis of basalt-wool fibers after leaching by (1)H$_2$O, (2) NaOH, (3) HCl and (4) H$_2$SO$_4$. The good solubility in acidic liquids is evident.

cell membrane-fiber adhesion, adsorption of functionally or structurally essential biological components, transport of foreign substances to target organisms, catalytic transformation of biochemical intermediates, etc (Jolicoer *et al.*, 1987).

Crocidolite asbestos efficiently binds and catalyzes the oxidation of different PAH. This may be a mechanism whereby inhaled asbestos enhances the incidence of lung cancer induced by cigarette smoke, which contains benzo(a)pyrene (Graceffa *et al.*, 1987). Synthetic chrysotile containing Mg, Ni or Co were used in animal experiments to estimate their carcinogenic potency. Fibers containing Ni or Co were more carcinogenic than Mg-Chrysotile (Visilieva *et al.*, 1989).

Exposures of low levels of O$_3$ have also results in enhanced pulmonary retention of inhaled asbestos fibers. Ozone of ambient air can impair clearance of fibrogenic and carcinogenic materials, such as asbestos, from the lungs (Pinkerton *et al.*, 1989).

Exposure to amphibole asbestos fibers could explain the occurrence of most mesothelioma cases in Canada, other inorganic fibers, including chrysotile seem to have much less ability to induce mesothelioma. Fibrous tremolite, an amphibole contaminant of many industrial minerals including chrysotile, may be the causative agent in many of mesothelioma cases the Quebec asbestos mining region (McDonald *et al.*, 1989).

EMPIRICAL FIBER TOXICITY MODEL

Asbestos and other mineral fibers induce predominantly two types of cancers: mesotheliomas and bronchogenic carcinomas. Fiber size is an important factor in the carcinogenic activity of these substances. The specific fibrous form allows the fibers to penetrate easily into cytoplasm where they meet reducing agents and generate reactive species on their surface. But also fiber durability and the mentioned surface properties of fiber are important properties affecting carcinogenicity. Evidence exists that asbestos is a complete carcinogen, an initiator and a promoter. Multiple mechanisms must be operative to explain the diverse effects of mineral fibers. Although e.g. asbestos is inactive as gene mutagens, there is clear evidence that it induces chromosomal mutations. The fibers are phagocytized by the cells and accumulate in the perinuclear region of the cells. When the cell undergoes mitosis, the physical presence of the fibers interferes with chromosome segregation and results in anaphase abnormalities (Barrett *et al.*, 1989).

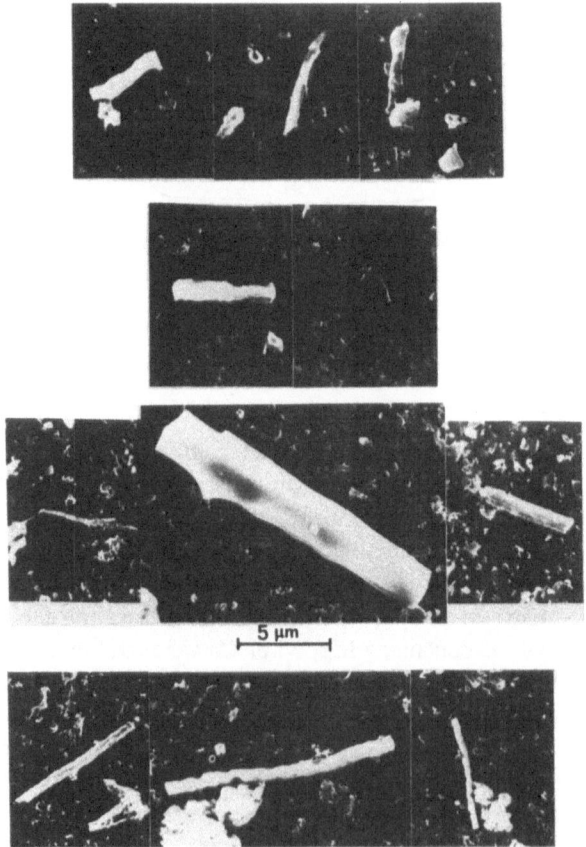

Figure 8. Scanning electron micrographs of man made mineral fibers after 3 year exposure in a rabbit tissue. The solubility and corrosion could be well demonstrated.

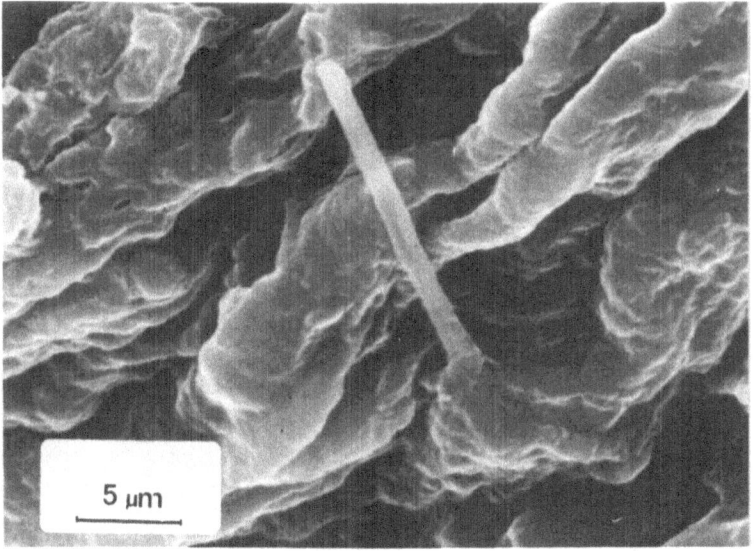

Figure 9. Scanning electron micrograph of an asbestos fiber deposited in the human lung tissue

TABLE 1. Proposed fibre toxicity tests
Preparation of homogeneous respirable fiber samples

Screening Test No. 1
Determination of physico-chemical properties
Fiber size distributions; Fiber specific surface;
Fiber persistency (physical and chemical degradation)
Fiber electric charge (Zeta-potential)
Peroxidation (Reactive Oxygen Metabolite Mesearuments)

Screening Test No. 2
Biophysical animal tests. - *In vitro* cell tests
Determination of the experimental lung clearance kinetics (rats)
(original fibers and chemically leached fibers)

Screening Test No. 3
Biological animal tests
Determination of the carcinogenic potency by instillation experiments (rats)

END Test
Animal inhalation test for carcinogenicity determination (rats)

After deposition of fibers in the lung tissue (Fig. 9) an interaction with cells can start. Our knowledge about the pathogenesis and etiology of the carcinogenic effect of fibers is still insufficient. The same is true regarding the biochemical and molecular-biological processes that occur inside the cell during this interaction.

On the other hand, as already mentioned, correlations do exist between some physico-chemical properties of mineral fibers and their tested tumor rates, e.g. by animal instillation. These correlations can be applied as a "black box model" for a preliminary, fast and cheap prediction of carcinogenic potency in the case of new types of mineral fiber: the carcinogenic effect (CE) is a function of different parameters - fiber size (FS), fiber persistency (FP), zeta potential (ZP), hydroxyl radical production (HRP), etc.

$$CE = f(FS + FP + ZP + HRP + ...)$$

CONCLUSIONS

In our opinion, it should be possible to develop and standardize a step-wise screening test procedure for preliminary estimation of the carcinogenic potency of new types of mineral fibers (see Table 1). In investigations planned in Germany, we will try to follow these research directions with the goal being to develop cheaper toxicity screening tests for new MMMF.

REFERENCES

Barrett,J.C., Lamb,P.W. and Wiseman,R.W. (1989) Multiple mechanisms for the carcinogenic effects of asbestos and other mineral fibers. *Environ. Hlth Perspect.* **81**:81-89.

Bignon,J., Peto,J. and Saracci,R. (1989) Non-occupational exposure to mineral fibres. IARC Scientific Publications **90**, Lyon, pp. 1-529.

Case,B.W. *et al* (1986) Asbestos effects on superoxide production. *Environ. Res.* **39**:299-306.

Graceffa,P. and Weitzman,S.A. (1987) Asbestos catalyzes the formation of the 6-oxobenzo(a)pyrene radical from 6-hydroxy-benzo-[a]-pyrene. *Arch. Biophys.* **257**:481-484.

Jolicoer,C. and Pisson,D. (1987) Surface physico-chemical studies of chrysotile asbestos and related minerals. *Drug Chem. Toxicol.* **10**:1-47.

Korkina,L.G., Suslova,T.B., Cheremisina,Z.P. and Velichkovski,B.T. (1988) Catalytic properties of asbestos fibers and their biological activity. *Studia Biophysica* **2**:99-104.

Light,W.G. and Wei,E.T. (1977) Surface charge and asbestos toxicity. *Nature* **265**:537-539.

McDonald,J.C., Amstrong,B., Case,B., Doell,D., McCaughey,T.E., McDonald,A.D. and Sebastien,P. (1989) Mesothelioma and asbestos fiber type. *Cancer* **63**:1544-1547.

Pinkerton,K.L., Brody,A.R., Miller,F.J. and Crapo,J.D. (1989) Exposure to low levels of ozone results in enhanced pulmonary retention of inhaled asbestos fibers. *Am. Rev. Respir. Dis.* **140**:1075-1081.

Pott,F. (1978) Some aspects of the dosimetry of the carconigenic potency of asbestos and other fibrous dust. *Straub-Reinhalt Luft.* **38**:486-489.

Pott,F. (1987) The fiber as a carcinogenic agent. *Zbl. Bakt. Hyg. B* **184**:1-23.

Scheule,R.K., Holian,A. (1989) I$_g$G specifically enhances chrysotile asbestos-stimulated superoxide anion production by alveolar macrophage. *Am. J. Respir. Cell Mol. Biol.* **1**:313-318.

Spurny,K.R. *et al* (1983) On the chemical changes of asbestos fibers and MMMFs in biologic residence and in the environment. *Am. Ind. Hyg. Assoc. J.* **11**:833-845.

Stanton,M.F., Layard,M., Tegeris,A., Miller,E., May,M. and Kent,E.J. (1977) Carcinogenesis of fibrous glass: pleural response in relation to fiber dimension. *J. Natl. Cancer Inst.* **58**:587-603.

Vasilieva,L.A., Pylev,L.N., Vezencev,A.I. and Smolikov,A.A. (1989) The carcinogenic activity of the synthetic chrysotile asbestos with different sizes of fibers and chemical composition. *Exp. Oncology* (Russ) **11**:26-29.

Walton,W.H. and Coppock,S.M. (1987) Man-made mineral fibres in the working environment. *Ann. Occup. Hyg.* **31 B**:517-834.

FIBRE SIZE AND CHEMISTRY EFFECTS *IN VITRO* AND *IN VIVO* COMPARED

R.C.Brown, J.A.Hoskins, E.A.Sara, C.E.Evans and K.J.Cole
MRC Toxicology Unit
Woodmansterne Road
Carshalton
Surrey SM5 4EF. U.K.

INTRODUCTION

The biological activities of mineral fibres including their ability to cause disease are primarily related to their physical form. Fibres with widely different chemical compositions and composed of crystalline or amorphous materials can all cause the same biological effects if the fibres are within certain size ranges (Chamberlain & Brown, 1978, Stanton & Layard, 1978, Brown *et al.*1978, Pott *et al.*, 1987; Donaldson *et al.*, 1989). However to cause disease such fibres must persist *in situ* in animals or man for a period probably related to life span and therefore shorter in animals than man. In *in vitro* systems with a time of exposure only rarely exceeding a few days even very soluble fibres can exert an effect even if that effect is still dependent on fibre dimension.

The chemical composition of the fibre is an obvious determinant of solubility but the role of chemistry as a further determinant of activity is far from certain. Throughout this volume the catalytic activities of some types of fibre are described but even so it is not obvious that such effects are primary determinants of pathogenesis. Although the composition of the fibres may contribute to the production of, for example, biologically damaging free radicals (see chapters by Aust, Pezerat, Yano etc. in this volume) it is not clear that such effects actually cause disease *in vivo*. The best evidence that a particular activity is responsible for pathogenesis is the absence of that activity in a closely related but harmless analogue or conversely the absence of pathogenicity in a material lacking the suspect short term activity.

If the composition of the fibre is important then it might be possible to create fibres with chemical compositions minimising pathogenicity. This possibility alone would make an examination of the effect of composition on activity worthwhile.

We have been studying the activities of mineral fibres and attempting to correlate these activities with both chemical composition and size. In particular we have been modifying the surface of fibres to obtain populations which differ in their apparent chemical composition but retain similar size distributions. We have also been studying the activities of thermally degraded ceramic fibres in which partition of the components of the originally vitreous materials causes changes in both bulk and surface chemistry.

MATERIALS AND METHODS

Histopaque 1119 (sodium diatrizoate-Ficoll mixture (s.g. 1.119), poly-D-lysine and poly-L-lysine (molecular weight 130,000) and other biochemicals were obtained from Sigma, Poole, Dorset; the pentapeptide gly-arg-gly-asp-ser (GRGDS) was obtained from both Sigma and Boehringer, Lewes, Sussex. Octyldimethylchlorosilane and octadecyldimethylchlorosilane were obtained from The Aldrich chemical Co., Gillingham, Dorset.

Tissue culture medium, serum and sterile plastic-ware were from Flow Laboratories, Irvine, Scotland, and Gibco Europe, Paisley, Scotland. Fibronectin depleted serum, prepared by affinity chromatography on gelatin-sepharose, and rat plasma fibronectin were a kind gift from Dr R.Davies (MRC Toxicology Unit).

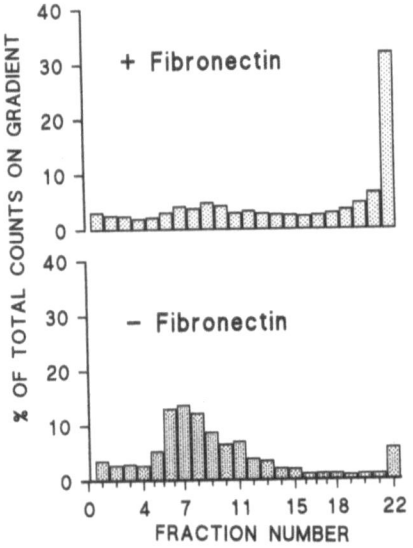

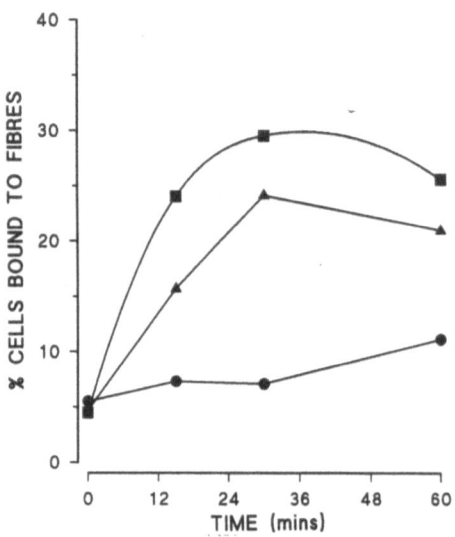

Figure 1. The time course of binding of amosite fibres to ^{3}H-Thymidine labelled V79-4 cells, in Minimum Essential Medium (MEM) containing 10% fibronectin depleted serum or the same with the addition of 50 μg/ml human plasma fibronectin. The proportion of cells attached to the fibres was estimated by adding samples of cell suspension to density gradients 15 minutes after mixing cells and fibres. Upon centrifugation cells attached to fibres migrate to the bottom of the gradient.

Figure 2. The attachment of V79-4 cells to Standard grade ceramic fibres (squares) and the same fibres after exposure to 1200°C (triangles) and 1400°C (●) for two weeks. The proportion of cells bound to the fibres is estimated by totalling the proportion of radioactivity in the last 2 fractions on density gradients prepared as in figure 1.

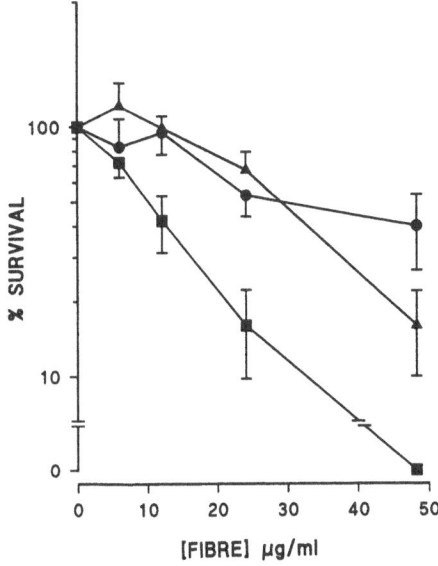

Figure 3. The cytotoxicity of the three ceramic fibre samples whose attachment to cells is reported in figure 2. The surviving fraction of V79-4 cells mixed with fibres at the concentrations shown was estimated in a clonogenic assay.

The UICC asbestos sample of amosite was chosen for modification and derivatisation was carried out as described previously (Sara *et al.*, 1990, Brown *et al.*, 1990). Milled amosite was prepared in an agate mill to avoid any contamination with metal. Fibres were coated with polylysine by suspension in a 0.1 mg/ml aqueous solution of the polypeptide with gentle sonication. After 10 minutes the fibres were sedimented by centrifugation and the coated fibres washed once with water. All centrifugation steps during fibre preparation were carried out at 16,000g for more than 10 min in an attempt to avoid losing the finer fibres at the meniscus and tube sides. When sterile fibres were required samples were weighed and autoclaved dry before use (121°C for 15 minutes in sealed tubes).

Immediately before use in an experiment fibres were suspended by brief sonication in complete medium to give a stock fibre concentration of 2 mg/ml. Where the fibre was extremely hydrophobic it was first suspended in acetone as described previously (Sara *et al.*, 1990). The stock fibre suspension was added to cell suspensions such that the final concentration of fibre was 200 μg/ml and, when used, the acetone concentration did not exceed 1%: a concentration that did not affect fibre/cell interaction or cell viability.

V79-4 and C3H10T1/2 cells were cultured as described previously (Chamberlain and Brown, 1978; Brown *et al.*, 1983).

Cell/fibre interaction was estimated using radio-labelled cells prepared from cultures exposed to medium containing 2 μCi/ml ³H-thymidine for at least two population doubling times. A single cell suspension (approximately 7×10^5 cells/ml) was prepared from these labelled cultures by trypsinisation. Ten millilitre aliquots of a cell suspension were transferred to silanised flasks in a shaking water bath at 37°C. After temperature and gas phase equilibration fibre suspension samples were added. The flasks were shaken at

100 strokes/minute: a shaking rate found in preliminary experiments to cause minimal cell damage.

Histopaque 1119 was diluted with phosphate buffered saline (PBS) to final concentrations of 60-100% Histopaque in nine even dilution steps. Stepwise gradients were constructed in 12 ml conical bottomed disposable centrifuge tubes with 1 ml aliquots of each dilution and a top layer of 1 ml PBS. One millilitre samples of the cell/fibre suspensions were removed at timed intervals from the flasks and layered onto the top of these gradients which were then centrifuged at 1,100g for five minutes at room temperature.

The spread of radioactivity through the gradient was measured by taking sequential 0.5 ml aliquots which were placed in plastic insert vials, 4 ml 'Liquiscint' (National Diagnostic, Manville, NJ, USA) was added and the mixture counted in a liquid scintillation counter.

Cell survival in the presence and absence of amosite and its derivatives was determined as described previously (Chamberlain & Brown, 1978). The results are presented in Figure 1 as a percentage survival compared with that in untreated controls. Both surface silylation treatments reduced the toxicity of the asbestos.

In life-time animal studies male LACP rats were injected intrapleurally with the three dust samples. The fibres were suspended in DMSO with sonication and this suspension was diluted 1:1 with normal phosphate buffered saline. Each rat was lightly anaesthetised with diethyl ether and a 20 gauge needle inserted into the pleural cavity. Twenty mg of each fibre in 0.5 ml vehicle or vehicle alone was then injected. The rats were allowed food and water *ad libitum* and allowed to live until either dying or showing distress. Autopsy and histological examination of any tissues was by usual methods.

RESULTS AND DISCUSSION

The primary interaction between fibres and cells, in particular the time taken for the interaction between amphibole fibres and cells in suspension, both of which have negatively charged surfaces, is in excess of 15 minutes and may approach an hour (Evans *et al.*, 1983; Sara *et al.*,1990; Brown *et al.*, 1990). Using the simple method described above the effect of fibre type, medium composition and environmental conditions on cell/fibre interaction have been investigated. We have shown that when serum is depleted of fibronectin it supports less fibre/cell interaction and on readdition of this protein interaction is restored (figure 1); additional fibronectin can increase the adhesion of less adherent cell lines. The role of fibronectin in fibre cell adherence is also supported by the blocking effect of the peptide GRGDS which is believed to represent the binding site on fibronectin that binds to the cell "RGD" receptors, and this is discussed fully in Brown *et al.*, (1990). We have previously shown that surface-modification of asbestos by glutaraldehyde coupled protein can reduce the rate of fibre/cell interaction but not longer term cytotoxicity (Evans *et al.* 1983). In figure 2 the modification of ceramic fibres by conditions similar to those encountered in use is also shown to reduce the rate of fibre/cell interaction and also to reduce the overall extent of this interaction. The work with amosite asbestos supported the idea that fibre size was the determinant of *in vitro* toxicity and *in vivo* pathogenicity. However the work with the ceramic fibres has shown that as well as reducing short term interaction with fibres overall cytotoxicity is also affected (figure 3) the results of *in vivo* injection experiments with these fibres is still awaited.

In an attempt to examine the role of fibre chemistry in determining pathogenic activity we have modified fibres in a way designed to alter their surface physicochemical properties and to be sufficiently resilient to withstand long term residence in experimental animals. Such alterations have to consist of covalent additions to the surface with the added group or groups having different properties to those of the parent fibre and being incapable of biodegradation. We have modified UICC Amosite by treatment with octyldimethylchlorosilane or octadecyldimethylchlorosilane (C_8 and C_{18}). This treatment alters the chemical structure of the surface but should not affect fibre size distribution: to allow for any selective loss of fibres during centrifugation a sample of fibre was "sham derivatised".

The progressive alteration of the surface has reduced both rate of interaction with cells and overall cytotoxicity, the derivatised fibres were also less tumourigenic when injected intrapleurally into rats. Of the rats injected with the sham-derivatised amosite and the C_8 material 4/10 developed large compressing mesotheliomas; two of which in the C_8 fibre group developed early. The tumours were frank and were the probable causes of death. With the C_{18} fibre only 1/10 rats developed a very small tumour on the pleural diaphragm. This was a preliminary experiment and so the difference in tumour incidence is difficult to analyse. It is also possible that the hydrophobic fibres agglomerated in some way in the pleural space and were thus unavailable for interaction with mesothelial cells. This type of objection to injection experiment is alluded to in the paper by Pott in this present volume, a true analysis of tumourigenicity awaits the use of this type of fibre in inhalation experiments.

When a fibre interacts with the surface of a cell we have demonstrated that, for negatively charged fibres, fibronectin (or vitronectin) binds to the surface and this then interacts with "RGD" receptors on the cell surface. Negatively charged fibres interact more rapidly and presumably with positively charged domains on the cell surface (Brown *et al.* 1990). The consequence of either type of interaction is an alteration in the ability of the cell to respond to stimulation by primary messengers and form functional junctions with other cells. Non-fibrous particles do not exert these effects. Chemical alteration of the surface prevents or slows fibre/cell interaction and also affects *in vivo* pathogenesis. We are currently designing alterations to surface properties which increase the avidity of fibre/cell interaction.

The production of free radicals and other chemically catalysed activities of fibres are not dependent on particle morphology and thus if this type of activity is responsible for tumourigenicity then isometric particles should also cause tumours. The fibre morphology can only be involved by affecting clearance from the lungs or pleural cavities. Hypotheses based on this suggestion will prove extremely difficult to test.

We believe that size, chemical activity and the ability to resist clearance are all determinants of *in vivo* activity and that, at present, it would be unwise to regard any of these as predominant.

REFERENCES

Brown,R.C., Carthew,P., Hoskins,J.A., Sara,E. and Simpson,C.F. (1990a) Surface modification can affect the carcinogenicity of asbestos. *Carcinogenesis* 11:1883-1885.

Brown,R.C., Poole,A. and Fleming,G.T.A. (1983) The influence of asbestos dust on the oncogenic transformation of C3H10T1/2 cells. *Cancer Letts.* 18:221-227.

Brown,R.C., Chamberlain,M., Griffiths,D.M. and Timbrell,V. (1978) The effect of fibre size on the *in vitro* biological activity of three types of amphibole asbestos. *Int. J. Cancer* **22**:721-727.

Chamberlain,M. and Brown,R.C. (1978) The cytotoxic effects of asbestos and other mineral dust in tissue culture cell lines. *Br. J. Exp. Path.* **59**:183-189.

Donaldson,K., Brown,G.M., Brown,D.M., Bolton,R.E. and Davis,J.M.G. (1989) Inflammation-generating potential of long and short fibre amosite asbestos samples. *Br. J. Indust. Med.* **46**:271-276.

Evans,P.E., Brown,R.C. and Poole,A.(1983) Modification of the *in vitro* activities of amosite asbestos by surface derivatization. *J. Toxicol. Environ. Hlth* **11**:535-544.

Pott,F., Ziem,U., Reiffer,F.J., Huth,F., Ernst,H. and Mohr,U. (1987) Carcinogenicity studies on fibres, metal compounds and some other dusts in rats. *Exp. Path.* **29**:129-152.

Sara,E., Brown,R.C., Evans,C.E., Hoskins,J.A. and Simpson,C.F. (1990) Interaction of Amosite and Surface Modified Amosite with a Chinese Hamster Lung Cell Line. *Environ. Hlth Perspect.* **85**:101-105.

Stanton,M.F. and Layard,M. (1978) The carcinogenicity of fibrous minerals. In, "Workshop on Asbestos: Definitions and Measurement Methods", C.C.Gravatt, P.D.LaFleur. & K.F.J.Heinreich, eds. National Bureau of Standards Special Publication 506. Washington DC. p. 143.

INFLAMMATION AND IMMUNOMODULATION CAUSED BY SHORT AND LONG AMOSITE ASBESTOS SAMPLES

K.Donaldson, *S.Szymaniec, X.Y.Li, D.M.Brown, G.M.Brown

Institute of Occupational Medicine
Roxburgh Place
Edinburgh, Scotland

* Ludwig Hertzfeldt Institute for Immunology
and Experimental Therapy
Polish Academy of Sciences
Czerska 12, 53-114 Wroclaw
Poland

ABSTRACT

In order to try and understand the processes underlying carcinogenesis in the peritoneal cavity following deposition of fibrous dusts, we instilled long and short fibre samples of amosite asbestos into the peritoneal cavity of mice and assessed some aspects of the inflammatory and immune response. In previous studies we have demonstrated that the long fibre sample causes mesotheliomas following intraperitoneal instillation while the short fibre sample is virtually inactive in this respect. There was a dramatic inflammatory response to a very low dose of long fibres and no effect with the same dose of short fibres. In animals immunised with Sheep Red Blood Cells, treatment with the long fibres caused a suppression of systemic humoral immunity, as assessed by the numbers of anti-body-producing cells in the spleen; splenic T-cell responses were also depressed. The short fibre sample caused much smaller effects in inhibiting these immune responses. Long fibre amosite also caused a greater release of the cytokines Interleukin-1 and TNF from alveolar macrophages than did the short fibre amosite sample. Inflammation and immunosuppression may play a role in the pathogenic processes following peritoneal cavity deposition of fibres that lead to mesothelioma; cytokine release by macrophages may also play a role in these processes.

INTRODUCTION

Fibre length is a major descriptor of the pathogenicity of fibrous dusts (Davis, 1989) although surface chemistry of fibres also has a role to play. We have used two samples of asbestos differing in fibre length and demonstrated that, when delivered by inhalation, the long fibre sample has the ability to cause more lung fibrosis and more lung tumours. The long fibre sample also produces more mesotheliomas when injected into the peritoneal cavity than does the short fibre sample (Davis *et al.*, 1986).

Great interest has been shown in the role of leukocytes in pathological change and in the responses of the mesothelial surfaces (both pleural and peritoneal) to asbestos. We therefore investigated the effects of the long and short fibres following intraperitoneal exposure on the two key leukocyte-mediated defence systems - the immune and inflammatory responses. We injected long and short fibre samples of amosite asbestos into the peritoneal cavity of mice and assessed general measures of immune response as well as the local inflammatory response. We also assessed the ability of the two different fibre types to stimulate the release of cytokine from macrophages *in vitro*.

MATERIALS AND METHODS

Dust samples

The dust samples used in the present study were those used previously in our laboratory (Davis *et al.*, 1986; Donaldson *et al.*, 1989). These samples were obtained from the same batch of South African amosite. The two samples did not differ elementally or crystallographically. The dimensions of the two samples are shown in Figures 1 and 2. From these figures it is evident that the samples do not differ substantially in their diameters but are quite different in the length with very few fibres longer than 10 μm in the short fibre sample. The UICC standard sample of chrysotile asbestos was also used.

Animals

Syngeneic C57Bl6 mice of 12-16 weeks were used throughout. Rats of the HAN/Wistar strain were used as a source of alveolar macrophages.

Inflammation Assay

Inflammatory potential of the two fibre samples was assessed by injecting 5 μg of dust, in a 0.5 ml suspension, into the peritoneal cavity of mice. Two days later the peritoneal cavity was lavaged with 3 x 2 ml volumes of phosphate buffered saline to retrieve the free cell population; the total number of cells was estimated and a differential cell count was made in Giemsa-stained preparations.

Assessment of the systemic humoral immunity

The ability of splenic lymphocytes to mount a humoral immune response was measured as the number of lymphocytes able to produce specific antibody following immunisation with Sheep Red Blood Cells (SRBC). On day 0 mice were exposed to fibre as described above, by intraperitoneal injection. On day 3 mice were immunised with an intraperitoneal injection of 0.2 ml of a 10% solution of SRBC. On day 7 the mice were killed, their spleens removed aseptically, and the proportion of splenocytes producing antibody to SRBC was determined using a haemolytic plaque assay as described in detail

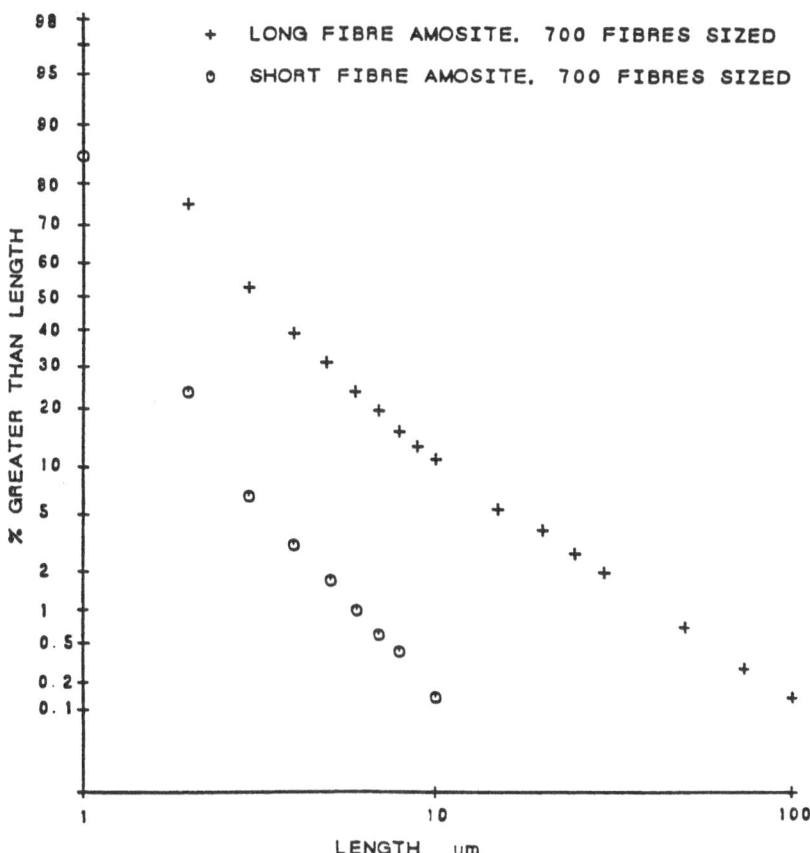

Figure 1. Length characteristics of the long and short fibre amosite samples. Results from 700 sized fibres.

in Szymaniec and co-workers (1990). Plaque-forming cells (PFC) were expressed as either PFC per spleen or PFC per 10^6 spleen cells.

Mitogenic responses of splenic T cells

   The status of T cells in fibre-exposed mice was ascertained by exposing to fibre and, 3 days later, measuring the splenic T lymphocyte responses to the T cell mitogen phytohaemagglutinin (PHA). Spleens were removed aseptically and mechanically dispersed before bringing into microtitre plate wells and assessing the mitogenic response to PHA, added at 2.5 and 5 μg/ml. Plates were then incubated for 2 days and pulse labelled with tritiated thymidine. Proliferation was measured by liquid scintillation counting 16 hours later. The method is described in detail in Hannant and co-workers (1985).

Production of Interleukin 1 and Tumour Necrosis Factor

   Rat alveolar macrophages were obtained by bronchoalveolar lavage from normal rats and brought into culture. Long or short fibre amosite asbestos at 25 μg/ml was added to the cultures and these were incubated for 24 hours. The supernatants were collected, centrifuged to clarity and stored at -20°C. Interleukin 1 (IL-1) was assayed using

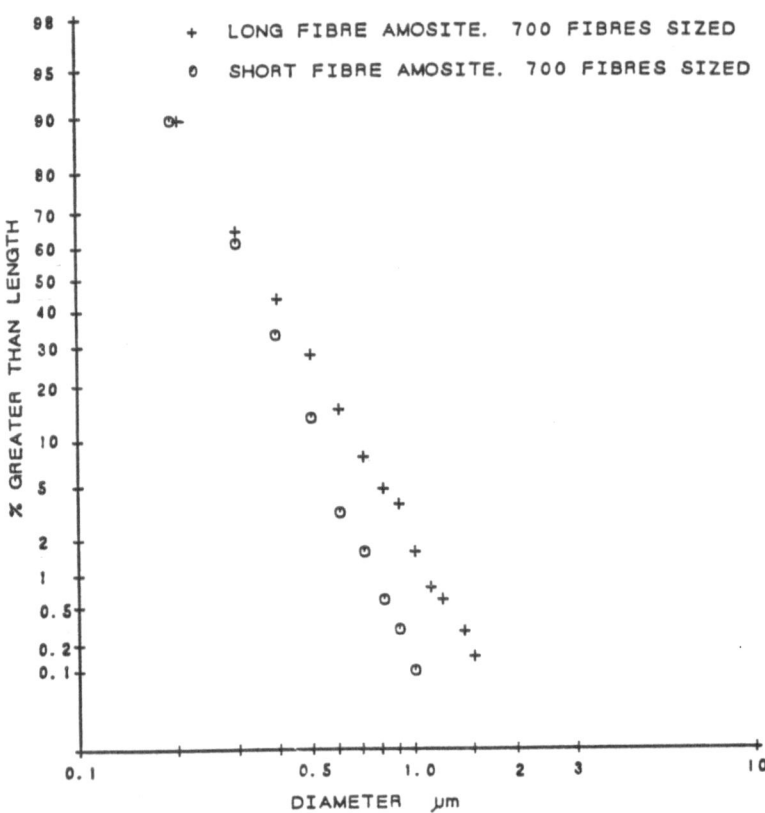

Figure 2. Diameter characteristics of the long and short fibre amosite asbestos samples. Results from 700 sized fibres.

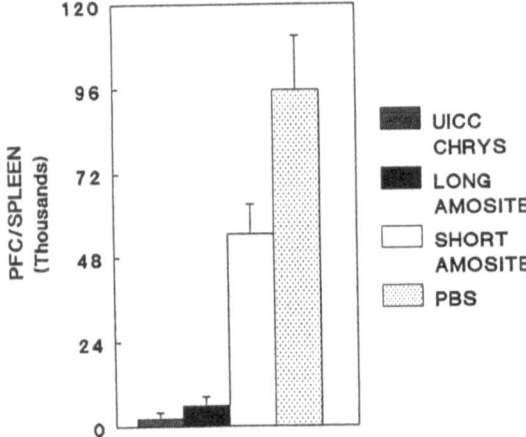

Figure 3. Numbers of anti-SRBC plaque-forming cells (PFC) in the spleens of mice injected intraperitoneally with 2.5 mg of the indicated dusts. Results represent the mean and standard deviation from 5 mice in a single experiment. Significantly reduced PFC with all dusts compared to PBS, and with long and UICC chrysotile compared to short amosite (P < 0.01).

TABLE 1. The peritoneal inflammatory response to 5 μg of long or short fibre amosite instilled 2 days previously. Significant increase in the numbers of macrophages and neutrophils with long amosite treatment (P < 0.001)

AMOSITE	MACROPHAGES	NEUTROPHILS
SHORT	*4.65(0.56)	0.20(0.02)
LONG	23.25(3.56)	13.28(3.71)

*CELLS x10^6 (mean($\pm$s.d.))

enhancement of the sub-optimal PHA mitogenesis of C57BL6 mouse thymocytes as described in detail in Kusaka and co-workers (1990). Tumour Necrosis Factor (TNF) was assayed using the TNF-sensitive cell line L929. By using standard cytokine preparations in the same assay systems we are able to determine the absolute amounts of IL-1 and TNF in the different samples of supernatant.

Statistical Analysis

Data from repeat experiments was analysed by analysis of variance and significance of differences between treatment groups was tested using a T test.

RESULTS

Inflammatory potential of the long and short amosite samples

As shown in Table 1 the long fibre sample caused substantially more inflammation in the mouse peritoneal cavity than the short fibre sample.

Immunomodulatory activity of the long and short fibre samples

The short fibre sample caused a reduction in the splenic antibody response of the same order of magnitude as that caused by exposure to the UICC chrysotile sample, when they were injected at equal mass (Figure 3). In contrast, the short fibre amosite sample suppressed splenocyte responses much less, although it did cause a significant reduction in PFC compared to the saline control.

Dose response of the humoral immunosuppression caused by long fibre amosite

When the PFC response was measured in mice exposed to lower levels of the long amosite it became clear in 3 separate experiments that, at very low doses (5 μg) of dust, there was still profound immunosuppression which appeared greater at the lower doses (Figure 4).

T-cell responses in the spleens of mice exposed to long and short fibre

As shown in Figure 5, intraperitoneal instillation of both the short and long fibres caused a decrease in T cell responses to PHA with the short fibre sample. The long fibre

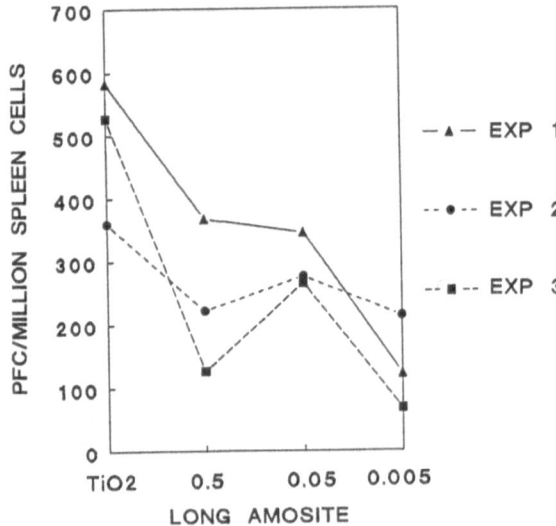

Figure 4. Numbers of anti-SRBC PFC in the spleens of mice injected intraperitoneally with 0.5 mg of TiO_2, 0.5, 0.05 or 0.005 mg of long amosite. Results are the means from 3 separate experiments. Significant reductions with all treatments compared to TiO_2 ($P < 0.01$).

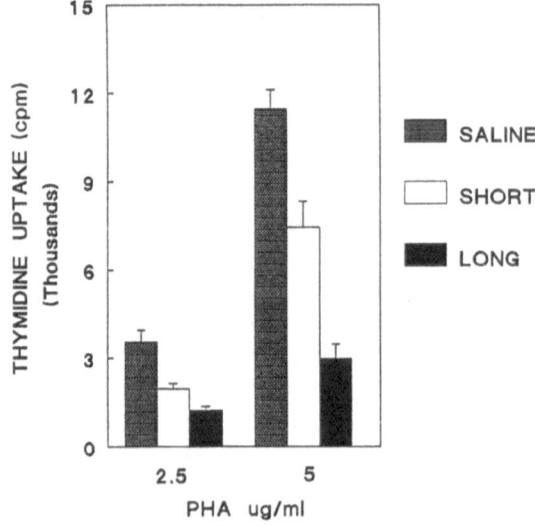

Figure 5. PHA-stimulated proliferation of splenocytes from mice injected intraperitoneally with the indicated dusts or saline. Results are the mean and standard deviation from 3 mice in each group in a single experiment. Significantly reduced proliferation with short and long amosite compared to saline and long compared to short amosite ($P < 0.05$).

sample caused substantially more suppression of T cell response than did the short fibre sample.

Release of cytokine by macrophages exposed to long and short fibre *in vitro*

Figures 6 and 7 reveal that the long fibre sample had more ability than the short fibre sample to stimulate the release of cytokine when alveolar macrophages were the indicator cells.

DISCUSSION

We set out to determine whether there was any difference in the ability of a long and a short fibre amosite asbestos sample to affect the two main body defence systems of immunity and inflammation. The two samples have greatly different abilities to cause lung fibrosis and cancer when inhaled by rats with the long fibre sample being considerably more carcinogenic and fibrogenic than the short fibre sample (Davis *et al.*, 1986).

We used the reaction of the mouse peritoneal cavity to assess these effects because it represents a more reliable site than the lung to deposit a known quantity of a fibrous dusts by instillation and it has been used as a site to study fibre carcinogenesis (*e.g.* Bolton *et al.*, 1982). It is, however, very different to the lung, having no clearance mechanism and a macrophage population which differs in several important respects. It is questionable, therefore, whether studies on the peritoneal cavity reveal information applicable to an understanding of events inside the lung itself. They should, however, aid understanding of the mesothelioma response found after intraperitoneal instillation.

Fibrous dusts deposited in the peritoneal cavity give rise to a fibrogenic (Donaldson, 1982) and carcinogenic (Bolton *et al.*, 1982) response and the ability of fibrous dusts to cause mesotheliomas in the peritoneal cavity following instillation has been used as a bio-assay for fibre carcinogenesis and to investigate some of the determinants of fibre carcinogenesis. When the long and short fibre amosite samples were injected into the rat peritoneal cavity in a previous study, there was a high proportion of mesothelioma in the long fibre-treated rats but little response in those animals treated with the short fibre (Davis *et al.*, 1986). We sought therefore to further understand the effect of instillation of these dusts into the peritoneal cavity on the immune and inflammatory responses.

The present study revealed that the long fibre sample was very active in causing inflammation in the peritoneal cavity and that in long fibre-treated rats there was an inhibition of a humoral immune response and impaired T cell function in the spleen. The short fibre sample was able to cause a small amount of inflammation but was many times less active than the long fibre sample. The short fibre sample also reduced both the B- and T-cell responses but this was modest in comparison to the long fibre sample.

We have reported previously that the non-carcinogenic dusts quartz and TiO_2 caused inflammation in the mouse peritoneal cavity but that this was substantially less than that caused by fibrous dusts (Donaldson *et al.*, 1988). We have also shown that TiO_2 and quartz can reduce the PFC response in the spleen in experiments similar to those shown here (Szymaniec *et al.*, 1990). However, the ability of these dusts to cause this immunosuppressive effect was markedly less than that of UICC chrysotile asbestos.

The peritoneal cavity has its own macrophage population and is lined by mesothelial cells. Macrophages are an important cell type in the initiation and execution of both inflammatory and immune responses and we have shown macrophage activation in the peritoneal cavity following injection of asbestos (Donaldson *et al.*, 1982). It is assumed

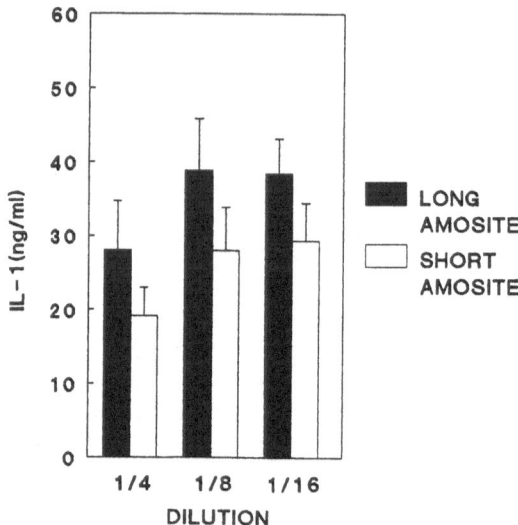

Figure 6. Production of Interleukin-1 (IL-1) by alveolar macrophages exposed *in vitro* to long or short amosite asbestos. Results are mean and standard deviation obtained from 3 separate experiments. Significantly increased IL-1 at all dilutions (P < 0.05).

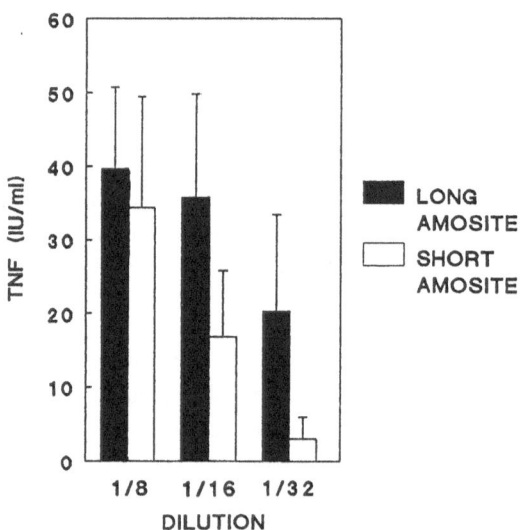

Figure 7. Production of Tumour Necrosis Factor (TNF) by alveolar macrophages exposed to long or short amosite asbestos *in vitro*. Results are mean and standard deviation from triplicate wells in 2 separate experiments. Significantly increased production with long amosite compared to short amosite at 1/16 and 1/32 dilutions (P < 0.05).

that, following instillation of asbestos, the macrophages phagocytose the fibres and this is an important event in leading to inflammation, immunomodulation and, ultimately, pathological change. The other cell type which could be directly affected are the mesothelial cells. One potential factor in inflammation and immunomodulation following fibre deposition and phagocytosis is the production of macrophage cytokines. These have powerful pro-inflammatory activity and also the ability to stimulate lymphocytes. The increased production of cytokine by alveolar macrophages treated with the long fibre *in vitro* demonstrated here, suggests one route whereby inflammation could be caused by the long fibres. It was notable that TNF showed a greater difference in response between the two fibre types than was shown by IL-1. This suggests that TNF could be more important than IL-1 in the development of fibre-associated pathology. The production of these particular cytokines is unlikely to be a primary cause of the immunosuppression that is seen with the long fibre sample. It does, however demonstrate that the long fibre sample can stimulate secretion; the production of other substances such as prostaglandin, with well-identified immunosuppressive activity, is therefore possible.

In general terms the layout of the experiment on plaque-forming cells would be expected to enhance immune responses since the antigen is deposited into the peritoneal cavity where inflammation is proceeding. In inflammatory sites the up-regulation of Ia on the surface of the macrophages should favour the immune response and the increased recruitment of macrophages should also aid the transfer of antigen to the lymph nodes for induction of the immune response. It is all the more surprising therefore that with long amosite, where the inflammation is greatest, there is profound suppression of immune response. We have, however previously reported that macrophages from the peritoneal cavity, following instillation of asbestos, release a factor(s) that inhibits lymphocyte response (Donaldson *et al.*, 1984). This factor is likely to be released locally but may have some function in the systemic immunosuppression described here for the long amosite fibres.

Inflammation and immunity are likely to be important in the fibrogenic and carcinogenic process. Leukocytes play an important role in fibrosis through the release of growth factors which aid in the accumulation, proliferation and synthetic activities of fibroblasts. In the case of carcinogenesis, growth factors may be important and many tumour promoters are inflammogens. The immune response, through surveillance for transformed cells, is considered to be important in preventing neoplasia; cancer arises in immunosuppressed individuals.

We would suggest that the activities that we have demonstrated here in the highly carcinogenic, long fibre sample *i.e.* the ability to cause inflammation in combination with an ability to suppress the immune response, could be important in the carcinogenic and fibrogenic properties of this preparation in the peritoneal cavity. Following deposition of a bolus of dust containing long fibres, it can be assumed that there will be an intense inflammatory response with the release of macrophage cytokines; there is also likely to be immunosuppression. These events may play a key role in the subsequent pathological response.

These changes to the inflammatory and immune response may also be of general relevance to fibre carcinogenesis following inhalation although direct extrapolation from the peritoneal cavity to the lung is perilous. However inflammation (Begin *et al.*, 1986; Donaldson *et al.*, 1988) and immunosuppression (Kagan *et al.*, 1977; Hannant *et al.*, 1985) have been reported in workers and animals exposed to asbestos. The processes that underlie carcinogenesis in the lung are likely to be fundamentally the same as those that occur during carcinogenesis in the pleural and peritoneal cavities.

ACKNOWLEDGEMENT
Research funded by the Colt Fibre Research Foundation.

REFERENCES

Begin,R., Bisson,G., Boileau,R. and Masse',S. (1986) Assessment of disease activity by Gallium-67 scan and lung lavage in the pneumoconioses. *Sem. Resp. Med.* **7**:275-280.

Bolton,R.E., Davis,J.M.G., Donaldson,K. and Wright,A. (1982) Variations in the carcinogenicity of mineral fibres. *Ann. Occup. Hyg.* **26**:569-582.

Davis,J.M.G. (1989) Mineral fibre carcinogenesis: Experimental data relating to the importance of fibre type, size, deposition and dissolution. In, "Non-occupational exposure to mineral fibres", eds. Bignon,J, Peto,J. and Saracci,R. IARC Scientific Publication No 90, Lyon. pp. 33-45.

Davis,J., Bolton,R.E.,Donaldson,K., Jones,A.D. and Smith,T. (1986) Pathogenicity of long versus short fibre samples of asbestos administered to rats by inhalation and intraperitoneal injection. *Br. J. Exp. Pathol.* **67**:415-430.

Donaldson,K. (1982) The effects of asbestos injection on the peritoneal macrophage population of the mouse. PhD Thesis. University of Edinburgh.

Donaldson,K., Davis,J.M.G. and James,K. (1982) Characteristics of peritoneal macrophages induced by asbestos injection. *Environ. Res.* **29**:414-424.

Donaldson,K., Davis,J.M.G. and James,K. (1984) Asbestos-activated peritoneal macrophages release a factor(s) which inhibits lymphocyte mitogenesis. *Environ. Res.* **35**:104-114.

Donaldson,K., Bolton,R.,E and Brown,D.M. (1988) Inflammatory cell recruitment as a measure of mineral dust toxicity. *Ann. Occup. Hyg.* **32**:299-305.

Donaldson,K., Bolton,R.E., Jones,A.D., Brown,G.M., Slight,J., Cowie,H. and Davis,J.M.G. (1988) Kinetics of the bronchoalveolar leukocyte response in rats following exposure to equal airborne mass concentrations of quartz, chrysotile asbestos or titanium dioxide. *Thorax* **43**:525-533.

Donaldson,K., Brown,G.M., Brown,D.M., Bolton,R.E. and Davis,J.M.G. (1989) The inflammation-generating potential of long and short fibre amosite asbestos samples. *Br. J. Indust. Med.* **46**:271-276.

Hannant,D., Donaldson,K. and Bolton,R.E. (1985) Immunomodulatory effects of mineral dust. 1.Effects of intraperitoneal dust inoculation on splenic lymphocyte function and humoral immune responses *in vivo. J. Clin. Lab. Immunol.* **16**:81-85.

Kagan,E., Solomon,A., Cochrane,J.C., Bressner,E.I., Gluckman,J., Rocks,P.H. and Webster,I. (1977) Immunological studies of patients with asbestosis. 1.Studies of cell-mediated immunity. *Clin. Exp. Immunol.* **28**:261-267.

Kusaka,Y., Cullen,R.T. and Donaldson,K. (1990) Immunomodulation in mineral dust-exposed lungs: stimulatory effect and interleukin-1 release by neutrophils from quartz-elicited alveolitis. *Clin. Exp. Immunol.* **80**:293-298.

Szymaniec,S., Donaldson,K., Brown,D.M., Chladzynska,M., Jankowska,E. and Polikowska,H. (1990) Antibody producing cells in the spleens of mice treated with pathogenic mineral dust. *Br. J. Indust. Med.* **46**:724-728.

FIBER DIMENSIONS AND MESOTHELIOMA: A REAPPRAISAL OF THE STANTON HYPOTHESIS

Agnes B. Kane

Department of Pathology and Laboratory Medicine
Brown University
Providence, RI 02912, U.S.A.

ABSTRACT

Fiber dimensions are postulated to be critical factors in the toxicity, fibrogenicity, and carcinogenicity of asbestos fibers. Recent *in vitro* experiments have provided evidence that the chemical composition of mineral fibers, especially surface iron content, is important in catalyzing the formation of highly reactive hydroxyl radicals that may cause acute toxicity, lipid peroxidation, and DNA damage. We have reexamined the roles of fiber length in the acute toxicity of crocidolite asbestos fibers *in vitro* and *in vivo* and in the induction of mesotheliomas in mice. Native UICC crocidolite asbestos fibers were separated into long and short fiber preparations by differential centrifugation. Both long and short fiber preparations stimulated the production of reactive oxygen species by elicited mouse peritoneal macrophages. Whether compared on the basis of equal mass, fiber number, or surface area, both long and short fiber preparations were toxic to macrophages. *In vitro* toxicity was prevented by the iron chelator, deferoxamine, or by exogenous superoxide dismutase or catalase. A single intraperitoneal injection of long crocidolite asbestos fibers caused deposition of fibers on the mesothelial surface at sites of lymphatic stomata, while short fibers were cleared to regional lymph nodes. Only the long fiber preparation caused an intense inflammatory reaction, local production of superoxide anions, and mesothelial cell injury. Similar to *in vitro* toxicity, mesothelial cell injury *in vivo* was ameliorated by deferoxamine or PEG-conjugated superoxide dismutase or catalase. If lymphatic clearance was prevented by daily repeated injections, short crocidolite asbestos fibers (but not titanium dioxide particles) accumulated at the mesothelial surface and stimulated an inflammatory reaction with local production of superoxide anions and injury to adjacent mesothelial cells. We tested whether repeated injections of short crocidolite asbestos fibers would prevent lymphatic clearance and produce mesotheliomas. Mice were injected weekly with equal numbers of native, long, or short crocidolite asbestos fiber preparations. After 22-60 weekly injections, 37.5% of mice injected with native crocidolite asbestos fibers developed malignant mesotheliomas. In contrast, 50.0% of mice injected with short fibers and 23.5% of mice injected with long fibers developed tumors. In summary, both long and short crocidolite asbestos fibers are toxic *in vitro* via an oxidant-dependent mechanism. *In vivo*, short fibers are also toxic and carcinogenic if lymphatic clearance is prevented.

INTRODUCTION

Asbestos fibers are divided into two categories: serpentine or chrysotile asbestos and amphiboles. Chrysotile asbestos consists of curly, white fibers and accounts for 90-95% of the asbestos used commercially. Amphiboles are long, straight fibers and include crocidolite, amosite, anthophyllite, and tremolite. Crocidolite and amosite were used extensively in the shipyards during World War II and anthophyllite was used to a limited extent in Finland. Although tremolite is not used commercially, this amphibole is a common contaminant of some deposits of sand, talc, and chrysotile asbestos (Antman & Aisner, 1987).

The usual route of exposure to asbestos fibers is inhalation. Workers exposed to asbestos have an increased risk of developing these lung diseases: effusions, parietal pleural plaques, diffuse fibrosis of the visceral pleura, diffuse interstitial fibrosis, bronchogenic carcinoma, and malignant mesothelioma. People exposed to asbestos fibers in the vicinity of asbestos factories or from household exposure to asbestos workers also have an increased risk of developing malignant mesothelioma. The health risks of exposure to asbestos in place and in urban environments in general are unknown. Since these exposures are usually lower than those occurring in the workplace, the major concern is an increased risk of developing bronchogenic carcinoma or malignant mesothelioma, not asbestosis which usually develops after prolonged exposure to high levels of fibers (Churg & Green, 1988).

The mechanisms underlying the development of these various lung diseases after inhalation of asbestos fibers are unknown. Fiber geometry (Stanton *et al.*, 1981), chemical composition (Monchaux *et al.*, 1981), and persistence in the target tissue (Pott, 1987) have been identified as properties of asbestos fibers relevant to production of disease. Fiber dimensions are a critical factor in inhalation, deposition, and clearance of fibers in the lungs. Only thin fibers penetrate into the lower respiratory tract. Curly chrysotile fibers are trapped more readily at bronchial bifurcations than straight amphibole fibers. Once deposited on the bronchial epithelium or in the alveolar spaces, short fibers are cleared more readily than long fibers. Long, straight amphibole fibers, especially, are thought to penetrate to the pleura although the route of fiber transport from the alveoli to the pleural linings is unknown. After fibers become trapped in the lung parenchyma or pleural lining, their chemical composition and crystalline structure influence their persistence. Amphibole fibers persist unchanged, perhaps for decades while chrysotile fibers fragment into shorter fibers and lose surface magnesium ions (Craighead, 1987). An extensive series of experiments by Wagner (1973) and Stanton and co-workers (1981) have established that fibers less than 0.25 μm in diameter and more than 8 μm long, regardless of their chemical composition, are more effective in inducing mesotheliomas after direct intrapleural or intraperitoneal injection. Subsequent studies have shown that fiber length is an important determinant of the biologic reactivity of asbestos fibers as tested in a variety of *in vitro* and *in vivo* model systems (reviewed by Harington, 1981 and Dunnigan, 1984).

Recent *in vitro* studies have identified reactive oxygen metabolites as potential mediators of asbestos cytotoxicity, fibrogenicity, and carcinogenicity. All types of asbestos fibers have the potential to catalyze the production of hydroxyl radicals as demonstrated in cell-free systems. This reaction is inhibited by coating fibers with iron chelators such as deferoxamine (Weitzman & Graceffa, 1984; Gulumian & Van Wyk, 1987). In various *in vitro* and *in vivo* models, deferoxamine-coated fibers are less toxic. These observations emphasize the importance of iron content of fibers: Fe^{+2} and Fe^{+3} ions can catalyze the formation of hydroxyl radicals by a modified Haber-Weiss or Fenton reaction. On the basis of these recent observations in *in vitro* models of asbestos toxicity, the Stanton hypothesis has been modified: incomplete or "frustrated phagocytosis" of long fibers stimulates generation of oxidants by the target cell (Mossman & Marsh, 1989). These oxidants, including H_2O_2 and $O_2^{\cdot-}$, serve as substrates for the iron-catalyzed generation of hydroxyl radicals that ultimately kill the target cell (Goodglick & Kane, 1986; Shatos *et al.*, 1987).

The importance of surface reactivity in the ability of asbestos fibers to catalyze the formation of hydroxyl radicals prompts a reassessment of the Stanton hypothesis. Most investigators who have studied the role of fiber length in asbestos toxicity and carcinogenicity produced short fibers by milling native asbestos ores or the UICC standardized asbestos fiber preparations. As reviewed by Dunnigan (1984), milling alters the surface reactivity of asbestos fibers. These alterations may decrease the ability of fibers to catalyze the formation of hydroxyl radicals. Therefore, we have prepared short and long fiber preparations from UICC crocidolite asbestos fibers by differential centrifugation. We have re-examined the cytotoxicity of these short and long fiber preparations *in vitro* and *in vivo*. The potential of short asbestos fibers to produce mesotheliomas after repeated exposures was also tested. These experimental observations will be summarized and compared with previous studies designed to test Stanton's hypothesis. On the basis of these *in vitro* and *in vivo* models, a new hypothesis incorporating the roles of fiber length and reactive oxygen species in the development of malignant mesotheliomas will be proposed.

EXPERIMENTAL EVIDENCE

A murine model system has been developed to study the acute and chronic effects of crocidolite asbestos fibers after direct intraperitoneal injection. In this model system, the clearance and deposition of fibers in the peritoneal lining will be described. Next, the acute reactions of the mesothelium to deposition of fibers will be characterized. After a single injection of crocidolite asbestos fibers, there is accumulation of macrophages and localized injury to mesothelial cells at sites of fiber deposition. We will summarize evidence from our *in vitro* and *in vivo* models that acute asbestos toxicity is mediated by reactive oxygen species. The ability of short crocidolite asbestos fibers to inflict oxidant-induced injury *in vitro* and *in vivo* will be tested. Injury to the mesothelium produced by a single injection of asbestos fibers is repaired by proliferation within 14-21 days. After weekly repeated injections of 200 μg of native UICC crocidolite asbestos fibers, dysplastic mesothelial cells appear after 12 weeks; malignant mesotheliomas develop after 30-60

TABLE 1. Characteristics of Asbestos Fiber Preparations

Sample	# of fibers per mg x 10^9	% of fibers > 5 μm	*in vitro* dose (μg)	*in vivo* dose(μg)
Mixed fibers	2.9	8.8	50	200
Long fibers	1.2	27.6	120	480
Short fibers	4.6	1.1	30	120

weeks. The ability of repeated injections of short asbestos fibers to produce mesotheliomas will be tested in this model system.

UICC crocidolite asbestos fibers were separated by a series of differential centrifugation steps and characterized by transmission electron microscopy as described previously (Moalli *et al.*, 1987; Goodglick & Kane, 1990). The number of fibers, lengths, and doses used are summarized in table 1.

Fiber Clearance and Deposition in the Peritoneal Lining

Peritoneal macrophages are an initial target of asbestos fibers injected into the peritoneal space. Fibers are rapidly phagocytized by the free peritoneal macrophages and cleared after 7-15 days (Macdonald & Kane, 1986). We have followed the kinetics of particle clearance from the free peritoneal macrophage population between 1 to 30 days after a single injection of 100 μg of crocidolite asbestos fibers. As summarized in Macdonald and Kane (1986), during the first week, 61-68% of the free peritoneal macrophages collected by saline lavage contained asbestos fibers. The number of fibers per cell ranged from 24.5 - 49.0. After 15 days, only 11% of the free peritoneal macrophages contained asbestos fibers; the mean number of fibers per cell was between 0.7 - 5.7.

During this same time period, clusters of fibers and macrophages accumulate on the peritoneal lining, especially on the inferior surface of the diaphragm and in lymphoid aggregates in the mesenteries called milky spots. These locations are the sites for clearance of any foreign particles introduced into the abdominal cavity (Courtice & Simmonds, 1954). In contrast to microorganisms or spherical mineral particles, asbestos fibers are not efficiently cleared by lymphatics and persist at the mesothelial surface causing localized inflammation and injury to mesothelial cells. We studied the localization of asbestos fibers on the inferior surface of the diaphragm using a combination of stereomicroscopy and scanning electron microscopy. Clusters of asbestos fibers appeared along the musculotendinous junction and radiated out over the muscular region of the diaphragm after a single injection of 200 μg of UICC crocidolite asbestos fibers. These sites are the location of lymphatic openings or stomata (Courtice & Simmonds, 1954). We confirmed this using scanning electron microscopy; the diameter of lymphatic stomata was measured as 10.7 $\pm$ 2.3 μm.

We hypothesized that asbestos fibers accumulated at these sites because the clearance of fibers longer than 8-12 μm was retarded by the size of lymphatic stomata. In order to test this hypothesis, we fractionated UICC crocidolite asbestos into preparations enriched for short and long fibers by differential contrifugation: 90.6% of the short fiber preparation is < 2.0 μm long; 60.3% of the long fiber preparation is > 2.0 μm long. Mice were injected with 200 μg of each of these preparations and the distribution of fibers determined by fiber counting after tissue digestion, stereomicroscopy, and scanning electron microscopy. Using the technique of bleach digestion, we recovered 1.8×10^6 fibers from the diaphragm after injection of native UICC crocidolite asbestos, 9.6×10^5 after injection of long asbestos fibers, and only 3.0×10^3 fibers after injection of short asbestos fibers. Microscopic examination confirmed the results of counting fibers after tissue digestion: no fiber clusters were seen on the diaphragmatic surface after injection of short fibers using the dissecting stereomicroscope or scanning electron microscope. Histologic examination revealed large numbers of short fibers in abdominal lymph nodes. These observations provide an anatomic basis for the Stanton hypothesis: fibers longer than 8 μm are trapped at sites of lymphatic stomata while shorter fibers are cleared to regional lymph nodes. We hypothesize that long asbestos fibers are trapped at the mesothelial lining at sites of lymphatic stomata and provoke injury and proliferation of mesothelial cells leading to the development of mesothelioma (Moalli *et al.*, 1987). These mesothelial reactions to asbestos fibers will be described next.

Mesothelial Cell Injury and Repair

Trapping of long asbestos fibers at lymphatic stomata places these fibers in close proximity to the mesothelial lining. Injury to adjacent mesothelial cells was documented by scanning electron microscopy, uptake of trypan blue, and recovery of lactate dehydrogenase activity in the peritoneal lavage fluid. This injury peaked at 3 days after a single injection and subsided after 7 days (Moalli *et al.*, 1987).

We hypothesize that reactive oxygen metabolites released from macrophages attempting to phagocytize asbestos fibers trapped at the mesothelial surface mediate this injury. This hypothesis is supported by the following evidence. First, macrophages are the predominant inflammatory cell found in clusters of asbestos fibers. These macrophages persist at sites of asbestos fiber deposition for several months. This persistent inflammatory response is not produced by injection of toxic silica particles or nontoxic titanium dioxide particles. Second, macrophages exposed to asbestos fibers *in vitro* (Goodglick & Kane, 1986) or *in vivo* (Goodglick & Kane, in press) release H_2O_2 and O_2^-. After *in vivo* exposure, reduction of NBT can be visualized directly at sites of fiber deposition on the surface of the diaphragm (Goodglick & Kane, 1990).

Asbestos toxicity to macrophages *in vitro* can be prevented by addition of exogenous superoxide dismutase or catalase (Goodglick & Kane, 1986; Shatos *et al.*, 1987). *In vivo*, intraperitoneal injection of these enzymes conjugated to polyethylene glycol (PEG) to prolong their half-life also decreased mesothelial cell injury caused by crocidolite asbestos fibers (Goodglick & Kane, in press).

Other inflammatory stimuli (zymosan) or inert particles (TiO_2) also trigger the release of H_2O_2 and O_2^- from macrophages; however, these agents are not toxic *in vitro* or *in vivo* (Goodglick & Kane, 1986; Moalli *et al.*, 1987). It is hypothesized that iron present in crocidolite asbestos fibers catalyzes the formation of highly reactive hydroxyl radicals, OH $\cdot$, by a modified Fenton reaction. Coating asbestos fibers with the ferric iron chelator, deferoxamine, decreases toxicity *in vitro* (Goodglick & Kane, 1986), and *in vivo* (Goodglick & Kane, in press).

The mesothelium is capable of regeneration following localized injury. We studied the restoration of the mesothelial lining after injury induced by asbestos fibers using scanning electron microscopy and autoradiography after pulse-labelling with ^{3}H-thymidine (Moalli *et al.*, 1987). Immature mesothelial cells can be identified at the periphery of

TABLE 2. Inflammation, NBT Reduction and Injury Caused by a Single Intraperitoneal Injection of Mineral Particles

Treatment	Albumin (µg)	NBT Reduction	Trypan Blue Staining
Saline	588	–	–
Thioglycollate	1938	–	–
TiO$_2$	567	–	–
Long Asbestos Fibers	2255	+ + +	+ + +
Short Asbestos Fibers	1625	+	+

NBT reduction and trypan blue staining were scored as follows (Goodglick & Kane, 1990):

– no NBT reduction or trypan blue staining,

+ < 10% of cells showing NBT reduction or trypan blue staining,

+ + 10-25% of cells showing NBT reduction or trypan blue staining,

+ + + 25-50% of cells showing NBT reduction or trypan blue staining.

fiber clusters 3-7 days after injection of native or long asbestos fibers. After 14-21 days, the surface of the fiber clusters is covered by a single layer of cuboidal mesothelial cells. These morphologic observations are confirmed by autoradiography: a peak of ^{3}H-thymidine labelling is seen 7 days after injection of long asbestos fibers. After a single injection of asbestos fibers, the surface mesothelium is restored by regeneration. After repeated weekly injections of asbestos fibers, there are repeated episodes of mesothelial cell injury and proliferation as fibers continue to accumulate around lymphatic stomata.

Short Fibers are Toxic *in vivo* if Lymphatic Clearance is Impaired

We next compared the extent of inflammation, NBT reduction by macrophages surrounding fiber clusters, and injury caused by a single or multiple injections of mineral particles. In these experiments, equal numbers (5.8 x 10^8) of TiO$_2$ particles or asbestos fiber after the last injection, peritoneal lavage fluid was collected and analyzed for albumin content as an index for the severity of inflammation (Moalli *et al.*, 1987). NBT reduction was used to assess O$_2$$^-$ production *in situ* at sites of fiber clusters and Trypan blue staining was used to monitor the extent of injury (table 2). After a single injection of TiO$_2$ particles or short asbestos fibers, there is no accumulation of particles or fibers on the mesothelial surface and minimal injury. In contrast, long asbestos fibers are trapped around lymphatic stomata, stimulate NBT reduction at these sites, and cause cell injury. Asbestos fibers, regardless of their length, elicit a more intense inflammatory reaction than an equal number of TiO$_2$ particles, confirming previous studies by Donaldson *et al.* (1990) using short and long preparations of amosite asbestos fibers.

In contrast, after repeated injections of short asbestos fibers, aggregates of macrophages and fibers impair lymphatic clearance. Inflammatory cells and fibers accumulate on the mesothelial lining, stimulate NBT reduction, and cell injury as summarized in table 3.

TABLE 3. Inflammation, NBT Reduction and Injury Caused by
Multiple Injections of Mineral Particles

Treatment	Albumin (μg)	NBT Reduction	Trypan Blue Staining
Saline	768	–	–
TiO$_2$	904	–	–
Long Asbestos Fibers	2407	+++	+++
Short Asbestos Fibers	2657	+++	+++
TiO$_2$ + Short Fibers	1925	+	+

NBT reduction and Trypan blue staining scored as in table 2.

TABLE 4. Induction of Mesotheliomas After Weekly Intraperitoneal Injections of Crocidolite Asbestos Fibers

Sample	Latency (wks.)	with tumors
Mixed fibers	36-63	37.5
Long fibers	35-51	23.5
Short fibers	54-66	50.0

In contrast, after repeated injections of TiO$_2$ particles followed by a single injection of short asbestos fibers, lymphatic clearance is not impaired, short fibers do not accumulate on the mesothelial lining, and mesothelial cells are not injured.

On the basis of this *in vivo* model, we conclude that both length and chemical composition are important determinants of acute asbestos toxicity after direct intraperitoneal injection of crocidolite asbestos fibers. If lymphatic clearance is impaired by an intense inflammatory reaction, short fibers accumulate on the mesothelial lining and produce acute injury. Finally, after weekly repeated injections, short fibers induce mesotheliomas, although the latent period is prolonged, as summarized in table 4.

DISCUSSION

Fiber dimensions are critical parameters for the deposition, clearance, translocation, and retention of mineral fibers after inhalation. Stanton and his colleagues demonstrated that long, thin mineral fibers, regardless of their chemical composition, were more likely to induce mesotheliomas than short fibers after direct intrapleural or intraperitoneal injection. Long fibers have also been shown to produce acute alterations in cell culture models, including cell death (Bey & Harington, 1971), inhibition of growth (Chamberlain & Brown, 1978), production of chromosomal damage (Hesterberg & Barrett, 1955), and transformation (Hesterberg & Barrett, 1984). Recently, the chemical composition of fibers, especially as it relates to surface reactivity (Bonneau *et al.*, 1986), has been linked to generation of hydroxyl radicals. We have re-examined the role of fiber length in acute cytotoxicity using *in vitro* and *in vivo* models. In contrast to many of the earlier studies of *in vitro* cytotoxicity, both long and short preparations of crocidolite asbestos fibers stimulated release of H_2O_2 from elicited mouse peritoneal macrophages. During the same time period, long and short fibers inflicted oxidant-induced injury as monitored by loss of the mitochondrial membrane potential. Acute toxicity was prevented by adding exogenous superoxide dismutase or catalase to the culture media or by coating the fibers with the iron chelator deferoxamine (Goodglick & Kane, 1990).

The conflicting evidence obtained in this model system in comparison with previously published studies is due to major differences in the experimental protocols. The most important difference is the method of fiber preparation (Dunnigan, 1984). Previous investigators prepared short fibers by milling, while this protocol separated short and long fibers by centrifugation. Other major differences are the range of doses, differences in fiber type (amphiboles vs. serpentine asbestos), and time of exposure. In addition, different target cell populations and different indices of toxicity were used (Bey & Harington, 1971; Chamberlain & Brown, 1978; Hesterberg & Barrett, 1985; Kaw, 1982). In our *in vitro* model, acute toxicity was monitored in quiescent cultures of elicited mouse peritoneal macrophages using dye exclusion or uptake assays (Goodglick & Kane, 1986). In other models, inhibition of growth in proliferating cultures was used as an index of toxicity. Both of these indices monitor the biologic reactivity of mineral fibers; however, the mechanisms responsible for acute cell death and inhibition of growth by fibers may differ. More importantly, which index of toxicity is more relevant for the *in vivo* effects of asbestos fibers? Additional studies are needed to determine the mechanisms of acute toxicity using different target cell populations in the lungs.

The mesothelium is an important target of asbestos fibers, especially as the site for development of malignant mesotheliomas. The usual model system used to induce these tumors in rodents is direct intrapleural or intraperitoneal injection of a single dose of fibers (10-40 mg). The usual endpoint in this model is production of mesotheliomas after 1-2 years (reviewed by Harington, 1981). We have modified this model and use weekly intraperitoneal injections of lower doses of fibers in mice. At early time points, clearance of this lower dose of fibers (100-200 μg) is not impaired, while a single large implant is walled off by fibrous tissue in the omentum (Courtice & Simmonds, 1954). In our model, we have described the pattern of deposition and mesothelial reactions to asbestos fibers. Short asbestos fibers and spherical mineral particles are cleared through lymphatic channels that open between mesothelial cells on the inferior surface of the diaphragm, while the clearance of long fibers is limited by the diameter of the lymphatic openings or stomata (10.7 $\pm$ 3.2 μm). Within 24-72 hours after a single injection of long asbestos fibers, acute injury was demonstrated at sites of fiber deposition around lymphatic stomata. Inflammatory cells, predominantly macrophages recently elicited from the circulation, accumulated around fiber clusters and were shown to release superoxide anions *in situ*. Similar to our *in vitro* model of acute toxicity to macrophages, acute mesothelial cell injury *in vivo* was decreased by injections of superoxide dismutase or catalase conjugated to polyethylene glycol or by injection of deferoxamine-coated asbestos fibers.

Our observations confirm Stanton's hypothesis: long fibers are trapped at the mesothelial lining because their clearance is limited by the size of lymphatic stomata. We observed acute mesothelial cell injury at sites of fiber deposition that depends on local production of oxidants. We propose the following working hypothesis for the development of mesotheliomas by asbestos fibers (modified from Moalli *et al.*, 1987):

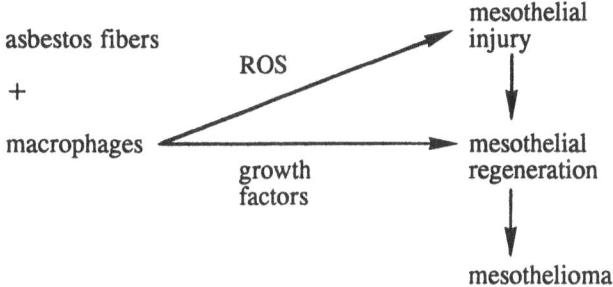

In our model system, acute injury depends on fiber length after a single injection of equal numbers of short or long fibers. However, after multiple injections of short fibers, lymphatic clearance is impaired leading to accumulation of short fibers at the same sites where long fibers are trapped. Under these conditions, short fibers stimulate oxidant production and localized mesothelial cell injury. Therefore, we conclude that short fibers, as tested in our *in vitro* and *in vivo* models are capable of inducing acute cell injury mediated by oxidants at the site of interaction between the fibers and target cells. Before this conclusion can be accepted, other interpretations of these experimental studies must be considered. The preparation of short fibers used in these experiments is contaminated by long fibers: 1.1% of the fibers are longer than 5 μm compared to 27.6% of the long fiber preparation. Is this injury caused by clustering of short fibers at the mesothelial lining or by a few long fibers that contaminate this preparation? This question is especially important in testing the ability of short fibers to induce mesotheliomas. Previous studies also report induction of mesotheliomas by short fibers (Kolev, 1982; Wagner *et al.*, 1984; Davis *et al.*, 1986). As in our experiments reported here, the latency period before tumors were induced was prolonged, but some animals did develop mesotheliomas after a single injection of short fibers. Purity of the fiber preparation was also a confounding factor in these studies.

Additional questions should be raised about impaired lymphatic clearance in this model system. As judged by histopathologic examination, repeated injections of short fibers, but not of TiO$_2$ particles, overloaded lymphatic channels. Lymphatic clearance should be quantitated under these conditions. Is lymphatic overload produced in the pleural lining after inhalation of asbestos fibers? If pleural plaques form around lymphatic stomata in the pleural lining, does this retard lymphatic clearance leading to pleural effusions? Is localized lymphatic overload in the mesothelial lining related to accumulation of fibers and development of mesotheliomas (Bolton *et al.*, 1983)? Further studies, including animal models and measurement of fiber burdens in the human pleura, are required to answer these questions and to assess whether large numbers of short fibers may cause human disease as suggested previously by Churg & Wiggs (1984).

ACKNOWLEDGEMENTS

This research was supported by grants from the National Institutes of Health (R01 ES 03721 and R01 ES 03189).

REFERENCES

Antman,K. and Aisner,J. (1987) "Asbestos-Related Malignancy", Grune & Stratton, Orlando, FL.

Bey,E. and Harington,J.S. (1971) Cytotoxic effects of some mineral dusts on Syrian hamster peritoneal macrophages. *J. Exp. Med.* **133**:1149-1169.

Bolton,R.E., Vincent,J.H., Jones,A.D., Addison,J. and Beckett,S.T. (1983) An overload hypothesis for pulmonary clearance of UICC amosite fibres inhaled by rats. *Br. J. Indust. Med.* **40**:264-272.

Bonneau,L., Marlard,C. and Pezerat,H. (1986) Studies on surface properties of asbestos. *Environ. Res.* **41**:268-275.

Chamberlain,M. and Brown,R.C. (1978) The cytotoxic effects of asbestos and other mineral dust in tissue culture cell lines. *Br. J. Exp. Pathol.* **59**:183-189.

Churg,A. and Green,F.H.Y., eds. (1988) "Pathology of Occupational Lung Disease", Igaku-Shoin, N.Y.

Churg,A. and Wiggs,B. (1984) Fiber size and number in asbestos-induced mesothelioma. *Am. J. Pathol.* **115**:437-442.

Craighead,J.E. (1987) Current pathogenetic concepts of diffuse malignant mesothelioma. *Human Pathol.* **18**:544-577.

Courtice,F.C. and Simmonds,W.J. (1954) Physiological significance of lymph drainage of the serosal cavities and lungs. *Physiol. Rev.* **34**:419-448.

Davis,J.M.G., Addison,J., Bolton,R.E., Donaldson,K., Jones,A.D. and Smith,T. (1986) The pathogenicity of long versus short fibre samples of amosite asbestos administered to rats by inhalation and intraperitoneal injection. *Br. J. Exp. Pathol.* **67**:415-430.

Donaldson,K., Brown,G.M., Brown,D.M., Bolton,R.E. and Davis,J.M.G. (1989) Inflammation generating potential of long and short fibre amosite asbestos samples. *Br. J. Indust. Med.* **46**:271-276.

Dunnigan,J. (1984) Biological effects of fibers: Stanton's hypothesis revisited. *Environ. Hlth Perspect.* **57**:333-337.

Goodglick,L.A. and Kane,A.B. ( ) The role of fiber length in crocidolite asbestos toxicity *in vitro* and *in vivo*. In, "Proceedings of the VIIth International Conference on the Pneumoconioses", Pittsburgh, August 23-26, 1988, (in press).

Goodglick,L.A. and Kane,A.B. (1990) Cytotoxicity of long and short crocidolite asbestos fibers *in vitro* and *in vivo*. *Cancer Res.* **50**:5153-5163.

Goodglick,L.A. and Kane,A.B. (1986) The role of reactive oxygen metabolites in crocidolite asbestos toxicity to macrophages. *Cancer Res.* **46**:5558-5566.

Gulumian,M. and Van Wyk,J.A. (1987) Hydroxyl radical production in the presence of fibres by a Fenton-type reaction. *Chem. Biol. Inter.* **62**:89-97.

Harington,J.S. (1981) Fiber carcinogenesis: epidemiologic observations and the Stanton hypothesis. *J. Natl. Canc. Inst.* **67**:977-989.

Hesterberg,T.W. and Barrett,J.C. (1985) Induction by asbestos fibers of anaphase abnormalities. *Carcinogenesis* **6**:473-475.

Hesterberg,T.W. and Barrett,J.C. (1984) Dependence of asbestos- and mineral dust-induced transformation of mammalian cells in culture on fiber dimension. *Cancer Res.* **44**:2170-2180.

Kaw,J.L., Tilkes,F. and Beck,E.G. (1982) Reaction of cells cultured *in vitro* to different asbestos dusts of equal surface area but different fibre length. *Br. J. Exp. Pathol.* **63**:109-115.

Kolev,K. (1982) Experimentally induced mesothelioma in white rats in response to intraperitoneal administration of amorphous crocidolite asbestos. *Environ. Res.* **29**:123-133.

Macdonald,J.L. and Kane,A.B. (1986) Identification of asbestos fibers within single cells. *Lab. Invest.* **55**:177-185.

Moalli,P.A., Macdonald,J.L., Goodglick,L.A. and Kane,A.B. (1987) Acute injury and regeneration of the mesothelium in response to asbestos fibers. *Am. J. Pathol.* **128**:425-445.

Monchaux,G., Bignon,J., Jaurand,M.-C., Lafuma,J., Sebastien,P., Masse,R., Hirsch,A. and Goni,J. (1981) Mesotheliomas in rats following inoculation with acid-leached chrysotile asbestos and other mineral fibers. *Carcinogenesis* **2**:229-236.

Mossman,B.T. and Marsh,J.P. (1989) Evidence supporting a role for active oxygen species in asbestos-induced toxicity and lung disease. *Environ. Hlth Perspect.* **81**:91-94.

Pott,F. (1987) Problems in defining carcinogenic fibers. *Ann. Occup. Hyg.* **31**:799-802.

Shatos,M.A., Doherty,J.M., Marsh,J.P. and Mossman,B.T. (1987) Prevention of asbestos-induced cell death in rat lung fibroblasts and alveolar macrophages by scavengers of active oxygen species. *Environ. Res.* **44**:103-116.

Stanton,M.F., Layard,M., Tegeris,A., Miller,E., May,M., Morgan,E. and Smith,A. (1981) Relation of particle dimensions to carcinogenicity in amphibole asbestoses and other fibrous minerals. *J. Natl. Canc. Inst.* **67**:965-975.

Wagner,J.C., Berry,G. and Timbrell,V. (1973) Mesotheliomata in rats after inoculation with asbestos and other minerals. *Br. J. Cancer* **28**:173-185.

Wagner,J.C., Griffiths,D.M. and Hill,R.J. (1984) The effect of fibre size on the *in vivo* activity of UICC crocidolite. *Br. J. Cancer* **49**:453-458.

Weitzman,S.A. and Graceffa,P. (1984) Asbestos catalyzes hydroxyl and superoxide radical release from hydrogen peroxide. *Arch. Biochem. Biophys.* **228**:373-376.

ACUTE PULMONARY EFFECTS OF INHALED WOLLASTONITE FIBERS ARE DEPENDENT ON FIBER DIMENSIONS AND AEROSOL CONCENTRATIONS

David B. Warheit, Kimberly A. Moore, Michael C. Carakostas and
Mark A. Hartsky

Central Research and Development
Haskell Laboratory for Toxicology and Industrial Medicine
E.I. du Pont de Nemours and Co.
Newark, DE 19714, U.S.A.

INTRODUCTION

Occupational exposure to asbestos fibers has been associated with the development of pulmonary fibrosis (i.e., asbestosis), bronchogenic carcinoma and mesothelioma. Therefore, the commercial use of asbestos is likely to be curtailed or limited in the near future and mineral fiber substitutes are currently being promoted to fill the void. Wollastonite fibers are natural acicular calcium silicate minerals which have been proposed as alternatives for asbestos in applications such as brake linings, wallboard and insulation materials. Wollastonite fiber diameters are generally in the 1 - 10 μm range with an average diameter of 3.5 μm (Vu, 1988).

There is a paucity of information regarding the pulmonary effects of wollastonite fiber inhalation. Recently, a 2-year inhalation study was carried out by the National Toxicology Program to assess the fibrogenicity and carcinogenicity of inhaled wollastonite fibers. In this study, groups of male Fischer 344 rats were exposed to 10 mg/m^3 (360 fibers/cc) of wollastonite for 12 or 24 months; these were compared to crocidolite asbestos and sham-exposed controls. Wollastonite exposure did not result in any fibrogenic or carcinogenic effects in rats (Adkins et al., 1989). However, the significance of these results is unclear as the dimensions of the fibers generated during the exposure were not measured; thus, the respirable nature of the sample could not be confirmed. In a recent study, the pulmonary effects of respirable dust from three commercially produced calcium silicate insulation materials were investigated in exposed rats. No pulmonary lesions were attributed to calcium silicate exposure. The authors concluded that the dust samples were harmless to rats at the doses tested (Bolton et al., 1986).

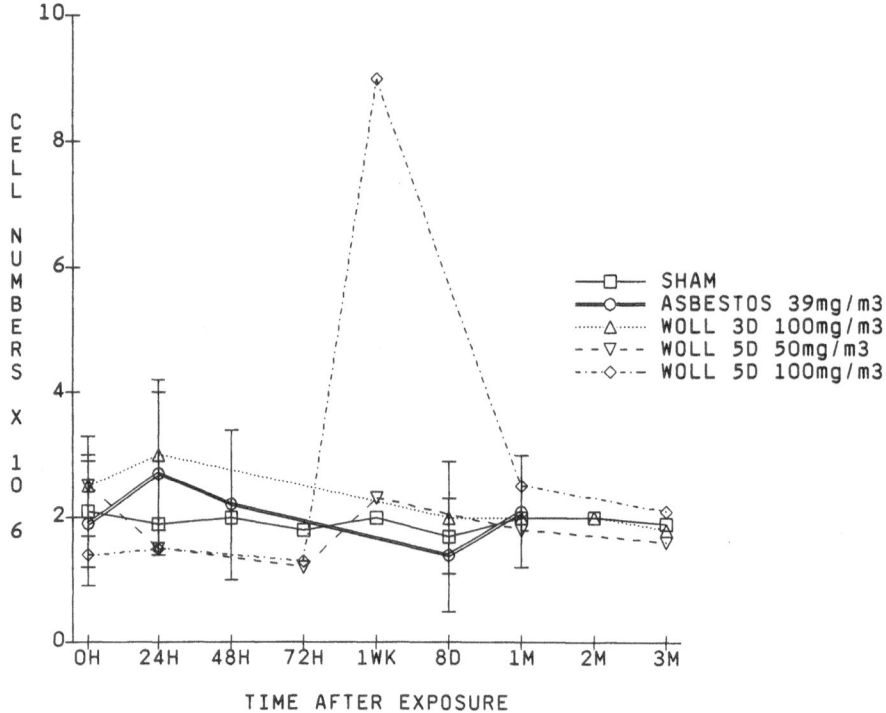

Figure 1. The total no. of cells recovered by BAL.

A short-term inhalation bioassay in rats has been developed to gauge the potential for inhaled particles to produce chronic lung injury in humans (Warheit *et al.*, 1991a). The efficacy of this predictive screen is predicated on the idea that the development of particle-induced pulmonary disease is correlated with four interdependent general factors:

1) the propensity of an inhaled material to cause lung cell injury (*i.e.*, general cytotoxicity);

2) the tendency for inhaled particles to produce ongoing inflammation;

3) particle-induced diminution of pulmonary macrophage clearance functions (*i.e.*, as measured by morphologic characteristics, phagocytosis and/or chemotaxis);

4) persistence or reduced clearance of the inhaled material in the lung (Bowden & Adamson, 1984).

The presence of these four factors has been described independently or in various combinations in previous reports of chronic lung disease with fibrosis (Lugano *et al.*, 1982; Dauber *et al.*, 1982; Bowden & Adamson, 1984; Reiser & Last, 1986). However, they have not been collectively used as predictive indicators of chronic lung disease.

In the present study, the pulmonary effects of short-term, high-dose inhalation exposures to wollastonite fibers at different fiber dimensions and fiber concentrations (numbers) were assessed in exposed rats. Rats exposed to crocidolite asbestos fibers served as positive controls. The results from biochemical and cellular studies suggest that inhalation of wollastonite fibers presents a reduced inhalation hazard when compared to crocidolite asbestos fibers. In addition, the method of fiber aerosol generation, the fiber size (aerodynamic) and the aerosol concentration and corresponding fiber number play important roles in producing wollastonite-related acute lung injury.

144

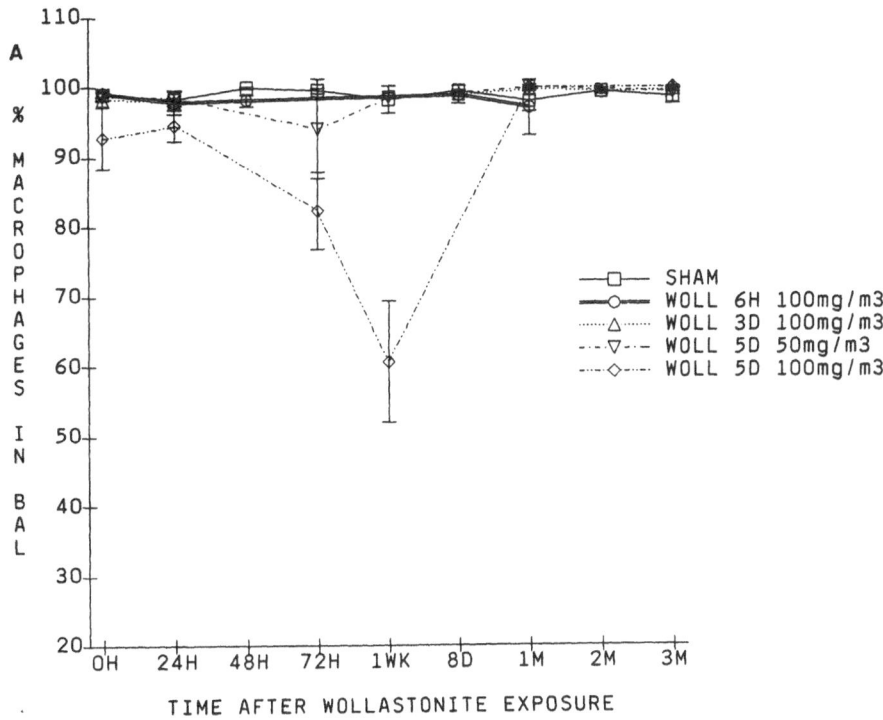

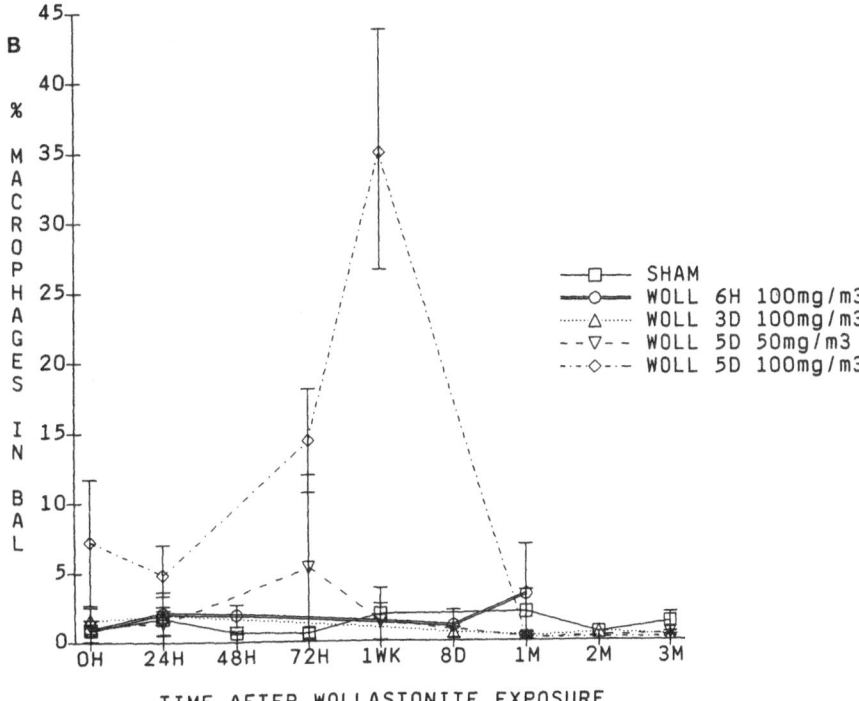

Figure 2. A) Pulmonary macrophage differential percentages in BAL of wollastonite-exposed rats. **B)** Granulocyte differential percentages in BAL of wollastonite-exposed rats.

MATERIALS AND METHODS

General Experimental Design

Groups of male Crl:CDBR rats (8 weeks old, Charles River Breeding Laboratories, Kingston, New York) were used to assess the pulmonary effects of short-term, high-dose aerosol exposures to either crocidolite asbestos or wollastonite fibers. Rats were exposed to design concentrations of 40 (asbestos), 50, or 100 mg/m³ (wollastonite) for 6 hours, or for 3 or 5 days (6 hrs/day). Following exposures, fiber-exposed animals and aged-matched sham controls were evaluated at 0, 24, and 48 hrs, 1 week, 8 days, or 1 month post exposure.

Fiber Preparations

The crocidolite asbestos fibers were obtained from NIEHS, and were derived from a UICC sample previously characterized (Warheit et al., 1984c). Wollastonite NYAD-G fibers were obtained from NYCO (Willsboro, NY). A preparation of fibers was measured using scanning electron microscopy (SEM) techniques; wollastonite fiber diameters ranged from 0.2 - 3.0 µm. Mean fiber diameters of the crocidolite sample were 0.15 µm.

Inhalation Exposure

Animals were placed in cylindrical polycarbonate or stainless steel holders equipped with conical nose pieces. The restrainers were inserted into face plates on the exposure chambers such that the nose of each animal protruded into the chamber. Atmospheres of crocidolite asbestos or wollastonite fibers were generated with a K-tron bin feeder equipped with twin feed screws. The dust was metered into a plastic funnel which transferred the test material into a microgenerator. Baffles were inserted into the generation apparatus for the 5-day wollastonite fiber experiments and served to reduce the MMAD from 5.8 to 2.6 µm (see Table 1). High pressures of air swept the test material into a polycarbonate and glass transfer tube into the exposure chamber. Chamber concentrations of wollastonite fibers were maintained by controlling the dust-feed rate into the generation apparatus, or by varying the air-flow rate.

For gravimetric analysis, samples of atmospheric crocidolite asbestos or wollastonite fibers were taken from the animal breathing zone at approximately 30-minute intervals by drawing calibrated volumes of chamber atmosphere through pre-weighed glass-fiber filters. Filters were weighed on a Cahn 26 automatic electrobalance. The atmospheric concentrations of asbestos or wollastonite fibers were determined from

TABLE 1. Fibre Exposure - specifications and dose

Fibre	Duration of exposure	MMAD µM	Mean gravimetric conc	Mean no fibres/cc	Retained fibre no fibres(x10⁶) /g lung tissue
Crocidolite	6h	2.2	41 ± 8	12,830	1.6±0.2
Wollastonite	6h	5.8	105 ± 14	476	—
Wollastonite	3d	5.8	118 ± 24	227	3.9±0.8
Wollastonite	5d	4.3	59 ± 21	123	—
Wollastonite	5d	2.6	114 ± 20	835	8.9±0.1

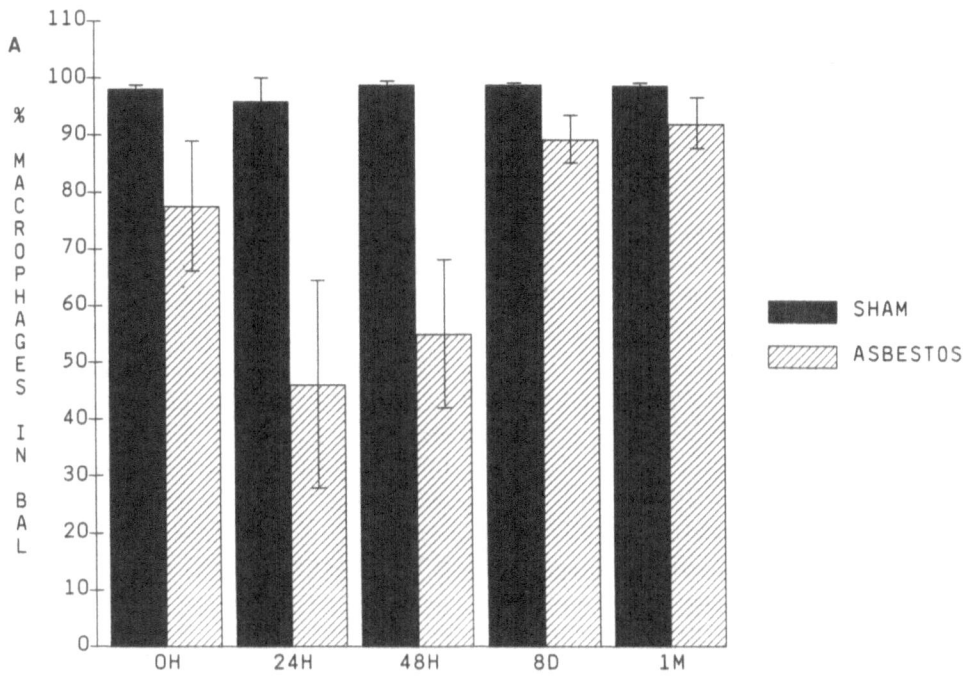

TIME AFTER 6 HOUR ASBESTOS EXPOSURE

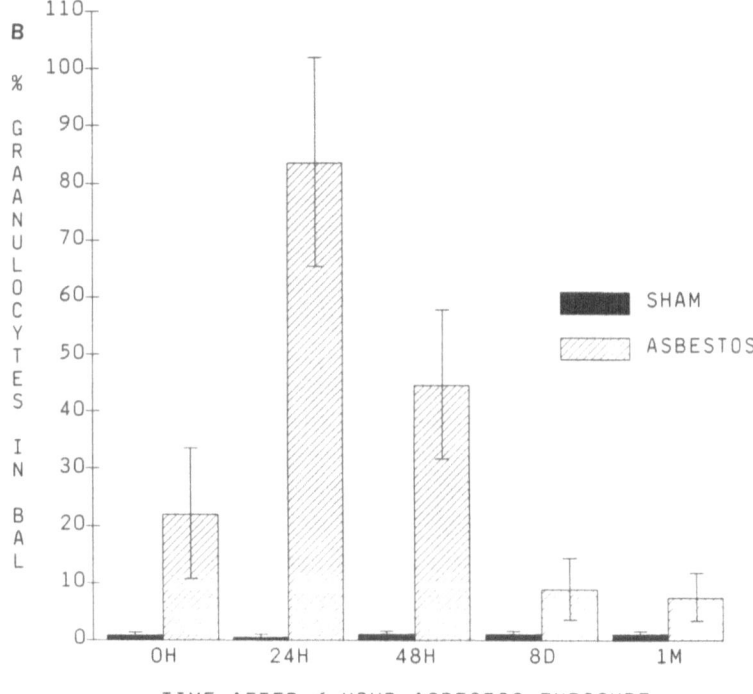

TIME AFTER 6 HOUR ASBESTOS EXPOSURE

Figure 3. A) Pulmonary macrophage differential percentages in BAL of crocidolite asbestos-exposed rats. B) Granulocyte differential percentages in BAL of crocidolite asbestos-exposed rats.

the filter weight differentials before and after sampling. Particle size measurements of airborne particles in the test chamber were determined with a Sierra cascade impactor and reported as mass median aerodynamic diameter (MMAD) and percent of particles less than 10 μm aerodynamic diameter. Fiber counts were carried out according to the NIOSH 7400B method.

Pulmonary Lavage and Preparation of Macrophage Monolayers

Bronchoalveolar lavage procedures were conducted according to methods previously described (Warheit *et al.*, 1983, 1984a). Briefly, CD rats were euthanized by intraperitoneal injections of sodium pentobarbital (Nembutal). The trachea, airways and lungs were gravity infused with a warmed (37°C), Ca^{++} and Mg^{++} free, phosphate-buffered saline solution. This bronchoalveolar lavage (BAL) procedure was carried out 6 times or until 50 ml of fluid were collected from each animal.

Lavaged fluids recovered from sham and dust-exposed rats were centrifuged at 250 x g, and the supernatant was removed and concentrated for biochemical studies (see below). The cell pellet was resuspended in Eagles Minimal Essential Medium (Eagles MEM F-11; pH= 7.2, GIBCO, Grand Island, NY) supplemented with penicillin and streptomycin. Cell numbers and viability were quantified using a hemocytometer and trypan blue solution. Cell differential counts were carried out on Diff-Quik stained (American Scientific Co.) cytocentrifuge preparations. A minimum of 500 cells per slide were counted for differential analysis.

Biochemical Assays on Bronchoalveolar Lavage Fluid Lavage fluid from the first two washes was centrifuged and the supernatant concentrated 10X in an Amicon concentrator (cut-off MW = 10,000). All biochemical assays were performed on concentrated BAL fluid at 30°C, using a semi-automated clinical chemical analyzer (Encore II, Baker Instruments, Allentown PA). Lavage fluid protein was measured using a commercially available reagent kit based on Coomassie Blue dye binding (QuanTtest, Quantimetrix, Hawthorne, CA). Lactate dehydrogenase (LDH) and alkaline phosphatase (ALP) were measured using commercially available reagent kits (Baker Instruments, Allentown, PA). All refrigerated lavaged samples containing enzymes and proteins were stable for a minimum period of 24 hours.

Macrophage Cell Culture and Phagocytosis of Iron Particles

Pulmonary macrophage cell culture and phagocytosis assays have been previously reported (Warheit *et al.*, 1983, 1984a). Briefly, macrophages recovered from sham and fiber-exposed rats were incubated for 45 minutes in a CO_2 incubator at 37°C (mean plating efficiency = 90%) to allow the cells to adhere to the culture dish surface. Subsequently, the monolayers were rinsed vigorously in Eagles MEM to remove non-adherent cells and either placed in fixative or further processed for phagocytosis studies.

For the phagocytic assay, a suspension of carbonyl iron particles was incubated with normal rat serum for 1 hour at 40°C and sonicated to reduce aggregations of particles. The iron particles ranged in diameter from 0.4 to 2.0 μm (mean particle size = 1.1 μm). A final concentration of 1.75 mg/ml (mass per volume in the culture dish = 204 μg/cm^2) was added to the monolayers. Phagocytosis of iron particles was carried out at 37°C in the CO_2 incubator for 1 hour. The coverslips were then rinsed in Eagles MEM to removed residual carbonyl iron and fixed in Karnovsky's solution before processing for scanning electron microscopy (SEM), as described below.

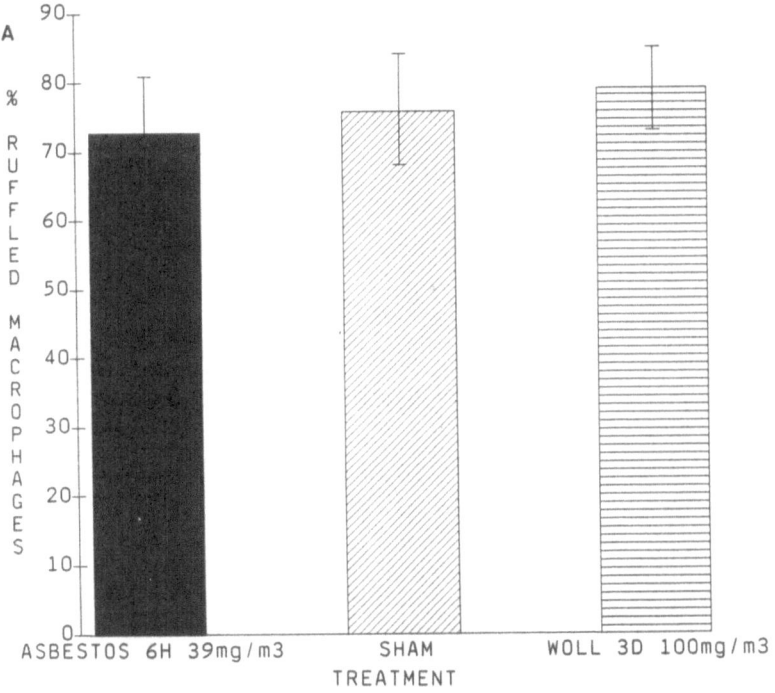

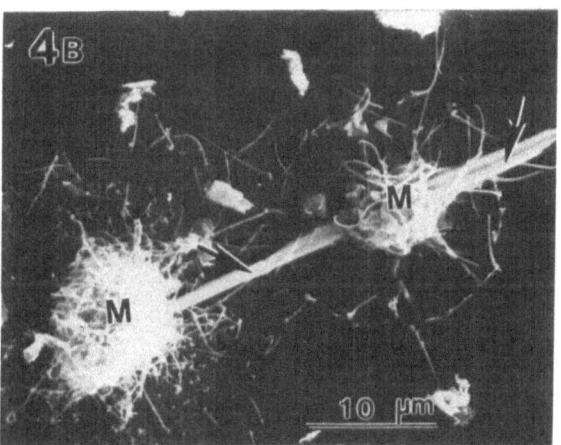

Figure 4. A) The percentages of ruffled macrophages recovered from fiber-exposed rats B) Morphology of pulmonary macrophages recovered from asbestos and wollastonite-exposed rats.

Scanning Electron Microscopy (SEM)

Specimens were prepared according to standard methods and examined using a JEOL 840 scanning electron microscope.

Studies of morphology and *in vitro* phagocytosis were implemented using SEM examination of randomly selected cultured cells as previously described (Warheit *et al.*, 1983). Cells were categorized according to their ruffled membranes or unruffled, smooth surface characteristics. For *in vitro* phagocytosis of carbonyl iron (CI) beads, the numbers

TABLE 2. Protein and LDH in BAL fluid recovered from asbestos and wollastonite exposed rats

a) BAL protein values (as percent of sham control)

Treatment	Post recovery time		
	24h	1 wk or 8 days	1 month
sham	100	100	100
Asb 6h (MMAD 2.2 μm)	250 ± 90	221 ± 23*	222 ± 51*
Woll 3d (MMAD 5.8 μm)	89 ± 4	113 ± 34	72 ± 9
Woll 5d (MMAD 4.3 μm)	175 ± 15	98 ± 24	76 ± 7
Woll 5d (MMAD 2.6 μm)	341 ± 128	653 ± 207*	63 ± 15

b) BAL LDH values (as percent of sham control)

Treatment	Post recovery time		
	24h	1 wk or 8 days	1 month
sham	100	100	100
Asb 6h (MMAD 2.2 μm)	362 ± 91*	180 ± 22	341 ± 31*
Woll 3d (MMAD 5.8 μm)	65 ± 25	88 ± 14	82 ± 21
Woll 5d (MMAD 4.3 μm)	317 ± 45*	154 ± 71	96 ± 32
Woll 5d (MMAD 2.6 μm)	749 ± 59*	662 ± 131*	73 ± 22

* $p < 0.05$

of interiorized CI particles were counted in individual macrophages using secondary and backscattered electron-imaging modes as described above. The capacity of macrophages to phagocytize carbonyl iron particles in vitro was assessed by categorizing the cells as phagocytic (i.e., > 1 internalized particles), or nonphagocytic (i.e., 0 endocytosed particles). The morphological and *in vitro* phagocytosis data are represented respectively as percentages of ruffled and phagocytic macrophages.

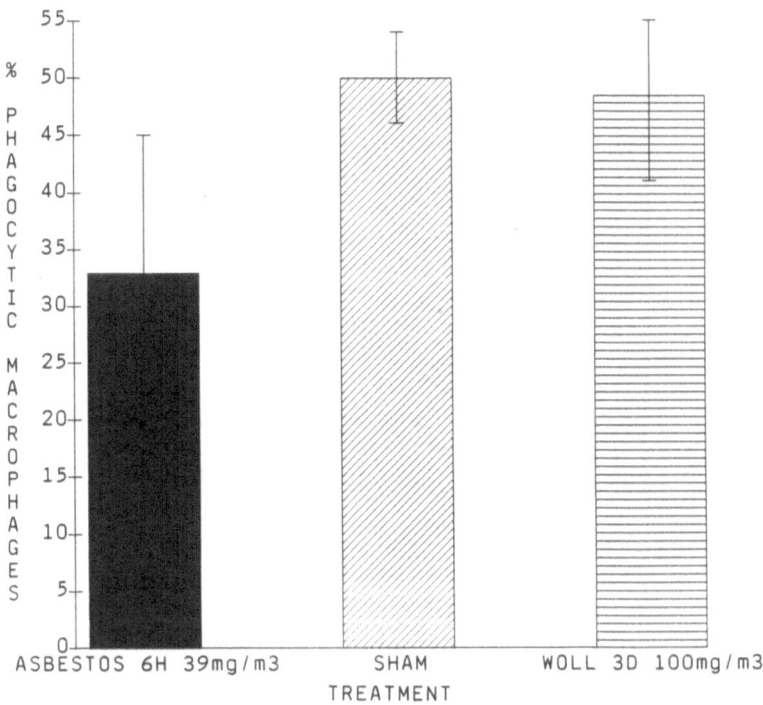

Figure 5. Changes in phagocytic capacities of macrphages recovered from fiber-exposed rats.

Lung Perfusion, Dissection and Tissue Preparation

The lungs of wollastonite or asbestos fiber-exposed and corresponding sham control rats were prepared for light or electron microscopy immediately following the termination of aerosol exposures. Animals were anesthetized by intraperitoneal injections of sodium pentobarbital. After anesthesia, lungs of rats were fixed either through the vasculature (vascular perfusion) or intratracheally (airway infusion). In both cases the trachea was exposed and clamped to prevent lung collapse. For intratracheal fixation a small incision was made below the clamp and a 19-gauge butterfly catheter (Abbott Labs., North Chicago, IL) was secured into the trachea. The catheter was connected to a reservoir located 15 cm above the thorax of the animal. The trachea and lungs of the rat were perfused by gravity flow of Karnovsky's fixative (1% formaldehyde-1% glutaraldehyde in cacodylate buffer, pH 7.2, 350 mOsm) at a pressure of 15 cm H_2O. After 15 min, with the trachea still clamped to prevent collapse, the lungs were removed from the chest cavity and immersed in fresh fixative for 48 hr.

Sagittal sections of the left lung were made with a razor blade. Tissue blocks were dissected from upper, middle and lower regions of the lung, and were subsequently prepared for light microscopy (paraffin embedded, sectioned, and hematoxylin-eosin stained), as well as for scanning and transmission electron microscopy (see below).

Scanning Electron Microscopy

Following completion of intratracheal fixation and wet dissection, tissue blocks were rinsed in a cacodylate-buffered sucrose solution and dehydrated through a series of ethanol solutions. The SEM samples were critical-point-dried, using carbon dioxide as the transitional fluid. The dried tissue blocks were mounted onto carbon stubs and then

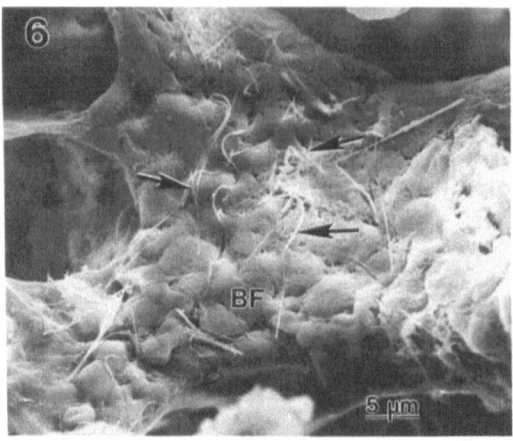

Figure 6. Crocidolite fibers at alveolar duct bifurcations.

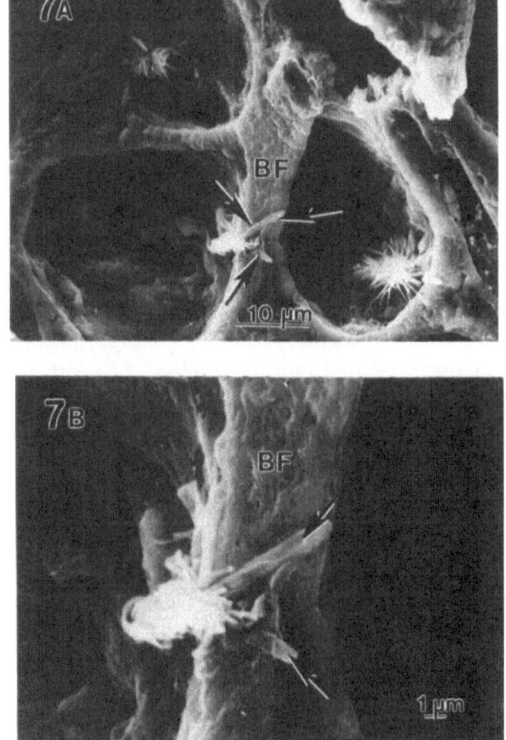

Figure 7. Wollastonite fibers in distal lung regions.

TABLE 3. Alkaline phosphatase in BAL fluids recovered from asbestos and wollastonite exposed rats

BAL alkaline phosphatase as percent of sham control

Treatment	Post recovery time		
	24h	1 wk or 8 days	1 month
sham	100	100	100
Asb 6h (MMAD 2.2 μm)	181 ± 80	172 ± 35	173 ± 38
Woll 3d (MMAD 5.8 μm)	128 ± 57	188 ± 46	133 ± 34
Woll 5d (MMAD 4.3 μm)	37 ± 13	54 ± 22	53 ± 7
Woll 5d (MMAD 2.6 μm)	392 ± 133	296 ± 145	29 ± 6

dissected with a razor blade until several functional units, i.e., terminal bronchioles and their corresponding alveolar ducts were apparent in a dissecting microscope. This assured the easy identification of these anatomic units in the SEM (Hill & Plopper, 1979; Warheit *et al.*, 1984b). Following dissection, the specimens were gold-coated and placed into a JEOL 840 electron microscope. Tissue prepared for SEM was used to investigate particle-induced cellular responses and pulmonary macrophage clearance responses at bronchoalveolar junctions.

Statistical Analyses

For analysis, each of the means of experimental values were compared to their corresponding sham control values for each time point. A one-way analysis of variance (ANOVA) and Bartlett's test were calculated for each sampling time. When the F-test from ANOVA was significant, the Dunnett test was used to compare means from the control group and each of the groups exposed either to silica or carbonyl iron. Significance was judged at the 5% probability level. The data were subsequently normalized and are represented as percent of sham control values for that experiment.

RESULTS AND DISCUSSION

Chamber Atmosphere Analysis and Retained Dose

The mass median aerodynamic diameters (MMAD) for wollastonite and crocidolite asbestos are summarized in Table 1. Fiber sizes ranged from 2.6 - 5.8 μm in the wollastonite fiber experiments and the variability was dependent upon the method of aerosol generation. In addition, selected samples of fixed lung tissue from wollastonite and asbestos-exposed rats were digested with a hypochlorite solution (Warheit *et al.*,

1991b) and the numbers of retained fibers (i.e., dose) were calculated. The results are enumerated in Table 1.

Analyses of Cellular Constituents in BAL

Increased numbers of cells were recovered 1 week post-exposure from the lungs of 5-day wollastonite-exposed rats at 100 mg/m^3. Otherwise, exposures to wollastonite or asbestos fibers did not significantly alter the total numbers of cells recovered by bronchoalveolar lavage (figure 1). The viability of cells recovered in fiber-exposed rats was generally greater than 95% and was not significantly different from sham controls at any post-exposure time period.

Cell differential analyses of lavaged cells recovered from wollastonite fiber-exposed rats demonstrated dose-dependent decreased percentages of macrophages in BAL fluids (figure 2A), and this corresponded to increases percentages of granulocytes in the lavageate during the early post-exposure recovery periods (i.e., 0 - 1 week post-exposure) (figure 2B). These results indicated that exposures to wollastonite for 5 days at 50 or 100 mg/m^3 provoked acute pulmonary inflammatory responses which were characterized by an influx of granulocytes (primarily neutrophils). This effect was a transient one and was not evident at the 1 week post-exposure period, or at any other subsequent time period.

Cell differential analysis of cells recovered from the lungs of asbestos-exposed rats demonstrated that a 6-hr inhalation exposure to crocidolite produced a transient PMN and eosinophil inflammatory response which returned to near normal levels within 8 days after exposure (figures 3A and B).

Enzyme and Protein Analyses in BAL Fluid

Significant increases in BAL fluid protein levels were measured 24 hrs and 1 week after a 5-day (100 mg/m^3) exposure to wollastonite fibers and had not returned to control levels by the 1 month post-exposure time period (Table 2). Protein values in BAL fluids of asbestos-exposed rats were still increased 8 days and 1 month after the end of exposures (Table 2).

Extracellular LDH in lung lavage fluids was considered to be a sensitive indicator of pulmonary cytotoxicity. Transient increases in LDH values were measured in rats exposed to wollastonite fibers for 5 days. However, no significant increases in LDH levels were measured by 1 month after exposure. In contrast, LDH values in BAL fluids of crocidolite-exposed animals were still increased 1 month after termination of exposures (Table 2).

Increased levels of alkaline phosphatase (ALP) were measured immediately after a 5-day exposure to wollastonite but this result was not statistically significant. No other differences in BAL fluid ALP levels were detected between sham- and fiber-exposed rats at any time period (Table 3).

Pulmonary Macrophage Studies

SEM techniques were utilized to assess morphologic characteristics and *in vitro* phagocytic capacities of lavage-recovered PM exposed to crocidolite asbestos or wollastonite fibers. The percentages of ruffled macrophages recovered from fiber-exposed rats were not significantly different from sham controls at any post-exposure period (figures 4A and 4B). The *in vitro* phagocytic capacities of macrophages recovered from asbestos fiber-exposed animals were decreased compared to wollastonite- or sham-exposed control rats (figure 5).

Scanning Electron Microscopy of Fiber-Exposed Lung Tissue

Examination of lung tissue by SEM revealed that inhaled crocidolite asbestos fibers deposited preferentially at alveolar duct bifurcations (figure 6). Alternatively, few of the wider (5.8 μm) wollastonite fibers deposited in distal lung regions (figures 7A and B). In previous studies, a mass standard was utilized for quantification of both particle and fiber exposures. However, it now appears that utilization of a gravimetric standard without corresponding fiber numbers may be misleading for assessing the relative toxicities of different fiber-types. In the studies presented here, a gravimetric concentration of 41 mg/m^3 crocidolite asbestos was equivalent to a fiber number of 12,800, whereas a gravimetric concentration of 114 mg/m^3 wollastonite was correlated with 835 fibers/cc. Thus, it seems reasonable to conclude that the toxicity of asbestos fibers in general (i.e., both serpentine and amphiboles) is probably related to the increased numbers of fibers per gravimetric concentration (e.g., 12800 *vs.* 835). At similar gravimetric concentrations there is an order of magnitude difference in the numbers of fibers presented to the peripheral regions of the lung. As a consequence it is conceivable that, on a fiber number basis, asbestos may not be significantly more toxic than any other fiber.

The results of this study demonstrated that a 6-hr inhalation exposure to crocidolite asbestos fibers in rats (41 mg/m^3, 12,800 f/cc) produced a transient PMN and eosinophil inflammatory response which returned to near normal levels within 8 days after exposure. However, noncellular, biochemical indices of inflammation persisted through a 1 month post-exposure period as increases in BAL LDH and protein were measured at most time periods post-exposure ($p < 0.05$). In contrast, wollastonite exposure produced transient pulmonary inflammatory responses and corresponding increases in lavage fluid parameters only when the MMAD was sufficiently small (i.e., 2.6 μm) and the exposure concentration exceeded 500 f/cc. We conclude that the method of fiber aerosol generation, the fiber size (aerodynamic), the aerosol concentration and corresponding fiber number as well as exposure duration are critical factors in producing wollastonite-related acute lung injury. Furthermore, the results suggest that wollastonite fibers present a reduced inhalation hazard when compared to asbestos fibers.

REFERENCES

Adkins,B. Jr., McConnell,E.E. and Hall,L. (1989) Carcinogenicity studies of wollastonite in rats. *Toxicologist* 9:212A.

Bolton,R.E., Addison,J., Davis,J.M.G., Donaldson,K., Jones,A.D., Miller,B.G. and Wright,A. (1986) Effects of the inhalation of dusts from calcium silicate insulation materials in laboratory rats. *Enivron. Res.* 39:26-43.

Bowden,D.H. and Adamson,I.Y.R. (1984) The role of cell injury and the continuing inflammatory response in the generation of silicotic pulmonary fibrosis. *J. Pathol.* 144:149-161.

Dauber,J.H., Rossman,M.D. and Daniele,R.P. (1982) Pulmonary fibrosis: Bronchoalveolar cell types and impaired function of alveolar macrophages in experimental silicosis. *Environ. Res.* 27:226-236.

Hill,L.H. and Plopper,C.G. (1979) Use of large block embedding for correlated LM, TEM and SEM characterization of pulmonary airways at known anatomic location. *Proc. Southeast Elect. Microsc. Soc.* 1:24.

Lugano,E.M., Dauber,J.H. and Daniele,R.P. (1982) Acute experimental silicosis. Lung morphology, histology, and macrophage chemotaxin secretion. *Am. J. Pathol.* 109:27-36.

Reiser,K.M. and Last,J.A. (1986) Early cellular events in pulmonary fibrosis. *Exp. Lung Res.* **10**:331-355.

Vu,V.T. (1988) Health Hazard Assessment of Non-asbestos Fibers. *USEPA report.*

Warheit,D.B., Hill,L.H. and Brody,A.R. (1983) Pulmonary macrophage phagocytosis: quantification by secondary and backscattered electron imaging. *Scan. Electron Microsc.* **1**:431-437.

Warheit,D.B., Hill,L.H. and Brody,A.R. (1984a) Surface morphology and correlated phagocytic capacity of pulmonary macrophages lavaged from the lungs of rats. *Exp. Lung Res.* **6**:71-82.

Warheit,D.B., Chang,L.Y., Hill,L.H., Hook,G.E.R., Crapo,J.D. and Brody,A.R. (1984b) Pulmonary macrophage accumulation and asbestos-induced lesions at sites of fiber deposition. *Am. Rev. Respir. Dis.* **129**:301-310.

Warheit,D.B., Hill,L.H. and Brody,A.R. (1984c) *In vitro* effects of crocidolite asbestos and wollastonite on pulmonary macrophages and serum complement. *Scan. Electron Microsc.* **II**:919-926.

Warheit,D.B., Overby,L.H., George,G. and Brody,A.R. (1988) Pulmonary macrophages are attracted to inhaled particles through complement activation. *Exp. Lung Res.* **14**:51-66.

Warheit,D.B., Carakostas,M.C., Hartsky,M.A. and Hansen,J.F. (1991a) Development of a short-term inhalation bioassay to assess pulmonary toxicity of inhaled particles: Comparisons of pulmonary responses to carbonyl iron and silica. *Toxicol. Appl. Pharmacol.* **107**:350-368.

Warheit,D.B., Hwang,H.C. and Achinko,L. (1991b) Assessments of lung digestion methods for recovery of fibers. *Environ. Res.*, in press.

TOXICOLOGICAL ASPECTS OF THE PATHOGENESIS
OF FIBER-INDUCED PULMONARY EFFECTS

G. Oberdörster and B. E. Lehnert[*]

The University of Rochester
Environmental Health Sciences Center
Rochester, NY 14642, U.S.A.

[*]Los Alamos National Laboratory
Life Sciences Division
Los Alamos, NM 87545, U.S.A.

INTRODUCTION

The potential for inhaled asbestos fibers to induce adverse pulmonary effects depends on numerous and diverse factors: their deposition characteristics in the respiratory tract, their retention kinetics and clearance pathways, their interaction with airway and alveolar free cells, their interactions with specific target cells, and their ability to stimulate the release of cellular mediators and thereby influence intercellular communications. Figure 1 depicts a scheme of the hypothetical pathogenic sequence for the effects of fibers in the respiratory tract. It illustrates the movement of fibers within the respiratory tract, specific responses to mediators and several recognized biological responses to fibers, in particular asbestos fibers, *i.e.*, bronchial carcinoma, pleural mesothelioma and lung fibrosis. It should be noted that while many of the steps outlined in figure 1 have received considerable experimental support, several others remain controversial and continue to be in need of further experimental examination. In this report, we address some of the more prominent questions about fiber-lung responses that need to be investigated in future studies. One important aspect in fiber toxicology relates to the use and interpretation of animal data for extrapolation of effects in humans. Species specific differences in respiratory tract dosimetry and in basic cellular responses have to be considered for risk assessment, as will be pointed out in the following overview.

DOSIMETRY

Fiber Deposition

During exposure to airborne fibers, the dose deposited in the respiratory tract is a critical parameter that governs the ability of fibers to induce adverse effects. Bronchial carcinoma occurs in response to fibers in the conducting airways and mesothelioma and pulmonary fibrosis result from fibers deposited in the alveolar region. An important determinant of the deposited dose is the mode of breathing, *i.e.*, nasal *vs.* mouth-breathing. Figure 2 shows the difference in the deposition of inhaled fibers when humans breathe either through the nose or through the mouth, as predicted by Yu (1990, personal communication). The deposition of long and thin fibers in the extrathoracic airways (nasopharynx and oropharynx) decreases significantly during mouth-breathing as compared to nasal breathing. This means that during mouth-breathing not only will a larger dose of fibers be delivered to the lower respiratory tract, but also that a different fiber fraction, with regard to length and diameter, will penetrate to the deeper lung: the nose effectively retains these fibers. This is of particular importance for exposures in the workplace where increased physical activity may result in more mouth-breathing.

Unlike humans, laboratory animals such as the rat, are obligatory nose breathers. Inhalation studies conducted in these animals will necessarily lead to regional deposition of certain fiber fractions different from that in humans. The nose of the rat retains a significant fraction of inhaled fibers (figure 3), protecting the tracheobronchial and alveolar regions. This difference in respirability between rats and humans has to be taken into account when planning rat inhalation studies with fibers. Reported negative outcomes of such studies with respect to tumor induction (Davis *et al.*, 1978; Wagner *et al.*, 1985;

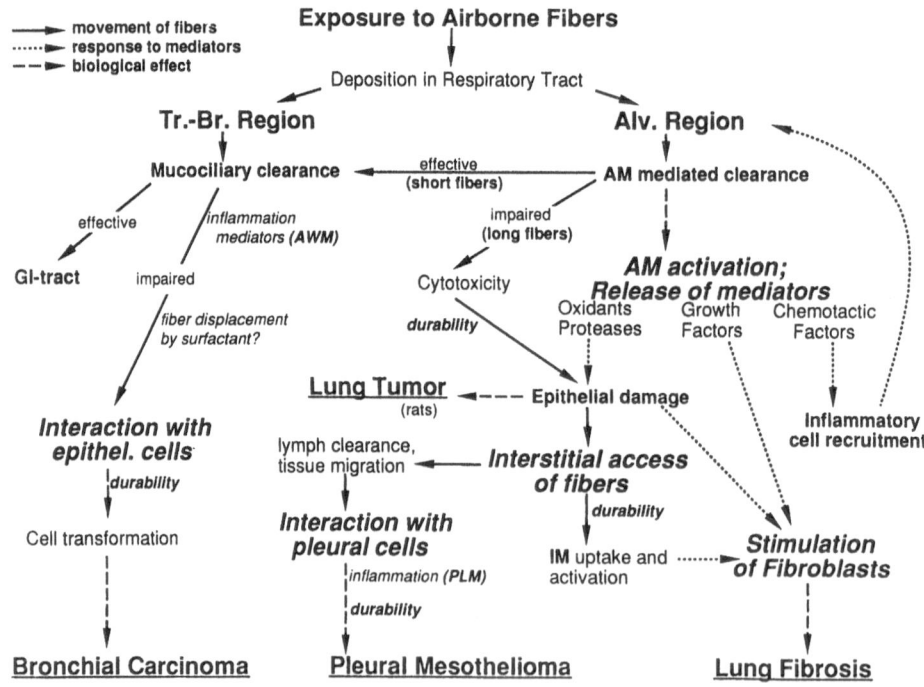

Figure 1: Hypothetical Pathogenic Sequence for Effects of Fibers in the Respiratory Tract.

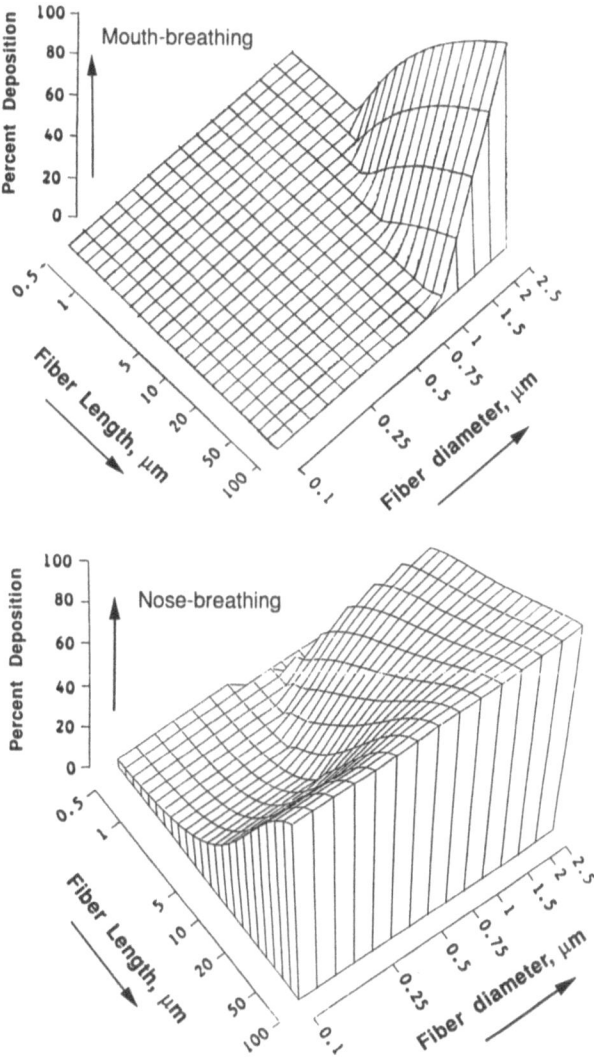

Figure 2: Predicted Deposition of Inhaled Fibers in Human Extrathoracic Airways (Yu, 1990).

Wagner *et al.*, 1987; Muhle *et al.*, 1987) may not only be a matter of the deposited dose, they may also reflect the selective deposition of a fiber size which may be less potent in inducing carcinogenic effects. This poses a possible dilemma when trying to use data from fiber inhalation studies in rats for extrapolation to humans. In order to interpret and extrapolate results from rat studies in a meaningful way, a careful analysis and comparison of the dosimetry in the respiratory tract of both species needs to be performed.

Fiber Clearance

The clearance kinetics of fibers is another important parameter to consider in respiratory tract dosimetry. As indicated in figure 1, elimination of fibers from the alveolar region is quite effective for short fibers but becomes increasingly less efficient for longer fibers. Figure 4 outlines some basic mechanisms responsible for the clearance of fibers after deposition in either the conducting airways or the alveolar region. These consist of physical removal and transfer processes and of chemical mechanisms involving dissolution of certain types of fibers in lung cells and in the pulmonary interstitium. For the conducting airways, the most important clearance mechanism is the movement of fibers up the mucociliary escalator towards the pharynx, with subsequent transfer to the GI tract. Fibers are transported either in translocating alveolar macrophages (AM) or in airway macrophages that may phagocytize them (Gehr *et al.*, 1990; Lehnert *et al.*, 1990), or as free fibers.

Phagocytosis of fibers by alveolar macrophages (AM) is the most prominent defense mechanism leading to elimination of particles from the alveolar region. However their inability to phagocytize fibers beyond a certain length is a limiting factor for the effective physical removal of fibers from this region (Drew *et al.*, 1987; Morgan & Holmes, 1982; Morgan *et al.*, 1978). Long fibers which are phagocytized by AM may not be fully engulfed and one or both ends of the fiber may protrude possibly inducing irreversible damage and cell death. The less effective clearance of long fibers from the lung can also be deduced from results reported by Timbrell (1982) in which he showed that the size distribution of fibers retained in the lungs of an anthophyllite worker included more long fibers than that of the inhaled dust (figure 5). Fibers not cleared from the alveolar region by AM via the mucociliary escalator may penetrate into the interstitium. The recently reported action of the surfactant layer to actively move spherical particles toward the epithelial surface (Schürch *et al.*, 1990) may also play a role for the

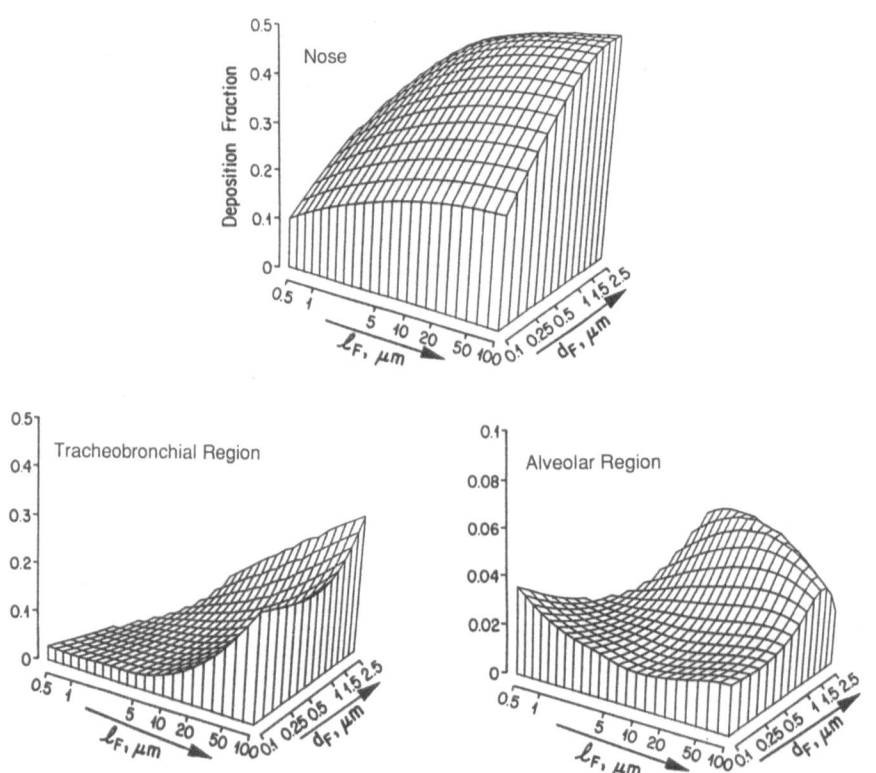

Figure 3: **Predicted Deposition of Inhaled Fibers in the Respiratory Tract of the Rat (Asgharian & Yu, 1989).**

a) *Physical mechanisms*:

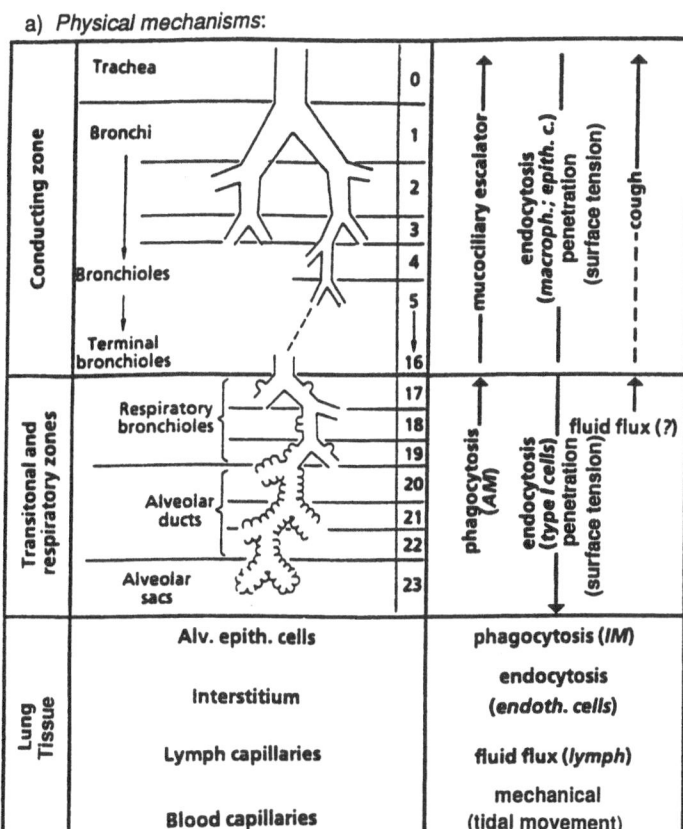

b) *Chemical mechanisms*: **Dissolution**

Figure 4: Clearance Mechanism for Fibers in the Lower Respiratory Tract.

translocation of non-phagocytized fibers by facilitating their contact with epithelial cells and their subsequent movement into the interstitium.

Normal functioning of the mucociliary escalator is an obvious prerequisite for the physical removal of fibers from conducting airways. However, humans are often exposed to pollutants that may damage this mechanism, for example, total mucociliary clearance rates in smokers are significantly lower, even if they are symptom-free, and become increasingly impaired as bronchitis develops (Vastag *et al.*,1986, figure 6). This decreased clearance of fibers will increase the possibility of fiber contact with epithelial cells of the conducting airways, thereby increasing the dose of fibers to target cells in this region. The well known synergism between smoking and exposure to asbestos fibers on the induction of bronchial carcinoma in humans (Hammond *et al.*, 1979) may be, partly or entirely, based on this dosimetric mechanism (Oberdörster, 1989). In contrast, animals exposed solely to fibers in a long-term inhalation study will not necessarily develop an impairment of mucociliary clearance; this may be another important difference to consider when extrapolating from such studies to humans.

With impaired mucociliary clearance both long and short fibers will have a prolonged retention time in the conducting airways; should we therefore consider the importance of short fibers in carcinogenesis? The widely held view that fibers shorter than

TABLE 1. Net Surface Charge/Superoxide Anion-Oxidant Production by Macrophages *in vitro*

Fiber-Particle Type (Net charge)	O_2^-/Oxidant Production	Macrophage Type	Investigators
Chrysotile A&B (+)	Yes	GP AM	
Crocidolite (-)	Yes	GP AM	
Anthophyllite (-)	Yes	GP AM	Hatch et al., 1980
Fiberglass (-)	No	GP AM	
Glass beads (-)	No	GP AM	
Crocidolite 9 (-)	Yes	M PM	Goodglick & Kane, 1986
Chrysotile (+)	Yes	GP AM	
Anthophyllite (-)	No	GP AM	Roney & Holian, 1986
Crocidolite (-)	No	GP AM	
Amosite (-)	No	GP AM	
Crocidolite (-)	Yes	H, R AM	Hansen & Mossman, 1987
Latex Microspheres	No	GP AM	
Latex Microspheres (-)	No	GP AM	
Latex Microspheres (+)	No	GP AM	
Talc (-)	No	GP AM	Scheule & Holian, 1989
Metallic Aluminium (+)	No	GP AM	
Silica (-)	No	GP AM	
Chrysotile (+)	Yes[*]	GP AM	[*]IgG-Mediated Enhancement
Chrysotile (+)	Yes	R AM	Petruska et al., 1990
Crocidolite (-)	Yes	R AM	

AM: alveolar macrophage; GP: guinea pig; H: hamster; R: rat; M PM: mouse peritoneal macrophages

5 μm only have a low or no carcinogenic potential is based on results from *in vitro* and *in vivo* studies (Stanton et al., 1981; Pott, 1978). If, in the *in vivo* studies, normal clearance prevented the shorter fibers from being retained and becoming biologically active then their possible tumorigenic potential may not have been detectable. Indeed, recent experiments by Kane and coworkers (1990) indicated that clogging of lymphatic clearance pathways in the peritoneal cavity of rats did lead to the same inflammatory and carcinogenic response from both short and long crocidolite fibers. If the importance of the clearance function for the pathogenesis of asbestos induced carcinoma by short fibers can be confirmed, then proposed standards defining a biologically active fiber based on a certain length, diameter, and aspect ratio may have to be reconsidered (Kane, 1991).

CELLULAR MECHANISMS

As indicated in figure 1, a central role in the pathogenesis of fiber induced effects is the activation of AM with the ensuing release of mediators. These mediators can induce a variety of cellular responses, such as damage to the epithelial layer (oxidants, proteases), stimulation of epithelial and interstitial cells (growth factors) and recruitment of additional inflammatory cells (chemotactic factors). The physicochemical features relating to influence on the functional status of AM remain unclear. Table 1 shows data regarding the potential role the net surface charge or geometry of a particle may play in stimulating the production of oxidant radicals by AM. These show no definitive relationships and experimental results from different laboratories often conflict, even when AM from the same animal species and, presumably, the same type of particle preparations, were examined for an oxidant response.

It is now well recognized that particle stimulated AM are capable of producing a wide variety of both pro-mitogenic and anti-mitogenic factors that can influence the proliferative activities of a wide range of cell types. Some cytokines serve to both inhibit and stimulate the proliferation of target cells as a function of their relative concentrations and the co-presence of other cytokine types, *inter alia* (tables 2 and 3). Under healthy, homeostatic conditions where normal tissue structure and function are maintained, it is reasonable to assume that the constitutive production and relative abundance of the various cytokines in the lung are well balanced and finely tuned. Yet it is clear that a broad range of cellular responses may occur under conditions of abnormal cytokine production, *e.g.*, fibroblast proliferation. Such responses, given the diversity of actions of many of the cytokines *in vivo*, are currently difficult to completely predict.

For example, particle-associated stimulation of tumor necrosis factor-α (TNF-α) production by AM would be expected to favor the development of fibrosis and perhaps

TABLE 2. Macrophage-Derived Promitogenic Factors*

Cytokines	Target Cells
Macrophage-Derived Pneumocyte Growth Factor (MPGF)	Type II pneumocytes
Transforming Growth Factor-a (TFG-a)	Epithelial cells, fibroblasts, lymphocytes
Basic Fibroblast Growth Factor (bFGF)	Epithelial cells, endothelial cells
Platelet-Derived Growth Factor (PDGF)	Fibroblasts
Tumor Necrosis Factor-a (TNF-a)	Fibroblasts, epithelial cells
Interleukin-1 (IL-1)	Fibroblasts, hemopoietic origin cells
Colony Stimulating Factor for Granulocytes and Macrophages (CSF-G/M)	Macrophages and granulocytes
Colony Stimulating Factor for Granulocytes (CSF-G)	Hematopoietic origin cells
Interferons (IFN)	Lymphocytes, fibroblasts, epithelial cells
Insulin-like Growth Factors (IGFs)	Broad spectrum
Fibronectin	Fibroblasts

*Adapted with modification from Kelly, 1990.

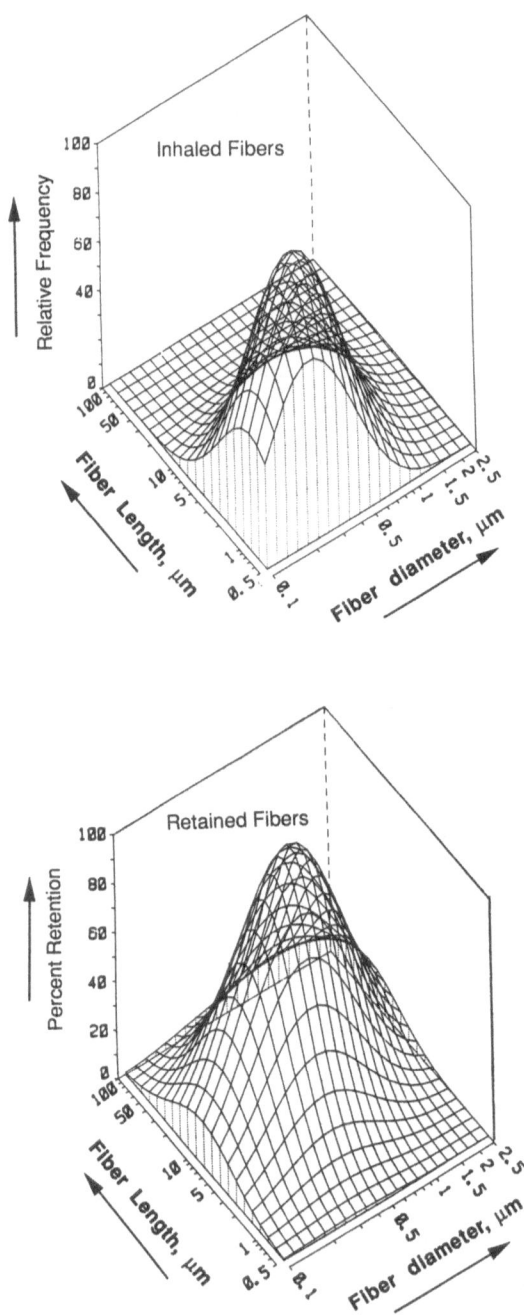

Figure 5: Clearance of anthophyllite Fibers
from the Human Lung (Timbrell, 1982).

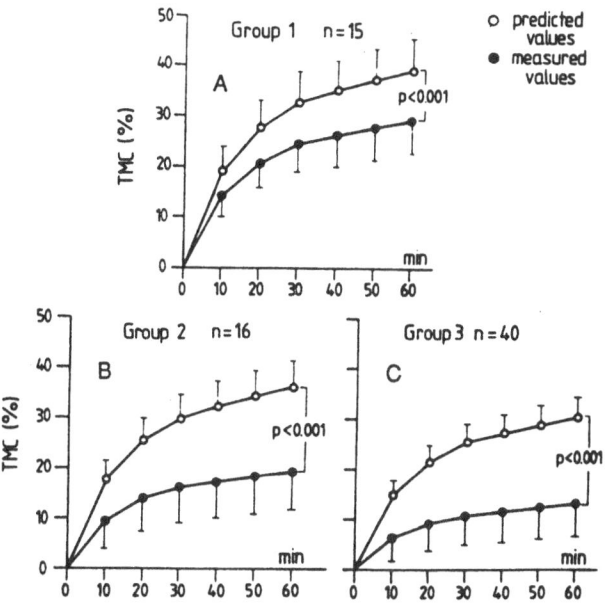

Figure 6: Total Mucociliary Clearance Rates in Smokers. A: symptom free, B: Simple chronic bronchitis, C: Obstructive chronic bronchitis (Vastag *et al.*, 1986).

epithelial cell hyperplasia (table 2), whereas this same cytokine may serve to protect against lung cancer (table 3). This might have important pathogenetic consequences for adverse pulmonary effects related to high inhaled concentrations of particles. For example, the incidence of lung cancers in coal workers with pneumoconiosis is not significantly elevated (Merchant *et al.*, 1986) and may even be lower than in those not suffer-

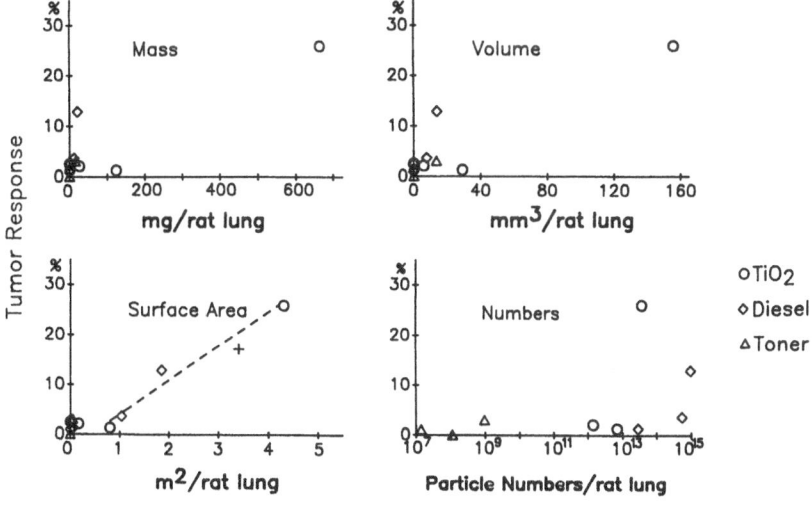

Figure 7: Parameters of Retained Particles in the Rat Lung and Tumor Induction: Long-Term Inhalation Studies in Rats with Different Particulate Material. + indicates study with carbon black reported by Heinrich (1990) (Oberdörster & Yu, 1991).

TABLE 3. Macrophage-Derived Anti-mitogenic Factors*

Cytokine	Target Cells
Basic Fibroblast Growth	Cancer cells
Transforming Growth Factor-a (TGF-a)	Cancer cells
Transforming Growth Factor-b (TGF-b)	Fibroblasts, epithelial cells, lymphocytes
Interleukin-1 (IL-1)	Cancer cells
Tumor Necrosis Factor-a (TNF-a)	Cancer cells (tumoricidal)
Interferons (INF)	Fibroblasts

*Adapted with modification from Kelly, 1990.

TABLE 4. Examples of Particle/Fiber Effects on Cytokine Production

Particle/Fiber Type or Condition	Effect	Investigators
Chrysotile B (Rat)	Increases in a FGF by AM	Lemaire *et al.*, 1986
Canadian Chrysotile (Sheep)	Increased lavagable fibronectin	Begin *et al.*, 1986
Silica (Rats)	Increased IL-1 by AM	Struhar *et al.*, 1989
Silica (Human) Diamond Dust	Increased IL-1 by monocytes No increase in IL-1	Schmidt et al., 1984
Chrysotile A (Rat) Silica Latex Microspheres	Increased TNF by AM Increased TNF by AM No increase in TNF	Dubois *et al.*, 1989
Silica (Rat) Titanium Dioxide	Increased IL-1, TNF by AM No increase in IL-1, TNF increases transiently	Driscoll *et al.*, 1990
Asbestosis Silicosis Coal Workers' Pneumoconiosis	In all conditions: Increased fibronectin and AMDGF (IGF-1) by AM	Rom *et al.*, 1987
Coal Workers' Pneumoconiosis	Increased TNF by AM	Lassalle *et al.*, 1990
Coal Workers' Pneumoconiosis	Increased TNF by blood monocytes	Borm *et al.*, 1988

ing from this disease (Parkes, 1982). This is despite the fact that pneumoconiosis might be associated with pulmonary particle overload and this overload has also been associated with increased incidences of tumors in laboratory rats (Lee *et al.*, 1986; Mauderly *et al.*, 1987). Conceivably, the enhanced production of the TNF-α cytokine that occurs in coal workers' pneumoconiosis, table 4, may bring about the demise of emerging transformed lung cells.

Although information about the abilities of some types of particles to stimulate the elaboration of cytokines by AM continues to mount (table 4), specific information about how a particle's physical features, *e.g*, size, geometry, and surface characteristics, affect the production of cytokines has not been systematically obtained. Information about the cytokines that may be produced by AM in response to a given type of particulate material, and the kinetics and relative abundance of cytokine production by particle-stimulated AM, is currently nonexistent. Thus, while cytokine production by particle-stimulated AM undoubtedly figures prominently in mediating the pathological responses to fibers, specific details about the complex roles of the cytokines and the overall relationships of their production by AM in terms of the properties of inhaled particles remain to be elucidated.

One major important factor underlying the effects of cytokines is the functional integrity of the alveolar epithelial cell layer. The tight junctions of the alveolar epithelium normally prevent the rapid diffusion of larger-sized proteins (MW > 10,000) across this layer. The junctions of damaged epithelial cells, however, become leaky and proteins, including cytokines, can readily penetrate and subsequently interact with interstitial cells such as fibroblasts. For example, Mangum and coworkers (1991) reported recently that PDGF could permeate type II cells in mono-culture layers only after the cells had been damaged which has obvious implications for the situation *in vivo*.

A second mechanism with regard to the effects of cytokines relates to their stimulatory effect on cell proliferation. Injured type I epithelial cells are readily replaced by proliferating type II cells and this evidently is supported by growth factors released from activated AM (Brandes & Finkelstein, 1990). An increased proliferation rate implies an increasing probabliity that transformed cells will emerge (Moolgavkar & Knudson, 1981). The same mechanism may be responsible for the increased lung tumor incidence observed in long-term rat inhalation studies after the animals had been exposed to high concentrations of different types of particles, including asbestos fibers (Lee *et al.*, 1986; Mauderly *et al.*, 1987; Davis, 1991). Tumors induced in such studies are typically in the periphery of the lung, often apparently originating from type II cells. This may be an important difference with respect to tumors observed in humans after exposure to asbestos since the human tumors are more likely to be of bronchogenic origin. However, peripheral lung tumors (adenocarcinoma) are also found in asbestos-exposed humans (Suzuki & Selikoff, 1986).

Another aspect of the significance of the alveolar epithelial layer relates to the interstitial passage of fibers across the epithelium (figure 1). The importance of this event leading to a direct interaction of particles with interstitial cells, such as the interstitial macrophage (IM), has been discussed by several authors (Adamson *et al.*, 1989; Bowden *et al.*, 1989; Oberdörster *et al.*, 1990) and may be the main mechanism underlying particle-induced fibrosis. For example, long-term inhalation studies with high concentrations of relatively benign dusts could induce both fibrosis as well as lung tumors in rats (Lee *et al.,*, 1986; Mauderly *et al.*, 1987). These effects are thought to be non-specific and are referred to as being "lung overload" related (Morrow, 1988). One hallmark of such overload is a severe retardation of AM-mediated particle clearance with subsequent accumulation of particles in the alveolar space and increased penetration and sequestration of the particles in the interstitium. This interstitialization of particles is possibly a key event leading to the interaction of particles with IM followed by the release of cytokines, which in turn can stimulate other cells, such as fibroblasts. The relevance for such events for certain types of spherical particles and also for fibers is shown by examples of particle/fiber effects on cytokine production by macrophages, as indicated in table 4.

PARTICLE PROPERTIES

The foregoing raises the question as to which particle parameters can be correlated with the production of pro- and anti-mitogenic cytokines. We need to establish a correlation between pathological endpoints such as carcinoma or fibrosis and the particle parameters. This in turn may indicate the involvement of specific cytokines in fiber-induced pathogenicity.

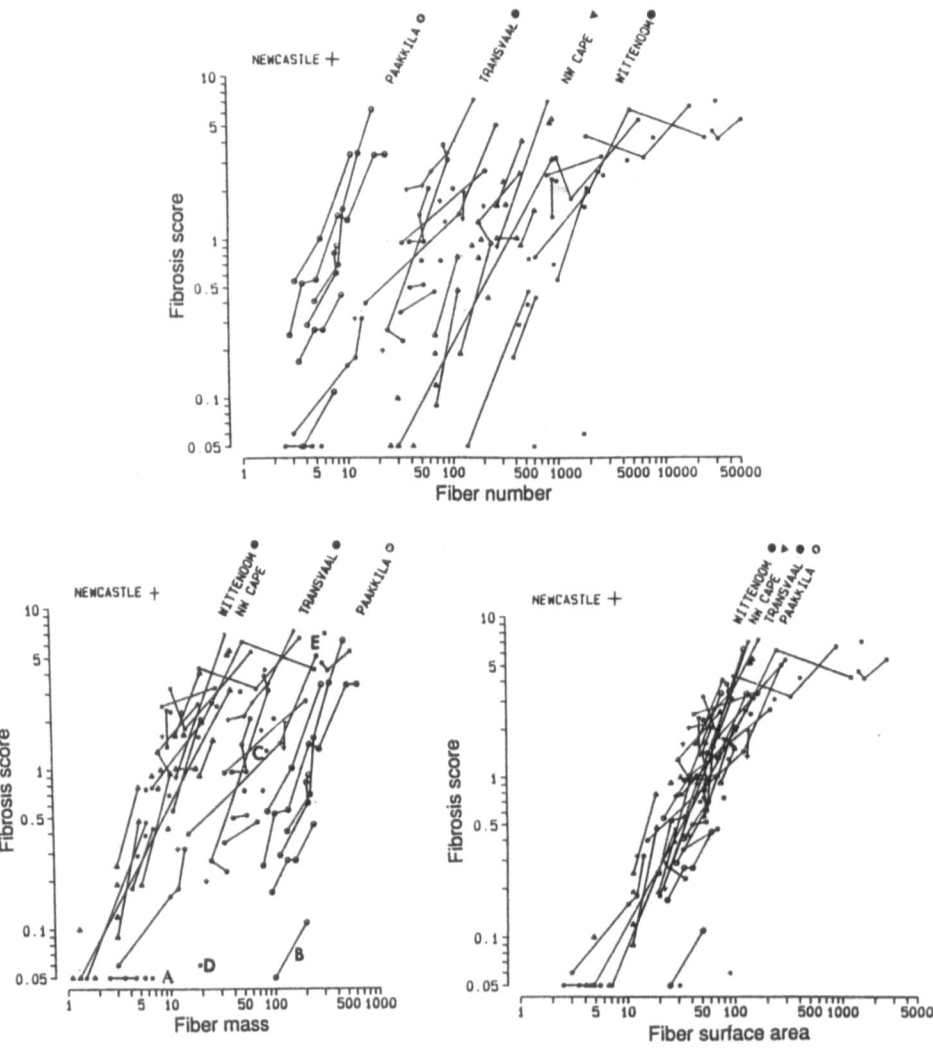

Figure 8: Characteristics of Retained Fibers in Human Lung and Degree of Lung Fibrosis (Timbrell, 1988).

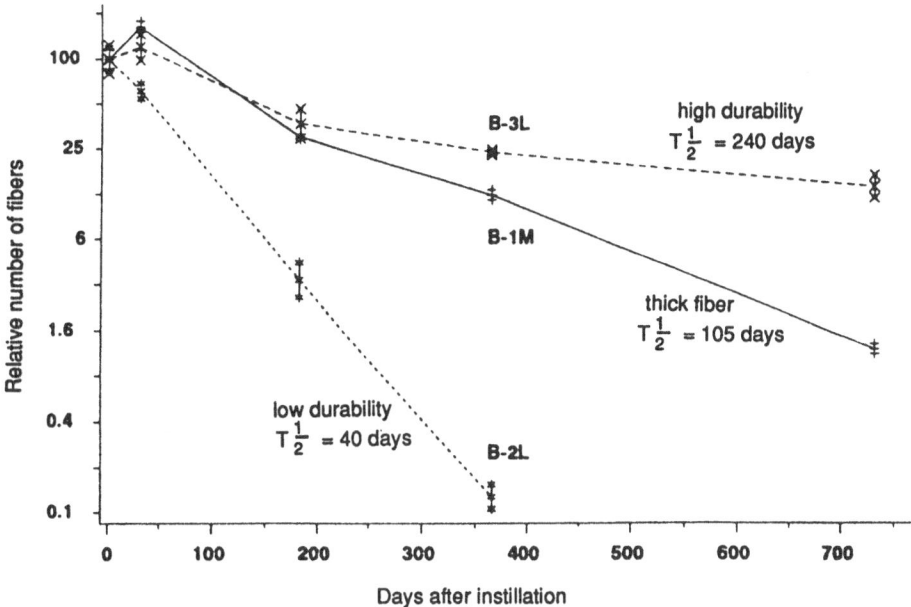

Figure 9: Pulmonary Clearance of Glass Fibers in Rat Lung (Bellmann *et al.*, 1990).

Surface area

The carcinogenic effect observed in chronic high exposure inhalation studies with "nuisance" particles in rats, for example, has been associated with a "particle overload" effect (see above). In these "particle overload" studies only the surface area of the retained materials showed good correlation with the carcinogenic response (Oberdörster & Yu, 1990) (figure 7). This in spite of the fact that the retained particles were of markedly dissimilar types, *i.e.*, TiO_2, diesel, toner. In contrast to the surface area, expressing the retained particle load as mass, volume or numbers of particles did not reveal a correlation with the lung tumor response. This suggests that a "critical" surface area of the retained particle mass may exist below which no significant tumor response is to be expected. The relationship between surface area of retained particles and the biological response was strengthened recently by Heinrich (1991) who reported that a one-year inhalation exposure of rats to 6 mg/m³ carbon black resulted in a lung tumor induction in 17% of the animals. The surface area of the retained mass of carbon black, which was equivalent to 3.5 m², and the tumor response fits well into the plot in figure 7.

It is surprising that the surface area of the various particulate materials used in the studies correlated so well with the carcinogenic response. This might be coincidental, but one can hypothesize that it is the surface area of the particles that interacts with AM receptors. In turn this may induce AM to elaborate mediators that favor the proliferation/transformation of sensitive target cells. The release of these mediators and their interactions with specific target cells would then be proportional to the particulate surface area in this hypothetical scenario. Whether a similar correlation exists for fibrous particles has not been investigated. Very likely, other properties of fibers in addition to their surface area may play a role. The relationship between surface area and tumor response described above pertains only to particles of relatively benign dusts. Toxic particles, like SiO_2, would not fit into this scheme since Muhle and coworkers (1989) found a significant lung tumor response in rats at much lower lung burdens of SiO_2 than would be predicted on the basis of surface area of a benign dust.

The surface area of fibers has also been found to correlate with a biological effect. Inhaled asbestos fibers of different types and origins produce a fibrotic response which correlates best with their surface area rather than with their mass or their numbers (Timbrell, 1988, figure 8). Surface characteristics, including the charge of asbestos fibers, were also found by other investigators to be of importance for biological effects (Klockars *et al.*, 1990).

Studies with 20 nm TiO_2 ultrafine particles have shown a highly increased pulmonary inflammatory reaction compared to bigger (~200 nm) "benign" TiO_2 particles with one eighth their specific surface area (Oberdörster *et al.*, 1991). Coffin and coworkers (1989) have speculated similarly that it may be the extremely large internal surface of erionite fibers as compared to crocidolite that makes the erionite such a potent inducer of mesothelioma. On the other hand, Wagner (1988) could not induce mesothelioma in rats with erionite if the fiber length was less than 5 μm. While we can only speculate on underlying basic molecular mechanisms, it appears to be useful to focus future research on the significance of surface characteristics of fibers with respect to cytokine release and the subsequent induction of adverse effects.

Fibre Durability

Another important aspect in fiber toxicology is the durability of fibers in the target tissue (figure 1). It is generally accepted that a fiber of low durability is less potent at inducing a carcinogenic response because during the long latency period required from exposure to the manifestation of tumors a less durable fiber will be dissolved. For example, Bellman and coworkers (1990) determined that a specially manufactured glass fiber of low durability (high solubility in tissue) was cleared from the lungs of rats with a biological half-time of 40 days. In contrast, they found that glass fibers of high durability (low solubility in tissue), but otherwise of the same diameter and length, were retained in rat lungs with a 6-fold longer biological half-time of 240 days (figure 9). When these

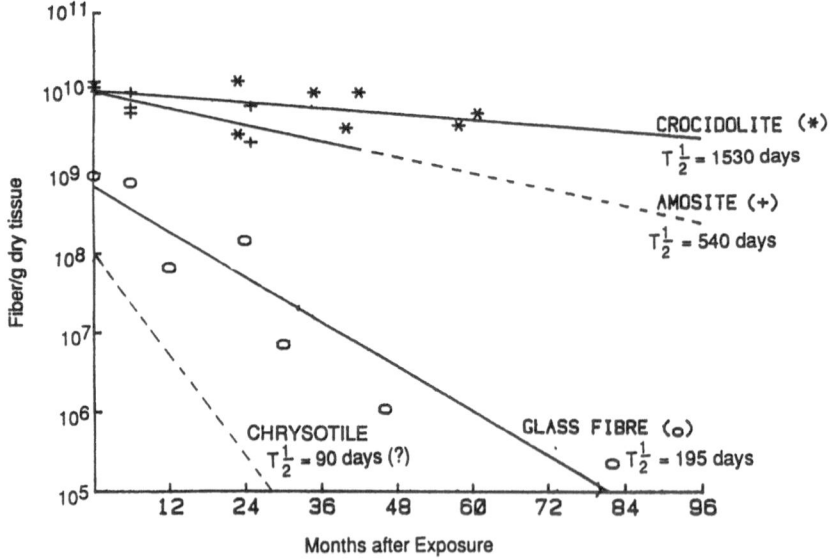

Figure 10: Pulmonary Clearance of Different Inhaled Fibers in Baboons (after Rendall, 1988).

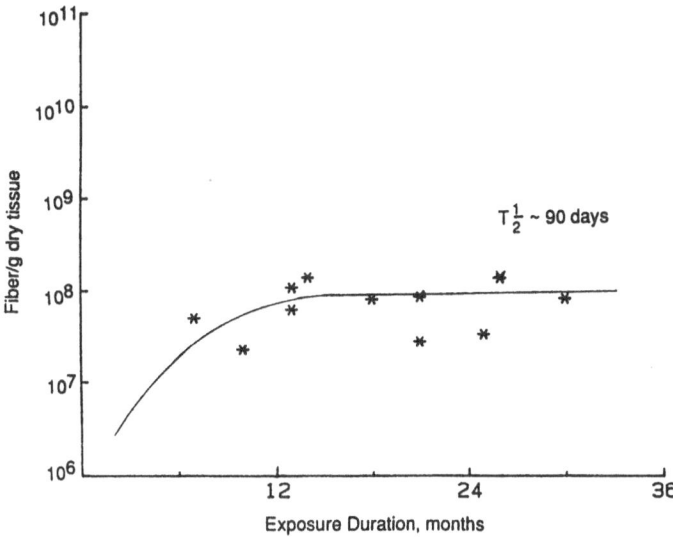

Figure 11: Accumulation of Inhaled Chrysotile in Lungs of Baboons During Continuous Exposure (after Rendall, 1988).

fibers were administered in an i.p. carcinogenicity test, only the highly durable glass fiber type induced tumors in the abdominal cavity. No tumors were induced by the more soluble glass fibers (Pott *et al.*, 1991). Morgan and coworkers (1982) reported that glass fibers > 10 μm long were dissolved faster in lung tissue *in vivo* than glass fibers < 10 μm long. They attributed this to the fact that the shorter fibers were phagocytized by AM and therefore dissolved less rapidly than the longer fibers in lung tissue since the solubility of glass increases with pH.

Although readily dissolved fibers have only a short retention time, it can be argued that constant replenishment during a chronic inhalation exposure, will lead to a steady state dose of fibers in the respiratory tract and, therefore, may eventually also induce adverse responses. This could be especially true for effects in the conducting airways since some fibers are directly deposited in this region or are translocated there during clearance from the alveolar region. Contact with epithelial target cells in the conducting airways could occur shortly after inhalation while that with target cells in the pleural space (mesothelioma) requires a possible lengthy translocation phase. Thus, it remains conceivable that even a low durability fiber may cause effects at epithelial sites of the respiratory tract but not at pleural sites. Possibly, chrysotile fibers might fit into this category since they are known to induce lung cancer and fibrosis in exposed humans but are believed to be of relatively low durability in tissue and have a very low potency for inducing mesothelioma in humans. Those cases where mesothelioma have occurred in chrysotile-exposed workers may be attributable to contamination of chrysotile with more durable fibers, like tremolite (Churg *et al.*, 1984; Weill *et al.*, 1990).

The low durability of chrysotile fibers has been inferred from studies in which the pulmonary retention has been estimated from the analysis of tissue samples by light and electron microscopy. For example, Rendall and coworkers (1988) determined in baboons the pulmonary retention of crocidolite, amosite, chrysotile and of a glass fiber after exposure to these different fibers for about 3 years. They found pulmonary retention half-times of 1530 and 540 days for the durable asbestos fibers, crocidolite and amosite respectively (figure 10). The glass fiber was cleared more rapidly with a retention half-time of 195 days. They were unable to determine the half-time for chrysotile since all of the chrysotile-exposed monkeys died during the exposure period. However, based on the data they did obtain about the accumulation of chrysotile fibers in the lung during the exposure (figure 11), it can be estimated that steady state conditions of fiber numbers were reached within 12 months of exposure. Their data is consistent with a very short biological half-time of about 90 days (figure 11) provided that their method for determining chrysotile fibers in lung tissue was sensitive enough to detect the fibers equally after the different times of exposure. Chrysotile fibers tend to split longitudinally into ever finer fibrils, which can affect their reliable detection and the accurate assessment of dose as well as their durability or solubility in tissues. A final judgment on the durability of chrysotile awaits further study, even though several authors refer to chrysotile as being of low durability (Davis, 1984; James *et al.*, 1989). Additionally, the surface area concept is potentially important for chrysotile because of its propensity to split longitudinally, thereby increasing the fiber surface area available for cellular interaction.

Even if fibers of low durability under chronic exposure conditions still have a tumorigenic potential, this potential has to be much lower than that of a durable fiber. Therefore, one characteristic which should be emphasized in the search for a new fibrous material as replacement for asbestos should be its low durability in tissue. Muhle and coworkers (1991), for example, reported that wollastonite appears to be highly soluble in the rat lung while Pott and coworkers (1989) found that its tumorigenic potential after i.p. injection was very low. Further studies are needed to confirm these findings.

Once fibers have reached the pulmonary interstitium they may be cleared via lymphatic channels to regional lymph nodes and to the pleura (figure 1). This pathway has been found to be limited to fibers up to about 16 μm in length and a small fraction of fibers may even appear in the postnodal lymph (Oberdörster *et al.*, 1988) and spread to extra-pulmonary body compartments via lympho-hematogenous translocation. Direct migration of interstitial fibers towards the pleura has also been suggested, supported possibly by the tidal movement of ventilation. Once the fibers have reached the pleura they may interact with pleural macrophages (PLM, figure 1) (Lehnert *et al.*, 1988). Interaction of fibers with pleural mesothelial cells may also stimulate these cells, which in turn then release chemotactic factors to attract fibroblasts (Kagan *et al.*, 1990). This finding leads to an attractive hypothesis about a correlation between initial fibrotic and subsequent tumorigenic effects. Stimulation of fibroblasts in turn could not only result in increased collagen production but also cause the release of growth factors which stimulate other target cells, *e.g.*, epithelial cells (Everett *et al.*, 1990).

TUMORIGENESIS AND EXPERIMENTAL ASSESSMENT

Some investigators (Kuschner, 1984; Kagan, 1985) have suggested that the production of collagen could be a pre-condition for fiber-induced tumors. Others do not hold this view (Pott, 1991). While it may be conceivable that local attraction and stimulation of fibroblasts occurs at the site of subsequent tumor induction, a causal relationship between these two events has yet to be established. At least some lung tumor cell types produce factors that stimulate fibroblast proliferation and extracellular matrix deposition (see Pelton & Moses, 1990). Thus, it appears possible that fibrosis associated with the *in situ* appearance of lung tumors may in some cases be a result of local factors elaborated by the neoplastic cells, as opposed to the emergence of transformed cells in pre-existing fibrotic sites. Other mechanistic events possibly involved in fiber-induced transformation of cells include DNA damage by oxygen radicals, a direct effect of fibers on chromosomes and dividing cells (Jaurand *et al.*, 1991) and mutational events and genetic damage induced by fibers (Hei *et al.*, 1991).

The events outlined in the preceding paragraphs may also be involved in the induction of bronchogenic carcinoma, *i.e.*, interaction of epithelial cells of the conducting airways with fibers (figure 1). Existing inflammatory conditions may also contribute to the tumorigenic response. Fiber durability may be of less importance for effects in the conducting airways, compared to the pleural space. Yet the "minimum" durability of a fiber to induce bronchial carcinoma is not known. Any impairment of the mucociliary escalator would increase the dose of fibers in the bronchial region. Long-term inhalation experiments with fibers in healthy laboratory animals may not show bronchogenic tumors because the animals are selected to be healthy and do not have an impaired clearance in this region. Rather, tumors observed in such animal studies are of peripheral origin, in some cases due to a high fiber load in the alveolar region, *e.g.*, 5-6 mg of amosite in the rat lung (Davis, 1991).

The question arises from this discussion whether the issue of central bronchial carcinoma (humans) *vs* peripheral lung tumors (rats) is really a matter of impaired bronchial clearance (bronchitis) *vs* impaired alveolar clearance (overload)? If indeed differences in the tumorigenicity of fibers between humans and rats can be based on such secondary events, the next question is deciding what is a relevant animal model for demonstrating and detecting the carcinogenic potential of a fibrous particulate material? Inhalation studies in rats may not be unequivocally recommended as the ultimate test methodology. Differences in inhalability and respirability between rats and humans exist, as discussed above, and negative outcomes with respect to tumor induction in chronic animal inhalation studies have been reported (Davis *et al.*, 1978; Wagner *et al.*, 1985, 1987; Muhle *et al.*, 1987). Therefore, such inhalation studies need to be carefully designed (Berstein *et al.*, 1991). Perhaps an animal model with impaired mucociliary clearance may be better suited for studying a bronchial carcinogenic potential. Studying all newly developed fibers for such a potential would be prohibitively expensive if they were subjected to a chronic inhalation test. Should intratracheal instillation or injection studies be used instead? Possibly, a tiered system starting with other tests (injection and instillation) would be better suited to the screening of unknown fibers. Therefore, the initial question should not be whether to estimate a tumorigenic risk by animal inhalation or injection/instillation studies, but rather, which of the available simpler tests could be used for screening.

Intrapleural and intraperitoneal injection and intratracheal instillation should be considered: each method has advantages and disadvantages. For example, erionite is known to induce pleural mesothelioma in humans and experimental animals by inhalation, yet 500 μg erionite instilled intratracheally at 15 weekly intervals into rats did not show such a tumor response (Pott, personal communication), whereas i.p. injected erionite gave a positive tumor response (Pott *et al.*, 1987). Likewise, a recent long-term inhalation study in hamsters and rats with refractory ceramic fibers (Bernstein *et al.*, 1991; Hesterberg *et al.*, 1991) clearly revealed their tumorigenic potency with respect to pleural mesothelioma, whereas intratracheal instillations of these fibers by the same investigators did not show this response. Intraperitoneal injection, on the other hand again showed a positive outcome. I.p. injected fibers can immediately interact with serosal target cells. The long translocation period needed to reach serosal cells in the pleural space is excluded. Therefore, the durability of a fibrous material would have to be tested if it shows a positive response in the i.p. test and one is concerned about induction of mesothelioma. A significant tumor response in the i.p. test may also be viewed as a predictor for lung carcinogenicity in general; such a view may be acceptable if it can be shown that there is a positive correlation between these two endpoints. Such a correlation has been suggested for asbestos fibers, (Pott *et al.*, 1990).

The advantages and disadvantages of each of these tests, including chronic inhalation, have to be carefully considered when making a decision as to their use. The recent study with refractory ceramic fibers mentioned in the previous paragraph showed a significant response of pleural mesothelioma after inhalation in hamsters and in rats, and abdominal tumors after intraperitoneal injection in both species, whereas intratracheal instillation (three times within one week) did not induce pleural mesothelioma. This result may favor the use of the intraperitoneal test as a predictive tool, however, questions of the dose in either test are of prime importance and have to be carefully evaluated. The intraperitoneal test will reveal the potential of a fiber to induce a tumor response primarily at mesothelial sites, provided a non-specific effect caused by a mechanism related to the aforementioned overload effect of particles in the lung can be excluded. While the i.p. test will show the potential of a fiber to cause these effects *in vivo*, its potential to do so after inhalation *in vivo* depends very much on the ability of a fiber to reach mesothelial sites, which includes most of all the durability of a fiber in the pulmonary tissue. The complex *in vivo* behaviour of inhaled fibers, involving dosimetry, translocation, durability and molecular biology after cellular interaction together determine the potency of a fiber to induce adverse effects including local inflammation, fibrosis, immune reactions and tumors.

ACKNOWLEDGEMENT

This paper is based partly on work supported by NIH grants ESO4872 and ESO1247.

REFERENCES

Adamson,I.Y.R., Letourneau,H.L. and Bowden,D.H. (1989) Enhanced macrophage-fibroblast interactions in the pulmonary interstitium increases fibrosis after silica injection to monocyte-depleted mice. *Am. J. Pathology* **134**, No. 2:411-418.

Asgharian,B. and Yu,C.P. (1988) Deposition of inhaled fibrous particles in the human lung. *J. Aerosol Med.* 1:37-50.

Begin,R., Bisson,G., Lambert,R., Cote,Y., Fabi,D., Martel,M., Lamoureux,G., Rola-Pleszczynski,M., Boctor,M., Dalle,D. and Masse,S. (1986) Gallium-67 uptake in the lung of asbestos exposed sheep: Early association with enhanced macrophage-derived fibronectin accumulation. *J. Nucl. Med.* 27:538-544.

Bellmann,B., Muhle,H. and Pott,F. (1990) Study on the durability of chemically different glass fibers in lungs of rats. *Zbl. Hyg.* **190**:310-314.

Bernstein,D.M., Fleissner,H., Bouvier,C., Hesterberg,T. and Mast,R. (1991) An inhalation model for evaluation of fiber oncogenicity in rodents. In, "Mechanisms in Fibre Carcinogenesis", eds. Brown,R.C., Hoskins,J.A. and Johnson,N.F., NATO ASI Series, Plenum Press, New York, (this volume).

Borm,P.J.A., Meijers,J.M.M. and Swaen,G.M.H. (1990) Molecular epidemiology of coal worker's pneumoconiosis: Application to risk assessment of oxidant and monokine generation by mineral dusts. *Exp. Lung Res.* **16**:57-71.

Bowden,D.H., Hedgecock,C. and Adamson,I.Y.R. (1989) Silica-induced pulmonary fibrosis involves the reaction of particles with interstitial rather than alveolar macrophages. *J. Pathology* **158**:73-80.

Brandes,M.E. and Finkelstein,J.N. (1990) The production of alveolar macrophage-derived growth-regulating proteins in response to lung injury. *Toxicology Letters* **54**:3-22.

Churg,A., Wiggs,B., Depaoli,L., Stevens,B. and Kampe,B. (1984) Lung asbestos content in chrysotile workers with mesothelioma. *Am. Rev. Respir. Dis.* **130**:1042-1045.

Coffin,D.L., Peters,S.E., Palekar,L. D. and Stahel,E.P. (1989) A study of the biological activity of erionite in relation to its chemical and structural characteristics. In, "Biological Interaction of Inhaled Fibers and Cigarette Smoke", ed. Wehner,A., Columbus, OH, Battelle Press, pp. 313-323.

Davis,J. (1984) The pathology of asbestos related disease. *Thorax* **39**:801-808.

Davis,J.M.G. (1991) The carcinogenicity of mineral fibres in experimental animals: An overview. In, "Mechanisms in Fibre Carcinogenesis", eds. Brown,R.C., Hoskins,J.A. and Johnson,N.F., NATO ASI Series, Plenum Press, New York, (this volume).

Davis,J.M.G., Addison,J., Bolton,R.E., Donaldson,K. and Jones,A.D. (1978) Inhalation and injection studies in rats using dust samples from chrysotile asbestos prepared by a wet dispersion process. *Br. J. Exp. Pathol.* **67**:113-129.

Drew,R.T., Kuschner,M. and Bernstein,D.M. (1987) The chronic effects of exposure of rats to sized glass fibres. *Ann. Occup. Hyg.* **31** (4B):711-729.

Driscoll,K.E., Lindenschmidt,R.C., Maurer,J.K., Higgins,J.M. and Ridder,G. (1990) Pulmonary response to silica or titanium dioxide: Inflammatory cells, alveolar macrophage-derived cytokines and histopathology. *Am. J. Respir. Cell Mol. Biol.* **2**:381-390.

Dubois,C.M., Bissonnette,E. and Rola-Pleszczynski,M. (1989) Asbestos fibers and silica particles stimulate rat alveolar macrophages to release tumor necrosis factor: An autoregulatory role of leukotriene B_4. *Am. Rev. Respir. Dis.* **139**:1257-1264.

Everett,M.M., King,R.J., Jones,M.B. and Martin,H.M. (1990) Lung fibroblasts from animals breathing 100-percent oxygen produce growth factors for alveolar type-II cells. *Am. J. Physiol.* **259** (4):L247-L254.

Gehr,P., Schürch,S., Berthiaume,Y., Im Hof,V. and Geiser,M. (1990) Particle retention in airways by surfactant. *J. Aerosol Med.* **3**:27-44.

Geiser,M., Im Hof,V., Gehr,P. and Cruz-Orive,L.M. (1988) Histological and stereological analysis of particle deposition in the conductive airways of hamsters. *J. Aerosol Med.* **1**:197-198.

Goodglick,L.A. and Kane,A.B. (1986) Role of reactive oxygen metabolites in crocidolite asbestos toxicity to mouse macrophages. *Cancer Res.* **46**:5558-5566.

Hammond,E.C., Selikoff,I.J. and Seidman,H. (1979) Asbestos exposure, smoking and death rates. *Ann. NY Acad. Sci. USA* **330**:473-492.

Hansen,K. and Mossman,B.T. (1987) Generation of superoxide (O_2^-) from alveolar macrophages exposed to asbestiform and nonfibrous particles. *Cancer Res.* **47**:1681-1686.

Hatch,G.E., Gardner,D.E. and Menzel,D.B. (1980) Stimulation of oxidant production in alveolar macrophages by pollutant and latex particles. *Environ. Res.* **23**:121-136.

Hei,T.K., He,Z.Y., Piao,C.Q., Waldren,C.A. and Hall,E.J. (1991) The mutagenicity of asbestos fibers: Cellular and molecular characterization. In, "Mechanisms in Fibre

Carcinogenesis", eds. Brown,R.C., Hoskins,J.A. and Johnson,N.F., NATO ASI Series, Plenum Press, New York, (this volume).

Heinrich,U. (1991) Presentation at Workshop on Health Assessment Document for Diesel Emissions, Research Triangle Park, July, 1990.

Hesterberg,T.W., Mast,R., McConnell,E.E., Chevalier,J., Bernstein,D.M., Bunn,W.B. and Anderson,R. (1991) Chronic inhalation toxicity of refractory ceramic fibers in Syrian hamsters. In, "Mechanisms in Fibre Carcinogenesis", eds. Brown,R.C., Hoskins,J.A. and Johnson,N.F., NATO ASI Series, Plenum Press, New York, (this volume).

Jaurand,M.C., Endo Capron,S., Levy,F., Renier,A., Saint-Etienne,L. and Khenang,L. (1991) Genotoxic effects of fibres. In, "Mechanisms in Fibre Carcinogenesis", eds. Brown,R.C., Hoskins,J.A. and Johnson,N.F., NATO ASI Series, Plenum Press, New York, (this volume).

Jones,A.D., Vincent,J.H., McIntosh,C., McMillan,C.H., Addison,J. (1989) The effect of fibre durability on the hazard potential of inhaled chrysotile asbestos fibres. *Exper. Pathol.* **37**:98-102.

Kagan,E. (1985) Current perspectives in asbestosis. *Ann. Allergy* **54**:465-473.

Kagan,E., Kuwahara,M., Gersten,D. and Diglio,C. (1990) Asbestos induced pleural fibrosis: Role of a mesothelial cell-derived chemoattractant for lung fibroblasts. Presented at Sixth International Colloquium on Pulmonary Fibrosis, Stowe, VT, October, 1990.

Kane,A.B. (1991) Fiber dimensions and mesothelioma: A reappraisal of the Stanton hypothesis. In, "Mechanisms in Fibre Carcinogenesis", eds. Brown,R.C., Hoskins,J.A. and Johnson,N.F., NATO ASI Series, Plenum Press, New York, (this volume).

Kelly,J. (1990) State of the Art: Cytokines of the lung. *Am. Rev. Respir. Dis.* **141**:765-788.

Klockars,M., Hedenborg,M. and Vanhala,E. (1990) Effect of two particle surface-modifying agents, polyvinylpyridine-N-oxide and carboxymethylcellulose, on the quartz and asbestos mineral fiber-induced production of reactive oxygen metabolites by human polymorphonuclear leucocytes. *Arch. Env. Health* **45**:8-14.

Kuschner,M. (1984) Peer review: Pathogenicity of MMMF in contrast to natural fibres. In, "Biological Effects of Man made Mineral Fibres". Proceedings of a WHO/IARC Conference, Vol. 2. WHO Regional Office for Europe, pp. 367-369.

Lassalle,P., Gosset,P., Aerts,C., Fournier,E., Lafitte,J.J., Degreef,J.M., Wallaert,B., Tonnel,A.B. and Voisin,C. (1990) Abnormal secretion of interleukin-1 and tumor necrosis factor α by alveolar macrophages in coal worker's pneumoconiosis: Comparison between simple pneumonconiosis and progressive massive fibrosis. *Exp. Lung Res.* **16**:73-80.

Lee,K.P., Henry,N.W. III, Trochimowicz,H.J. and Reinhardt,C.F. (1986) Pulmonary response to impaired lung clearance in rats following excessive TiO_2 dust deposition. *Environ. Res.* **41**:144-167.

Lehnert,B.E., Valdez,Y.E., Sebring,R.J., Lehnert,N.M., Saunders,G.C. and Steinkamp,J.A. (1990) Airway intra-luminal macrophages: Evidence of origin and comparisons to alveolar macrophages. *Am. J. Respir. Cell Mol. Biol.* **3**:377-391.

Lehnert,B.E., Dethloff,L.A. and Valdez,Y.E. (1988) Leukocytic responses to the intrapleural deposition of particles, particle-cell association and the clearance of particles from the pleural space compartment. *J. Toxicol. Environ. Hlth* **24**:41-66.

Lemaire,I., Beaudoin,H., Masse,S. and Grondin,C. (1986a) Alveolar macrophage stimulation of lung fibroblast growth in asbestos-induced pulmonary fibrosis. *Am. J. Pathol.* **122**:205-211.

Lemaire,I., Beaudoin,H. and Dubois,C. (1986b) Cytokine regulation of lung fibroblast proliferation: Pulmonary and systemic changes in asbestos-induced pulmonary fibrosis. *Am. Rev. Respir. Dis.* **134**:653-658.

Mangum,J.B., Everitt,J.E., Bonner,J.C., Moore,L.R. and Brody,A.R. (1991) Co-culture of primary pulmonary cells to model alveolar injury and translocation of proteins. *In vitro: Cellular and Developmental Biology*, in press.

Mauderly,J.L., Jones,R.K., Griffith,W.C., Henderson,R.F. and McClellan,R.O. (1987) Diesel exhaust is a pulmonary carcinogen in rats exposed chronically by inhalation. *Fund. Appl. Tox.* **9**:208-221.

Merchant,J.A., Taylor,G. and Hodous,T.K. (1986) Coal-workers pneumoconiosis and exposure to other carbonaceous dusts. In, "Occupational Respiratory Diseases", ed. Merchant,J.A. DHHS (NIOSH), Wash., D.C., Publ. No. 86-102, pp. 329-384.

Moolgavkar,S.H. and Knudson,A.G. (1981) Mutation and Cancer: A model for human carcinogenesis. *J. Natl. Cancer Inst.* **66**(6):1037-1052.

Morgan,A., Holmes,A. and Davison,W. (1982) Clearance of sized glass fibres from the rat lung and their solubility *in vivo*. *Ann. Occup. Hyg.* **25**(3):317-331.

Morgan,A., Talbot,R.J. and Holmes,A. (1978) Significance of fibre length in the clearance of asbestos fibres from the lung. *Brit. J. Ind. Med.* **35**:146-153.

Morrow,P.E. (1988) Possible mechanisms to explain dust overloading of the lungs. *Fund. Appl. Tox.* **10**:369-384.

Muhle,H., Bellmann,B. and Pott,F. (1991) Durability of various mineral fibers in rat lungs. In, "Mechanisms in Fibre Carcinogenesis", eds. Brown,R.C., Hoskins,J.A. and Johnson,N.F., NATO ASI Series, Plenum Press, New York, (this volume).

Muhle,H., Mermelstein,R., Bellmann,B. and Morrow,P.E. (1989) Lung tumor induction upon long-term low level inhalation of crystalline silica. *Am. J.Indust. Med.* **15**:343-366.

Muhle,H., Pott,F., Bellmann,B., Takenaka,S. and Ziem,U. (1987) Inhalation and injection experiments in rats to test the carcinogenicity of MMMF. *Ann. Occup. Hyg.* **31** (4B):755-764.

Oberdörster,G. (1989) Combined effects of tobacco smoke and asbestos fibers in the lung: Synergism or increased dose to target sites? In, "Biological Interaction of Inhaled Mineral Fibers and Cigarette Smoke", ed. Wehner,A.P. Columbus, OH, Battelle Press, pp. 195-209.

Oberdörster,G., Ferin,J., Gelein,R., Soderholm,S.C. and Finkelstein,J. (1991) Role of the alveolar macrophage in lung injury: Studies with ultrafine particles.*Exp. Hlth Perspect.*, (in press).

Oberdörster,G. and Yu,C.P. (1990) The carcinogenic potential of inhaled diesel exhaust: A particle effect? *J. Aerosol Sci.*, **21**:S397-S401.

Oberdörster,G., Morrow,P.E. and Spurny,K. (1988) Size dependent lymphatic short-term clearance of amosite fibers in the lung. *Ann. Occup. Hyg.* **32**, Suppl. 1:149-156.

Parkes,W.R. (1982) Pneumoconiosis due to coal and carbon. In, "Occupational Lung Disorders", Chapter 8: Second Ed., Butterworths, London.

Pelton,R.W. and Moses,H.L. (1990) The b-type transforming growth factor: Mediators of cell regulation in the lung. *Am. Rev. Resp. Dis.* **142**:S31-S35.

Petruska,J.M., Marsh,J., Bergeron,M. and Mossman,B.T. (1990) Brief inhalation of asbestos compromises superoxide production in cells from bronchoalveolar lavage fluid. *Am. J. Respir. Cell Mol. Biol.* **2**:129-136.

Pott,F. (1978) Some aspects on the dosimetry of the carcinogenic potency of asbestos and other fibrous dusts. *Staub-Reinhalt Luft* **38**, Nr. 12:486-489.

Pott,F. (1991) Neoplastic findings in experimental asbestos studies and conclusions to fiber carcinogenesis in humans. *Ann. NY Acad. Sci.*, (in press).

Pott,F., Roller,M., Ziem,U., Reiffer,F.J., Bellmann,B., Rosenbruch,M. and Huth,F. (1989) Carcinogenicity studies on natural and man-made fibres with the intraperitoneal test in rats. In, "Non-occupational Exposure to Mineral Fibres", eds. Bignon,J., Peto,J and Saracci,R. IARC, Lyon, (IARC Scientific Publication No. 90). pp. 173-179.

Pott,F., Schlipköter,H.W., Roller,M., Rippe,R.M., Germann,P.G., Mohr,U. and Bellmann,B. (1990) Carcinogenicity of glass fibres with different durability. *Zbl. Hyg.* **189**:563-566.

Pott,F., Roller,M., Rippe,R.M., Germann,P.G. and Bellmann,B. (1991) Tumours by the intra-peritoneal and intrapleural routes and their significance for the classification of mineral fibers.In, "Mechanisms in Fibre Carcinogenesis", eds. Brown,R.C., Hoskins,J.A. and Johnson,N.F., NATO ASI Series, Plenum Press, New York, (this volume).

Rendall,R.E. (1988) The retention and clearance of inhaled glass fibres and different varieties of asbestos by the lung. Dissertation, Faculty of Medicine, University of Witwatersrand, Johannesburg, South Africa.

Rom,W.N., Bitterman,P.B., Rennard,S.I., Cantin,A. and Crystal,R.G. (1987) Characterization of the lower respiratory tract inflammation of non-smoking individuals with interstitial lung disease associated with chronic inhalation of inorganic dusts. *Am. Rev. Respir. Dis.* **136**:1429-1434.

Roney,P.L. and Holian,A. (1989) Possible mechanism of chrysotile asbestos-stimulated superox-ide anion production in guinea pig alveolar macrophages. *Toxicol. Appl. Pharm.* **100**:132-144.

Scheule,R.K. and Holian,A. (1989) IgG specifically enhances chrysotile asbestos-stimulated superoxide anion production by the alveolar macrophages. *Am. J. Cell Mol. Biol.* **1**:313-318.

Schmidt,J.A., Oliver,C.N., Lepe-Zuniga,J.L., Green,I. and Gery,I. (1984) Silica-stimulated monocytes release fibroblast proliferation factors identical to Interleukin 1: A potential role for interleukin 1 in the pathogenesis of silicosis. *J. Clin. Invest.* **73**:1462-1472.

Schürch,S., Gehr,P., Im Hof,V., Geiser,M. and Green,F. (1990) Surfactant displaces particles toward the epithelium in airways and alveoli. *Respir. Physiol.* **80**:17-32.

Stanton,M.F., Layard,M., Tegeris,A., Miller,E., May,M., Morgan,E. and Smith,A. (1981) Relations of particle dimension to carcinogenicity in amphibole asbestoses and other fibrous minerals. *JNCI* **67**, No. 5:965-975.

Struhar,D.J., Harbeck,R.J., Kawada,H. and Mason,R.J. (1989) Increased expression of class II antigens of the major histocompatibility complex on alveolar macrophages and alveolar type II cells and interleukin-1 (IL-1) secretion from alveolar macrophages in an animal model of silicosis. *Clin. Exp. Immunol.* **77**:281-284.

Suzuki,Y. and Selikoff,I.J. (1986) Pathology of lung cancer among asbestos insulation workers. *Fed. Proc.* **45**:744.

Timbrell,V. (1982) Deposition and retention of fibres in the human lung. *Ann. Occup. Hyg.* **26**, No. 1-4:347-369.

Timbrell,V., Ashcroft,T., Goldstein,B., Heyworth,F., Meurman,L.O., Rendall,R.E.G., Reyn-olds,J.A., Shilkin,K.B. and Whitaker,D. (1988) Relationships between retained amphi-bole fibres and fibrosis in human lung tissue specimens. *Ann. Occup. Hyg.* **32**, Suppl. 1:323-340.

Vastag,E., Matthys,H., Zsamboki,G., Köhler,D. and Daikeler,G. (1986) Mucociliary clearance in smokers. *Eur. J. Resp. Dis.* **68**:107-113.

Wagner,J.C. (1988) Significance of the fibre size of erionite. Presented at VIIth International Pneumoconiosis Conference, Pittsburgh, PA (August, 1988).

Wagner,J.C., Griffiths,D.M. and Munday,D.E. (1987) Experimental studies with palygorskite dusts. *Br. J. Indust. Med.* **44**:749-753.

Wagner,J.C., Skidmore,J.W., Hill,R.J. and Griffiths,D.M. (1985) Erionite exposure and mesotheliomas in rats. *Br. J. Cancer* **51**: 727-7430.

Weill,H., Abraham,J.L., Balmes,J.R., Case,B., Churg,A., Hughes,J., Schenker,M. and Sebastien,P. (1990) Health Effects of Tremolite (Official Statement of the American Thoracic Society). *Am. Resp. Dis.* **142**:1453-1458.

DURABILITY OF VARIOUS MINERAL FIBRES IN RAT LUNGS

H. Muhle, B. Bellmann, *F. Pott

Fraunhofer Institute of Toxicology and Aerosol Research
D-3000 Hannover, F.R.G.

* Medical Institute for Environmental Hygiene at the Heinrich Heine
University, D-4000 Dusseldorf, F.R.G.

ABSTRACT

The durability of crocidolite, glass fibres, and wollastonite in rat lungs was
examined after intratracheal instillation. Experiments were based on the assumption that
thin ($< 2 \mu$m), long ($> 5 \mu$m) and durable fibres are of special importance for the carcinogenic potency of these materials. After serial sacrifices up to two years after administration of the test materials, the retained fibres in lungs were analysed by scanning or
transmission electron microscopy. Retention half times of between 10 and more than 500
days were calculated for the various fibre types investigated.

The results show that the chemical composition of mineral fibres is an essential
determinant of the durability of these materials in lungs. A correlation between the durability of fibres in lungs and results for carcinogenic potency in the intraperitoneal test
(i.p.) shows the plausibility of an interrelationship.

INTRODUCTION

The persistence of mineral fibres in susceptible tissues like the bronchial epithelium or the serosa is thought to be of principal importance for fibre-induced tumours
(Davis, 1986; Pott *et al.*, 1989). Persistence is defined as the long-term residence of
fibres at the same location and persistent fibres have to be durable in biological tissue.
The persistence of fibres is terminated by physical migration, dissolution or disintegration. Physical migration itself is influenced by the shape of the fibres and the lung burden.

For long fibres ($> 5 \ \mu$m) physical transport of entire particles is very low (Morgan, 1980). Furthermore, after inhalation exposure of insoluble particles leading to a retained mass of about 1 mg per rat lung, a considerable prolongation of the alveolar retention has been observed (Vincent *et al.*, 1985; Muhle *et al.*, 1990).

The disappearance of fibres from the lung is influenced by various parameters. We have tried to establish experimental conditions which enable conclusions on the durability of fibres to be drawn. For a comparative investigation of the durability of various fibres, it is desirable to obtain fibre samples with a substantial fraction of long fibres ($> 5 \ \mu$m) to eliminate or minimise the impact of the physical migration *e.g.* after uptake by macrophages. In addition, the animals were treated with similar fibre masses and partly similar size distributions.

This paper supplements prior publications (Bellmann *et al.*, 1987, 1990).

The supposed easier determination of the durability of fibres *in vitro* discloses a considerable weakness. Morgan and Holmes (1986) stated that it is very difficult to reproduce the intra- and extra-cellular environments of the lung *in vitro* and therefore, extrapolation of *in vitro* solubility to persistence in animals or humans should be treated with caution until reliable analogous data are available from both systems.

MATERIALS AND METHODS

Fibres

The fibres investigated are listed in table 1. The chemical composition of the glass fibres is presented in table 2.

The glass fibres and wollastonite were pretreated by milling to shorten the fibres. For characterisation of the samples used, fibres were suspended in water, briefly sonicated (< 1 min) and filtered onto a Nucleopore filter. Fibre number and size distribution were determined from TEM or SEM photos by means of a graphics tablet and computer supported data analysis. Magnification was chosen so that both the longest fibres and the thinnest fibres could be measured sufficiently precisely. For most fibre types the range of these dimensions was so wide that two or even three different magnifications had to be used to satisfy these requirements (see table 3). The same size of object field was analysed for each magnification. To avoid double counting, different fibre length limits were set for the counts at each magnification. The size distribution approximated to log-normal for all fibre types and was classified by the limits for length and diameter of 10, 50 and 90% of the fibre number.

The results of the fibre characterisation are shown in table 3. The mass fraction of critical fibres and the number of critical fibres per mass are presented according to a definition of fibre length L $> 5 \ \mu$m, fibre diameter D $< 2 \ \mu$m and a ratio of L/D > 5. To demonstrate the significance of the fibre diameter, in addition both parameters are listed according to a fibre definition of L $> 5 \ \mu$m, D $< 5 \ \mu$m and L/D >5. For the analysis of the fibre sizes the only criterion used was a ratio of L/D > 5 to present the overall spectrum of fibres.

Design of animal experiment

Two milligrams of fibres suspended in 0.3 ml of 0.9% NaCl (0.4 ml for glass fibres) solution were instilled intratracheally into 200-220 g female Wistar rats (Central Institute for Laboratory Animal Breeding, Hannover). At the dates specified in table 4,

TABLE 1. Fibrous materials used, their origin and pretreatment

Fibrous material	Origin/Producer	Pretreatment
Crocidolite	UICC	
Glass fibre B1	Bayer AG	Milling:120 min knife mill 3 min ball mill
Glass fibre B2	Bayer AG	Milling:60 min knife mill 0.5 min ball mill
Glass fibre B3	Bayer AG	Milling:60 min knife mill 0.7 min ball mill
Wollastonite 1	India[a]	Milling [b]
Wollastonite 2	Eternit, Kapelle	Milling by a knife mill Sizing by sedimentation in water [b]
Wollastonite 3 NYAD G	NYCO	Milling:15 min knife mill

[a] Distributed by Osthoff-Petrosch K.G, Hamburg, FRG, Type D-1
[b] Preparation by Dr. K. Spurny

three to six rats of each group were sacrificed. The lungs were subjected to low-temperature ashing, the ash was suspended in acetic acid or water (for wollastonite) and filtered on Nucleopore filter (pore size = 0.2 μm, 0.4 μm for glass fibres). The lungs of each animal were analysed as a separate sample. As indicated in Table 4, for fibres with short residence times only samples of the earlier sacrifice dates were analysed, whereas for persistent fibres samples up to two years after treatment were investigated. The total number of fibres per lung was determined from the weight of the sample used and the filter area investigated. For calculation of the clearance kinetics of the fibres, an exponential curve was fitted to data from individual animals.

RESULTS

In figure 1 the relative number of fibres in the lung ash of the different sacrifice dates is shown. The fibre number 1 or 2 days after treatment was used as 100% and only data from those fibres longer than 5 μm are presented.

The calculated half-times are shown in table 4. For up to two years almost no change in crocidolite fibre numbers was found. For the glass fibre B2, which contains a high calcium oxide content and which is thinner than the glass fibre B1, a half-time of only 38 days was observed. After 1 year, only 0.1% of those fibres which were found 1 day after treatment were still in the lungs. The thicker, but chemically identical, glass fibre B1 showed a half-time of 107 days. Glass fibre B3, which had a similar size distribution but different composition to fibre B2, had a half-life of 238 days. All three wollastonite samples investigated showed very short half-lives of about 10 days.

TABLE 2. Chemical composition of fibres examined (%)

Fibrous material	SiO_2	B_2O_3	Na_2O	Al_2O_3	$FeO+Fe_2O_3$	BaO	ZnO	K_2O	CaO	MgO	Others
Glass fibre 1	60.7	3.3	15.4	-	0.2	-	-	0.7	16.5	3.2	
Glass fibre 2	60.7	3.3	15.4	-	0.2	-	-	0.7	16.5	3.2	
Glass fibre 3	58.5	11.0	9.8	5.8	0.1	5.0	3.9	2.9	3.0	-	
Wollastonite [a]	49-55	-	-	0.1-0.7	0.1-0.6	-	-	-	42-48	0.1-0.8	<1.5

[a] From IARC (1986)

TABLE 3. Characterization of initial fibrous materials

Fibrous material	Mass fraction of critical fibres		Number of critical fibres/ng		Fibre length[c] (μm)			Fibre diameter[c] (μm)			Instrument & mag.
	a	b	a	b	10% <	50% <	90% <	10% <	50% <	90% <	
Crocidolite	30.4	30.4	46	46	0.5	1.2	3.1	0.06	0.13	0.59	TEM 6400x
Glass fibre B1	14.9	70.1	2.7	6.7	6.7	10.7	21.1	0.88	1.68	2.49	SEM 500x-1500x
Glass fibre B2	86.5	89.9	58	58	2.3	6.0	14.7	0.26	0.51	0.88	SEM 500x-1500x
Glass fibre B3	92.0	97.0	23	23	1.6	5.6	20.7	0.15	0.34	0.82	SEM 500x-1500x
Wollastonite 1	10.2	81.6	3.1	6.1	4.2	8.4	18.4	0.64	1.14	2.29	SEM 1000x
Wollastonite 2	3.5	14.9	1.5	1.9	3.1	5.6	13.2	0.39	0.71	1.35	SEM 500x-2000x
Wollastonite 3	3.4	28.1	0.7	1.2	2.1	5.4	22.2	0.32	0.72	2.34	SEM 200x-2000x

a L > 5 μm, D < 2 μm, L/D > 5/1
b L > 5 μm, D < 5 μm, L/D > 3/1
c L/D > 5/1

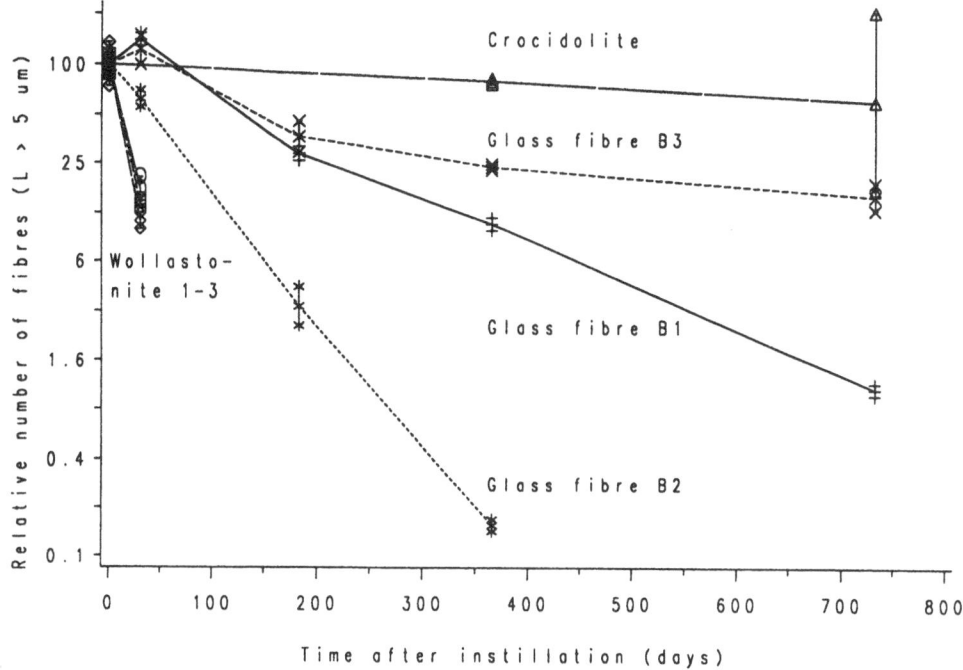

Figure 1. Decrease of number of fibres in the lung ash after intratracheal instillation (Fibre definition: length > 5 μm, diameter < 2 μm, aspect ratio > 5:1)

TABLE 4. Half-time and 95 % confidence limit of lung clearance of fibres
(based on fibre numbers L > 5 μm)

Fibrous material	Half-time (95% CL) (days)	Sacrifice dates used for calculation (days)	Carcinogenic in intraperitoneal test	
			Dose (10^9 fibres)	Result
Crocidolite	1000 (215->1000)	1, 365,730	0.05	+
Glass fibre B1	107 (98-119)	1, 182, 365, 730	0.5	−
Glass fibre B2	38 (35-41)	1, 182, 365	1	−
Glass fibre B3	238 (183-340)	1, 182, 365, 730	0.15	+
Wollastonite 1	11 (9-13)	2, 30	0.3	−
Wollastonite 2	10 (7-15)	2, 30	0.05	−
Wollastonite 3	12 (9-18)	2, 30	[a]	

[a] not investigated

DISCUSSION

With regard to the potential carcinogenicity of long fibres the fibre fraction longer than 5 μm was selected for the evaluation of the fibre clearance from lungs. For the calculation of the mass fraction of critical fibres used for the treatment, a diameter of fibres < 2 μm was applied. One of the reasons for this selection is that only these fibres can reach the deep lungs of the experimental animals by the inhalation route.

Figure 1 and table 4 demonstrate considerable differences between the fibre types with regard to their persistence in lungs. Crocidolite fibres, which were used in the experiment as a prototype of a very durable fibre, showed the expected results. The size distribution of the glass fibres B2 and B3 were relatively similar; therefore the pronounced difference in the calculated half-lives in lungs is due to solubility differences. The longer half-life of glass fibre B1 compared to fibre B2, which are chemically identical, is obviously influenced by the larger diameter of glass fibre B 1.

All three wollastonite fibre types used were fairly soluble probably because of the very high content of calcium. These data support our previous statement that a high calcium content promotes the solubility in the rat lung (Bellmann et al., 1987).

In the intraperitoneal test only crocidolite and glass fibre B3 were considered as carcinogenic (Pott et al., 1987, 1990; Rittinghausen et al., this volume). This coincides very nicely with the long persistence of these two fibre types in lungs. These data support the hypothesis that long, thin and durable fibres are capable of inducing tumours.

REFERENCES

Bellmann,B., Muhle,H., Pott,F., König,H., Klöppel,H. and Spurny,K. (1987) Persistence of man-made mineral fibres (MMMF) and asbestos in rat lungs. *Ann. Occup. Hyg.* **31**:693-709.

Bellmann,B., Muhle,H. and Pott,F. (1990) Study on the durability of chemically different glass fibres in lungs of rats. *Zbl. Hyg.* **190**:310-314.

Davis,J.M.G. (1986) A review of experimental evidence for the carcinogenicity of man-made vitreous fibres. *Scand. J. Work Environ. Hlth* **12**:Suppl. 1, 12-17.

IARC (1987) Silica and some silicates. IARC monographs on the evaluation of the carcinogenic risk of chemicals to humans, Vol. **42**, . World Health Organisation International Agency on Cancer, Lyon.

Morgan,A. (1980) Effect of length on the clearance of fibres from the lung and on body formation. In, "Biological effects of mineral fibres", ed. Wagner,J.C. IARC, Lyon, pp. 329-335.

Morgan,A., Holmes,A. (1986) Solubility of asbestos and man-made mineral fibres *in vitro* and *in vivo*: its significance in lung disease. *Environ. Res.* **39**:475-484.

Muhle,H., Bellmann,B., Creutzenberg,O., Fuhst,R., Kilpper,R., Koch,W., MacKenzie,J.C., Morrow,P., Mohr,U., Takenaka,S. and Mermelstein,R. (1990) Subchronic inhalation study of toner in rats. *Inhal. Toxicol.* **2**:341-359.

Pott,F., Ziem,U., Reiffer,F.-J., Huth,F., Ernst,H. and Mohr,U. (1987) Carcinogenicity studies on fibres, metal compounds, and some other dusts in rats. *Exp. Path. (Jena)* **32**:129-152.

Pott,F., Roller,M., Ziem,U., Reiffer,F.J., Bellmann,B., Rosenbruch,M. and Huth,F. (1989) Carcinogenicity studies on natural and man-made fibres with the intraperitoneal test in rats. In, "Non-occupational exposure to mineral fibres", eds. Bignon,J., Peto,J. and Saracci,R. IARC Science Publication No. 90. International Agency for Research on Cancer, Lyon, pp. 173-179.

Pott,F., Schlipköter,H.-W., Roller,M., Rippe,R.M., Germann,P.-G., Mohr,U. and Bellmann,B. (1990) Kanzerogenität von Glasfasern mit unterschiedlicher Beständigkeit. *Zbl. Hyg.* **189**:563-566.

Rittinghausen,S., Ernst,H., Muhle,H., Fuhst,R. and Mohr,U. (1990) Histopathological analysis of tumour types after intraperitoneal injection of mineral fibres in rats. This volume.

Vincent,J.H., Johnston,A.Y., Jones,A.D., Bolton,R.E. and Addison, J. (1985) Kinetics of deposition and clearance of inhaled mineral dusts during chronic exposure. *Br. J. Indust. Med.* **42**:707-715.

HEALTH-RELATED ASPECTS OF THE HEATING OF REFRACTORY CERAMIC FIBRE

J. Young, J.J. Laskowski, R. Acheson and S.D. Forder

Department of Applied Physics and Materials
Materials Research Institute
Sheffield City Polytechnic
Sheffield S1 1WB, U.K.

ABSTRACT

Concern has arisen that inhalation of cristobalite-containing dust, derived when aluminosilicate ceramic fibre devitrifies during high-temperature use, may pose a health hazard. This paper identifies the types of fibres concerned and critically reviews recent studies of the devitrification products, the occurrence and significance of devitrification in industrial hygiene terms and the results of initial *in vitro* studies of fibre toxicity. Necessary developments of these studies are argued and identified.

INTRODUCTION

Refractory ceramic fibres are used extensively in insulation, fire protection, refractory and engineering applications due to their advantages of lower thermal capacity, improved resistance to thermal shock, good overall high-temperature performance and cheaper installation and maintenance costs when compared with more traditional materials (Griffiths, 1986). Concern, though, has arisen that changes in their physical and chemical properties, occurring during these high-temperature applications, may pose a health hazard through the subsequent release and inhalation of cristobalite-containing dust (Gantner, 1986), Holroyd *et al.*,1988)).

This paper focusses on the occurrence of cristobalite in refractory ceramic fibres and the associated material and health questions. The main types of commercially-available, refractory ceramic fibre, and the phases, phase changes and microstructures observed when materials are used at high operating temperatures are identified. Results of

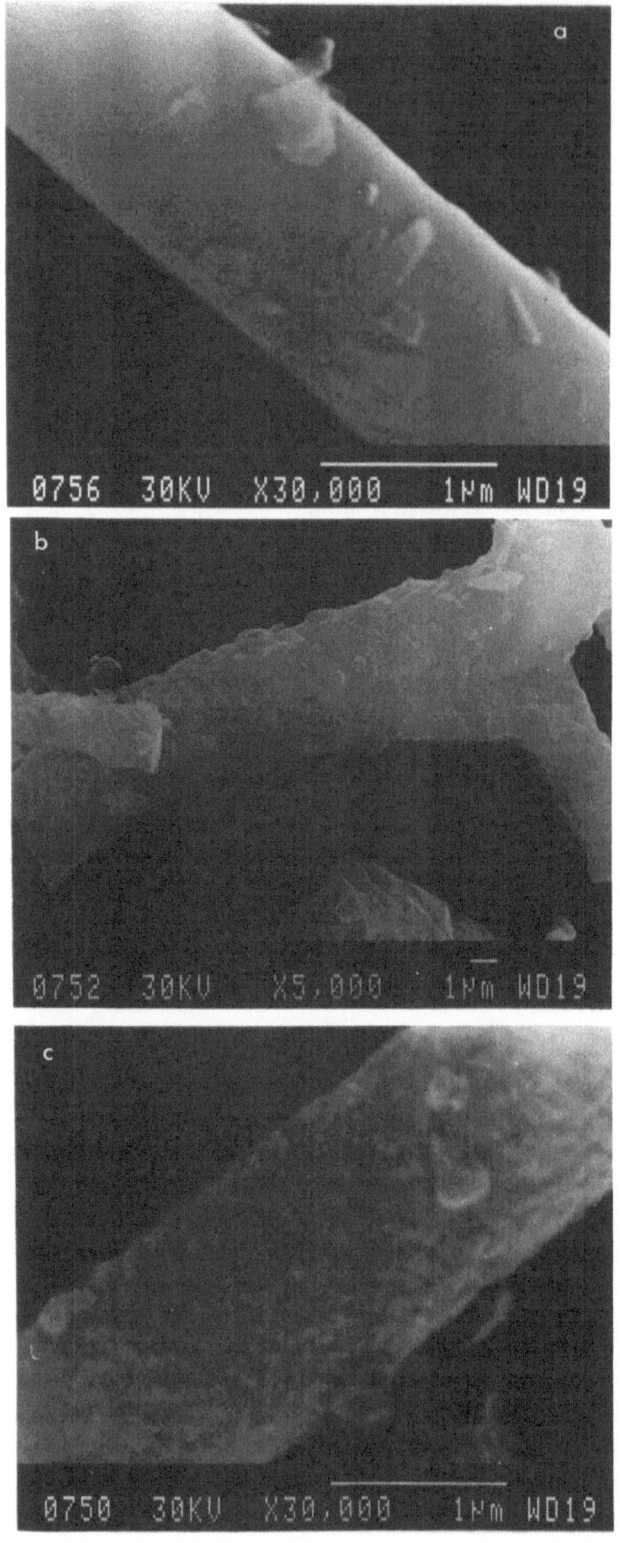

Figure 1. The gross effects of heating on the surface structure of fibres.
(a) unexposed fibre; (b) fibre exposed at 1200°C; (c) fibre exposed at 1400°C.

studies concerning the occurrence of cristobalite in furnace-exposed linings, of dust release from furnace-exposed materials, the industrial hygiene of furnace stripping operations and the *in vitro* toxicity testing of furnace-exposed materials are presented and critically discussed. Necessary further work is identified.

REFRACTORY CERAMIC FIBRES

The Materials

Amorphous, aluminosilicate, refractory ceramic fibre wools are normally manufactured by blowing or spinning high-temperature melts of alumina and silica. As shown by table 1, the "standard-grade" material has a classification, or indicative maximum service, temperature of about 1260°C whereas materials with increased alumina content or additional oxide components have higher classification temperatures. These types of fibre show a relatively wide distribution of fibre diameters with arithmetic-mean fibre diameters of typically 2 - 3 μm. Dependent on their intendent application they may be supplied with or without surface coatings, and as-blown, as opposed to spun, fibre contains a significant proportion of unfiberised material or shot which may be removed by washing (Manufacturer's data, Young, in press). These materials are used for high-temperature (>800°C) insulation, fire protection, refractory and engineering applications and bulk fibre is further processed to provide secondary products in blanket, board, module, textile, paper, coatable etc forms.

In addition to the aluminosilicate refractory ceramic fibres identified, a range of refractory ceramic fibres wools with higher classification temperatures are also available. These are produced by non-melt methods and include polycrystalline mullite fibre with a classification temperature of 1650°C and mean fibre diameter of 2 - 3.5 μm (Fibermax from Carborundum Resistant Materials Ltd), and microcrystalline, high-alumina fibre with a classification temperature of 1600°C and mean fibre diameter of 3 μm (Saffil from ICI Chemicals and Polymers Ltd). As these materials do not form cristobalite they will not be considered further.

These materials, their manufacture and application, are further reviewed by Dickson (1979) and Young (in Press).

Phase Changes and Devitrification

On heating aluminosilicate refractory ceramic fibre of approximately 45 mass-% alumina and 55 mass-% silica, the initially vitreous fibre shows a glass transition at 840 °C and then, on further heating, the formation of mullite at 980°C and the onset of the formation of cristobalite at about 1050°C. With further increases in temperature, the concentration of mullite remains essentially constant while the cristobalite concentration increases to a maximum about 1200°C and then reduces to zero by about 1400°C with its conversion to a viscous, silica-rich liquid phase (Belyakova *et al.*,1981; Hickling *et al.*., 1981;, Jager *et al.*, 1984; Holroyd *et al.*, 1988).

The gross effects of heating on the surface structure of these fibres are illustrated by figures 1a, 1b and 1c for unexposed fibre, fibre exposed at 1400°C and fibre exposed at 1400°C respectively. The unexposed fibre has the smooth surface expected of the original glassy material whereas the fibre exposed at 1200°C shows surface crystallisation, and the fibre exposed at 1400°C shows mullite crystallites "floating" in a sea of silica-rich supercooled liquid.

TABLE 1. Composition and classification temperatures of selected aluminosilicate fibre products

Grade of Product	Concentration / mass-%				Classification temperature °C
	Al_2O_3	SiO_2	Cr_2O_3	ZrO_2	
Standard[*]	43 - 47	53 - 57	-	-	1260
High-Duty[*]	52 - 56	44 - 48	-	-	1400
Chrome[*]	40 - 45	53 - 57	2 - 3	-	1450
Zirconia[*]	32 - 36	45 - 48	-	18.5 - 20.5	1425
R-bulk[**]	47	52.8	-	-	1260

[*] Morganite Thermal Ceramics Ltd.
[**] Carborundum Resistant Materials Ltd.

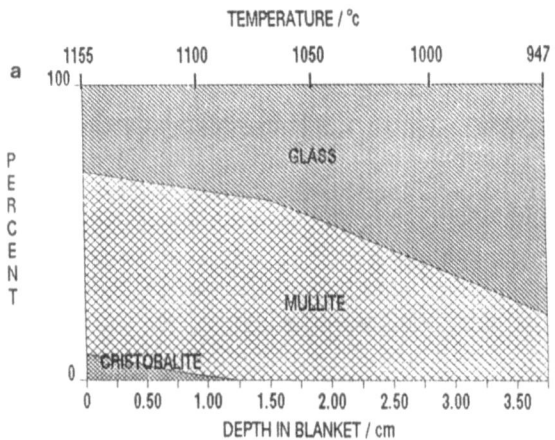

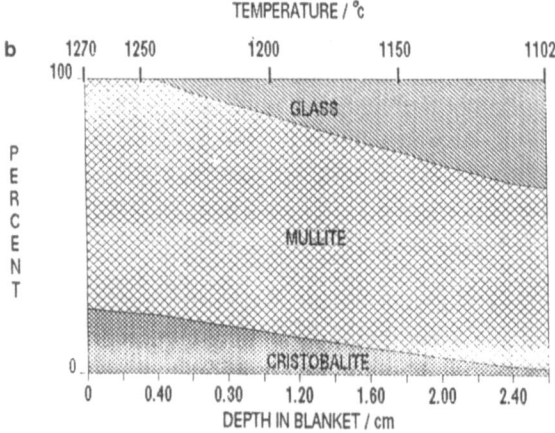

Figure 2. Depth profiles for cristobalite and mullite in furnace-exposed blankets. (a) standard grade; (b) high-duty grade.

With reference to blankets formed from this fibre, blanket shrinkage increases significantly with temperature above 1000°C, and above 1300°C the fibres sinter together to form a locked and brittle fibre structure (Hickling *et al.*, 1981). This shrinkage, and the sintering caused by the presence of the silica-rich liquid phase above 1300°C, has caused the fibre manufacturers to limit the classification temperature of "standard grade" material to 1260°C. Methods to increase this classification temperature, as indicated in table 1a, involve increasing the concentration of alumina with an associated reduction in the silica-rich liquid content or adding further oxides, eg chromia or zirconia, to stabilise the cristobalite phase to higher temperatures (Olds *et al.*, 1980).

The surface structure of devitrified fibres (figure 1b) has been identified (Gaodu *et al.*, 1977) as comprising mullite crystallites standing in relief due to rapid crystallisation and the shrinking back of the remaining vitreous silica-rich phase. This suggests that the cristobalite phase would subsequently form in the bulk of the fibre.

A question which concerns cristobalite is the non-formation of the low-temperature or α-form normally expected at room temperature (Young *et al.*, 1989). Analysis of X-ray diffraction traces indicates that the high-temperature or ß-form is observed at room temperature and, therefore, that the expected displacive transformation from high- to low-temperature forms has been prevented. Similarly, we report that the devitrification of "zirconia-grade" refractory ceramic fibre results in the formation at room temperature of the high-temperature, tetragonal form of zirconia rather than, as expected, the monoclinic room-temperature phase.

HEALTH-RELATED STUDIES OF FIBRE DEVITRIFICATION

Simulated Furnace Exposures

Holroyd and co-workers (1988) report depth profiles for the occurrence of cristobalite and mullite in furnace-exposed blankets of standard and high-duty grade refractory ceramic fibre. As indicated by figures 2a and 2b for furnace-exposed standard and high-duty grade blankets respectively, cristobalite formed in both types of blanket above a threshold temperature of about 1080°C. The same authors also reported the occurrence of cristobalite in an industrial, tube furnace lining. Rea and co-workers (1988) reported studies of dust release from furnace-exposed standard, high-duty and zirconia grade blankets. Samples were exposed for a range of temperatures and times, subjected to a controlled mechanical test and the resulting dust and blanket components were separated and weighed. Figures 3a, 3b and 3c, for standard, high-duty and zirconia grade materials respectively, indicate that all three materials show significant levels of dust release at all exposure temperatures and times investigated, and that there are significant increases in dust release with temperature at and above 1200°C. Released dust was analysed in terms of respirable fibre content and fibre dimensions and the results are summarised in table 2.

Industrial Hygiene Study

Gantner (1986) carried out industrial hygiene studies during the stripping of refractory ceramic fibre linings from experimental, heat-treatment furnaces. He found that cristobalite formed within the linings to a depth determined by a threshold temperature for cristobalite formation, he observed the release of significant levels of cristobalite-containing dust, and found, for personal and area samples and for respirable and total dust, that measured exposure levels were invariably in excess of calculated Threshold Level Values

TABLE 2. Summary data for released fibrous dust

Sample	Percentage Respirable Fibre	Geometric Mean Diameter (μm)	Geometric Mean Length (μm)
Standard grade 1270°C; 94 dy	86	1.0	11.6
High-duty grade 1400°C; 25 dy	86	1.3	14.2
Zirconia grade 1400°C; 25 dy	55	2.2	19.4

for cristobalite. He concluded, "there appears to be a significant exposure risk to cristobalite dust that is released during removal of ceramic fiber insulation that has been subjected to high temperatures."

In vitro Toxicity Testing

Initial *in vitro* studies of the toxicity towards V79-4 cells and macrophage-like cells of furnace-exposed standard, high-duty and zirconia grade fibres, suggest that the heat-treated, devitrified fibres containing cristobalite are less toxic towards the cells than the original, as-manufactured materials (R.C. Brown, private communication, 1989).

DISCUSSION

These studies have been pursued due to concern that cristobalite dust, formed when refractory ceramic fibre is used in high-temperature applications, may be released and inhaled with subsequent risk of silicosis.

Studies of these materials (section 2) indicate that cristobalite forms as a devitrification product when refractory ceramic fibres of alumina and silica composition are used in the temperature range from about 1050 to 1400°C. The quantity of cristobalite decreases as the alumina content increases to 72 mass-%, the composition of mullite. Addition of stabilising oxides, for example zirconia or chromia, stabilises the cristobalite to higher temperatures and prevents the formation of the silica-rich liquid phase. It is noted that cristobalite forms at service temperatures below the classification temperatures of these materials.

While the detailed bulk and surface microstructure of devitrified aluminosilicate fibres has not been fully determined, it has been suggested that mullite crystals form on the surface of the fibre enclosing the cristobalite. The mullite crystals may provide a protective layer which prevents the toxicity of the cristobalite towards cells.

The cristobalite product formed in the fibres does not transform to the low-temperature form when cooled to room temperature but rather the high-temperature form persists. While the non-formation of low-temperature cristobalite requires explanation, the question arises as to whether this high temperature form of cristobalite has a similar toxicity to cells as the low-temperature form. It is noted that the devitrification of zirco-

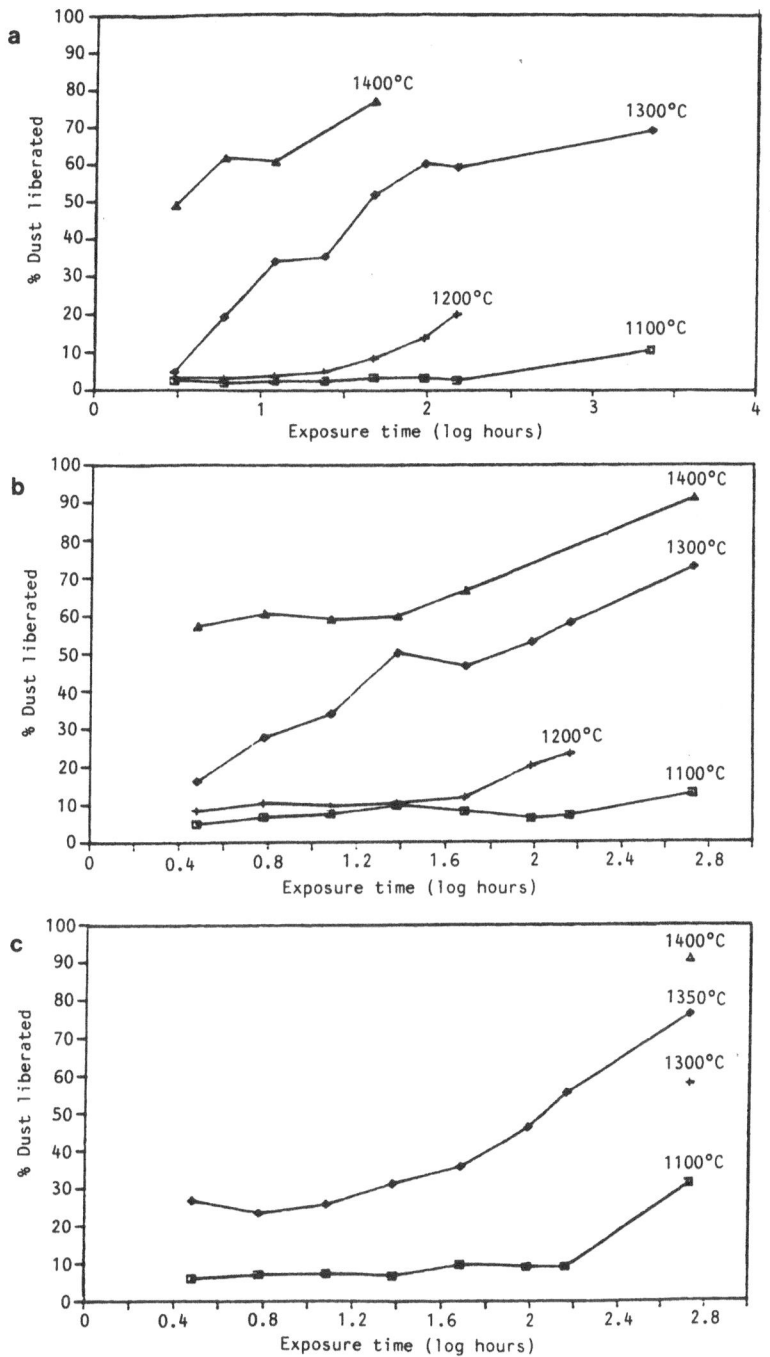

Figure 3. Dust release from furnace-exposed blankets. (a) standard grade; (b) high-duty grade; (c) zirconia grade.

nia-grade aluminosilicate fibre gives the tetragonal, high-temperature form of zirconia in addition to the high-temperature form of cristobalite.

The simulated furnace exposures, and the industrial hygiene studies of Gantner (1986), indicate that cristobalite forms in aluminosilicate fibrous materials when they are exposed in service to temperatures exceeding a threshold value of about 1050°C. The hygiene study and the dust release experiment indicated that furnace-lining stripping can provide a significant risk of exposure to cristobalite-containing dust.

The initial *in vitro* results, suggesting that devitrified, cristobalite-containing, aluminosilicate fibres are less toxic to cells than the original materials, are contrary to what was expected. While the aim is to repeat this experiment and to separately test the full range of materials concerned, potential explanations for a reduced toxicity could be the enclosure of cristobalite in a mullite surface coat or the production of high-temperature rather than low-temperature cristobalite.

The programme of work currently in progress aims to: (a) provide a more complete description of the regions of temperature, time and fibre composition corresponding to the formation of cristobalite devitrification product, (b) describe more clearly the detailed microstructure of devitrified fibres and explain the non-formation of low-temperature cristobalite, (c) undertake an industrial hygiene study of a furnace stripping operation, notably to study the collected airborne dust products, and (c) to repeat and extent the toxicity testing experiment as well as testing devitrified materials *in vivo* in animals. The results of these experiments will be reported in due course.

ACKNOWLEDGEMENTS

The studies at Sheffield City Polytechnic are funded by the Health and Safety Executive of the United Kingdom and the work is undertaken in collaboration with the MRC Toxicology Unit, Carshalton, Surrey UK and the HSE Occupational Medicine and Hygiene Laboratories, London, UK.

REFEERENCES

Belyakova,N.P., Kutukov,V.F., Ustyantsev,V.M. and Trebnikova,M.G. (1981) Phase transition in materials based on high-alumina fibre. (Translation) *Inorg. Mater.* **17**:948-951.

Dickson,R.R. (1983) Ceramic fibres - a growing refractory product. *Indust. Miner., Lond.*, November, pp. 23-31.

Gaodu,A.N., Pitak,N.V., Volfson,R.E. and Drizheruk,M.E. (1977) Crystallisation in heated aluminosilicate fibers. (Translation) *Inorg. Mater.* **13**:1802-1804.

Gantner,B.A. (1986) Respiratory hazard from removal of ceramic fiber insulation from high temperature industrial furnaces. *Am. Ind. Hyg. Assoc. J.* **47**:530-534.

Griffiths,J. (1986) Synthetic mineral fibres. *Indust. Miner., Lond.*, September, pp. 20-43.

Hickling,H., Thomas,D.H. and Briggs,J. (1981) High temperature behaviour of alumino-silicate ceramic fibres. *Sci. Ceram.* **11**:397-403.

Holroyd,D., Rea,M.S., Young,J. and Briggs,G. (1988) Health-related aspects of the devitrification of aluminosilicate refractory fibres during use as a high-temperature furnace insulant. *Ann. Occup. Hyg.* **32**:171-178.

Jager,A., Stadler,Z. and Wernig,J. (1984) Investigations on microstructural changes undergone by ceramic fibres at elevated temperatures, particularly as regards the formation of cristobalite. *Ber. Dt. Keram. Ges.* **61**:143-147.

Olds,L.E., Miller,W.C. and Pallo,J.M. (1980) High temperature alumino-silicate fibres stabilised with Cr_2O_3. *Ceramic Bull.* **59**:739-741.

Rea,M.S., Young,J. and Briggs,G. (1988) Dust release and crystallisation in furnace-exposed aluminosilicate fibre blankets. Proc. Meeting of Aerosol Society, Bournemouth, U.K. pp. 241-246.

Young,J. (1989) The non-formation of α-cristobalite in devitrified commercial-grade aluminosilicate refractory ceramic fibre. *Br. Ceram. Trans. J.* **88**:58-62.

Young,J. (1990) Man-made mineral fibres. In, "Mineral fibres and Health", Liddell,D. and Miller,K., eds. Chapter 4, CRC Press.

ANALYSIS OF POLYCYCLIC AROMATIC HYDROCARBONS ON VAPOR-GROWN CARBON FIBERS

T.C. Pederson, C.A. Powell, J. Santrock, L. Rosenbaum, J. Siak,
G.G. Tibbetts[1] and R.L. Alig[2]

Biomedical Science and [1]Physics Departments
General Motors Research Laboratories
Warren, MI, U.S.A.

[2]General Motors Delco Products Division
Dayton OH
U.S.A.

ABSTRACT

The ability to grow carbon fibers by pyrolysis of gas phase hydrocarbons in the presence of catalytic iron particles holds the promise of new technologies for low-cost production of novel fibers with improved mechanical properties for use in existing or new manufacturing applications. As part of a research effort to investigate methods for producing vapor-grown carbon fibers, a series of fiber samples were analyzed for their content of polycyclic aromatic hydrocarbons (PAH) which are an undesirable by-product of many industrial pyrolysis processes. Adsorbed hydrocarbons were extracted and analyzed by both capillary gas chromatography (GC/FID & GC/MS) and the Ames bacterial mutation assay. PAH compounds were readily detected in most carbon fiber extracts. The highest concentrations (1% - 5% sample mass) were detected on fibers produced by a continuous flow-through reactor. Much lower concentrations (<0.2%) were found on fibers produced by batch processes. Most prominent PAH components, containing 2-6 aromatic rings, were identified with available reference compounds. Use of an aluminum-clad column also permitted analysis of high-molecular-weight-PAH (C26-C34) which contributed from 1% to 80% of the total PAH in various samples. Distinctly different patterns of PAH isomer distribution were found in these samples which can be correlated with changes in the proportionate contributions of individual PAH compounds to the mutagenic activity of the total extract. The extent of exposure to PAH from handling fibers and the likely significance of an association between genotoxic chemicals and potentially respirable fibers require further consideration, but these results have demonstrated the importance of continuing efforts to minimize production of PAH and their retention on vapor-grown carbon fibers.

INTRODUCTION

A process for the production of carbon fibers by pyrolysis of vapor phase hydrocarbons in the presence of iron-based catalytic particles has been recently developed at General Motors Research Laboratories (Tibbetts, 1990).

These fibers possess advantageous properties, including high tensile strength, high thermal conductivity, and low electrical resistivity, which derive from their unique concentric structure of lamellar carbon in a form analogous to the basal plane of graphite. This technology offers the possibility of a low cost source of a non-mineral fiber for use in fiber/polymer composites and development of other advanced materials.

Many industrial carbon products, most notably the reinforcing carbon blacks, are produced by pyrolytic carbonization of natural gas or petroleum hydrocarbons. An undesirable by-product of most pyrolysis processes is the formation of polycyclic aromatic hydrocarbons, PAH, which include numerous individual compounds that have been shown to be animal carcinogens (IARC, 1983). PAH may also be intermediate structures in carbonization processes such as formation of soot in flames (Howard & Longwell, 1983). Condensed phase pyrolysis of PAH generates intermediate-scale carbonized products that ultimately form the three-dimensionally ordered form of graphite (Lewis & Singer, 1988). Although the iron particle-catalyzed growth of carbon fibers may involve fundamentally different mechanisms for carbon condensation, the conditions employed are very analogous to those associated with formation of PAH.

Both anecdotal and epidemiological evidence acquired over the past two centuries have established a clear correlation between industrial exposures to high concentrations of PAH and excess cancer incidence, principally cancer of the skin or the lung (IARC, 1984a, 1984b, 1984c; Kipling & Cook, 1984). The biological consequences of exposure to PAH adsorbed to the surface of respirable fibers could be significantly greater than the separate effects of genotoxic chemicals and the possible cellular pathology associated with the fiber. In fact, a co-carcinogenic effect between asbestos and the mineral oils used in textile plants has been suggested in explanation of the much greater risk of lung cancer among textile workers than among miners exposed to the same asbestos (Sebastien et al., 1989). Consequently, the issue addressed in these studies is the biological and chemical characterization of the extractable pyrolysis products on vapor-grown carbon fibers.

PRODUCTION OF A NOVEL CARBON FIBER FROM VAPOR PHASE HYDROCARBONS

A schematic diagram of the pyrolysis process and the modes of operation employed in the development of methods for growing carbon fibers are shown in Fig. 1. Fibers have been produced from a variety of hydrocarbon sources including methane, hexane, benzene and naphthalene. Initial experiments were conducted in a small batch-type device, GMR Type I, where magnetite particles adsorbed to a ceramic substrate

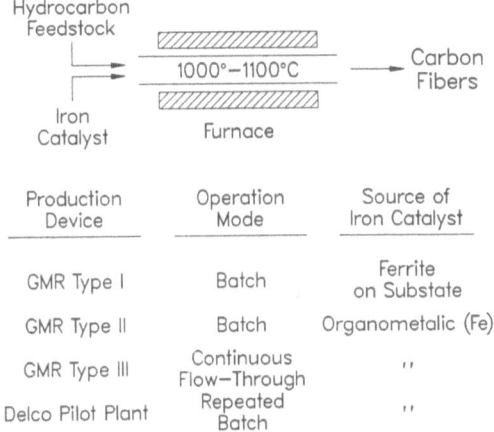

Production Device	Operation Mode	Source of Iron Catalyst
GMR Type I	Batch	Ferrite on Substate
GMR Type II	Batch	Organometalic (Fe)
GMR Type III	Continuous Flow—Through	''
Delco Pilot Plant	Repeated Batch	''

Figure 1. A schematic representation of the process used for vapor-grown carbon fibers and the differing modes of operation used to prepare the samples described in this study.

served as the catalyst for fiber growth (Tibbets *et al.*, 1987; 1988). The catalyst has also been generated from an organo-iron complex dissolved in the hydrocarbon feedstock to produce a suspended iron particle. The latter process has permitted production of fibers in a continuous process, GMR Type III. Vapor-grown carbon fibers have also been produced for applications development in a pilot scale device at GM Delco Products Division.

The electron micrographs shown in Fig. 2 illustrate the wide range of fiber dimensions obtained with varying procedures for vapor-grown carbon fibers. All fibers originate as thin filaments which grow from iron-based particles of nanometer dimensions. Control of temperature and hydrocarbon concentrations can first promote filament lengthening followed by a thickening of existing filaments to ultimately yield fibers several

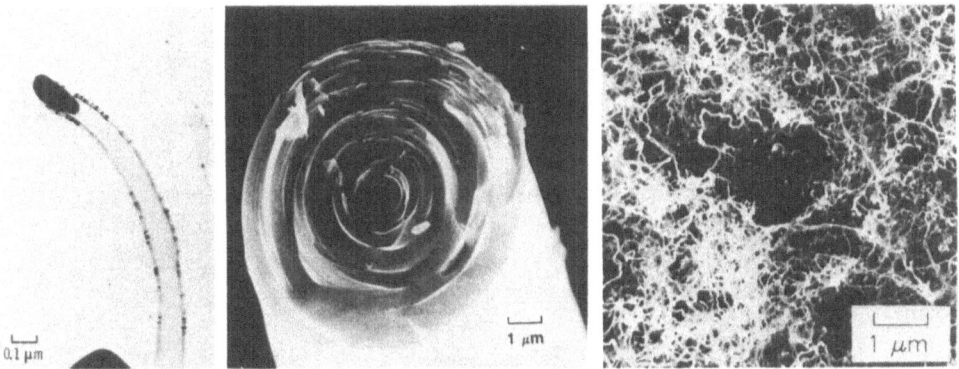

Figure 2. Electron micrographs of the fiber products produced by vapor phase pyrolysis of hydrocarbons in the presence of catalytic iron particles. Left: a view of a newly formed filament attached to the catalytic iron particle. Center: a transverse cross section of a thickened fiber from a batch process. Right: a meshwork of small diameter fibers produced by the continuous flow process.

TABLE 1. Summary of Experimental Methods

— procedures —	— techniques —	——— conditions ———
Extraction of Fiber Samples	Soxhlet Apparatus	benzene-ethanol, 4:1(v/v), 15-16 hr, ~500 cycles
Salmonella/S9 Mutation Assays	Plate-Incorporation	strain TA100, 5 ml/plate rat liver S9 (arochlor pretreated)
PAH Analysis by Capillary Gas Chromatography	GC/FID	Al-clad column, 30-400°C, methyl-5%-phenyl silicone, on-column injection, 2,2'-difluorobiphenyl as internal standard
	GC/MS	split mode injection, 30-320°C, CI & positive ion detection

millimetres in diameter and several centimetres long. The continuous process, GMR Type III, produces fibers of much smaller diameters and they exit the reactor as an entangled mesh rather than as discrete fibers. The fibers produced by the Delco pilot plant are similar to those from the continuous process, but this reactor was operated with a much lower linear flow rate which increased transit time for the pyrolyzing hydrocarbon feedstock and permitted the fibers to form and remain as an intermeshed plug within the heated reactor. Elevated temperatures (1000°-1100°C) were maintained for some period of time following the cessation of fiber formation.

Fifteen samples of vapor-grown carbon fibers were obtained from the four modes of fiber production described in Fig. 1 using a wide variety of feedstocks, operating conditions and collection methods. Two types of carbon black were also acquired for comparative purposes. One sample, designated CB1 is a thermal black (Type MT) produced by pyrolysis of natural gas in the absence of combustion (Mantell, 1968) which should be somewhat analogous to the conditions used to produce the carbon fibers. The second sample, CB2, is a reinforcing rubber-grade carbon black (Type ISAF) produced by a furnace process.

MUTAGENIC ACTIVITY OF EXTRACTS FROM VAPOR-GROWN CARBON FIBERS

The methods used in these studies for analysis of extractable constituents on carbon fibers are briefly summarized in Table 1. The benzene-ethanol azeotrope was used

TABLE 2. Carbon Pyrolysis Products Extracted for Analysis

Sample Description	Label	Extractable Organics	Mutagenic Activity[a]	
		mg/g sample	per μg extract	per mg sample
Vapor-Grown Fibers				
production device				
GMR Type I	PF1	2.7	< 1.4	> 5
GMR Type II	PF2	11	18	195
GMR Type III	PF3a	40	77	3100
"	PF3b	69	89	6100
"	PF3c	61	97	5900
"	PF3d	45	63	2800
"	PF3e	44	84	3700
"	PF3f	44	101	4400
"	PF3g	45	54	2400
"	PF3h	33	52	1750
"	PF3i	17	68	1150
"	PF3j	17	57	980
Delco Pilot Plant	DPFa	4.1	15	63
"	DPFb	1.1	< 3	< 5
"	DPFc	3.1	11	34
Commercial Carbon Products				
Carbon Black, type MT	CB1	2.6	57	147
Carbon Black, type ISAF	CB2	1.5	1.3	2

[a] Number of *Salmonella* revertants

as an extracting solvent because of its demonstrated ability to effectively extract both neutral aromatic compounds and more polar components (Williams and Chock, 1989; Siak *et al.*, 1980). The bacterial mutation assays were performed in accordance with the general procedures recommended by Maron and Ames (1983). Mutagenic activities of the fiber extracts and of a series of reference PAH compounds were all measured under identical assay conditions with a range of concentrations sufficient to demonstrate a linear dose response.

The amounts of material extracted from the carbon fiber and carbon black samples varied from 0.15% to 7% of the sample weights. Table 2 lists the amount of extractable material obtained from each sample and the mutagenic activity of each extract in the Salmonella/S9 assay. The extracts exhibited little or no activity in the absence of the rat liver S9 enzymes. When mutagenic activity is expressed as activity/mg of original sample, the carbon fibers produced by the GMR Type III reactor, which operates in a

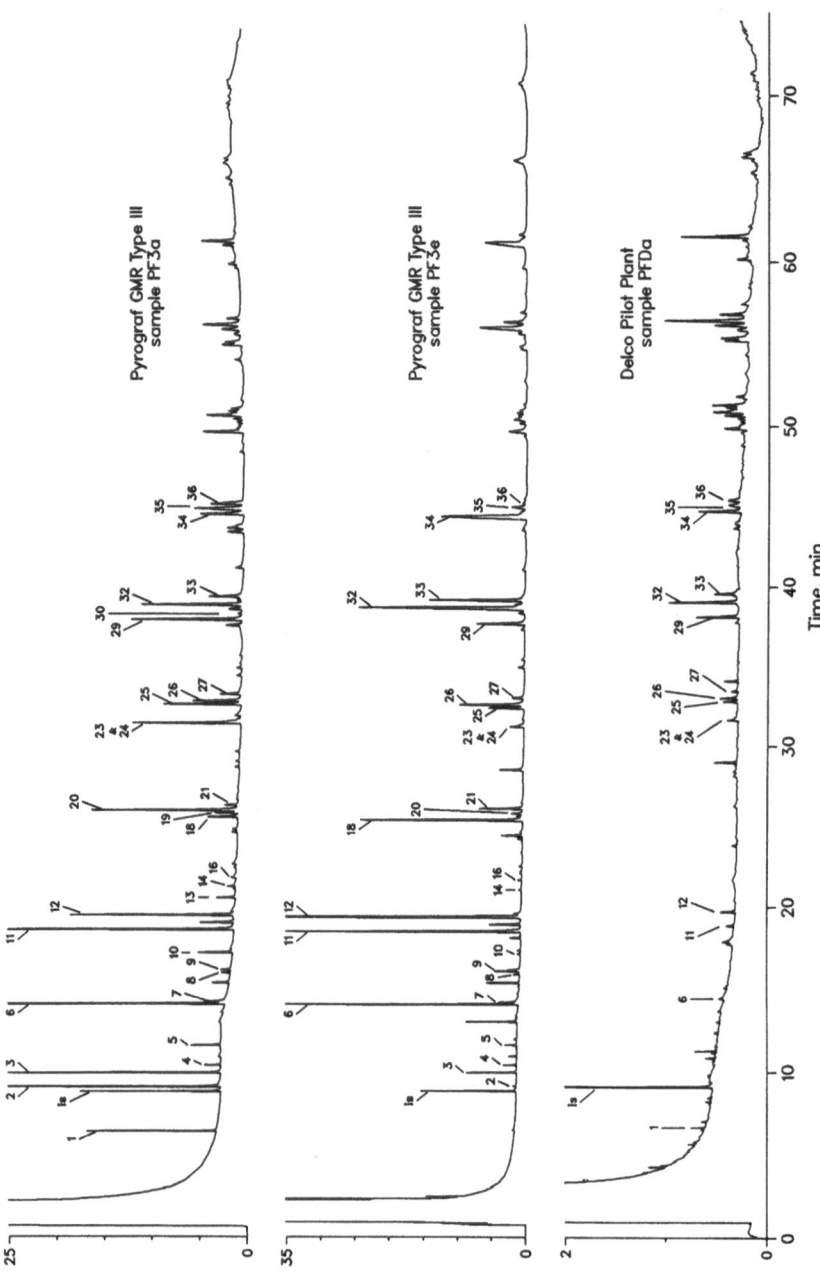

Figure 3. GC/FID chromatograms of three extracts which exemplify the wide range of PAH compounds in carbon fiber extracts. They include the high molecular weight components which elute starting at about 48 minutes in the temperature program. The numbered peaks, identified in Table 3, co-elute with PAH reference compounds.

TABLE 3. Identification and quantitation of PAH from vapor-grown fibers by GC/FID & GC/MS

#	GC/MS mol wt	Identification	µg/g sample PF3a	PF3e
1	128	Naphthalene	378	
2	154	Biphenyl	1642	20
3	152	Acenaphthalene	887	241
4	154	Acenaphthene	76	81
5	166	Fluorene	137	80
6	178	Phenanthrene	2506	1883
7	178	Anthracene	48	99
	204	(1-Phenylnaphthalene)	116	261
8	230	o-Terphenyl	58	72
9	190	4-H-Cyclopenta[def]phenanthrene	79	227
10	204	2-Phenylnaphthalene	206	39
11	202	Fluoranthene	2118	2599
	202	(Acephenanthrylene)	264	342
12	202	Pyrene	1303	6770
13	230	m-Terphenyl	158	
14	230	p-Terphenyl	64	33
16	216	Benzo[b]fluorene	44	51
	226	(Benzo[ghi]fluoranthene)	50	276
	226	- unidentified -	80	56
18	226	Cyclopenta[cd]pyrene	297	2025
19	228	Benz[a]anthracene	206	40
20	228	Chrysene/Triphenylene	1453	124
21	228	3,4-Dihydrocyclopenta[cd]pyrene	117	533
23-4	252	Benzo[-]fluoranthenes, (b,j & k)	1370	227
	252	(Benzo[a]fluoranthene)	74	
25	252	Benzo[e]pyrene	837	536
26	252	Benzo[a]pyrene	525	633
27	252	Perylene	255	128
	276	- unidentified -	187	71
29	276	Indeno[1,2,3-cd]pyrene	1271	791
30	278	Dibenz[-]anthracenes, (a,c & a,h)	35	
32	276	Benzo[ghi]perylene	1172	3405
33	276	Anthanthrene	402	1402
	302	M.W. 302 complex	452	199
34	300	Coronene	669	2879
35	302	Dibenzo[a,e]pyrene & others	916	144
36	302	Dibenzo[a,l]pyrene & others	594	391
		High mol wt compounds	6251	8723

	PF3a	PF3e
Σ as % of total GC hydrocarbons	98%	94%
Σ as % of total extract weight	69%	81%

continuous flow-through mode, are much more mutagenic than the other samples. The much lower values for fibers produced by the batch process modes (GMR Type I & II, and the Delco pilot plant), and for the carbon black samples, correspond with the lower amounts of extractable material, but expression of activity as revertants per mg extractable mass does not indicate a uniform mutagenic potency for these extracts. A better assessment of similarities or differences among these extracts can be made by comparisons of mutagenicity with PAH content and the estimated contributions of identifiable PAH mutagens.

ANALYSES OF PAH CONTENT IN FIBER EXTRACTS

The PAH content of each extract was analyzed by the gas chromatographic methods described in Table 1. The use of an aluminum-clad capillary column, tolerant of temperatures over 400°C, extends the range of GC analysis to include much larger and less volatile PAH. The low-temperature on-column injection technique in combination with an internal standard yields a uniform response for PAH with the FID detector (Grob & Grob, 1978). Chromatograms from three of the fiber extracts are shown in Fig. 3. These analyses enabled the identification and quantitative determination of thirty or more individual PAH compounds in most extracts, ranging in size from naphthalene ($C_{10}H_8$) through the dibenzopyrenes ($C_{24}H_{18}$), plus varying amounts of high molecular weight components presumed to be large (C_{26}-C_{34}) PAH compounds. The identities of most the components, other than the high molecular weight compounds, were assigned with use of available reference compounds, published values for retention indices, and determination of parent-ion molecular weights from GC/MS analyses. Table 3 lists all the compounds that were identified in most of the extracts and compares the amounts of each component in extracts of samples PF3a and PF3e.

Although the identified PAH are quite similar in extracts of PF3a and PF3e, the proportionate contributions vary in a systematic manner. In PF3a, there is more fluoranthene than pyrene, more cyclopenta[cd]pyrene than chrysene, more benzofluoranthenes than benzo[a]pyrene, and more indeno[cd]pyrene than benzo[ghi]perylene. In PF3e the inverse is true in each case. These patterns were also evident in the PAH content of many of the other extracts. Further analysis demonstrated that at least two kinds of processes were contributing to significantly different PAH distributions. One distribution difference was evident in the varying proportionate amounts of low and high molecular weight PAH fractions, which is an understandable consequence of fractionation by thermal gradients within the apparatus for formation and collection of fibers. The second difference was identifiable with the two previously described patterns of isomer distribution within the C_{16}, C_{18}, C_{20}, C_{22}, and C_{24} families of PAH isomers. These differences could presumably have considerable effects on the mutagenic or carcinogenic properties of the extracts.

EFFECT OF PAH DISTRIBUTION PATTERNS ON MUTAGENIC ACTIVITY

In Table 4, comparisons of mutagenic activity are made among extracts that are grouped together according to similarities in both isomer pattern distributions and the proportionate amounts of high and low molecular weight fractions. A division among eleven of the carbon fiber samples was made on the basis of a Type I or Type II isomer

TABLE 4. Comparative mutagenicity of extracts with differing PAH distribution patterns

Extracts Classification by PAH distribution	Characteristic mutagens[a]						C_{24}-PAH 302 mol wt	Other PAH		Mutagenicity Revertants per µg of 4-6 ring PAH
	Revertants per µg[b]							low mol wt 2-3 ring	high mol wt (C_{26}–C_{34})	
	FLT	PYR	CPcdP	CHR	BxFTs	BaP				
	35	5	700	(160)	(100)	940	?			
	% of 4-6 ring PAH							% of total hydrocarbon		
Type I isomer ratios										
PF2	46	25	0.6	6.2	2.7	0.9	1.7	19	2	90
PF3a	14	9	2.0	9.7	9.2	3.5	7.0	23	22	210
PF3b	19	7	4.2	12.7	8.7	2.7	7.3	27	14	200
group #1 PF3c	15	18	2.0	7.1	5.6	4.4	3.9	9	20	180
PF3f	14	17	1.6	9.5	5.6	4.1	4.6	7	20	185
Type II isomer ratios										
PF3d	11	26	8.4	0.7	0.9	2.9	2.6	5	32	135
PF3e	11	29	8.5	0.5	1.0	2.9	2.4	7	23	155
group #2 PF3i	10	31	5.2	0.4	0.8	3.9	2.8	3	27	115
PF3j	11	31	5.3	0.4	0.7	3.6	2.7	5	27	110
group #3 PF3g	15	42	2.7	0.9	0.9	2.3	2.2	40	3	95
PF3h	14	46	3.5	0.5	0.8	2.1	1.4	46	1	110
Other extracts										
PFDa	1.7	0.7	-0-	-0-	2.0	4.6	14.7	0.3	79	155
PFDc	13	19	2.2	2.3	2.6	5.7	2.5	15	5	130
CB1	4.4	15	2.9	0.2	1.0	3.1	6.0	0.5	44	185

[a] PAH abbreviated as FLT = fluoranthene, PYR = pyrene, CPcdP = cyclopenta[cd]pyrene, CHR = chrysene/triphenylene
BxFTs = benzo[-]fluoranthenes (bj & k) and BaP = benzo[a]pyrene.

[b] The specific mutagenicity of each PAH compound. Values in parentheses are weighted estimates for the combined activity of isomer mixtures.

distribution pattern which affects the relative content of several characteristic mutagens. Four of the five extracts with a Type I isomer distribution, group #1, have similar PAH distributions and also very similar values for mutagenic activity expressed as revertants per milligram of 4-6 ring PAH. The other extract with a Type I isomer distribution, PF2, had a much higher proportionate amount of both fluoranthene and pyrene than any other extract which presumably explains a mutagenic activity that is about half that of the other Type I extracts. The extracts with Type II isomer ratios can be divided into groups #2 and #3 based on contrasting content of low and high molecular weight fractions. The mutagenic activity of these two groups are both significantly lower than that of group #1. The three remaining extracts in Table 4 have values for mutagenic activity per total 4-6 ring PAH that are similar to the other extracts, even though the PAH distributions in these samples are quite different.

Although the differing PAH distributions do evidently have some effect on values for mutagenic activity per total 4-6 ring PAH, more distinctive consequences are evident when comparing the estimated contributions of individual mutagens to the total activity of each extract. Such comparisons are shown in Fig. 4. Within each of the three groups of extracts defined in Table 4, the estimated contributions from individual mutagens were quite similar and, therefore, average values are shown for each group. In general, these summations demonstrate that the principle PAH mutagens identified in carbon fiber extracts account for the majority of the total mutagenic activity.

The variations between extracts in proportionate contributions of individual mutagens may be of greater significance than differences in total mutagenicity with regard to the carcinogenic potential of these extracts. A comparative ranking of most PAH carcinogens, as weak, moderate, or strong, can be derived from periodic reviews and other studies (IARC, 1983; Nagao & Suigimura, 1978; LaVoie et al., 1979, Wood et al., 1980). Benzo[a]pyrene, a strong carcinogen and a strong mutagen, is indicated to be a prominent mutagen in all extracts, but its contribution ranges from 10% to 40% of the total activity. In the extract of PFDa, where the C_{20}, C_{22}, and C_{24} isomers are the predominant constituents of the 4-6 ring PAH, the contribution from ideno[cd]pyrene (weak carcinogen. moderate mutagen) is equal to that of benzo[a]pyrene.

The mutagenicity profiles for extracts with a Type II PAH isomer distribution pattern, groups #2 and #3, can be clearly distinguished from extracts with a Type I distribution by the much larger proportionate contribution of cyclopenta[cd]pyrene (weak carcinogen, strong mutagen) whereas, the contributions of chrysene (weak carcinogen, moderate mutagen) and the benzofluoranthenes (moderate carcinogen, moderate mutagen) appear significantly greater for extracts with a Type I isomer distribution. The extract of carbon fiber sample PF2, like the four extracts in group #1, has a Type I isomer distribution, but fluoranthene and pyrene, which have little or no carcinogenic activity, are indicated to be the components with the largest contribution to total mutagenic activity.

A number of high molecular weight PAH compounds, ranging from C_{26} to C_{30} have been tested for carcinogenic activity and several reportedly exhibit low or moderate activity (Dipple et al., 1984; Lee et al., 1981; Dai Qianhuan, 1985). Since the amounts high molecular weight PAH varied greatly among the extracts included in this study, some assessment can be made of the mutagenic activity from these components. Extracts of PFDa and CB1 have the highest proportionate content of the high molecular weight components. The amounts of 4-6 ring PAH and high molecular weight PAH are about equal in the extract of CB1 and the contribution of known mutagens from the 4-6 ring fraction evidently accounts for less than half of the total activity which suggests that the high molecular weight components might be responsible for much of the remaining activity. However, the extract of PFDa contains about four times as much of the high molecular weight components as 4-6 ring PAH, yet the identifiable mutagens in the 4-6 ring fraction account for about 80% of the mutagenicity in this extract. Similarly, all four carbon fiber extracts included in group #2 have considerable amounts of the high molecular weight components (27% to 32% of total hydrocarbon), yet the contribution of identifiable mutagens accounts for 85% of the total activity. Thus, there is little evidence that these components, presumed to be C_{26} to C_{32} PAH, contribute a significant proportion of mutagenic activity of carbon fiber extracts.

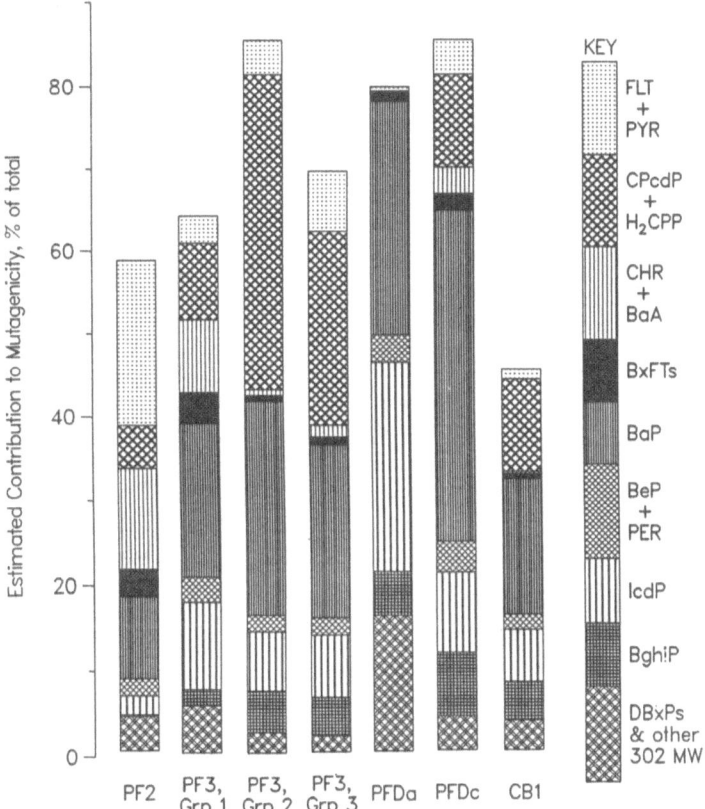

Figure 4. The estimated contributions to measured mutagenic activity by identified PAH mutagens. Extracts within each of three groups, designated Grp 1, 2, & 3 and defined in Table 4, exhibit common activity profiles. Abbreviations for individual PAH and their mutagenic activities were partially described in Table 4 and also include H_2CPP = 3,4-dihydrocyclopenta[cd]pyrene (70 rev't/μg), BaA = benz[a]anthracene (65 rev't/ug), BeP = benzo[e]pyrene (90 rev't/μg), PER = perylene (80 rev't/μg), IcdP = indeno[1,2,3-cd]pyrene (280 rev't/μg), BghiP = benzo[ghi]perylene (40 rev't/μg), and DBxPs = dibenzo[-]pyrenes. The estimates of the mutagenicity for the combined 302 mol wt PAH isomers are based on an arbitrary value of 100 rev't/μg.

SUMMARY AND CONCLUSIONS

These studies have demonstrated that polycyclic aromatic hydrocarbons are formed under the conditions used for production of vapor-grown carbon fibers and may remain as contaminants adsorbed to the surface of the fibers.

The amounts and mutagenic activities of these pyrolysis by-products were compared in a series of fiber samples produced by several procedures. The findings and implications of these efforts are summarized below.

- Fiber samples from the flow-through reactor had the greatest amounts of PAH, ranging from 1% to 5% of total sample weight, and the highest mutagenic activity in the Salmonella/S9 assay. The much lower PAH concentrations and mutagenicity present in samples produced by the pilot plant reactor appear comparable to those of commercial carbon black samples.

- Distinctly different patterns of PAH isomer distribution affecting prominent mutagenic PAH were noted, but mutagenic activity of these extracts remained closely correlated with the total amount of 4-6 ring PAH.

- Use of an aluminum-clad capillary GC column, and a temperature program extending to 400RC, permits analysis of high molecular weight PAH (C26-C34). These components contributed from 1% to 80% of the total PAH in the fiber extracts, but did not produce a demonstrable increase in mutagenic activity.

- The lower PAH content of fibers from the pilot plant reactor was presumably a consequence of retaining the fiber within the heated reactor following the cessation of fiber formation. Inferences drawn from these operations suggest that practical means exist for elimination of PAH from the vapor-grown carbon fibers.

The health-effects issues not addressed by these studies concern primarily the properties of the fibers themselves. There is an obvious need to characterize potential and actual exposure to respirable forms of vapor-grown carbon fibers that may be encountered during production, processing, or use. Since the vapor phase growth of carbon filaments is dependent on the catalytic properties of small iron-based particles, the biological significance of the fiber's iron content (*i.e.* bioavailability) can be examined. Finally, a comparative evaluation of vapor-grown carbon fibers can be made in appropriate models (*in vivo* or *in vitro*) of fiber associated pathology.

REFERENCES

Dai Qianhuan (1985) Di-Region Theory - New Conception for Quantitative Structure-Carcinogenic Activity Relationship and Mechanism of Chemical Carcinogenesis. In *Polynuclear Aromatic Hydrocarbons: Mechanisms Methods and Metabolism* (M. Cooke and Dennis, A.J., eds) pp. 1045-1073.

Dipple,A., Moschel,R.C., Bigger,C.A.H. (1984) *Polynuclear Aromatic Carcinogens. in Chemical Carcinogens*, Second Edition (C.E. Searle, ed.) American Chemical Society, Washington, DC, Vol. 1, pp. 41-163.

Grob,K. and Grob,K., Jr. (1978) On-Column Injection on to Glass Capillary Columns. *J. Chromatogr.* 151:311-320.

Howard,J.B., Longwell,J.P. (1983) Formation Mechanisms of PAH and Soot in Flames. In *Polynuclear Aromatic Hydrocarbons: Formation, Metabolism and Measurement* (M. Cooke and Dennis, A.J., eds) pp. 27-62.

IARC (1983) Chemical, Environmental and Experimental Data. *IARC Monographs on the Evaluation of Carcinogenic Risk of Chemicals to Humans*, Vol. 32, Polynuclear Aromatic Hydrocarbons, Part 1, Lyon.

IARC (1984a) Carbon Blacks; 1. Chemical and Physical Data. *IARC Monographs on the Evaluation of Carcinogenic Risk of Chemicals to Humans*, Vol. 33, Polynuclear Aromatic Hydrocarbons, Part 2, Lyon.

IARC (1984b) Idustrial Exposures in Aluminum Production, Coal Gasification, Coke Production and Iron and Steel Founding. *IARC Monographs on the Evaluation of Carcinogenic Risk of Chemicals to Humans*, Vol. 34, Polynuclear Aromatic Hydrocarbons, Part 3, Lyon.

IARC (1984c) Carbon Blacks; 1. Bitumens, Coal-Tars and Derived Products, Shale Oils and Soots. *IARC Monographs on the Evaluation of Carcinogenic Risk of Chemicals to Humans*, Vol. 35, Polynuclear Aromatic Hydrocarbons, Part 4, Lyon.

Kipling,M.D. and Cooke,M.A. (1984) Soots, Tars, and Oils as Causes of Occupational Cancer. In *Chemical Carcinogens*, Second Edition (C.E. Searle, ed.) American Chemical Society, Washington, DC, Vol. 1, pp. 165-174.

LaVoie,E., Bedenko,V., Hiorota,N., Hecht,S.S. and Hoffmann,D. (1979) A comparison of the Mutagenicity, Tumor-Initiating Activity and Complete Carcinogenicity of Polynuclear Aromatic Hydrocarbons. in Polynuclear Aromatic Hydrocarbons, *Third International Symposium on Chemistry and Biology, Carcinogenesis and Mutagenisis* (P.W. Jones and P. Leber, eds) Ann Arbor Science Publ., Inc., Ann Arbor, MI, pp. 705-721.

Lee,M.L., Novotny,M.V. and Bartle,K.D. (1981) Appendix 5: Polycyclic Aromatic Hydrocarbons That Have Been Tested for Carcinogenicity. In *Analytical Chemistry of Aromatic Compounds*, Academic Press, New York, pp 441-449.

Lewis, I.C., Singer, L.S. (1988) Thermal Conversion of Polynuclear Aromatic Compounds to Carbon. In *Polynuclear Aromatic Compounds* (L.B. Ebert, ed.) American Chemical Society, Washington, DC, pp. 269-285.

Maron, D.M., Ames,B.N. (1983) Revised Methods for the Salmonella Mutagenicity Test. *Mutation Res.* 113:173-215.

Mantell, C.L. (1968) *Carbon Black Carbon and Graphite Handbook*, John Wiley and Sons, Inc., New York, Chapter 6, pp. 76-105.

Nagao, M., Suigimura, T. (1978) Mutagenesis: Microbial Systems. In *Polycyclic Hydrocarbons and Cancer* (H. Gelboin and P. Ts'o, eds.) Academic Press, New York, Vol. 2:pp. 99-121.

Sebastien, P., McDonald, J.C., McDonald, A.D., Case, B., Harley, R. (1989) Respiratory Cancer in Chrysotile Textile and Mining Industries: Exposure Inferences from Lung Analysis. *Br. J. Ind. Med.* 46:180-187.

Siak, J.-S., Chan, T.L., Lee, P.S. (1980) Characterization of Diesel Particulate Exposure., in Health Effects of Diesel Engine Emissions: Proceedings of an International Symposium, (W.E. Pepelko, R.H. Danner and N.A. Clark, eds) EPA-600/9-80-057a, Vol.1, pp. 245-259.

Tibbetts, G.G. (1990) Carbon Fibers from Vapor Phase Hydrocarbons. SAE Technical Paper Series No. 901036.

Tibbetts, G.G., Devour, M.G., Rodda, E.J. (1987) An Adsorption-Diffusion Isotherm and its Application to the Growth of Carbon Filaments on Iron Catalyst Particles. *Carbon* 25:367-375.

Tibbetts, G.G., Rodda, E.J. (1988) High Temperature Limit for the Growth of Carbon Filaments on Catalytic Iron Particles. *Mat. Res. Soc, Symp. Proc.* 111:49-52.

Williams, R.L., Chock, D.P. (1980) Characterization of Diesel Particulate in Health Effects of

Diesel Engine Emissions: Proceedings of an International Symposium, (W.E. Pepelko, R.H. Danner and N.A. Clark, eds) EPA-600/9-80-057a, Vol.1, pp. 3-32.

Wood, A.W., Levin, W., Chang, R.L., Huang, M.-T., Ryan, D.E., Thomas, P.E., Kumar, S., Koreeda, M., Akagi, H., Ittah, Y., Dansette, P., Haruhiko, Y., Jerina, D.M., Conney, A.H. (1980) Mutagenicity and Tumor-Initiating Activity of Cyclopenta[cd]pyrene and Structurally Related Compounds. *Cancer Res.* **40**:642-649.

INDUCTION OF INFLAMMATION AND FIBROSIS AFTER EXPOSURE TO INSOLUBLE AND ISOMETRIC PARTICLES

R. Mermelstein, H. Muhle[1] and P. Morrow[2]

Environmental Health & Safety, Xerox Corporation (W843)
Joseph C. Wilson Center for Technology
Webster, New York 14580, USA

[1]Fraunhofer-Institut fur Toxikologie und Aerosolforschung
Nikolai-Fuchs Strasse 1
3000 Hannover 61, FRG

[2]University of Rochester
Rochester
New York 14642, USA

ABSTRACT

Several published reports show that long-term inhalation of large quantities of non-fibrous, innocuous particles such as TiO_2, carbon black, petroleum coke, toner, and other polymeric materials can result in a spectrum of pulmonary changes. Exposure of F344 rats to 0, 1, 4 or 16 mg/m^3 of test toner for 6 hours/day, 5 days/week resulted in inflammatory changes determined by broncho-alveolar lavage at the two highest exposure levels after 15 months. Minimal to mild pulmonary fibrosis was detected in 22% of the group at 4 mg/m^3, while mild to moderate fibrosis was observed in 92% of the high exposure group at the termination of the study (25.5 months). In another experiment, F344 rats were exposed to the same test toner at 40 mg/m^3 for 3-months and observed for 15-months post-exposure. The inflammatory changes, detected after exposure, persisted during the entire period of observation. In both of these experiments, the observed effects were interpreted in terms of "lung overloading". The presence of large quantities of insoluble, non-fibrous particles, generally considered to be innocuous, leads to an inflammatory response in the lungs. Depending on the duration and magnitude of the insult, the long-term presence of the foreign material in the lungs can lead to histopathological changes which included lipoproteinosis, fibrosis, and in some cases neoplasia.

INTRODUCTION

Numerous reports show that inhalation of fibers can result in pulmonary dysfunction and pathological lesions which include fibrotic and neoplastic changes. We believe it is important to recognize that the same effects may result from the long-term, high level, inhalation of innocuous or relatively benign isometric particles. This possibility should be considered in the design and interpretation of the results of inhalation investigations involving insoluble particles.

Chronic high level inhalation of the "nuisance dust", titanium dioxide, by rats, was reported, to result in a spectrum of pulmonary changes including fibrosis and lung cancer (Lee *et al.*, 1985). Long-term, high level, inhalation exposures to other innocuous and insoluble and isometric particles can lead to a persistent inflammatory process, which is accompanied or followed by alveolar lipoproteinosis, fibrosis and finally neoplasia. If this is a generic process, then for health protection reasons it would appear to be prudent to prevent the occurrence of the first rather than the last serious end-point.

In the present report, we document our findings with respect to the induction of inflammation and fibrosis upon exposure to insoluble and isometric particles and provide preliminary information, on a dosimetric approach which appears to correlate lung burdens with the onset of pulmonary fibrosis in rats. Similar results were obtained upon inhalation of volcanic ash and petroleum coke particles.

MATERIALS & METHODS

Test toner was specially prepared for these studies. Toner is a pigmented plastic powder with density of about 1.2 g/cm³, and a softening range of 85 - 100°C. The powder contained about 90% random co-polymer (CAS 25213-39-2) and 10% high purity, medium color furnace type carbon black (CAS 7440-44-0). The polymer was composed of styrene and 1-butyl-methacrylate in the ratio of 58:42, and its weight average molecular weight was approximately 70,000 Da. The mass median aerodynamic diameter (MMAD) of the test toner was about 4.0 μm (geometric standard deviation of 1.5 μm). The toner was about 35% respirable according to the 1984 ACGIH criterion. The combined level of volatile impurities and residual monomer, measured by capillary gas chromatography was less than 0.1%. Both extracts of the carbon black and samples of the test toner were inactive in the Salmonella- mutagenicity assay (Butler *et al* 1982).

In the chronic inhalation study of toner in rats, SPF, female F344 rats were exposed for 6 hr/day, 5 days/week for up to 24 months, to toner at 0, 1 (low), 4 (medium), and 16 mg/m³ (high), or TiO_2 at 5 mg/m³, or SiO_2 at 1 mg/m³, by inhalation, and then kept for an additional 6 weeks in filtered air. The test toner was enriched in fine particles such that the ACGIH respirable aerosol concentrations were 0, 0.35, 1.4 and 5.6 mg/m³. Two materials, titanium dioxide and crystalline silicon dioxide were used as negative and positive controls for fibrogenicity. The TiO_2 'Bayertitan T', (Bayer A.G. D-3150 Krefeld), 99.5% rutile, density of 4.3 mg/m³, and SiO_2 'DQ-12' (Bergbauforschung, D4300 Essen), density 2.6 mg/m³, 87% crystallinity as Quartz by X-ray crystallinity, respirable concentrations were 3.87 and 0.74 mg/m³, respectively.

In the toner inhalation and recovery study in rats, female F344 rats were exposed to one of three toner concentrations; control (0 mg/m³), low (10 mg/m³) or high (40 mg/m³) for 13 weeks, followed by five 13-week observation intervals. The response

TABLE 1. Chronic and recovery studies: key findings

Resp. Exposure conc. (ACGIH) (mg/m³)	Lung Burden (mg/g lung)		Pulmonary response **Chronic study** (end of study: 25.5 Mos.)
0.35	0.21		None
1.40	1.80		Symptoms of overloading Slight decrease in clearance/increase in retention Slight chronic inflammation Limited, very slight-slight fibrosis: 22% of group
5.60	15.0		Extensive symptoms of overloading Decrease in clearance/increase in retention Chronic inflammation Decrease in pulmonary function Increase in lung weight Slight-moderate fibrosis in 92% of group

Recovery study

	Lung Burden (mg/g lung) End Exp.	Lung Burden (mg/g lung) Post Exp.	End exposure	15 Mos. post exposure
14.0 (1 wk) 3.5 (13 wks)	0.28 0.40	0.11 0.12	Slight inflammatory response Slight inflammatory response Slight decrease in clearance	Return to normal within 90 days Return to normal within 180 days Persistent clearance impairment
14.0 (13 wks)	3.01	2.65	Symptoms of overloading Moderate inflammatory response Decrease in clearance	Extensive symptoms of overloading Persistent inflammatory response Persistent decrease in clearance Very slight fibrosis in 2/5 animals

215

TABLE 2. Principal Lung-Histopathological Findings

	Number of rats investigated	Number of rats with primary lung tumors	Number and type of tumor — Benign	Number and type of tumor — Malignant	Lipo-proteinosis	Fibrotic foci	Peribronchiolar granulatomatous foci
Air only	100	3	2(A)	1(AC)	-	1(s)	-
Toner Low	100	1	1(A)	-			
Toner Med	100	0	-	-		20(s)	
Toner High	100	3	2(A)	1(AC)		72(s) 19(m)	
TiO$_2$	100	2	1(A)	1(AC)	1(vs)	4(vs) 1(s)	
SiO$_2$	100	19**	3(A) 4(KSCT)	11(AC) 1(ASC) 1(SCC)	61(s) 47(m)	1(vs) 13(s) 85(m)	15(s) 78(m) 2(se)

Abbreviations: A = adenoma; AC = adenocarcinoma; ASC = adenosquamous carcinoma; KSCT = keratinizing cystic squamous cell tumour; SCC squamous cell carcinoma; vs = very slight; s = slight; m = moderate; se = severe
** P<0.01

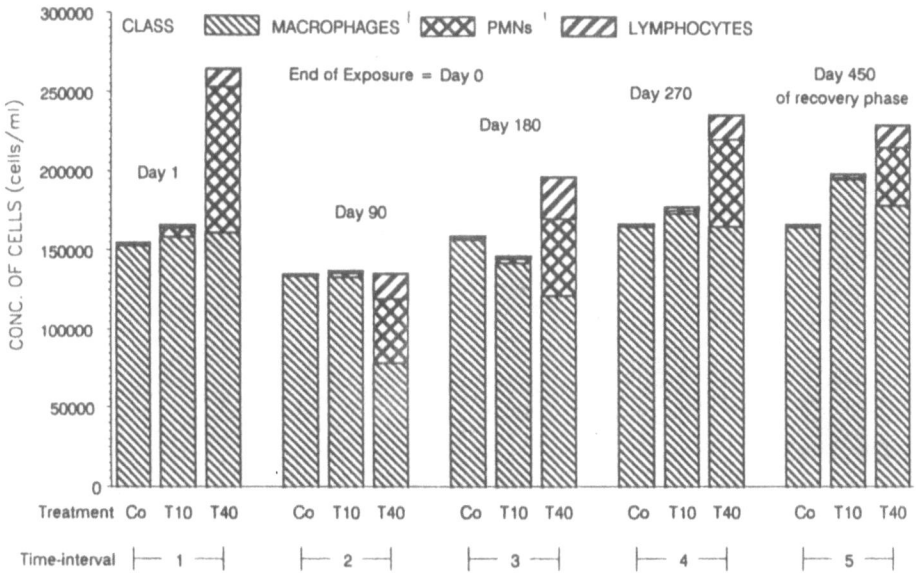

Figure 1. Differential cell count. Recovery study from a 90 day subchronic exposure.

parameters evaluated were: lung weight, lung retention, alveolar clearance, broncho-alveolar lavage (BAL), pulmonary function and histopathology. Grading of the degree of fibrosis was done according to the Wagner scale using Masson-trichrome stained tissue. The details of the broncho-alveolar lavage and clearance and retention procedures have been described (Muhle *et al.*, 1991; Bellmann *et al.*, 1991).

RESULTS & DISCUSSION

A.   Chronic Inhalation Study

The objective of the chronic inhalation study was to determine the biological effects of long-term inhalation of toner. Two control materials were employed, because in our view, the use of air only controls would be an inadequate basis for assaying the effects of a relatively non-toxic particle which is retained in the lungs for extended periods of time.

The general results are shown below while the significant pulmonary responses are summarized in Table 1. (Bellmann *et al.*, 1991; Muhle *et al.*, 1990b). There was no detectable effect of toner exposure on survival and causes of death, body weight, and food consumption. Clinical chemistry and hematology parameters were essentially unaffected. There was no evidence of systemic or upper respiratory tract toxicity. The only significant change was a consistent increase in the weight of the lungs and associated lymph nodes in animals exposed to the highest concentration of toner. There was no detectable fibrosis in the toner low group which is the most relevant exposure level with respect to potential human exposures. Lung fibrosis of minimal to mild degree was observed in 22% of the rats in the toner middle exposure group, while mild to moderate degree of fibrosis was detected in the lungs 92% of the animals in the toner high exposure group. There was a significant increase in the frequency of lung tumors in the rats exposed to crystalline silica (Table 2).

TABLE 3. Mean level of control and levels normalized to controls of LDH, ß-Glucuronidase and protein in the lavagate at various times during the chronic toner inhalation study in rats

Groups	Exposure (months)		
	15	21	24
LDH			
CONTROL (U/l)	25	35	39
Control	1.00	1.00	1.00
Toner low	0.92	0.74	0.85
Toner medium	1.36	1.06	1.36
Toner high	5.68**	5.63**	5.64**
TiO_2	1.16	0.71	1.26
SiO_2	16.24**	13.83**	13.36**
ß-Glucuronidase			
CONTROL (U/l)	0.15	0.20	0.18
Control	1.00	1.00	1.00
Toner low	0.70	0.70	0.83
Toner medium	1.26	1.20	2.38
Toner high	13.46**	16.35**	28.30**
TiO_2	1.00	0.70	1.89
SiO_2	53.06**	46.35**	55.50**
Protein			
CONTROL (U/l)	108	114	146
Control	1.00	1.00	1.00
Toner low	1.06	1.10	1.12
Toner medium	1.13	1.19	1.14
Toner high	2.59**	3.05**	2.54**
TiO_2	1.04	0.99	1.86
SiO_2	5.67**	5.48**	4.36**

** Significant compared to controls (P < 0.01 Dunnett test)

B.   Toner Inhalation & Recovery

The objective of the toner inhalation and recovery study in rats was to investigate (as a function of exposure conditions), whether some of the alterations observed in the chronic inhalation study in rats were reversible, persist, or show a delayed occurrence. In general the results paralleled those of the chronic study. The significant pulmonary responses are summarized in Table 1. A minimal to mild degree of fibrosis was detected in all the toner high exposure group 15 months post-exposure. A large influx of PMN's, a moderate increase in lymphocytes as well as an increase in lavaged cell concentration were noted in the toner high exposure group. These responses were still elevated 9 months and 15-months post exposure.

C.   Study Interpretations

The comparable survival and organ weights between the air only control and the three toner exposure groups indicates the absence of gross chronic toner toxicity within the limits examined in the study. However, the consistently increased weight of the lungs and lung associated lymph nodes in the animals exposed to the highest toner concentration provided an early indication that the respiratory system was the target organ for toner.

The comparable lung tumor frequencies observed for the three different toner

218

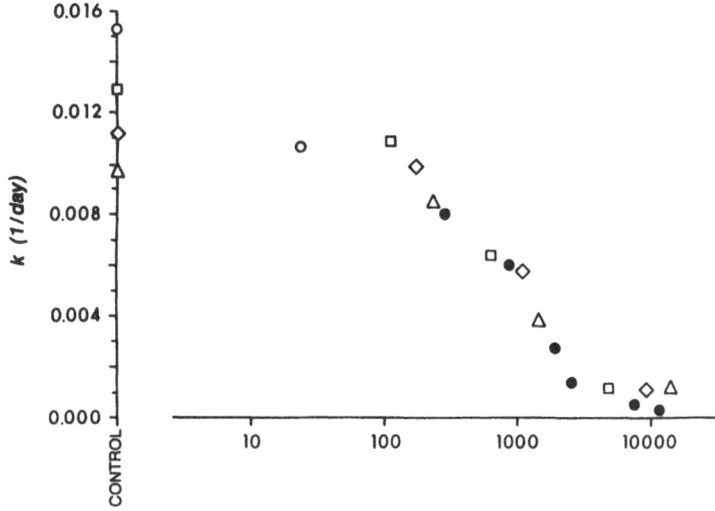

Figure 2. Retained toner in lungs (μg).

exposure groups, the air-only and, TiO_2 controls, demonstrates that toner is not tumorigenic for the respiratory system.

The responses observed (Table 3) in the biochemical and cytological BAL parameters in the toner high exposure group were similar to those observed with mineral dusts and diesel exhaust, and are indicative of an inflammatory process (Henderson *et al.*, 1985). Increases in the numbers of polymorphonuclear leukocytes (PMN), and increased lysosomal enzyme release have been observed following both intratracheal instillation and inhalation exposure to various dusts. Further, inhaled quartz and diesel exhaust exposure resulted in elevated protein levels, increased numbers of PMN's, as well as increased levels of cytoplasmic and lysosomal enzymes. The elevated levels of LDH in the lavagate are indicative of cell damage, while increased levels of total protein suggest that there was transudation across the alveolar-capillary barrier. The increase in LDH, ß-GLU and

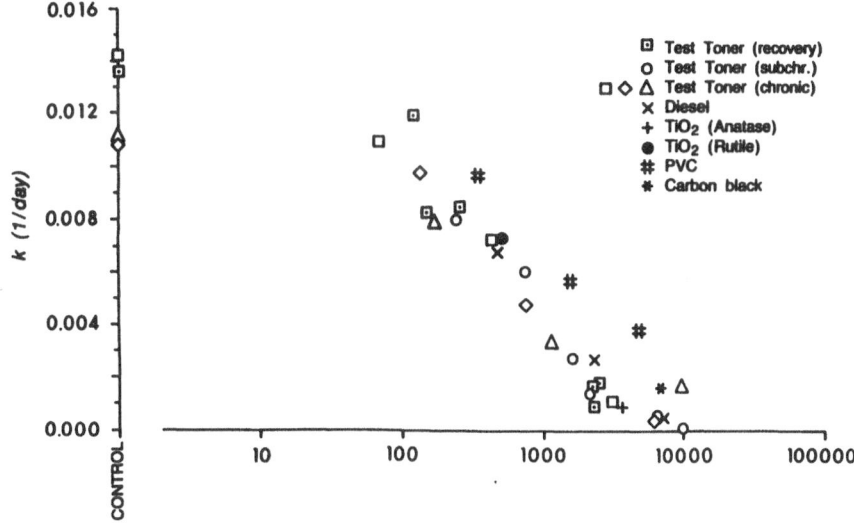

Figure 3. Retained dust volume in lungs (nl).

219

TABLE 4. Subchronic Inhalation Study of Toner-2 in Rats[a]

Exposure group Toner-2 (mg/m³)	Lung weight (g/lung) Mean ± SD	Toner retention (µg/lung) Mean ± SD	Alveolar clearance halftime (days)	PMN (Percent)	LDH (U/l) Mean ± SD	ß-Gluc. (U/l) Mean ± SD
Control	0.709±.033		66	1.1	32±6	0.10±.04
Toner Low-1	0.699±.067	74 ±58	69	1.4	34±8	0.09±.04
Toner Med-4	0.677±.047	131 ±31	68	1.2	28±3	0.10±.03
Toner High-16	0.700±.053	795±123	112**	8.9*	42±9	0.17±.05
Toner Very high-64	0.704±.042	4135±675	525**	36.3**	137±23**	1.29±.23**
Control Toner-1 High-16	0.711±.048	735±135	119**	7.2*	43±11	0.17±.09
Toner High-12d		161±64		7.2*		
Toner Very High-12d		945±205	105**	25.0**	55±8*	0.32±.06*

[a] Results after 90 days of exposure, except clearance, days 14-91 post-exposure; *P <0.05; **P <0.01

TABLE 5. Generic Response in Particle Inhalation Studies

Test material	TiO$_2$	Volcanic Ash	Fly Ash	Petroleum-Coke	Diesel	PVC	Toner
RESPONSE							
Increased lung weight	+	+	+	+	+	+	+
Disproportional increase test material retention	?	+	+		+	+	+
Decreased dust clearance		+	+		+	+	+
Chronic inflammatory process		+		+	+	+	+
Changes in pulmonary mechanics		+			+		+
-Increase no. part. laden macrophages	+	+	+	+	+	+	+
-Increased particles in interstitium	+	+	+	+	+	+	+
-Septal thickening	+	+			+	+	+
-Lipoproteinosis	+	+			+		+
-Fibrosis	+	+		+	+	?	+
-Tumors	+	?		+	+	No	No

The responses specifically reported as present are designated by the (+). Equivocal findings reported are indicated by (?). Reported negative findings are indicated by (No), whereas, responses not specifically reported by the investigators are left blank.

221

TABLE 6. Some characteristics of lung overload

Chronic inflammatory process

Decreased macrophage mediated alveolar clearance

Disproportional increase in lung retention of particles

Changes in pulmonary ventilation and elastic responses

Increased lung weight

Increased number of particle laden macrophages

Increased number of particles in the interstitium

Alveolar septal thickening

Alveolar lipoproteinosis

Alveolar fibrosis

Increased incidence of lung tumors

protein seen at the toner high exposure level at 15, 21 and 24 months were very similar to that observed upon 6-months high level of exposure to diesel exhaust (Henderson *et al*, 1988).

The slight increase in PMNs seen in both the toner middle and TiO_2 exposure groups, was not accompanied by appreciable changes in the level of cytoplasmic or lysosomal enzymes. The comparable response for all of the measured parameters at 15,

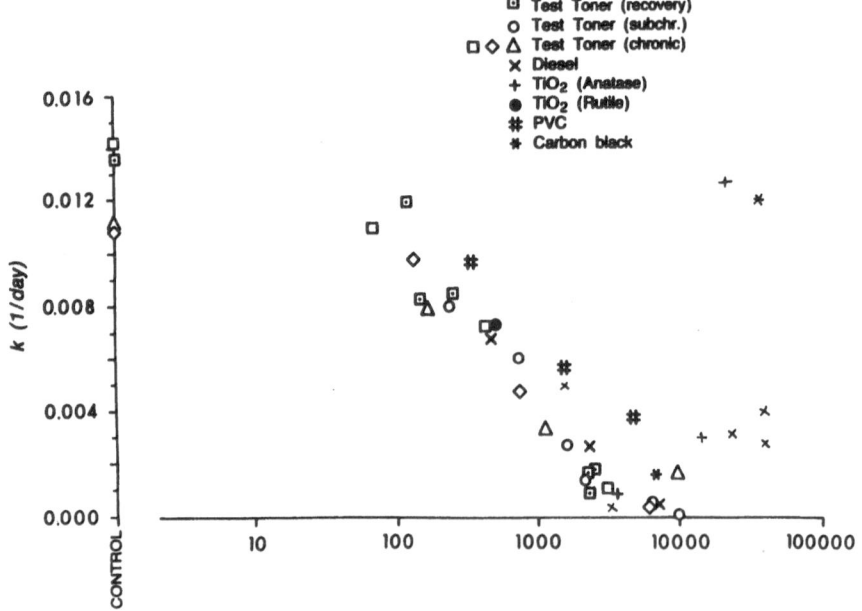

Figure 4. Retained dust volume in lungs (nl).

21 and 24 months of exposure indicate the persistent nature of the inflammatory process for both SiO_2 and toner high exposed groups. We believe that this is not a late occurring response, and it is likely that similar results would have been obtained, had the measurements been made earlier in the study. The changes observed in the crystalline silica group; specifically the decrease in total cell number is due to the known cytotoxicity, while the large increase in the number of PMN, and levels LDH and ß-glucuronidase are consistent with the reported inflammatory response of crystalline silica (Henderson *et al*, 1985). The observed response (LDH, ß-GLU) was 3-4 times higher in the silica than in the toner high exposure groups. If we merely consider that there was 10-18 times more test toner than crystalline silica retained in the lungs at comparable times, then the inflammatory potency of the silica is calculated to be 30-72 times higher.

In interpreting the results of these investigations, it is important to note that, at the highest concentration of toner the rats had lung burdens greater than that which would have been predicted from lung burdens found at the lowest concentration of toner used in this investigation. The observed lung burden ratio for the low:medium:high exposure levels was 1:9:72 at the termination of the study the ratio should be 1:4:16 on theoretical grounds. Similarly, in the "recovery study", at the end of the exposure; the observed lung burden ratio in the low:high exposure groups was 1:7.5; and at the end of the investigation 1:22, compared to the theoretical 1:4 ratio. This disproportionality is primarily associated with the retarded clearance of particles from the lungs upon prolonged exposure. The relationship between the rate of alveolar macrophage mediated particle clearance and the retained pulmonary burden of toner is illustrated in Figure 1. Impaired macrophage mediated alveolar clearance becomes apparent between 100-1000 μg of retained toner dust in the lungs (Figure 1).

Similar observations, consisting of disproportional retention, decreased alveolar clearance, increased lung weight and inflammatory changes at the highest exposure level, were obtained in the sub-chronic inhalation study of an alternate toner formulation (toner-2), containing a different type and level of carbon black and a quaternary ammonium salt (Table 4). These results indicate that the prior findings are not restricted to a single toner formulation but may be typical for relatively benign materials.

D.   Extension to Other Materials and "LUNG OVERLOAD"

The non-specific nature of the responses obtained in toner studies raised the question : Do all particulates, benign or otherwise, produce a common spectrum of pulmonary responses, with the more highly toxic materials simply eliciting their common responses at lower doses, along with any material-specific effects?

A summary of several inhalation toxicology studies published during the last decade is shown in Table 5. There is a striking similarity in findings reported in these studies, with those from the toner studies. In the Fischer 344 rat, significant changes are associated with a lung burden of about 1 mg of a persistently retained dust per gram of lung. At about 10 mg of dust per gram of rat lung, dust clearance appears to virtually cease.

It would seem logical to compare the effects of insoluble dusts of differing density is the volume rather than the mass of material retained in the lungs (Morrow, 1988; Morrow & Mermelstein, 1988). Impaired macrophage mediated alveolar particle clearance becomes apparent between a pulmonary burden of 100-1000 nl of retained dust in the lungs. The similarity in experimental observations reported in the inhalation evaluation of particles varying in chemical composition and acute toxicity suggests that there is likely to be a common feature or cause to explain their behaviour. Our view is

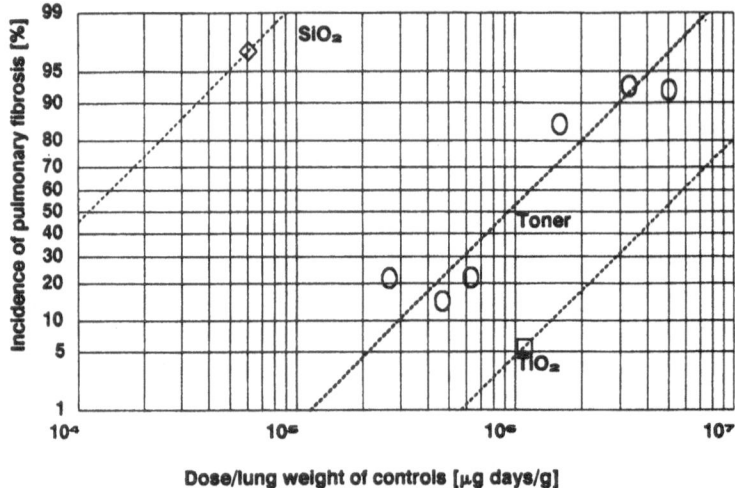

Figure 5. Log-probability plot of the dose (time integrated retained mass) of toner per gram of lungs of concurrent controls and the incidence of pulmonary fibrosis.

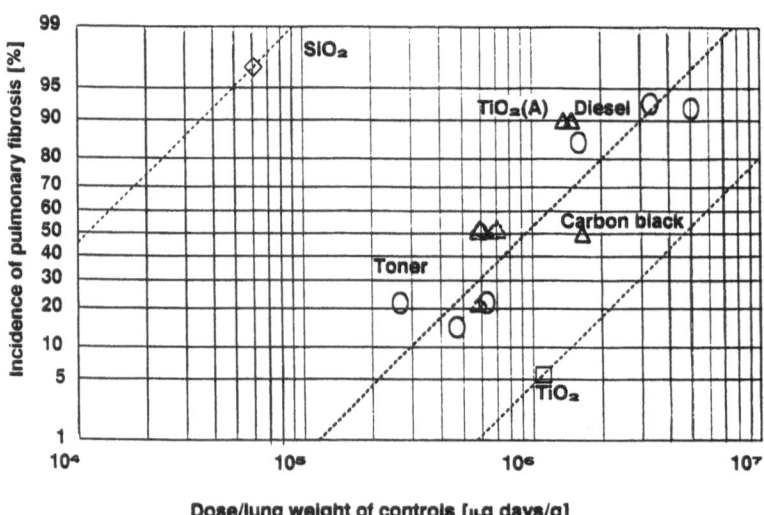

Figure 6. Log-probability plot of the dose (time integrated retained mass) of toner per gram of lungs of concurrent controls and the incidence of pulmonary fibrosis. The filled triangles represent the incidence of fibrosis when diagnosed from H & E stained lung tissue slides, while the open triangles indicate projected results from the specially stained Masson-trichrome lung tissue slides.

TABLE 7. Inhalation Study of Diesel Exhaust, Carbon Black & TiO$_2$ in Wistar Rats

Exposure group	Total dust conc. (mg/m^3)	Retained mass (mg/lung)	Retained volume (μl/lung)	Alveolar clearance halftime (days)	Inflammation
Control air	0	0	0	61	No
Diesel exhaust	0.8$\pm$0.5	0.95$\pm$0.25	0.47	109 [**]	Mild
Diesel exhaust	2.4$\pm$1.2	4.7 $\pm$1.2	2.3	292 [**]	Moderate
Diesel exhaust	6.8$\pm$1.6	14.4$\pm$2.3	7.2	3050 [**]	Severe
Carbon black	7.4$\pm$1.5	13.7$\pm$2.0	6.9	472 [**]	Moderate
TiO$_2$ (Anatase)	7.2$\pm$1.2	14.2$\pm$2.2	3.7	1222 [**]	Severe

Results after 4.5 months of exposure, 95 hrs/week; [**]$P < 0.01$

Data from: Mihle et al., (1990b) Dust overloading of lungs, *J. Aerosol Med.* (In press).

that lung overload is responsible for and provides the mechanistic basis to explain the observed behaviour of these different particles in the lungs.

The term "lung overload" refers to the generic changes observed in response to excessive amounts of any dust retained in the lungs for a prolonged time interval. This phenomenon should be considered similar to other toxicological overload situations where the normal detoxification pathways are saturated. A comprehensive discussion of the possible mechanisms to explain lung overloading has been previously published Morrow (1988). However, at dust volumes exceeding 10,000 nl. the previously described relationship is no longer valid. At these levels, the breathing pattern of the animals is altered and the clearance behaviour becomes unpredictable.

Additional insights are gained from a recently reported experiment at the Fraunhofer Institute for Toxicology in which diesel engine exhaust is being evaluated at three exposure levels while carbon black and TiO$_2$ are each being evaluated at a single exposure level Muhle et al., (1990b). In this investigation a daily exposure interval of 19 hours is used. The interim results after 4.5 months of exposure indicate evidence for lung overload. The relationship between alveolar clearance and retained dust in the lungs for these three materials follows the previously described behaviour, and the early data points fall on the same line as described for toner particles previously.

E. Dosimetry

Kuschner (1987) reported that lung cancers induced by asbestos exposure are preceded by fibrosis. He cited a study by Suzuki and Selikoff (1986), involving 356 cases of lung cancer in asbestos insulation workers, where all but one case showed interstitial fibrosis, with the majority being graded as moderate to severe fibrosis. The recognition of the widespread and generic effects of lung overloading, coupled by the report of Kuschner (1987) that lung carcinoma associated with asbestos exposure is a consequence of scarring within the lung, focussed our attention to the importance of this problem. It appeared to us that if these processes are indeed related, then one should be able to prevent the occurrence of the more serious end-points by proper recognition of earlier pulmonary changes.

We calculated the cumulative dose (time integrated retained mass) of toner in the various investigations which resulted in a diagnosis of fibrosis and displayed the results on a log-probability plot (Figure 2). The diagnosis of fibrosis was made from Masson-tri-chrome stained histopathology slides of the lungs. The results clearly show that there is an obvious correlation between the dose of material in the lungs and the onset of pulmonary fibrosis.

It is instructive to compare the pulmonary behaviour of the test toner with that of TiO_2 the nuisance dust and SiO_2 (quartz) the material used as positive control for fibroge-nicity (Figure 2). For determining the relative fibrogenic potency of these three materials, the assumption of a common fibrogenic mechanism was made. Thus the slopes of the dose response curves are assumed to be parallel, and the slope of the toner line can be applied to the single point TiO_2 and SiO_2 measurements. This assumption appears to be most reasonable for toner and TiO_2, since the fibrogenic properties of both dusts have been demonstrated only under conditions of lung overloading (Muhle et al., 1990b; Lee et al., 1985). While the methodology applied indicates that the fibrogenic potency of TiO_2 and toner are comparable (1:5), it has not taken into account the almost four-fold differ-ence in their densities. In contrast, quartz is fibrogenic at much lower lung burdens than the other two materials, so that expectations would be for a steeper dose response curve. Consequently, the assumption made for our estimations only, is very likely to grossly underestimate the fibrogenic potency of crystalline silica. Using a point-to-point basis for comparison, the respective doses producing 100 percent (or 97%) incidence leads to the same potency ratio of 1:103 for toner versus quartz. Furthermore,, the use of a 100% response point for SiO_2 which was already found after the first serial sacrifice at 9-months, does not take into account the distinct likelihood that a lower SiO_2 dose would produce the same 100% incidence. Hence all the methods of determining relative fibroge-nicity of SiO_2 at our disposal are prone to underestimate its potency. In summary, the fibrogenic potency of toner and TiO_2 are comparable, while that of the silica is at least two orders of magnitude higher.

If various dusts retained in the lungs for excessive time intervals exert their effects leading to pulmonary fibrosis by a common mechanism, then these materials also should follow the above dosimetric relationship. Accordingly, we calculated the cumulative dose of diesel exhaust particles, carbon black and titanium dioxide in the lungs, respectively, and predicted at what time interval (serial sacrifice) fibrosis should be detectable. The preliminary results for this ongoing investigation indicate, that the relationship between the cumulative dose of these three different materials and the onset of pulmonary fibrosis also follows the above described relationship. Two obvious caveats should be noted. First; in this investigation, Wistar, not Fischer rats were used, and the two strains of animals differ in size. Further, the diagnosis of fibrosis in this study was made on H & E stained lung tissue slides, rather than the more sensitive specially stained Masson tri-chrome slides. It is expected that with Masson trichrome stained slides, higher incidence of fibrosis would be detected at the same time interval, or fibrosis may be detected earlier in the study. Restaining of the slides with the special stain is currently underway. This approach is being extended to other dusts and fibers. Validation of the above concept may eventually lead to a more rigorous specification of the permissible human exposure limits for various dusts.

SUMMARY

The information presented, clearly illustrates that neither the development of pulmonary fibrosis nor the onset of lung tumors is unique to fibers. Since high pulmonary doses of both benign and isometric particles and fibers may results in a spectrum of pulmonary changes/lesions, this possibility must be kept in mind in the design and interpretation of the inhalation investigation of such materials. Obviously this is an important point to keep in mind in the discussion of the possible mechanisms in fiber carcinogenesis.

A dosimetric approach has been developed which correlates the dose, time integrated retained quantity material, with the onset of pulmonary fibrosis in rats. Extension of this relationship to other dusts and extrapolations of the results to humans are underway. Validation of these concepts may permit, for the first time, a scientific basis and specification of permissible human exposure limits for respirable dusts.

REFERENCES

Bellmann,B., Muhle,H., Creutzenberg,O., MacKenzie,J., Morrow,P. and Mermelstein,R. (1991) Lung clearance and retention of toner, utilizing a tracer technique during chronic inhalation exposure in rats . *Fundament. Appl. Toxicol.* (in press)

Butler,M.A., Evans, D.L., Giammarise,A.T., Kiriazides,D.K., Marsh,D., McCoy,E.C., Mermelstein,R., Murphy,C.B. and Rosenkranz,H.S. (1982) Application of the Salmonella assay to carbon blacks and toners. In Polynuclear aromatic hydrocarbons. Seventh international symposium on formation, metabolism and measurement. Cooke,M.W. and Dennis,A.J. (eds.) Battelle Press. pp. 225-241.

Henderson,R.F., Benson,J.M., Hahn,F.F., Hobbs,C.H., Johns,R.K., Mauderly,J.L., McClellan,R.O. and Pickerell,J.A. (1985) New approaches for the evaluation of pulmonary toxicity: Bronchoalveolar lavage fluid analysis, *Fundament. Appl. Toxicol.* **5**:451-458.

Henderson,R.F., Pickrell,J.A., Jones,R.K., Sun,J.D., Benson,J.M., Mauderly,J.L. and McClellan,R.O. (1988) Response of rodents to inhaled diluted diesel exhaust: Biochemical and cytological changes in bronchoalveolar lavage fluid and in lung tissue. *Fundament. Appl. Toxicol.* **11**:546-565.

Kuschner,M. (1987) The effects of MMMF on animal systems: Some reflections on their pathogenesis. *Ann. Occup. Hyg.* **31**:791-797.

Lee,K.P., Trochimowicz,H.J. and Reinhardt C.F. (1985) Pulmonary Response of Rats Exposed to Titanium Dioxide (TiO$_2$) by Inhalation for Two-Years. *Toxicol. Appl. Pharmacol.* **79**:179-192.

Morrow,P.E. (1988) Possible mechanisms to explain dust overloading of the lungs. *Fundament. Appl. Toxicol.* **10**:369-384.

Morrow,P.E. and Mermelstein,R. (1988) Chronic Inhalation Toxicity Studies: Protocols and Pitfalls. In: "Inhalation Toxicology. The Design and Interpretation of Inhalation Studies and Their Use in Risk Assessment". Springer-Verlag. pp. 103-117.

Muhle,H., Bellmann,B., Creutzenberg,O., Fuhst,R., Kilpper,R., Koch,W., MacKenzie,J.C., Morrow,P., Mohr,U., Takenaka,S. and Mermelstein,R. (1990a) Subchronic Inhalation study of toner in rats. *Inhal. Toxicol.* **2**:341-359.

Muhle,H., Creutzenberg,O., Bellmann,B., Heinrich,U. and Mermelstein,R. (1990b) Dust Overloading of the Lungs: Investigations of Various Materials, Species Differences and Irreversibility of Effects. *J. Aerosol Med.* **3**: Suppl. 1, 111-128.

Muhle,H., Bellmann,B., Creutzenberg,O., Dasenbrock,C., Ernst,H., Kilpper,R., MacKenzie,J.C., Morrow,P., Mohr,U., Takenaka,S. and Mermelstein,R. (1991) Pulmonary response to toner upon chronic inhalation exposure in rats. *Fundament. Appl. Toxicol.* (in press).

Morrow,P.E., Muhle,H., Mermelstein,R. (1991) Chronic inhalation study findings as a basis for proposing a new occupational dust exposure limit. *J. Am. College Toxicol.* (in press).

Suzuki,Y and Selikoff,I.J., (1986) Pathology of lung cancer among asbestos insulation workers. *Feder. Proc.* **45**:744.

CHEMICAL AND PHYSICAL PROPERTIES

2. Minerology

ASSOCIATION OF TREMOLITE HABIT WITH BIOLOGICAL POTENTIAL: PRELIMINARY REPORT

R.P. Nolan, A.M. Langer, G.W. Oechsle, J. Addison[1]
and D.E. Colflesh[2]

Environmental Sciences Laboratory
Brooklyn College of the City University of New York
Brooklyn, New York, U.S.A.

[1]Institute of Occupational Medicine, Ltd.
8 Roxburgh Place
Edinburgh EH8 9SU, Scotland

[2]School of Medicine
State University of New York at Stony Brook
Stony Brook, New York, U.S.A.

INTRODUCTION

Tremolite is an amphibole mineral which occurs naturally in three distinct morphological forms or mineral habits. It may occur as asbestos, splintery fibres or in massive crystalline deposits. Mineral habit is determined at the time of its crystallization and conversion to a different habit would require re-crystallization. The massive crystalline deposits, on crushing may yield elongated fibrous-looking particles referred to as acicular cleavage fragments (Langer *et al.*, 1979). Tremolite's ability to form these three habits imparts it with a range of physico-chemical properties.

The health effects from the occupational exposure to tremolite has been evaluated epidemiologically among vermiculite and talc workers (Reger & Morgan, 1990). Case reports have associated both pleural calcification and mesothelioma with environmental exposure to tremolite in Turkey, Cyprus and Greece. Experimental studies using various routes of administration in different animal species and *in vitro* assays have been used to compare tremolite specimens. These studies clearly indicate that tremolite possesses a range of biological activities, which should be expected considering the range of physico-chemical properties associated with the mineral's different habits of crystallization. The preliminary results of a comparison of several tremolite specimens which produced this range of biological activities are described in this paper.

EPIDEMIOLOGICAL AND CLINICAL STUDIES OF POPULATIONS INDUS-TRIALLY EXPOSED TO TREMOLITE

The pulmonary hazards of fibrous tremolite exposure were evaluated among a group working in a vermiculite exfoliating plant. Twelve cases of benign pleural effusion occurred in the group over a 12 year period. The cause for this unusually high number of cases could not be established and a study was designed to determine the prevalence of pulmonary abnormalities in these workers. The vermiculite ore used in the plant was primarily obtained from Montana, which was reported to contain at least 0.006% - 0.41% fibrous tremolite (Banks, 1980; Lockey et al., 1984). The study protocol involved a review of the industrial hygiene data, administration of a modified American Thoracic Society (ATS) respiratory questionnaire, physical examination and determination of single-breath diffusion capacity.

The study population consisted of 513 current employees with a history of exposure to vermiculite containing tremolite. The control group was made up of employees who were unexposed to dust. About 44% were current smokers and 20% ex-smokers with no significant differences among the exposure groups. The medical component of the cross-sectional epidemiological study was correlated with exposure by job category, cumulative fibre exposure and time from first exposure to the vermiculite containing fibrous tremolite.

A statistically significant association was found between cumulative fibre exposure and symptoms, i.e., shortness of breath with wheezing, dyspnea on exertion and pleuritic chest pain. The radiographic survey involved 501 of the 513 individuals in the exposed group. Pleural and/or parenchymal changes were found in 4.4%. Eleven showed costo-phrenic angle blunting, and eleven showed pleural abnormalities. The mean cumulative exposure for an abnormal X-ray was 8.74 ± 11.71 (f/ml) x yr.

The health effects of exposure to vermiculite containing tremolite on miners and millers in Montana has been studied (McDonald et al.,(1986a,b; Amandus et al., 1987a,b,c). The two studies were carried out separately, though in parallel, and each involved a cohort mortality study and a cross-sectional radiographic survey. Slightly different criteria were used to define each cohort and the McDonald cohort contained 406 men with 165 deaths and the Amandus cohort contained 575 men with 161 deaths.

Each research group used historical air samples to estimate an exposure index for each member of the cohort. The older measurements were all made with the midget impinger and a conversion was made from million particles per cubic foot (mppcf) to approximate the fibres per milliliter (f/ml). The exposures in the dry mill, before the installation of dust control equipment in 1964 were estimated by McDonald and co-workers (1986a) at ≈ 100 f/ml and Amandus and co-workers (1987a) at ≈ 168 f/ml and 1965 to the close of the dry mill in 1974 ≈ 20 f/ml and ≈ 33 f/ml respectively. These were the highest exposures except for sweeping the floor in the dry mill which was $\approx 20\%$ higher.

The McDonald cohort has an SMR for total mortality of 1.17 with 23 respiratory cancer cases (SMR = 2.45) and 4 mesotheliomas (3 pleural and 1 peritoneal). The SMR for total mortality of the Amandus cohort was 1.10 with 20 respiratory cancer cases (SMR = 2.23) and 2 mesotheliomas. The lung cancer SMR for more than 20 years since hire and all exposure levels for the McDonald and Amandus cohorts were 2.42 and 2.79 respectively. Both cohorts had an SMR of ≈2.5 for non-malignant respiratory disease.

The mortality experienced in Montana was compared to miners and millers of vermiculite in the Enoree region of South Carolina where the ore contains trace amounts of fibrous tremolite (McDonald et al., 1988). This cohort was made up of 194 men. For men with >15 years since onset of exposure, 51 deaths had occurred. The SMR for total mortality was 1.17, which was similar to Montana. All the air samples, except one, analyzed by phase contrast microscopy (using the NIOSH reference method) had mean concentrations <0.01f/ml (N=58). The analysis of the same samples by analytical transmission electron microscopy were higher and the mean concentrations ranged from 0.01 - 0.32 fibres >5 μm in length/ml. The average fibre diameter was 1.1 μm with an average length of 12.7 μm. Less than half the fibres sized had elemental compositions consistent with the tremolite-actinolite series. Of the 51 deaths, four were from respiratory cancer (SMR = 1.21), 3 of 4 lung cancers in lowest exposure group <1(f/ml) x yr with no mesotheliomas or pneumoconiosis deaths.

ENVIRONMENTAL EXPOSURE IN TURKEY, CYPRUS AND GREECE

A series of tumours of the lung and pleura in southeast Turkey have been studied retrospectively in the Diyarbakir Chest Hospital (Yazicioglu et al, 1980). Between 1968 and 1976, 177 malignant lung cancers and 44 pleural tumours were admitted to the chest hospital. The geographic distribution of the tumours suggested an etiology due to asbestos exposure. The incidence of pleural neoplasms and pulmonary tumours were 11.4 times higher and 2.5 times higher respectively in areas where asbestos was extensively used compared to areas with no asbestos.

Of the 44 mesotheliomas found in the group of 221 malignant tumours, ten (6 females, 4 males) occurred among the 20-40 year old age group. The early age at which these tumours appeared, and the near equal distribution between the sexes, indicates the exposure was environmental and began in early childhood. Interestingly, many cases of benign pleural effusion were found as well. A similar observation was later reported (Lockey et al., 1983) among the workers in a vermiculite exfoliation plant in the United States.

A radiographic cross-sectional survey of 7,000 individuals revealed that 6.6% (461) had pleural thickening and calcification, of the total survey group 1.5% (103) had interstitial pulmonary fibrosis. By 65 years of age, 50% of this group would have radiographic evidence of pleural abnormalities. The exposure continues from birth as long as the individual remains in this environment, which may explain why the changes occur so early and are very extensive. The clinical effects of these childhood exposures are found years later in individuals who moved away from these asbestos areas. This feature was noted by Constantopoulus and co-workers (1987b) in their observations pertaining to geographic distribution of plaques and tumours in Greece.

The 1978 report attributed the malignant tumours, particularly those of the pleura, to be numerous outcrops of tremolite asbestos, which was used locally to make a whitewash or stucco for the walls, floors and roofs of the houses (Yazicioglu *et al.*, 1980). The whitewash contained fibrous tremolite and the non-fibrous minerals talc, chlorite and antigorite/lizardite. Although occasional chrysotile fibres were found in the environment, the investigators attributed the pleural reactions, pulmonary fibrosis, and the malignant tumours of the lung and pleural in the Cermik region to the tremolite asbestos.

Asbestos related diseases of the chest, including mesothelioma, have also been reported in the small Anatolian Village of Caparkayi in Turkey (Baris *et al.*, 1988a, b). Four cases of pleural mesothelioma were reported in a population of 425 over a three year period. All of the tumours occurred in women between 26 and 40 years of age. Again, the tumours occurred at a young age and in women indicating the non-occupational nature of the exposure.

In a radiographic cross-sectional survey of 167 individuals over 20 years old from the village, 63 abnormalities were found. Due to a migration of the younger people, the village population has a higher percentage of older individuals than would generally be expected (51% are over 20 years). Approximately 15% of the 167 individuals surveyed had calcified pleural plaques, interlobar fissure thickening and/or diffuse interstitial fibrosis.

Although no asbestos mine is near the village, a commonly used white stucco was described by Baris and co-workers (1988b) as "rich in tremolite asbestos including some very fine fibre." The report indicates that the high incidence of mesothelioma and some of the pleural and parenchymal abnormalities in the village are associated with exposure to tremolite fibres.

Located in the central mountains of the island of Cyprus is a large chrysotile mine which has been in commercial operation since 1904. Initially it was thought that the site would provide an opportunity to study human mesothelioma from exposure only to chrysotile. The first mesothelioma identified was found in a women who had never worked in the mine, although she lived in a village nearby. The lung tissue in this case contained asbestos bodies and amphibole asbestos. Fourteen cases in total were reported between the onset of the study in 1969 and March, 1986. No tremolite was found in the chrysotile specimen taken from the mine, although tremolite and chrysotile were found to be present in an environmental dust sample taken from the roof eaves of the houses (McConnochie *et al.*, 1987).

Of the 14 cases of mesothelioma identified, 7 were confirmed by a panel of pathologists. In one case the diagnosis was in doubt, and in another the individual may have had prior exposure to asbestos in South Africa. Eight cases were either chrysotile miners or the wives of miners, and two were residents of mining communities, leaving only two cases with environmental exposure. Examination of the lung tissue burden of both human and sheep (from within 5 miles of the mine) identified chrysotile and tremolite. The tremolite found was in a form that included long, thin fibres having a similar size distribution to crocidolite.

Further study has positively identified 13 cases of mesothelioma, 5 of which occurred in persons unconnected with the local asbestos mine. A stucco used in the region contained fine fibrils of chrysotile and long, thin tremolite fibres. The reports indicate that the distribution of tumours and the naturally occurring tremolite asbestos, particularly its use as stucco, suggests the mine is not the major source of disease (McConnochie *et al.*, 1989).

Six deaths occurring from malignant pleural mesothelioma have been reported among residents in the villages of Milea, Metsovo, Anilio and Votonosi in northwest

Greece (Constantopoulos *et al.*, 1987a). These six deaths occurred among seven mesotheliomas diagnosed in the region. Three mesotheliomas occurred in males and four in females. Bilateral pleural plaques, pleural thickening, restrictive lung function and mesotheliomas constitute the cluster of disease referred to as Metsovo lung (Constantopoulos *et al.*, 1985). Before 1940 virtually all of the inhabitants of the area painted their homes with a whitewash containing tremolite asbestos (Langer *et al.*, 1987). Exposure to the tremolite asbestos contained in the whitewash has been associated with Metsovo lung. A hypothesis for expecting such findings to be regional and occur in other parts of the world has been proposed (Constantopoulos *et al.*, 1987b).

EXPERIMENTAL ANIMAL STUDIES

The carcinogenic properties of four tremolite specimens were determined through single intrapleural injections into Syrian golden Hamsters (Smith, 1974; Smith *et al.*, 1979). The model was validated by determining the carcinogenicity of the four commercial asbestos varieties. The greatest proportion of tumours were produced with crocidolite, although, all the commercial asbestos specimens tested produced tumours, extensive pleural adhesions and densely fibrotic lesions. Two of the four tremolite specimens produce no tumours.

These results were compared to a specimen of tremolite talc which contained 50% fibrous tremolite, 35% talc (about 25% of which was itself fibrous or rolled), 10% antigorite and 5% chlorite. By light microscopy the fibres ranged from 2.5-16.5 μm in length with an average length of 5.7 μm and diameter of 1.6 μm. Intrapleural injection of a 25 mg dose of tremolite talc produced no tumours in 50 animals. An identical mass of three chrysotile specimens, one of which was tested both heated and unheated, produced $\approx 18 \pm 2$ tumours, and 1 mg injection of UICC crocidolite produced 4% tumours (Smith, 1974).

Three additional specimens, a non-asbestos tremolite (#275), a western tremolitic talc and a tremolite asbestos, were evaluated using the same bioassay (Smith *et al.*, 1979) The specimens were found to be 95%, 90% and 95% tremolite respectively. The non-asbestos tremolite produced no tumours at doses of 10 and 25mg while the western tremolitic talc and tremolite asbestos produced $\approx 2.5\%$ and 24% tumours respectively at 10 mg and 21% and 50% tumours at the 25 mg dose respectively.

Both of the tremolite specimens which produced no tumours were from the same geological locale, although, the tremolitic talc specimen evaluated in the earlier experiment contained a significant proportion of both platy and fibrous talc, antigorite and chlorite. Extensive pleural fibrosis was produced with tremolite asbestos, less with western tremolitic talc, and very slight pleural fibrosis with the two specimens that produced no tumours. In these experiments tumour production correlated with fibrosis (Kuschner, 1987).

Two tremolite asbestos specimens, both from the same lot, were evaluated by Stanton and co-workers (1981) using a pleural implantation model in rats. The fibre diameters of these specimens were distinctly smaller than the diameters of the specimen reported by Smith, 1974. A 40 mg dose of tremolite 1 and 2, containing 55.2 x 10^6 and 27.7 x 10^6 fibres >8.0 μm in length and <0.25 μm in diameter respectively, produced a tumour incidence of 22/28 and 21/28 respectively. For both of these specimens the model predicted a 100% probability of tumour induction.

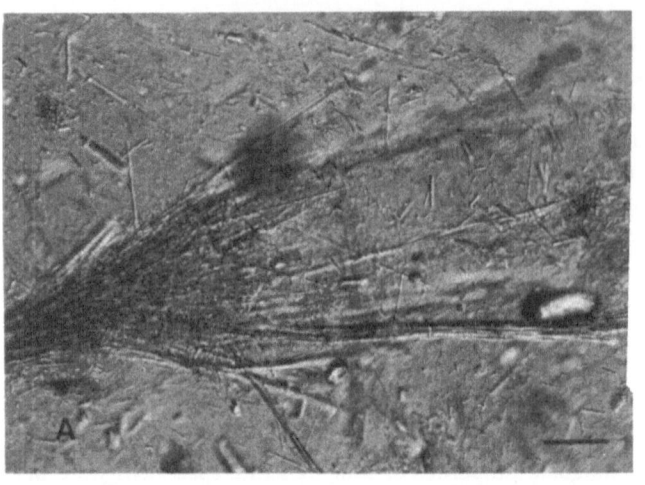

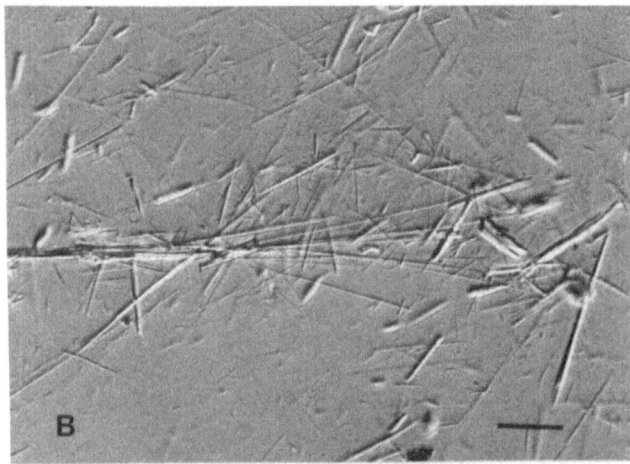

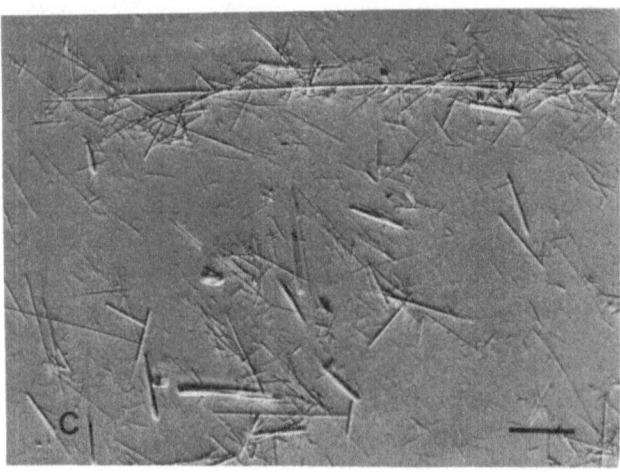

Figure 1: Light photomicrographs of tremolite asbestos using
Hoffman Interference Optics except (f), which was taken using
a Zeiss bright field lens:
(a) Whitewash from **Metsovo, Greece**; (b) South **Korea**;
(c) Beneficiated from vermiculite deposit near **Libby**, Montana

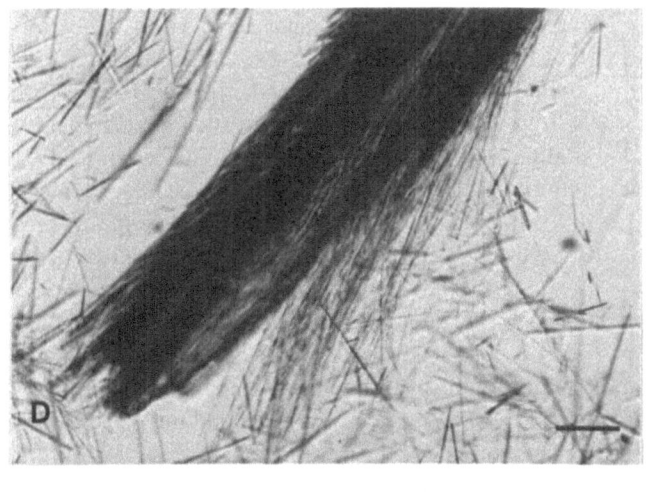

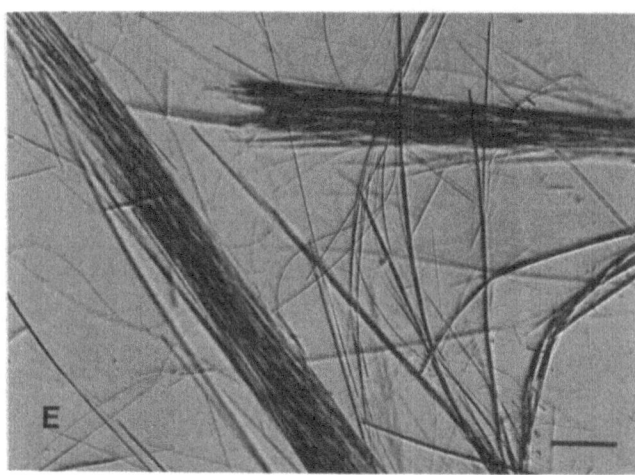

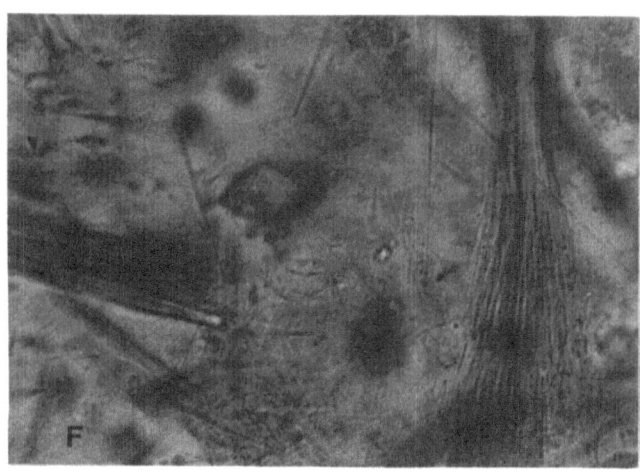

(d) **Swansea**, Wales; (e) **Jamestown**, California;
(f) **Inyo County**, California. (Bar represents 50 μm)

By intrapleural injection into rats Wagner (Wagner, 1982; Wagner *et al.*, 1982) evaluated three tremolite specimens while using either UICC crocidolite or Super-Fine Asbestos (SFA) chrysotile as positive controls. One tremolite specimen was beneficiated from a California tremolitic talc deposit by froth flotation. The starting material was $\approx 62\%$ talc and $\approx 38\%$ tremolite, and after froth flotation the material enriched to $\approx 95\%$ tremolite (the remaining minerals were talc and calcium carbonate). The 20 mg dose injected intrapleurally, containing $1,020 \times 10^6$ fibres (3.3% of which were $> 8\ \mu m$ in length and $< 1.5\ \mu m$ in diameter) which produced no tumours in thirty-one rats. The positive control, 20 mg of SFA chrysotile, produced twenty mesotheliomas in thirty-two rats.

A second tremolite specimen from Greenland, which contained 960×10^6 fibres (with no fibres $> 8\ \mu m$ in length and $< 1.5\ \mu m$ detected) in a 20 mg dose produced no tumours in 48 rats. A highly fibrous tremolite asbestos specimen from Korea contained $3,100 \times 10^6$ fibres (with 35% being $> 8\ \mu m$ in length and $< 1.5\ \mu m$ in diameter) in a 20 mg dose. This specimen produced 14 mesotheliomas in forty-seven rats. Due to a poor survival rate the positive control in this experiment, UICC crocidolite, produced only two mesotheliomas.

The tremolite asbestos from Korea has produced two mesotheliomas, two adenomas and sixteen carcinomas in thirty-nine animals using inhalation as the route of administration. The dust cloud contained $\approx 10\ mg/m^3$ with $\approx 1,600$ fibres $> 5\ \mu m$ in length/ml. At the end of the exposure of 7 hrs/day for a total of 224 days in a 12 month period, the mean tremolite lung burden of four rats was 10.8 mg at the end of the exposure period. Additionally intraperitoneal injection of 25 mg of the tremolite asbestos from Korea produced tumors in twenty-seven out of twenty-nine rats (Davis *et al.*, 1985).

Recently, six tremolite specimens were studied by the intraperitoneal injection of a 10 mg dose in thirty rats (Davis, 1990; Davis *et al.* in press). Three were tremolite asbestos specimens from California, Swansea and Korea and contained 121×10^6, 8.0×10^6 and 48×10^6 fibres $> 8\ \mu m$ in length and $< 0.25\ \mu m$ in diameter per dose, which was 0.9%, 0.4% and 0.6% of the total fibres, and produced 100%, 97% and 89% tumours respectively. The mean survival time was less than 428 days for the animals in these three exposure groups. The splintery fibres from Italy contained 58×10^6 fibres per dose $> 8\ \mu m$ in length, and approximately 1.7% were $< 0.25\ \mu m$ in diameter and produced 67% tumours with a mean survival time of 755 days. Although the non-asbestos specimens from Carr Brae and Shinness contained no fibres $> 8\ \mu m$ in length and $< 0.25\ \mu m$ in diameter, the dose did contain 134×10^6 and 17×10^6 fibres per dose ($> 8\ \mu m$) in length respectively, and produced 12% and 5.6% tumours respectively with too few tumours for a mean survival time to be calculated for the lifetime of the animals. Using the intraperitoneal model, the incidence of tumours ~10% with a 10 mg dose indicates it is unlikely that the non-asbestos tremolite from Shinness or Dornie would cause tumours by inhalation. (Ilgren & Wagner, 1991).

MATERIALS AND METHODS

Origin of the mineral specimens:
● Mixture of tremolite and richterite asbestos from **Libby**, Montana, U.S.A. obtained from W. Banks, Bureau of Mines, United States Department of the Interior. Epidemiology studies of the health effects of industrial exposure by Lockey *et al.*, 1983;

Lockey et al., 1984; McDonald et al., 1986a,b; Amandus et al., 1987a,b,c.

- Tremolite asbestos from **Inyo** County, California obtained from G.J. Gill of Cypress Industrial Minerals, Denver, Colorado, U.S.A.
- Tremolite asbestos from **Metsovo**, Greece used as a whitewash and associated with pleural mesothelioma (Langer et al., 1987).
- Mixture of tremolite asbestos and splintery fibres, Udaipur District, Rajastan State, **India**. obtained from Ward Scientific, Rochester, New York, U.S.A.
- Non-asbestos **respirable** tremolite from **Gouverneur**, New York, U.S.A. obtained from C.S. Thompson, R.T. Vanderbilt and Company, Norwalk, Connecticut, U.S.A. The same specimen (Sample No. 275) was used in experimental animal studies by Smith et al. 1979.
- Non-asbestos tremolite from **Gouverneur**, New York, U.S.A. used in NTP feeding studies, 1990 (Campbell et al., 1979, Campbell et al., 1986).
- Non-asbestos tremolite found in a vermiculite deposit in the Enoree region of **South Carolina** (McDonald et al., 1988).

Respirable fractions of the following specimens were characterized mineralogically and used in experimental animal studies (Wagner et al., 1982, Davis et al., 1985, Davis, 1990, Addison and Davis, in press, Davis et al., in press).

- Tremolite asbestos, **Jamestown**, California, U.S.A.
- Tremolite asbestos, **South Korea**.
- Tremolite asbestos, of unknown geological locale, obtained from **Swansea** Laboratory, Wales, United Kingdom.
- Splintery fibres, **Ala di Stura**, near Turin, Northern Italy.
  UICC Reference Asbestos Specimens
- Crocidolite from the Cape Province, Republic of South Africa.
- Amosite from the Transvaal, Republic of South Africa.
- Anthophyllite from Paakila, Finland.
- Chrysotile A from Zimbabwe.
- Chrysotile B, a blend of chrysotile ores from eight Canadian mines.
  The positive control for membranolysis was:
- Quartz, a commercially available silica flour referred to as Min-U-Sil 15 was obtained from the Pennsylvania Glass and Sand Company, Pittsburgh, PA.
  The inhibitors of membranolytic activity, and their sources, are as follows:
- Poly (2-vinylpyridine-N-Oxide); (2-PVPNO) was obtained from Polyscience Inc, Washington, PA. The weight average molecular weight, of the 2-PVPNO polymer, as determined by light scattering, was 276,000.
- Pyridoxal 5-Phosphate, 1.5 H_2O was obtained from Sigma Chemical Company, St. Louis, MO.

Human Membranolytic Model

The ability of each of the mineral specimens to alter the permeability of a population of human erythrocytes was determined quantitatively. The HC_{50} is the concentration of particles (given in mg/ml) required to lyse 50% of the erythrocytes in a suspension containing 1.8×10^8 cells/ml (see Nolan et al., 1981 for details).

Hydrogen Bonding Determinations

The ability of a mineral specimen's surface to bind 2-PVPNO was determined by adding various concentrations of the mineral to 20 ml of 50 μg/ml 2-PVPNO. The

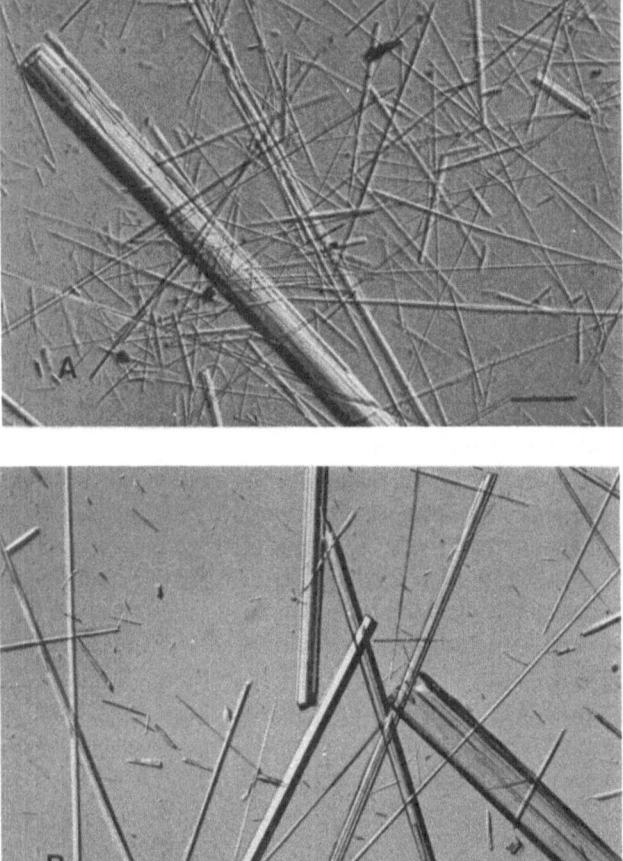

Figure 2. Light photomicrographs of splintery fibres using a) Hoffman interference optics and b) a Zeiss bright field optics:
a) Udaipur District, Rajastan State, India.
(Bar represents 50 μm)
b) **Ala di Stura, near Turin, Italy.** (Bar represents 100 μm)

suspension of mineral and polymer were sonicated using the same conditions as those for the membranolysis experiments and allowed to stand at room temperature for 60 minutes. The suspensions were vortexed occasionally to resuspend the mineral particulates. After an hour, the 2-PVPNO bound to the mineral surface was separated from the free 2-PVPNO by centrifugation at Fc $\approx$ 10,000 x g. The concentration of 2-PVPNO in the clear supernatant was determined by the absorbence of the polymer at 260nm. The amount of 2-PVPNO bound to a given amount of mineral was determined by this method.

Zeta Potential Measurements

The surface charge was approximated by determining the zeta potential using a commercially available instrument from Zeta-Meter, Inc., Long Island City, New York.

240

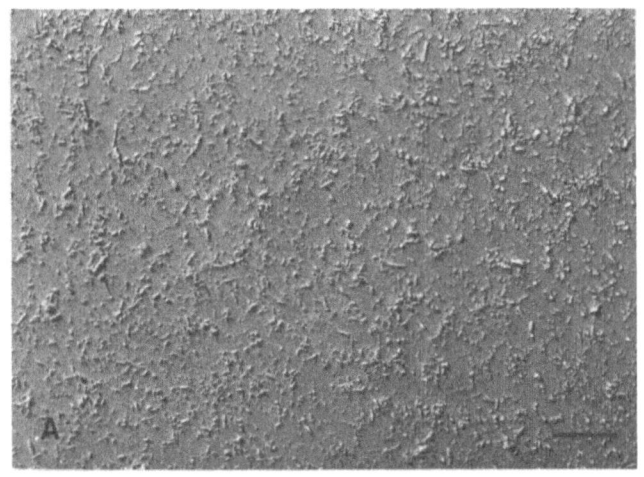

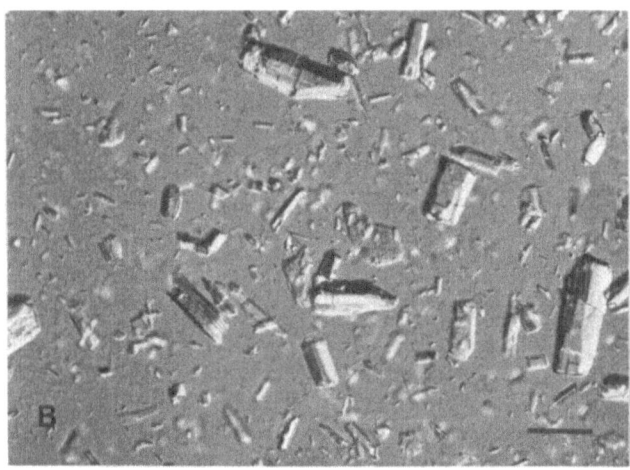

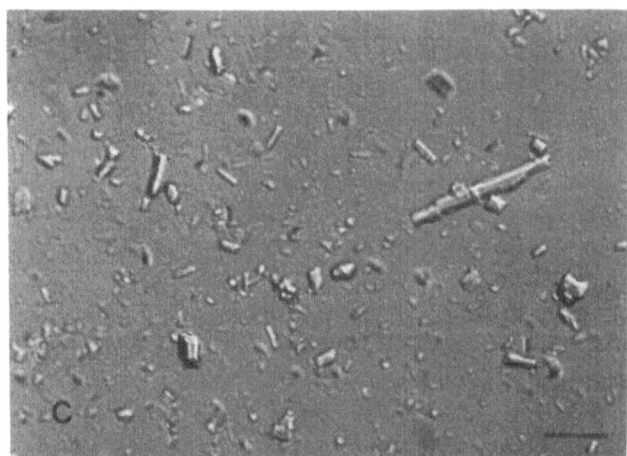

Figure 3. Light photomicrographs of non-asbestos
tremolite specimens using Hoffman interference optics:
a) **Gouverneur, New York** used by Smith *et al.*, 1979.
b) **Gouverneur, New York** used in NTP feed study, 1990.
c) Beneficiated from a vermiculite deposit, **Enoree region, South
Carolina**. (Bar represents 50 μm)

TABLE 1: Gross characteristics of minerals specimens

Metsovo, Greece: Fibres visible with unaided eye. Fibre lengths up to 1 mm, fibre widths < 0.5 mm. Fibres noted to protrude from matted clumps. Fibre color is gray-white.

Korea: Fibres visible with unaided eye. Fibre lengths up to 2 mm, fibre widths <1 mm. Few fibres visible. Much granular powder present associated with fine, matted clumps. Fibre color is light-tan.

Libby, Montana: Fibres visible with unaided eye. Fibre lengths up to 2 mm, fibre widths up to 0.5 mm. Specimen composed almost entirely of clumps of fibre. Fibre color is gray.

Swansea, Wales: Fibres visible with unaided eye. Fibre lengths up to 1 cm, fibre widths up to 1 mm. Fibre visible associated with granular powder. Fibre occurs in matted clumps. Fibre color is gray-white.

Jamestown, California: Fibres visible with unaided eye. Fibre lengths up to 2.5 cm, fibre widths up to 3 mm. Specimen composed almost entirely of fibre. Fibres appear kinked, curled and splayed. Fibre color is predominately white.

Inyo County, California: Fibres visible with unaided eye. Fibre lengths up to 2 mm, fibre widths up to 0.5 mm. Fibres are present in matted clumps. Fibre color is white.

India: Fibres visible with unaided eye. Fibre lengths up to 1 cm, widths up to 1 mm. Fibres visible as individual bundles and as components of clumps. Fibre color is gray.

Ala di Stura, Italy: Fibres visible with unaided eye. Fibre lengths up to 1.5 cm, widths up to 2.0 mm. Specimen composed almost entirely of fibres. Fibres form straight, bladed crystals. Fibre color is light-gray.

Gouverneur, N.Y., Respirable: No fibres visible in specimen. Specimen consists of granular, white powder.

Gouverneur, N.Y., NTP: No fibres visible. Specimen consists of fine, white powder of small particle size.

Enoree Region, South Carolina: No fibres visible. Specimen consists of granular powder. The powder is grey in color.

Note: Some of these characteristics were better observed with the aid of a 10 power hand lens.

Each specimen was suspended at a concentration of 100 to 250 mg per liter in veronal buffer without saline at pH 7.4. The particles were dispersed by ultrasound at 50 Watts of power for five minutes and allowed to cool back to room temperature. The electrophoretic mobility (EM) was then measured by determining the time (in seconds) required for 10 different particles to migrate 160 μm in an electric field. The field strength was varied from 10 to 20 V/cm to allow a tracking time of ≈ 3 seconds. Once the electrophoretic mobility (given in microns per sec/volts per cm) is determined, the zeta potential (ZP) can be calculated using the Helmholtz-Smoluchowski equation:

$$ZP = \frac{4\pi V_t}{D_t} \times EM$$

where:

EM = Electrophoretic Mobility at actual temperature; V_t = Viscosity of the suspending liquid at temperature, D_t = Dielectric constant of the suspending liquid at temperature;
ZP = Zeta Potential in electrostatic units.

TABLE 2. Characteristics of the Mineral Specimens by Polarized Light Microscopy

Metsovo, Greece: Asbestos fibre bundles. Fibres with splayed ends, polyfilamentous, and parallel extinction. Asbestos constitutes about 50-75% of specimen. Tremolite cleavage fragments are present, possibly constituting up to half of specimen. Other silicate minerals present. Figure 1a.

Korea: About 50% asbestos fibre (may be even less). Splintery fibres abundant ($\approx 25\%$). Cleavage fragments present as alteration products. Figure 1b.

Libby, Montana: About 90% asbestos. Polyfilamentous bundles of fibres. Parallel to near-parallel extinction, actinolite, other silicates fragments present. Figure 1c.

Swansea, Wales: About 75%-85% asbestos fibre, polyfilamentous bundles, splayed ends, parallel extinction. Some cleavage fragments ($\approx 10\%$), and splintery fibres ($\approx 5\%$). Figure 1d.

Jamestown, California: Over 90% asbestos fibre (polyfilamentous, parallel extinction, splayed ends, etc.) Some cleavage fragments ($\approx 3\%$), splintery fiber ($\approx 2\%$), trace of chlorite (?). Figure 1e.

Inyo County, California: Asbestos fibres constitute ~15-20% of specimen. Asbestos fibres exhibit polyfilamentous character, splayed ends, parallel extinction. Tremolite cleavage fragments present associated with carbonate minerals, talc, other silicates. Figure 1f.

India: Asbestos fibre constitutes $\approx 90\%$ of specimen. Asbestos fibres exhibit polyfilamentous character, splayed ends, parallel extinction. Some cleavage fragments observed, non-fibrous silicates, some splintery fibres. Figure 2a.

Ala di Stura, Italy: About 90% laths (splintery fibre), no asbestos present, all laths have parallel extinction, laths display extreme length: width ratio. Figure 2b.

Respirable, Gouverneur, N.Y.: Non-fibrous fine powder, aggregates of particles, no asbestos present, small cleavage fragments observed in talc, some talc fibres visible. Figure 3a.

NTP, Gouverneur, N.Y.: Tremolite cleavage fragments, some talc, no asbestos present, some other silicates. Figure 3b.

Enoree Region, South Carolina: Tremolite cleavage fragments, some other silicates, no asbestos present, some elongate cleavage fragments noted (aspect ratio ~10:1). Figure 3c.

Mineralogical Characterization of the Specimens

Bulk powders were available for all the specimens examined (Table 1). An aliquot of each was characterized by polarized light microscopy (Table 2). Specimens found to contain polyfilamentous fibre bundles with splayed ends exhibiting parallel extinction, were classified as asbestos (Figure 1, see Langer *et al.* this volume). Invariably, these specimens contained fibres with high aspect ratios and very narrow diameters. The specimens in which the fibres exhibited the optical properties consistent with single crystals and oblique extinction in polarized light were classified as non-asbestos (Figure 2 and 3). These specimens invariably contained very few fibres with narrow diameters.

The specimen from India contained both polyfilamentous bundles with parallel extinction, which were classified as asbestos, and high aspect ratio fibres with oblique extinction which were classified as splintery fibres, which are not asbestos. The fibres were constituted of single crystals, of larger diameter than amphibole asbestos fibrils. The specimen from **Ala di Stura** contained predominantly splintery fibres (see Figure 2). The asbestos specimens contained variable amounts of cleavage fragments and/or splintery fibres and other impurities (Table 2). The four non-asbestos specimens contained no polyfilamentous asbestos fibre bundles.

Each specimen was examined by continuous scan X-ray diffraction and transmission electron microscopy using energy dispersive spectrometry. The complete results will be published elsewhere. Each specimen, with the exception of Enoree and Inyo County, produced a diffractogram in which the major crystalline mineral phase present was an amphibole.

Membranolytic Activity of the Mineral Specimens

The membranolytic activity of the tremolite asbestos specimens varied from 1.14 $\pm$ 0.19 mg/ml for Jamestown, California to 3.06 $\pm$ 1.06 for Metsovo, Greece, a factor of ~3-fold. The average HC_{50} for the six specimens is 1.94 $\pm$ 0.75 mg/ml. The HC_{50} of the specimen from India was approximately equal to Min-U-Sil 15 (Nolan *et al.*, 1981) while the specimen from Italy contained long, rigid fibres which could not be pipetted. The non-asbestos tremolite specimens from Gouverneur, New York ranged from an HC_{50} of 1.02 $\pm$ 0.22 mg/ml for the respirable tremolite to 5.21 $\pm$ 0.11 mg/ml for the NTP specimen. The UICC reference asbestos amphiboles varied from 1.76 $\pm$ 0.28 mg/ml for amosite to 4.05 $\pm$ 1.22 mg/ml for crocidolite a factor of 2.3 fold. The two UICC chrysotile specimens had any average HC_{50} of 0.07 mg/ml which was ~28-fold and ~45-fold more active than the average for the tremolite asbestos specimens or the average for the UICC asbestos amphiboles respectively (Table 3).

Effect of Inhibitors of Membranolytic Activity and a Comparison of Surface Properties

The activity of all of the asbestos and non-asbestos tremolite specimens were inhibited by 2-PVPNO, a hydrogen bonding polymer. The dose response inhibition of non-asbestos tremolite, Gouverneur, NY and tremolite asbestos, Korea were compared to quartz. The inhibition of lysis occurred in a similar concentration range to quartz. The ability of these three specimens to hydrogen bond 2-PVPNO correlated with their membranolytic activity (Figure 4).

The zeta potential of the two non-asbestos tremolite specimens from Gouverneur, NY averaged –13.4 $\pm$ 4.1 mV, while six tremolite asbestos specimens were found to have an average zeta potential of –44.9 $\pm$ 10.4 mV (Table 3). The non-asbestos tremolite from Gouverneur, NY is primarily prismatic with very few elongated or acicular cleavage fragments and the zeta potential is 70% less than for the asbestos analogues. The difference in zeta potential indicates the charge density on the surface of the non-asbestos tremolite maybe lower than on its asbestos analog. The anion transport inhibitor, pyridoxal 5-phosphate, which is known to block positively charged sites on the erythrocyte membrane was a more effective inhibitor for the non-asbestos tremolite than for tremolite asbestos or quartz (Figure 5). These differences may be due to the non-asbestos tremolite having a lower number of ionized sites on its surface.

DISCUSSION AND CONCLUSIONS

Eleven specimens thought to be in the tremolite-actinolite series were selected for study and compared to the UICC asbestos reference specimens and quartz. For all but two of the eleven specimens either epidemiological and/or experimental animal studies were available to indicate their biological potential. The habit of each tremolite specimen was

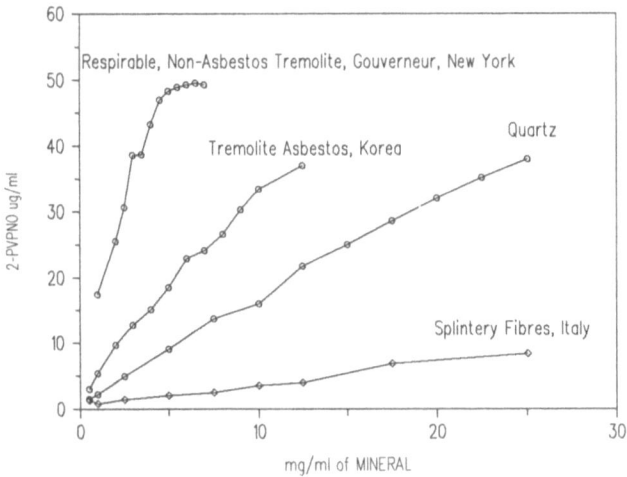

**Figure 4. A comparison of the ability of three tremolite speci-
mens having different habits to bind the hydrogen binding
polymer, 2-PVPNO.**

characterized as asbestos, splintery fibres or massive on the basis of their optical characteristics when examined by polarized light microscope (PLM). The splintery fibres and
the massive form are non-asbestos, (for the criteria used see Langer *et al.*, this volume).
The PLM observations were supported by the determination of each specimen's selected
area electron diffraction characteristics, internal structure and morphology using transmission electron microscopy.

The specimens from Metsovo, Korea, Swansea, Jamestown, Inyo and India contain asbestos in the tremolite-actinolite series. Only about half of the asbestos fibres,
analyzed by energy dispersive spectroscopy, from the vermiculite mine near Libby,
Montana had an elemental composition consistent with that series. The other half of the

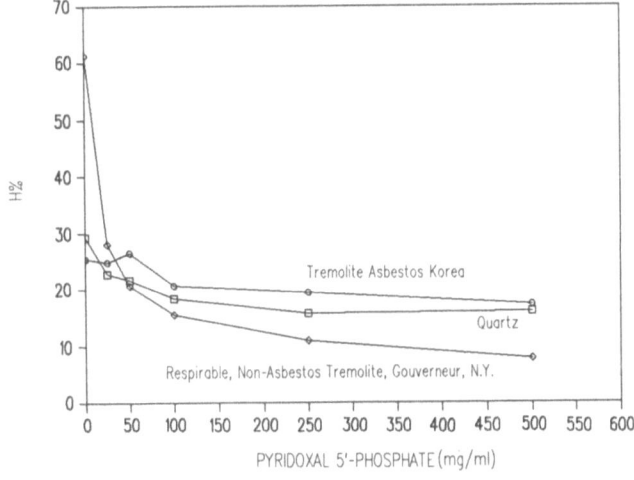

**Figure 5. The effect of the anion transport inhibitor, pyridoxal
5-phosphate on the membranolytic activity of an asbestos and
non-asbestos tremolite specimens compared to quartz.**

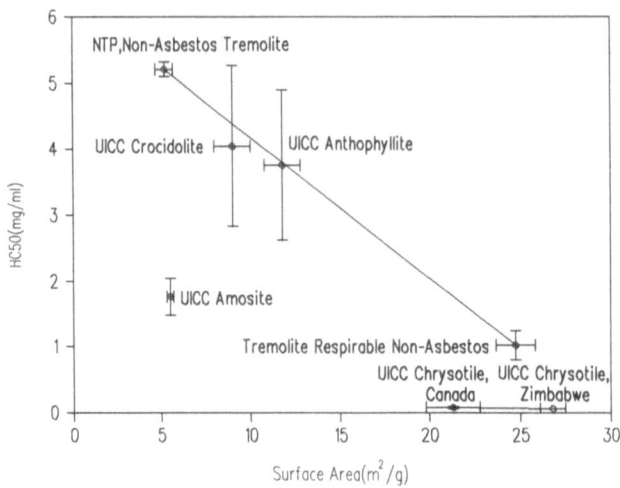

Figure 6. A comparison of the HC_{50}'s *vs* surface areas for two non-asbestos tremolite specimens and the UICC reference asbestos specimen. Values are given as mean ± standard deviation.

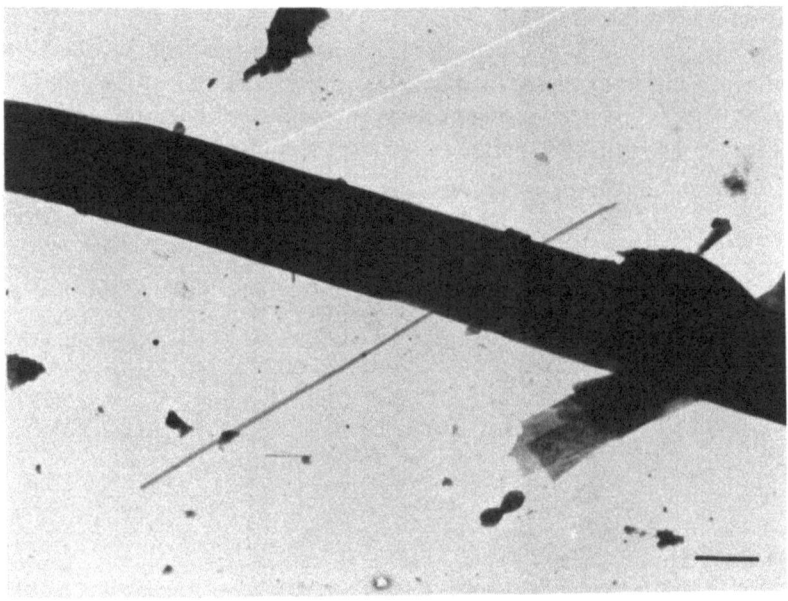

Figure 7. A transmission electron photomicrograph of the specimen from Ala di Stura, Italy. The large fibre is > 3 μm in diameter with an elemental composition consistent with tremolite-actinolite series. The smaller fibre ($\sim$17.5 μm in length and $\sim$0.´5 μm in diameter) is a magnesium iron silicate which is consistent with the cummingtonite-grunerite series.

amphibole asbestos fibres contained either sodium or sodium and potassium (Nolan & Langer, unpublished data). The fibres from Libby found to be present in the tremolite-actinolite series were predominately actinolite asbestos (Moatamed *et al.*, 1986). The specimen from Rajastan, India contained a mixture of two fibers, tremolite asbestos and splintery fibres having oblique extinction.

The Ala di Stura specimen contained predominately splintery fibres with oblique extinction. These non-asbestos fibres have diameters which vary as a function of length, *i.e.*, the long fibres tend to have larger diameters (Wylie, 1988). Also, the specimen contains a sub-population of iron magnesium silicate fibres with high aspect ratios and diameters < 0.25 μm (Figure 7) (Nolan & Langer, unpublished data). This sub-population of fibres could have contributed to the high proportion of late occurring mesotheliomas after intraperitoneal injection in rats (Davis, 1990; Davis *et al.*, in press).

The two specimens from Gouverneur, New York and the Enoree region of South Carolina had optical characteristics consistent with non-asbestos tremolite (Figure 3). No iron was detected in the two specimens from Gouverneur, New York when examined by energy dispersive spectrometry. The tremolite specimen from the Enoree region contained barely detectable iron, indicating that the three specimens have elemental compositions consistent with tremolite.

The concentration of mineral required to lyse half the erythrocytes in suspension, the HC_{50} for the seven amphibole asbestos specimens ranged from 1.14 ± 0.19 mg/ml for Jamestown to 3.06 ± 1.06 mg/ml for Metsovo with an average of 2.07 ± 0.76 mg/ml (Table 3). On average, the UICC reference amphibole asbestos specimens were less active, requiring 3.19 ± 1.25 mg/ml to lyse half the erythrocytes. The two UICC chrysotile specimens were considerably more active than any specimen studied.

The range of HC_{50}'s for the two non-asbestos tremolite specimens from Gouverneur, New York were from 1.02 ± 0.22 mg/ml for the respirable tremolite used by Smith *et al.*, 1979 to 5.21 ± 0.11 mg/ml for the specimen used in the NTP feeding studies. The respirable specimen was more active, and the NTP specimen less active, than any of the amphibole asbestos specimens studied. The two specimens differ in particle size distribution which is reflected in each having different surface areas per unit mass (Table 3). The surface area of the respirable non-asbestos tremolite is ~5 fold greater than the NTP specimen and also ~5 fold more membranolytic per unit mass.

A plot of the HC_{50}'s vs. surface area compares the two non-asbestos specimens to the UICC reference asbestos standards (Figure 6). The anthophyllite, crocidolite and the two non-asbestos tremolite specimens had similar HC_{50}'s per unit surface area. Although the amosite specimen had a surface area comparable with the NTP tremolite, the amosite was almost 3 fold more membranolytic. The UICC chrysotile specimens from Zimbabwe and Canada had HC_{50}'s of 0.06 ± 0.00 mg/ml and 0.08 ± 0.02 mg/ml and surface areas of 26.8 ± 0.7 m^2/g and 21.3 ± 1.5 m^2/g respectively. Although the surface area of the Zimbabwe specimen is comparable with the respirable non-asbestos tremolite, the chrysotile specimen is 17 fold more membranolytic. The membranolytic activity per unit surface area was greatest for chrysotile > > amosite > anthophyllite $\approx$ crocidolite $\approx$ non-asbestos tremolite were approximately the same. Although Smith and co-workers (1974) produced mesotheliomas, extensive pleural adhesions and dense fibrotic lesions by intrapleural injection of the UICC reference amphibole asbestos specimens, crocidolite and anthophyllite, comparable doses of the respirable non-asbestos tremolite produced no tumours and very slight fibrosis (Smith *et al.*, 1979).

The hydrogen bonding polymer, 2-PVPNO, inhibited the membranolytic activity of all the tremolite specimens studied. The dose-response inhibition of non-asbestos

TABLE 3. Comparison of the membranolytic activity, surface area, and zeta potential of tremolite asbestos, non-asbestos tremolite and amphibole asbestos reference samples.

Specimen	HC_{50} (mg/ml)	Surface Area Determined by N_2 Adsorption (m^2/g)	Zeta Potential (mv)
Tremolite-Actinolite Asbestos			
Metsovo, Greece	3.06 ± 1.06		−45.6
Korea	1.38 ± 0.47		−56.3
Libby, Montana	2.07 ± 1.72		−44.3 ± 2.3
Swansea, Wales	2.53 ± 0.05		−56.7
Jamestown, California	1.14 ± 0.19		−32.9
Inyo, California	1.48 ± 0.69		−33.6
India	2.81 ± 0.33		
Non-Asbestos Tremolite-Actinolite			
Respirable, Gouverneur, NY	1.02 ± 0.22	24.76 ± 1.09	−13.3 ± 5.8
NIEHS, Gouverneur, NY	5.21 ± 0.11[2]	5.20 ± 0.50	−13.5 ± 2.3
Ala di Stura, Italy	NP[1]		
UICC Reference Asbestos Specimens			
Crocidolite	4.05 ± 1.22	9.8 ± 1.0[3],	8.3 ± 1.0[4]
Amosite	1.76 ± 0.28	4.0 ± 0.1[3],	5.7 ± 0.3[4]
Anthophyllite	3.76 ± 1.14	11.8 ± 1.0[4]	
Chrysotile, Canada	0.08 ± 0.02	21.3 ± 1.5[4]	
Chrysotile, Zimbabwe	0.06 ± 0.00	26.8 ± 0.7[4]	

[1] NP- Not Pipettable
[2] Duplicate Determinations
[3] Campbell *et al.*, 1980.
[4] Timbrell, 1969.

tremolite from Gouverneur, New York and tremolite asbestos from Korea occur over a concentration range approximately the same as a standard quartz specimen. Although the two tremolite specimens have similar HC_{50}'s, the respirable non-asbestos tremolite bound significantly more polymer than the tremolite asbestos specimen (Figure 4).

At a physiological pH of 7.4, six asbestos specimens had an average zeta potential of −44.9 ± 10.4 mV while the two non-asbestos tremolites averaged −13.4 ± 0.1 mV. Surface charge heterogeneity has been previously reported between amphibole cleavage fragments and their asbestos analogues (Schiller *et al.*, 1980). The quartz surface charge can be reduced by blocking the ionized surface silanol groups and this reduced the mineral's membranolytic activity (Nolan *et al.*, 1981). One site on the erythrocyte membrane which would have the appropriate chemistry to bind an anion would be the receptor site for anion transport on the band 3 protein (Cabantchik *et al.*, 1978). The site specific anion transport inhibitor, pyridoxal 5-phosphate, binds to the erythrocyte membrane and inhibited the membranolytic activity of the non-asbestos tremolite, tremolite asbestos and quartz (Figure 5). Of the three minerals studied, the non-asbestos tremolite had the lowest zeta potential, i.e. surface charge, which may account for pyridoxal-5-phosphate being such an effective inhibitor of its membranolytic activity.

ACKNOWLEDGEMENTS

We wish to acknowledge support under a Contract from the United States Consumer Products Safety Commission, The Asbestos Institute of Canada and the R.T. Vanderbilt Company. RPN was supported by a Fellowship from the Stony Wold-Herbert Fund.

REFERENCES

Addison,J. and Davies,J.M.G. (1991) A comparison of the carcino-genicity of six tremolites using the intrapleural injection assay in rats: Interim report. VIIth International Pneumoconiosis Conference, Pittsburgh, Pennsylvania, August, 1988. (in press)

Amandus,H.E., Wheeler,R., Jankovic,J. and Tucker,J. (1987a) The morbidity and mortality of vermiculite miners and miller exposed to tremolite-actinolite: Part I Exposure estimates. *Am. J. Indust. Hyg.* **11**:1-14.

Amandus,H.E. and Wheeler,R. (1987b) The morbidity and mortality of vermiculite miners and millers exposed to tremolite-actinolite: Part II Mortality. *Am. J. Indust. Hyg.* **11**:15-26.

Amandus,H.E., Althouse,R., Morgan,W.K.C., Sargent,N. and Jones,R. (1987c) The morbidity and mortality of vermiculite miners and millers exposed to tremolite-actinolite: Part III Radiographic findings. *Am. J. Indust. Hyg.* **11**:27-37.

Banks,W. (1980) Asbestiform and/or Fibrous Minerals in Mines, Mills, and Quarries. U.S. Department of Labor, Mine Safety and Health Administrastion. Informational Report IR 1111.

Baris,Y.I., Artvinli,M., Sahin,A.A., Bilir,N., Kalyoncu,F. and Sebastien,P. (1988) Non-occupational asbestos related chest diseases in a small Anatolian village. *Br. J. Indust. Med.* **45**:841-842.

Baris,Y.I., Bilir,N., Artvinli,M., Sahin,A.A., Kalyoncu,F. and Sebastien,P. (1988) An epidemiological study in an Anatolian village environmentally exposed to tremolite asbestos. *Br. J. Indust. Med.* **45**:838-840.

Cabantchik,Z.I, Knauf,P.A. and Rothstein,A. (1978) The anion transport system and the red blood cell. The role of membrane proteins evaluated by the use of probes. *Biochem. Biophys. Acta (Biomembr. Rev.)* **515**:239-302.

Campbell,W.J., Huggins,C.W. and Wylie,A.G. (1980) Chemical and physical characterization of amosite, chrysotile, crocidolite, and non-fibrous tremolite for oral ingestion studies by the National Institute of Environmental Health Sciences. Bureau of Minerals Report of Investigations RI 8452. United States Department of Interior.

Campbell,W.J., Steel,E.B., Virta,R.L. and Eisner,M.H. (1979) Relationship of mineral habit to size characteristics for tremolite cleavage fragments and fibres. Bureau of Mines Report of Investigation RI 8367. United States Department of the Interior.

Constantopoulos,S.H., Malamou-Mitsi,V.D., Gouvdevenos,J.A., Papthanasiou,M.P., Pavlidis,N.A. and Papadimitriou,C.S. (1987a) High incidence of malignant pleura mesothelioma in neighboring villages of northwestern Greece. *Respiration* **51**:266-271.

Constantopoulos,S.H., Langer,A.M., Saratzis,N. and Nolan,R.P. (1987b) Regional findings in Metsovo lung. *Lancet* i:452-453.

Davis,J.M.G. (1990) The biological effects of mineral fibres: What we know in 1989? Proceedings of AIA 6th International Colloquium on Dust Measurement Technique and Strategy. Jersey, Channel Islands. pp. 12-22.

Davis,J.M.G., Addison,J., Bolton,R.E., Donaldson,K., Jones,A.D. and Miller,B.G. (1985) Inhalation studies on the effect of tremolite and brucite dust. *Carcinogenesis* 6:667-674.

Davis,J.M.G., Addison,J., McIntosh,C., Miller,B.G. and Niven,K. (1991) Variations in the carcinogenicity of tremolite dust samples of differing morphology In: The Proceedings of the Third Wave of Asbestos Disease: Exposure to asbestos in place. Public Health Control, Landrigan, P.J., Kazemi, H. (eds). (in press)

Ilgren,E.B. and Wagner,J.C. (1991) Background incidence of mesothelioma: animal and human evidence. *Regul. Toxicol. Pharmacol.* 13:133-149.

Kuschner,M. (1987) A review of experimental studies on the effects of MMMF on animal systems. *Ann. Occup. Hyg.* 31, No. 4B:791-797.

Langer,A.M., Rohl,A.N., Wolff,M.S. and Selikoff,I.J. (1979) Asbestos, fibrous minerals and acicular cleavage fragments: Nomenclature and biological properties. In: "Dust and Disease." Dement,J.A. and Lemen,R.A. (eds), Pathotox Publishers, pp. 1-22.

Langer,A.M., Nolan,R.P., Constantopoulos,S.H. and Moutsopoulos,H.M. (1987) Association of Metsovo lung and pleural mesotheliomas with exposure to tremolite-containing whitewash. *Lancet* :965-967.

Langer,A.M., Nolan,R.P. and Addison,J. (1991) Distinguishing between amphibole asbestos and elongated cleavage fragments of their non-asbestos analogues. In: "Nato Advanced Research Workshop on Mechanisms of Fibre Carcinogenesis". Brown,R.C., Hoskins,J.A. and Johnson,N.F. Plenum Press. (this volume)

Lockey,J., Jarabek,A.M., Carson,A., McKay,R., Harber,P., Khoury,P., Morrison,J., Wiot,J., Spitz,H. and Brooks,S.M. (1983) Pulmonary hazards of vermiculite exposure. In: "Health Studies Related to Metal and Non-metal Mining". Wagner,W.L., Rom,W.N. and Merchant,J.A. (eds) Butterworth Publishers. Boston London. pp. 303-315.

Lockey,J., Brooks,S.M., Jarabek,A.M., Khoury,P.R., McKay,R.T., Carson,A., Morrison,J.A., Wiot,J.F. and Spitz,H.B. (1984) Pulmonary changes after exposure to vermiculite contaminated with fibrous tremolite. *Am. Rev. Respir. Dis.* 129:952-958.

Moatamed,F., Lockey,J.E. and Parry,W.T. (1986) Fibre contamination of vermiculites: A potential occupational and environmental health hazard. *Environ. Res.* 41:207-218.

McConnochie,K., Simonato,L., Mavrides,P., Christofides,P., Pooley,F.D. and Wagner,J.C. (1987) Mesothelioma in Cyprus: the role of tremolite. *Thorax* 42:342-347.

McConnochie,K., Simonato,L., Mavrides,P., Christofides,P., Mitha,R., Griffiths,D.M. and Wagner,J.C. (1989) Mesothelima in Cyprus. In: "Non-Occupational Exposure to Mineral Fibres", Bignon,J., Peto,J. and Saracci,R. (eds) IARC Scientific Publication No. 90, Lyon, France. pp. 411-419.

McDonald,J.C., McDonald,A.D., Armstrong,B. and Sebastien,P. (1986a) Cohort study of mortality of vermiculite miners exposed to tremolite. *Br. J. Indust. Med.* 43:436-444.

McDonald,J.C., Sebastien,P. and Armstrong,B. (1986b) Radiological survey of past and present vermiculite miners exposed to tremolite. *Br. J. Indust. Med.* 43:445-449.

McDonald,J.C., McDonald,A.D., Sebastien,P. and Moy,K. (1988) Health of vermiculite miners exposed to trace amounts of fibrous tremolite. *Br. J. Indust. Med.* 45:630-634.

National Toxicology Program. Toxicology and Carcinogenesis Studies of Tremolite in F344/N Rats (Feed Studies). Technical Report Series No. 277. March 1990.

Nolan,R.P., Langer,A.M., Harington,J.S., Oster,G. and Selikoff,I.J. (1981) Quartz hemolysis as related to its surface functionalities. *Environ. Res.* **26**:503-520.

Reger,R. and Morgan,W.K.C. (1990) On talc, tremolite and tergiversation. *Br. J. Indust. Med.* **47**:505-507.

Schiller,J.E., Payne,S.L. and Khalafalla,S.E. (1980) Surface charge heterogenicity in amphibole cleavage fragments and asbestos fibers. *Science* **204**:1530-1532.

Smith,W.E. (1974) Experimental Studies on Biological Effects of Tremolite Talc on Hamsters. In: "Proceedings of the Symposium on Talc". Washington, D.C. Information Circular 8639, U.S. Bureau of Mines, pp. 43-48.

Smith,W.E., Herbert,D.D., Sobel,H.J. and Marquet,E. (1979) Biological tests of tremolite in hamsters. In: "Dusts and Disease". Dement,J.A. and Lemen,R.A. (eds) Pathotox Publisher, pp. 335-339.

Stanton,M.F., Layard,M., Tegeris,A., Miller,E., May,M., Morgan,E. and Smith,A. (1981) Relationship of particle dimension to carcinogenecity in amphibole asbestos and other fibrous minerals. *J. Natl. Cancer Inst.* **67**:965-975.

Timbrell,V. (1969) Characteristics of the International Union Against Cancer Standard Reference Samples of Asbestos. Inter-national Conference on Pneumoconiosis. Johannesberg, Republic of South Africa. pp. 17-25.

Wagner,J.C. (1982a) Health hazards of substitutes. In: "Proceedings of the World Symposium on Asbestos". Canadian Asbestos Information Centre. Quebec. pp. 244-266.

Wagner,J.C., Chamberlain,M., Brown,R.C., Berry,G., Pooley,F.D., Davies,R. and Griffiths,D.M. (1982b) Biological Effects of Tremolite. *Br. J. Cancer* **45**:352-360.

Wylie,A.G. (1988) Relationship between the growth habit of asbestos and the dimensions of asbestos fibers. *Mining Engineering* :1036-1039.

Yazicioglu,S., Ilcayto,R., Balci,K., Sayli,B.S. and Yorulmaz,B. (1980) Pleural mesotheliomas and bronchial cancers caused by tremolite dust. *Thorax* **35**:564-569.

DISTINGUISHING BETWEEN AMPHIBOLE ASBESTOS FIBERS AND ELONGATE CLEAVAGE FRAGMENTS OF THEIR NON-ASBESTOS ANALOGUES

A.M. Langer, R.P. Nolan and J. Addison[1]

Environmental Sciences Laboratory
Brooklyn College
Brooklyn, NY 11210, U.S.A.

[1]Institute of Occupational Medicine Ltd.
8 Roxburgh Place
Edinburgh EH8 9SU, Scotland

INTRODUCTION

In 1986, a letter of correspondence to the New England Journal of Medicine (Germine, 1986) communicated that 2-4% tremolite asbestos was present in a crushed carbonate marble, marketed as a sand to be used in children's sand boxes. Analysis of a specimen of this sand by the Environmental Sciences Laboratory for the US Consumer Products Safety Commission found that although tremolite was present in the amounts stated, it was not asbestos but rather common tremolite which, upon crushing, yielded generally blocky, prismatic cleavage fragments (see Langer & Nolan, 1987 and Figures 1,2,3).

The criteria used by Langer and Nolan (1987) to distinguish asbestos fibre from cleavage fragments included data which appeared in the mineralogical literature over the past decade or more. These data indicated that a mineral occurred as a monoclinic amphibole asbestos when it: possessed anomalous optical properties (*e.g.*, parallel extinction); was composed of fibrils rendering it polyfilamentous; its fibrils were curvilinear and splayed, and possessed in greater or lesser degree fibril parting along (100) twin and (010) planes; its fibrils displayed extreme length and narrow diameter, *etc.* (Figures 4,5). None of these properties nor characteristics were observed for the tremolite found in the play sand (compare Figures 1 and 4). Langer and co-workers (1987) had described tremolite asbestos in a whitewash in Greece as an agent of pleural disease there (Constantopoulos *et al.*, 1987).

In his reply to Langer and Nolan (1987) who characterized the tremolite in play sand as non-asbestos, Germine (1987) suggested that the properties used to distinguish the two habits were commercially constrained "mineralogic abstractions." He stated that the criteria used by us "were not consistent with those used in (our) past work" (ca. 1975). Since that time, more data had become available on the nature of these minerals. The data presented here form the scientific basis for the distinction between asbestos and non-asbestos tremolite.

DEFINITIONS USED IN THE ASBESTOS STANDARD - BY OCCUPATIONAL SAFETY AND HEALTH ADMINISTRATION

The first OSHA Standard for asbestos regulation in the United States was promulgated in 1972. It included a broad definition of asbestos and a specific criteria for determining which fibres in the workplace were to be assayed. At the time of the first standard, asbestos was defined mineralogically as one of several silicate minerals with the following characteristics:

- Fibre bundles are composed of "hair-like" (filiform) fibrils, each with a large length-to-width ratio;
- Fibre bundles are polyfilamentous, that is, composed of fibril strands which may be easily separated by hand;
- Fibres are chemically durable and flexible and may be woven like organic fibres;
- Fibres possess diameter-dependent tensile strength.

Fibres which exhibited these properties were said to display the asbestos habit.
The OSHA asbestos standard included five amphiboles (Table 1). However, it also included morphological specifications for counting asbestos fibres in airborne dust samples, i.e., asbestos was a fibre of a length ≥ 5 μm and aspect ratio $> 3:1$. The portion of the 1972 standard regarding length and aspect ratio was based on both practical and theoretical considerations:

- The 5 μm size length limit for asbestos fibre counting in a work environment was based on several factors. Firstly, it eliminated the requirement for the microscopist to distinguish short fibre from non-fibrous particles, especially important since the fibre standard (fibre/ml) replaced the total particle standard (millions of particles/ft^3). Secondly, counting of fibres greater than 5 μm in length improved the precision of the determination by a single microscopist, and especially among different microscopists analyzing the same specimen. Additionally, the fibres used in textile mills, the principal work site from which the standard was developed, produced dust containing long fibers when carded and spun. Although **short** fiber, < 5 μm in length, was the predominant component in the air, it constituted a small component of the total dust assayed by light microscopy at 100x magnification. Lastly, the then existing theory of the etiology of asbestosis was based on the disintegration of asbestos bodies over time with the release of silicic acid. Asbestos bodies were noted to form on long fibres (≥ 5 μm) (Beattie, 1961).
- The 3:1 aspect ratio was also established as part of a counting strategy, principally to eliminate "particulates" and fibre clumps from environmental assays. Again, this form limitation improved both the precision and accuracy of fibre counting on air filters. The use of the aspect ratio was not introduced to define asbestos.

These latter elements currently remain in the standard. With time, many interpreted these counting criteria as part of the definition of what asbestos was as a material.

Figure 1. Cleavage fragments of tremolite recovered from a carbonate play sand from New York State. Fragments are visualized by interference light microscopy. Fragment **A** is ≈ 75 μm in length with an aspect ratio of $\approx 4.4{:}1$. Its surface topography displays the characteristic angular relationship anticipated for amphibole cleavage planes. Fragment **B** is ≈ 17 μm in length with an aspect ratio of $\approx 3{:}1$. Fragment **C** is also typical of the tremolite fragments recovered from the sand, very nearly equant in aspect ratio. None of the fragments, A, B or C, are respirable. Scale as marked. Note that all the fragments are "single crystals."

PROBLEMS EMERGING AS THE RESULT OF DEFINITIONS

The regulatory definition of asbestos used by OSHA included elements of an occupational environmental survey strategy. This mattered little in that the strategy was formulated for use in workplaces where **asbestos** was manipulated. However, as the workplace levels decreased and monitoring expanded to environments where asbestos and non-asbestos materials might be found, problems emerged. The definition of asbestos was too broad and many non-asbestos minerals began to be assayed as asbestos, by OSHA's definition. These minerals possessed some of the morphological characteristics similar to asbestos and had the same mineral name. Additionally, the environmental assay was to be carried out with a phase-contrast microscope which was both resolution limited and incapable of acquiring data useful for fiber identification. Non-asbestos mineral fragments, by regulatory fiat, became asbestos.

To further complicate the issue, the hypothesis was advanced that morphology was the primary determinant of biological activity. It therefore did not matter whether the object counted by light microscopy was actually asbestos or not in that if the mineral fragment possessed a fibrous **morphology**, it should be counted as asbestos. Tremolite asbestos and tremolite cleavage fragments, if morphologically similar, became synonymous. The asbestos standard was now expanded to include elongate, non-asbestos, mineral fragments.

TABLE 1. Asbestos Minerals Cited in the 1972 OSHA Standard.

Asbestos Mineral Cited	Mineral Name	Counting Criteria; Analytical Problems
Amosite	Grunerite-Cummingtonite	Requirements for counting in workplace: >5 μm length; aspect ratio $>3:1$; <3 μm diameter; <100 μm length. *n.b.* Grunerite-Cummingtonite is **not regulated**.
Crocidolite	Riebeckite	Requirements for counting are same as for amosite. *n.b.* Riebeckite is **not regulated**.
Chrysotile	Chrysotile	Requirements for counting are same as for amosite. Is always asbestos.
Actinolite	Actinolite	Requirements for counting are same as for amosite. Same name for asbestos fibre and mineral, which may occur with **non-asbestos habit**. Possible to include cleavage fragments as "asbestos" in filter analysis, especially with phase contrast microscopy.
Anthophyllite	Anthophyllite	Requirements for counting are same as for amosite. Problems of distinguishing asbestos and non-asbestos varieties are the same as for actinolite.
Tremolite	Tremolite	Requirements for counting are same as for amosite. Problems of distinguising asbestos and non-asbestos varieties are the as for actinolite.

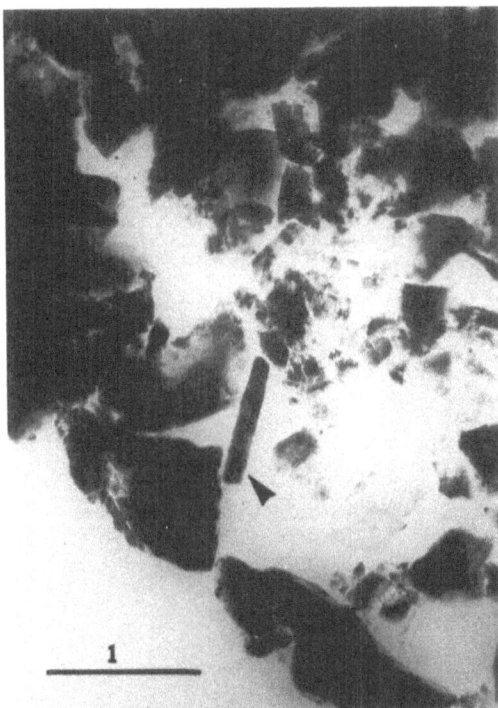

Figure 2. Transmission electron micrograph of a water-fractionated fine-size fraction of the carbonate play sand. Arrowhead indicates a tremolite particle within the carbonate martrix, ≈ 0.9 μm in length, ≈ 0.15 μm in width (AR $\approx$ 6:1). About 18% of the fragment population displays aspect ratios $\geq$ 5:1; about 1.5% $\geq$ 10:1. Most of the respirable tremolite fragments are less than 5 μm in length.

STATE OF THE ART IN IDENTIFICATION OF ASBESTOS; BIOLOGICAL DIFFERENCES

Since 1972 when the United States asbestos standard was set, new mineralogical and crystallographical studies of the asbestos minerals and their non-asbestos analogues have contributed to the modern understanding of their structure and properties. These studies have helped to explain the anomalous optical properties of the monoclinic amphibole asbestos minerals and their relationship with their non-asbestos analogues. These forms can be distinguished.

Data accumulated which suggested that amphibole asbestos, and its non-asbestos analogues, possessed very different biological potentials (Wagner *et al.*, 1982; Addison and Davis, 1988; Nolan *et al.*, 1991). Therefore, it is important to distinguish among them. The question arose as to the criteria and methods which could distinguish between different forms of the same mineral.

TABLE 2: Comparison of the Physical Characteristics Displayed by Tremolite Asbestos from Metsovo, Greece and Non-Asbestos Tremolite found in a New York Play Sand.

Property	Tremolite Asbestos (Metsovo White-Wash)	Tremolite Cleavage Fragments (NYS Play Sand)
By Polarized Light Microscopy		
Fibre, on Light Optical Level,	Polyfilamentous; curvilinear; exhibits splayed ends; exhibits parallel to near parallel extinction; characteristic amphibole cleavage are observed for some few associated fragments present. Large fibrils show (100) and (010) surfaces.	Optically continuous single fragment; straight-edges along length and termination; angular extinction; amphibole cleavage discernable; no polyfilamentous bundles observed. Occasionally, some striations parallel to long axis.
Aspect Ratio	Avg. = 10.9:1, ≥20:1, 18%; Log-normal distribution towards high aspect ratios (≥10:1).	Avg. = 3.7:1, ≥20:1, 0%; Log-normal distribution towards low aspect ratios (≤5:1).
By Transmission Electron Microscopy		
Fibre/Fibril on Sublight Level	Tends toward long and thin. Parallel sides to fibril. Polyfila-mentous when in fiber.	Tends toward short and wide. Irregular stepped-sides, edges and ends.
Fibril Structure	Twins (100) frequent; closely spaced, <1.0 μm.	Twins rare; spacing > >1.0 μm.
Selected Area Electron Diffraction (SAED)	Common Diffraction Nets: (001) x (010) (001) x (100) (103) x (010) Diffraction spots common. No Kikuchi lines.	Common Diffraction Nets:* (110) x (1̄1̄1̄) (020) x (1̄1̄1̄) (131) x (1̄1̄1̄) Diffraction spots common. Kikuchi lines common.

* Note: Most of the play sand fibres were too thick to permit passage of electrons. They are optically opaque on the TEM level

258

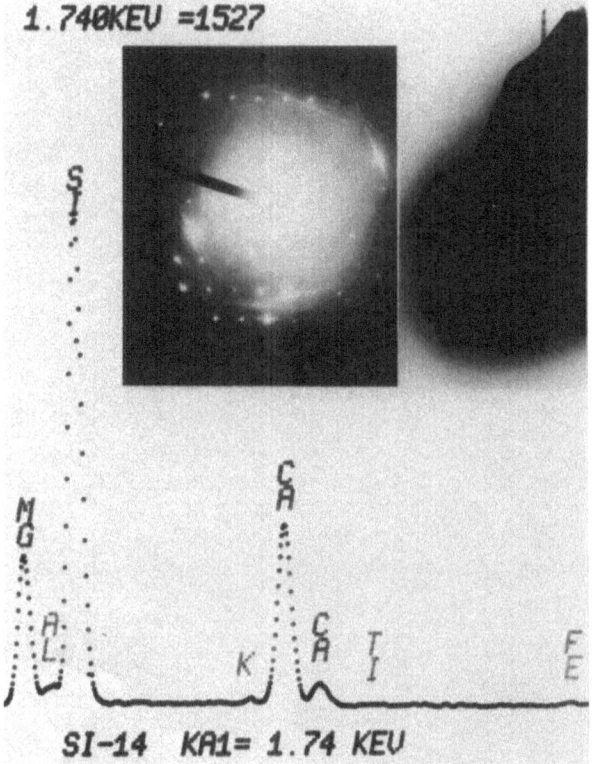

1.740KEV =1527

SI-14 KA1= 1.74 KEU

Figure 3. Transmission electron micrograph, energy dispersive X-ray spectrum (EDXS) and selected area electron diffraction (SAED) pattern obtained on tremolite fragment (upper right). The fragment's length is ≈ 6 μm, its width ≈ 3.4 μm (AR $\approx 1.8{:}1$), the SAED pattern shows Kikuchi lines indicating a thick fragment. The Laue net is about (131 x ($\bar{1}\bar{1}\bar{1}$)) indicating the particle lies near the (110) cleavage plane. The EDXS indicates a chemistry consistent with tremolite.

THE NATURE OF THE AMPHIBOLE ASBESTOS MINERALS

Anomalous Properties of Asbestos

On the level of electron microscopic examination, the amphibole asbestos minerals display anomalous characteristics which were reported in the early 1970s (see, *e.g.*, Skikne *et al.*., 1971; Langer *et al.*., 1974). Fibres 0.20 μm in diameter, and greater, produce selected area electron diffraction patterns which included reflections forbidden for the space group symmetry, and spatial periodicities which were not accountable for by the amphibole unit cell dimensions. Twinning and defects were suspected as the causes of these anomalies (Langer *et al.*, 1974). It was also noted that tilting of amphibole asbestos fibers in the electron beam, at angles of up to 20°, failed to produce new symmetry nets (Skikne *et al.*., 1971; Seshan, 1976). This behavior was not observed for common amphibole minerals. It should be noted that thick fibres ($>>200$ Å in diameter) gave rise to electron effects in the crystal, *e.g.*, multiple reflections, which produced "uninterpretable" diffraction patterns. These patterns do not resemble those obtained with X-rays

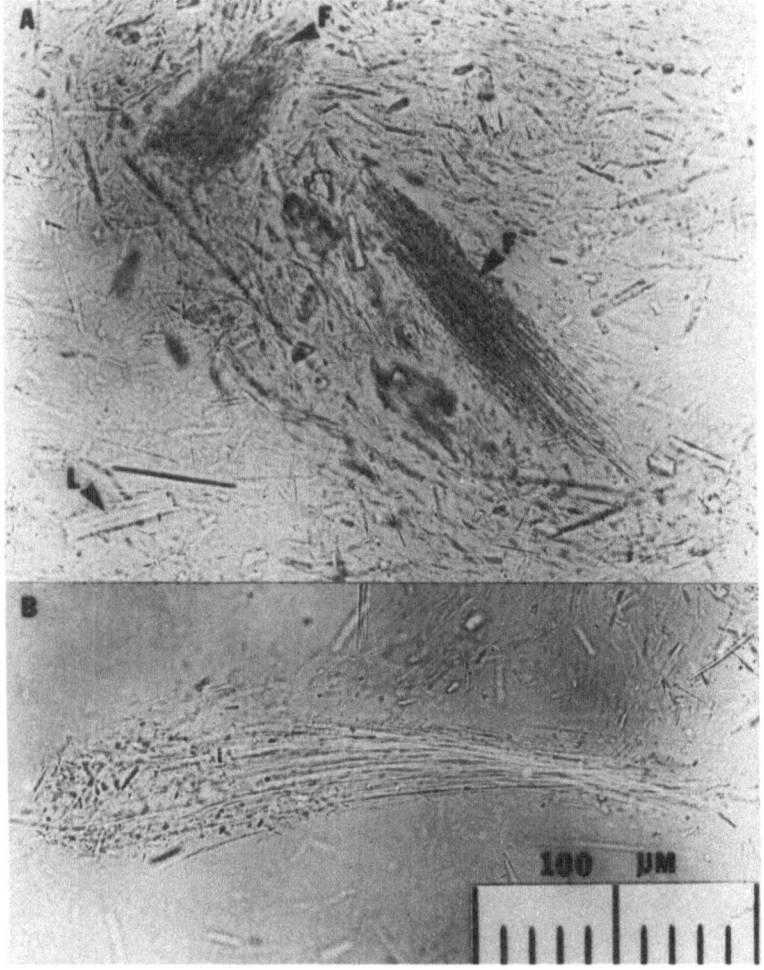

Figure 4. Tremolite asbestos, as visualized in plane polarized light. The tremolite asbestos is a major mineral component of whitewash (stucco) from Metsovo, Greece. (A) Fibrous polyfilamentous bundles of fibrils(F, arrow-heads) and cleavage fragments or laths (L, arrowhead) are present. (B) Fibre bundle as viewed between partially crossed Nicols. The fibre bundle is clearly polyfilamentous, curvilinear and splayed open at its ends. The unit fibrils have aspect ratios exceeding 100:1. Fibre extinction is parallel. Scale as marked in (B). Compare with cleavage fragment in Figure 1.

(Whittaker, 1979). Only where very thin particles are available can diffraction data be obtained which yield interpretable patterns (Whittaker, 1979).

Nature of Twins

Twins, considered a form of crystal defect, are defined as differently oriented structural lattices (which form multiple individuals) contained within the same crystal. A number of specific symmetry operations produce these differing orientations and individuals. At the boundary of these differently oriented lattices, there exists a layer of atoms which is common to each lattice, and in which the atoms fit exactly into the structures of each individual. When a single crystal contains these differently oriented lattices it is said to be twinned.

In monoclinic amphiboles, the twin is defined by the following symmetry operation: a mirror reflection of the same individual across a plane; the reflection plane is the (100), and it is referred to as the twin plane. It is important to add the following: the individual twins which are reflected across this plane may be faulted or offset (crystallographically referred to as translated) so that instead of the anticipated shared atom between these individual lattices, preserving crystal bond cohesive forces, the atoms in the facing twin planes occupy sites which set up repulsive forces in the crystal. This causes a reduction in cohesion. These faulted twin planes are therefore planes of structural weakness, where failure will tend to occur. Failure of this type will occur at a lower energy input than required by the cleavage process for a single, defect-free, crystal. This failure is referred to as parting rather than cleavage.

Twinning accounts, in part, for the asbestos fibre's insensitivity to tilt in the electron beam, where change in crystal orientation, unexpectedly, does not produce new patterns of reflections. These are defined by crystallographers as reciprocal lattice projections which satisfy the Laue equations for diffraction (Buerger, 1958). This behavior of amphibole asbestos fibre/fibril in the electron beam has been explained by the presence of microscopic twins. Crushed fragments of non-asbestos amphiboles very rarely display these characteristics.

Asbestos Fibril Defects

Since the mid-1970s, electron beam studies of amphibole asbestos varieties have found: a high frequency of twinning in amosite parallel to the (100) plane (Hutchison *et al.*, 1975); a high frequency of (100) twining in amosite and crocidolite (Champness *et al.*, 1976; Harlow *et al.*, 1985); the presence of "faults" along adjacent (100) twin surfaces in tremolite asbestos (Seshan & Wenk, 1976); and the presence of a high density of translated "faults" along the (100) in crocidolite (Crawford, 1980). Seshan and Wenk compared the frequency of twinning in calcic amphibole asbestos with its non-asbestos analogue. Non-asbestos tremolite, and common hornblende, exhibit the (100) twin only rarely. Harlow and co-workers (1985) compared the reflection twin (100) periodicity in amosite and grunerite. Amosite twin periodicity ranged from 0.004 μm to 0.02 μm ($\approx$ 40-200 Å), whereas grunerite twin periodicity was rare and ranged from 1.0 μm to 100 μm. Differences in density of planes of failure produces thin, bladed, elongate fibrils of amosite on comminution, as compared to low aspect ratio, thick fragments of grunerite. The twin plane separation, termed parting, is a lower energy process than is required for amphibole cleavage.

Comparison of Selected Area Electron Diffraction (SAED) Patterns

Harlow and co-workers (1985) also showed that amosite generated SAED patterns corresponding to particles lying on or near (100), whereas grunerite particles generally were either too thick to permit penetration of the electron beam which was accelerated at 100 kV, or produced Kikuchi lines when penetrated, or showed patterns indicating most particles lie near or at (110), the prominent amphibole cleavage plane. Dorling and Zussman (1987) have suggested that the (100) plane may be a growth surface for amphibole asbestos (see Figures 3 and 5B).

It should be noted that Kikuchi lines represent a special condition of coherent electron diffraction which produces a pattern of lines rather than discrete reflection spots.

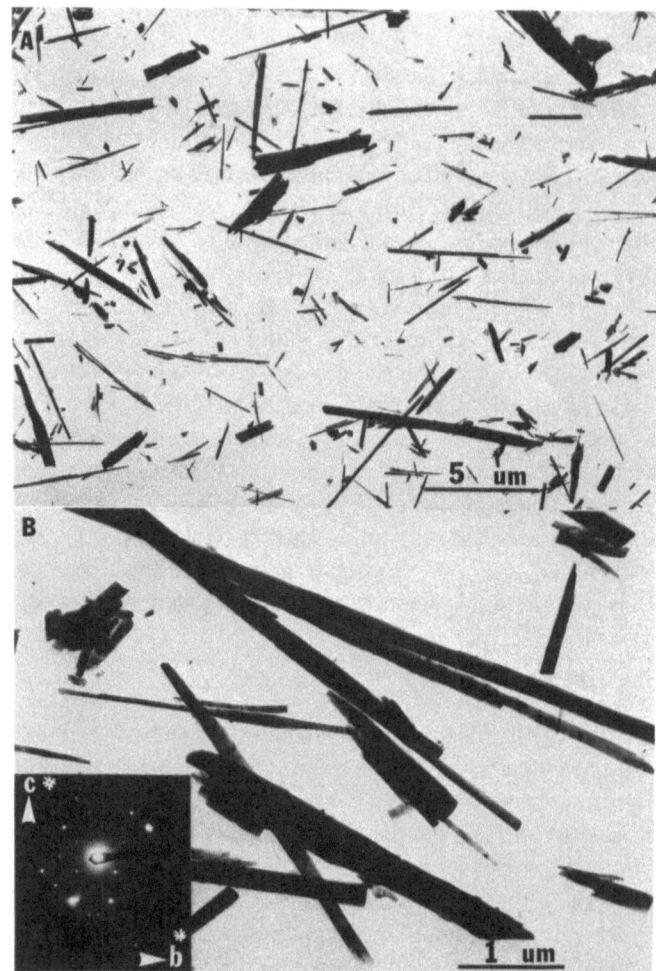

Figure 5. Tremolite asbestos, fine fraction as visualized by TEM. (A) High aspect ratio fibres/fibrils predominate the particle population. Fragments are less common. (B) Tremolite asbestos fibrils at higher magnification display greater aspect ratios. SAED pattern (Insert B) obtained on a thin fibril, displays a ≈(001) x (010) net. The fibril is thin and is lying close to the (100) twin plane. Note thin fibrils ≈0.1 μm in width.

It is commonly observed for thick crystals before complete beam absorption (Van der Biest & Thomas, 1976 and Fig.3) in SAED.

Chain-Width Errors

In 1973, Chisholm reported the presence of anomalies in several amphibole minerals. Rather than the anticipated double-chain groups forming the *b* dimension of the unit cell (≈18Å), the chain was formed of six planar-linked tetrahedra, producing a *b*-cell dimension of ≈27Å. These were termed Wadsley defects, a crystallographical term for similar structural anomalies noted in other minerals (Chisholm, 1973; Veblen *et al.* 1977). Wadsley defects, producing compositionally anomalous (010) planes, were noted with some frequency in specimens of both amosite and crocidolite by Champness and co-workers (1976). Crawford (1980) noted the presence of single (pyroxene), double (amphibole) and triple (pyrobole) chains in UICC crocidolite. The Wadsley defects were both continuous and structurally staggered along the *a*-axis.

Random Orientation of Fibrils

Franco and co-workers (1977) added to the understanding of the structural complexity of the amphibole asbestos minerals. Transmission micrographs obtained on crocidolite showed that unit fibrils were randomly oriented with respect to their *ab* planes. Only the *c*-axes of the fibrils were in common alignment (Franco *et al.*, 1977). Using a preparation technique to reduce artifact formation, disoriented fibrils were also observed by Crawford (1980). He also observed the presence of pore spaces between some juxtaposed fibrils. This azimuthal, completely random orientation of unit fibrils which make up the asbestos fibre, was also shown to exist for actinolite asbestos, based on the polarized light microscopy studies by Wylie (1979).

Properties Produced by Structural Defects

Fibres of the monoclinic amphibole asbestos varieties, made up of fine submicroscopic fibrils, display parallel extinction when examined by polarized light microscopy. The fibres, composed of polyfilamentous bundles of fibrils, are often splayed on their ends and display curvilinear shape along the fibre axis. Indices of refraction measured on these fibres tend to show unanticipated values.

Twinning of or in large individual fibrils of monoclinic amphibole asbestos (≥ 0.5 μm diameter) produces gross structural modification, i.e., an anomalous orthorhombic symmetry for each fibril. Three indices of refraction define each of these fibrils, but again, values are unanticipated ($n'_{\alpha1}$, $n'_{\beta1}$, $n'_{\gamma1}$) and the fibre extinction is parallel rather than oblique (Wylie, 1979; Dorling and Zussman, 1987). Crocidolite fibre bundles, where unit fibrils are rotated with respect to each other and/or are small in diameter (≤ 0.5 μm) produce fibres which display almost **uniaxial** symmetry, i.e., only **two** indices of refraction ($n'_{\alpha1}$, $n'_{\gamma1}$) and parallel extinction. These characteristics have been observed for fibre bundles of tremolite asbestos, actinolite asbestos and amosite.

In contrast, the monoclinic amphibole cleavage fragments display both the morphology and optical properties anticipated for the specific mineral regardless of the presence or absence of twinning. When lying on a flat surface the asbestos fibril lies principally on or near the (100) twin plane and frequently on the (010) plane. Irregular planes dominate for crocidolite (Whittaker, 1979). The fibril therefore generates diffraction patterns which are significantly different from those generated by cleavage fragments which lie at or near the (110) cleavage plane. Therefore, **populations** of fragments/fibres may be distinguished as common amphibole or as asbestos on the submicroscopic level, by a determination of their selected area electron diffraction characteristics.

Comminution of asbestos produces a dust of fine particle size, with narrow diameter fibres/fibrils, high particle number per unit mass of dust, and high surface area. These properties provide an understanding of why asbestos has a distinctly different health hazard evaluation compared to its crushed common amphibole analogue.

THE AMPHIBOLE ASBESTOS FIBRE

By the end of the 1970s, a pattern emerged which was formulated on data obtained on studies of many forms of common amphiboles and their asbestos analogues. Amphibole asbestos fibres, of all varieties, were found to consist of polyfilamentous bundles of azimuthally disoriented, fibrils. Many forms showed that each fibril was complexly twinned on (100) surfaces, and that these twins were commonly offset by faults. Some fibrils possessed chain-width errors, Wadsley defects, on (010) surfaces. These structural characteristics helped explain a number of properties unique to the monoclinic amphibole asbestos minerals which are not observed for their normal amphibole counterparts:

- For the most part, asbestos fibrils are easily separable, in part, because of translocation along the (100) twin plane, producing a much reduced cohesion. The disorientation of the fibril bundle is thought to contribute to this property. Non-asbestos analogues must be crushed to yield elongate particles;

- Mechanical manipulation of asbestos fibre produces many long, thin fibres/fibrils rapidly, as compared to identical manipulation of common amphiboles. Vigorous mechanical manipulation of non-asbestos analogues also induces failure across the fibre axis, producing short, equant fragments as compared to thin, elongate asbestos. Different physical processes are involved in the size reduction of non-asbestos amphiboles than for the disaggregation of asbestos , (in the absence of other binder minerals).

- Amphibole asbestos fibrils express planar elements (100) > > (010) > (110) > (hkl), produced by failure along the twin, defect and cleavage planes, as compared to cleavage fragments expressed planes (110) > > (010) > (100) > (hkl). The plane of cleavage dominates the expressed surfaces of crushed normal amphiboles;

The size distributions and aspect ratios for populations of asbestos fibres/fibrils show them to be mostly smaller in diameter and greater in aspect ratio than non-asbestos analogues;

- Optical properties of the monoclinic amphibole asbestos fibre are "anomalous" (indices of refraction of amphibole asbestos fibres have different values than their normal amphibole single crystals or cleavage fragments analogues);

- The monoclinic amphibole asbestos fibre bundle, made of very narrow diameter fibrils (< 0.2 μm), usually possesses a uniaxial optical indicatrix, displaying two refractive indices, rather than three; fibres display parallel extinction, rather than oblique extinction anticipated for the crystal symmetry, when viewed by polarized light microscopy (Heinrich, 1965; Wylie, 1979; Dorling & Zussman, 1987). It should be noted that tremolite asbestos was originally misidentified as anthophyllite because of the observed parallel extinction (see Heinrich, 1965);

- The different surfaces produced upon comminution, impart on amphibole asbestos fibrils a crystallographic orientation which in an electron beam differs significantly from the orientations of cleavage fragments. The different Laue zones of symmetry can be determined by SAED. This results in monoclinic amphibole asbestos fibrils, exhibiting reciprocal diffraction nets near or at, *e.g.*, (001)x(010), (001)x(100), (103)x(010), *etc.*, as compared to cleavage fragments which exhibit diffraction nets near or at (110)x(111), (020)x(111), (131)x(111), *etc.* (*e.g.*, see Lee *et al.*, 1978).

These characteristics will vary slightly as a function of mineral type and geological occurrence. In some instances, for a single isolated particle, it may be impossible to distinguish an acicular cleavage fragment from asbestos fibril.

The Current Definition

The OSHA-NIOSH definition of asbestos is a composite of mineralogical terms enmeshed in a light microscopy strategy for particle counting in the workplace. It does not uniquely identify or define asbestos. The play sand analysis is an illustration of how this broad definition led to the misidentification of cleavage fragments as asbestos. The play sand consists of cleavage fragments rather than asbestos fibre. The analytical results are summarized in Table 2. Phase contrast optical microscopy techniques and the current definition are insufficient to distinguish elongate prismatic or acicular cleavage fragments from asbestos, or many other fibrous particles.

DISCUSSION AND CONCLUSIONS

Monoclinic amphibole asbestos fibre from non-asbestos cleavage fragment can be distinguished by polarized light microscopy and by transmission electron microscopy. Study of the predominant diffraction nets by TEM requires a **population** of particles and a determination of the principal Laue zones of symmetry displayed by the diffracted objects. The asbestos fibril tends to part along planes uncommon for its non-asbestos counterpart. Its resulting diffraction nets therefore generally differ. Sub-microscopic respirable particles may thus be distinguished on a population basis. These data are too recent to be reflected in the scientific basis upon which the first regulatory asbestos standard was promulgated. A review of the OSHA asbestos standard in the United States to evaluate the merit of including some of the more recent data should be considered.

The OSHA asbestos standard is used to regulate **asbestos minerals**. In the mining and milling environment, crushing of rocks generate fragments of normal amphiboles which conform to the OSHA-NIOSH definition of asbestos. If the asbestos standard is to regulate asbestos only, then polarized light microscopy is required to identify and distinguish between the minerals present. The identification is required for effective monitoring.

The New York State play sand, which reportedly contained tremolite asbestos, did not. Rather it contained tremolite cleavage fragments. Due to no asbestos being present, there can be no asbestos risk associated with its use. The amount of respirable tremolite dust in the specimens studied, of length and diameter considered biologically important, exists in the hundreds of ppm range. Based on known properties thought to contribute to the biological potential of **mineral fibre**, the risk associated with the use of the sand is considered to be so quantitatively different from asbestos that any comparison is not justified.

ACKNOWLEDGEMENTS

We wish to acknowledge support under a Contract from the United States Consumer Products Safety Commission. RPN was supported by a Fellowship from the Stony Wold-Herbert Fund. The writing of this manuscript was made possible by a grant from The Asbestos Institute of Canada and the R.T. Vanderbilt Company.

REFERENCES

Addingley,C.F. (1966) Asbestos dust and its measurements. *Ann. Occup. Hyg.* **9**:73-82.

Addison,J. and Davis,J.M.G. (1988) A comparison of the carcinogenicity of six tremolites using the Intrapleural Injection Assay in Rats: Interim Report. *VIIth International Pneumoconiosis Conference*, Pittsburgh, PA, August 23-26, 1988.

Beattie,J. (1961) The asbestosis body. In: "Inhaled Particles and Vapours", Proc. Intl. Symposium BOHS. Davies,C.N. (ed), Pergamon Press, Oxford. pp. 434-442.

Buerger,M.J. (1958) *X-ray Crystallography*. J. Wiley & Sons, Inc., New York. 531 p.

Champness,P.E., Cliff,G. and Lorimer,G.W. (1976) The identification of asbestos. *J. Microscopy* **108**:231-249.

Chisholm,J.E. (1973) Planar defects in fibrous amphiboles. *J. Mater. Sci.* **8**:475-483.

Constantopoulos,S.H., Malamou-Mitsi,V.D., Goudevenos,J.A., Papathanasiou,M.P., Pavlidis,N.A. and Papadimitriou,C.S. (1987) High indicence of malignant pleural mesothelioma in neighboring villages in northwest Greece. *Respiration* **51**:226-271.

Crawford,D. (1980) Electron microscopy applied to studies of the biological significance of defects in crocidolite asbestos. *J. Microscopy* **120**:181-192.

Dorling,M. and Zussman,J. (1987) Characteristics of asbestiform and non-asbestiform calcic amphiboles. *Lithos* **20**:469-489.

Franco,M.A., Hutchison,J.L., Jefferson,D.A. and Thomas,J.M. (1977) Structural imperfection and morphology of crocidolite (blue asbestos). *Nature* **266**:520-521.

Germine,M. (1986) Asbestos in play sand. Correspondence. *N. Engl. J. Med.* **315**:891.

Germine,M. (1987) Asbestos in play sand. Reply. *N. Engl. J. Med.* **316**:882.

Harlow,G.E., Kimball,M.R., Dowty,E. and Langer,A.M. (1985) Observations on amosite/grunerite dusts. *Proc. 2nd Intl. Congr. Appl. Mineralogy in the Minerals Industry.* Park,W.C., Hausen,D.M. and Hagni,R.D. (eds), pp. 1147-1157.

Heinrich,E.W. (1965) *Microscopic Identification of Minerals*. McGraw Hill, New York, p. 414 (see p. 247-248).

Hutchison,J.L., Irusteta,M.C. and Whittaker,E.J.W. (1975) High resolution electron microscopy and diffraction studies of fibrous amphiboles. *Acta. Cryst.* **A31**:794-801.

Langer,A.M. and Nolan,R.P. (1987) Asbestos in play sand. Correspondence. *N. Engl. J. Med.* **316**:882.

Langer,A.M., Mackler,A.D. and Pooley,F.D. (1974) Electron microscopical investigation of asbestos fibres. *Environ. Hlth Perspect.* **9**:63-80.

Langer,A.M., Nolan,R.P., Constantopoulos,S.H. and Moutsopoulos,H.M. (1987) Association of Metsovo lung and pleural mesothelioma with exposure to tremolite-containing whitewash. *Lancet* **i**:965-976.

Lee,R.J., Lally,J.S. and Fisher,R.M. (1978) Identification and counting of mineral fragments. In: *Workshop on Asbestos: Definitions and Measurement Methods.* *NBS Special Publ.* **506**:387-402.

Nolan,R.P., Langer,A.M., Oechsle,G.W., Addison,J. and Colflesh,D.E. (1991) Association of tremolite habit with biological potential: Preliminary report. In *Mechanisms of Fibre Carcinogenesis*. Brown,R.C., Hoskins,J.A. and Johnson,N.F. (this volume).

Seshan,K. (1976) Explanation for the insensitivity to tilts of electron diffraction patterns of amphibole asbestos fibres. *Environment. Res.* **14**:46-58.

Seshan,K. and Wenk,H.R. (1976) Identification of faults in asbestos minerals and application to pollution studies. *Proc. Electron Microscope Soc. Amer.*, 34th Ann. Mtg. Bailey,G.W. (ed.) Claitor's Publ., Baton Rouge, LA. pp. 616-617.

Skikne,M.I., Talbot,J.H. and Rendall,R.E.G. (1971) Electron diffraction patterns of UICC asbestos samples. *Environment. Res.* **4**:141-145.

Wagner,J.C., Chamberlain,M., Brown,R.C., Berry,G., Pooley,F.D., Davies,R. and Griffiths,D.M. (1982) Biological effects of tremolite. *Br. J. Cancer* **45**:352-360.

Whittaker,E.J.W. (1979) Mineralogy, chemistry and crystallography of amphibole asbestos. In: "Short Course in Mineralogical Techniques of Asbestos Determination". Ledoux,R.C. (ed.) Mineralogical Association of Canada.

Van der Biest,O. and Thomas,G. (1976) Fundamentals of Electron Microscopy. In: "Electron Microscopy in Mineralogy". Wenk,H.R. (ed.) Springer-Verlag, Berlin. pp. 18-51.

Veblen,D.R., Busek,P.R. and Burnham,C.W. (1977) Asbestiform chain silicates: New minerals and structural groups. *Science* **198**:359-366.

Wylie,A.G. (1979) Optical properties of fibrous amphiboles. *Ann. NY Acad. Sci.* **330**:611-619.

ASBESTIFORM MINERALS ASSOCIATED WITH CHRYSOTILE FROM THE WESTERN ALPS (PIEDMONT - ITALY): CHEMICAL CHARACTERISTICS AND POSSIBLE RELATED TOXICITY

Antonella Astolfi, Bice Fubini, Elio Giamello
and Marco Volante

Dipartimento di Chimica Inorganica, Chimica Fisica
e Chimica dei Materiali
Università di Torino
Via Pietro Giuria 7
10125 Torino, Italy

Elena Belluso and Giovanni Ferraris

Dipartimento di Scienze della Terra
- Sezione di Mineralogia e Cristallografia
Università di Torino
Via Valperga Caluso 37
10125 Torino, Italy

ABSTRACT

Two new asbestiform minerals, balangeroite (Balangero, Italy) and carlosturanite (Val Varaita, Italy), have been investigated from the standpoint of their potential toxicity. Their characteristics have been compared with chrysotile and antigorite with which they are always associated. EPR spectra of the minerals revealed the presence of paramagnetic ions, including Fe^{3+} and Mn^{2+}, in different crystal configurations. The presence of Fe^{2+} was shown by the enhancement of the Fe^{3+} signals after grinding in air. Balangeroite is the richer in both Fe^{2+} and Fe^{3+} which are in magnetic interaction with each other, unlike antigorite, in which only isolated and weakly interactive ions are present. Profound modifications of the structures are found upon standing in solutions mimicking the biological environment. This suggests that these minerals may interact in a number of ways *in vivo*. Taking into account both their form and chemical composition they may be regarded as potentially carcinogenic components of fibrous minerals extracted from the Piedmont mines.

INTRODUCTION

Two new asbestiform silicate minerals have recently been found in serpentinite rocks from the Western Alps (Piedmont, Italy): balangeroite at Balangero (Compagnoni *et al.*, 1983) and carlosturanite in Val Varaita (Compagnoni *et al.*, 1985; Mellini *et al.*, 1985). A subsequent systematic field exploration of the Piedmont Western Alps resulted in the discovery of two new localities for balangeroite and twenty-one for carlosturanite (Belluso & Ferraris, in press). Carlosturanite has also been reported from Taberg (Sweden) by Mellini and Zussman (1986).

Both minerals occur as brown centimetric fibres, macroscopically very similar to long-fibre chrysotile with which they have always been mistaken. The fibres consist of bundles of single fibres which are randomly distributed around the fibre axis (Fig. 1). Single fibres (Fig. 2) have cross-sections of a few hundred Ångströms and are usually intergrown with chrysotile and other fibrous minerals at submicroscopic level. The new minerals are clearly distinguishable by their chemical and physical properties.

Balangeroite

Balangeroite occurs as brown, rigid and brittle xyloid fibres. Apart from the other physical properties reported by Compagnoni and coworkers (1983), the best way to identify balangeroite is by diffraction; e.g., its XRD powder pattern exhibits strong reflections for d_{hkl} at 9.59, 6.77, 3.378, 3.278 and 2.714 Å.

Balangeroite is monoclinic with a strong pseudo-orthorhombic symmetry (Ferraris *et al.*, 1987) and its crystal structure can be described as an octahedral framework with channels occupied by a chain of silicate tetrahedra (Fig. 3a). In the unit cell there are four formula units with the ideal composition:

$$M_{21}O_3(OH)_{20}(Si_4O_{12})_2$$

where M (octahedral) is Mg, Fe(II) and Fe(III) (major cations); Mn, Al; Ca, Cr, Ti (traces).

Carlosturanite

The macroscopic characteristics of carlosturanite are similar to those of balangeroite. The brown colour is slightly lighter although samples from Taberg (Mellini & Zussman, 1986) are green. Also carlosturanite can only be reliably identified by diffraction. The XRD powder pattern of carlosturanite is, however, similar to that of chrysotile and so care must be taken to record the low angle reflection with d_{hkl} of 18.02 Å. Other strong reflections have d_{hkl} at 7.17, 3.595, 3.397, 2.562 Å.

Carlosturanite is monoclinic (Mellini *et al.*, 1985) and its crystal structure preserves the octahedral sheet of serpentine, while in the tetrahedral sheet rows of $[Si_2O_7]^{6-}$ groups are substituted by rows of $[(OH)_6.H_2O]^{6-}$ groups (Fig. 3b). In the unit cell there are two formula units with the ideal formula:

$$M_{21}T_{12}O_{28}(OH)_{34} \cdot H_2O$$

where M (octahedral) = Mg (major cation), Fe, Ti, Mn with traces of Cr and Ca; T (tetrahedral) = Si (mainly) and Al. Carlosturanite can be regarded as a Si-poor and H_2O-rich serpentine.

Both fibrous minerals contain paramagnetic ions in low oxidation states (e.g. Mn(II), Fe(II)) which upon contact with air or in a biological medium may give rise to active oxygen species. Their toxicity to humans is unknown as their association with chrysotile and antigorite (another serpentine) inhibits the assessment of the individual toxicity of each rock component. No *in vivo* or *in vitro* studies have been performed so far. Several epidemiological studies have been carried out on the workers from the Balangero mine (Rubino *et al.* 1979a,b; Berrino *et al.* 1983; Piolatto *et al.*, in press), which revealed asbestosis and an excess of laryngeal cancer and pleural mesothelioma. An investigation of the non-occupationally exposed population of the village is underway.

In order to investigate the possible role of each individual fibre in the overall toxicity of the mineral dust, the chemical properties of balangeroite, chrysotile, carlosturanite and antigorite have been compared. The various fibres extracted from the rocks have been characterised by XRD, electron microscopy and chemical analysis in order to check their purity.

The present paper reports some preliminary results showing the presence and localisation within the fibres of paramagnetic ions detected by Electron Paramagnetic Resonance (EPR). Taking into account the proposal that iron is implicated in asbestos related toxicity (Mossman *et al.*, 1988; Pezerat *et al.*, 1989), much care has been devoted to the detection of Fe^{2+} and Fe^{3+} ions in different crystal configurations. Both these ions may play a role in the production of active oxygen species, via redox cycles. Fe^{2+} at the surface partially reduces oxygen to O_2^- initiating a redox chain and yielding $OH\cdot$ in aqueous media (Zalma *et al.*, 1987). The ferric ion may also be active in $OH\cdot$ production, but requires the presence of H_2O_2 and, in some cases, a reducing agent such as ascorbate (Kennedy *et al.*, 1989).

To evaluate any reaction which may occur upon inhalation of the fibres, small amounts were incubated at 37 °C in the dark in phosphate buffered saline (PBS). Since the fate of an inhaled particle may involve phagocytosis by alveolar macrophages, saline media containing H_2O_2 and ClO^-, to mimic the environment within the macrophage, were also used. The solids were then re-examined by EPR to detect modifications due either to redox reactions or to the release of components into the media.

MATERIALS AND METHODS

Balangeroite was collected from the Ponte del Diavolo (Lanzo-Torino); carlosturanite from Colle Sampeyre (Cuneo); the two chrysotile samples from Alpe Praiet and Punta Sbaron (Lemie-Torino) and antigorite from the Rio Milanese (Sampeyre-Cuneo).

A long preliminary preparation of the rock specimens was necessary before analysis of the minerals because of their asbestiform morphology and natural association with other minerals. Small bundles of fibres were separated from the matrix rock with the aid of tweezers and then the bundles were freed from impurities with the aid of the optical microscope. Separated fibres were powdered as finely as possible in acetone suspension to avoid loss, mixed with Vaseline oil and spread on the specimen support grid. Finally, the samples were submitted to X-ray diffraction for identification.

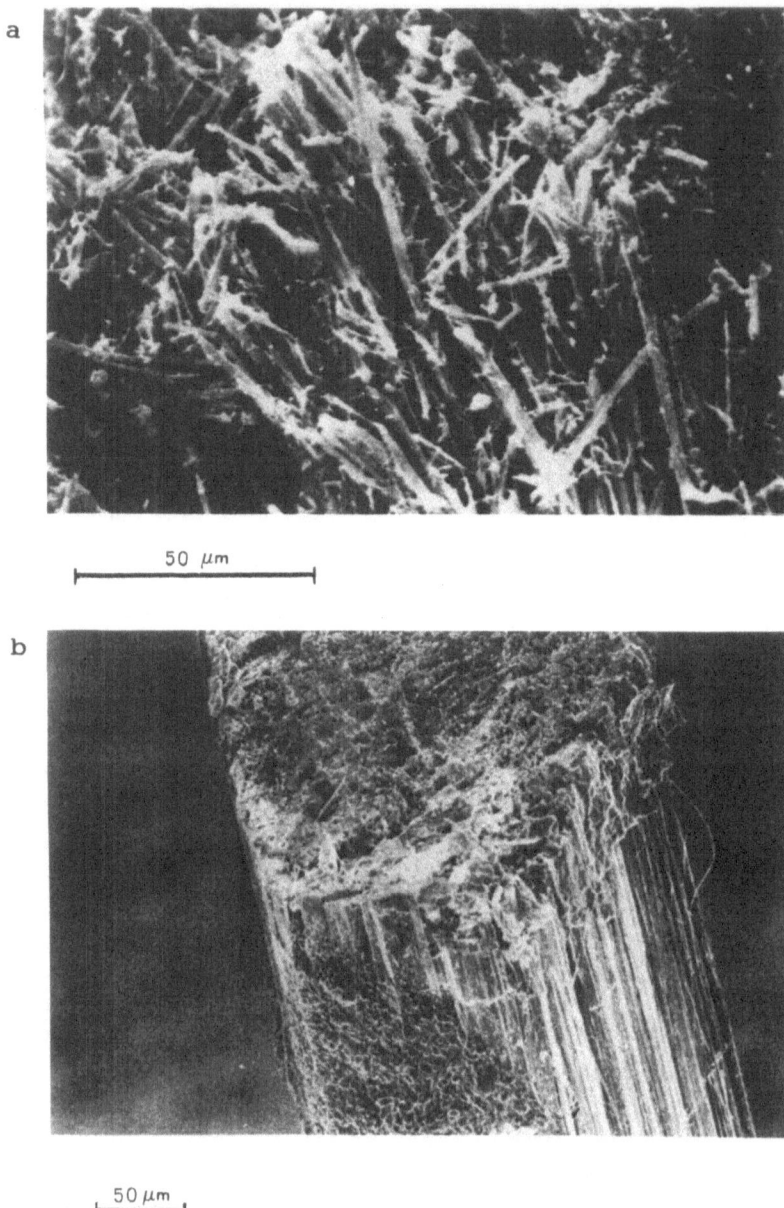

a

50 μm

b

50 μm

Figure 1. Scanning electron microscope (SEM) images of bundles of single fibres of balangeroite (a) and carlosturanite (b).

Powder X-ray diffraction and chemical analysis of the samples was carried out using:

- a Nonius Guinier-Lenné camera for powder XRD with CuK_α radiation.
- an ARL SEMQ microprobe.
- a Cambridge S-360 SEM with EDS 860-500 Link System.
- a Philips 400T electron microscope with EDAX 707.

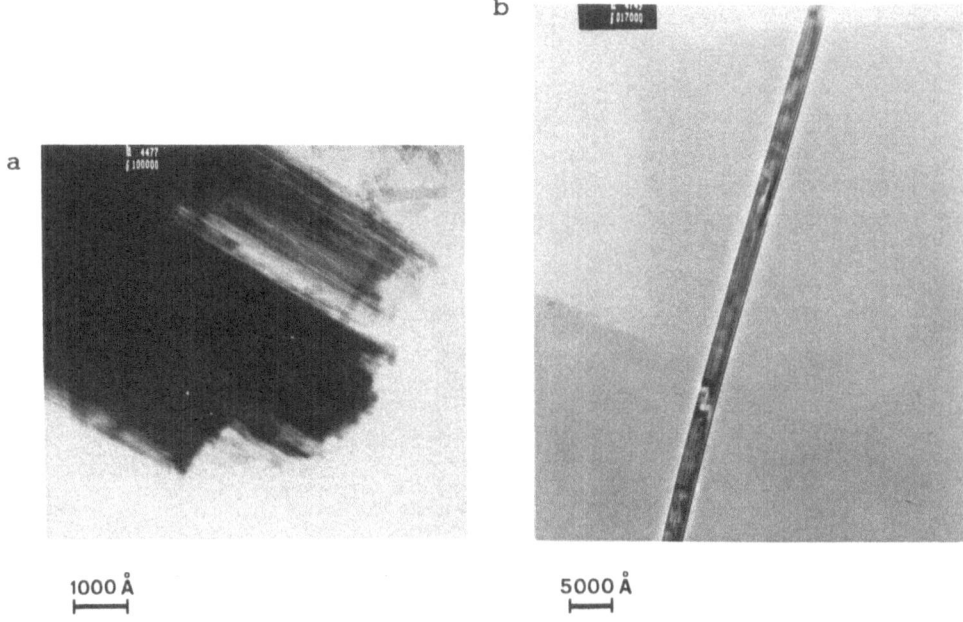

a

b

1000 Å

5000 Å

Figure 2. Transmission electron microscope (TEM) images of single fibres of balangeroite (a) and carlosturanite (b).

Electron Paramagnetic Resonance spectra were recorded at 77 K on a Varian E 109 using the appropriate cells. All spectra are shown with the magnetic field increasing from left to right and at the same scale: a corresponding scale for the g values is also given, together with the amplification used during recording. The EPR spectra of the minerals was recorded before and after grinding to determine the presence and abundance of paramagnetic ions and to study the development of the Fe^{3+} spectra arising through oxidation of Fe^{2+} exposed to the air during grinding.

In order to provide evidence for their possible modification *in vivo* the fibres were kept for 15 days in the dark at 37 °C in the following solutions: i) Dulbecco and Vogt PBS; ii) PBS plus H_2O_2 (3.5%); iii) PBS plus NaClO (0.02 M). The solids were then washed, vacuum dried and transferred into the EPR cell.

RESULTS

Characteristics of the EPR spectra of the new fibrous minerals. Comparison with UICC standards.

The EPR spectra of balangeroite, carlosturanite, chrysotile and antigorite are shown in Fig. 4-7. To distinguish the characteristic features of the minerals from those of occasional contaminants spectra from at least two separate samples were examined. No significant variations were found between different samples of each mineral with the exception of chrysotile. In this mineral remarkable differences in the intensities of the spectral lines were found. Figure 6 shows two chrysotile spectra, samples A and B, of rock from two different locations: sample A was associated with calcite. The main feature of the spectra in Fig. 4-7 is the very intense broad band given by both balangeroite and carlosturanite which contrasts with the lower intensity bands of chrysotile and antigorite

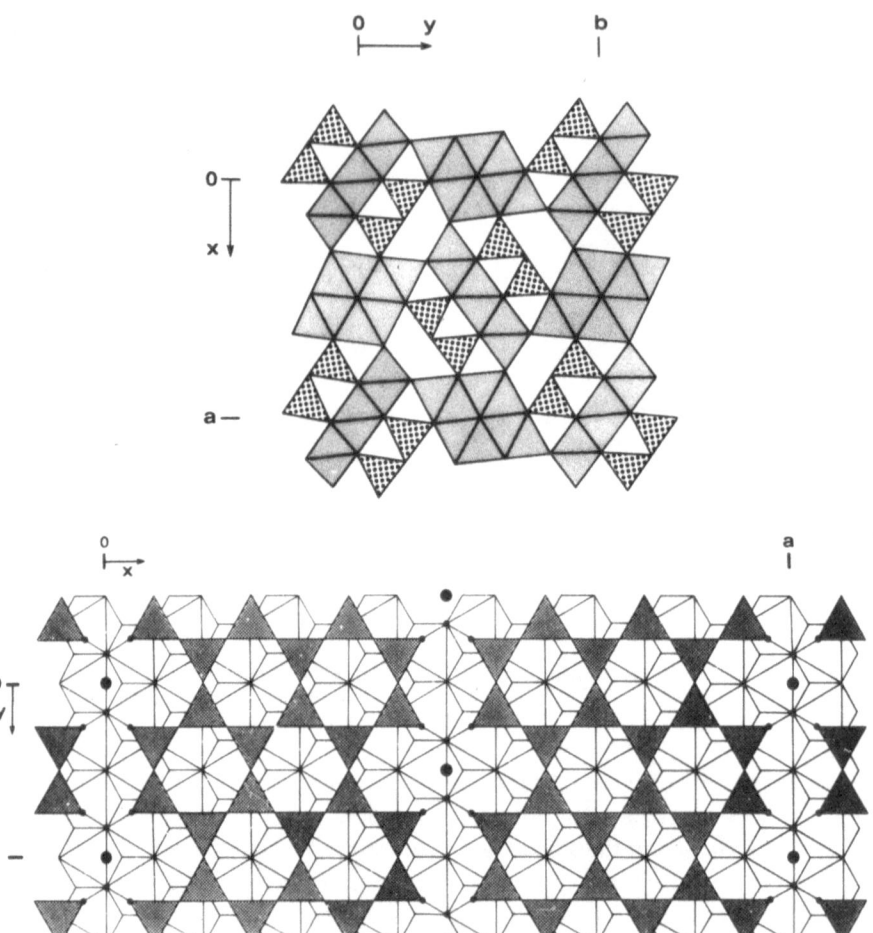

Figure 3. a) Projection along the z axis (fibre axis) of the crystal structure of balangeroite referred to a pseudo-orthorhombic cell. Chains of silicate tetrahedra (large dots) lie within the channels defined by Mg-octahedra (small dots) gathered in groups of three and four rows running along the fibre axis. **b) Projection along the z axis of the crystal structure of carlosturanite.** The fibre axis is parallel to the y axis. A brucite octahedral sheet is coupled with a phyllosilicate tetrahedral sheet where [010] rows of silicate tetrahedra are substituted with hydroxyl anions (small filled circles) and water molecules (large filled circles).

which are resolved into well defined different components. The sextet of lines centred at $g = 2$ in the spectra of antigorite and chrysotile A are due to dispersed Mn^{2+} ions. We assume, on the basis of chemical composition, that all other resonances are due to Fe^{3+} centres isolated or in magnetic interaction with other paramagnetic ions of the same or different kind.

The spectrum of UICC crocidolite (Fig. 5c) closely resembles those of balangeroite (Fig. 4a,b) and carlosturanite (Fig. 5a,b) each consisting of a broad intense signal. Spectra of integral fibres show fine structure superimposed on the broad band which is lost on grinding. Chrysotile and antigorite have spectra (Fig. 6a and 7a) similar to that of

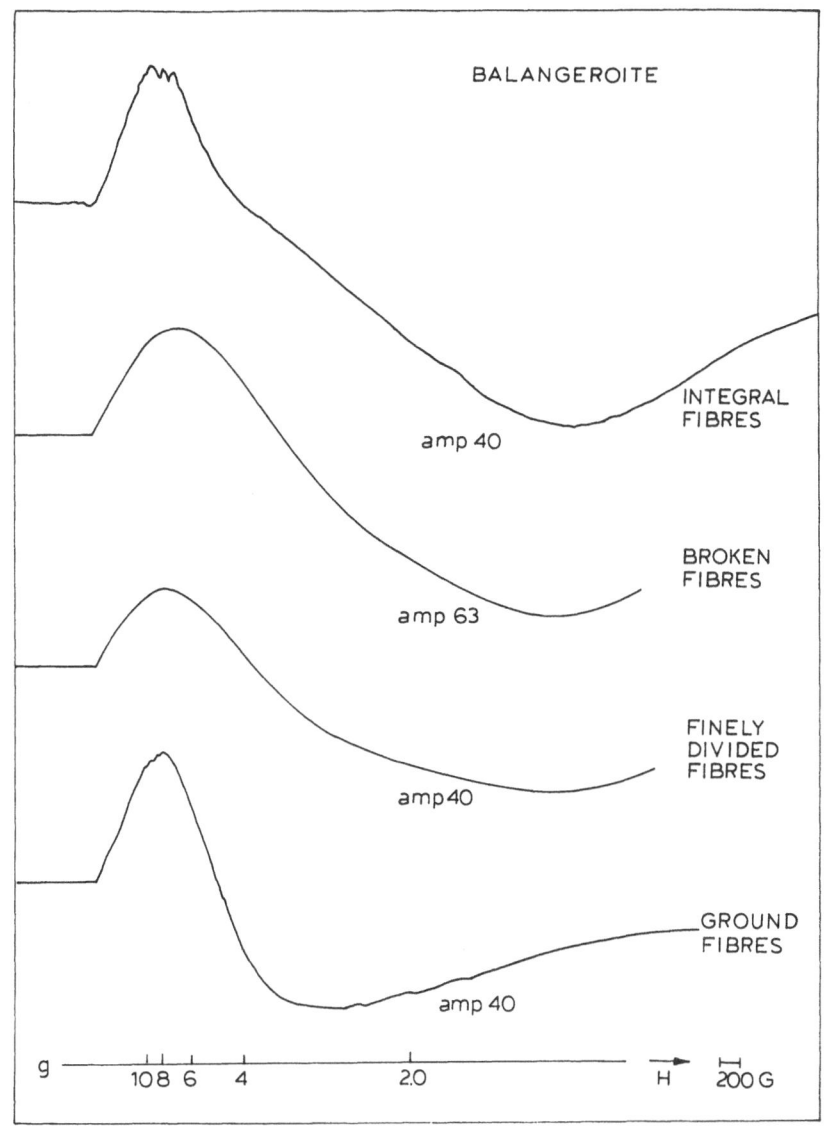

Figure 4. EPR spectra of balangeroite after progressive fragmentation of the fibres.

UICC chrysotile (Fubini *et al.*, 1991, this volume). The spectrum of UICC chrysotile has a broad band superimposed on single lines. This band is less intense in chrysotile B and absent from antigorite and chrysotile A. The intensity of this broad component increases when the spectrum is recorded at room temperature, which may be indicative of interionic magnetic interactions involving ferric iron (Friebele *et al.* 1971).

The effect of grinding

Samples which remain as long fibres after extraction from the rocks are modified on grinding because fresh surfaces are brought into contact with the air when oxidation of newly exposed ions may occur. The spectra of balangeroite after progressive comminu-

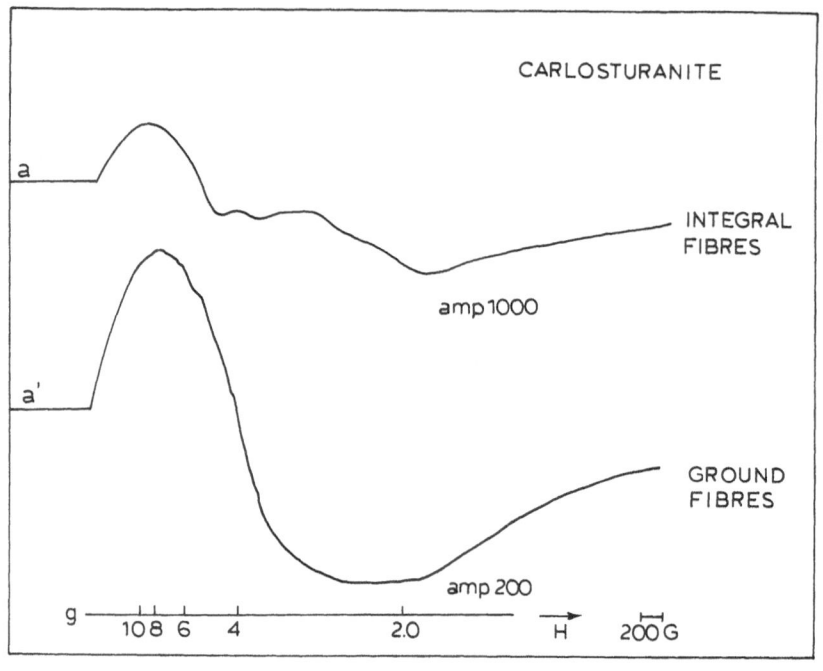

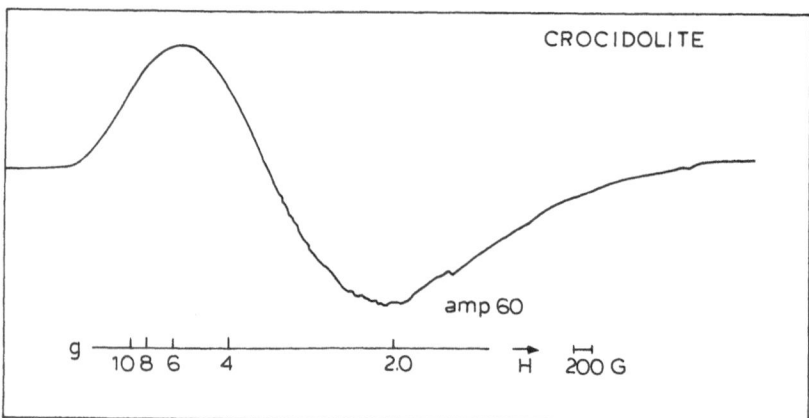

Figure 5. EPR spectra of carlosturanite fibres before (spectrum a) and after grinding (spectrum a') and of UICC crocidolite.

tion and then grinding is shown in Fig. 4 . Increasing fragmentation causes disappearance of the low field spectral bands and a shift of the whole spectrum towards a lower field. Grinding changes both linewidth and position of the band with displacement of the centroid to lower field. The spectrum of carlosturanite (Fig. 5a and b) shows an increase in intensity and a shift to lower field after grinding and the disappearance of the lines superimposed on the broad signal from the starting material. With chrysotile A and antigorite grinding produces a small signal centred at $g = 2$ superimposed on the sextet of lines. The spectrum of chrysotile B, in contrast, is considerably changed by grinding, probably because the mineral contains more iron ions.

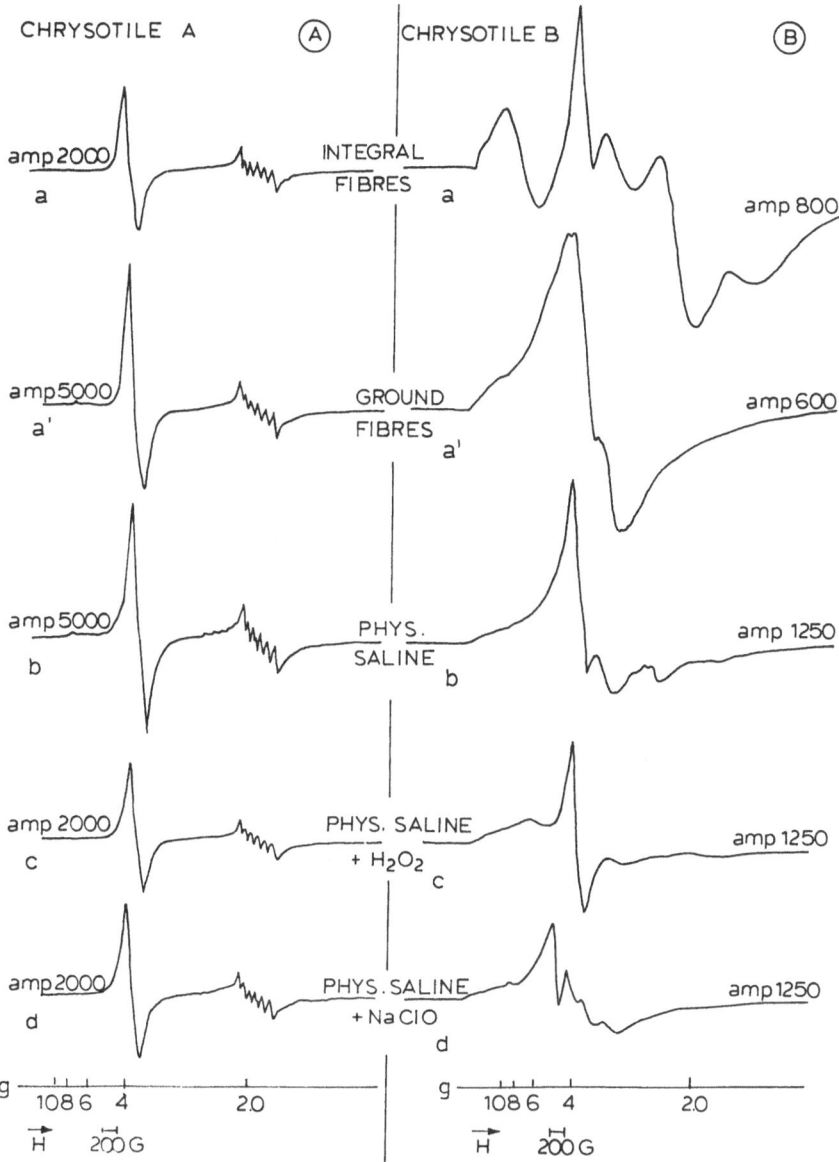

Figure 6. EPR spectra of chrysotile A and B. Spectrum a: integral fibres; spectrum b: kept in phosphate buffered saline (PBS); spectrum c: kept in PBS with added H_2O_2; spectrum d: kept in PBS with added NaClO.

Evidence of paramagnetic ions in various configurations

The EPR signals shown in Figs. 4 and 5 are very broad and asymmetric. This structure can be due to various phenomena. There can be mutual magnetic interactions between similar ions (Fe^{3+}-Fe^{3+}) or dissimilar ions (Fe^{3+}-Fe^{2+}, Mn^{2+}); the possible presence of a magnetically ordered (or partially ordered) lattice; additional strong distortions of the crystal electric field felt by paramagnetic ions, e.g. those at surfaces. In balangeroite and carlosturanite the number of different octahedral sites for Fe^{3+} or other cations is at least 21 and 11 per unit cell respectively. Each site will give a slightly different contribution to the overall spectrum: assignment of bands is clearly not easy in such cases. The similarity of the spectra of the two new fibrous minerals with that of crocidolite, whose spectrum is mainly ascribed to the simultaneous presence of iron in the two oxidation states in relatively high concentration, suggests the presence of similar arrangements in the fibres under study.

The spectra of crysotile A, with only three independent sites, and antigorite, are similar. Each consists of a relatively narrow signal at g = 4.3 and a sextet of lines centred near g = 2. A similar spectrum has been reported for one sample of Canadian chrysotile (Sharrock, 1982). The signal at g = 4.3 is characteristic of ferric ions in a tetrahedral or slightly distorted octahedral field. In the former case it is attributed to an iron ion substituting for silicon in the silicate tetrahedra, in the latter to an iron ion substituting for magnesium in the brucite layer. It is well known that the ratio between these two configurations varies from one mineral to another. The sextet of lines is due to

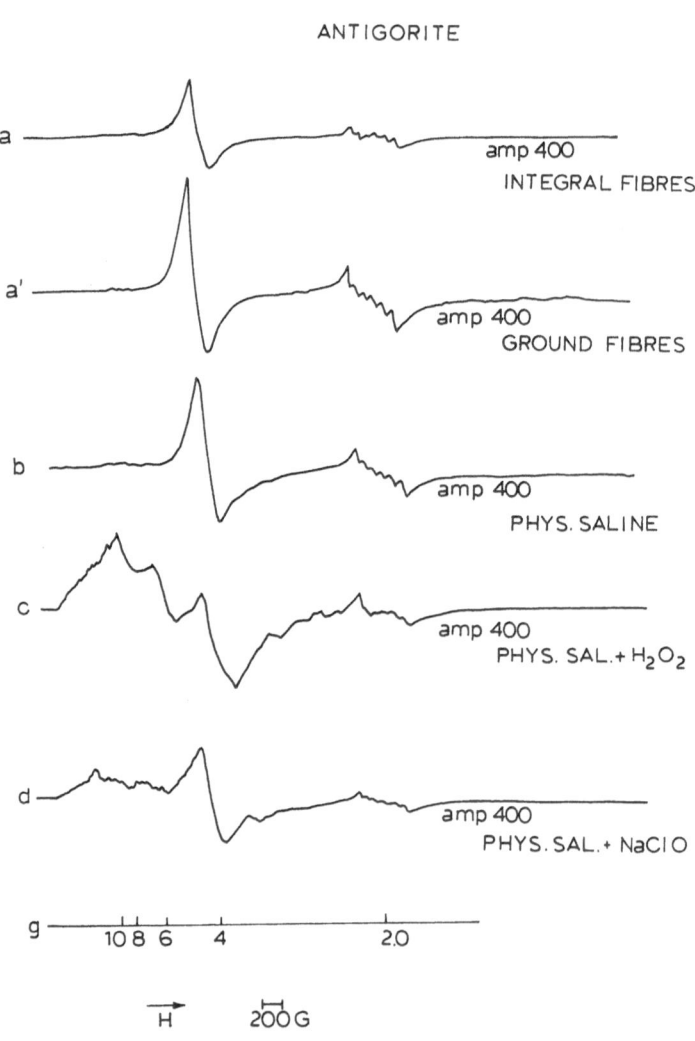

Figure 7. EPR spectra of antigorite. Details as in Figure 6.

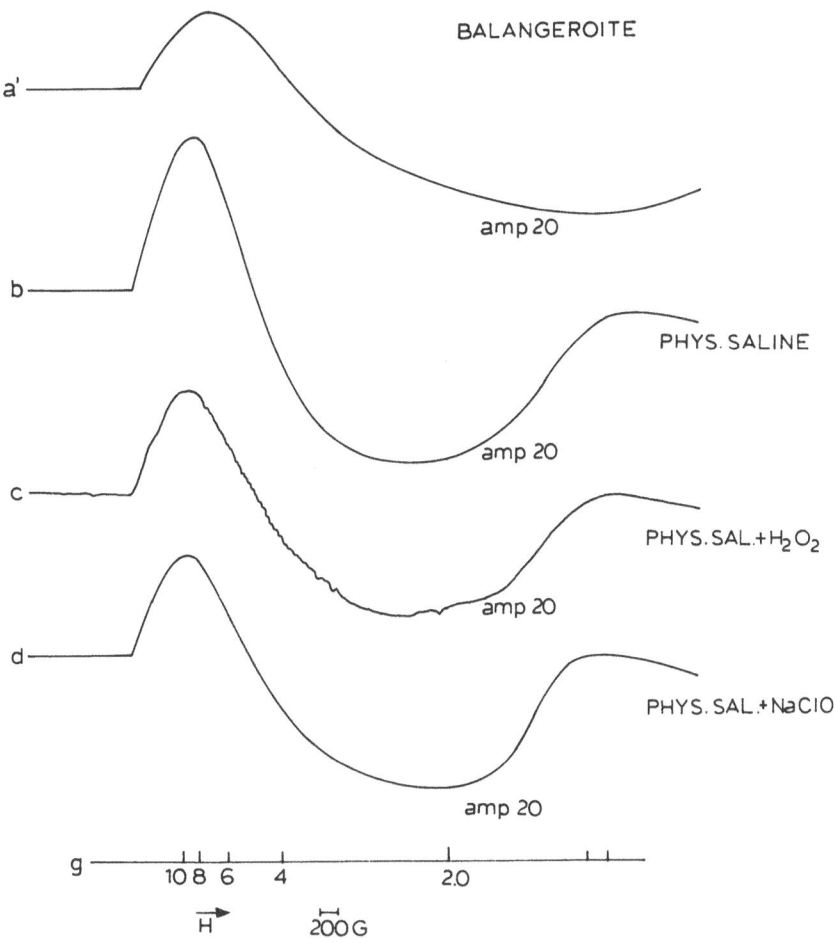

Figure 8. EPR spectra of balangeroite kept in various solutions. Spectrum a': starting material; other features as in Figure 6.

manganese substituting for magnesium in the brucite layer; the relatively small linewidth indicates that the manganese is well dispersed in the lattice at a concentration below 1%. The absence of broad bands in the spectra of Chrysotile A and antigorite shows that these minerals are poorer in iron than chrysotile B. This confirms the large variation in the amounts of iron substituting for magnesium from one chrysotile to another.

The effect of immersion in aqueous solutions: mimicking the biological environment

The spectra from fibres left standing at 37 °C for two weeks in PBS (b), PBS with added with hydrogen peroxide (c), or PBS with added hypochlorite (d) are compared with those of the starting materials in Fig. 6-9. The spectrum of balangeroite is a broad signal due to magnetic interactions between the various paramagnetic ions. After immersion the overall features are still the same, but variations in shape, symmetry and the g value of the centroid indicate that some ions have undergone chemical modification. In physiological solutions balangeroite undergoes a slow fragmentation, the resulting material

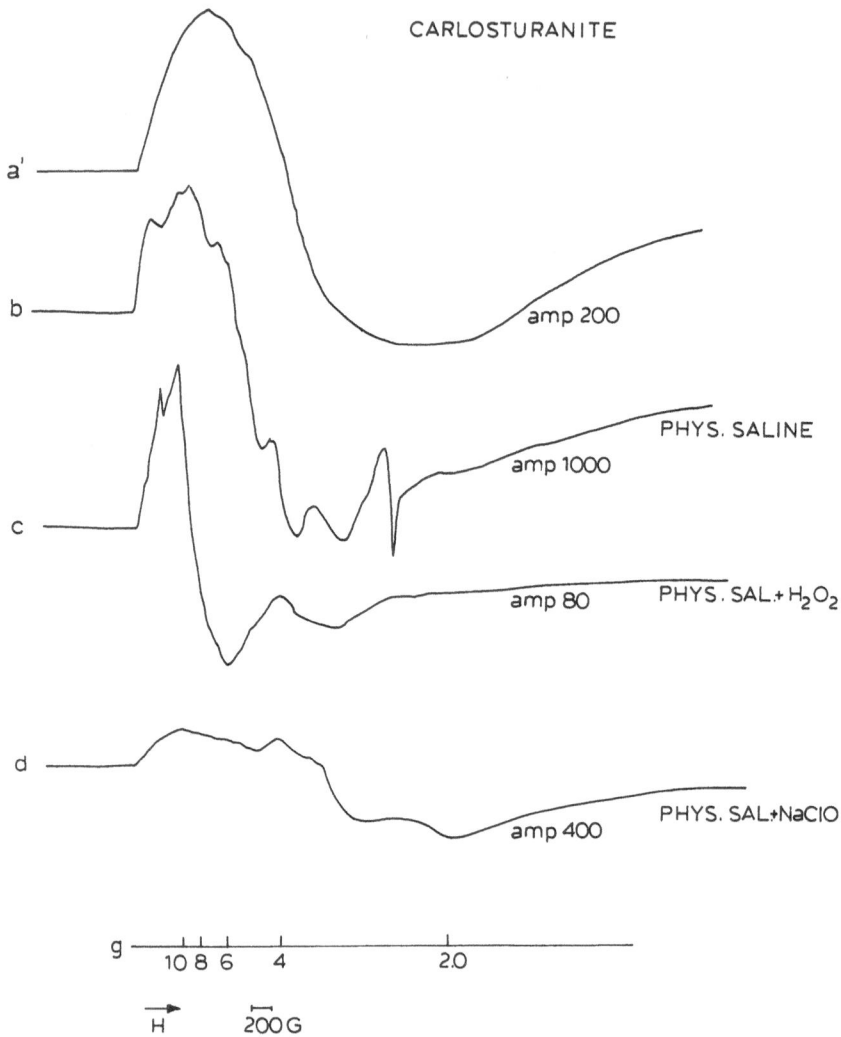

CARLOSTURANITE

a'

b — amp 200

PHYS. SALINE

amp 1000

c

amp 80   PHYS. SAL.+ H_2O_2

d

PHYS. SAL+NaClO

amp 400

g ———— 10 8 6   4   2.0

H   200 G

Figure 9. EPR spectra of carlosturanite kept in various solutions. Features as in Figure 8.

consisting of much smaller fibrils and appearing as a brownish powdery material. Leaching with hydrochloric acid completely eliminates the brown colour and leaves a still fibrous white material which contains no iron. XRD shows that the original structure is preserved.

Carlosturanite (Fig. 9) is profoundly changed by immersion. A constant depletion of iron ions occurs in PBS: the broad signal in the spectrum decays but several new lines appear with an overall decrease in intensity. In contrast the presence of H_2O_2 produces an increase in signal intensity. Most spectral features are changed: two new bands appear at $g = 10$ and $g = 4.2$, both characteristic of Fe^{3+} in distorted octahedral positions. These resonance lines are also visible in the sample treated with NaClO but are less intense and the overall intensity is less than that of the starting material. No visible changes were detected in the shape of the fibres.

Chrysotile A (Fig. 6A b,c,d) is virtually unchanged by the saline solutions whereas chrysotile B (Fig. 6B b ,c ,d) undergoes profound modification. Some spectral features of the starting material, before grinding, reappear after standing in the saline solution, indicating a solubilisation of Fe^{3+} brought to the surface by grinding. Differences in the spectral features in Fig. 6c and d indicate that different redox reactions have occurred in the two cases. The presence of iron therefore imparts a particular reactivity to the mineral.

Antigorite (Fig.7) is little affected by immersion in the various solutions. No significant variations in the overall intensity of the spectra were found and no broad bands appeared indicating an extreme dilution of all the paramagnetic ions. The most remarkable feature of the spectra are the bands at $g = 6$ and $g = 10$ characteristic of isolated Fe^{3+} which only show up with the H_2O_2 solution, indicating some oxidation of Fe(II) to Fe(III).

DISCUSSION

The role of iron in asbestos toxicity

The main hypothesis of this research is that the chemical functionalities implicated in asbestos toxicity have to be sought in the chemistry of iron and its involvement in the production of active oxygen species (AOS). This is in agreement with what has recently been proposed (Mossman *et al.* 1988, Pezerat *et al.* 1989, Zalma *et al.* 1987). We have looked for the presence of Fe(II) or other ions in a low oxidation state and examined their oxidation in different media. Besides the well known Fenton reaction which occurs in presence of H_2O_2, various other reactions may be envisaged in which iron produces AOS directly from atmospheric oxygen (Pezerat *et al.* 1989). It has recently been shown that the initiation of lipid peroxidation by Fe^{2+} and H_2O_2 is mediated by an oxidant which requires both Fe^{2+} and Fe^{3+} (Minotti & Aust, 1987). Pairs of these two ions may be present at the surface of some asbestos minerals and there is the possibility that similar reactions could also occur at the solid liquid interphase. We will therefore focus our interest on the presence of Fe^{2+}. Fe^{2+} does not give an EPR spectrum under our experimental conditions. However, we can show evidence for the presence of this ion either through its interaction with Fe^{3+} (broad band) or by the enhancement of the signal due to Fe^{3+} ions produced by oxidation of Fe^{2+}. Although no biological data are available, we also consider that Mn^{2+} can act as a possible redox site.

New asbestiform minerals: balangeroite and carlosturanite.

The chemical formulae of the two new minerals show that they contain relatively large amounts of Fe^{2+}, Fe^{3+} and Mn^{2+}. For this to affect their toxicity, however, requires that those ions exist at, or can be brought to, the surface of the solid where the redox reaction involving AOS will occur in a biological medium.

Our results show:
i)   the presence of relatively large amounts of Fe^{2+} and Fe^{3+}, similar to those found in crocidolite;
ii)   that changes occur on grinding which indicate that oxidation is taking place at the newly formed surfaces;
iii)   a decrease in EPR signal intensity and other spectral changes after immersion in a

physiological solution which partially mimics the macrophage environment (H_2O_2 or ClO^-). These changes imply that redox reactions occur at the mineral surface and there is substantial leaching of metal ions into the media. The relevance of both these findings to the potential reactivity of these minerals *in vivo* is obvious.

Chrysotile and antigorite from the Western Alps

The chrysotile specimens examined differed in their iron content. Iron substitutes for magnesium in the matrix and marked differences in the composition of rock samples is found. The samples also underwent modification in physiological solutions. Antigorite - whose toxicity has never been reported to our knowledge - behaves in physiological solutions in a similar manner to chrysotile A. No significant changes could be measured in PBS and only minor ones in oxidising solutions. The two main features of the spectrum, those due to isolated Fe^{3+} and Mn^{2+}, are unchanged. These ions, therefore, must be away from the surface and unable to react with molecules in biological media.

Effect of H_2O_2 and ClO^-

The assignment of each spectral component to a defined ion in its crystallographic environment is required for the interpretation of the chemical reactions brought about independently by H_2O_2 and NaClO. It is noteworthy that the two oxidants act in different ways. Hydrogen peroxide carries out redox reactions at various sites and creates new surface arrangements of ions, whereas ClO^-, which suppresses part of the spectra, probably oxidises and assists solubilisation of surface ions. Since both H_2O_2 and ClO^- are present within the macrophage, the effects of both these species will have to be taken into account in any discussion of mineral toxicity.

CONCLUSIONS

Chrysotile asbestos from the Western Alps contains two recently described fibrous silicates: balangeroite and carlosturanite. A preliminary examination of this asbestos suggests a potential toxicity for these fibres *in vivo* but not for the other associated mineral - antigorite. Balangeroite has an EPR spectrum very similar to crocidolite and has a higher iron content, both Fe(II) and Fe(III), than carlosturanite. Considering its association with chrysotile in the Balangero mine it may well be that the excess of cancer in workers at this mine should be related to its presence. The use of EPR spectra to illustrate the modifications that occur when the minerals are kept in physiological solutions, mimicking the biological environment of the inhaled fibre, turns out to be an effective tool for assessing their reactivity. The changes, both in the intensity and in the relevant lines of the spectra, monitors the release of ions into the medium and the formation of new potentially active sites on the mineral surface.

ACKNOWLEDGEMENTS

We would like to dedicate this paper about minerals from Balangero to the memory of Dr. Primo Levi, chemist and novelist, who worked as an analyst in the mine during hard times and reported on it in the short novel "Nickel" in his book "The Periodic Table", Shocken Books, New York, 1984.

REFERENCES

Belluso,E. and Ferraris,G. (1991) New data on balangeroite and carlosturanite from Alpine serpentinites, *Eur. J. Mineral.*, in press.

Berrino,F., Crosignani,P., Aresini,G.A., Guerrieri,M.C. and Vigliani,F.G. (1983) Mortality study of asbestos workers at the Balangero chrysotile mine. In, "VIth International Pneumoconiosis Conference", Bochum, Sept. 20-23, 1983.

Compagnoni,R., Ferraris,G. and Fiora,L. (1983) Balangeroite, a new fibrous silicate related to gageite from Balangero, Italy. *Am. Mineral.*, **68**:214-219.

Compagnoni,R., Ferraris,G. and Mellini,M. (1985) Carlosturanite, a new asbestiform rock-forming silicate from Val Varaita, Italy. *Am. Mineral.*, **70**:767-772.

Ferraris,G., Mellini,M. and Merlino,S. (1987) Electron-diffraction and electron microscopy study of balangeroite and gageite: Crystal structure, polytypism, and fiber texture, *Am. Mineral.* **72**:382-391.

Friebele,E.J., Wilson,L.K., Dozier,A.W. and Kinser,D.L. (1971) Antiferromagnetism in an oxide semiconducting glass. *Phys. Stat. Sol.*, **45**:323-331.

Fubini,B., Bolis,V., Giamello,E. and Volante,M. (1991) Chemical functionalities at the broken fibre surface relatable to free radicals production. This volume.

Kennedy,T.P., Dodson,R., Rao,N.V., Ky,H., Hopkins,C., Baser,M., Tolley,E. and Hoidal,J.R. (1989) Dusts causing pneumoconiosis generate OH⁻ and produce hemolysis by acting as Fenton Catalysts. *Arch. Biochem. Biophys.* **269**:359-364.

Mellini,M., Ferraris,G. and Compagnoni,R. (1985) Carlosturanite: HRTEM evidence of a polysomatic series including serpentine. *Am. Mineral.* **70**:773-781.

Mellini,M. and Zussman,J. (1986) Carlosturanite (not "picrolite") from Taberg. *Sweden Min. Mag.* **50**:675-679.

Minotti,G. and Aust,S.D. (1987) The requirement for iron(III) in the initiation of lipid peroxidation by iron(II) and hydrogen peroxide *J. Biol. Chem.* **262**:1098-1104.

Mossman,B.T., Marsh,J.P., Shatos,M.A., Doherty,J., Gilbert,R. and Hill,S. (1988) Implication of active oxygen species as second messengers of asbestos toxicity. In, "Asbestos Toxicity", eds. Fisher,G.L. and Gallo,M.A., Dekker, New York, pp. 157-181 and references therein.

Pezerat,M., Zalma,R., Guignard,J. and Jaurand,M.C. (1989) Production of oxygen radicals by the reduction of oxygen arising from the surface activity of minerals fibres. *IARC Scient. Publications* **90**:100-111.

Piolatto,G., Negri,E., La Vecchia,C., Pira,E., Decarli,A., Peto,J. (1991) An update of cancer mortality among crysotile asbestos miners in Balangero, Northern Italy. *Brit. J. Ind. Med.*, in press.

Rubino,G.F., Newhouse,M., Murray,R., Scansetti,G., Piolatto,G. and Aresini,G. (1979a) Radiologic changes after cessation of exposure among chrysotile asbestos miners in Italy. *Ann. New York Acad. Sci.* **330**:157-162.

Rubino,G.F., Piolatto,G., Newhouse,M.L., Scansetti,G., Aresini,G.A. and Murray,R. (1979b) Mortality of chrysotile asbestos workers at the Balangero Mine, Northern Italy. *Brit. J. Ind. Med.* **36**:187-194.

Sharrock,P. (1982) Chrysotile asbestos fibres from Quebec: electron magnetic resonance identification. *Geochim. Acta* **46**:1311-1315.

Zalma,R., Bonneau,L., Jaurand,M.C., Guignard,J. and Pezerat,M. (1987) Production of hydroxyl radicals by iron solid compounds. *Toxicol. Environ. Chem.* **13**:171-188; Formation of oxy-radicals by oxygen reduction arising from the surface activity of asbestos. *Canad. J. Chem.* **65**:2938-2341.

MECHANISMS OF PATHOGENESIS
1. Genotoxic effects

MECHANISMS OF FIBRE GENOTOXICITY

Marie-Claude Jaurand

Laboratory of Cellular and Molecular Toxicology
CHU Henri Mondor, 94010 Creteil
CEDEX, France

INTRODUCTION

Neoplastic transformation is a multistep process involving the arrangement, amplification, mutation or deletion of specific genes. Several mechanisms may account for these gene modifications; they include point mutations and chromosome mutations (see Barrett, this volume). Genotoxic agents may be responsible for primary or secondary events which play a critical role in cell transformation by the production, not only of point mutations, but also of heritable changes through gene modifications and rearrangements. Agents that interact with DNA structure, chromosomes or even interfere with DNA metabolism are potentially genotoxic.

The first studies investigating the mechanisms of fibre carcinogenesis appear to be those reported in 1975 (Lavappa *et al.*,1975; Sincock & Seabright, 1975). These described chromosome damage following the *in vitro* treatment of rodent cells with asbestos. No structural chromosome damage was detected with glass fibres or glass powder (Sincock & Seabright,1975). Later, an absence of mutagenic effect in bacteria was reported when asbestos or glass fibres were tested in the Ames' test (Chamberlain & Tarmy, 1977). Several questions arise from these early results: what is the relevance of the clastogenic potency of fibres? To what extent is it possible to compare the effects of different fibre types? What is the importance of particle shape to the observed effects? These questions are now topical issues for various reasons including the production of new fibres and the need to evaluate their toxicity in comparison to that of asbestos fibres.

This review will summarise the published results and will suggest a mechanism of action of fibres. Moreover, an attempt will be made to place these hypotheses in an *in vivo* context.

Inhalation is the normal route of human exposure to aerosols. Fibres are deposited in the airways, at different levels, depending on their size (Bernstein *et al.*, 1991). Fibres must interact first with the surface layer covering the respiratory airways and then with the epithelial cells and interstitial cells. Deposition in the alveoli is much higher for small isometric particles or thin fibres (Gross & Braun, 1984; Schlesinger, 1985) than for larger particles or thicker fibres. In the alveolar region, interaction with alveolar macrophages and epithelial cells may occur. Fibres can be translocated towards the pleura. A deposition of inhaled particles in the subpleural part of the lung has been reported (Morgan *et al.*, 1977), and fibres have been found in the human parietal pleura (Bignon *et al.*, 1980; Sebastien *et al.*, 1980).

During their deposition, size selection and translocation, fibres encounter different cell types. From the types of fibre-associated cancers, the genotoxicity of fibres should occur mainly in bronchial cells and mesothelial cells. In this context, the carcinogenicity, at the organ level, will depend, among other factors, on the number of fibres reaching the "target" cells and on the defence mechanisms and immunological processes operating at these sites.

I. GENOTOXICITY OF FIBRES

Results reported in the literature on the genotoxicity of fibres have recently been reviewed (Jaurand, 1989; 1991). An attempt will be made here to compare the different literature findings.

Effects on DNA

It should be remembered that few data are available on DNA breakage. Breakage of DNA in mammalian cells treated with asbestos has not been found (Fornace 1982; Mossman, 1983; Harris *et al.*, 1985) with direct methods of detecting single strand breaks (Kohn & Grimek-Ewig, 1973). Breakage of bacteriophage DNA in an acellular system did not reach high levels in the absence of tobacco smoke (Jackson *et al.*, 1987).

These results may indicate that, under the experimental conditions used, asbestos or asbestos-treated cells did not produce single strand breaks, or that other types of DNA damage occurred (Jaurand, 1991). However, from other work, it is known that asbestos and asbestos-treated cells produce molecular species such as oxygen derivatives that might damage DNA. This has been demonstrated in acellular systems where DNA damage (Kasai & Nishimura, 1984) or guanine hydroxylation (Leanderson *et al.*, 1988, 1989) have been observed in the presence of hydrogen peroxide and asbestos or man made mineral fibres (MMMF). Moreover, a high rate of DNA strand breaks was obtained by Jackson and co-workers (1987) when hydrogen peroxide and asbestos were incubated with bacteriophage DNA.

In our laboratory, using chrysotile-treated mesothelial cells and the method of nucleoid formation, we have been unable to show DNA strand breaks probably for technical reasons resulting from the interaction between DNA and fibres during cell lysis (Renier *et al.*, unpublished data).

DNA damage by asbestos fibres in cultured cells has been found through the use of indirect methods. Thymidine incorporation has been observed by Libbus and others (1989) after treatment of rat embryo cells with crocidolite fibres. DNA repair, as assessed by the unscheduled DNA synthesis technique (UDS), has been found in rat pleural

TABLE 1. *In vitro* mutagenicity of fibres

Cells[*]	Fibres	Mutagenicity[**]	Reference
Bacteria	Asbestos[§]	-	Chamberlain &
	Glass fibres	-	Tarmy, 1977
CHL	AsbestosE	[+][°]	Huang *et al.*, 1978
			Huang 1979
Adult rat	Chrysotile	-	Reiss *et al.*, 1982
liver	Crocidolite	-	"
	Amosite	-	"
Adult rat	Chrysotile	-	Reiss *et al.*, 1983
liver	+ B[a]P	> B[a]P	"
	+ MNNG	= MNNG	"
Bacteria	Richterite	+	Cleveland, 1984
SHE	Chrysotile	-	Oshimura *et al.*, 1984
	Crocidolite	-	"
H Lymph	Crocidolite	-	Kelsey *et al.*, 1986
	Erionite	-	"
CHO	Crocidolite	-	Kenne *et al.*, 1986
	+ B[a]P	= B[a]P	"
Hybrids AL	Crocidolite	+	Hei *et al.*, 1991

[*]CHL: Chinese hamster lung; CHO: Chinese hamster ovarian; H Lymph: human
lymphoblastoid; SHE: Syrian hamster embryo
[**]B[a]P: benzo[a]pyrene; MNNG: *N*-methyl-*N'*-nitro-nitrosoguanidine
[§]amosite, crocidolite and chrysotile
[°]weak effect

mesothelial cells (RPMCs) treated with chrysotile or crocidolite fibres (Renier *et al.*, 1990). In contrast, rat hepatocytes did not exhibit repair following treatment with chrysotile fibres (Denizeau *et al.*, 1985). These different UDS responses could be due to the nature of the cells used in the two experiments. Different cell types may have different repair efficiency (Hanawalt, 1990); in addition DNA damage could result from secondarily formed molecules whose production may depend on cell type. It is interesting to note that UDS enhancement is not directly associated with phagocytosis since another fibre type, a short attapulgite sample, was ingested by RPMCs without the detection of UDS enhancement and also chrysotile fibres were ingested by rat hepatocytes (Fleury *et al.*, 1983). The UDS method has also been applied to C3H10T1/2 and V79 cells treated with erionite fibres. This treatment allowed the detection of DNA repair of the damage produced by the fibres (Poole *et al.*, 1983).

Until now, the mutagenicity of asbestos was questionable because few studies were positive (Table 1). However, recently Hei and others (1990) have reported mutage-

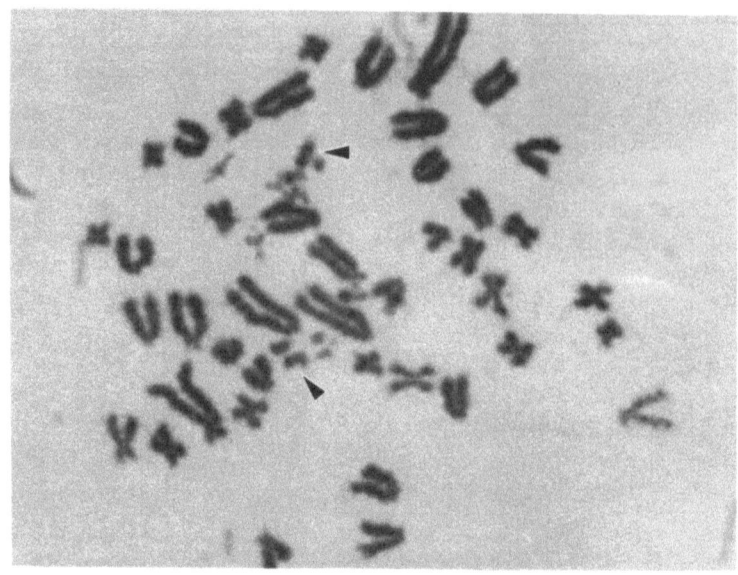

Figure 1. Metaphase from a RPMC culture treated with 1 μg/cm² of chrysotile fibres for 48 hours and showing different abnormalities including fragments (arrowheads).

nicity from the action of crocidolite on human/hamster cell hybrids; these authors found mutations resulted from deletions. Therefore the lack of mutagenicity in bacterial systems (Chamberlain & Tarmy, 1977) might have been due to the fact that the mutation did not allow observation of the phenotype reversion. Another difference between bacteria and mammalian cells in regard to their reactivity towards fibres is the inability of bacteria to ingest fibres. If the occurrence of mutation needs the presence of fibres in the cell, this inability to phagocytose could also account for the lack of effect.

Chromosomal Abnormalities

Unrepaired DNA damage may impair the process of DNA synthesis and result in the formation of structural chromosome abnormalities (CAs) or of sister chromatid exchanges (SCEs). Several authors have studied the formation of CAs (including the formation of micronuclei) in asbestos-treated cells (Table 2). Experiments have been performed using either human or rodent cells. It appears that rodent cells exhibited CAs in all the studies while human cells did not. Figure 1 shows an example of CAs produced in RPMCs by chrysotile fibres. It should be noted that the levels of chromosomal damage obtained in the different systems, while significant, were moderate. Also fewer occurred, even at the highest experimental dose, than was the case with polycyclic aromatic hydrocarbons (Jaurand *et al.*, 1986).

Some studies have been carried out with human cells but with the exception of one study using human lymphocytes (Valerio *et al.*, 1983), no or few structural chromosome aberrations have been reported. Two studies have been carried out with human mesothelial cells (Lechner *et al.*, 1985; Olofsson & Mark,1989). These studies reported numerical chromosomal abnormalities, but according to Olofsson and Mark (*ibid.*), no increased frequency of chromosome breakage or structural aberration was detected following

TABLE 2. Observation of *in vitro* chromosome damage by fibres (including micronuclei)

Cells[*]	Chrysotile	Crocidolite	Others[**]	Reference
SHE	+			Lavappa et al., 1975
CHO-K1	+	+	-GF, - GP	Sincock & Seabright, 1975
CHL		+		Huang et al. 1978
V79-4			-GF 110T, +GF 110R	Brown et al., 1979
CHO	+			Babu et al. 1980
CHO-K1	+	+	+GF 100, -GF 110	Sincock et al. 1982
H fibro (2)	-	-	-GF 100, -GF 110	"
H lymphoid (2)	-	-	-GF 100, -GF 110	
H lymphocytes	+	+		Valerio et al. 1983
SHE	+	+	+GF 100	Oshimura et al. 1984
HMC			[+]°Amosite§	Lechner et al. 1985
RPMC	+			Jaurand et al. 1986
CHO		+	+Erionite, -Qz	Kelsey et al. 1986
RTC	+[***]	+[***]		Hesterberg et al. 1987
HBE	[+]E			Kodama et al. 1987
V79	+	+	+Erionite	Palekar et al. 1987
HMC	-	-	-Amosite	Olofsson & Mark 1989

[*] Symbols: see Table 1 and HBE: human bronchial epithelial; H fibro: human fibroblasts; RPMC: rat pleural mesothelial cell; V79: Chinese hamster lung cells.

[**]   GF: glass fibres; T: total; R: respirable; GP: glass powder; Qz: quartz

[***] Micronuclei

° Weak effect

§ No breakage has been mentioned. Dicentrics have been found.

treatment of four mesothelial cell lines with 100 μg/ml of chrysotile or amphibole fibres. Lechner and co-workers (1985) found only the formation of di-centric chromosomes. The differences exhibited between rodent cells and human cells are of great interest. Sincock and co-workers (1982) had previously mentioned that the human cell lines tested in their experiments did not form CAs. The authors suggested that this could have been partly due to the greater proficiency of human cells to repair DNA damage as compared with rodent cells. This has also been demonstrated by Hanawalt (1990) while studying repair of cyclobutane pyrimidine dimers. Thus, the results observed with human mesothelial cells might be explained by such differences in repair capabilities. It should be added here that fibroblasts from xeroderma pigmentosum patients were more sensitive to asbestos than fibroblasts from normal subjects (Yang et al., 1984), in agreement with an involvement of DNA repair in the maintenance of cell viability.

TABLE 3. Observation of *in vitro* aneuploidy and polyploidy in cells exposed to fibres (including binuclei formation)

Cells[*]	Chrysotile	Crocidolite	Others[**]	Reference
CHO-K1	+	+	[+]GF -GP	Sincock *et al.*, 1975
CHL		+		Huang *et al.*, 1978
V79-4		+	-Min-U-Sil	Price-Jones *et al.*, 1980
CHO-K1	+	+	+GF 100, -GF 110	Sincock *et al.* 1982
H fibro (2)	-	-	-GF 100, -GF 110	"
H lymphoid (2)	-	-	-GF 100, -GF 110	"
RPMC	+[**]	+[**]		Jaurand *et al.* 1983
H lymphocytes	-	-		Valerio *et al.* 1983
SHE	+	+	+GF 100	Oshimura *et al.* 1984
HMC			+Amosite	Lechner *et al.* 1985
H fibro	+			Vershave & Palmer 1985
RPMC	+			Jaurand *et al.* 1986
CHO		+	+Erionite, -Min-U-Sil	Kelsey *et al.* 1986
CHO		+		Kenne *et al.* 1986
RTE	+	+		Hesterberg *et al.* 1987
HBE	+[**,§]			Kodama *et al.* 1987
V79	+	+	+Erionite	Palekar *et al.* 1987

[*] Symbols: see Tables 1 and 2
[**] Binuclei
[§] No numerical change

Few experiments have been carried out with fibres other than asbestos, but erionite produced CAs in rodent cells. In four studies carried out with glass fibres, CAs were found with code 100 fibres and with a respirable fraction of code 110 fibres (Brown *et al.*, 1979; Sincock *et al.*,1982; Oshimura *et al.*, 1984).

Aneuploidy has been described in different cell types and species after treatment with asbestos, erionite or glass fibres (Table 3). In most experiments, asbestos-treated cells, except lymphocytes, exhibited aneuploidy and/or polyploidy. This has been observed with rodent cells as well as with human cells. Erionite and glass fibres also produced these effects in cells, but not lymphocytes. The lack of effect on lymphocytes might be due to the absence of fibre ingestion by these cell types.

It has been mentioned above that phagocytosis should be an important parameter in the genesis of genetic damage by fibres. While it does not appear that DNA damage is necessarily associated with phagocytosis, aneuploidy could easily be dependent on phagocytosis. Aneuploidy may be a consequence of the interaction between fibres and components of the mitotic spindle or chromosomes. There are several observations that may account for the formation of aneuploid cells as a result of phagocytosis. It should first be remembered that phagocytosis is defined as an ingestion process which is followed by a degranulation of lysosomes in the phagocytic vacuole. This has been observed *in vitro* in macrophages and in RPMCs treated with chrysotile fibres; moreover, mesothelial cells in pleural explants can take up fibres (unpublished data and Jaurand *et al.*, 1979a). Internalised fibres have been found in the perinuclear region in Syrian hamster embryo (SHE) cells and in tracheal cell lines (Hesterberg *et al.*, 1986, 1987 and Fig. 2). Electron microscopic studies have revealed that, in interphase cells, fibres are sometimes observed crossing the nucleus (Fig. 3) or located near or in the nucleus (Johnson & Davies, 1983; Jaurand, 1991 and Fig. 4). However, the fibres still seem to be surrounded by a membrane. When fibres are present within cells they may interact with the cell components during mitosis. Fibres have been observed in the different phases of the mitosis of RPMCs (Jaurand *et al.*, 1983; Jaurand, 1991). It has been suggested that interaction with the mitotic spindle may produce chromosome mis-segregation and thus result in the formation of aneuploid cells (Barrett *et al.*, 1987; 1990). Abnormal anaphases characterised by the presence of lagging chromosomes and bridges have been observed in SHE cells treated with crocidolite (Hesterberg & Barrett, 1985), human fibroblasts treated with chrysotile (Vershave & Palmer, 1985) and V79 cells treated with chrysotile, crocidolite or erionite (Palekar *et al.*, 1987). Using scanning electron microscopy, chrysotile and crocidolite have been found to be associated in cells with chromosomes (Wang *et al.*, 1987).

These different pieces of evidence suggest that aneuploidy most likely results from the interaction between fibres and subcellular structures; a phenomenon that can occur following fibre ingestion and which is associated with cell division. Whether such events may take place after fibre inhalation will be considered later.

The involvement of aneuploidy in a multistep model of carcinogenesis is discussed elsewhere in this book (Barrett, 1991). It will be mentioned here that aneuploidy has been characterised in SHE cells transformed by chrysotile fibres: a non-random trisomy of chromosome 11 has been reported (Oshimura *et al.*, 1986). Aneuploidy has also been found in mesothelioma cells from asbestos induced rat peritoneal tumours (Libbus & Craighead, 1988) as well as in mesotheliomas in patients who had been exposed to asbestos (Tiainen *et al.*, 1989). Rat pleural mesothelial cells treated *in vitro* with chrysotile fibres exhibited a modal chromosome number of 43 after several treatments with chrysotile fibres: trisomy 1 partly accounted for this aneuploidy (Jaurand *et al.*, 1991). It is interesting to note that other authors have mentioned this type of numerical abnormality in mesothelioma cells derived from pleural tumours induced in rats by chrysotile or erionite (Palekar *et al.*, 1989). It should also be mentioned that trisomy 1 is associated with spontaneous peritoneal mesotheliomas in the Fisher rat (Walker *et al.*, 1991). Further research is necessary to understand the relevance of these chromosomal modifications in the neoplastic transformation of rat mesothelial cells and to determine to what extent they are specific to the effect of fibres.

Unlike chromosome breakage, aneuploidy following treatment with fibres does not seem species specific. However, an absence of aneuploid cells has been reported with human lymphocytes (Sincock *et al.*, 1982; Valerio *et al.*, 1983). One possible explanation, is the lack of fibre ingestion by lymphocytes, as suggested above. In Valerio's experiment, unlike others carried out with lymphocytes, CAs were obtained. This absence of aneuploid cells should reinforce the hypothesis of a clastogenic effect mediated via secondary molecules, possibly formed by monocytes as the method used to prepare leukocytes does not seem to eliminate monocytes (Moorhead *et al.*, 1960). Recently we have

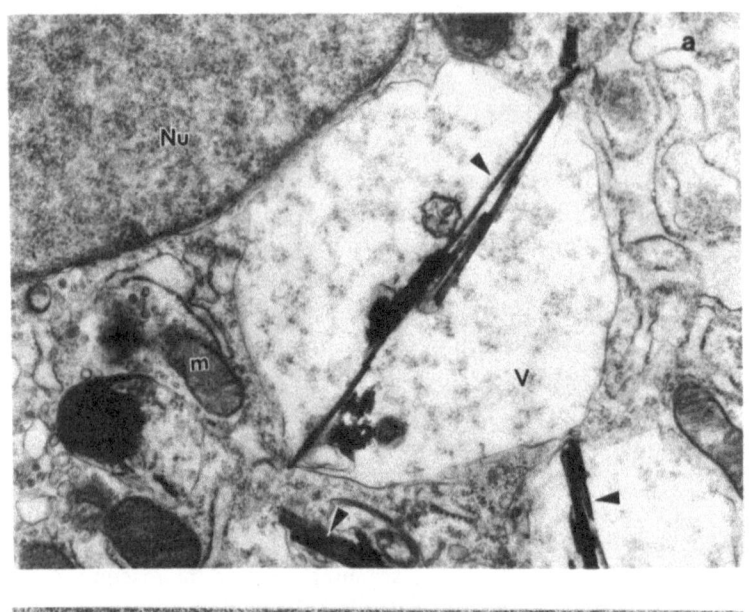

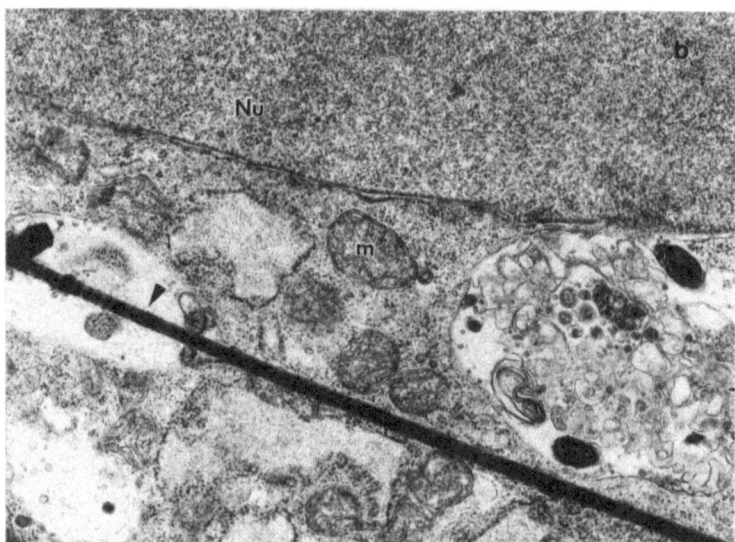

Figure 2. Transmission electron microscopy of an ultra thin section of a RMPC from a culture treated with 1 μg/cm^2 of asbestos fibres (arrowheads). Fibres are located in a phagocytic vacuole near the nucleus. m: mitochondria, Nu: nucleus; V: vacuole characteristic of treatment with chrysotile. a: chrysotile (X21,200); b: crocidolite (X25,200).

studied the production of sister chromatid exchanges in human lymphocytes treated with different sorts of silica (Pairon *et al.*, 1989). A weak effect was observed with tridymite but only when monocytes were present in the incubation media. It has been suggested that this effect was mediated by secondary molecules resulting from particle ingestion by monocytes. However, additional experiments are necessary to determine the relevance of these hypotheses.

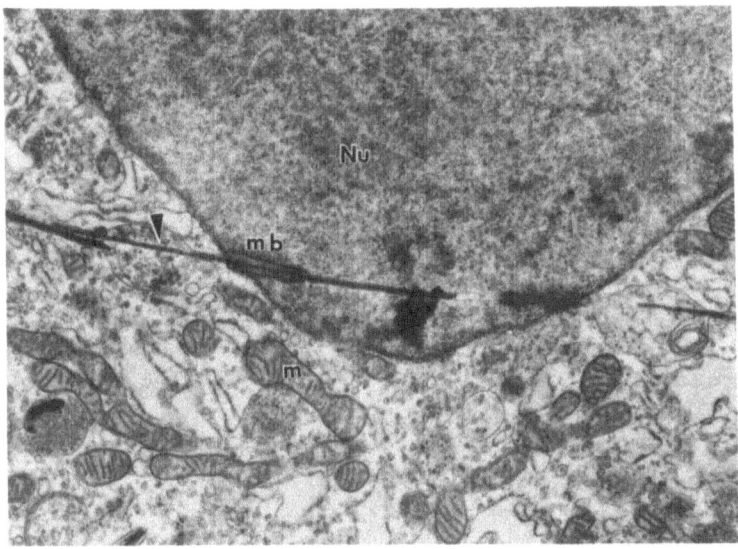

Figure 3. Transmission electron microscopy of an ultra thin section of a RPMC from a culture treated with 1 μg/cm^2 of chrysotile fibres. A fibre is seen crossing the nucleus but surrounded by the nuclear membrane (mb) (X15,000). Symbols: see Figure 2.

Sister Chromatid Exchanges (SCEs)

Experiments resulting in both the enhanced production of SCEs and the absence of effect have been reported (Table 4). In most experiments, no effect was observed with crocidolite, the fibre type most often tested. There is no clear explanation to account for the lack of consistency. It could be suggested that the differences were related to several parameters including differences in cell cycle kinetics (Kelsey *et al.*, 1986) or in the culture conditions. It seems that the production of SCEs is strongly dependent on the culture conditions, especially on the nature of the serum used in the medium. The formation of SCEs has been well studied with the promoter *O*-tetradecanoyl-phorbol-13-acetate (TPA) where oxygen derivatives, either directly or indirectly, are likely to be involved in the production of SCEs (Emerit & Lahaud-Maghani, 1989). Contradictory results have also been reported (Loveday & Latt, 1979; Gentil *et al.*, 1980; Thompson *et al.*, 1980; Schwartz *et al.*, 1982). According to Nagasawa & Little (1981) the differences could be attributed to the nature of the serum and they recommend that heat inactivated serum should be used. In addition, if oxygen derivatives are involved in the production of SCEs, the level of superoxide dismutase in serum might modulate the result (Baret & Emerit, 1983). Moreover, cultured cells have to be incubated for about 48 hours, *i.e.*, over two cycles to allow the detection of SCEs. Therefore the repair mechanisms and the nature of the culture medium could be important factors modulating the expression of SCEs.

Other Genetic Effects

It is remarkable that asbestos fibres can facilitate the transfer of foreign genes into recipient cells. Dubes and Mark (1988) have reported the transfection of viral RNA into mammalian cell cultures by several types of asbestos, but this effect was not asbestos specific since other non-fibrous particles exerted the same effect. Moreover, Appel and co-workers (1988) have transfected the proto-oncogene coding for the p53 protein into monkey Cos cells and observed expression of the protein. Chrysotile fibres were as active as the calcium phosphate which is usually used in gene transfer experiments, but one sample of glass fibres failed to achieve transfection.

The significance of these experiments is unknown because of the lack of information on the general mechanism of transfection. However, they do indicate that several types of particles can facilitate the uptake of nucleic acids and their integration in the cell genome.

In vivo Genotoxicity

Few studies have been carried out to study the effects of fibres on the production *in vivo* of genotoxic effects. Rita and Reddy (1986) have reported that chrysotile fibres failed to produce abnormal chromosomes in germ cells of mice fed orally with the fibres. However, it is not known if the fibres reached the germ cells. Other data is needed to determine whether fibres are able to induce germinal mutations.

II. FIBRE CHARACTERISTICS RELEVANT TO GENOTOXICITY

From the results published in the literature on the mechanisms of action of fibres, it appears that several parameters may play a role in the carcinogenic process. The parameters are the size (length and diameter) and the physico-chemical properties (charge, interaction with biological molecules, ability to split, chemical stability, surface reactivity).

Fibre Dimensions:

Fibre size is an important parameter that influences the deposition, retention and clearance of fibres. The size is also of importance in the toxicological process and fibres of different size have different cell toxicity and genotoxicity as demonstrated by *in vitro* tests (Brown *et al.*, 1979; Tilkes & Beck, 1983; Hesterberg & Barrett, 1984; Brown *et al.*, 1986; Jaurand *et al.*, 1987). Sincock *et al.* (1982) have reported investigations on the genotoxicity of glass fibre of different dimensions using different cell types. Chromosomal aberrations were detected with code 100 fibres when rodent cells were exposed to 10 μg/ml while no effect was observed with the thicker code 110 fibres.

It is generally reported that long fibres are more toxic than short fibres. Fisher (1989) has studied the SCEs of JM 104 and JM 100 glass fibres, a fine and an ultrafine fraction produced by a colloid chemical method. It was concluded that a given number of the short fibres had less effect than did the longer fibres.

Two papers have reported studies performed with JM 100 and JM 110 fibres. Brown and co-workers (1979) tested both glasses including respirable fractions of these samples using cytotoxicity and genotoxicity assays. The results were compared on the basis of weight and number of fibres. On a per weight basis, there was little or no cytotoxicity using code 110 fibres but the respirable fractions gave increased biological activity. Code 100 fibres exhibited a greater cytotoxicity than code 110. However, the number of fibres contained in a given weight was not the same in the different samples and the conclusions changed when the results were compared on the basis of the number of fibres. Thus, it appears that a given number of code 110 fibres is more toxic than the same number of code 100 fibres. When only the fibres more than 10 μm long were taken

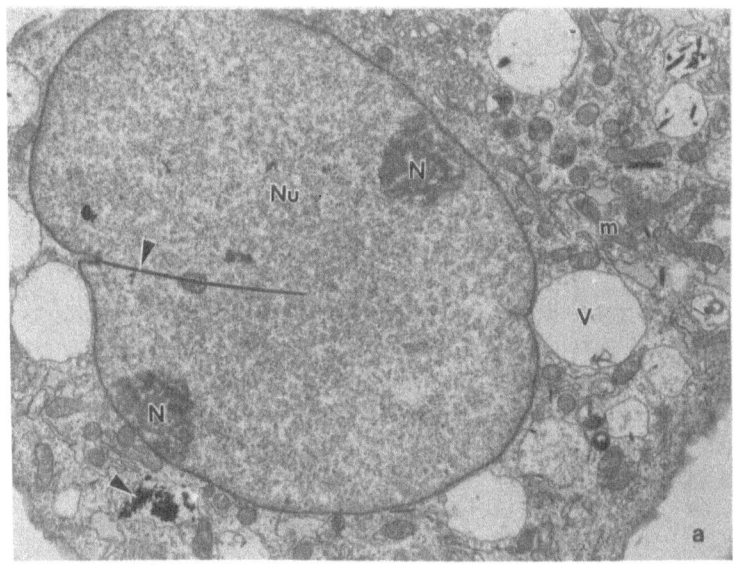

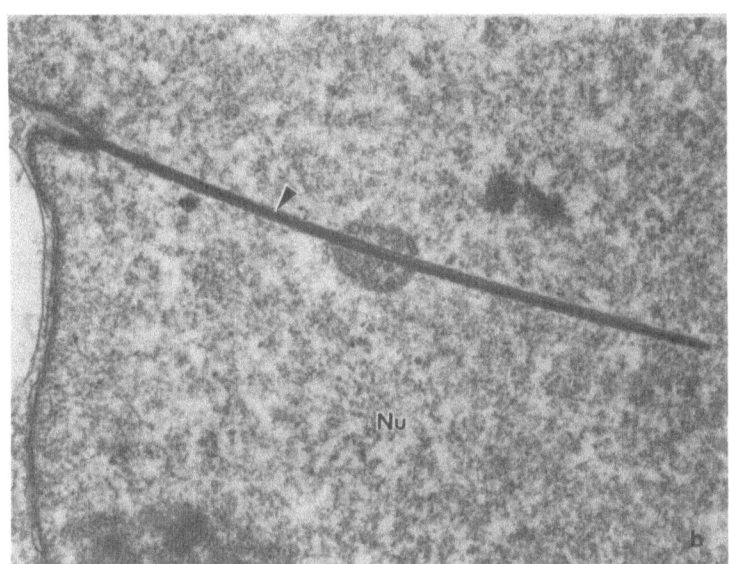

Figure 4. Transmission electron microscopy of an ultra thin section of a RPMC from a culture treated with 1 μg/cm² of chrysotile fibres. Symbols: see Figure 2. N: nucleolus. a: X6,900; b: X32,500.

into consideration the same trend was found. A similar effect of fibre sizes has been found by Hesterberg and Barrett (1984) using a transformation assay with SHE cells.

Several studies have been carried out with milled versus unmilled fibres. Milling fibres results in a modification of the fibre dimensions but may also modify the physico-chemical properties (Langer *et al.*, 1978; Langer, 1985). Ririe and co-workers (1985) have reported that milling reduced the cytotoxicity of code 100 to rat tracheal epithelial cells. Hesterberg and Barrett (1984) and Hesterberg and co-workers (1986) have reported that milling code 100 glass fibres resulted in a reduction of the genotoxicity as well as the cytotoxicity and transformation of SHE cells.

From these experiments it seems that the higher carcinogenic potency of long fibres compared to short fibres is also found in genotoxicity assays. Moreover, they emphasise how difficult it is to compare quantitatively the activity of different samples which differ in their size distribution, number of fibres on a per weight basis, chemical and/or physical properties.

Fibre Chemistry

Few data are available on the dependence of genotoxicity on fibre durability, in one investigation the effects of chrysotile and phosphorylated chrysotile on the *in vitro* production of CAs and aneuploidy in RPMC were compared (Jaurand *et al.*, 1988). Phosphorylated chrysotile was found to be chemically more durable in short-term assays, after ingestion by macrophages, than native chrysotile. However, the level of chromosome abnormalities was not significantly different (Jaurand *et al.*, 1988).

The effects of the production of oxygen derivatives has not been properly tested in terms of genotoxicity. Their production by asbestos or some samples of MMMF has been shown in acellular systems (Weitzman & Graceffa, 1984; Zalma *et al.*, 1987), but no attempt has been made to correlate this effect to the level of DNA or chromosome damage in cells. However, the hydroxylation of guanine in DNA, mentioned above (see Effects on DNA) may be one of the events related to carcinogenesis (Floyd, 1990). It could be one of the stages directly or indirectly involved in the multistep process proposed for fibre carcinogenesis (Barrett, 1991). Further experiments are necessary to understand the importance of this physico-chemical property.

III. CAN INHALED FIBRES BE GENOTOXIC?

The genotoxicity of fibres has mainly been demonstrated *in vitro*. Such studies have shown the mutagenic potency of fibres and the question which now arises is whether these effects can occur *in vivo*.

As mentioned at the beginning of this paper, inhaled fibres are deposited in the respiratory airways where they may interact with different types of epithelial cells. In the rat, fibre deposition is mainly observed at the bifurcation of the bronchus (Brody & Hill, 1982). After deposition in the airways, they pass through the epithelium and reach the interstitial compartment (Brody & Hill, 1982). According to Lee *et al.* (1981b), fibres can migrate through direct penetration or via transfer of phagocytes to the lymphatic system and blood circulation. The fibre surface can be modified by biological molecules such as surfactant or proteins that have the ability to be adsorbed at the fibre surface: adsorption could modify, at least transiently, the fibre reactivity (Jaurand *et al.*, 1979b; Wallace *et al.*, 1985).

If aneuploidy is an important stage in the multistep process of carcinogenesis by fibres, two events have to occur to allow the formation of aneuploid cells. First, the fibres should be ingested by the target cell; second, the cells should undergo mitosis.

It seems that bronchial cells are the main target of inhaled fibres. In certain studies, an association between lung adenocarcinoma and asbestos exposure has been found (Mollo *et al.*, 1990). Experimental studies have shown that epithelial cells may take up asbestos (Mossman *et al.*, 1977; Lee *et al.*, 1981a; Brody & Hill, 1982; Woodworth *et al.*, 1983; Roggli & Brody, 1984); this is an important step in the production of genotoxicity.

TABLE 4. Observation of SCEs in cells treated *in vitro* with fibres

Cells[*]	Chrysotile	Crocidolite	Others	Reference
CHO		+	-Amosite	Livingston *et al.* 1980
V79-4		-		Price-Jones *et al.* 1980
CHO-K1	-	-	-GF 100, -GF 110	Casey 1983
H fibro	-	-	-GF 100, -GF 110	
H lymphocytes	-	-	-GF 110	
C3H 10T1/2	-	-	+radiation> radiations	Hei *et al.* 1985
CHO		-	+Erionite	Kelsey *et al.* 1986
RPMC		+	-Attapulgite	Achard *et al.* 1987
Chinese hamster			+GF 100	Fisher 1989

[*] Symbols: see Tables 1 and 2; C3H 10T1/2: mouse fibroblasts

In addition, after inhalation of chrysotile an increased incorporation of thymidine has been reported in epithelial and interstitial cells (Brody & Overby, 1989) which most likely indicates a cell proliferation.

In contrast to bronchogenic carcinoma, mesothelioma is more specifically related to fibre exposure, even though it is thought that mesotheliomas do not always result from an exposure to fibres (Bignon & Brochard, 1991). The major problem in interpreting the genotoxicity results in relation to *in vivo* effect is the lack of sufficient data concerning the rate of translocation of fibres from the respiratory airways towards the pleural mesothelial cells. According to the mutational hypothesis, fibres should be present in mitotic cells to exert their effect. Sebastien and co-workers (1980) and Bignon and co-workers (1980), for example, have shown that short chrysotile fibres are present in human parietal pleural tissue.

How relevant is the presence of short fibres, which are thought to be weakly carcinogenic? There are at least three answers. First, although short fibres are much less efficient than long fibres on a per number basis they do have some activity. This has been shown in several *in vitro* experiments. Secondly, after intrapleural inoculation, short chrysotile fibres retain some carcinogenic potency when compared to longer fibres (Jaurand *et al.*, 1987; Kane, 1991) and the reduced activity could partly be due to the rapid migration from the pleural space towards other regions. This suggestion is in agreement with recent data from Kane (1991) who has reconsidered the potency of fibres according to their size and found that, if retained at the peritoneal site, short fibres will produce some mesotheliomas. Finally, it can be assumed that short fibres found in human decades after inhalation are derived from longer fibres. Thus, even if fibres are not seen in the mesothelial cells they could be present in the pleura. Moreover, we have observed that fibres are ingested by mesothelial cells from pleural explants. Another intriguing point is that in animal experiments, where the fibres were administered intra-tracheally, fibres have not been detected in the pleural fluid or in the pleural cells (Lee *et al.*, 1981a;

Oberdoerster *et al.*, 1983). It is possible that few fibres reach the pleural space and these are very quickly retranslocated towards the stomata present at the surface of the parietal pleura (Wang, 1975), or towards the lung. In effect, after intrapleural inoculation, fibres migrate to the lung via the lymphatic system.

The other question regarding the mutational hypothesis concerns cell proliferation. One can imagine that cells divide because of epithelial regeneration following injury resulting from the presence of fibres. However, although this process may occur experimentally, it is unlikely to be relevant to the human situation if one assumes that a small number of fibres will reach the pleura. Other processes can be suggested which produce cell division. Growth factors can be produced by other cell types as a consequence of the inflammatory process. Three observations should be mentioned. According to Dodson and Ford (1985), pleural mesothelial cells show *in vivo* morphological changes after intratracheal instillation of amosite fibres in guinea pigs, possibly resulting from the effect of circulating factors. Bryks and Bertalanffy (1971) reported an increased tritiated thymidine labelling index after intratracheal instillation of chrysotile in rats. More recently Coin and co-workers (1991) also reported increased thymidine incorporation in the DNA of mouse pleural mesothelial cells *in vivo* after inhalation of chrysotile fibres. These results suggest that either a few fibres or circulating factors might stimulate mesothelial cells after inhalation. It has been reported that mesothelial cells release chemotactic factors for polymorphonuclear neutrophils (Antony *et al.*, 1989) and these cells might in turn produce stimulating factors.

In the relationship between *in vitro* and *in vivo* studies the results obtained *in vitro* indicate that under certain experimental conditions, high transformation rates can be obtained (Jaurand *et al.*, 1990). Since the total number of fibres determined by transmission electron microscopy (TEM) in the UICC/A sample of chrysotile is about 10^7 per μg (Jaurand *et al.*, 1987) an incubation with 1 μg/cm^2 or 25 μg per 25 cm^2 flask will represent about 2.5×10^8 fibres or, on a surface basis, about 80 fibres per cell. With crocidolite there are about four times fewer fibres per unit mass.

Inhalation, both in human and in rodents, produces low rates of mesothelioma. For example, Lee *et al.* (1981) exposed rats to a cumulative dose of 7.1×10^6 fibres.ml^{-1}.h of amosite (optical count); Wagner and co-workers (1985) exposed rats to 2.9×10^6 fibres.ml^{-1}.h of crocidolite and Mühle and co-workers (1987) to 2.0×10^6 fibres.ml^{-1}.h of crocidolite (SEM counts) or 0.24×10^6 fibres.ml^{-1}.h of chrysotile. Smith and co-workers (1987) have exposed rats and hamsters to 9.4×10^6 crocidolite fibres (scanning electron microscopy determination). In these experiments where 35 to 60 animals per group were exposed only one mesothelioma was observed in a rat (Smith *et al.*, 1987). Assuming that a rat inhales 240 mls of air per minute, it would inhale during its "working" life a total of 1.4×10^4 times the hourly dose, *i.e.*, about 10^{10} to 10^{11} fibres. Although about 1% of these fibres will be deposited deep in the lung, an unknown percentage will reach the pleura. The number reaching any mesothelial cell should be lower than that used in an *in vitro* incubation.

It is difficult to compare inhalation studies to those *in vitro*, and it is not the aim of this discussion to derive a conversion factor. But from the calculations of fibre numbers, it appears that the differences between *in vitro* responses and the *in vivo* tumour yield could at least partly result from the different doses applied to, or received by, the cells. However, other parameters have to be taken into consideration to account for the low rate of mesotheliomas generally observed by inhalation. These parameters are tissue specificity and species specificity. The data reported here indicate a genotoxic potential of fibres that seems to be modulated by the fibre's physical characteristics. The mechanisms whereby fibres are genotoxic need certain *in vivo* events to take place after inhalation. Genotoxicity may result in cell transformation via a multistep process but this does not imply that a tumour must arise. The mutated cell has to undergo several other steps related to neoplastic transformation and to overcome other tissue and organ reactions including immunological surveillance to form first a tumour and secondly a detectable neoplasm.

REFERENCES

Achard,S., Perderiset,M. and Jaurand,M.C. (1987) Sister chromatid exchanges in rat pleural mesothelial cells treated with crocidolite, attapulgite, or benzo-3,4-pyrene. *Br. J. Indust. Med.* **44**:281-283.

Antony,V.B., Owen,C.L. and Hadley,K.J. (1989) Pleural mesothelial cells stimulated by asbestos release chemotactic activity for neutrophils *in vitro*. *Amer. Rev. Resp. Dis.* **139**:199-206.

Appel,J.D., Fasy,T.M., Kohtz,S., Kohtz,J.D. and Johnson,E.M. (1988) Asbestos fibers mediate transformation of monkey cells by exogenous plasmid DNA. *Proc. Natl. Acad. Sci. U.S.A.* **85**:7670-7676.

Babu,K.A., Lakkad,B.C., Nigam,S.K., Bhatt,D.K., Karnik,A.B., Thakore,K.N., Kashyap,S.K. and Chatterjee,S.K. (1980) *In vitro* cytological and cytogenetic effects of an Indian variety of chrysotile asbestos. *Environ. Res.* **21**:416-422.

Baret,A. and Emerit,I. (1983) Variation of superoxide dismutase levels in fetal calf serum. *Mutat. Res.* **121**:1293-297.

Barrett,J.C. (1991) Asbestos carcinogenicity: a mutational hypothesis. (this volume).

Barrett,J.C., Lamb,P.W. and Wiseman,R.W. (1990) Hypotheses on the mechanisms of carcinogenesis and cell transformation by asbestos and other mineral dusts. In "Health Related Effects of Phyllosilicates", NATO ASI Series, vol. **G21**, Bignon, J., ed., Springer-Verlag, Berlin-Heidelberg, pp. 281-307.

Barrett,J.C., Oshimura,M., Tanaka,N. and Tsutsui,T. (1987) Genetic and epigenetic mechanisms of presumed nongenotoxic carcinogens. In, "Nongenotoxic Mechanisms in Carcinogenesis", Banbury Report 25, Butterworth, B. and Slaga, T.J., eds., Cold Spring Harbor Press, New York, pp. 311-324.

Bernstein,D.M., Fleissner,H., Bouvier,C., Hesterberg,T. and Mast,R. (1991). An inhalation model for evaluation of fiber oncogenicity in rodents. this volume).

Bignon,J. and Brochard,P. (1991) Animal and cell models for understanding and predicting fibre-related mesothelioma in man. (this volume).

Bignon,J., Jaurand,M.C. and Sebastien,P. (1980) Interaction between mineral fibres and pulmonary cells. In, "The Lung in its Environment", Bonsignore, G. and Cumming, G., eds., Plenum Press, New York and London, pp. 321-342.

Brody,A.R. and Hill,L.H. (1982) Interstitial accumulation of inhaled chrysotile asbestos fibers and consequent formation of microcalcifications. *Am. J. Pathol.* **109**:107-114.

Brody,A.R. and Overby,L.H. (1989) Incorporation of tritiated thymidine by epithelial and interstitial cells in bronchiolar-alveolar regions of asbestos-exposed rats. *Am. J. Pathol.* **134**:133-140.

Brown,R.C., Chamberlain,M., Davies,R., Gaffen,J. and Skidmore,J.W. (1979) *In vitro* biological effects of glass fibers. *J. Environ. Pathol. Toxicol.* **2**:1369-1383.

Brown,G.M., Cowie,H., Davis,J.M.G. and Donaldson,K. (1986) *In vitro* assays for detecting carcinogenic mineral fibres: a comparison of two assays and the role of fibres size. *Carcinogenesis* **7**:1971-1974.

Bryks,S. and Bertalanffy,F.D. (1971) Cytodynamic reactivity of mesothelium. *Arch. Environ. Hlth* **23**:469-472.

Casey,G. (1983) Sister chromatid exchange in cell kinetics in CHO-K1 cells, human fibroblasts and lymphoblastoid cells exposed *in vitro* to asbestos and glass fibers. *Mutat. Res.* **116**:369-377.

Chamberlain,M. and Tarmy,E.M. (1977) Asbestos and glass fibers in bacterial mutation tests. *Mutat. Res.* **43**:159-164.

Cleveland,M.G. (1984) Mutagenesis of *Escherichia coli* (CSH50) by asbestos (41954). *Proc. Soc. Exp. Biol. Med.* **177**:343-346.

Coin,P.G., Moore,L.B., Roggli,V. and Brody,A.R. (1991) Pleural incorporation of ^{3}H-TdR after inhalation of chrysotile asbestos in the mouse. ATS meeting, (to be published).

Denizeau,F., Marion,M., Chevalier,G. and Cote,M. (1985) Inability of chrysotile asbestos fibers to modulate the 2-acetylaminofluorene-induced UDS in primary cultures of rat hepatocytes. *Mutat. Res.* **155**: 83-90.

Dodson,R.F. and Ford,J.F. (1985) Early response of the visceral pleura following asbestos exposure: an ultrastructural study. *J. Toxicol. Environ. Hlth* **15**:673-686.

Dubes,G.R. and Mack,L.R. (1988) Asbestos-mediated transfection of mammalian cell cultures. *In vitro Cell Dev. Biol* **24**:175-182.

Emerit,I. and Lahoud-Maghani,M. (1989) Mutagenic effects of TPA-induced clastogenic factor in Chinese hamster cells. *Mutat. Res.* **214**:97-104.

Fisher,A.B. (1989) Induction of sister chromatid exchanges by fibrous dusts alone and in combination with other xenobiotics in Chinese hamster cells. In, "Effects of Mineral Dusts on Cells", NATO ASI Series, vol. H3, Mossman, B.T., and Bégin, R.O., eds., Springer-Verlag, Berlin, pp. 149-156.

Fleury,J., Cote,N.G., Marion,G., Chevalier,G. and Denizeau,F. (1983) Interaction of asbestos fibers with hepatocytes: An ultrastructural study. *Toxicol. Lett.* **19**:15-22.

Floyd,L.A. (1990) The role of 8-hydroxyguanine in carcinogenesis. *Carcinogenesis* **11**:1447-1450.

Fornace,A.J. Jr. (1982) Detection of DNA single-strand breaks produced during the repair of damage by DNA-protein cross-linking agents. *Cancer Res.* **42**:145-149.

Gentil,A., Renault,G. and Margot,A. (1980) The effect of the tumour promoter 12-0-tetradecanoyl-phorbol-13-acetate (TPA) on UV- and MNNG-induced sister chromatid exchanges in mammalian cells. *Int. J. Cancer* **26**:517-521.

Gross,P. and Braun,D.C. (1984) Toxic and biomedical effects of fibers. Asbestos, talc, inorganic fibers, man-made vitreous fibers, and organic fibers. In "Man-Made Vitreous Fibers", Chapter 5, Noyes Publications, Park Ridge, NJ, USA. pp. 143-222.

Hanawalt, P. (1990) Selective DNA repair in expressed genes in mammalian cells. In "Mutation and the Environment, Part A", Mendelsohn,M.L. and Albertini,R.J., eds., Wiley Liss, Inc., New York, USA. pp. 213-222.

Harris,C.C., Lechner,J.F., Yoakum,G.H., Amstad,P., Korba,B.E., Gabrielson,E., Graftstrom,R., Shamsuddin,A. and Trump,B.F. (1985) *In vitro* studies of human lung carcinogenesis. In "Carcinogenesis, vol. 9", Barrett, J.C. and Tennant, R.W., eds., Raven Press, New York, pp. 257-269.

Hei,T.K., He,Z.Y., Piao,C.Q., Waldren,C.A. and Hall,E.J. (1991) The mutagenicity of asbestos fibers: cellular and molecular characterization. (this volume).

Hei,T.K.J., Geard,C.R., Osmak,R.S. and Travisano,M. (1985) Correlation of *in vitro* genotoxicity and oncogenicity induced by radiation and asbestos fibres. *Br. J. Cancer* **52**:591-597.

Hesterberg,T.W. and Barrett,J.C. (1984) Dependence of asbestos- and mineral dust-induced transformation of mammalian cells in culture of fibre dimension. *Cancer Res.* **44**:2170-2180.

Hesterberg,T.W. and Barrett,J.C. (1985) Induction by asbestos fibers of anaphase abnormalities: mechanism for aneuploidy induction and possibly carcinogenesis. *Carcinogenesis* **6**:473-476.

Hesterberg,T.W., Butterick,C.J., Oshimura,M., Brody,A. and Barrett,J.C. (1986) Role of phagocytosis in Syrian hamster cell transformation and cytogenetic effects induced by asbestos and short and long glass fibers. *Cancer Res.* **46**:5795-5802.

Hesterberg,T.W., Ririe,D.G., Barrett,J.C. and Nettesheim,P. (1987) Mechanisms of cytotoxicity of asbestos fibres in rat tracheal epithelial cells in culture. *Toxicol. in vitro* **1**:59-65.

Huang,S.L. (1979) Amosite, chrysotile and crocidolite asbestos are mutagenic to Chinese hamster lung cells. *Mutat. Res.* **68**:265-274.

Huang,S.L., Saggioro, D., Michelman,H. and Malling,H.V. (1978) Genetic effects of crocidolite asbestos in Chinese hamster lung cells. *Mutat. Res.* **57**:225-232.

Jackson,J.H., Schraufstatter,I.U., Hyslop,P.A., Vosbeck,K., Sauerheber,R., Weitzman,S.A. and Cochrane,C.G. (1987) Role of oxidants in DNA damage: hydroxyl radical mediates the synergistic DNA damaging effects of asbestos and cigarette smoke. *J. Clin. Invest.* **80**:1090-1095.

Jaurand,M.C. (1989) Particulate state carcinogenesis: a survey of recent studies on the mechanisms of action of fibres. *IARC Sci. Pub.* **90**:54-73.

Jaurand,M.C. (1991) Mechanisms of action of fibres in Carcinogenesis. In "Asbestos Related Cancer", Sluyser,M., ed., Ellis Horwood Ltd., (in press).

Jaurand,M.C., Bastie-Sigeac,I., Bignon,J. and Stoebner, P. (1983) Effect of chrysotile and crocidolite on the morphology and growth of rat pleural mesothelial cells. *Environ. Res.* **30**:255-269.

Jaurand,M.C., Fleury,J., Monchaux,G., Nebut,M. and Bignon,J. (1987) Pleural carcinogenic potency of mineral fibers (asbestos, attapulgite) and their cytotoxicity on cultured cells. *J. Natl. Cancer Inst.* **79**:797-804.

Jaurand,M.C., Kaplan,H., Thiollet,J., Pinchon,M.C., Bernaudin,J.F. and Bignon,J. (1979a) Phagocytosis of chrysotile fibers by pleural mesothelial cells in culture. *Am. J. Pathol.* **94**:529-532.

Jaurand,M.C., Kheuang,L., Magne,L. and Bignon,J. (1986) Chromosomal changes induced by chrysotile fibres and benzo-3,4-pyrene in rat pleural mesothelial cells. *Mutat. Res.* **169**:141-148.

Jaurand,M.C., Magne,L. and Bignon, J. (1979b) Inhibition by phospholipids of haemolytic action of asbestos. *Br. J. Indust. Med.* **36**:113-116.

Jaurand,M.C., Renier,A., Gaudichet,A., Kheuang,L., Magne,L. and Bignon,J. (1988) Short-term tests for the evaluation of potential cancer risk of modified asbestos fibers. *N.Y. Acad. Sci.* **534**:741-753.

Jaurand, M.C., Saint Etienne, L., Van der Neeren,A., Endo-Capron,S., Renier,A. and Bignon,J. (1990) Neoplastic transformation of rodent cells. Cell and molecular aspects of fiber carcinogenesis. "Current Communication in Molecular Biology." Cold Spring Harbor, No. 2, (in press).

Johnson,N.F. and Davies,R. (1983) Effects of asbestos on the P388D1 macrophage like cell line: preliminary ultrastructural observations. *Environ. Hlth Perspect.* **51**:109-117.

Kane,A. (1991) Fiber dimensions and mesothelioma: A reappraisal of the Stanton hypothesis. (this volume).

Kasai,M. and Nishimura,S. (1984) DNA damage induced by asbestos in the presence of hydrogen peroxide. *Gann* **75**:841-844.

Kelsey,K.T., Yano,E., Liber,H.L. and Little,J.B. (1986) The *in vitro* genetic effects of fibrous erionite and crocidolite asbestos. *Br. J. Cancer* **54**:107-114.

Kenne,K., Ljungquist,S. and Ringertz,N.R. (1986) Effects of asbestos fibers on cell division, cell survival, and formation of thioguanine-resistant mutants in Chinese hamster ovary cells. *Environ. Res.* **39**:448-464.

Kodama,Y., Hesterberg,T.W., Maness,S.C., Iglehart,J.D. and Boreiko,C.J. (1987) Asbestos-induced cytogenetic effects in cultured human bronchial epithelial cells. *Proc. Am. Soc. Cancer Res.* **28**:85.

Kohn, K.W. and Grimke-Ewig,R.A. (1973) Alkaline elution analysis, a new approach to the study of DNA single-strand interruptions in cells. *Cancer Res.* **33**:1849-1853.

Langer,A.N., Wolff,N.S., Rohl,A.N. and Selikoff,I.J. (1978) Variation of properties of chrysotile asbestos subjected to milling. *J. Toxicol. Environ. Hlth* **4**:173-188.

Langer,A. (1985) Physico-chemical properties of minerals relevant to biological activities: state of the art. In "*in vitro* Effects of Mineral Dusts", NATO ASI Series, Series G: Ecological Sciences, vol. 3, Beck,E.G., Bignon,J., eds., Springer-Verlag, Berlin, pp. 9-24.

Lavappa,K.S., Fu,N.R. and Epstein,S.S. (1975) Cytogenetic studies on chrysotile asbestos. *Environ. Res.* **10**:165-173.

Leanderson,P., Söderkvist,P., Tagesson,C. and Axelson,O. (1988) Formation of 8-hydroxy-deoxyguanosine by asbestos and man made mineral fibres. *Br. J. Indust. Med.* **45**:309-311.

Leanderson,P., Söderkvist,P. and Tagesson,C. (1989) Hydroxy radical mediated DNA base modification by man made mineral fibres. *Br. J. Indust. Med.* **46**:435-438.

Lechner,J.F., Tokiwa,T., La Veck,M., Benedict,W.F., Banchs-Schlegel,S., Yeager,M., Banerjee,A. and Harris,C.C. (1985) Asbestos-associated chromosomal changes in human mesothelial cells. *Proc. Natl. Acad. Sci. U.S.A.* **82**:3884-3888.

Lee,K.P., Barras,C.E., Griffith,F.D. and Waritz,R.S. (1981a) Pulmonary response and transmigration of inorganic fibers by inhalation exposure. *Am. J. Pathol.* **102**:314-323.

Lee,K.P., Barras,C.E., Griffith,F.D., Waritz,R.S. and Lapin,C.A. (1981b) Comparative pulmonary responses to inhaled inorganic fibers with asbestos and fiberglass. *Environ. Res.* **24**:167-191.

Libbus,B.L. and Craighead,J.E. (1988) Chromosomal translocations with specific breakpoints in asbestos-induced rat mesotheliomas. *Cancer Res.* **48**:6455-6461.

Libbus,B.L., Illenye,S.A. and Craighead,J.E. (1989) Induction of DNA strand breaks in cultured rat embryo cells by crocidolite asbestos and assessed by nick translation. *Cancer Res.* **49**:5713-5718.

Livingston,G.K. Rom,W.N. and Morris,M.V. (1980) Asbestos induced sister chromatid exchanges in cultured Chinese hamster ovarian fibroblast cells. *J. Environ. Pathol. Toxicol.* **4-2**:373-382.

Loveday,K.S. and Latt,S.A. (1979) The effect of a tumor promoter, 12-0-tetradecanoyl-phorbol-13-acetate (TPA), on sister chromatid exchange formation in cultured Chinese hamster cells. *Mutat. Res.* **67**:343-348.

Mollo,F., Piolatto,G., Bellis,D., Adrion,A., Delsedime,L., Bernardi,P., Pira,E. and Ardissone,F. (1990) Asbestos exposure and histologic cell types of lung cancer in surgical and autopsy series. *Int. J. Cancer* **46**:576-580.

Moorhead,P.S., Nowell,P.C., Mellman,J., Battips,D.M. and Hungerford,D.A. (1960) Chromosome preparations of leukocytes cultured from human peripheral blood. *Exptl. Cell Res.* **20**:613-616.

Morgan,A., Evans,J.C. and Holmes,A. (1977) Deposition and clearance of inhaled fibrous minerals in the rat. Studies using radioactive tracer techniques. In "Inhaled Particles, vol. IV", Walton, W.H., ed., Pergamon Press, Oxford, UK. pp. 259-272.

Mossman,B.T. (1983) *In vitro* approaches for determining mechanisms of toxicity and carcinogenicity by asbestos in the gastrointestinal and respiratory tracts. *Environ. Hlth Perspect.* **53**:155-161.

Mossman,B.T., Kessler,J.B., Ley,B.W. and Craighead,J.E. (1977) Interaction of crocidolite asbestos with hamster respiratory mucosa in organ culture. *Lab. Invest.* **36**:131-139.

Mühle,H., Pott,F., Bellmann,B., Takenaka,S. and Ziem,U. (1987) Inhalation and injection experiments in rats to test the carcinogenicity of MMMF. *Ann. Occup. Hyg.* **31**:755-764.

Nagasawa,H. and Little,J.B. (1981) Factors influencing the induction of sister chromatid exchanges in mammalian cells by 12-*O*-tetradecanoyl-phorbol-13-acetate. *Carcinogenesis* **2**:601-607.

Oberdoerster,G., Ferin,J., Marcello,N.L. and Meinhold,S.H. (1983) Effect of intrabronchially instilled amosite on lavagable lung and pleural cells. *Environ. Hlth Perspect.* **51**:41-48.

Olfosson,K. and Mark,J. (1989) Specificity of asbestos-induced chromosomal aberrations in short-term cultured human mesothelial cells. *Cancer Genet. Cytogenet.* **41**:33-39.

Oshimura,M., Hesterberg,T.W. and Barrett,J.C. (1986) An early non-random karyotypic change in immortal Syrian hamster cell lines transformed by asbestos: trisomy of chromosome 11. *Cancer Genet. Cytogenet.* **22**:225-237.

Oshimura,M., Hesterberg,T.W., Tsutsui,T. and Barrett,J.C. (1984) Correlation of asbestos-induced cytogenetic effects with cell transformation of Syrian embryo cells in culture. *Cancer Res.* **44**:5017-5022.

Pairon,J.C., Jaurand,M.C., Kheuang,L., Brochard,P. and Bignon,J. (1990) Sister chromatid exchanges in human lymphocytes treated with silica. *Br. J. Indust. Med.* **47**:110-115.

Palekar,L.D., Eyre,J.F. and Coffin,D.L. (1989) Chromosomal changes associated with tumorigenic mineral fibers. In, "Biological Interaction of Inhaled Mineral Fibers and Cigarette Smoke", Welner,A.P., ed., pp. 355-372.

Palekar,L.D., Eyre,J.F., Most,B.M. and Coffin,D.L. (1987) Metaphase and anaphase analysis of V79 cells exposed to erionite, UICC chrysotile and UICC crocidolite. *Carcinogenesis* **8**:553-560.

Poole,A., Brown,R.C., Turner,J.C., Skidmore,J.W. and Griffiths,D.M. (1983) *In vitro* genotoxic activities of fibrous erionite. *Br. J. Cancer* **47**:697-705.

Price-Jones,M.J., Gubbings,G. and Chamberlain,M. (1980) The genetic effects of crocidolite' asbestos; a comparison of chromosome abnormalities and sister chromatid exchanges. *Mutat. Res.* **79**:331-336.

Reiss,B., Solomon,S., Tong,C., Levenstein,M., Rosenberg,S.H. and Williams,G.M. (1982) Absence of mutagenic activity of three forms of asbestos in liver epithelial cells. *Environ. Res.* **27**:389-397.

Reiss,B., Tong,C., Telang,S. and Williams,G.M. (1983) Enhancement of benzo[a]pyrene mutagenicity by chrysotile asbestos in rat liver epithelial cells. *Environ. Res.* **31**:100-104.

Renier,A., Levy,F., Pilliere,F. and Jaurand,M.C. (1990) Unscheduled DNA synthesis in rat pleural mesothelial cells treated with mineral fibres or benzo-[a]-pyrene. *Mutat. Res.* **241**:361-367.

Ririe,D., Hesterberg,T.W., Barrett,J.C. and Nettesheim,P. (1985) Toxicity of asbestos and glass fibers for tracheal epithelial cells in culture. In, "*In vitro* Effects of Mineral Dusts", Beck,E. and Bignon,J., eds. NATO ASI Series, **G3**, , Springer Verlag, Berlin, Heidelberg, pp. 177-184.

Rita,P. and Reddy,P.P. (1986) Effect of chrysotile asbestos fibers on germ cells of mice. *Environ. Res.* **41**:139-143.

Roggli,V.L. and Brody,A.R. (1984) Changes in numbers and dimensions of chrysotile asbestos fibers in lungs of rats following short-term exposure. *Exptl. Lung Res.* **7**:133-147.

Schlesinger,R.B. (1985) Comparative deposition of inhaled aerosols in experimental animals and humans: a review. *J. Toxicol. Environ. Hlth* **15**:197-214.

Schwartz,J.L., Banda,N.J. and Wolff,S. (1982) 12-*O*-tetradecanoyl phorbol-13-acetate (TPA) induces sister chromatid exchanges and delays in cell progression in Chinese hamster ovary and human cell lines. *Mutat. Res.* **92**:393-409.

Sebastien,P., Janson,X., Gaudichet,A., Hirsch,A. and Bignon,J. (1980) Asbestos retention in human respiratory tissues: comparative measurements in lung parenchyma and in parietal pleura. In, "Biological Effects of Mineral Fibres", Wagner, J.C., ed., IARC Scientific Publication No. 30, INSERM Symposia Series, vol. 92, pp. 237-246.

Sincock,A. and Seabright,M. (1975) Induction of chromosome changes in hamster cells by exposure to asbestos fibres. *Nature* **257**:56-58.

Sincock,A.M., Delhanty,J.D.A. and Casey,G. (1982) A comparison of the cytogenetic response to asbestos and glass fibre in Chinese hamster and human cell lines. *Mutat. Res.* **101**:257-268.

Smith,D.M., Ortiz,L.W., Archuleta,R.F. and Johnson,N.F. (1987) Long-term health effects in hamsters and rats exposed chronically to man-made vitreous fibres. *Ann. Occup. Hyg.* **31**:731-754.

Tiainen,M., Tammilehto,L., Rautonen,J., Tuomi,T., Mattson,K. and Knuutila,S. (1989) Chromosomal abnormalities and their correlations with asbestos exposure and survival in patients with mesotheliomas. *Br. J. Cancer* **60**:618-626.

Thompson,L.H., Baker,R.M., Carrano,A.V. and Brookman,K.W. (1980) Failure of the phorbol ester 12-*O*-tetradecanoyl-phorbol-13-acetate to enhance sister chromatid exchange, mitotic segregation, or expression of mutations in Chinese hamster cells. *Cancer Res.* **40**:3245-3251.

Tilkes,F. and Beck,E.G. (1983) Influence of well-defined mineral fibers on proliferating cells. *Environ. Hlth Perspect.* **51**:275-279.

Valerio,F., DeFerrari,M., Ottagio,L., Repetto,E. and Santi,L. (1983) Chromosomal aberrations induced by chrysotile and crocidolite in human lymphocytes *in vitro. Mutat. Res.* **122**:397-402.

Vershave,L. and Palmer,P. (1985) On the uptake and genotoxicity of UICC Rhodesian chrysotile A in human primary lung fibroblasts. *Naturwissenchaften* **72**:326-327.

Wagner,J.C., Skidmore,J.W., Hill,R.J. and Griffiths,D.M. (1985) Erionite exposure and mesotheliomas in rats. *Br. J. Cancer* **51**:727-730.

Walker,C., Everitt,J., Bermudez,E. and Funaki,K. (1991) Involvement of chromosome 1 in rat mesothelial cell transformation. *The Toxicologist*, (in press).

Wallace,W.E., Vallyathan,V., Keane,M.J. and Robinson,V. (1985) *In vitro* biologic toxicity of native and surface-modified silica and kaolin. *J. Toxicol. Environ. Hlth* **16**:415-424.

Wang,N.S. (1975) The preformed stomas connecting the pleural cavity and the lymphatics in the parietal pleura. *Am. Rev. Resp. Dis.* **111**:12-20.

Wang,N.S., Jaurand,M.C., Magne,L., Kheuang,L., Pinchon,M.C. and Bignon,J. (1987) The interactions between asbestos fibers and metaphase chromosomes of rat pleural mesothelial cells in culture. *Am. J. Pathol.* **126**:343-349.

Weitzman,S.A. and Graceffa,P. (1984) Asbestos catalyzes hydroxyl and superoxide radical generation from hydrogen peroxide. *Arch. Biochem. Biophys.* **228**:373-376.

Woodworth,C.D., Mossman,B.T. and Craighead,J.E. (1983) Interaction of asbestos with meta-plastic squamous epithelium developing in organ cultures of hamster trachea. *Environ. Hlth Perspect.* **51**:27-33.

Yang,L.L., Kouri,R.E. and Curren,R.D. (1984) Xeroderma pigmentosum fibroblasts are more sensitive to asbestos fibers than are normal human fibroblasts. *Carcinogenesis* **5**:291-294.

Zalma,R., Bonneau,L., Jaurand,M.C., Guignard,J. and Pezerat,H. (1987) Formations of oxy-radicals by oxygen reduction arising from the surface of asbestos. *Can. J. Chem.* **65**:2338-2341.

ASBESTOS CARCINOGENICITY: A MUTATIONAL HYPOTHESIS

J. Carl Barrett

Laboratory of Molecular Carcinogenesis
National Institute of Environmental Health Sciences
P. O. Box 12233, Research Triangle Park
North Carolina 27709, U.S.A.

INTRODUCTION

One approach to understanding the carcinogenic process is to define the molecular alterations involved in the evolution of a malignant cell and to elucidate the mechanisms by which carcinogenic substances can affect these alterations in target cells. With the identification of the genes involved in carcinogenesis, this approach is becoming increasingly useful. This can be illustrated with asbestos-induced neoplasms, mesotheliomas and lung cancers.

There is increasing evidence for the involvement of two classes of genes, oncogenes and tumor suppressor genes, in carcinogenesis (Table 1). Oncogenes are the activated forms of a family of normal cellular genes, proto-oncogenes. When proto-oncogenes are activated by a variety of mutational events, including point mutation, chromosome translocation, and gene amplification, the activated oncogenes produce positive, proliferative signals for the cell (Boyd & Barrett, 1990). In contrast, tumor suppressor genes negatively regulate cell growth and are frequently inactivated or lost in tumors. These genes are inactivated by chromosome loss, chromosome deletion, and recombination, as well as by point mutations (Barrett & Wiseman, 1987). A family of proto-oncogenes with 40 to 50 members has been identified, but only 10 to 12 of these genes have been shown to be altered in human tumors (Bishop, 1987). The size of the tumor suppressor gene family is unknown, but an estimate of 20 to 50 genes is reasonable (Boyd & Barrett, 1990). One of the major lines of evidence for the existence of tumor suppressor genes is the finding of non-random deletions or losses of heterozygosity of specific chromosomal regions in human tumors (Klein, 1987; Weinberg, 1989; Skuse & Rowley, 1989). Although alterations of two oncogenes are occasionally observed in some human tumors (Land *et al.*, 1983), multiple allelic deletions (of putative tumor suppressor

TABLE 1

Proto-oncogenes	Tumor Suppressor Genes
1. Involved in cellular growth. and differentiation	1. Function unknown but possibly involved in cellular growth and differentiation (negative regulators of cell growth?)
2. Family of genes exists	2. Family of genes exists
3. Must be activated in cancers (qualitatively or quantitatively)	3. Must be inactivated, lost or mutated in cancers
4. Mutational activation by point mutation, chromosome translocation or gene amplification	4. Mutational inactivation by chromosome loss, chromosome deletion, point mutation, somatic recombination or gene conversion
5. Little evidence for involvement in hereditary cancers	5. Clear evidence for involvement in hereditary cancers

TABLE 2. Examples of multiple tumor suppressor genes in specific tumors

Multiple chromosomes show losses of heterozygosity in human tumors

Colon	-	Chromosomes 5, 17p, 18q and others
Lung	-	Chromosomes 3p, 11p, 13q, and 17p
Breast	-	Chromosomes 1p, 1q, 3p, 11p, 13q, and 17p
Bladder	-	Chromosomes 9q, 11p, 17p and others

Multiple chromosomes suppress tumorigenicity following microcell transfer

Endometrial	-	Chromosomes 1, 6 and 9
Lung adenocarcinoma	-	Chromosomes 3 and 11
Fibrosarcoma	-	Chromosomes 1 and 11

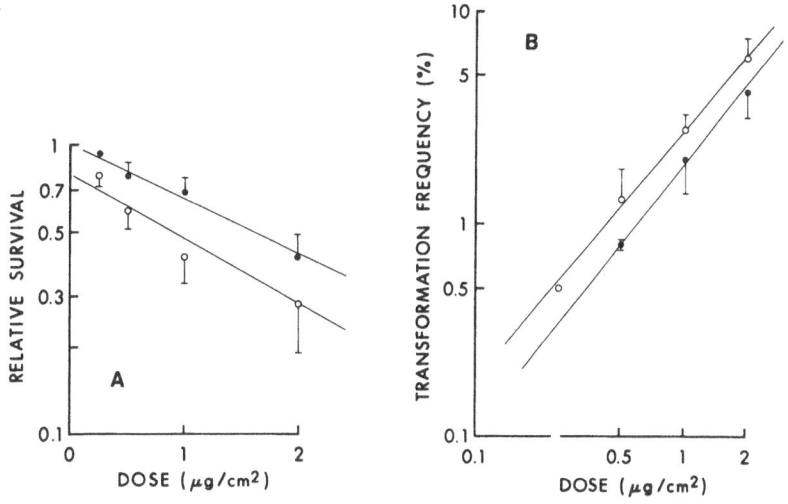

Figure 1. Effects of different doses of chrysotile (o) and crocidolite (●) asbestos on the relative survival (A) and the morphological transformation frequency (B) of Syrian hamster embryo cells in culture (bars, ±SE). (Reprinted, with permission, from Hesterberg & Barrett 1984).

genes) are frequently found in most adult solid cancers (Fearon & Vogelstein, 1990; Callahan, 1989; Weston *et al.*, 1989) (Table 2). Furthermore, functional evidence for multiple tumor suppressor genes can be demonstrated by transfer of different normal human chromosomes into tumor cells with resultant suppression of tumorigenicity (Yamada *et al.*, 1990; Oshimura, 1990). These findings emphasize the importance of chromosomal losses and deletions in the development of human cancers.

Unlike most carcinogens, asbestos fibers are not directly electrophilic, do not form adducts with DNA, are inactive in the Ames Salmonella test, and fail to induce mutations at specific genetic loci in mammalian cells (Chamberlain & Tarmy, 1977; Barrett, 1987; Jaurand, 1989). Although asbestos is not active as a gene mutagen in a variety of test systems, there is now clear evidence that it induces chromosomal mutations (aneuploidy and aberrations) in a wide variety of mammalian cells (reviewed in Jaurand, 1989 and Barrett *et al.*, 1991). We reported several years ago that asbestos, like other chemical carcinogens, can heritably induce neoplastic alterations of treated cells in culture and proposed a mutational basis involving chromosomal changes for this effect (Hesterberg & Barrett, 1984). In this report, I will briefly review this work and discuss its significance in light of recent findings of the molecular genetics of mesotheliomas and lung cancer.

MECHANISMS OF ASBESTOS INDUCED CELL TRANSFORMATION

Jaurand and coworkers (Patrour *et al.*, 1985; Jaurand *et al.*, 1989) have shown that asbestos induces transformation and chromosomal changes in rat mesothelial cells in culture. Lechner and co-workers (1985) also reported that asbestos fibers alter the growth properties of human mesothelial cells, and this is associated with chromosomal changes in the treated cells. Extensive studies from our laboratory have shown that asbestos and other mineral dusts induce neoplastic transformation of Syrian hamster embryo cells in culture. These studies have provided possible insights into the mechanism of mineral dust carcinogenicity and the role of fiber dimensions in this process (Barrett *et al.*, 1990).

We observed that chrysotile and crocidolite asbestos induce morphological transformation of Syrian hamster embryo cells that is indistinguishable from cell transformation induced by other carcinogens (Hesterberg & Barrett, 1984). The morphologically
transformed cells progress and become tumorigenic after multiple passages, again similar
to the findings with other carcinogens (Barrett & Ts'o, 1978). The frequency of morphologically transformed cells increases linearly with increasing dose of asbestos (Fig. 1).
Three other laboratories also have shown that asbestos fibers transform Syrian hamster
embryo cells in culture (DiPaolo et al., 1983; Mikalsen et al., 1988; A. Tu, personal
communication).

One of the more important (and intriguing) factors that influences mineral dust
carcinogenicity is fiber dimension. Stanton et al. (1981) have shown in classical experiments that the mesothelioma-inducing activity of mineral dusts is related to fiber size, i.
e., long, thin fibers are more active than short, thick fibers. Similar to the induction of
mesotheliomas in vivo, cell transformation by mineral fibers is dependent on fiber size
(Hesterberg & Barrett, 1984). Code 100 fiberglass (average length = 9. 5 μm) is as
active as crocidolite asbestos of a similar length (Hesterberg & Barrett, 1984). Milling of
the fiberglass reduces the fiber length to 2.2 μm and decreases the transforming activity
by 10-to 20-fold; further reduction to < 1 μm in length eliminates the transforming activity at the doses tested (Fig. 2). Thus, the fiber length dependence of cell transformation
parallells the results of mesothelioma induction in vivo.

In order to explore the mechanism of the asbestos-induced cell transformation, we
examined the mutagenic effects of asbestos fibers in the same cell system (Oshimura

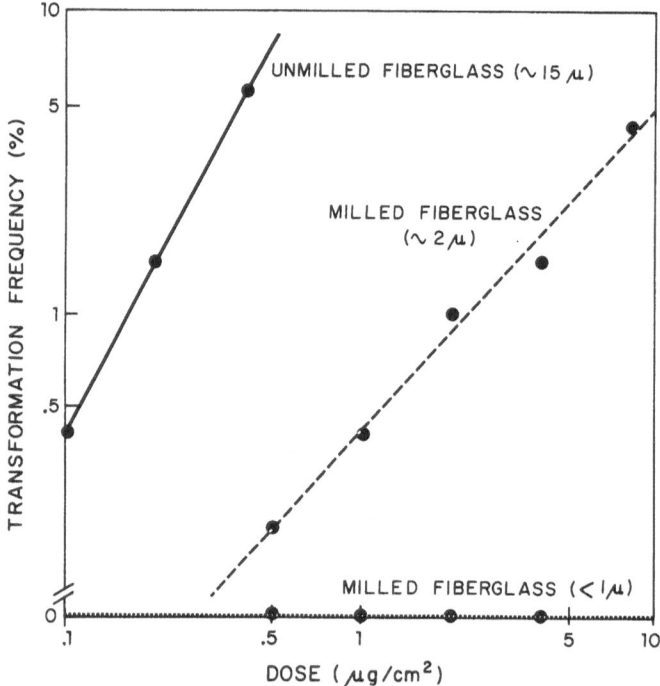

Figure 2. Effects of different doses of unmilled code 100 glass fibers
(—) or milled code 100 (--) on transformation frequency of SHE cells
in culture. (Reprinted, with permission, from Hesterberg & Barrett
1984.)

312

TABLE 3. Cytogenetic effects of 2 $\mu g/cm^2$ of various mineral dusts on Syrian hamster embryo cells *in vitro*

Treatment	Transformation frequency[a]	Aneuploid cells[b] (%)	Chromosome aberrations[c]	Cells with micronuclei [d] (%)	Tetraploid cells[d,e] (%)	Binucleated cells (%)
Control	0	1.7	1	0.3	5.0	0.3
Chrysotile	6.2	12.5	5	2.6*	33.0*	25.0*
Crocidolite	4.6	9.0*	4	1.1*	14.0*	11.2*
Fiberglass[f]	3.0	7.0*	4	3.0*	20.0*	18.4*
Milled Fiberglass	0.0	2.0	1	0.5	6.0	0.4
Alpha quartz	0.0	3.0	1	0.5	5.0	0.3

[a] Cited from Hesterberg & Barrett (1984). The transformation frequency was calculated by dividing the number of morphologically transformed colonies by the total number of colonies examined X 1000

[b] This represents the percentage of metaphases that contained a near-diploid number of chromosomes.

[c] Percentage of metaphases containing the following aberrations: chromatid breaks, isochromatid breaks, chromosome fragments, chromatid exchanges or dicentric chromosomes.

[d] For each treatment group 1000 cells were scored.

[e] Cells with a tetraploid (4N=88) or near tetraploid (70-100) number of chromosomes.

[f] This fiberglass was obtained from Johns Manville (code 100) and processed as described previously (Hesterberg & Barrett, 1984).

* Statistically significant from the control, $p < 0.05$, Fisher's exact test.

These data were reproduced from Oshimura *et al.*, 1984, with permission

et al., 1984). Consistent with other studies with mammalian cells, asbestos fibers do not induce gene mutations in the cells under conditions that induce cell transformation. In contrast, asbestos fibers induce chromosomal abnormalities in the treated cells, and the production of chromosomal mutations correlates with the transforming activity of the different fiber types (Table 3). Chromosome mutations induced by the fiberglass are fiber-length dependent. The mineral fibers induce numerical chromosome changes, both aneuploid cells in the near-diploid range and polyploid cells in the tetraploid range. A low level of structural chromosome alterations are also induced. The induction of aneuploidy appears to be the most significant chromosome change in the transformation process. This conclusion is based on the finding of a non-random chromosome change, an extra copy of chromosome 11, in the transformed cells (Oshimura *et al.*, 1986). Trisomy 11 is the sole

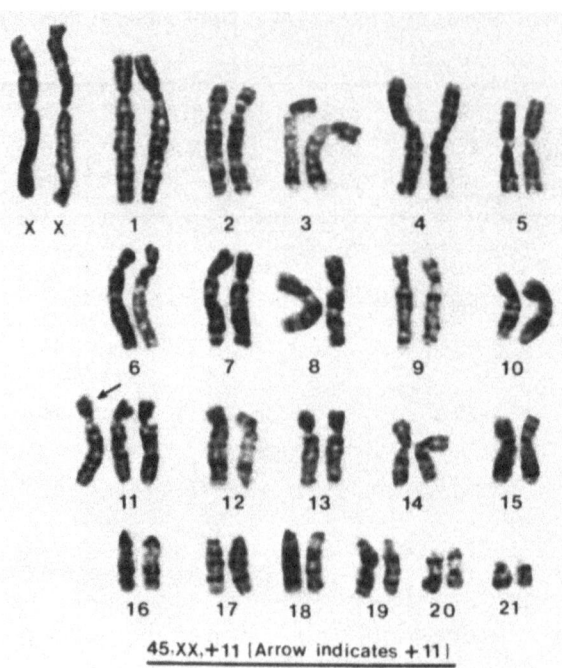

45.XX,+11 [Arrow indicates +11]

Figure 3. G-banded karyotype showing trisomy of chromosome 11 in the. asbestos-induced Syrian hamster embryo cell line (10W). Arrow indicates the extra chromosome. (Reprinted, with permission, from Oshimura & Barrett, 1985).

karyotypic change found in several of the asbestos-induced cell lines (Fig. 3). Aneuploidy induction by asbestos is random so the finding of a non-random change in the transformed cells indicates that this change is involved in the transformation process.

We have proposed a mechanism for asbestos-induced aneuploidy, i.e., losses and gains of individual chromosomes (Hesterberg & Barrett, 1985; Hesterberg *et al.*, 1986). In collaboration with Dr. Arnold Brody, we showed that crocidolite asbestos fibers are taken up by the cells within 24 hr after treatment by phagocytosis (Hesterberg *et al.*, 1986); the intracellular fibers accumulate around the perinuclear region of the cells 24 to 48 hr after exposure. When the cells undergo mitosis, the physical presence of the fibers results in interference with chromosome segregation. Analysis of chrysotile-exposed cells in anaphase (Hesterberg & Barrett, 1985) reveals a large increase in the number of cells with anaphase abnormalities (Fig. 4) including lagging chromosomes, bridges, and sticky chromosomes. Asbestos fibers are observed in the mitotic cells and appear, in some cases, to interact directly with the chromosomes. Using ultrastructural analysis, Wang *et al.* (1987) have observed asbestos fibers apparently interacting with metaphase chromosomes after treatment of rat mesothelial cells in culture. We have proposed that the physical interaction of the asbestos fibers with the chromosomes or structural proteins of the spindle apparatus causes mis-segregation of chromosomes during mitosis, resulting in aneuploidy (Hesterberg & Barrett, 1985). These findings provide a mechanism, at the chromosomal level, by which asbestos and other mineral fibers can induce cell transformation and cancer. Malignant human and rat mesotheliomas are highly aneuploid (Gibas

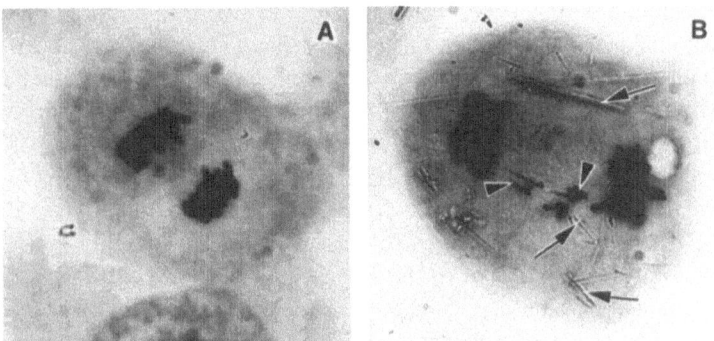

Figure 4. A normal (A) and an abnormal (B) anaphase from crocidolite asbestos-treated Syrian hamster embryo cells. Note the asbestos fibers (arrows), some of which appear to be associated with displaced chromosomes (arrowheads) in the abnormal anaphase. (Reprinted, with permission, from Hesterberg & Barrett, 1985.)

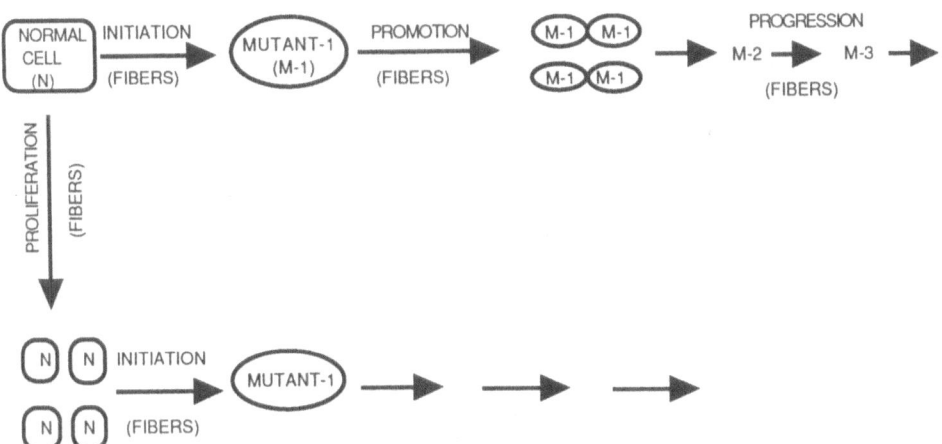

Figure 5. The multistep process of carcinogenesis results from multiple genetic changes required for a normal cell to evolve into a malignant cell. Asbestos fibers may initiate this process by a chromosomal mutation. However, additional mutations are required for clinically evident neoplasms. These secondary mutations are more likely to arise if the initiated (mutant-1) cells are stimulated to proliferate, which may result from fiber-related promotional mechanisms. Tumor promotion results in clonal expansion of the initiated cells, but preneoplastic cells must progress and acquire additional mutations. Asbestos fibers may influence this progression by either inducing chromosomal mutations in the intermediate cells and/or by stimulating cellular proliferation. Asbestos fibers may also stimulate proliferation of the normal cells thereby increasing the rate of either asbestos-induced or spontaneous mutations. Therefore, asbestos fibers can influence the carcinogenic process either at early or late stages by both genetic and epigenetic mechanisms. Synergism with other agents (e. g., cigarette smoke) can be explained by asbestos acting at one step and the synergistic carcinogen acting at a different step in this multistep process.

et al., 1986; Stenman *et al.*, 1986; Tiainen *et al.*, 1989). It is possible that asbestos exposure leads to the generation of these aneuploid cells *in vivo* as well as *in vitro*. Asbestos-induced, immortal hamster cells have an alteration in a cellular senescence gene. Re-introduction of a specific human chromosome (chromosome 1) corrects this defect and results in restoration of the normal cellular phenotype (Sugawara *et al.*, 1990). The region involved has now been mapped to 1q23 (Futreal, P.A. & Barrett, J.C., unpublished). Interestingly, this region is also frequently altered in human mesotheliomas (Tiainen *et al.*, 1989).

Two factors may contribute to the fiber-size dependence of asbestos- induced cell transformation and chromosome damage. Fiber length appears to affect the phagocytosis of fibers as well as the ability of intracellular fibers to induce cytogenetic damage and the resultant transformation (Hesterberg *et al.*, 1986).

ASBESTOS FIBERS AND MULTISTEP CARCINOGENESIS

We have proposed a mutational hypothesis that explains in part the carcinogenic activity of fibers (Barrett *et al.*, 1989). Since human and rodent mesotheliomas and lung cancers exhibit non-random karyotypic changes and carcinogenic fibers induce chromosomal mutations in treated cells in culture, it is reasonable to propose that fiber induction of chromosomal alterations play some role in asbestos carcinogenesis. Since fibers also induce heritable alterations in the growth control of mesothelial cells treated in culture and a chromosome mutational basis for these alterations has been proposed, it is also plausible to propose that this is a mechanism for the initiation of human mesothelioma formation *in vivo*. These data also provide a mechanistic explanation for fiber size dependence of mesothelioma induction (Hesterberg & Barrett, 1984). However, as previously proposed (Barrett *et al.*, 1989), asbestos fibers are likely to operate through multiple mechanisms.

Carcinogenesis is a multistep process involving multiple genetic changes for the transformation of a normal cell into a malignant cell (Fig. 5). Fibers may induce mutational changes either early or late in the carcinogenic process. Asbestos acts at an early stage in mesothelioma formation (Peto, 1985) and the proposed mutational hypothesis can explain this effect. However, for chromosomal mutations to arise in mesothelial cells, these normally quiescent cells must be stimulated to proliferate. Asbestos fibers may so stimulate normal cells, either by directly killing cells and inducing regenerative hyperplasia or by stimulating the production of paracrine growth factors generated by macrophages or other distant cells stimulated by fibers and acting at a distance. Asbestos exposure also increases the incidence of lung cancer in smokers and nonsmokers (Hammond *et al.*., 1979). A synergistic effect of asbestos exposure and smoking is observed and asbestos appears to affect a late stage in lung cancer (Peto, 1985). One explanation for this effect is that asbestos-induced mutations in smoking-induced pre-neoplastic cells increases the progression phase of lung cancer by inducing chromosomal mutations (Fig. 5).

In addition to the genetic mechanisms emphasized in this review, asbestos fibers may also influence the carcinogenic process by epigenetic mechanisms (Barrett *et al.*, 1990). Asbestos fibers may promote spontaneous and fiber-induced initiated cells by stimulating clonal expansion of these cells (Fig. 5). Thus, asbestos probably operates through multiple genetic and epigenetic mechanisms necessary for neoplastic development (Barrett *et al.*, 1989).

REFERENCES

Barrett,J.C. (1987) Relationship between mutagenesis and carcinogenesis. In, "Mechanisms of Environmental Carcinogenesis: Role. of Genetic and Epigenetic Changes", Barrett,J.C. ed. Boca Raton, CRC Press, Vol. I. pp. 129-142.

Barrett,J.C., Lamb,P.W. and Wiseman,R.W. (1989) Multiple mechanisms for the carcinogenic effects of asbestos and other mineral fibers. *Environ. Hlth Perspect.* **81**:81-89.

Barrett,J.C. (1991) Role of chromosome mutations in asbestos-induced cell transformation. In, "Cellular and Molecular Fiber Carcinogenesis", Brinkley,B., Lechner,J., Harris,C. eds. Cold Spring Harbor Laboratory, New. York, Cold Spring Laboratory Press, in press.

Barrett,J.C., Lamb,P.W. and Wiseman,R.W. (1990) Hypotheses on the mechanisms of carcinogenesis and cell transformation by asbestos and other mineral dusts. In, "Health related effects of phyllosilicates", Bignon,J., ed. NATO ASI Series, Vol. **G21**, Springer-Verlag. Berlin-Heidelberg, pp. 281-307.

Barrett,J.C. and Ts'o,P.O.P. (1978) Evidence for the progressive nature of neoplastic transformation *in vitro. Proc. Natl Acad. Sci. USA* **75**:3761-3765.

Barrett,J.C., Everitt,J. and Walker,C., (1991) Possible cellular and molecular mechanisms for asbestos carcinogenicity. *Am. J. Ind. Med.*, in press.

Barrett,J.C. and Wiseman,R.W. (1987) Cellular and molecular mechanisms of multistep carcinogenesis: relevance to carcinogen risk assessment. *Environ. Hlth Perspect.* **76**:65-70.

Bishop,J.M. (1987) The molecular genetics of cancer. *Science* **235**:305-311.

Boyd,J.A. and Barrett,J.C. (1990) Tumor suppressor genes: possible functions in the negative regulation of cell proliferation. *Molec. Carcinogen.* **3**:, in press.

Callahan,R. (1989) Genetic alterations in primary breast cancer. *Breast Cancer Res. Treat.* **13**:191-203.

Chamberlain,M. and Tarmy,E. M. (1977) Asbestos and glass fibers inbacterial mutation tests. *Mutat. Res.* **43**:159-164.

DiPaolo,J.A., DeMarinis,A.J. and Doniger,J. (1983) Asbestos and benzo(a)pyrene synergism in the transformation of Syrian hamster embryo cells. *Pharmacology* **27**:65-73.

Fearon,E.R.and Vogelstein,B. (1990) A genetic model for colorectal tumorigenesis. *Cell* **61**:759-767.

Gibas,Z., Li,F.P., Antman,K.H., Bernai,S., Stahel,R. and Sandberg,A.A. (1986) Chromosome changes in malignant mesothelioma. *Cancer Genet. Cytogenet.* **20**:191-201.

Hammond,E.C., Selikoff,I.J.and Seidman,H. (1979) Asbestos exposure, cigarette smoking and death rates. *Ann. NY Acad. Sci.* **330**:473-490.

Hesterberg,T.W. and Barrett,J.C. (1984) Dependence of asbestos- and mineral dust-induced transformation of mammalian cells in culture on fiber dimension. *Cancer Res.* **44**:2170-2180.

Hesterberg,T.W. and Barrett,J.C. (1985) Induction by asbestos fibers of anaphase abnormalities: mechanism for aneuploidy induction and possibly carcinogenesis. *Carcinogenesis* **6**:473-475.

Hesterberg,T.W., Butterick,C.J., Oshimura,M., Brody,A.R. and Barrett,J.C. (1986) Role of phagocytosis in Syrian hamster cell transformation and cytogenetic effects induced by asbestos and short and long glass fibers. *Cancer Res.* **46**:5795-5802.

Jaurand,M.C. (1989) A particulate state carcinogenesis: Recent data on the mechanisms of action of fibers. *IARC Publications* **90**:54-73.

Klein,G. (1987) The approaching era of the tumor suppressor genes. *Science* **238**:1539-1545.

Land,H., Parada,L.F. and Weinberg,R.A. (1983) Cellular oncogenes and multistep carcinogenesis. *Science* **222**:771-778.

Lechner,J.F., Tokiwa,T., LaVeck,M., Benedict,W.F., Banks-Schlegel,S., Yeager Jr.,H., Banerjee,A. and Harris,C.C. (1985) Asbestos-associated chromosomal changes in human mesothelial cells. *Proc. Natl Acad. Sci. USA* **82**:3884-3888.

Mikalsen,S.O., Rivedal,E. and Sanner,T. (1988) Morphological transformation. of Syrian hamster embryo cells induced by mineral fibres and the alleged enhancement of benzo(a)pyrene. *Carcinogenesis* **9**:891-899.

Oshimura,M. (1990) Lessons learned from studies on tumor suppression by microcell-mediated chromosome transfer. *Environ. Hlth Perspect.*, in press.

Oshimura,M., Hesterberg,T.W. and Barrett,J.C. (1986) An early, nonrandom karyotypic change in immortal Syrian hamster cell lines transformed by asbestos: trisomy of chromosome 11. *Cancer Genet. Cytogenet.* **22**:225-237.

Oshimura,M., Hesterberg,T.W., Tsutsui,T. and Barrett,J.C. (1984) Correlation of asbestos-induced cytogenetic effects with cell transformation of Syrian hamster embryo cells in culture. *Cancer Res.* **44**:5017-5022.

Patrour,J.J., Bignon,J. and Jaurand,M.C. (1985) *In vitro* transformation of rat pleural mesothelial cells by chrysotile and/or benzo-a-pyrene. *Carcinogenesis* **6**:523-529.

Peto,J. (1985) Problems in dose-response and risk assessment: the example of asbestos. Risk Quantitation and Regulatory Policy, *Banbury Report.* **19**:89-101.

Skuse,G.R., Rowley,P.T. (1989) Tumor suppressor genes and inherited predisposition to malignancy. *Semin. Oncol.* **16**:128-137.

Stanton,M.F., Layard,M., Tegeris,A., Miller,E., May,M., Morgan,E. and Smith,A. (1981) Relation of particle dimension to carcinogenicity in. amphibole asbestoses and other fibrous minerals. *J. Natl Cancer Inst.* **67**:965-975.

Stenman,C., Olofsson,K., Mansson,T., Hagmar,B. and Mark,J. (1986) Chromosomes and chromosomal evolution in human mesotheliomas as reflected in sequential analysis of two cases. *Hereditas* **105**:233-239.

Sugawara,O., Oshimura,M., Koi,M., Annab,L. and Barrett,J.C. (1990) Induction of cellular senescence in immortalized cells by human chromosome 1. *Science* **247**:707-710.

Tiainen,M., Tammilehto,J., Rautonen,J., Tuomi,T., Mattson,K. and Knuutila,S. (1989) Chromosomal abnormalities and their correlations with asbestos exposure and survival in patients with mesothelioma. *Br. J. Cancer* **60**:618-626.

Wang,N.S., Jaurand,M.C., Magne,L., Kheuang,L., Pinchon,M.C. and Bignon.J. (1987) The interactions between asbestos fibers and metaphase chromosomes of rat pleural mesothelial cells in culture. *Am. J. Pathol.* **126**:343-349.

Weinberg,R.A. (1989) Oncogenes, anti-oncogenes, and the molecular bases of multistep carcinogenesis. *Cancer Res.* **49**:3713-3721.

Weston,A., Willey,J.C., Modali,R., Sugimura,H., McDowell,E.M., Resau,J., Light,B., Haugen,A., Mann,D.L., Trump,B.F. and Harris,C.C. (1989) Differential DNA sequence deletions from chromosomes 3, 11, 13, and 17 in squamous-cell carcinoma, large-cell carcinoma, and adenocarcinoma of the human lung. *Proc. Natl Acad. Sci. USA* **86**:5099-5103.

Yamada,H., Wake,N., Fujimoto,S., Barrett,J.C. and Oshimura,M. (1990) Multiple chromosomes carrying tumor suppressor activity for a uterine endometrial carcinoma cell line identified by microcell mediated chromosome transfer. *Oncogene* **5**:1141-1147.

THE MUTAGENICITY OF MINERAL FIBERS

Tom K. Hei, Zhu Y. He, Chang Q. Piao and *Charles Waldren

Center for Radiological Research
College of Physicians & Surgeons
Columbia University
New York, U.S.A.

*Dept. Radiation Biology & Radiology
Colorado State University
Fort Collins
Colorado, U.S.A.

INTRODUCTION

Asbestos fibers are carcinogenic to both man and experimental animals. At present the cellular and molecular mechanisms for fiber carcinogenesis, however, are not entirely clear. Various *in vitro* and *in vivo* studies have identified several factors that may influence the biological effects of mineral fibers. Fibers that are long and thin are more carcinogenic than fibers that are short and thick (Stanton *et al*, 1977). *In vitro* studies using oncogenic transformation as an endpoint have demonstrated that asbestos fibers can induce malignantly transformed foci in certain rodent cells (Hesterberg *et al*, 1984) and that asbestos fibers, in combination with either benzo[a]pyrene (Brown *et al*, 1983) or radiation (Hei *et al*, 1985, 1989) can synergistically enhance the oncogenic transforming activity of these agents. Although several types of asbestos fibers have been shown to induce chromosomal aberrations and sister chromatid exchanges in both rodent and human cells (Jaurand *et al*, 1986; Lechner *et al*, 1985), reports on the mutagenicity of asbestos fibers have largely been negative. Huang *et al* (1978) reported the only positive mutagenic studies of asbestos fibers, using Chinese hamster lung cells at the hypoxanthine guanine phosphoribosyl transferase (HGPRT) locus with 6-thioguanine as the selective agent.

While the use of gene markers such as the HGPRT located on essential, monosomic chromosomes provides an accurate and convenient measure for agents that induced mainly gene mutations, the incidence of chromosomal mutations, such as large deletions and non-disjunctional process, is underestimated if these mutations extend into vital genes. In the present studies, mutations induced by graded doses of crocidolite fibers are examined using the A_L hybrid cell system. The increased sensitivity of this assay is due to the larger genomic targets in the A_L cells for mutation than conventional gene markers.

MATERIALS AND METHODS

The human-hamster hybrid cell line (A_L), developed by Puck and coworkers (1971) was used in these studies. In addition to the standard set of hamster chromosomes, these hybrid cells contain a single copy of human chromosome 11. Several cell surface antigen markers have been identified on these cells and have been regionally mapped on chromosome 11. These include the genes for a_1, a_2 and a_3 surface antigens. Normal rabbit serum was used as a source of complement and specific poly- and monoclonal antisera against a_1 and a_2 antigenic marker were produced as described (Waldren *et al*, 1979). Cells were maintained in Ham's F12 medium supplemented with 8% heat-inactivated fetal bovine serum (Hyclone Laboratories, Logan, UT), 2x normal glycine (2×10^{-4} M) and 25 μg/ml gentamycin.

UICC standard reference crocidolite fibers were used in these studies. The compositional analysis, size distribution and preparation of the fibers have been described previously (Timbrell *et al*, 1968; Hei *et al*, 1984). Exponentially growing A_L cells were treated with graded doses of the fibers in 5 ml of culture medium for 24 hours. Following treatment, the cultures were washed twice with buffered salt solution, trypsinized, counted and replated at a density such that approximately 5×10^4 viable cells were included. Corresponding dishes were plated at lower densities to determine survival.

The asbestos treated cultures were incubated for one week before mutagenesis testing began as described (Hei *et al*, 1988a). This expression period permitted the surviving cells to multiply to the point at which the progeny of the mutated cells no longer contain lethal amounts of the surface antigens. Briefly, aliquots containing 5×10^4 cells were plated into each of six 60 mm diameter dishes in 2 ml culture medium. The dishes were incubated for 2 hours to allow for cell attachment and subsequently challenged with 0.6% antisera together with 1.5% freshly thawed complement. Cultures were incubated for 7-8 days at which time they were fixed, stained and counted. The cultures were tested each week for 2 consecutive weeks to ensure full expression of the mutants. The induced mutant fractions remained fairly constant for several weeks. Mutation frequencies were determined as the number of surviving colonies divided by the total number of cells plated after correction for any non-specific killing due to complement.

To assay for induction of HGPRT⁻ mutants, asbestos treated cultures as described above were replated into 100 mm diameter dishes at a density of 1×10^5 cells/dish (Hei *et al*, 1988b). Each dish contained 12 ml of culture medium together with 40 μM 6-thioguanine. A total of 30 dishes for mutant selection and 12 dishes/dose point for plating efficiency were plated. After incubation for 9-10 days, all dishes were fixed and stained as described above.

To analyze DNA from asbestos induced and spontaneously-derived HGPRT⁻ mutants, they were isolated by cloning and high molecular weight DNA was then

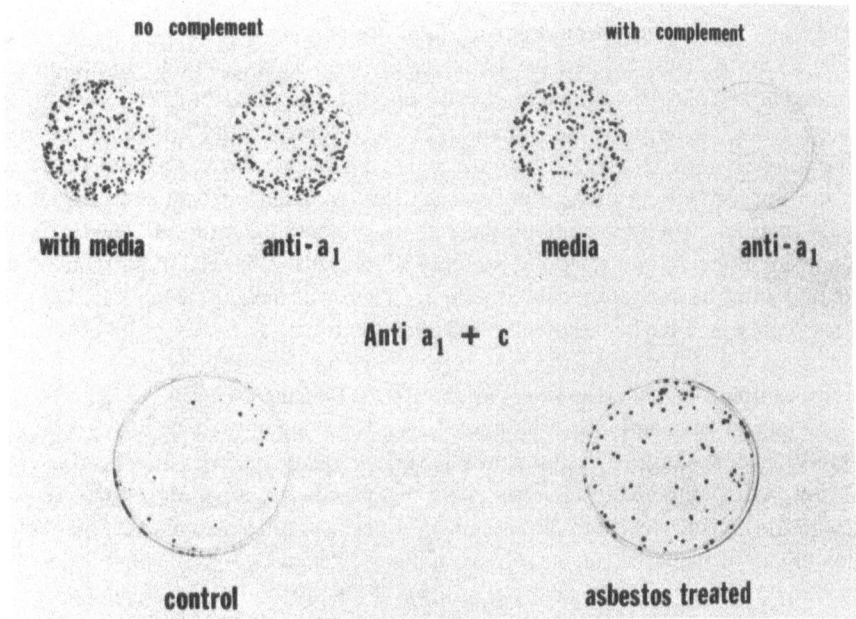

no complement · with complement

with media · anti-a_1 · media · anti-a_1

Anti a_1 + c

control · asbestos treated

Figure 1. Characterization of asbestos induced a_1^- mutants.

prepared according to standard technique. The DNA was then digested with Pst I restriction endonuclease and fractionated by agarose gel electrophoresis. The blotting was done overnight in 20x SSC buffer onto a nitrocellulose sheet. After baking for 2 hours at 80°C, the blot was pre-hybridized, then hybridized with ^{32}P-labelled HGPRT cDNA probe (pHPT12) as described previously (Fuscoe *et al*, 1986). The filter was then washed and developed according to standard techniques.

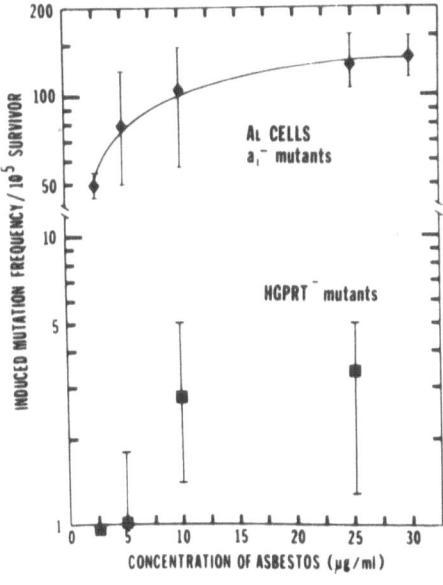

Figure 2.

Characterization of A_L mutation assay at the a_1 locus

Wild type A_L cells express the surface antigen markers a_1 and a_2. In the presence of complement and specific antiserum against either the a_1 or a_2 antigens, wild type A_L cell system either in the presence (upper right) or absence (upper left) of complement when 300 A_L cells were challenged with the antiserum against the a_1 surface antigen. The average non-specific killing of A_L cells by the 1.5% complement treatment ranged from essentially none to 10% depending on the batches of complement used. The 0.6% antibody treatment alone did not affect the viability of the cells. When 5×10^4 normal A_L cells were treated with the antiserum and complement, most of the cells were killed except a few which represented the background mutant fraction (lower left).

Mutation induction by asbestos fibers at the HGPRT locus

The mutant fraction induced by graded doses of crocidolite fibers in A_L cells at the a_1 and HGPRT loci examined is shown in Figure 2. Pooled data from three to five experiments showed that at the HGPRT locus, there was no consistent pattern in dose response relationship for mutant yield through the range of concentrations examined. Although the absolute number of induced mutants (observed minus background) at the high dose groups was not zero, there was no statistically significant difference in mutant induction between the highest and lowest doses studied.

The spontaneous HGPRT$^-$ mutant fraction in the A_L cells used ranged from 0.16 to 2.3 mutants per 10^5 survivors. When 6-thioguanine resistant mutants were cloned, expanded in cultures and subsequently tested in medium containing aminopterin and hypoxanthine, none of them survived, indicating that they were true HGPRT$^-$ mutants.

Mutation induction by crocidolite fibers at the a_1 locus

In contrast to the HGPRT locus, crocidolite fibers induced a dose dependent mutagenesis at the a_1 locus from the same treated population over the same concentration ranges. At the lowest fiber concentration examined (2.5 μg/ml, surviving fraction = 0.87), while the HGPRT$^-$ mutation induction by crocidolite fiber was indistinguishable from the spontaneous level, the mutant fraction at the a_1 locus was at least 50 fold higher. The average number of pre-existing a_1^- mutants per 10^5 survivors ranged from 57 to 150. This background mutant fraction represented the accusal of a_1^- cells in the population over a long period of time before these experiments were begun. Fluctuation analysis showed that the rate of loss of spontaneous a_1 marker amounted to approximately 7×10^{-6}/cell/doubling time (Waldren et al, 1986). Since the rate of spontaneous loss of the a_1 marker was low compared to the frequency of induced mutations, the background was subtracted to give the induced frequencies at each asbestos dose.

Cellular and molecular analysis of mutants

To determine the proportion of a_1^- mutants that had also lost the a_2 marker gene located on the long arm of the human chromosome 11, selected a_1^- mutants from asbestos treated and control dishes were cloned, expanded in culture and subsequently tested for their sensitivity to antisera against either the a_1 or a_2 marker genes. Of the 16 asbestos-induced a_1^- mutants that had been analysed thus far, 13 of them had lost both the a_1 and a_2 markers (82% versus 18% that only lost the a_1 gene). Similar results were also obtained with spontaneous mutants derived from control dishes. Recent studies using Southern blotting techniques probed to other human gene markers on human chromosome 11, indicated that most of the fiber-induced mutants still retained substantial regions of human chromosome 11 (data not shown).

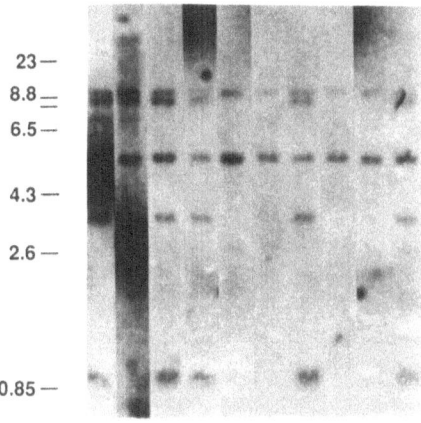

1 2 3 4 5 6 7 8 9 10

23 —

8.8 —
6.5 —

4.3 —

2.6 —

0.85 —

Figure 3.

Analysis by Southern blots of the HGPRT⁻ mutants disclosed that almost all of the mutants generated from the asbestos treated cultures showed some alteration of the gene (23/24). In contrast, spontaneous mutants showed either no detectable change at the HGPRT sequences or partial deletions. Figure 3 shows representative results from several experiments in which Pst I restriction fragments from mutants were separated by electrophoresis, transferred to nitrocellulose and hybridized to ^{32}P-labelled pHPT12. The full length Chinese hamster HGPRT cDNA probe hybridized to six fragments of 8.8, 8.2, 7.6, 5.2, 3.2 and 0.85 kb in DNA from normal A_L cells (lane 1). In some samples, a weakly hybridizing fragment that corresponded to 2.6 kb could also be identified. The fragments that corresponded to size 5.2 and 2.6 kb represent pseudogenes (Fuscoe et al., 1986). Two of the three spontaneous mutants shown in Figure 3 showed partial deletions (lanes 3&4) while a third one showed no detectable change in the HGPRT sequences (lane 2). Patterns from asbestos induced mutants are shown in lanes 5 through 10. Four of the mutants (lanes 5,6,8 and 9) apparently have lost 4 out of the 5 functional HGPRT sequences and thus represent major deletion of the gene, while two of the mutants (lanes 7 & 10) showed only partial deletions with the 7.6 kb fragment missing.

SUMMARY

Results of the present studies indicate that crocidolite fibers are highly mutagenic to mammalian cells. It is shown that, while few mutants are induced at the HGPRT locus, at equivalent fiber concentrations there is a 50 fold increase in mutant induction at the a_1 locus of human chromosome 11. While the use of gene markers such as the HGPRT located on essential, monosomic chromosomes, provides an accurate and convenient measure for agents that produce mainly gene mutations, chromosomal mutations such as large deletions, and non-disjunctional process may be underestimated. Deletions at the HGPRT locus much larger than 40 kilobases are not seen because they extend into vital genes, which results in the cell's death (Waldren, 1979; Thilly, 1985). The utilization of the A_L hybrid cells for mutagenesis study has advantage over conventional assay systems because the whole human chromosome 11 can act as a target for the mutagen. Since this chromosome is not essential for the viability of the cell, all kinds of mutations can be scored. The fact that earlier studies failed to detect mutation by asbestos fibers at the HGPRT or ouabain loci may be a consequence of large gene deletions not compatible with the survival of the induced mutants, as is the case for mutations by various radiations including X-rays. These larger kinds of damages cannot be ignored since they have been implicated in a number of human mutational diseases and in the activation of oncogenes (Croce, 1986).

The A_L cells have been in culture for almost two decades and represents a stable hybrid cell line. While the majority of a_1^- mutants, either induced by fibers or spontaneously derived, have also lost the a_2 marker gene, most of them still retain regions of the human chromosome 11 (Waldren, unpublished data). Hence, mutagenesis in the A_L cells arises not as a propensity of the hamster cells to lose human chromosome.

Although the mechanism of asbestos induced genetic damages are not known, previous studies undertaken to elucidate the cellular and molecular mechanism(s) of interaction between radiation and asbestos fibers have demonstrated that oxygen radicals may be an important intermediate in the observed synergistic interaction (Hei & Kushner, 1987). Reactive oxygen species have been shown to be mutagenic in both the Ames's bacterial assay and in several mammalian gene loci as well. The fact that asbestos has not been shown to be mutagenic in bacterial assay systems highlights the importance of fiber-cell interaction. The multilocus type of damage induced by asbestos fibers is in agreement with the proposed oxy-radical mechanism of fiber carcinogenesis since recent studies using the polymerase chain reaction technique have suggested that reactive oxygen species induce mostly deletions in mammalian cells (Hsie *et al*, 1990).

ACKNOWLEDGEMENT

The authors would like to thank Drs Howard Libermann and Gregg Freyer for helpful discussions. This work was supported by Grant CA-49062 from the National Cancer Institute.

REFERENCES

Brown,R.C., Poole,A. and Fleming,G.T.A. (1983) The influence of asbestos dust on the oncogenic transformation of C₃H 10T1/2 cells. *Cancer Letters* **18**:221-227.

Croce,C.M. (1986) Chromosome translocations and human cancer. *Cancer Res.* **46**:6019-6023.

Fuscoe,J.C., Ockey,C.H. and Fox,M. (1986) Molecular analysis of X-ray induced mutants at the HPRT locus in V-79 Chinese hamster cells. *Int. J. Radiat. Biol.* **49**:1011-1020.

Hei,T.K., Geard,C.R., Osmak,R. and Travisano, M. (1985) Correlation of *in vitro* genotoxicity and oncogenicity induced by radiation and asbestos fibers. *Br. J. Cancer* **52**:591-597.

Hei,T.K. and Kushner,S. (1987) Radiation and asbestos fibers: interaction and possible mechanism. In, "Anticarcinogenesis and radiation protection", eds. Cerutti,P., Nygarrd,O. and Simic,M. ,Plenum Press, N.Y. pp. 345-348.

Hei,T.K., Hall,E.J. and Waldren,C.A. (1988a) Mutation induction by neutrons as determined by an antibody complement mediated cell lysate system. I. Experimental observation. *Radiation Res.* **115**:281-291.

Hei,T.K., Chen,D., Brenner,D. and Hall,E.J. (1988b) Mutation induction by charged particles of defined LET. *Carcinogenesis* **9**:1233-1236.

Hesterberg,T.W. and Barrett,J.C. (1984) Dependence of asbestos and mineral dust induced transformation of mammalian cells in culture on fiber dimension. *Cancer Res.* **44**:2170-2180.

Hsie,A.W., Xu,W., Yu,Y., Sognier,M.A. and Hrelia,P. (1990) Molecular analysis of reactive oxygen species induced mammalian gene mutation. *Terato. Carcin. Mut.* **10**:115-124.

Huang,S.L., Saggioro,D., Michelmann,H. and Malling,H.V. (1978) Genetic effects of crocidolite asbestos in CHO cells. *Mutat. Res.* **57**:225-232.

Jaurand,M.C., Kheuang,L., Magne,L. and Bignon,J. (1986) Chromosomal changes induced by chrysotile fibers or benzo(3,4)pyrene in rat pleural mesothelial cells. *Mutat. Res.* **169**:141-145.

Lechner,J.K., Tokiwa,T., LaVeck,M., Benedict,W.F., Bankschlegel,S., Yeager,H., Barnegee,J.A. and Harris,C.C. (1985) Asbestos associated chromosomal changes in human mesothelial cells. *Proc. Natl Acad. Sci. USA* **82**:3884-3889.

Puck,T.T., Wuchier,P., Jones,C. and Kao,F.T. (1971) Genetics of somatic mammalian cells: Lethal antigens and genetic markers for study of human linkage groups. *Proc. Natl Acad. Sci. USA* **68**:3102-3106.

Stanton,M.F., Layard,M. and Tegeris, A. (1977) Carcinogenesis of fibrous glass: pleural response in the rat in relation to the fiber dimension. *J. Natl. Cancer Inst.* **58**:587-604.

Thilly,W.G. (1985) Dead cells don't form mutant colonies: a serious source of bias in mutation assays. *Environ. Mutat.* **7**:255-258.

Timbrell,V., Gilson,J.C. and Webster,I. (1986) UICC standard reference samples of asbestos. *Int. J. Cancer* **3**:406-410.

Waldren,C., Jones,C.C. and Puck,T.T. (1979) Measurement of mutagenesis in mammalian cells. *Proc. Natl Acad. Sci. USA* **76**:1358-1362.

Waldren,C., Correll,L., Sognier,M.A. and Puck,T.T. (1986) Measurement of low levels of X-ray mutagenesis in relation to human disease. *Proc. Natl Acad. Sci. USA* **83**:4839-4843.

CHROMOSOMAL DAMAGE AND GAP JUNCTIONAL INTERCELLULAR COMMUNICATION IN MESOTHELIOMA CELL LINES AND CULTURED HUMAN PRIMARY MESOTHELIAL CELLS TREATED WITH MMMF, ASBESTOS AND ERIONITE

Kaija Linnainmaa, Katarina Pelin-Enlund, Kaarina Jantunen, Esa Vanhala, Timo Tuomi, Jim Fitzgerald[1], Hiroshi Yamasaki[1]

Institute of Occupational Health
Topeliuksenkatu 41 a A, SF00250
Helsinki, Finland

[1] International Agency for Research on Cancer
Lyon, France

ABSTRACT

Mesothelioma tumour cells exhibit karyotypic changes involving different chromosomes. Gap junctional intercellular communication (GJIC) is reduced in the tumour cells compared to normal mesothelial cells. Human primary mesothelial cells were treated with defined fractions of various mineral fibres for acute chromosomal aberration studies and for effects on GJIC. No dose response was observed in the induction of chromosomal aberrations by the fibres tested. Some doses of rock wool, thin and coarse glass wool, wollastonite, amosite and erionite caused significant increases in the number of aberrant cells. A major or persistent diminution of GJIC was not affected by any of these mineral fibres.

INTRODUCTION

Exposure to asbestos fibre is the primary causal factor in the etiology of malignant mesothelioma (Wagner *et al.*, 1960); association to asbestos exposure has been documented in 80 to 100% of all mesothelioma cases (Parkes, 1982). Animal studies indicate that asbestos and other mineral fibres can act as complete carcinogens, but there is also evidence that they possess a tumour promoting activity. Cell transformation has shown to be dependent on fibre dimension. According to hypotheses, the physical presence of fibres inside the cells interferes with chromosome segregation during the mitosis and results in chromosomal abnormalities (Barrett *et al.*, 1989). Asbestos and glass fibres have shown to induce chromosomal aberrations and aneuploidy in a wide variety of rodent cell assays (Brown *et al.*, 1979; Sincock *et al.*, 1982; Oshimura *et al.*, 1984; Hesterberg & Barrett, 1985; Jaurand *et al.*, 1986), but less data are available with human cells (Sincock *et al.*, 1982; Linnainmaa *et al.*, 1986) and conflicting results have been reported.

Cultured human mesothelial cells provide a relevant model for mechanistic studies of fibre carcinogenicity, since they are the cells from which mesothelioma originates. Primary mesothelial cells can be obtained from pleural effusions from non-cancerous donors and short term cultures can be established as demonstrated by Lechner and co-workers (1985).

We have been examining the etiology and cellular mechanisms of fibre-induced mesothelioma by establishing and characterising malignant mesothelioma cell lines from tumour tissue or pleural fluids from patients with asbestos associated mesothelioma. These collaborative studies have included histological, morphological and karyotypic characterisation of the established cell lines, as well as studies on oncogene and growth factor expression and studies on the gap junctional intercellular communication (GJIC).

Since such studies indicated a possible role of chromosomal aberrations and inhibited GJIC in the genesis of mesothelioma in asbestos-exposed persons, we examined the effects of mineral fibres on structural chromosomal aberrations and on GJIC of cultured human mesothelial cells. The fibre types in our studies have included amosite and chrysotile asbestos, erionite, wollastonite and man-made-mineral fibres (MMMF), glass wool and rock wool. Titanium dioxide was used as a non-fibrous control dust. In addition to the effects of different fibre types, the role of fibre size has been addressed in these experimental studies.

In this chapter we summarise our results on the chromosome and GJIC studies both from the *in vitro* genotoxicity experiments with primary human mesothelial cell cultures and from the studies on our mesothelioma tumour cell lines.

STUDIES ON CHROMOSOMAL EFFECTS

Fibre samples

Different size fractions of MMMF were prepared from rock wool and glass wool by liquid milling and sedimentation. The lengths and widths of all the fibres included in the study were defined by electron microscopy and the number of fibres per unit weight was calculated. The fibre fractions were defined as follows: thin glass wool (median length 3.8 μm, width 0.21 μm), coarse glass wool (median length 13.0 μm, width 0.71 μm), thin rock wool (median length 6.1 μm, width 0.29 μm), coarse rock wool (median length 25.0 μm, width 1.1 μm), milled rock wool (no fibres), amosite (median length 2.3 μm, width 0.18 μm), chrysotile (median length 1.0 μm, width 0.09 μm), erionite (median length 1.4 μm, width 0.12 μm) and wollastonite (median length 2.6 μm, width 0.38 μm). The number of fibres in one microgram of materials were as follows: chrysotile (8×10^6), erionite (1.4×10^6), amosite (0.4×10^6), thin glass wool (29×10^3), thin rock wool (11×10^3) coarse rock wool (5.8×10^3), coarse glass wool (4.7×10^3),

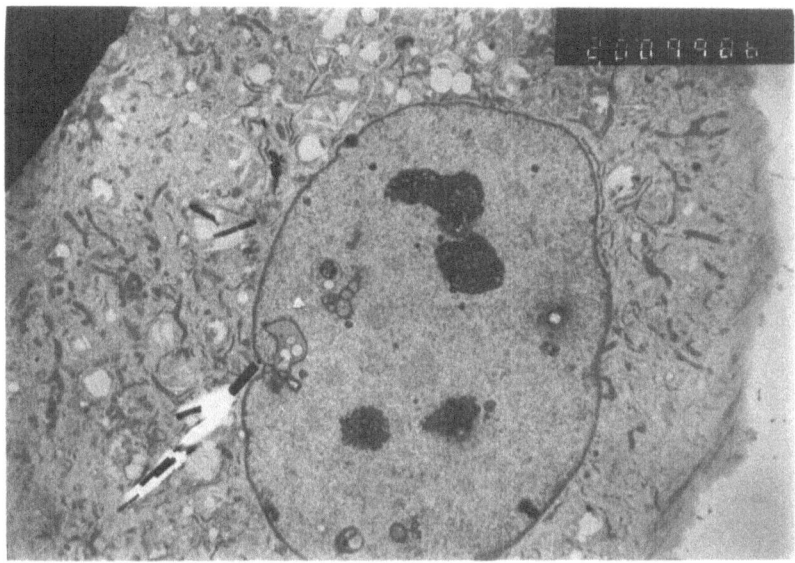

Figure 1. A transmission electron micrograph of amosite fibres inside a mesothelial cell. The fibres are located in the cytoplasm near the nucleus (magnification x 3000).

wollastonite (3×10^3) and milled rock wool (0.5×10^3). Titanium dioxide was used as an inert, non-fibrous control dust.

Induction of chromosomal aberrations

Phagocytosis of amosite, chrysotile and glass wool fibres was studied by scanning and transmission electron microscopy. The fibres were rapidly engulfed by the cells and within 24 hours after the exposure were seen almost exclusively in the cytoplasm, either in phagosomes or free (figure 1).

For the chromosomal aberration studies human mesothelial cells were treated with fibres for 48 hours. The cells used in the experiments originated from four different donors. In general, no dose response was observed in the induction of chromosomal aberrations by the various fibres tested. These results together with the numbers of fibres per unit area of the tissue culture flask are presented in Table 1. Significant increases in the number of aberrant cells were observed for thin rock wool (4 μg/cm^2), thin glass wool (2, 3 and 5 μg/cm^2), coarse glass wool (8 μg/cm^2), wollastonite (0.5 and 5 μg/cm^2), amosite (0.5, 1 and 2 μg/cm^2), and erionite (1 μg/cm^2). The positive results were observed with the higher fibre doses and no clear differences were noted between thin and coarse MMMF. Results indicate that mesothelial cells from different individuals vary in their sensitivity to the toxic and clastogenic effects of mineral fibres. Our present studies have not included aneuploidy analyses, but we have previously shown that both amosite asbestos and JM 100 glass fibres also induce aneuploidy (mainly hypodiploidy) in cultured human mesothelial cells at lower fibre doses than those at which they induce structural aberrations (Linnainmaa *et al.*, 1986). More precise studies on the induction of aneuploidy in human mesothelial cells are currently underway at our laboratory.

329

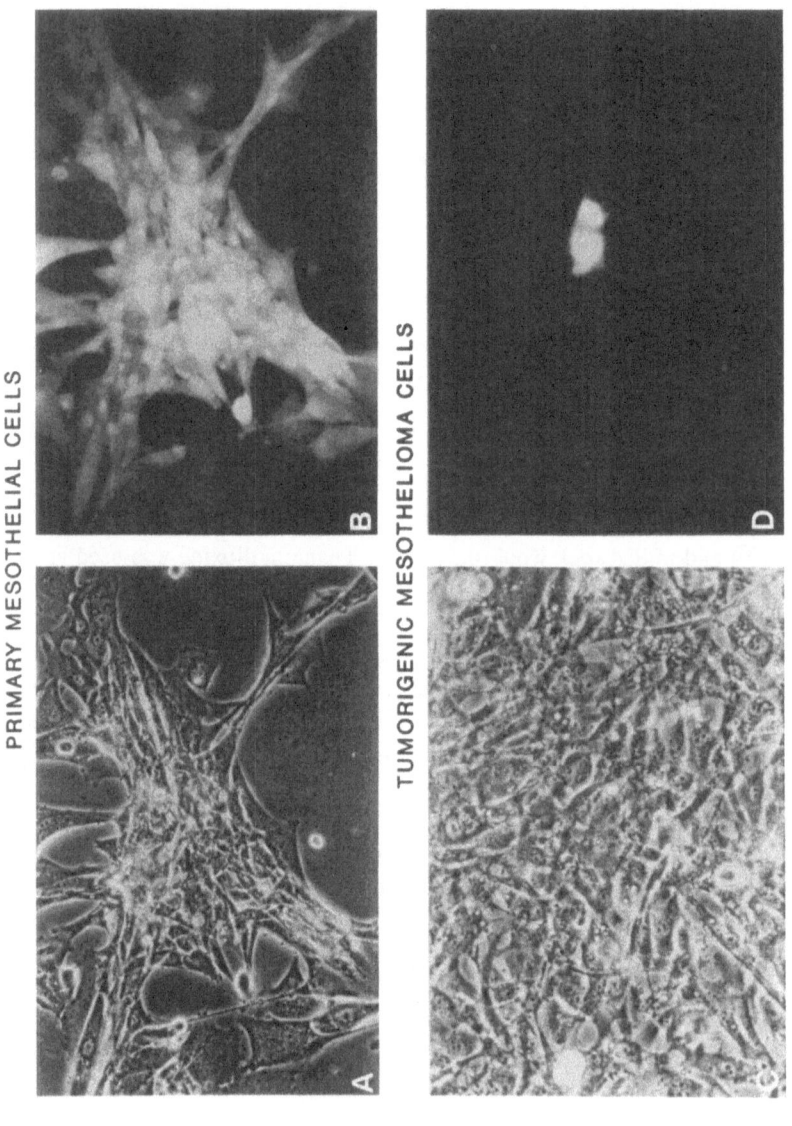

PRIMARY MESOTHELIAL CELLS

TUMORIGENIC MESOTHELIOMA CELLS

Figure 2 A) A phase contrast micrograph of a primary mesothelial culture. B) Communicating primary mesothelial cells after fluorescent dye-injection of one cell. C) A phase contrast micrograph of a tumorigenic mesothelioma cell culture. D) No communication between mesothelioma cells after fluorescent dye-injection of one cell.

Tumour cell line karyotypes

Several studies have indicated that human mesothelioma tumour cells exhibit complex structural and numerical abnormalities (Mark, 1978; Becher *et al.*, 1984; Gibas *et al.*, 1986; Stenman *et al.*, 1986; Bello *et al.*, 1987; Popescu *et al.*, 1988; Tiainen *et al.* 1988; Pelin-Enlund *et al.*, 1990). We have established and characterised seven immortal cell lines from tumour samples or pleural effusions from mesothelioma patients with a history of occupational asbestos exposure (Pelin-Enlund *et al.*, 1990). As reported by others, the karyotypes typically exhibit complex structural and numerical chromosome abnormalities. These involve different chromosomes, e.g., 1, 4, 5, 6, 7, 8, 9, 11, 12, 13, and 22. In addition, an excess of chromosome material of the short arm of chromosome 5 was consistently seen in six of the seven cell lines. These consistent clonal changes concerning chromosome 5 seem to be non-random and could be typical of malignant mesothelioma. Changes in chromosome 13, mainly monosomy, were also frequently noted. This may be of interest, because chromosome 13 carries the Rb gene locus which has been shown to play important role in small cell lung cancer (Harbour *et al.*, 1988), as well as in other cancers of mesenchymal origin (Friend *et al.*, 1987). Monosomy of chromosome 9 and rearrangements of chromosome 1 were also frequently noted which is in accordance with several other studies (Gibas *et al.*, 1986; Bello *et al.*, 1987; Popescu *et al.*, 1988; Tiainen *et al.*, 1988). Chromosome 1 has been shown to carry several proto-oncogenes, thus changes in chromosome 1 may have importance in a variety of malignancies; alterations have been found to be present in several different types of solid tumours (Sandberg & Turc-Carel, 1987; Friend *et al.*, 1987; Sandberg, 1982; Sandberg, 1983; Yunis, 1983). The most frequently found karyotypic changes in our mesothelioma cell lines are listed in Table 2.

The role of the consistent changes observed in chromosome 5 and 13 as well as 1, along with their possible association with oncogene activations or growth factor productions in the aetiology of asbestos-related malignant mesothelioma, remains to be solved. Mesothelioma cell lines have been shown to produce more platelet-derived growth factor (PDGF) than primary mesothelial cells in culture (Gerwin *et al.*, 1987). The gene for the PDGF-A chain is located in chromosome 7 and for the PDGF-B chain in chromosome 22. Partial trisomy of chromosome seven was noted in one of our cell lines and chromosome 22 was found to be monosomic in two of our cell lines. Whether the observed fibre-induced chromosomal changes contribute to the karyotypic abnormalities in mesothelioma tumour cells remains unknown.

STUDIES ON GAP JUNCTIONAL INTERCELLULAR COMMUNICATION

In normal tissues cells are tightly controlled by surrounding cells. Cancer cells, on the contrary, are uncontrolled and neglect the interaction of surrounding cells. Numerous studies suggest that the lack of this control associates with abnormal gap junctional intercellular communication (GJIC) in tumour cells. It has been shown that GJIC can be disturbed in the process of carcinogenesis by various agents including tumour promoters (Loewenstein, 1979 & 1981; Yamasaki *et al.*, 1988; Yamasaki, 1990). Since asbestos fibres have been shown to have tumour promoting activity, we decided to examine the GJIC in our mesothelioma tumour cell lines and find the effects of amosite and chrysotile asbestos, as well as those of thin and coarse glass fibres, on the GJIC of normal human mesothelial cells.

Effects of mineral fibres on GJIC in primary mesothelial cell cultures

Gap junctional intercellular communication in human mesothelial cells treated with 2 $\mu g/cm^2$ of chrysotile, amosite, thin glass wool and coarse glass wool was studied by scoring the number of dye-coupled cells after micro-injection of single cells with Lucifer Yellow CH. The two asbestos fibre types were largely ineffective in blocking GJIC acutely. Some decrease was observed after 24 hours, but this did not persist, even when the fibres had been completely phagocytised by the cells. For the glass fibres, both types appeared to exert some early effect on GJIC but recovery was evident in 24-48 hours (Table 3). Exposure of primary mesothelial cells to the tumour promoter TPA, on the other hand, revealed a rapid and complete inhibition of GJIC but by 24 hours this effect was reversed. This complete but reversible effect is also typical of the effect of TPA on rodent epithelial cells (Mesnil *et al.*, 1986; Fitzgerald *et al.*, 1980). Thus, no major or persistent inhibition of GJIC was caused by any of the mineral fibres under study.

GJIC in mesothelioma tumour cell lines

When the GJIC characteristics of normal human mesothelial cells and a panel of our mesothelioma tumour cell lines was studied, GJIC was found to be markedly reduced in the mesothelioma tumour cells as compared to normal human cells (Table 4, figure 2). Thus it appears that a permanent and heritable inhibition of GJIC is a feature of resultant mesotheliomas, whatever the mechanism for mineral fibre action is. Since exposure of the primary mesothelial cells to the asbestos and glass fibres revealed no significant persistent effect on GJIC over the 48 hours time period, it is possible that longer time periods may be required to see any effect. The fibres may act in the later stages of the multistage transformation process. Mechanisms involved in permanent loss of GJIC and perhaps the role of fibres in them in the case of mesothelioma, may be better understood when the genes and their locations in the chromosomes related to gap junction functions become known.

ACKNOWLEDGEMENTS

We thank Ms. Marjatta Vallas for her technical help. Part of the study was supported by the Finnish Work Environment Fund.

REFERENCES

Barrett,J.C., Lamb,P.W. and Wiseman,R.W. (1989) Multiple mechanisms for the carcinogenic effects of asbestos and other mineral fibers. *Environ. Hlth Perspect.* **81**:81-89.

Becher,R., Wake,N., Gibas,Z., Ochi,H. and Sandberg,A.A. (1984) Chromosome changes in soft tissue sarcomas. *J. Natl. Cancer Inst.* **72**:823-831.

Bello,M.J., Rey,J.A., Aviles,M.J., Arevalo,M. and Benitez,J. (1987) Cytogenetic findings in an effusion secondary from pleural mesothelioma. *Cancer Genet. Cytogenet.* **29**:75-79.

Brown,R.C., Chamberlain,M. and Skidmore,J.W. (1979) *In vitro* effects of man-made mineral fibres. *Ann. Occup. Hyg.* **22**:175-179.

Fitzgerald,D.J. and Murray,A.W. (1980) Inhibition of intercellular communication by tumor promoting phorbol esters. *Cancer Res.* **40**:2935-2937.

Friend,S.H., Horowitz,J.M., Gerber,M.R., Wang,X.-F., Vogenmann,E., Li,F.P. and Weinberg,R.A. (1987) Deletions of a DNA sequence in retinoblastoma and mesenchymal tumors: organization of the sequence and its encoded protein. *Proc. Natl. Acad. Sci. USA* **84**:9059-9063.

Gerwin,B.I., Lechner,J.F., Reddel,R.R., Roberts,A.B., Robbins,K.C., Gabrielson,E.W. and Harris,C.C. (1987) Comparison of production of transforming growth factor-ß and platelet derived growth factor by normal human mesothelial cells and mesothelioma cell lines. *Cancer Res.* **47**:6180-6184.

Gibas,Z., Li,F.P., Antman,K.H., Bernal,S., Stahel,R. and Sandberg,A.A. (1986) Chromosome changes in malignant mesothelioma. *Cancer Genet. Cytogenet.* **20**:191-201.

Harbour,J.W., Lai,S.-L., Whang-Peng,J., Gazdar,A.F., Minna,J.D. and Kaye,F.J. (1988) Abnormalities in structure and expression of the human retinoblastoma gene in SCL. *Science* **241**:353-357.

Hesterberg,T.W. and Barrett,J.C. (1985) Induction by asbestos fibers of anaphase abnormalities: mechanism for aneuploidy induction and possibly carcinogenesis. *Carcinogenesis* **6**:473-475.

Jaurand,M.C., Kheuang,L., Magne,L. and Bignon,J. (1986) Chromosomal changes induced by chrysotile fibres or benzo-3,4-pyrene in rat pleural mesothelial cells. *Mutat. Res.* **169**:141-148.

Lechner,J.F., Tokiwa,T., LaVeck,M., Benedict,W.F., BanksSchlegel,S., Yeager,H., Banerjee,A. and Harris,C.C. (1985) Asbestos-associated chromosomal changes in human mesothelial cells. *Proc. Natl. Acad. Sci. USA* **82**:3884-3888.

Linnainmaa,K., Gerwin,B., Pelin,K., Jantunen,K., Lechner,J.F. and Harris,C.C. (1986) Asbestos-induced mesothelioma and chromosomal abnormalities in human mesothelial cells *in vitro*. Third US-Finnish Joint Science Symposium, Frankfort, Kentucky.

Loewenstein,W.R. (1979) Junctional intercellular communication and the control of growth. *Biochim. Biophys. Acta* **605**:33-91.

Loewenstein,W.R. (1981) Junctional intercellular communication: the cell-to-cell membrane channel. *Physiol. Rev.* **61**:829-913.

Mark,J. (1978) Monosomy 14, monosomy 22 and 13q-; three chromosomal abnormalities observed in cells in two malignant mesotheliomas studied by banding techniques. *Acta Cytol. (Baltimore)* **22**:398-401.

Mesnil,M., Montesano,R. and Yamasaki,H. (1986) Intercellular communication of transformed and non-transformed rat liver epithelial cells. *Exp. Cell Res.* **165**:391-402.

Oshimura,M., Hesterberg,T.W., Tsutsui,T. and Barrett,J.C. (1984) Correlation of asbestos-induced cytogenetic effects with cell transformation of Syrian hamster embryo cells in culture. *Cancer Res.* **44**:5017-5022.

Parkes,W.R. (1982) Occupational lung disorders. Butterworth, London, 2nd ed.

Pelin-Enlund,K., Husgafvel-Pursiainen,K., Tammilehto,L., Klockars,M., Jantunen,K., Gerwin,B.I., Harris,C.C., Tuomi,T., Vanhala,E., Mattson,K. and Linnainmaa,K. (1990) Asbestos-related malignant mesothelioma: growth, cytology, tumorigenicity and consistent chromosome findings in cell lines from five patients. *Carcinogenesis* **11**:673-681.

Popescu,N.C., Chahinian,A.P. and DiPaolo,J.A. (1988) Non-random chromosome alterations in human malignant mesotheliomas. *Cancer Res.* **48**:142-147.

Sandberg,A.A. (1982) Chromosomal changes in human cancers: specificity and heterogeneity. In, "Tumor cell heterogeneity: origins and implications", eds. Owens,A.H.,Jr., Coffey,D.S. and Baylin,S.B. Academic Press, New York, pp. 367-397.

Sandberg,A.A. (1983) A chromosomal hypothesis of oncogenesis. *Cancer Genet. Cytogenet.* **8**:277-285.

Sandberg,A.A. and Turc-Carel,C. (1987) The cytogenetics of solid tumors. *Cancer* **59**:387-395.

Sincock,A.M., Delhanty,J.D.A. and Casey,G. (1982) A comparison of the cytogenetic response to asbestos and glass fibre in Chinese hamster and human cell lines. Demonstration of growth inhibition in primary human fibroblasts. *Mutat. Res.* **101**:257-268.

Stenman,G., Olofsson,M., Månsson,T., Hagmar,B. and Mark,J. (1986) Chromosomes and chromosomal evolution in human mesotheliomas as reflected in sequential analyses of two cases. *Hereditas* **105**:233-239.

Tiainen,M., Tammilehto,L., Mattson,K. and Knuutila,S. (1988) Non-random chromosomal abnormalities in malignant pleural mesothelioma. *Cancer Genet. Cytogenet.* **33**:251-274.

Wagner,J.C., Sleggs,C.A. and Marchand,P. (1960) Diffuse pleural mesotheliomas and asbestos exposure in the North-western Cape Province. *Br. J. Indust. Med.* **17**:260-271.

Yamasaki,H., Enomoto,K., Fitzgerald,D.J., Mesnil,M., Katoh,F. and Hollstein,M. (1988) Role of intercellular communication in the control of critical gene expression during multistage carcinogenesis. In, "Cell Differentiation, Genes and Cancer", eds. Kakunaga,T., Sugimura,L. Tomatis,H. and Yamasaki,T. IARC Science Publication No. 92, IARC, Lyon. pp. 57-75.

Yamasaki,H. (1990) Gap junctional intercellular communication and carcinogenesis. *Carcinogenesis* **11**:1051-1058.

Yunis,J.J. (1983) The chromosomal basis of human neoplasia. *Science* **221**:227-235.

CHRYSOTILE, CROCIDOLITE, AND ANTHOPHYLLITE FACILITATION OF TRANSFECTION OF CULTURED MOUSE CELLS BY POLYOMAVIRUS DNA

George R. Dubes

Department of Pathology and Microbiology
College of Medicine
University of Nebraska Medical Center
Omaha, Nebraska 68198-6495, U.S.A.

ABSTRACT

Three decades have elapsed since the discovery that the silicates talc and fuller's earth are potent facilitators of transfection of mammalian cells. Although the possibility of a role for silicate facilitation in silicate-induced carcinogenesis was suggested in 1971, testing for facilitation by the important carcinogen asbestos apparently did not occur until 1984. There then appeared reports showing asbestos facilitated transfection of cultured mammalian cells by various viral RNAs and plasmid and polyomavirus DNA.

Further results with polyomavirus DNA are now reported, using cultured mouse 3T6 cells as permissive host, plaques as indicators of transfection events, and three asbestos varieties. While all three varieties facilitated transfection anthophyllite was especially potent, much more so than calcium phosphate, a classic "insoluble" facilitator. The amount of facilitation increased markedly with anthophyllite concentration up to 2.5 mg/ml and was strong with both crude and partially purified viral DNA. The anthophyllite-facilitated polyomavirus DNA/3T6 cell complexes initiating the transfection events were formed rapidly (< 40 seconds). Anthophyllite also made the cells readily transfectible by viral DNA added later, even after intervening cell washing. The theory that asbestos facilitation of transfection and oncogene transfer plays an important role in asbestos-induced carcinogenesis is discussed.

INTRODUCTION

The carcinogenic process is thought to involve the acquisition by a cell of two or more somatic mutations resulting in interference with normal cell regulation. These somatic mutations can be gene mutations, chromosome mutations, or genome mutations, as defined by Rieger and co-workers (1976), or some combination of these kinds of mutation. This can result in the acquisition of oncogenes, the conversion of protooncogenes into oncogenes, or the inactivation or elimination of tumor suppressor genes.

A priori, there are two ways a somatic cell could acquire two or more such mutations:

(1) Two or more mutations in that cell or in its progenitor cells;

(2) Transfer of the second mutation (and the third mutation *et cetera*, if required) from a cell of a different lineage.

Accordingly, agents which are mutagenic or "transfectogenic" would be expected to be carcinogenic, provided they (a) can reach the necessary DNA and target cells in an active form and (b) are sufficiently durable in the tissue to provide the time necessary for the mutagenic or "transfectogenic" action. The carcinogenicity of many mutagenic agents is well known and based on numerous studies. By contrast, there has been little investigation of the possible connection between "transfectogenic" activity and carcinogenicity.

One of the major categories of "transfectogenic" agents for mammalian cells is that of the "insoluble" facilitators of transfection. The first such "insoluble" facilitator found, using viral RNA, was calcium phosphate, followed by magnesium phosphate, calcium carbonate, magnesium carbonate, aluminum oxide, magnesium fluoride, calcium sulfate, chromic oxide, cobaltic oxide, ferric oxide, fuller's earth, magnesium silicate (talc), nickelous oxide, and zinc sulfide (Dubes & Klingler, 1960 and 1961). Subsequent work, showed that the silicates kaolin (Engler and Tolbert, 1963) and bentonite (Dubes, 1972) are also "transfectogenic".

"Insoluble" facilitators are also "transfectogenic" for viral DNA, as was first shown by Ishikawa and Furuno (1967), who used kaolin and SV40 DNA. Later, Graham and Van Der Eb (1973) reported the "transfectogenic" activity of calcium phosphate for human adenovirus type 5 DNA. It is especially relevant here that calcium phosphate was subsequently used to facilitate transfection of mouse cells of the NIH 3T3 line by human oncogene DNAs, resulting in transformation of these cells (Krontiris & Cooper, 1981; Murray *et al.*, 1981). Thus, at least one "insoluble" facilitator is "transfectogenic" for oncogene DNA, with the consequent expression of that DNA in the transfected cell and an associated marked alteration in the growth behavior of that cell and its descendants. At the present time, there appears to be no reason for believing that calcium phosphate is unique among "insoluble" facilitators in being "transfectogenic" for oncogene DNA.

The strong "transfectogenic" activity of the silicates fuller's earth, talc, and kaolin prompted the following comment (Dubes, 1971), with regard to silicate-induced neoplasms: ".......but mechanisms involving the adsorption and carrying of nucleic acids, viral or cellular, with consequent disturbance of cellular regulatory mechanisms, do not seem to have been considered." However, as far as I am aware, the first tests for this activity in the important silicate carcinogen- asbestos - were not carried out until 1984. It was then found that anthophyllite asbestos (from the Powhatan Mining Company) strongly facilitated transfection with poliovirus RNA and cells of the chimpanzee liver cell line CLI (Dubes, 1985). This work was extended to include tests of standard UICC and NIEHS asbestos samples, a variety of viral RNAs, and three additional mammalian cell lines (KB, human carcinoma; eta, rhesus monkey kidney; NIH 3T3, mouse embryo)

336

(Dubes & Mack, 1986 and 1988). In summary, all thirteen samples of asbestos tested were found to be "transfectogenic", and asbestos was active for every viral RNA/host cell combination tested.

At the same time asbestos/plasmid DNA studies were conducted by Fasy, Johnson, and coworkers. These investigators demonstrated that asbestos has strong "transfectogenic" activity for plasmid DNAs on cultured monkey cells (Fasy *et al.*, 1986; Appel *et al.*, 1988). UICC Canadian chrysotile asbestos, the pSV2*neo* and p11-4 plasmids, and the COS-7 line of African green monkey kidney cells were used.

This present paper is based on experimental studies initiated with the purpose of testing for asbestos facilitation of transfection of mammalian cells by viral DNA and of acquiring basic information about such facilitation. Early results in this project have been reported previously (Unpublished and Dubes, 1988b).

MATERIALS AND METHODS

Virus

The LP147-2 strain of polyomavirus, kindly supplied by Dr. G. Kimura of Kyushu University, was used.

Cells

For host cells permissive for polyomavirus, the 3T6 cell line, derived originally from mouse embryos, was selected and a sample obtained from the American Type Culture Collection (ATCC), Rockville, Maryland. These were grown in plastic or glass usually using 96% (by volume) Eagle's Basal Medium (BME) with Hanks' salts but with NaHCO$_3$ at 20 mM and L-cystine at 0.20 mM. Penicillin G sodium at 50 units/ml and streptomycin sulfate at 50 μg/ml were used and this medium was suplemented by 4% (V/V) foetal bovine or calf serum. In some cases, other basal media were substituted for BME, with satisfactory results.

After several passages following receipt from ATCC, samples of the 3T6 cells were frozen and stored in liquid nitrogen. The early passage cells were satisfactory but after many passages the cells showed only relatively minor and transitory cytopathogenic effects (cpe) of the polyomavirus. When this occurred, early-passage-level 3T6 cells were retrieved from liquid-nitrogen storage.

Antisera and Control Sera

Samples of three different batches of polyclonal anti-polyomavirus rabbit sera were a generous gift of Professor Richard Consigli of Kansas State University, Manhattan, who gave their hemagglutinin inhibition titers as follows: batch 1, titer 2.56×10^4; batch 2, titer 8.00×10^5; batch 3, titer 6.4×10^6. Two normal rabbit sera were used as controls: one from CDC, Atlanta, Georgia; the other from M.A. Bioproducts, Walkersville, Maryland.

Electrophoresis

Some nucleic acid preparations were analyzed by submarine horizontal gel electrophoresis, using 1.0% agarose (BRL, Inc. or FMC BioProducts,) in tris-borate-EDTA (TBE) buffer (89 mM tris(hydroxymethyl)aminomethane, 89 mM boric acid, 2.0 mM EDTA, pH *circa* 8.3, Sigma Chemical Co., Saint Louis, Missouri). The conditions were

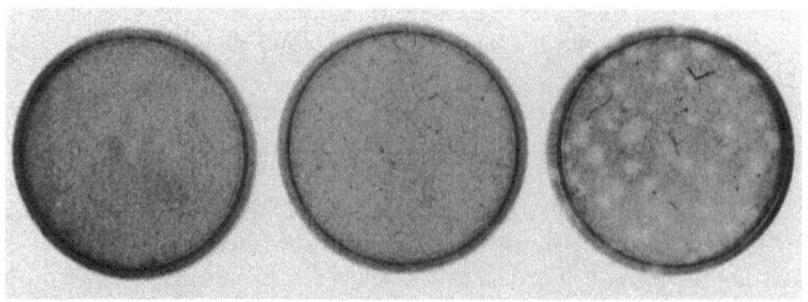

Figure 1. Asbestos Facilitation of Transfection by Polyomavirus DNA.
Mouse 3T6 cells were cultured in 60 x 15-mm Petri dishes until confluent, washed twice with Buffer M, inoculated with phenol-extracted polyomavirus DNA preparation at dilution 1:16 without asbestos (dish at left), anthophyllite asbestos at 2.5 mg/ml without DNA (middle), or the phenol-extracted polyomavirus DNA at 1:16 premixed with the anthophyllite at 2.5 mg/ml (right), 0.30 ml/culture, incubated at 37°C for 10 minutes, then agar-overlayered for plaquing. Neutral red overlay was added on day 9 post-inoculation. Photograph was taken two days later, when the plaque counts, left to right, were: 0; 0; 115.

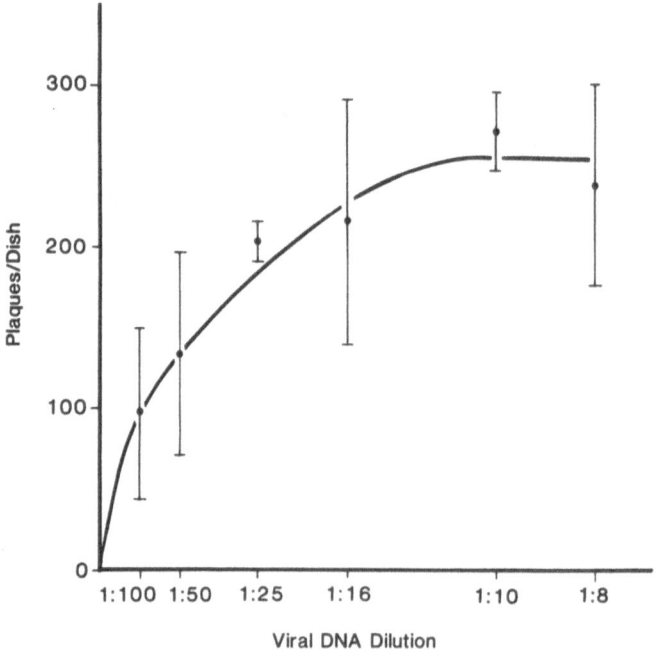

Figure 2. Asbestos-Facilitated Transfection as a Function of DNA Dilution Inoculated. Mouse 3T6 cell sheet cultures were washed twice with Buffer M, inoculated with phenol-extracted polyomavirus DNA preparation M36 at the dilutions indicated with anthophyllite constant at 2.5 mg/ml, 0.30 ml/culture, two cultures per inoculum, incubated at 37°C for 10 minutes, then agar overlayered for plaquing. Each point with bar represents arithmetic mean ± its standard error. The abscissa scale is linear for the concentration, or amount, of DNA inoculated.

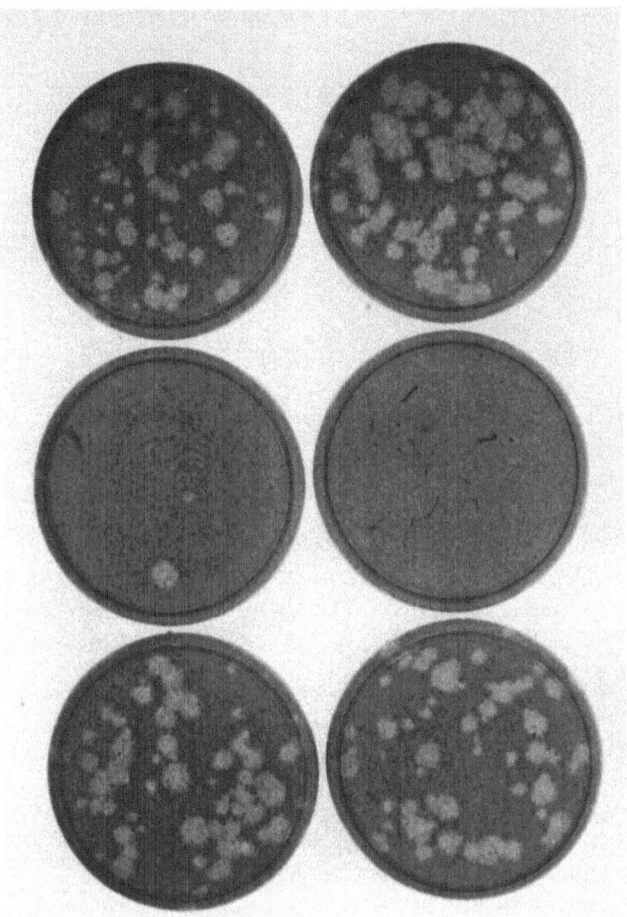

Figure 3. Verification: Pancreatic Deoxyribonuclease Inactivation of Transfectivity. Phenol-extracted polyomavirus DNA preparation M26 at 9:20 dilution was incubated with pancreatic deoxyribonuclease at 0 (control), 3, or 30 μg/ml at 37°C for 20 minutes. For the cold control without the enzyme, the polyomavirus DNA was left in the ice-water bath. Dilutions were then made so that DNA dilution in the inocula was constant at 1:16 and anthophyllite was constant at 2.5 mg/ml. The 37°C control was split into three parts, with one remaining without deoxyribonuclease and the other two containing terminally added enzyme at 0.42 and 4.2 μg/ml, respectively, as enzyme late-addition controls, having the same inoculum concentrations of pancreatic deoxyribonuclease as the experimentals. Immediately following these late additions, 3T6 cell sheet cultures were inoculated, 0.30 ml/culture, incubated at 37°C for 10 minutes, then agar-overlayered for plaquing. Photograph was taken on day 13 postinoculation. Left to right: Top, controls: cold and 37°C. Middle, experimentals: 3 and 30 μg/ml. Bottom, late-addition controls: 0.42 and 4.2 μg/ml. For the six cultures, plaque counts were, respectively: 80; 90; 2; 0; 69; 63.

TABLE 1. Asbestos Cytotoxicity on 3T6 Cells under Conditions of Transfection and Plaquing

Asbestos Variety	Asbestos concentration		
	1.0 mg/ml 14 $\mu g/cm^2$	1.5 mg/ml 21 $\mu g/cm^2$	2.5mg/ml 35 $\mu g/cm^2$
Anthophyllite (Pow)	None discerned	Not tested	Slight
Chrysotile (NIEHS short)	None discerned	Very slight	Moderate
Crocidolite (NIEHS)	Slight	Moderate	Strong

such that the relative mobilities of the three DNA forms of a given plasmid DNA or of a given papovavirus DNA would be: supercoiled > linear > open circular, *i.e.* nicked and relaxed (Mainwaring *et al.*, 1982). The gel was stained with ethidium bromide at 5.0 $\mu g/ml$ at 23°C for 15 minutes, destained in H_2O at 23°C for 15 minutes, and then examined or photographed under ultraviolet light.

Nucleic Acid Preparations

Polyomavirus stocks were produced by viral multiplication in 3T6 cell cultures at 36°C to 37.5°C. In some cases, the culture medium was modified BME alone, without serum. Virus stocks were harvested when the viral cpe was extensive by freezing the infected cells plus culture medium and storing at -20° C.

For most experiments, the nucleic acid preparations used were obtained by deproteinizing virus stock using extraction with phenol. The virus stock was first diluted 1:2 with Dulbecco's phosphate-buffered saline (PBS) without $CaCl_2$ and $MgCl_2$ but with 300 μM L-histidine and this dilution was then extracted at 0°C with an equal volume of water-saturated phenol. Residual phenol was removed by three extractions with benzene and traces of benzene flushed away by bubbling with nitrogen. Such nucleic acid preparations contain RNA and mouse cellular DNA in addition to the polyomavirus DNA. These preparations were stored at -20°, where they maintained high transfectivity for at least two years. Dilutions of these preparations are herein expressed relative to the original virus stock, *i.e.* the pre-phenol dilution 1:2 is included in the calculated overall dilution.

More recently, purified polyomavirus DNA preparations have been obtained using two commercial kits intended for the purification of plasmid DNAs from bacterial cells. It appeared that, because of the basic structural similarity (supercoiled, covalently closed, double-stranded DNA) of polyomavirus DNA to plasmid DNA, these kits might be applicable to the purification of polyomavirus DNA from mouse cells. These purified polyomavirus DNA preparations have been radiolabeled with [3][H]-thymidine, with the eventual purpose of using the radiolabel to follow the fate of the polyomavirus DNA in an asbestos/transfection experiment.

Although the DNA isolated by using the CircleprepKit (BIO 101, La Jolla, California) appeared suitable by electrophoresis it was unstable under the conditions used and degraded to low-molecular-weight products. The second kit used was the Nucleobond AX Plasmid Kit, with the anion-exchange cartridge AX-20 (Macherey-Nagel, Düren, FRG).

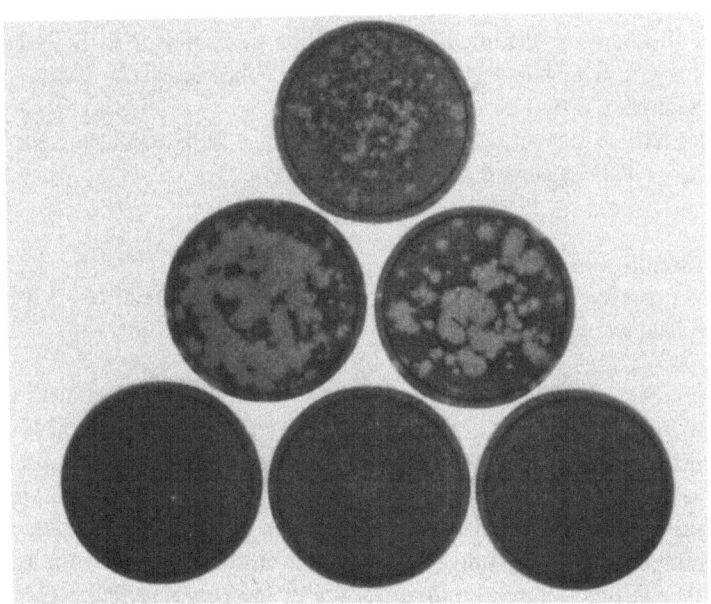

Figure 4. Verification: Anti-Polyomavirus Serum Blocking of Development of Plaques Initiated by Asbestos-Mediated Transfection by Viral DNA. The six 3T6 cell sheet cultures were all inoculated with viral DNA at 1:16 dilution and anthophyllite at 2.5 mg/ml, pre-mixed, 0.30 ml/culture, incubated at 37°C for 10 minutes, then agar-overlayered for plaquing, using six different agar-overlays. All six contained 4% by volume normal bovine calf serum, and all except the overlay for the top culture also contained 1% by volume rabbit serum, Five different rabbit sera were used: center-left, normal from CDC; center-right, normal from M.A. Bioproducts; bottom row, left to right, anti-polyomavirus batch numbers 1, 2, and 3. The agar-overlays added after the day of inoculation all contained only the 4% normal bovine calf serum and no rabbit serum. The photograph was taken on day 14 postinoculation. Plaque counts were: top, 129; center-left, 203; center-right, 106; bottom, 0, 0, and 0.

The standard protocol with this kit produces five fractions (flowthrough, washes 1-3, and eluate). Isopropanol precipitation produced visible pellets only for the flowthrough, wash #1, and eluate fractions. The pellet, visible or invisible, was then washed with 70% ethanol and dissolved in 25 μl pure sterile water. Storage was at 4°C. Such fractions have been analyzed by agarose gel electrophoresis and tested in asbestos/ transfection experiments. Eluate fractions appear to be stable at 4°C for at least six weeks.

Asbestos

Three varieties of asbestos previously tested for "transfectogenic" activity for viral RNAs (Dubes & Mack, 1988) were selected:

(1) Anthophyllite: Purchased Powminco Brand, purified, acid-washed, medium fiber from Powhatan Mining Company, Baltimore, Maryland via J.T. Baker Chemical

Company, Phillipsburg, . Identified as anthophyllite by Professor R. B. Nelson, Department of Geology, University of Nebraska-Lincoln (Dubes & Mack, 1988). Referred to below as anthophyllite (Pow)

(2) Chrysotile: NIEHS standard chrysotile of short range fibre length, log no. 031N

(3) Crocidolite: NIEHS standard, log no. 517V.

Calcium Phosphate and Kaolin

For comparison with asbestos, two classic "insoluble" facilitators were selected: calcium phosphate and kaolin. The commercial products used were:

$CaHPO_4 \cdot 2H_2O$, N.F., powder, catalogue no. 4265, control NBD, Mallinckrodt Chemical Works, Saint Louis, Missouri; kaolin, N.F., powder, catalogue no. 2242, lot 37126, Baker Chemical Co. In addition, calcium phosphate was also freshly precipitated by mixing solutions of Na_2HPO_4 and $CaCl_2$. Two sequences of addition were used for preparing viral DNA inocula containing fresh calcium phosphate. One sequence resulted in the precipitation of the calcium phosphate *before* the addition of the polyomavirus DNA, the other in precipitation *in the presence* of the polyomavirus DNA (Graham & Van Der Eb,1973) For each sequence, the components mixed were: (1) 618 μl H_2O; (2) 41 μl 200 mM Na_2HPO_4; (3) 41 μl 200 mM $CaCl_2$; (4) 700 μl viral DNA at 1:8 dilution. In one case, the sequence was 1234; in the other (coprecipitation), 1243. There was mixing after each addition. Precipitation of calcium phosphate occurred promptly after adding the 200 mM $CaCl_2$.

Plaquing

Infectivity of polyomavirus and transfectivity of its DNA were tested and titrated by plaquing on 3T6 cell sheet cultures in 60 x 15-mm Lux Contur Petri dishes, usually at 36°C to 37.5°C. Nutritional agar overlays were usually added on the day of inoculation and on days 4, 8, and 12 postinoculation. The last overlay contained neutral red as a vital stain. Plaques were often large enough for counting on day 13 postinoculation, but sometimes more time was required. Various compositions for the agar overlays were tested and the best ones, which were based on BME and Fischer's medium, were used.

The identity of the virus producing the plaques on the 3T6 cells was confirmed as polyomavirus by neutralization tests: Virus which had been incubated at dilution 1:10,000 with any one of the three anti-polyomavirus rabbit sera at dilution 1:50 at 37°C for 60 minutes was neutralized and produced no plaques. Virus after corresponding incubation with either of the two control rabbit sera produced hundreds of plaques, as expected.

Transfection Tests

In the standard procedure, the 3T6 cell sheet cultures were grown to confluence, and the culture medium was replaced with fresh medium 3-8 hours before inoculation. Just before inoculation, the cultures were washed twice with Buffer M (PBS without $CaCl_2$ but with 2.5 mM $MgCl_2$), usually 4 ml/wash, inoculated with the DNA, with or without asbestos or classic "insoluble " facilitator, 0.30 ml per culture, incubated at 37°C for 10 minutes, and then overlayered with agar medium for plaque development. This standard transfection procedure was varied, as required.

TABLE 2. Data Showing Asbestos Facilitation of Transfection by Polyomavirus DNA Prepared Using One Phenol Extraction

Experiment and 3T6 Cell Batch*	Polyomavirus DNA Preparation and Its Dilution in Inoculum ■	Plaque Counts on Individual Cultures**				
		Without Asbestos	With Asbestos			
			Crocidolite at 1.0mg/ml	Crocidolite at 1.5mg/ml	Chrysotile 1.5mg/ml	Anthophyllite 2.5mg/ml
M17	B9 at 1:8	0 0		5 7	0 0	
	B9 at 1:16	0 0		0 7	5 5	
M21	M21 at 1:16	0 0	3 7		2 6	
M28	M26 at 1:16	3 17	27 18			
M31	M26 at 1:16	14 3				55 87
M34	B9 at 1:16	0 0				38 12
	M21 at 1:16	0 0				48 93
	M26 at 1:16	15 0				ca 267 ca 400
M37	M37 at 1:16	3 5 1				22 27 20

*A different batch of 3T6 cell cultures was grown in Petri dishes for each experiment. Despite similarities in batch planting and handling, there were significant quantitative differences between cell batches in transfectibility (competence); e.g. with anthophyllite as facilitator, cell batch M34 was more transfectible than cell batch M31.

■Different DNA preparations had quantitatively different transfectivities; e.g., DNA preparation M26 produced more plaques than DNA preparations B9 and M21.

**Gaps in the table indicate tests not done.

343

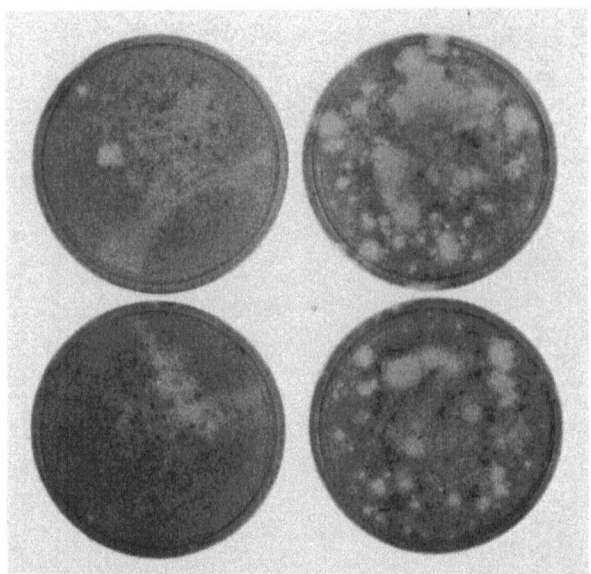

Figure 5. Asbestos Facilitation of Transfection by Tritiated Polyomavirus DNA Purified by the Nucleobond■AX Plasmid Kit: Run 2 Eluate Fraction. Polyomavirus-infected 3T6 cells were lysed *in situ* by Kit reagents, and the polyomavirus DNA was purified using an AX-20 cartridge and the Kit protocol. The Run 2 eluate fraction at 1:140 dilution without asbestos and with anthophyllite at 1.0 mg/ml was inoculated onto 3T6 cell sheet cultures. At day 14 postinoculation, plaque counts, top to bottom, were: left dishes, without asbestos: 1 and 1; right dishes, with asbestos: 95 and 66. (The larger light spot in the upper-left dish is an artifact, not a plaque.)

TABLE 3. Asbestos Faciliatation of Transfection by Polyomavirus DNA Purified using the Nucleobond■AX Plasmid Kit

Eluate Fraction of run number	Asbestos in Inoculum[*]	Plaque Counts[*]	
1	None (Control)	0	0
	Anthophyllite at 1.0 mg/ml	22	18
2	None (Control)	1	1
	Anthophyllite at 1.0 mg/ml	95	66

[*] Inoculum was 0.30 ml with eluate fraction at dilution 1:140 per culture;   thus, each plaque count is that derived from 2.1 μl of the undiluted eluate fraction.
[*] The two eluate fractions were inoculated onto different batches of 3T6 cells, possibly differing in transfectibility.

TABLE 4. Anthophyllite-Mediated Transfectivity of Five Fractions from Runs 2 and 3 with the Nucleobond■ AX Plasmid Kit

Run	Fraction	Age*	Dilution[§]	Plaque Counts	
2	Flowthrough	3 days	1:140	30	15
	Wash 1	3 days	1:140	22	23
	Wash 2	3 days	1:140	18	4■
	Wash 3	3 days	1:140	2	6
	Eluate	3 days	1:140	95	66
		46 days	1:140	54	110
3	Flowthrough	2.2 hours	1:140	54	69
	Wash 1	2.2 hours	1:140	45	28
	Wash 2	2.2 hours	1:140	9	10
	Wash 3	2.2 hours	1:140	4	2
	Eluate	2.2 hours	1:213	37	58
			1:560	62	41

* Age of the fraction when inoculated for determination of transfectivity; storage at 4°C.
[§] Inoculum was 0.30 ml of the fraction dilution indicated, anthophyllite was constant at 1.0 mg/ml, per culture. Each plaque count derives from the following volumes of undiluted fraction: 2.1 μl for dilution 1:140; 1.4 μl for 1:213; and 0.54 μl for 1:560. Note that these volumes are substantially less than the volume, 4.75 μL, of undiluted fraction analyzed per lane in Fig. 6.
■ Only 50% of the area of this culture was scorable for plaques.

Asbestos Cytotoxicity

At higher concentrations, the three varieties of asbestos were cytotoxic towards the 3T6 cells revealed by failure of some of the 3T6 cells to stain with the neutral red. These asbestos cytotoxicity results are summarized in Table 1. Asbestos cytotoxicity described as moderate or less did not interfere with identifying and counting polyomavirus plaques.

RESULTS

Asbestos Facilitation of Transfection by Polyomavirus DNA.

A typical result is shown in Fig. 1, where plaques were produced on the 3T6 cell sheet which had been inoculated with the mixture of the polyomavirus DNA and the anthophyllite asbestos. The chrysolite and crocidolite asbestos samples also facilitated transfection by polyomavirus DNA, but to a lesser extent. The results of several experiments showing asbestos facilitation are given in Table 2. For most of the control 3T6 cell sheet cultures inoculated with polyomavirus DNA without asbestos, there was no transfection and no plaques were produced; but for some controls, one, a few plaques or occasionally a "jackpot" number of 10-18 plaques, were produced. The distribution of plaque counts among the replicate control cultures indicated significant intra-batch inter-culture variation in transfectibility. Even so the mean plaque counts from transfection by inocula containing asbestos were usually many times higher than the mean counts without asbestos (Table 2).

The data of Experiment M17 of Table 2 suggest that, in the DNA dilution range 1:8 to 1:16, the system may be saturated with the DNA preparation. This suggestion was

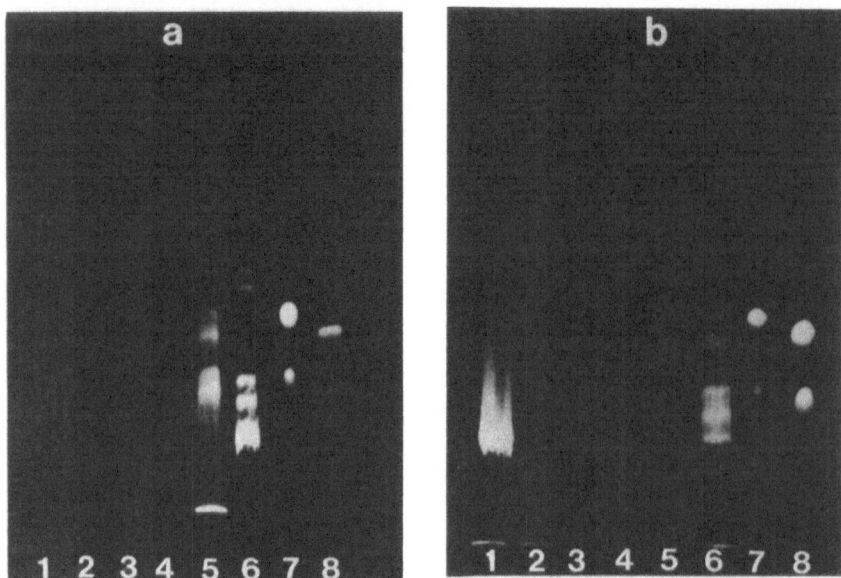

Figure 6. Agarose Gel Electrophoresis of Fractions from Runs 2-4 with the Nucleobond·AX Plasmid Kit versus DNA Standards for Reference. Gels were 1.0% Seakem·agarose, and wells were 4.1 mm long x 0.8 mm wide. Fraction loading was 4.75 μL per well, not including the 5% glycerol. The 16 well loadings were:

	Panel	
Lane	a	b
1	Run 2 flowthrough	Run 4 eluate
2	Run 2 wash #1	Run 3 flowthrough
3	Run 2 wash #2	Run 3 wash #1
4	Run 2 wash #3	Run 3 wash #2
5	Run 2 eluate	Run 3 wash #3
6	·DNA Hind III markers, 2.0 μg	Run 3 eluate
7	pBR322 DNA, 0.5 μg	pBR322 DNA, 0.5 μg
8	SV40 DNA, 0.5 μg	SV40 DNA, 0.5 μg

Runs 2 and 3 were from polyomavirus-infected cells, Run 4 from control uninfected cells. For the pBR322 DNA and SV40 DNA standards, the leading band is the supercoiled form, the trailing band the nicked (relaxed) form.

confirmed by an experiment in which the DNA dilution range 1:8 to 1:100 was tested, while holding asbestos concentration constant in the inocula (Fig. 2). Saturation level was reached at DNA dilution about 1:16.

In the experiments of Fig. 1 and 2 and Table 2, the DNA had been prepared by one phenol extraction. The asbestos-mediated infectivity of inocula prepared from such polyomavirus DNA preparations was rapidly and completely destroyed by incubation with pancreatic deoxyribonuclease (Fig. 3); similar verification was also obtained for the low infectivity found in controls without asbestos. It was also verified that the plaques produced by asbestos-facilitated transfection are indeed plaques of polyomavirus (Fig. 4).

TABLE 5. Rapidity of Formation of Polyomavirus DNA/3T6 Cell Complexes Initiating Transfection Events under Conditions of Anthophyllite Facilitation

Duration of Incubation[*]	Plaque Counts[*]	
40 sec	233	305
60 sec	215	209
5 min	252	309
10 min	290	246
17 min	222	269
20 min	ca 425	ca 469

[*]Duration of incubation of the inoculated 3T6 cell sheet cultures in 60 x 15 mm Petri dishes at 37°C before adding the agar-overlay and swirling. The inoculum and the washed cell sheets were pre-warmed at 37°C before inoculation.

[*]Inoculum was 0.30 ml of phenol-extracted polyomavirus DNA preparation M36 at dilution 1:16 with anthophyllite at 2.5 mg/ml per culture. Plaques were counted blind at day 13 postinoculation.

TABLE 6. Stability of Anthophyllite-Facilitated Polyomavirus DNA/3T6 Cell Complexes Initiating Transfection Events to Washing with Buffer

Number of buffer washes of the 3T6 cell sheet cultures after incubating them with viral DNA and asbestos	Plaque Counts[*]	
0	34	38
1	33	45

[*] The cell sheet cultures in 60 x 15-mm Petri dishes were first washed twice with Buffer M, 4 ml per wash per culture, inoculated with 0.30 ml of phenol-extracted polyomavirus DNA preparation M26 at dilution 1:16 with anthophyllite at 2.5 mg/ml per culture, incubated at 37°C for 10 minutes, then washed once with 4 ml Buffer M per culture or left unwashed, and finally agar-overlayered for plaquing.

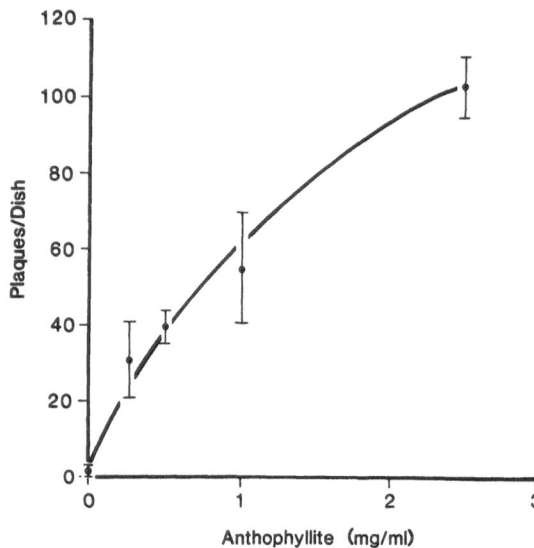

Figure 7. Transfection as a function of asbestos concentration. Confluent 3T6 cell sheet cultures were washed twice with Buffer M, inoculated with phenolextracted polyomavirus DNA preparation M36 at dilution 1:16 without asbestos or with anthophyllite asbestos at the concentrations indicated, 0.30 ml/culture, incubated at 37°C for 10 minutes, then agar-overlayered for plaquing. Each point with bar represents arithmetic mean ± its standard error; n = 2. The four anthophyllite concentrations tested correspond to 3.5; 7.1; 14; and 35 µg/cm².

Asbestos also facilitated transfection by polyomavirus DNA purified by the Nucleobond AX Plasmid Kit. Three purification runs (Runs 1-3) were made using this kit, as well as one control run (Run 4) using an uninfected 3T6 cell sheet culture, which was of the same cell batch as the infected culture used for Run 3. For Runs 1 and 2, the eluate fractions, expected to contain the polyomavirus DNA, were inoculated with and without asbestos onto 3T6 cell sheet cultures; the results are shown in Table 3. The four cultures in the experiment testing the Run 2 eluate fraction are shown in Fig. 5. It is clear that anthophyllite strongly facilitates transfection by polyomavirus DNA purified using this method.

Fractions of Runs 2-4 were examined by 1.0% agarose gel electrophoresis (Fig. 6). Reference materials were bacteriophage lambda DNA *Hind III* markers (Promega Corp, Madison, Wisconsin), plasmid pBR322 DNA [4,363 base-pairs (bp), Promega], and SV40 DNA (5,243 bp, BRL). The DNA of polyomavirus strain A2 has 5,297 bp (Salzman, 1986). Strain LP147-2, used here, presumably has a similar number of base-pairs; and the supercoiled and nicked forms of the DNA of this strain are thus expected to have similar electrophoretic mobilities to the corresponding forms of the reference SV40 DNA shown in the two lanes 8 of Fig. 6. Strong bands in or very near these expected locations were indeed found for both infected eluate fractions tested, but additional bands were also found (Fig. 6). Moreover, the eluate fraction from control Run 4 from unin-

TABLE 7. Facilitation by asbestos added after Polyomavirus DNA

First Inoculum[*]	First Incubation at 37°C[§]	Intervening Washes[■]	Second Inoculum[*]	Second Incubation at 37°C[§]	Plaque Counts
DNA & Asbestos	10 min	0	Buffer	10 min	93,83,52
(Reference)		3	Buffer	10 min	50,75,25
DNA	10 min	0	Asbestos	10 min	105,97,66
		3	Asbestos	10 min.	25,32, 7

[*] Each inoculum was 0.30 ml per culture. Where present, phenol-extracted viral DNA was at dilution 1:50 and asbestos was anthophyllite at 2.5 mg/ml.

[§] Inoculum was not removed before next step.

[■] Each wash was with 4 ml Buffer M per culture.

fected cells showed a slow-moving mass streaking upward (lane 1 of panel b). The polyomavirus DNA prepared using the Kit is therefore considered as *partially* purified; further steps, or a different purification method, would be necessary to obtain a purer product.

Figure 6 also shows that the great majority of the nucleic acid recovered from the infected cells is in the eluate fraction. Relatively small amounts were found in the flow-through and wash #1 fractions, and none was definitely discernible in washes #2 and #3. For Runs 2 and 3, these four fractions were also plated with asbestos to determine transfectivity. The results (Table 4) show unexpectedly high transfectivity for the flowthrough and wash #1 fractions. In the case of the wash #1 fractions, this transfectivity can possibly be ascribed to the very weak bands (Fig. 6) in or very near the expected positions for the supercoiled and nicked forms of polyomavirus DNA. Conceivably, small amounts of the supercoiled and nicked forms were adsorbed weakly enough onto the anion exchanger to be eluted by the pH 6.3 buffer with 1,300 mM KCl used to wash the cartridge. But for the flowthrough fractions, it is difficult to ascribe the transfectivity to the supercoiled and nicked forms of the polyomavirus DNA as no bands are discernible at the expected positions (Fig. 6).

The data of Table 4 also suggest (a) that, as tested with Run 3 eluate fraction, inoculum DNA at dilution 1:560 may still be at DNA saturation level for transfection; and (b) that, as tested with Run 2 eluate fraction, transfectivity may be stable at 4°C for at least six weeks.

Dependence of Asbestos Facilitation on Asbestos Concentration

The dilution of a phenol-extracted polyomavirus DNA preparation was held constant and anthophyllite concentration was varied in the inocula. The graphed results (Fig. 7) show the marked increase in anthophyllite facilitation with increasing anthophyllite concentration. The relationship is not linear, as the slope of the function continuously decreases.

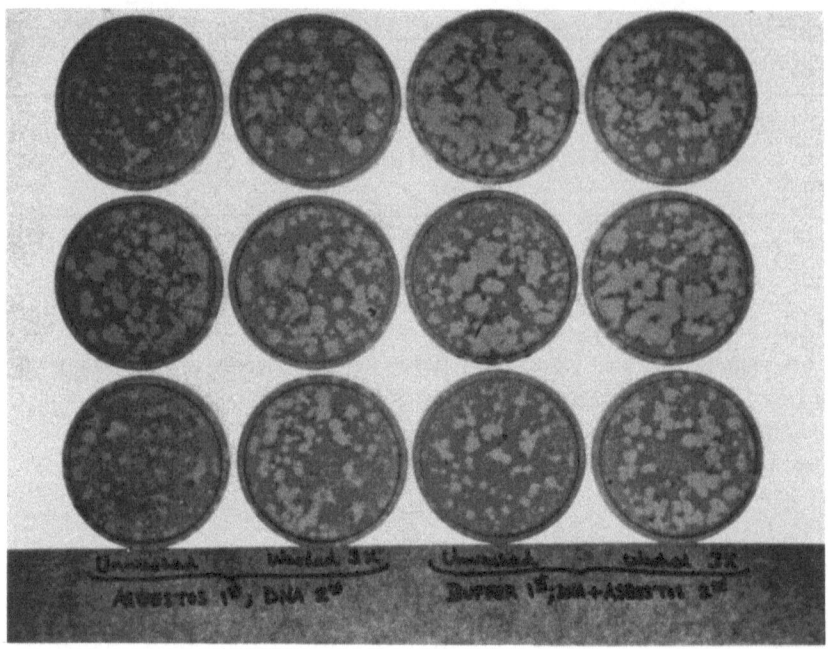

Figure 8. Asbestos can make 3T6 cells transfectible by Polyomavirus DNA added later.
Twelve 3T6 cell sheet cultures in 60-mm dishes were washed twice with Buffer M and then divided into four groups of three dishes each, which were then handled as follows:

Sequence of Handling Steps before Agar-overlayering

Trio[a]	First Inoculation[b]	First Incubation[c]	Washings[d]	Second Inoculation[b]	Second Incubation[c]
Left	Anthophyllite, 2.5 mg/ml	37° C, 10 min	None	Polyomavirus DNA, 1:50	37° C, 10 min
Left-Center	Anthophyllite, 2.5 mg/ml	37°C, 10 min	Three	Polyomavirus DNA, 1:50	37°C, 10 min
Right-Center	Buffer alone	37°C, 10 min	None	Polyomavirus DNA, 1:50 + anthophyllite, 2.5 mg/ml	37°C, 10 min
Right	Buffer alone	37°C	Three	Polyomavirus anthophyllite, 2.5 mg/ml	37°C

[a]Photograph was taken on day 13 postinoculation, when plaque counts, top to bottom, were: Left: 148; 145; 97. Left-Center: 88; 168; 151. Right-Center: ca 267; 218; 182. Right: 185; 178; 156.
[b]0.30 ml/culture.
[c]Inoculum not removed before next step.
[d]Each washing with 4 ml Buffer M per culture.

Rapidity of Formation of the Asbestos-Facilitated Polyomavirus DNA/3T6 Cell Complexes Initiating the Transfection Events

In the standard asbestos facilitation procedure used here, the inoculum of polyomavirus DNA plus asbestos is incubated with the 3T6 cell sheet culture at 37°C for 10 minutes before adding the agar-overlay and swirling. During this 10-minute period many polyomavirus DNA/cell complexes initiating transfection events are formed. *A priori*, these complexes could be direct or indirect, with or without asbestos fibres; an indirect complex could of course be a polyomavirus DNA molecule attached to an anthophyllite fibre, which in turn is attached to a cell.

To determine the rapidity of formation of the polyomavirus DNA/cell complexes initiating transfection, the duration of incubation of the inoculum with the cell sheet was deliberately varied; the results are shown in Table 5. The relatively high counts for duration 20 minutes probably represent merely the chance and the major conclusion from the results is that asbestos facilitation is independent of duration of incubation over the time range tested. Therefore, under conditions of anthophyllite facilitation, the polyomavirus DNA/cell complexes initiating transfection must be sufficiently firm in 40 seconds or less to withstand disruption by agar-overlayering and swirling.

Test of Washing Cell Sheet Cultures after Incubation with Inoculum of Polyomavirus DNA and Asbestos

The rapidity of formation of the polyomavirus DNA/cell complexes firm enough to withstand agar-overlayering and swirling suggests that these complexes may likewise persist through washing of the cell sheets with buffer and indeed washing did not significantly affect the number of transfection events (Table 6).

Asbestos Can Make Cells Transfectible by DNA Added Later

The effect of adding asbestos to the cells before the DNA was also tested. In the experiment reported in Fig. 8 asbestos incubated with the cells at 37°C for 10 minutes before adding the viral DNA was still strongly "transfectogenic" even when three intervening washes with buffer werwe used to remove loose or weakly attached asbestos. At the time of the washings, much of the asbestos was already tightly bound to the cells and this bound asbestos was probably responsible for the high transfectibility of the washed cells.

Adding the Viral DNA to the Cells First and The Asbestos Later

Asbestos was also found to be effective when added to the cells after the viral DNA, except when intervening washes were used to remove the loose or weakly attached viral DNA (Table 7). These results showed no significant inactivation of the viral DNA by these washed 3T6 cell sheet cultures over the 10-minute pretreatment at 37°C.

Asbestos versus Classic Facilitators for Transfection by Polyomavirus DNA

Anthophyllite, chrysotile, and crocidolite were compared with calcium phosphate and kaolin for facilitation of transfection of 3T6 cell sheet cultures by polyomavirus DNA (Table 8). The anthophyllite was found to be much more effective than calcium phosphate and at least as effective as kaolin. The crocidolite was also much more effective than calcium phosphate and possibly at low concentrations as effective as kaolin. The chrysotile had lower effectiveness, though at low concentrations may have been more effective than calcium phosphate (Experiment M51).

In this system, facilitation by calcium phosphate was relatively low regardless of whether it had been precipitated in the presence or absence of the viral DNA.

TABLE 8. Asbestos versus Classic Facilitators for Transfection by Polyomavirus DNA

Experiment and 3T6 Cell Batch	Facilitator in Inoculum[§]	Plaque Counts[*]		
M26	None (control)	0	0	
	Kaolin, 5.0 mg/ml	15	18	
	Chrysotile, 1.5 mg/ml	0	0	
	Crocidolite, 1.0 mg/ml	4	18	
M43	None (control)	17	2	
	Calcium Phosphate (Mall.), 2.5 mg/ml	20	[**]	
	Kaolin, 2.5 mg/ml	188	208	
	Anthophyllite, 2.5 mg/ml	197	414	
	Chrysotile, 2.5 mg/ml	19	17	
	Crocidolite, 2.5 mg/ml	90	47	
M51	None (control)	0	0	
	Calcium Phosphate (Mall.), 1.0 mg/ml	3	2	
	Kaolin, 1.0 mg/ml	14	13	
	Anthophyllite, 1.0 mg/ml	47	51	
	Chrysotile, 1.0 mg/ml	8	17	
	Crocidolite, 1.0 mg/ml	17	24	
P43	None (control)	0	0	0
	Calcium Phosphate (Mall.), 1.0 mg/ml	0	0	0
	Calcium Phosphate (f.p.), 1.0 mg/ml	0	0	0
	Calcium Phosphate (co-ppt.), 1.0 mg/ml	1	0	0
	Anthophyllite, 1.0 mg/ml	10	32	19

[§] Mall. = $CaHPO_4.2H_2O$ from Mallinckrodt; f.p. = freshly pre-formed from Na_2HPO_4 and $CaCl_2$ before adding viral DNA; co-ppt. = freshly formed in the presence of the viral DNA.

[*] 3T6 cell sheet cultures were first washed twice with Buffer M, 4 ml per wash per culture, then inoculated with 0.30 ml of phenol-extracted polyomavirus DNA preparation at dilution 1:16, except that dilution 1:50 was used for Experiment M51, and facilitator at concentration indicated, incubated at 37°C for 10 minutes, and then agar-overlayered for plaquing. Different viral DNA preparations were used: preparations M26, M36, M36, and M37 in Experiments M26, M43, M51, and P43, respectively.

[**] The second culture was lost.

DISCUSSION

The major significance of this work is the demonstration that asbestos can strongly facilitate transfection by oncogene-containing viral DNA. The polyomavirus DNA genome contains two overlapping oncogenes, PV-MT and PV-LT, as well as a third gene, PV-ST, which has also been considered as an oncogene: see, e.g., Dulbecco and Ginsberg (1988). These viral oncogenes, in contrast to those like *src* of retroviruses, do perform essential viral functions for viral replication in permissive cells. Thus, in asbestos-facilitated transfection of a permissive mouse 3T6 cell by polyomavirus DNA, oncogenes PV-MT, PV-LT, and PV-ST are presumably being expressed. In the future, it will be important to test for asbestos facilitation of transfection and transformation of *nonpermissive* host cells by oncogene-carrying viral DNAs and of suitable target cells such as mouse NIH 3T3 cells by *cellular* oncogene DNAs.

Experiments with tritiated pure polyomavirus DNA in its various forms (supercoiled, nicked, and linear) will be important for (a) quantitating asbestos facilitation of transfection as a function of concentration of the viral DNA, essentially free of other DNAs or RNA, (b) determining possible effects (competition for asbestos, sparing, interference, or cotransfection) of the presence of other DNAs and RNA, and (c) following the fate of the polyomavirus DNA by tracking the tritium radiolabel at various stages of the asbestos-facilitated transfection process.

Experiments with crude DNA-containing entities such as chromosomes, chromosome fragments, or in the case of polyomavirus or related papovaviruses the minichromosomes from the virions, will also be important. Indeed, when cells are lysed *in vivo* through the cytocidal activity of asbestos, the DNA released will be in such crude forms, which however in some cases may have been partially digested by lytic enzymes of the lysing cells. It is crucial to determine whether asbestos can facilitate "transfection" and, where appropriate, transformation by such crude DNA-containing forms.

A theory assigning an important role to the "transfectogenic" activity of asbestos in asbestos-induced carcinogenesis was recently presented (Dubes, 1988a). The major reason for the attractiveness of this theory is the high potential of such "transfectogenic" activity for transferring and "re-shuffling" oncogenes. The oncogenes would presumably arise from those asbestos-induced *chromosome* mutations which resulted in chromosomal relocation, deregulation, and overexpression of proto-oncogenes. Such oncogenes would be oncogenes by virtue of overproduction of the normal gene product rather than through production of a gene product altered through gene mutation. When asbestos is present together with a *gene* mutagen, such as benzo[a]pyrene from cigarette smoke, there is a new opportunity for the "transfectogenic" activity of asbestos, namely, the opportunity to transfer an oncogene simply by transferring the whole chromosome carrying the benzo[a]pyrene-induced mutant oncogene to another cell. This could generate trisomy and could lead to a malignant cell if the recipient already contained one or more other oncogenes, or had lost one or more active antioncogenes.

However despite this important role envisioned for the "transfectogenic" activity of asbestos in asbestos-induced carcinogenesis, it seems clear that "transfectogenic" activity is neither sufficient nor necessary for carcinogenesis. "Transfectogenic" activity is not sufficient, since many other "insoluble" facilitators of experimental transfection of cells *in vitro* are apparently noncarcinogenic, e.g. calcium phosphate and calcium sulfate. Clearly, such facilitators do not possess other properties (e.g., fibre geometry and bio-durability) critical for carcinogenesis. "Transfectogenic" activity is apparently also not necessary, since asbestos-induced chromosome mutagenesis should presumably be able to produce the multiple mutational steps needed for carcinogenesis.

Several facts and observations can be adduced in favor of the theory and increase its plausibility and perhaps credibility:

(1)   Asbestos is indeed a potent "transfectogen" for mammalian cells cultured *in vitro*, more potent even than calcium phosphate, which is commonly used experimentally for this purpose in molecular biology.

(2)   Asbestos is even strongly "transfectogenic" when it is added to the cells before the DNA, even when there are intervening washes of the cells.

(3)   Asbestos fibres do adsorb DNA (Misra *et al.*, 1983; Chang *et al.*, 1983; Touray *et al.*, 1987).

(4)   Asbestos fibres are taken up by many kinds of mammalian cells which can later undergo mitosis, but the intracellular asbestos fibres interfere with anaphase and apparently form intimate associations with some of the chromosomes (Hesterberg & Barrett, 1985; Wang *et al.*, 1987).

(5)   Many of these asbestos-laden cells are later killed by the asbestos, producing a "puddle" of cell debris including the chromosomes and asbestos fibres (Lipkin, 1980).

(6)   In the presence of asbestos, the formation of the transfection-initiating DNA/cell complex can be very rapid (< 40 seconds), at least with cells *in vitro*. This rapidity may leave little time for extracellular nucleolytic enzymes to destroy the DNA.

(7)   The asbestos fibres released from a lysing cell can undergo re-phagocytosis by other cells in the tissue, followed ultimately again by lysis (Lipkin, 1980); and many of these phagocytosis-lysis cycles can occur in the tissue, apparently over years (Suzuki, 1974), as the asbestos fibre of "preferred Stanton geometry" is very bio-durable in the cell and in its external environment post-lysis and very bio-persistent in the tissue. Thus a given asbestos fibre can in time apparently "pass through" many host cells and putatively would then have multiple opportunities for transferring chromosomes and chromosome fragments from one cell to another.

Several important experiments remain to be done to test the theory. These experiments should address the following questions:

(1)   Does asbestos facilitate transfection by "crude" DNA (i.e. chromosome fragments or whole chromosomes) carrying oncogenes?

(2)   Does asbestos facilitate cotransfection by "crude" DNA carrying two different oncogenes on separate chromosome fragments or on separate chromosomes?

(3)   Does asbestos facilitate transfection of cultured scavenger or putative target cells: mesothelial cells, macrophages, bronchial epithelial cells?

(4)   Does asbestos facilitate transfection of cells *in vivo*?

(5)   Put (1-4) above all together: Does asbestos facilitate transfection of the putative target cells *in vivo* by "crude" DNA carrying oncogenes (one oncogene or two or more different oncogenes)?

(6)   If the answer to (5) above is yes, then does this facilitated transfection result in production of mesothelioma or bronchogenic carcinoma?

ACKNOWLEDGMENTS

The author thanks Professor Richard Consigli of Kansas State University for the samples of the three anti-polyomavirus rabbit sera, Mr. George Pallas and Mr. Sergio Diaz for photography, and Mrs. Mary Davis for typing the manuscript.

REFERENCES

Appel,J.D., Fasy,T.M., Kohtz,D.S., Kohtz,J.D. and Johnson,E.M. (1988) Asbestos fibers mediate transformation of monkey cells by exogenous plasmid DNA. *Proc. Natl. Acad. Sci., U.S.A.* **85**:7670-7674.

Chang,M.J.W., Joseph,L.B., Stephens,R.E. and Hart,R.W. (1983) Adsorption of macromolecules on mineral fibers. *J. Am. Coll. Toxicol.* **2**:187-192.

Dubes,G.R. (1971) "Methods for Transfecting Cells with Nucleic Acids of Animal Viruses: A Review". Birkhäuser, Basel: pp. 27.

Dubes,G.R. (1972) The mechanism of transfection enhancement by bentonite. *Arch. Ges. Virusforsch.* **39**:13-25.

Dubes,G.R. (1985) Asbestos facilitates transfection. *Proc. Nebr. Acad. Sci.*, p. 14.

Dubes,G.R. (1988a) Asbestos and transfection. *Advances in Cell Culture* **6**:199-260.

Dubes,G.R. (1988b) Asbestos facilitates transfection by viral DNA. *Proc. Nebr. Acad. Sci.*, p. 13.

Dubes,G.R. and Klingler,E.A. (1960) Facilitation of infection with poliovirus `ribonucleic acid'. *Genetics* **45**:984-985.

Dubes,G.R. and Klingler,E.A. (1961) Facilitation of infection of monkey cells with poliovirus `ribonucleic acid' *Science* **133**:99-100.

Dubes,G.R. and Mack,L.R. (1986) Asbestos-mediated transfection by picornavirus RNAs. *In Vitro Cell Dev. Biol.* **22**:50A.

Dubes,G.R. and Mack,L.R. (1988) Asbestos-mediated transfection of mammalian cell cultures. *In Vitro Cell Dev. Biol.* **24**:175-182.

Dulbecco,R. and Ginsberg,H.S. (1988) *Virology*, 2nd ed., Lippincott, Philadelphia: p. 337.

Dulbecco,R. and Vogt,M. (1954) Plaque formation and isolation of pure lines of poliomyelitis viruses. *J. Exper. Med.* **99**:167-182.

Engler,R. and Tolbert,O. (1963) Reversible inhibition of poliovirus ribonucleic acid. *Arch. Ges. Virusforsch.* **13**:470-481.

Fasy,T.M., Andrews,P.W. and Johnson,E.M. (1986) Asbestos fibers mediate the uptake of DNA into primate cells in culture. *Fed. Proc.* **45**:1849.

Graham,F.L. and Van Der Eb,A.J. (1973) A new technique for the assay of infectivity of human adenovirus 5 DNA. *Virology* **52**:456-467.

Hesterberg,T.W. and Barrett,J.C. (1985) Induction by asbestos fibers of anaphase abnormalities: mechanism for aneuploidy induction and possibly carcinogenesis. *Carcinogenesis* **6**:473-475.

Ishikawa,A., Furuno,A. (1967) Some ionic factors affecting efficiency of infection of African green monkey kidney cultures with SV40 DNA in isotonic saline media. *Jpn. J. Microbiol.* **11**:229-232.

Krontiris,T.G., Cooper,G.M. (1981) Transforming activity of human tumor DNAs. *Proc. Natl. Acad. Sci., U.S.A.* **78**:1181-1184.

Lipkin,L.E. (1980) Cellular effects of asbestos and other fibers: correlations with in vivo induction of pleural sarcoma. *Environ. Hlth Perspect.* **34**:91-102.

Mainwaring,W.I.P., Parish,J.H., Pickering,J.D. and Mann,N.H. (1982) "Nucleic Acid Biochemistry and Molecular Biology", Blackwell, Oxford: p. 82.

Misra,V., Rahman,Q. and Viswanathan,R.N. (1983) Adsorption of nucleic acids on asbestos fibers *in vitro. Toxicol. Lett.* **15**:187-191.

Murray,M.J., Shilo,B.-Z., Shih,C., Cowing,D., Hsu,H.W. and Weinberg,R.A. (1981) Three different human tumor cell lines contain different oncogenes. *Cell* **25**:355-361.

Rieger,R., Michaelis,A. and Green,M.M. (1976) Glossary of Genetics and Cytogenetics, 4th ed., Springer-Verlag, Berlin: pp. 377-378.

Salzman,N.P., (ed.) (1986) The Papovaviridae, Vol. 1 The Polyomaviruses, Plenum, New York: p. 380.

Suzuki,Y. (1974) Interaction of asbestos with alveolar cells. *Environ. Hlth Perspect.* **9**:241-252.

Touray,J.-C., Baillif,P., Jaurand,M.-C., Bignon,J. and Magne,L. (1987) Etude comparative de l'adsorption d'acide deoxyribonucleique sur le chrysotile et le chrysotile phosphate (chrysophosphate). *Canad. J. Chem.* **65**:508-511.

Wang,N.S., Jaurand,M.-C., Magne,L., Kheuang,L., Pinchon,M.C. and Bignon,J. (1987) The interactions between asbestos fibers and metaphase chromosomes of rat pleural mesothelial cells in culture. *Am. J. Pathol.* **126**:343-349.

MECHANISMS OF PATHOGENESIS

2. Effects on Gene Expresssion

ASBESTOS MEDIATED GENE EXPRESSION IN RAT LUNG

Yvonne M. W. Janssen, Joanne P. Marsh, *Paul J. A. Borm, Piyawan
Surinrut, Kaaren Haldeman and Brooke T. Mossman

Department of Pathology
University of Vermont
Burlington
Vermont 05405, USA

*Department of Occupational Medicine
University of Limburg
Maastricht, The Netherlands

INTRODUCTION

The term "asbestos" refers to a family of hydrated silicates with an aspect (length
to diameter) ratio of $> 3:1$ as defined by regulatory policy in the United States. Physico-
chemical characteristics vary considerably amongst the various types of asbestos. Chrysotile asbestos [$Mg_6Si_4O_{10}(OH)_8$], the most common asbestos type, accounts for most of the
world's production of asbestos. Crocidolite asbestos [$Na_2Fe(III)_2Fe(II)_3Si_8O_{22}(OH)_2$], an
iron containing fiber, belongs to the amphibole family. It is of less industrial importance
than chrysotile but is more pathogenic than chrysotile in man in the causation of malignant mesothelioma (Mossman et al., 1990a).

List of abbreviations:
AOS, Active Oxygen Species; O_2^{-}, Superoxide Anion; ODC, Ornithine Decarboxylase;
Mn-SOD, Manganese-containing Superoxide Dismutase; CuZn-SOD, Copper-Zinc-
containing Superoxide Dismutase; GPX, Glutathione Peroxidase; CMFPBS, Calcium
and Magnesium-free Phosphate Buffered Saline.

Exposure to asbestos in the workplace is associated with the development of pulmonary fibrosis, bronchogenic carcinoma and mesothelioma. The mechanisms involved in the pathogenesis of these diseases are not fully understood at present. We have hypothesized that active oxygen species (AOS) are important mediators of asbestos-induced lung toxicity (reviewed in Mossman & Marsh, 1989). For example, AOS can be formed by redox reactions occurring on the surface of asbestos fibers in cell-free systems. (Zalma *et al.*, 1987; Weitzman & Graceffa, 1984). In the presence of divalent iron, a species naturally occurring in crocidolite asbestos, superoxide anion (O_2^-) and hydrogen peroxide (H_2O_2) can be converted to hydroxyl radical ($OH^·$) by the modified Haber-Weiss (Fenton) reaction. Hydroxyl radical reacts readily with lipids, proteins and DNA. Another important source of AOS are inflammatory cells. During phagocytosis of asbestos fibers, inflammatory cells evoke a respiratory burst with release of AOS into the extracellular environment. The generation of O_2^- in alveolar macrophages after asbestos exposure is related to its size and geometry. Longer, thinner asbestiform fibers generate more AOS than smaller fibers or chemically identical particulate analogs of asbestos (Hansen & Mossman, 1987). This phenomenon is attributed to the concept of "frustrated phagocytosis" which proposes that long asbestos fibers are not completely phagocytized thereby causing continuous formation of AOS.

Studies conducted by our laboratory and others indicate a causal role of AOS in asbestos-induced cell damage and fibrosis (reviewed in Mossman & Marsh, 1989). Cytotoxicity of pulmonary cells (fibroblasts, alveolar macrophages, epithelial cells) observed after *in vitro* exposure to asbestos is ameliorated by the addition of scavengers of AOS, implying that AOS are important mediators of asbestos-induced cell death. Moreover, inflammation, pulmonary damage and fibrosis, normally observed in rats after inhalation of crocidolite asbestos, are inhibited by the systemic administration of polyethylene glycol-conjugated catalase (Mossman *et al.*, 1990).

The role of asbestos in bronchogenic carcinoma (lung cancer) appears to be that of a tumor promoter (reviewed in Mossman *et al.*, 1983). Addition of asbestos to tracheal epithelial cells *in vitro* causes morphological and biochemical alterations identical to those seen with application of classical phorbol ester tumor promoters. These biologic effects include stimulation of $Na^+K^+ATPase$ in isolated plasma membranes, induction of cell division, changes in normal cell differentiation, release of tritiated arachidonic acid, production of AOS and increased activity of ornithine decarboxylase (ODC), a rate limiting enzyme in the biosynthesis of polyamines (Landesman & Mossman, 1982; Marsh & Mossman, 1988). These growth regulatory molecules are necessary for the initiation of cell division. Moreover, increased ODC activity is observed consistently after application of phorbol ester tumor promoters to mouse skin and human epidermal cells. Previously, we have shown that fiber geometry and length of asbestos fibers are important factors in initiating ODC activity in tracheal epithelial cells. (Marsh & Mossman, 1988). In addition, AOS cause increased gene expression and enzyme activity of ODC in these cells which are prevented by addition of antioxidants (Marsh & Mossman, 1990). This finding suggests a putative role for AOS in proliferation of epithelial cells of the respiratory tract which may occur during tumor promotion.

The studies described above indicate that AOS are important mediators of asbestos-induced cell injury, pulmonary fibrosis and cell proliferation intrinsic to the development of bronchogenic carcinoma.

There is controversy over the genotoxic potential of asbestos fibers. In the present study we examined gene expression in rat lung after short-term inhalation of crocidolite asbestos. We were specifically interested in whether asbestos altered the expression of genes involved in cellular defense, i.e., genes encoding antioxidant enzymes as well as genes associated with increases in cell proliferation, e.g., ODC. Using a rodent inhalation model showing increased uptake of tritiated thymidine by both epithelial cells and fibroblasts after brief exposure to crocidolite (Mossman *et al.*, 1986), we examined lung homogenates from these animals, using Northern blot analysis, for steady-state mRNA levels of the antioxidant enzymes, manganese-containing superoxide dismutase (Mn-SOD), copper-zinc containing SOD (CuZn-SOD), catalase and glutathione peroxidase (GPX), as well as ODC.

METHODS

Inhalation Protocol

Male Fischer 344 rats, weighing approximately 200-250 grams, were exposed to NIEHS crocidolite asbestos (7-10 mg/m^3 air for 6 hr/day, 5 days/wk for 10 days) as described previously (Mossman *et al.*, 1990b; Mossman *et al.*, 1990c). Asbestos fibers were generated using a modified Timbrell dust generator. Sham control animals were placed in dust-free chambers and handled identically. After 1, 3, 6 and 9 days of inhalation and 2 weeks after cessation of a 10 day exposure period, rats (N = 4/exposure group) were anaesthetized with an intraperitoneal injection of pentobarbital, the chest cavity was opened, and the lungs were perfused with heparinized calcium and magnesium-free phosphate buffered saline (CMFPBS) via the pulmonary artery until they appeared white. Subsequently, lungs were lavaged with CMFPBS (5 x 4 ml), flash-frozen in liquid nitrogen, and pulverized. Pulverized lung was stored at -80°C until extraction of RNA.

Reagents

Phenol was obtained from Anachemia (Rouses Point, NY), chloroform from Fisher Scientific (Fair Lawn, NJ), [α-^{32}P]dATP from Dupont (Wilmington DE), CMFPBS from GIBCO (Grand Island, NY) and pentobarbital from Fort Dodge Laboratories, Inc. (Fort Dodge, IA). A Prime-a-Gene labelling system was purchased from Promega (Madison, WI). All other reagents were obtained from Sigma Chemical Co. (St. Louis, MO). Rat Mn-SOD and CuZn-SOD cDNA probes were generously provided by Y.-S. Ho, Duke University, Durham, NC. A rat catalase cDNA probe was obtained from S. Furata (Japan) and a mouse GPX probe from G. Mullenbach, Emeryville, CA. ODC cDNA (pOD$_{48}$), cloned from a mouse lymphoma cell line, was the kind gift of Dr P. Coffino, University of California, San Francisco, CA.

Northern Blot Analysis

Total RNA was extracted from lung tissues according to the procedure of Chomczynski and Sacchi (1987). Purity and concentration were determined by measuring UV absorbance at 260 and 280 nm. Fifteen micrograms of total RNA was denatured and fractionated by electrophoresis on a 1.0% agarose-formaldehyde gel. After UV examination of RNA migration, RNA was transferred onto nitrocellulose filters (Schleicher and Shuell, Keene, NH) and subsequently baked in a vacuum oven for 2 hours at 80°C. Filters were hybridized (50% deionized formamide, 5X SSC, 5X Denhardt's solution, 50 μg/μl Salmon testes DNA, 0.1% sodium dodecyl sulfate) with radiolabelled cDNA probes overnight at 42°C (Sambrook, 1989). Filters were washed and visualized by exposure to Kodak X-Omat AR film (Rochester, NY) at -70°C using intensifying screens. In addition, radioactivity on filters was directly quantitated using a betascope blot analyzer (Betagen Corp., Waltham, MA). Whereas steady-state mRNA levels of antioxidant enzymes were examined at all time periods specified above, steady-state mRNA levels of ODC have thus far been examined at one time period (9 days of exposure).

RESULTS

Antioxidant Enzyme Gene Expression

Figure 1 shows steady-state mRNA levels of antioxidant enzymes in sham and asbestos-exposed rat lungs after 9 days of exposure. Northern blots of Mn-SOD (Figure 1A) revealed 5 distinct species of mRNA which have also been observed in human lung (Wispe *et al.*, 1989). Quantitation of blots showed that Mn-SOD was the antioxidant enzyme that was affected most dramatically after exposure to asbestos, whereas gene expression of CuZn-SOD appeared comparable in sham and crocidolite-exposed lungs. Increases in steady-state mRNA levels of GPX and catalase also occurred in lungs after inhalation of asbestos (Janssen *et al.*, manuscript in preparation). Results in general corre-

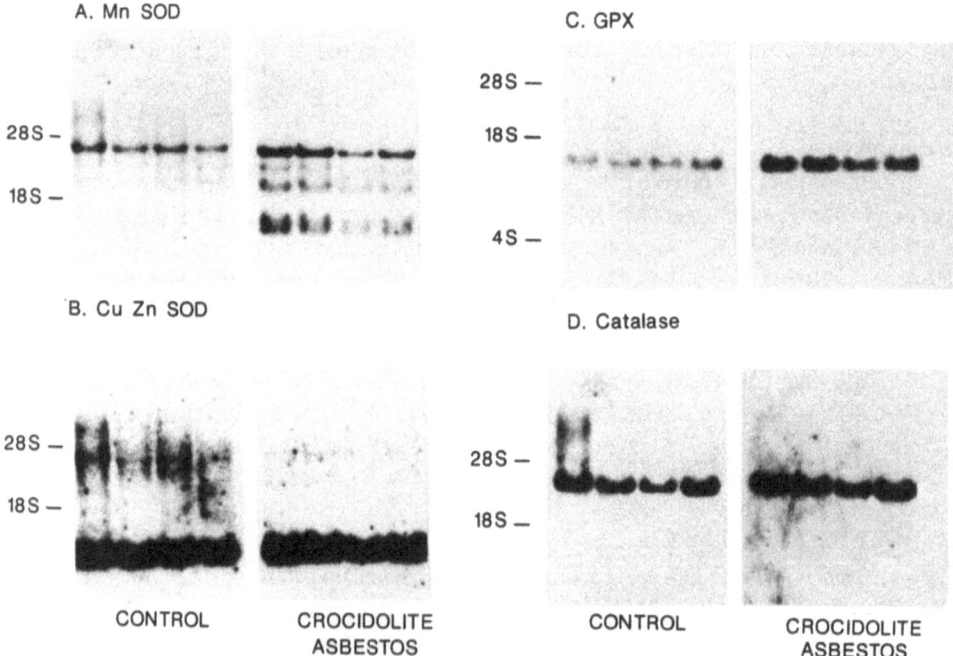

Figure 1. Northern blot analysis of Mn-SOD (A), CuZn-SOD (B), GPX (C) and catalase (D) in sham and asbestos-exposed rat lung (N=4/group) at 9 days. Total RNA was isolated by procedures described in the text and 15 μg of RNA fractionated on an agarose-formaldehyde gel, blotted onto nitrocellulose and hybridized to ^{32}P-labelled cDNA probes.

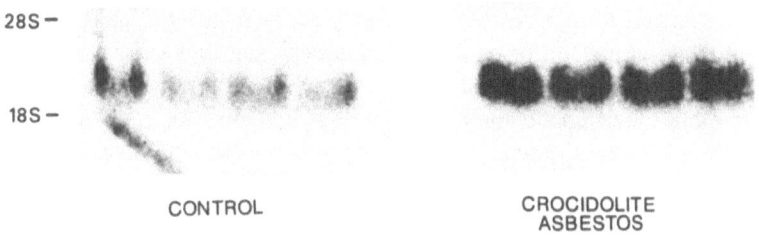

28S —

18S —

CONTROL   CROCIDOLITE
                           ASBESTOS

Figure 2. Northern blot analysis of mRNA for ODC in rat lungs at 9 days (N=4/group).
Total RNA was isolated by procedures described in the text and 30 μg of RNA fractionated on an agarose-formaldehyde gel, blotted onto nitrocellulose and hybridized to a ^{32}P-labelled cDNA probe for ODC.

lated with increases in antioxidant enzyme activities in crocidolite-exposed lungs as reported previously (Janssen *et al.*, 1990).

ODC Gene Expression

Figure 2 shows steady-state mRNA levels of ODC at 9 days after exposure to asbestos. Gene expression was increased in crocidolite-exposed rat lungs.

DISCUSSION

AOS appear to be involved as causative agents of acute lung injury and fibrosis after exposure to hyperoxia, paraquat, bleomycin, phorbol esters, NO_2 and ozone (reviewed in Heffner & Repine, 1989). In several of these models, the antioxidant enzyme system appears to be directly affected in response to oxidative stress. The increased activity of antioxidant enzymes under these circumstances might be an adaptive or protective response. For example, lung injury might ensue when this balance is disturbed by increased oxidative stress or when the antioxidant enzyme system is overwhelmed.

In our present study, we report that inhalation of high concentrations of asbestos results in altered expression of specific antioxidant enzyme genes in rat lung. However, these increases are not sufficient to protect the lung from subsequent injury and disease caused by asbestos exposure since rats develop acute pulmonary damage and diffuse interstitial fibrosis under these conditions (Mossman *et al.*, 1990b). Currently, we are exploring lower concentrations of asbestos in this inhalation model to determine whether threshold concentrations exist below which oxidant stress and antioxidant defense are in equilibrium, thereby protecting the lung from asbestos-induced damage. The possibility exists that antioxidant enzymes are altered only in certain cell types or compartments of the lung, leaving other components relatively unprotected. This phenomenon might account for a relative deficiency in antioxidant defense in certain areas resulting in focal cell damage and initiation of pulmonary fibrosis. Our future experiments are focused on identification of antioxidant enzyme proteins in certain cell types using immunocytochemistry in order to determine which cells are important in lung defense against mineral fibers.

Work presented here also indicates that genes involved in cell proliferation in lung are affected after short-term inhalation of asbestos. Whether this increase in gene expression occurs with other fibrogenic and carcinogenic materials at both low and high airborne concentrations of particulates is currently under investigation. The possibility that asbestos directly affects oncogenes involved in cell transformation is under exploration in a number of laboratories and results reported thus far have been inconsistent. However, our studies indicate a complex pattern of gene expression involving both genes participating in antioxidant defence as well as a gene which may be critical in tumor promotion by asbestos in lung.

ACKNOWLEDGEMENTS

The authors acknowledge the technical expertise of Laurie Sabens and James Rainville in preparation of this manuscript. This research was supported by grant R01 HL39469 and SCOR grant PHS 14212 from the National Heart, Lung and Blood Institute.

REFERENCES

Chomczynski,P. and Sacchi,N. (1987) Single-step method of RNA isolation by acid guanidinium thiocyanate-phenol-chloroform extraction. *Anal. Biochem.* **162**:156-159.

Hansen,K. and Mossman,B.T. (1987) Generation of Superoxide (O_2^{-}) from alveolar macrophages exposed to asbestiform and nonfibrous particles. *Cancer Res.* **47**:1681-1686.

Heffner,J.E. and Repine,J.E. (1989) Pulmonary strategies of antioxidant defense. *Am. Rev. Respir. Dis.* **140**:531-554.

Janssen,Y.M.W., Marsh,J.P., Absher,M., Borm,P.J.A. and Mossman,B.T. (1990) Increases in endogenous antioxidant enzymes during asbestos inhalation in rats. *Free Rad. Res. Comms.* **11**:53-58.

Janssen,Y.M.W., Marsh,J.M., Borm,P.J.A. and Mossman,B.T. Expression of antioxidant enzymes in rat lung after inhalation of crocidolite asbestos or cristobalite silica, (manuscript in preparation).

Landesman,J.M. and Mossman,B.T. (1982) Induction of ornithine decarboxylase in hamster tracheal epithelial cells exposed to asbestos and 12-O-tetradecanoylphorbol-13-acetate. *Cancer Res.* **42**:3669-3675.

Marsh,J.P. and Mossman,B.T. (1988) Mechanisms of induction of ornithine decarboxylase activity in tracheal epithelial cells by asbestiform minerals. *Cancer Res.* **48**:709-714.

Marsh,J.P. and Mossman,B.T. (1991) Role of asbestos and active oxygen species in activation and expression of ornithine decarboxylase in hamster tracheal epithelial cells. *Cancer Res.* **51**:167-173.

Mossman,B.T., Bignon,J., Corn,M., Seaton,A. and Gee,J.B.L. (1990a) Asbestos: scientific developments and implications for public policy. *Science* **247**:294-301.

Mossman,B.T., Marsh,J.P., Sesko,A., Hill,S., Shatos,M.A., Doherty,J., Petruska,J, Adler,K.B., Hemenway,D., Mickey,R., Vacek,P. and Kagan,E. (1990b) Inhibition of lung injury, inflammation, and interstitial pulmonary fibrosis by polyethylene glycol-conjugated catalase in a rapid inhalation model of asbestosis. *Am. Rev. Respir. Dis.* **141**:1266-1271.

Mossman,B.T., Janssen,Y.M.W., Marsh,J.P., Manohar,M., Garrone,M., Shull,S. and Hemenway,D. (1991c) Antioxidant defense mechanisms in asbestos-induced lung disease. *J. Aerosol Med.* **3(Suppl 1)**:S-75—S-81.

Mossman,B.T. and Marsh,J.P. (1989) Evidence supporting a role for active oxygen species in asbestos-induced toxicity and lung disease. *Environ. Hlth Perspect.* **81**:91-94.

Mossman,B.T., Gilbert,R., Doherty,J., Shatos,M.A., Marsh,J. and Cutroneo,K. (1986) Cellular and molecular mechanisms of asbestosis. *Chest* **89**:160S-161S.

Mossman,B., Light,W. and Wei,E. (1983) Asbestos: mechanisms of toxicity and carcinogenicity in the respiratory tract. *Ann. Rev. Pharmacol. Toxicol.* **23**:595-615.

Sambrook,J., Fritsch,E.F. and Maniatis,T. (1989) Molecular cloning: a laboratory manual. 2nd Ed., Cold Spring Harbor Laboratory Press, Cold Spring Harbor, NY.

Wispe,J.R., Clark,J.C., Burhans,M.S., Kropp,K.E., Korfhagen,T.R. and Whitsett,J.A. (1989) Synthesis and processing of the precursor for human mangano-superoxide dismutase. *Biochim. Biophys. Acta* **994**:30-36.

RAT PLEURAL CELL POPULATIONS: EFFECTS OF MMMF INHALATION ON

CYTOKINE mRNA EXPRESSION AND POPULATION CHARACTERISTICS

Klara Miller, B.N. Hudspith and C. Meredith

Immunotoxicology Department
British Industrial Biological Research Association
Woodmansterne Road, Carshalton
Surrey SM5 4DS, UK

INTRODUCTION

Because of the known dangers of asbestiform fibres, many man-made mineral fibres (MMMF) are being manufactured as replacements. Among substitute products in use today glass wool and rock/slag wool are probably "safer" than chrysotile asbestos as shown by epidemiological studies (Simonato *et al.*, 1987). However, little published human data exist for the refractory ceramic fibres developed to provide more efficient insulation particularly during high temperature uses. At present therefore, the biological activity of ceramic fibres may best be examined by cell culture and well-designed animal studies (Brown *et al.*, 1990). Such experiments need to take into account the possibility that differing mechanisms may be responsible for fibrosis and cancer. Transport of fibres to the serosal sites of the pleura is also poorly documented and several routes, i.e. lymphatic, systemic, direct translocation have been proposed, but none have been firmly established. Indeed, the mode of action whereby inhalation of inorganic fibres leads to pathogenic changes in pleural tissues is unknown, although there is a widespread assumption that the macrophage plays a central role via various inflammatory processes in initiating these events.

We have accordingly investigated the early cellular changes that short term inhalation of MMMF may induce in rat alveolar and pleural cell populations which could lead to fibrosis or pleural pathology. The hypothesis being examined is that inhalation of fibres induces changes in the ability of alveolar cells to synthesise and release immune inflammatory mediators. These mediators would then diffuse into the pleura and in turn activate pleural macrophages to release other mediators such as growth regulating proteins. A necessary adjunct of the investigation has been the determination of the alveolar and pleural macrophage phenotypes prior to assessing fibre-induced alterations in population characteristics.

MATERIALS AND METHODS

Exposure protocol

Inbred male PVG rats were exposed to glass fibres (John Manville Reference C100) refractory ceramic dust i.e. fibres plus particle (Kaowool; Morganite Ceramic Fibres) or specially prepared and sized refractory ceramic fibres (RCF1, RCF2; Carborundum, Niagara Falls, N.Y.size approx. 20 μm mean length and 1 μm diameter) via inhalation at mean mass concentrations between 10 and 30 mg/m³ 6 hours per day for 5 days. Animals were sacrificed and cells recovered 3, 30, 90 and 180 days post-exposure by bronchoalveolar (BAL) or pleural (PL) lavage. Eight rats per time point were compared with age-matched air exposed controls.

Collection of cells

Lavage of the pleural space was performed essentially as described by Oberdörster and co-workers (1983). The abdomen was opened, a cannula was inserted through the diaphragm and 8 ml of cold 0.9% saline (NS) installed into the thoracic spaces, after which several washes were performed. For BAL, lungs and trachea were removed, NS was installed through an intra-tracheal cannula and 6-8 washes were performed *ex situ*. Both the lavaged pleural and lung cells were kept on ice until centrifuged at 1200 rpm for 20 minutes and resuspended. Several cytocentrifuge preparations were made, one stained with "Diff-Quick" for cell differentiation, another for non-specific esterase activity using ß-naphthol butyrate as a substrate.

Phenotypic analysis

A panel of 5 rat monoclonal antibodies (ED1, ED2, MRC OX41, MRC OX42 and MRC OX4 (Class II)) were used to define macrophage subpopulations and follow the appearance of activation markers on these cells after MMMF exposure. The cytocentrifuge preparations were fixed in acetone for 10 minutes at -20°C, air dried and stored dry at -20°C until used. Staining was carried out at room temperature. Prior to staining endogenous peroxidase activity was blocked by flooding the slide with 3% hydrogen peroxide for 5 minutes. The slides were then rinsed in PBS and 50 μl of the monoclonal antibody was applied for 1 hour at dilutions we had previously established gave optimum staining. For ED1 and ED2 we used at a dilution of 1:500, OX41 was used at 1:400, OX42 and OX4 at 1:250. The slides were then washed in PBS and a biotinylated goat anti-mouse secondary antibody diluted 1 in 20 was applied for 30 minutes, after washing in PBS avidin-conjugated peroxidase reagent was applied for 20 minutes. Peroxidase activity was then visualised by a 10 minute incubation in a solution of 0.2 mg/ml 3-amino-9-ethylcarbazole (AEC) in 0.05 acetate buffer, pH 5.0 containing 0.01% H_2O_2. Slides were then washed in deionised water and counterstained with Meyers haematoxylin. At sites of antibody -antigen reactions a brown insoluble precipitate is formed.

ED1 and ED2 were purchased from Serotec UK, all other monoclonals were obtained from Sera Lab Ltd, UK. Reagents for the Avidin-Biotin staining procedure were purchased from Sigma UK.

TABLE 1. Distribution and staining pattern of rat macrophages by ED1 and ED2

	ED1	ED2
Peritoneal macrophages (Resident)	+++ (90%)	++ (45%)
Pleural macrophages	+++ (80%)	+++ (90%)
Alveolar macrophages	+++ (80%)	–
Lung "tissue" macrophages	+ (10%)	+++ (80%)

RNA Dot Blot and Hybridisation

Total cellular RNA was obtained from freshly lavaged, cultured and LPS-stimulated quiescent cell cultures by lysing adherent cells in the presence of 7.6 M guanidine hydrochloride (Cheley & Anderson, 1984). DNA in the lysate was sheared by repeated passage through a 25 g needle and 0.6 vol of 95% ethanol added. Following overnight storage at -20°C RNA was recovered by centrifugation at 13000 x g for 20 minutes. For dot-blotting total RNA was dissolved in a formaldehyde-buffer, blotted onto membranes

TABLE 2. Total and differential counts of pleural cells lavaged from rats exposed to ceramic fibres (RCF-1) at 19.6 mg/m^3 (3 days post-exposure)

	Control	Test
Total count (x 10^6)	2.65	2.80
% Cell type		
Macrophages	38.5	38.0
Lymphocytes	6.5	7.5
Neutrophils	17.5	16.0
Eosinophils	17.0	18.5
Mast cells	13.5	15.5
Others	7.0	4.5

and hybridised as described previously (Meredith *et al.*, 1990). The membranes were then exposed to X-ray film (Fuji) for 2-7 days at -70°C and the autoradiographs quantified by scanning video densitometry.

Scanning electron microscopy

Pleural and lung cells suspended in RPMI 1640 containing 10% FCS were added to glass cover slips (1 x 10^6 cells/ml), allowed to adhere for 2 h at 37°C, washed vigorously and cultured overnight. They were prepared for SEM as described previously (Miller & Kagan, 1976).

RESULTS AND DISCUSSION

The rat is the favoured animal in toxicological studies, but only recently have monoclonal antibodies delineating phenotypic subsets of macrophages in the rat become available. Two separate laboratories have now produced monoclonal antibodies that clearly demonstrate heterogeneity between different populations of macrophages (Robinson *et al.*, 1986; Dijkstra *et al.*, 1985).

MRC OX-41 and MRC OX-42 antibodies (Robinson *et al.*, 1986) although also labelling granulocytes recognise cell surface proteins that are synthesised specifically by macrophages. MRC OX-42 recognises an antigen associated with complement function, the iC3b-receptor. ED1 and ED2 antibodies (Dijkstra *et al.*, 1985) recognise mononuclear phagocytes exclusively, ED1 recognise a cytoplasmic antigen whereas ED2 binds to a membrane antigen. In agreement with published data, we find differences between alveolar and other macrophage populations. Cells recovered by broncho alveolar lavage (BAL) from PGV rats were found to be negative for ED2 but ED1 positive (as were peripheral blood monocytes); 90% were OX-41 positive and 20-30% were recognised by the OX-42 antibody (Figure 1a). Cells recovered by pleural lavage recognised all four antibodies, but differed from macrophage population obtained from the peritoneal serosal cavity in that only 50% of the peritoneal cells were recognised by the ED2 antibody (Figure 1b). It appears that pleural macrophages, although anatomically located in close proximity to the alveolar cell, share the phenotypic characteristic of a subset of peritoneal macrophages. Interestingly, in experiments in which interstitial macrophages were isolated from lung tissue 80% were found to be positive for ED2 whereas only 10% labelled with ED1. Thus all macrophage populations in the respiratory tract, the pulmonary alveolar macrophage in the peripheral air spaces and conducting airways, macrophages in lung tissues and macrophages in pleural space differ in their phenotypic expressions and show heterogeneity within their own population groups (Table 1). Although it was felt that alveolar cells might mature into ED2 positive cells after phagocytosis of fibres the inhalation of mineral fibres induced no alterations with respect to ED1 and ED2 in either alveolar or pleural macrophages.

An important difference between rat pleural and alveolar macrophages was the expression of MHC Class II (Ia) molecules on the surface membrane, said to correlate with the ability to present antigen and stimulate autologous lymphocyte proliferation. Approximately 50% of the pleural macrophages were positive for Class II when labelled with the MRC OX-4 monoclonal antibody whereas only 5% of alveolar macrophages expressed Class II antigen. During and immediately after fibre exposure there was an Class II positive cells increased by as much as six-fold and this was paralleled by similar

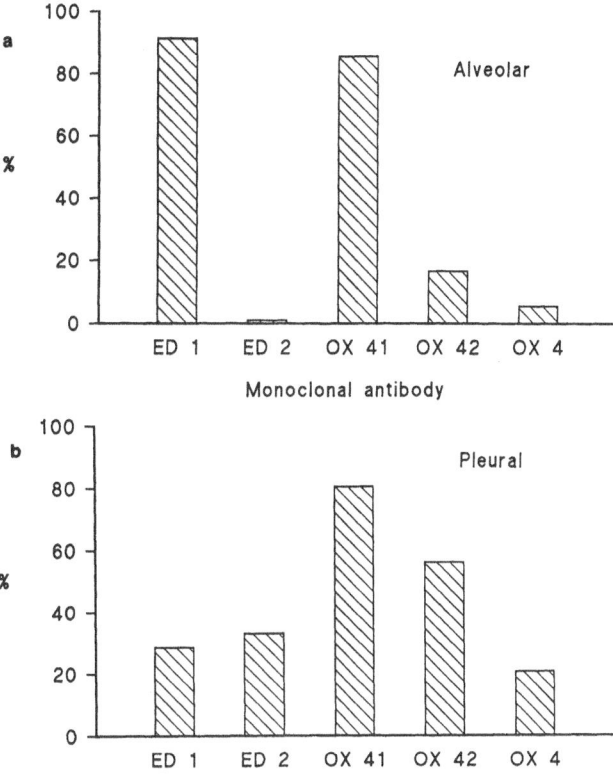

Figure 1a. Phenotypic labelling of cells recovered by bronchoalveolar lavage.
1b. Phenotypic labelling of cells recovered by pleural lavage.

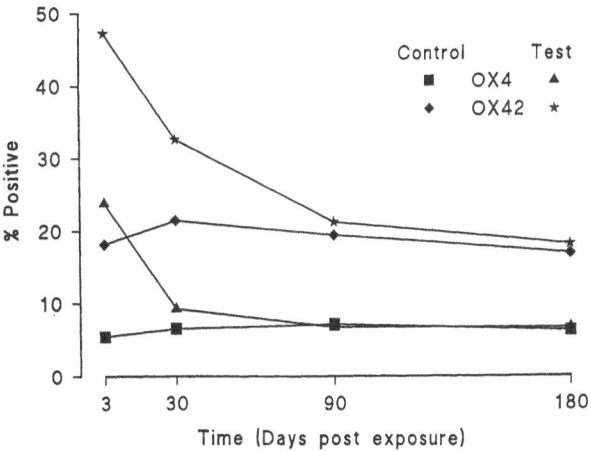

Figure 2. Expression of Class II and complement receptor (OX-42) in alveolar cells after exposure to MMMFs (composite graph).

increase in OX-42 and Class II expression by alveolar macrophages. The number of increase in OX-42 positive cells (Figure 2); this was found to be independent of the fibre type but increase in Class II expression was related to exposure dose. However, although the majority of alveolar cells contained phagocytosed fibres for up to 180 days, macrophages obtained 30 days post-exposure showed little increase in these activation markers. It is possible that expression of Class II and the complement receptor could have been induced by inflammatory stimuli. Although no significant increase in the cell population recovered by BAL during fibre exposure was evident, the number of neutrophils increased from a level of 2.2% ($\pm$ 2.5 E.M.) as seen in unexposed rats to 14.5% ($\pm$ 3.8%) in test animals 3 days post exposure. Again this rise in neutrophils declined with time and had returned to normal after 30 days post-exposure. It may that proteases or other mediators released by the neutrophils activated the macrophages to enable them to play a role in regulating potentially damaging inflammatory responses in the lung.

No alteration in Class II expression or any other phenotypic marker was evident in pleural cells during or after exposure to any of the MMMFs nor was there any increase in total cell numbers. As reported previously (Miller *et al.*, 1989) the recovered pleural cell population is a very heterogeneous one. Typical distribution from air or fibre exposed rats at 3 days post-exposure were 38% macrophages, 7% lymphocytes, 17% neutrophils, 18% eosinophils, and 14% mast cells (Table 2). In both air and fibre exposed animals there was an age related shift in the population from eosinophils to mast cells which increased to approximately 20%. The high numbers of inflammatory pleural cells found in this study have also been found by other investigators who make use of the rat model (Lehnert *et al.*, 1985), and raise the general question as to what extent pathological changes in the pleura may be influenced by such inflammatory cells. Serosal mast cells in particular have been shown to express very high levels of TNF-α, a multifunctional cytokine that can augment fibrosis and epithelial cell proliferation (Gordon *et al.*, 1990).

Pleural macrophages also differed from alveolar macrophages in that, whereas there was no discrepancy between morphology, and esterase staining with alveolar macrophages some pleural macrophages (as identified by the OX-4 marker) did not show any level of esterase activity. Such lack of esterase activity has previously been demonstrated (Lehnert *et al.*, 1988). No increase in Class II expression was observed in the pleural macrophages at any time post-exposure. However, after exposure to glass or refractory ceramic fibres adherent pleural cells appeared activated as evidenced by increased spreading and long cytoplasmic processes (Figure 3.). In earlier studies (Miller *et al.*, 1988) a few lavaged pleural cells when examined by scanning electron microscopy number were found to contain partly phagocytosed fibres as early as 3 days post-exposure after glass and ceramic fibres. However, when the number of fibres present in the pleural lavage supernatants were quantified they were very low (range: 2-16 fibres) and no increase was found 30 days post-exposure. We may therefore conclude that the possible role of short fibres in inducing pleural adherent cell activation is unlikely and postulate that activation of these cells could be mediated by cytokines released by fibre activated lung cells. Alveolar macrophage and pleural cell populations were therefore examined for fibre-induced alterations by assaying changes in the level of expression of specific cytokine genes (Miller & Hudspith, 1991).

Mononuclear phagocytes are adherent cells and *in vitro* assays to measure cytokines generally involve a 2 hour incubation in microtitre plates or plastic culture dishes to obtain pure macrophage monolayers. However adherence to plastic results in stimulation of IL-1 expression and secretion of the protein product (Danis *et al.*, 1990). In experiments using murine peritoneal macrophages we have found that adherence to culture dishes induced expressions at the level of mRNA of both IL-1 and IL-ß for at least 12 hours; by 18 hours the cells were quiescent with regard to transcription of IL-1, IL-6 and TNF (Meredith *et al.*, 1990). By contrast macrophages harvested from lymphoid organs, the peritoneal cavity, or the alveolar spaces do not express IL-1 mRNA (Fuhlbrigge *et al.*, 1987; Wewers & Heizyk, 1989). However, after stimulation with lipopolysaccharide (LPS) under appropriate *in vitro* conditions significant levels of these mRNAs can be detected.

We therefore examined the ability of alveolar and pleural macrophages for expression of IL-1 genes. We were unfortunately unable to examine the ability of the cells to express the fibrogenic cytokine TNF as no cDNA probe for rat TNF is available. Because most cytokine probes are of human or mouse origin, they are of limited use in the rat and depend strictly on the degree of homology for specific low-background hybridisation to occur. Using established techniques for mRNA extraction, then dot-blotting these samples and using appropriate genetic probes, macrophages harvested by bronchoalveolar lavage from fibre exposed rats showed no evidence of IL-1 gene expression. After these cells were cultured for 24 hours however, and then stimulated with LPS for 3 hours IL-1ß transcription increased significantly when compared to controls (Table 3). This finding of "priming" of alveolar cells has also been reported in a study into the effects of fibre inhalation in the context of antigen directed lymphoid activation (Hartmann *et al.*, 1984). No increase in IL-1 gene expression was found in pleural cells.

TABLE 3. Measurement of mRNA specific for Interleukin-1 (IL-1) and fibronectin (3 days post exposure; ceramic fibres, RCF1) expressed by alveolar macrophages

Culture conditions		mRNA		
Pre culture time	LPS (10 μg/ml) ± for 3hours	IL-α	IL-ß	Fibronectin
0 hrs				
Control	No	-	-	+
Test	No	-	-	+
24 hrs				
Control	Medium only	-	-	+
Test	Medium only	-	-	+
24 hrs				
Control	LPS	+	+	+
Test	LPS	+	+ + +	+

Figure 3. Scanning electron micrograph of pleural adherent cells obtained from rat exposed to ceramic dust (Kaowool) 3 days post-exposure.

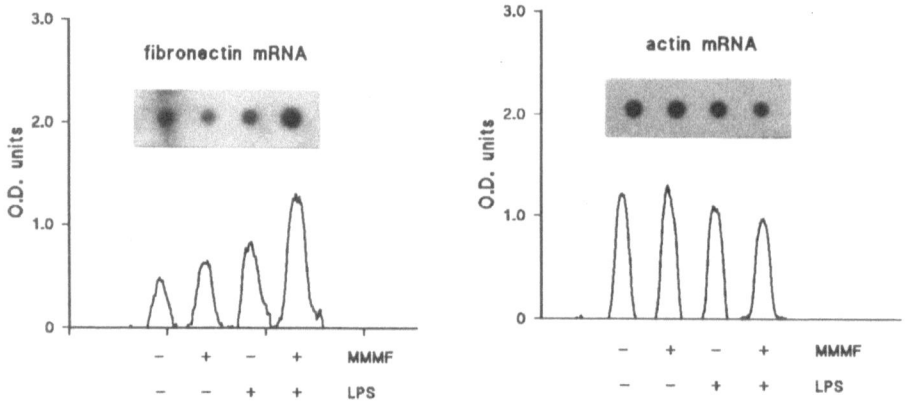

Figure 4. Expression of fibronectin and actin mRNA in pleural cells harvested 3 days post RCF1 exposure (19.6 mg/m^{3}M and cultured in the presence of lipopolysaccharide.

Another cytokine important in wound repair and inflammatory processes is fibronectin. Fibronectin is known to be expressed by mature mononuclear phagocytes including alveolar and peritoneal macrophages (Adachi *et al.*, 1988). No alteration in the level of fibronectin specific mRNA was found in alveolar cells from control or fibre exposed animals. However, pleural cells collected from test animals during or 3 days post exposure to ceramic fibres (RCF1) were found to express elevated levels of fibronectin mRNA after LPS stimulation (Fig 4.) whereas the expression of the house keeping gene actin was unaltered. No change in total number of cells or differential counts were observed and it is unlikely that the increased level of fibronectin expression is due to direct activation of the pleural cells by inhaled fibres. A more likely explanation is that alveolar or lung cells that come into contact and react directly with the inhaled fibres induce changes in the pleural cells via the release of cytokines or other mediators.

We would conclude that short-term inhalation is a promising method for investigating early cellular changes in the lung and that molecular biological analysis at the level of gene expression offers significant advantages over existing procedures for determining pathways that could be responsible for the fibrogenic, carcinogenic or promoting effects of inhaled fibres. However, in spite of familiarity with the rat, the use of a murine model would be far more productive. Firstly, the great majority of resident mouse pleural cavity cells consist of macrophages, and there are few mast cells present; secondly as stated earlier most cytokine probes are of human or mouse origin so that alterations in the expression of those cytokines important in the development of pulmonary fibrosis (i.e. TNF-α) can be easily measured (Piquet *et al.*, 1990) and thirdly, a valid murine model of fibre-induced pulmonary disease has already been established for some years (Bozelka *et al.*, 1983). The use of a murine model as demonstrated in recent studies by Piquet and co-workers would therefore have a number of advantages in monitoring the potential pathogenicity of inhaled man made mineral fibres.

ACKNOWLEDGEMENTS

We are grateful to the following for their gifts of cDNAs: P. LoMedico, Hoffman-La Roche (murine IL-1α); P. Gray, Genetech (murine IL-1ß); R. Hynes, MIT (rat fibronectin). This work was supported by the UK Health & Safety Executive to whom our thanks are due.

REFERENCES

Adachi,R., Yamuchi,K., Bernaudin,J.F., Fouret,P., Ferrans,V.J. and Crystal,R.G. (1988) Evaluation of fibronectin gene expression by *in situ* hybridization. *Am. J. Path.* **133**:193-203.

Bozelka,B.E., Sestini,P., Gaumer,H.R., Hammad,Y., Heather,C.J. and Salvaggio,J.E. (1983) A murine model of asbestosis. *Am. J. Path.* **112**:326-337.

Brown,R.C., Hoskins,J.A., Miller,K. and Mossman,B.T. (1990) Pathogenetic mechanisms of asbestos and other mineral fibres. *Molec. Aspects Med.* **11**:325-349.

Cheley,S. and Anderson,R. (1984) A reproducible microanalytical method for the detection of specific RNA sequences by dot-blot hybridisation. *Anal. Biochem.* **137**:15-19.

Danis,V.A., Kulecz,A.J., Nelson,D.S. and Brooks,P.M. (1990) Cytokine regulation of human monocyte interleukin-1 (IL-1) production *in vitro*. *Clin. Exp. Immunol.* **80**:435-443.

Dijkstra,C.D., Döp,E.A., Jopling,P. and Kroal,G. (1985) The heterogeneity of mononuclear phagocytes in lymphoid organs: distinct macrophage subpopulations in the rat recognized by monoclonal antibodies ED1, ED2 and ED3. *Immunol.* **54**:589-599.

Fulhbrigge,R.C., Chaplin,D.D., Kiely,J-M. and Unanue,E.R. (1987) Regulation of interleukin-1 gene expression by adherence and lipopolysaccharide. *J. Immunol.* **138**:3799-3802.

Gordon,J.R., Burd,P.R. and Galli,S.J. (1990) Mast cells as a source of multifunctional cytokines. *Immunol. Today* **11**:458-464.

Hartmann,D.P., Malagerogian,M., Ogisho,Y. and Kagan,E. (1984) Enhanced interleukin activity following asbestos inhalation. *Clin. Exp. Immunol.* **55**:643-650.

Lehnert,B.E., Valdez,Y.E. and Bomalski,S.H. (1985) Lung and pleural "free-cell responses" to the intrapulmonary deposition of particles in the rat. *J. Toxicol. Environ. Hlth* **16**:823-839.

Lehnert,B.E., Dethloff,L.A. and Valdez,Y.E. (1988) Leukocytic responses to the intrapleural deposition of particles, particle-cell associations and the clearance of particles from the pleural space compartment. *J. Toxicol. Environ. Hlth* **24**:41-66.

Meredith,C., Scott,M.P., Pekelharing,H. and Miller,K. (1990) The effect of Biostim (RU-41740) on the expression of cytokine mRNAs in murine peritoneal macrophages *in vitro*. *Toxicol. Lett.* **53**:327-337.

Miller,K. and Hudspith,B.N. (1991) Investigation into the effect of inhaled man-made mineral fibres on alvolear and pleural cell populations. *Human Exp. Toxicol.* **10**:83.

Miller,K. and Kagan,E. (1976) The *in vivo* effects of asbestos on macrophage membrane structure and population characteristics of macrophages. *J. Reticuloendothel. Soc.* **20**:159-171.

Miller,K., Lawrence,F. and Riley,A.R. (1989) Consequence of MMMF inhalation on lung and pleural cavity cell populations. In, "Effects of Mineral Dusts on Cells", eds. Mossman,B.T. & Begin,R.A. Springer-Verlag, Berlin, pp.321-328.

Oberdörster,G., Ferin,J., Marcello,N.L. and Meinhold,S.H. (1983) Effect of intrabronchially instilled amosite on lavagable lung and pleural cells. *Environ. Hlth Perspect.* **51**:41-48.

Piquet,P.L., Collart,M.A., Grau,G.E., Sappino,A.-P., Vasalli,P. (1990) Requirement of tumour necrosis factor for development of silica-induced pulmonary fibrosis. *Nature* **34**:245-247.

Robinson,A.P., White,T.M. and Mason,D.W. (1986) Macrophage heterogeneity in the rat as determined by two monoclonal antibodies, MRC OX-41 and MRC OX-42. *Immunol.* **57**:239-247.

Simonato,L., and 18 others (1987) The IARC historical cohort study of MMMF production workers in seven European countries: extension of the follow-up. *Ann. Occup. Hyg.* **31**:603-624.

GROWTH FACTOR AND GROWTH FACTOR RECEPTOR EXPRESSION IN TRANSFORMED RAT MESOTHELIAL CELLS

Cheryl Walker, Edilberto Bermudez, Wendy Stewart
and Jeff Everitt

Chemical Industry Institute of Toxicology
P.O. Box 12137, Research Triangle Park
NC 27709, U.S.A.

INTRODUCTION

Mesothelial cell transformation induced by asbestos and other mineral fibers has been shown to occur both *in vitro* (Paterour *et al.*, 1985) and *in vivo* (Craighead, 1987a; Mossman *et al.*, 1990). Transformation *in vivo* may occur through direct interaction of the fibers with target mesothelial cells or may occur through an indirect interaction of fibers with other cell populations (such as inflammatory cells) that produce cellular mediators, such as cytokines or oxygen radicals (Mossman *et al.*, 1989; Jaurand, 1989). Very little is known about the mechanism of transformation of mesothelial cells by either direct or indirect interactions with mineral fibers; a fact that has hampered the search for "safe" asbestos substitutes. An understanding of the biology of mesothelioma induction is critical in order to correctly assess the health effects associated with environmental and occupational exposure to natural and man-made fibers.

In man, transformation of normal mesothelial cells is accompanied by alterations in the production of and responsiveness to various growth factors, including platelet-derived growth factor (PDGF) (Gerwin *et al.*, 1987; Versnel *et al.*, 1988), epidermal growth factor (EGF) (Tubo & Rheinwald, 1987) and several hematopoietic cytokines (Demetri *et al.*, 1989). Of particular interest is the production of PDGF by transformed human mesothelial cells. Normal human mesothelial cells do not produce PDGF, but transformed cells express mRNA for both the A and B chains of PDGF and secrete a PDGF-like protein into the culture medium (Gerwin *et al.*, 1987; Versnel *et al.*, 1988). PDGF is also mitogenic for normal human mesothelial cells (Laveck *et al.*, 1988) and this has led to the suggestion that PDGF produced by human mesothelioma cells may be acting as an autocrine growth factor. However, PDGF has not been demonstrated directly to be mitogenic for transformed mesothelial cells (Gerwin *et al.*, 1990).

In rodents, the role of growth factors in mesothelioma is even less clearly understood. Rats develop mesothelioma both spontaneously (Tanigawa *et al.*, 1987) and in response to exposure to mineral fibers (Craighead, 1987a). Both normal and transformed rat mesothelial cells can be grown in culture (Jaurand *et al.*, 1981; Craighead *et al.*, 1987b), opening the way for the examination of changes in growth factor and growth factor receptor expression that occur during transformation. In this report, we characterize the expression of several growth factors and their cognate receptors in normal and transformed rat mesothelial cells.

EXPRESSION OF PDGF AND ITS RECEPTOR BY NORMAL AND TRANS-FORMED RAT MESOTHELIAL CELLS

Normal rat mesothelial cells were isolated from the intracostal region of the parietal pleura and cultured to produce cell strains under conditions previously described (Bermudez *et al.*, 1990). Strains of normal mesothelial cells can be isolated with a high frequency under these culture conditions, and nine independent rat mesothelial cell strains have been isolated to date (E. Bermudez, unpublished results). These cells co-express cytokeratin and vimentin and appear morphologically by both light and electron microscopy (Bermudez *et al.*, 1990) to be similar to mesothelial cells *in situ* (Figure 1). In two normal cell strains tested to date (NRM1 and NRM2), the cells were non-tumorigenic when late passage cells were injected into nude mice by subcutaneous and peritoneal inoculation (passage 33 and 81, respectively).

The level of expression of PDGF A and B in the normal rat mesothelial cell strain NRM2 was compared to the level of expression of these growth factors in a transformed cell line (II 45) derived from asbestos-induced rat mesothelioma (Craighead *et al.*, 1987b). As shown in Figure 2, at the level of RNA expression, no transcripts for PDGF A or B could be detected in either the normal or transformed rat cells. This finding contrasts to that for cells of human mesothelioma which express abundant PDGF A and B transcripts and secrete PDGF-like material into the culture medium (Gerwin *et al.*, 1987; Versnel *et al.*, 1988). The PDGF-receptor is a dimer composed of two types of subunits, termed α and ß (Hart *et al.*, 1988; Gronwald *et al.*, 1988; Matsui *et al.*, 1989; Seifert *et al.*, 1989). PDGF AA is recognized by the αα homodimer receptor molecule, but not αß or ßß receptors (Seifert *et al.*, 1989). In contrast, PDGF BB is recognized by all three isoforms of the receptor (Siefert *et al.*, 1989). Both normal and transformed rat mesothelial cells express transcripts for the B chain of the PDGF receptor, but no transcripts for the A chain of the PDGF receptor could be detected (Figure 2).

The presence of ß-type PDGF-receptors in both normal and transformed cells was confirmed by using a competition assay for the binding of PDGF BB homodimer. As shown in Figure 3, unlabeled PDGF BB but not PDGF AA or EGF, competes effectively for binding of ^{125}I-PDGF-BB to the normal rat mesothelial cell strain NRM2. Similar results were obtained for the II-45 cell line derived from the asbestos-induced rat mesothelioma. These transformed cells also bind ^{125}I-PDGF-BB, and PDGF BB but not PDGF AA or EGF, can compete for this binding. In contrast, Swiss 3T3 cells which express both α and ß type PDGF-receptors (Seifert *et al.*, 1989) show approximately a 50% inhibition of ^{125}I-PDGF binding in the presence of PDGF AA (data not shown). PDGF AA is not recognized by ß-type receptors, and the inability of PDGF AA to

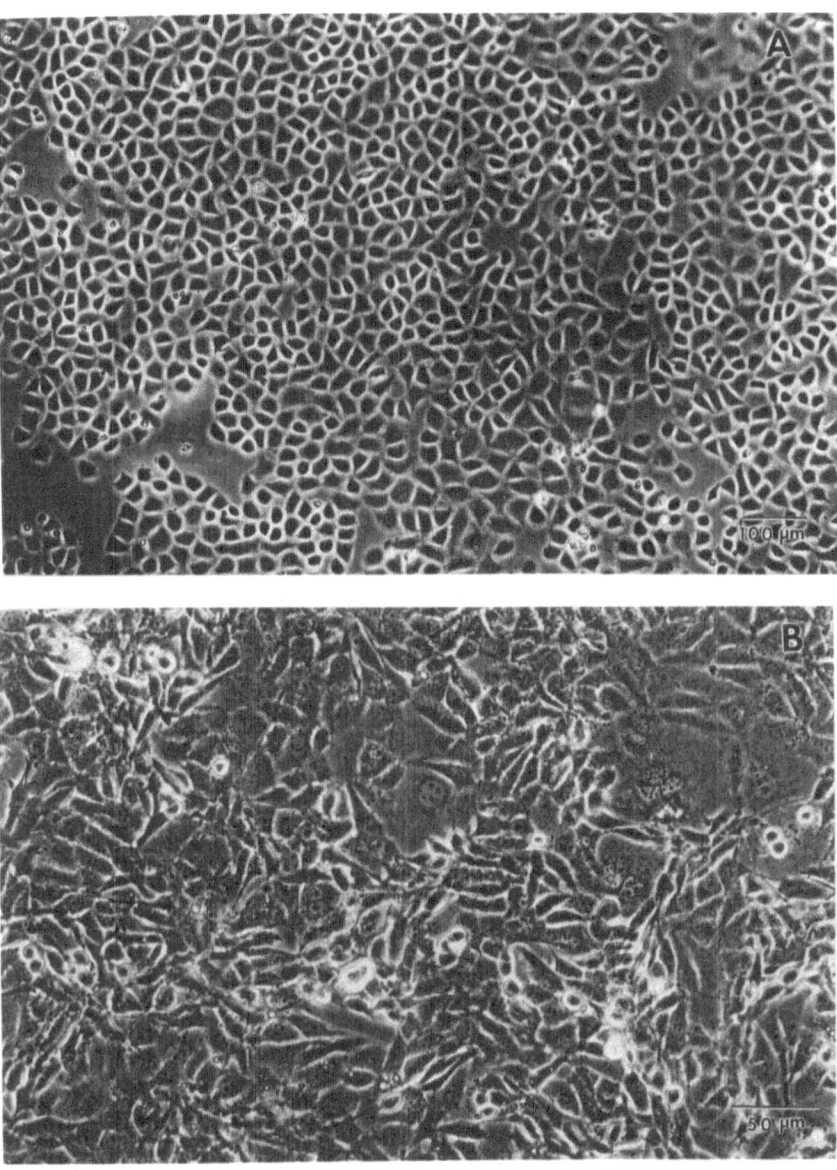

Figure 1. Phase contrast micrograph of normal and transformed rat mesothelial cells in culture. (A) NRM2, a normal mesothelial cell line derived from the parietal pleura. (B) II45, a cell line established from an asbestos-induced peritoneal mesothelioma.

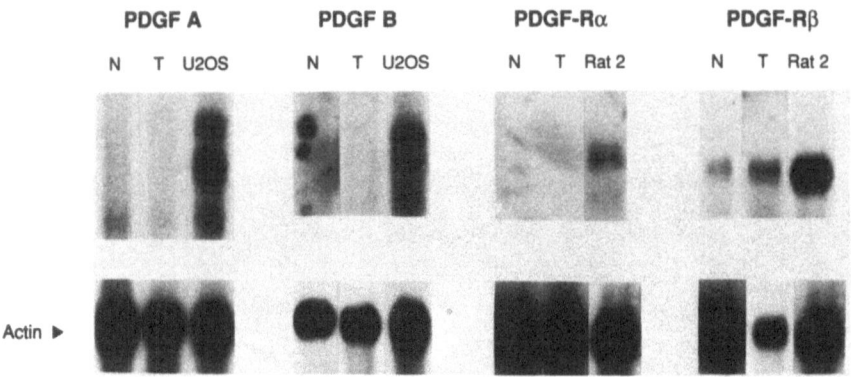

Figure 2. Northern blot analysis of poly (A)$^+$ RNA isolated from normal (N), tumor-derived cells (T), and cell lines U-2 OS and Rat 2 (used as positive controls). Radiolabeled probes for the A and B chains of PDGF and the α and ß subunits of the PDGF receptor were used to probe RNA blotted onto nitrocellulose (5 μg/lane). The blots were also probed with radiolabeled actin cDNA to indicate the integrity of the RNA and the relative loading between lanes.

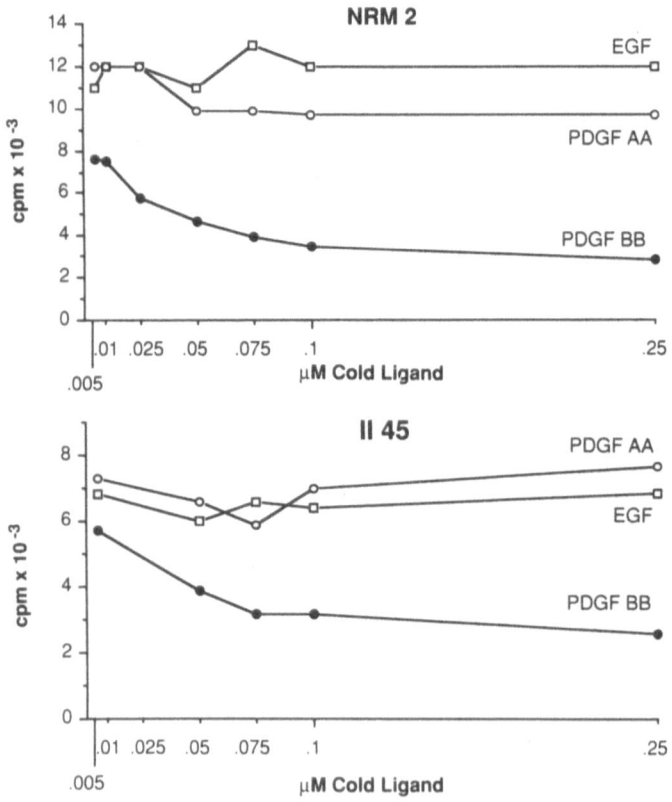

Figure 3. Competition assay for binding of PDGF BB. Normal (NRM2) and transformed (II-45) rat mesothelial cells were exposed to 0.5 nM ^{125}I-PDGF-BB in the presence of increasing concentrations of unlabeled PDGF-BB, PDGF-AA or EGF. Binding of ^{125}I-labeled ligand was determined after 3hr at 4°C.

displace the BB homodimer suggests that PDGF BB is being recognized by ß rather than α-type receptors. These results, therefore, are consistent with the RNA expression data indicating that only ß-type PDGF-receptors are expressed by the rat mesothelial cells.

OTHER GROWTH FACTORS EXPRESSED BY NORMAL AND TRANSFORMED MESOTHELIAL CELLS

In addition to PDGF, the expression of basic fibroblast growth factor (bFGF), transforming growth factor alpha (TGFα) and insulin like growth factor II (IGF-II) was compared in normal and transformed rat mesothelial cells. Transcripts for bFGF (6.3kb) were expressed by both normal and transformed rat mesothelial cells (Figure 4). In contrast, TGFα mRNA (4.5kb) was only expressed by the transformed rat mesothelial cells. IGF-II transcripts showed the opposite pattern of expression; the 3.5kb transcripts were expressed by the normal but not the transformed mesothelial cells. How the expression of TGFα by the mesothelioma cells, or expression of IGF-II by cultured normal cells, impacts cell growth remains to be determined. EGF is not mitogenic for normal rat mesothelial cells which only express low affinity EGF-receptors (Van der Meeren *et al.*, 1990), however, the effects of this growth factor on transformed rat mesothelial cells has not been investigated. TGFα, like EGF, binds to the EGF-receptor which mediates its effects on cell growth (Pike *et al.*, 1982; Massagué, 1983). The expression of EGF-receptors in transformed rat mesothelial cells has not been characterized, and studies to analyze the expression of EGF-receptors in these cells are currently in progress. IGF-II binds to the IGF-I receptor and this receptor appears to mediate the effects of this member of the insulin-like growth factor gene family. Interestingly, normal rat mesothelial cells express both insulin (Bermudez *et al.*, 1990) and IGF-I receptors (data not shown). Production of IGF-II by the normal rat mesothelial cells, therefore, may be related to establishment of

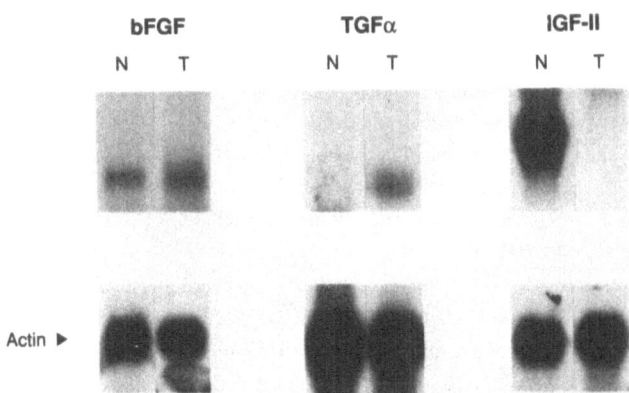

Figure 4. Northern blot analysis of poly (A)$^+$ RNA isolated from normal (N) and tumor-derived cells (T). Radiolabeled probes for bFGF, TGFα and IGF-II were used to probe RNA blotted onto nitrocellulose (5 μg/lane). The blots were also probed with radiolabeled actin cDNA to indicate the integrity of the RNA and the relative loading between lanes.

normal cells in tissue culture. Both TGFα and the insulin-like growth factor gene family represent good candidates for further study of the role of growth factors in mesothelial cell transformation.

SUMMARY

We have compared the expression of growth factors and their receptors in normal rat mesothelial cells and cells derived from asbestos-induced rat mesothelioma. PDGF receptors were expressed by both normal and transformed mesothelial cells, and cells expressing PDGF receptor mRNA specifically bound ^{125}I-PDGF, confirming the presence of PDGF receptors. However, no expression of either the A or B chain of PDGF was obtained in either cell type, in contrast to what has been observed in human mesothelioma (Gerwin et al., 1987; Versnel et al., 1988). Preliminary results indicate that normal mesothelial cells express receptors for insulin and IGF-I, and express IGF-II. In contrast, TGFα is expressed by transformed but not by normal rat mesothelial cells. Transcripts for bFGF, which are expressed by human mesothelial cells, were expressed by both normal and transformed cells. Taken together, these results suggest that TGFα and insulin-like growth factors, rather than PDGF, may potentially play a role in the genesis of mesothelioma in the rat.

REFERENCES

Bermudez,E., Everitt,J. and Walker,C. (1990) Expression of growth factor and growth factor receptor RNA in rat pleura mesothelial cells in culture. *Exp. Cell Res.* **190**:91-98.

Craighead,J.E. (1987a) Current pathogenetic concepts of diffuse malignant mesothelioma. *Human Pathol.* **18**:544-557.

Craighead,J.E., Akley,N.E., Gould,L.B. and Libbus,B.L. (1987b). Characteristics of tumors and tumor cells cultured from experimental asbestos-induced mesotheliomas in rats. *Am. J. Path.* **129**:448-462.

Demetri,G.D., Zenzie,B.W., Rheinwald,J.G. and Griffin,J.D. (1989) Expression of colony-stimulating factor genes by normal human mesothelial cells and human malignant mesothelioma cell lines *in vitro*. *Blood* **74**:940-946.

Gerwin,B.I., Lechner,J.F., Reddel,R.R., Roberts,A.B., Robbins,K.C., Gabrielson,E.W. and Harris,C.C. (1987) Comparison of production of transforming growth factor-ß and platelet-derived growth factor by normal human mesothelial cells and mesothelioma cell lines. *Cancer Res.* **47**:6180-6184.

Gerwin,B.I., Betsholtz,C., Linnainmaa,K., Pelin,K., Reddel,R., Gabrielson,E., Seddon,M., Greenwald,R., Harris,C.C. and Lechner,J. (1990) *In vitro* studies of human mesothelioma. In, "Biological Toxicology and Carcinogenesis of Respiratory Epithelium", eds. Thomasson,D.G. and Nettesheim,P., Hemisphere Publishing, Washington, DC.

Gronwald,R.G.K., Grant,F.J., Haldeman,B.A., Hart,C.E., O'Hara,P.J., Hagen,F.S., Ross,R., Bowen-Pope,D.F. and Murray,M.J. (1988) Cloning and expression of a cDNA coding for the human platelet-derived growth factor receptor: Evidence for more than one receptor class. *Proc. Natl. Acad. Sci. USA* **85**:3435-3439.

Hart,C.E., Forstrom,J.W., Kelly,J.D., Seifert,R.A., Smith,R.A., Ross,R., Murray,M.J. and Bowen-Pope,D.F. (1988) Two classes of PDGF receptor recognize different isoforms of PDGF. *Nature* **240**:1529-1534.

Jaurand,M.C., Bernaudin,J.F., Renier,A., Kaplan,H. and Bignon,J. (1981) Rat pleural mesothelial cells in culture. *In Vitro* **17**:98-106.

Jaurand,M.C. (1989) Particulate-state carcinogenesis: A survey of recent studies on the mechanisms of action of fibres. In, "Non-occupational exposure to Mineral Fibres", eds. Bignon,J., Peto,J. and Saracci,R., IARC Scientific Publications No. 90, Lyon, France. pp. 54-73.

Laveck,M.A., Somers,A.N.A., Moore,L.L., Gerwin,B.I. and Lechner,J.F. (1988) Dissimilar peptide growth factors can induce normal human mesothelial cell multiplication. *In Vitro Cell. Develop. Biol.* **24**:1077-1082.

Massagué,J. (1983) Epidermal growth factor-like transforming growth factor II. Interaction with epidermal growth factor receptors in human placenta membranes and A431 cells. *J. Biol. Chem.* **258**:13614-13620.

Matsui,T., Heidaran,M., Miki,T., Popescu,N., LaRochelle,W., Kraus,M., Pierce,J. and Aaronson,S. (1989) Isolation of a novel receptor cDNA establishes the existence of two PDGF receptor genes. *Science* **243**:800-803.

Mossman,B.T., Hansen,K., Marsh,J.P., Brew,M.E., Hill,S., Bergeron,M. and Petruska,J. (1989) Mechanisms of fibre-induced superoxide release from alveolar macrophages and induction of superoxide dismutase in the lungs of rats inhaling crocidolite. In, "Non-occupational Exposure to Mineral Fibres", eds. Bignon,J., Peto,J. and Saracci,R., IARC Scientific Publications No. 90, Lyon, France. pp. 81-92.

Mossman,B.T., Bignon,J., Corn,M., Seaton,A. and Gee,J.B.L. (1990) Asbestos: Scientific developments and implications for public policy. *Science* **247**:294-300.

Paterour,M.J., Bignon,J. and Jaurand,M.C. (1985) *In vitro* transformation of rat pleural mesothelial cells by chrysotile fibres and/or benzo[a]pyrene. *Carcinogenesis* **6**:523-529.

Pike,L.J., Marquardt,H., Todaro,G.J., Gallis,B., Casnellie,J.E., Bornstein,P. and Krebs,E.G. (1982) Transforming growth factor and epidermal growth factor stimulate the phosphorylation of a synthetic, tyrosine-containing peptide in a similar manner. *J. Biol. Chem.* **257**:14628-14631.

Seifert,R.A., Hart,C.E., Phillips,P.E., Forstrom,J.W., Ross,R., Murray,M.J. and Bowen-Pope,D.F. (1989) Two different subunits associated to create isoform-specific platelet-derived growth factor receptors. *J. Biol. Chem.* **264**:8771-8778.

Tanigawa,H., Onodera,H. and Maekawa,A. (1987) Spontaneous mesotheliomas in Fischer rats - A histological and electron microscopic study. *Tox. Path.* **15**:157-163.

Tubo,R.A. and Rheinwald,J.G. (1987) Normal human mesothelial cells and fibroblasts transfected with the Ejras oncogene become EGF-independent, but are not malignantly transformed. *Oncogene Res.* **1**:407-421.

Van Der Meeren,A., Levy,F., Renier,A., Katz,A. and Jaurand,M.C. (1990) Effect of epidermal growth factor on rat pleural mesothelial cell growth. *J. Cell. Physiol.*, in press.

Versnel,M A., Hagemeijer,A., Bouts,M.J., van der Kwast,T.H. and Hoogsteden,H.C. (1988). Expression of c-sis (PDGF B-chain) and PDGF A-chain genes in ten human malignant mesothelioma cell lines derived from primary and metastatic tumors. *Oncogene* **2**:601-605.

MECHANISMS OF PATHOGENESIS

3. Fibres and free radicals

THE SURFACE ACTIVITY OF MINERAL DUSTS AND THE PROCESS OF OXIDATIVE STRESS

H. Pezerat

Laboratoire de Réactivité de Surface et Structure
Université P. et M. Curie, CNRS, URA 1106
4 place Jussieu, 75252, Paris Cédex 05, France

INTRODUCTION

Any discussion of the mechanisms of carcinogenicity of a compound must include the possibility that that compound can be metabolised, with the formation of electrophilic species which are able to damage the genome either directly or indirectly. This approach to the mechanisms of carcinogenicity is valid for all carcinogenic compounds whether organic or inorganic, soluble or insoluble. When the compound is in the form of poorly soluble mineral particles then the electrophilic species must be examined principally in the context of the oxidising potential of surface sites. These sites are chemical entities at the boundary of the particles, capable of reacting with target molecules with only a reduced degree of freedom compared to molecules in solution.

Two modes of activity can be considered for such oxidising surface sites: either they react with biological molecules to liberate organic compounds able to act as carcinogens (indirect), or they react directly with the genome macromolecules to induce the first stage of mutation. In the first case, the activity can take place in the extracellular medium, but in the second case we must envisage the interactions occurring, probably during mitosis, after phagocytosis of the particle.

When considering the two situations it is essential to understand that the oxidising species are not present, in general, on the surface of environmental mineral particles. The situation is not very different from the one of benzo-[a]-pyrene, which *per se* is not a toxic molecule. It is only after several successive transformations that toxic metabolites appear. Such transformation steps are also essential to an understanding of the toxicity of mineral particles, but in this case they take place in an aqueous medium, the particles being in general hydrophilic and not lipophilic, with little contribution from enzyme systems.

THE FORMATION OF THE OXIDISING SPECIES

Various processes can be imagined which allow the appearance of oxidising species on the surface of particles of mineral dusts. In some cases it is possible to imagine an interaction between surface sites and H_2O_2 or O_2^- but only in phagocytic cells, or near them in the extracellular medium. For a process occurring inside the epithelial cells after phagocytosis, we would propose that the most interesting interactions are the ones between reducing surface sites and dissolved molecular oxygen present at constant concentration. We can summarise this interaction as follows:

$$S_R \xrightarrow[H_2O]{O_2} O^*$$

where S_R represents electron donor sites at the solid-liquid interface, and O^*, the set of all the activated species of oxygen, frequently bound to the particle surface. It is in this O^* group that we have the best candidates for triggering oxidative reactions in a biological medium. To understand the formation of electrophilic species, we must examine the ways in which reducing surface sites may occur in a biological medium.

The probability that mineral particles will have reducing surface sites in an aqueous medium depends on:

1) The presence of reducing cations such as Fe^{2+}, Cu^+, Sn^{2+}, or a metal compound (this last case is not discussed in the present paper), in the principal solid phase or in the contaminant phases.

2) The kinetics of the interactions between the particles and the aqueous medium, which conditions the emergence of these reducing sites at the interface.

3) The history of the particles taking into account the time elapsed since their grinding, any heat treatment, the previous contact of the particles with water, etc.

This last factor is related to the formation of an oxidised coating on the surface of the particles, a process whose speed depends on the relative humidity and the structure and composition of the solid phase. The thickness and the cohesion of this coating and also its kinetics of leaching or solubilisation in biological medium depend on the crystallochemical nature of the materials.

It is readily apparent that differentiating between various types of mineral particles is probably a task at least as difficult, or more so, than between species of organic molecule, particularly if we want to predict their probable surface oxidising activity in biological media. The paper by J. Fournier and others in this volume gives various details about the conditions under which the reducing sites and then the oxidising species appear on the surface of fibres or other inorganic materials.

THE NATURE OF THE OXIDISING SPECIES

Let us limit our investigations to the species derived from molecular oxygen which can appear in epithelial cells. The most frequently cited mechanism involves the three steps of oxygen reduction, with the appearance of hydroxyl radicals:

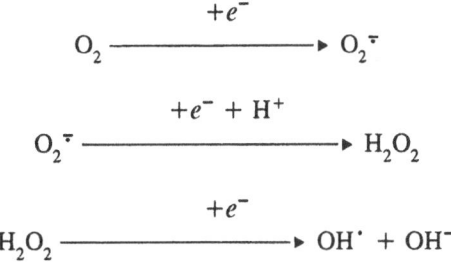

$$O_2 \xrightarrow{\quad +e^- \quad} O_2^{\cdot-}$$

$$O_2^{\cdot-} \xrightarrow{\quad +e^- + H^+ \quad} H_2O_2$$

$$H_2O_2 \xrightarrow{\quad +e^- \quad} OH^\cdot + OH^-$$

Only rarely, when the electrons are furnished by reducing sites, such as Fe(II), on the surface of the mineral particles, does such a mechanism seem to play a role. We are able to observe significant formation of $OH^\cdot$ only in the case of well dispersed reducing sites in an inert mineral matrix. One example is nemalite, $Mg(OH)_2$, containing 6 to 8% of Fe(II) in place of the Mg(II), after a light grinding. In this case Zalma (1988) has shown that the hydroxyl radical can represent about a quarter of the total oxidising species. The commonest case, observed with minerals containing Fe(II) and Fe(III), is the presence of a strong oxidising ability unrelated to the formation of hydroxyl radicals. Using, for example, formate as a target molecule for the oxidising species, we can obtain, with some mineral phases, appreciable formation of an adduct between the spin trap agent DMPO (5,5'-dimethyl-1-pyrroline-N-oxide) and the carboxylate radical anion ($CO_2^{\cdot-}$). We do not observe, in absence of formate, a comparable amount of the radical adduct with $OH^\cdot$. In this case, which is the most frequent observed, we consider that the probable oxidising species are the iron-oxo (ferryl) or the nickel-oxo species which are known for their strong oxidising power (Koppenol & Liebman, 1984).

Various symbols have been used to represent these strong oxidising species, such as Fe(IV)=O, Fe(V)=O, FeO_2^+, FeO^{2+}, in the case of the iron compounds, and various routes of formation have been proposed. The scheme below is the one hypothesised by Yamaguchi and others (1988).

$$M(II) \xrightarrow{\quad +O_2 \quad} \underset{\substack{\text{superoxo} \\ \text{species}}}{MO_2^{\cdot-}} \xrightarrow{\quad +e^- + H^+ \quad} M(III)\text{-}O_2^{(-)}\text{-}H$$

$$\underset{\text{oxo species}}{M(V)=O} \xleftarrow[\quad -H_2O \quad]{\quad +H^+ \quad}$$

This scheme allows us to note the appearance of two important and very different species, the oxo and the superoxo, which we can consider as representative of two groups of oxidising species that we symbolise as the A^* group and P^* group.

The A^* Group: We define this group of chemical oxidising species as those capable of triggering, with a high rate constant, a reaction involving the abstraction of an hydrogen atom from a target molecule such as formate, ethylic alcohol, acetone, etc.

$$A^* + RH \longrightarrow AH + R^\cdot$$

Hydroxyl radicals, the iron-oxo and the nickel-oxo species, for example, belong to this group. These species have a short life-time. They are not affected by SOD or catalase and they are able to hydroxylate various RH compounds, according to, for example, the following scheme:

$$Fe(V)=O \qquad Fe(III)$$
$$\diagup\!\!\!\!\diagdown$$
$$RH \qquad ROH$$

The hydroxylation of guanosine in DNA in the C-8 position by various Fe(II) containing materials is probably related to the presence of these A^* species. To detect these species we use a test in which mineral particles, such as asbestos fibres, are placed at $37°C$, without light, in an aerated buffer medium (pH=7.4), rich in phosphate (0.5M) to accelerate the leaching of the inorganic solid phases. We added a spin-trap agent (DMPO) and a target molecule for the oxidising species (Zalma *et al.* 1987, 1989; Zalma, 1988). With formate as a target molecule, the size of the e.s.r. signal from the adduct with the carboxylate species trapped by the DMPO, allows us to classify the materials according to their capacity to liberate A^* species.

$$A^* + HCO_2^- \longrightarrow CO_2^{\cdot-} + AH$$

$$CO_2^{\cdot-} + DMPO \longrightarrow (DMPO,CO_2^-)^\cdot$$

Being strongly oxidising, these A^* species, when they appear in the cells, can also participate in the metabolism of organic xenobiotics and contribute, by synergy, to the formation of carcinogenic metabolites. So, we can begin to imagine the meaning of the phenomenon of synergism at the level of molecular interactions between pollutants.

The P^* group: This group contains all the oxidising species, not belonging to the A^* group, which are however capable of triggering a lipid peroxidation with a high rate constant, for example some superoxide species:

$$M(III)O_2^{\cdot-} \xrightarrow{\qquad >\!\!=\!\!< \qquad} \underset{MO_2}{>\!\!-\!\!<^\cdot}$$

The members of the A^* subset are also able to trigger lipid peroxidation. Therefore, the union of the two disjoint subsets A^* and P^* represents all the oxidising species capable of triggering a lipid peroxidation. If we compare the P^* species with the A^* species, the first are weaker oxidising entities, much more susceptible to the classic biological defence systems such as SOD and catalase, but much easier to produce in a biological medium from a large variety of materials.

To produce the P^* species it is not necessary to have reducing sites such as Fe(II) on the solid. Fe(III), in some special coordination, is capable in the presence of oxygen, of triggering lipid peroxidation. Moreover, in the presence of H_2O_2 and O_2^-, and as a consequence of their probable role in phagocytic cells, these P^* species can arise bound to a great variety of cations (Mg^{2+}, Ti^{4+}, etc.). Their possible role in a carcinogenesis mechanism is indirect and implied by the presence of degradation products (aldehydes, peroxides, etc.) which result from lipid peroxidation (Ames, 1988). It is possible to refer to such a mechanism for tissues with weak defences i.e. those tissues without normal contact with xenobiotics.

We are putting the final touches to an experimental protocol which tests the capacity of solid phases to participate in a lipid peroxidation mechanism. However there are some final difficulties regarding the absorption of the degradation products on the solid phase.

The two groups A^* and P^* refer to two processes of oxidation. Others can be envisaged, but we consider that these are the two most important. The most interesting species in each group are not yet fully characterised, but with our tests of the oxidation of target molecules, we can define to what group the oxidising species belong and which appear with what mineral. As they appear in biological media, these oxidising species, including those bound to the surface of particles, are able to participate in certain reactions involving oxidative stress. They therefore appear to be good candidates to be considered as participants in various mechanisms of carcinogenesis.

MESOTHELIOMA AND LUNG CANCER INDUCED BY FIBRES: TWO DIFFERENT MECHANISMS

There is no reason to postulate the same mechanisms of carcinogenesis at the molecular level in two different tissues, which have different levels of defence, are at a different pH, etc.
On the contrary we have several reasons to postulate, for fibre-related cancers, two different mechanisms, one for mesothelioma, the other for lung cancer. These reasons, well known to the specialists, are the following:

- The difference in the latency times.
- The difference in the dose-effect relationship.
- The difference in the effects of synergy.
- The fact that all the fibrous mineral particles are carcinogenic in the mesothelial tissue if their retention time is sufficient, which is not the case in the lung.

The most spectacular example is that of erionite. This is very active in the mesothelial tissue and apparently not carcinogenic in the lungs (epidemiological and animal data). A second example is furnished in this volume by Hesterberg and others using ceramic fibres by inhalation in an animal experiment which resulted in a high level of mesothelioma and no lung cancer.

There is a high level of SOD and catalase in the lungs but these are probably at a lower level in the mesothelial tissue which is not normally in contact with xenobiotics. We can therefore postulate the necessity for very strong electrophilic species to induce a lung cancer but only a common and weak electrophilic species could induce mesothelioma. Before any propositions can be made for mechanisms at the molecular level, it is necessary to take into consideration and explain the "fibre effect".

THE FIBRE EFFECT

It is known that the clearance of small and isometric particles by the lymphatic fluid is much more efficient and rapid in the pleura or peritonea than in the pulmonary parenchyma. However the case of long and rigid fibres is peculiar, these particles stay trapped in the mesothelial tissue, unable to pass through the small diameter of the stomata (2-10 μm). Consequently the residence time necessary for oxidising activity to occur in the mesothelial tissue is - in general - insufficient for the isometric particles but sufficient for the fibres (probably several days or weeks). Thus the "fibre effect" appears essentially due to a clearance phenomenon. However, it is possible to induce a mesothelioma in animals without fibres and three examples are known, each involving a compound rich in iron. The first two examples used a crocidolite sample ground for a long time so that it contained no fibres. In the first case a large quantity was used (Kolev, 1982) and in the second case there was a weekly repetition of dosing (Kane, this volume). In the third example the authors used ferric saccharate with or without addition of nitrilotriacetic acid (NTA) and with 72 intraperitoneal injections (Okada et al., 1989). In all the three cases the authors have used compounds which are known to be active in lipid peroxidation. They compensated for the short retention time of their compounds either by giving a massive dose, or by repeated doses over several weeks.

If we now consider the human data, it appears that only with the principal asbestos varieties and the erionite fibres has there been established a relationship between mesothelioma and the exposure of various populations. Indeed, at this level, we must take into consideration the parameter of chemical durability (Spurny, this volume). If a fibrous morphology facilitates migration towards the pleura, it is also necessary for the fibres to resist the leaching effects of the biological medium.

Some authors (Hesterberg & Barrett, 1984; Hesterberg et al., 1986) have attempted to show a fibre effect in the transformation, in vitro, of chrysotile and glass fibres, using short fibres obtained by a long period of grinding. Other authors (Davis & Jones, 1988) have used in inhalation experiments in rats, long and short chrysotile fibres, the last being obtained by a sedimentation method. Both sets of experiments seem to give undeniably positive results with only the long fibres, but we have demonstrated (Zalma, 1988; Zalma et al., 1987; Zalma et al., 1989) that the two methods of obtaining the short fibres, applied to these particular materials, give particles where the surface properties are completely transformed. Heavily oxidised during the treatment, these short fibres are no longer able to generate the oxidising species A* like the long fibres. So, we consider that the fibre effect is a biological effect which, predominately, is concerned in the induction of mesothelioma.

THE MESOTHELIAL MECHANISM: a proposition

The fact that all fine, long and durable fibres, whatever their crystallo-chemical nature, induce mesotheliomas after an i.p. injection, implies the presence of relatively common oxidising species and does not imply specific cations in the solids. So we propose a mechanism involving some readily formed oxidising species, such as superoxide anions (O_2^-), bound to a surface cation and capable of triggering lipid peroxidation. The formation of these species is probably partially due to the secretion of O_2^- and H_2O_2 by phagocytic cells particularly for those materials with a weak redox surface activity (alumina, silica). Indeed in animal experiments phagocytic cells are recruited in great numbers to face the invasion of the tissue by the injection or implantation of fibres.

In humans, we must take into account the fact that few fibres are capable of reaching the mesothelial tissue. Consequently the probability that they will be able to initiate a mesothelioma will be greater with iron containing fibres, which are as capable of producing P^* species after reaction with O_2 as after reaction with O_2^- and H_2O_2. Indeed, it is relatively easy to obtain lipid peroxidation with the iron containing minerals. Several papers have emphasised the role of minerals such as crocidolite (Gulumian, 1987) or magnetite (Fontecave et al., 1987) in these reactions: the last compound is an important contaminant of chrysotile.

It is probable that the P^* species we are discussing can be represented by species such as Fe(III)-O_2^- or other more complex species. Many papers in recent years have been devoted to these types of studies (Yamaguchi et al., 1988; Minotti & Aust, 1987; Aust et al., 1985; Puntarulo & Cederbaum, 1988).

We now have a classic test for the peroxidation of linolenic acid detecting the malonaldehyde (MDA) produced by reaction with thiobarbituric acid (TBA). However some practical difficulties need to be resolved when these reactions are performed in the presence of solids with a strong capacity for absorption of some of the degradation products of the lipids and maybe also with a catalytic activity able to transform either the MDA or the MDA-TBA adduct. This test is not yet able to furnish a classification of the minerals according to their capacity to induce lipid peroxidation, in conditions similar to the ones present in the mesothelial tissue. It is probable that erionite presents some particularities, at a biological level, which enable it to amplify its carcinogenic potential. Indeed, erionite is capable of retaining the degradation products (aldehydes, peroxides, etc.) in its internal pores and of carrying these carcinogenic molecules inside the mesothelial cells after phagocytosis.

Using our tests, we have shown that the oxidised fibres of crocidolite, erionite and magnetite are not capable of producing the oxidising species A^*, but the three are capable of producing the P^* species leading to lipid peroxidation (Fournier et al, to be published). The hypothesis of a predominant role for the P^* species in the mechanism of mesothelioma induction is in agreement with the long latency period (the indirect mechanism with weak final carcinogens) and the fact that an individuals susceptibility, probably related to the level of their defences, seems to play a role in the distribution of this tumour in the exposed population.

LUNG CANCER MECHANISM: a proposition

We would like to propose a mechanism to explain the effects of fibres and inorganic particles which implicates the production of reactive oxidising species, capable of overwhelming the usual mechanisms of the body's defence. Therefore, we postulate a relationship between a mechanism for the induction of lung cancer and the appearance in the intracellular medium, i.e. at the moment of mitosis, of the A^* species, in general oxo species (Fe(V)=O, Ni(V)=O) bound to the surface of mineral particles. To characterise the capacity of mineral dusts to produce these species, we have used the test described above, which furnishes a classification of the mineral dusts according to their oxidising potency by a radical route. Some of the results concerning the compounds containing divalent iron are given by Fournier and others (this volume). This test is also appropriate for nickel compounds (Costa *et al.*, 1989a) with the proviso that the reducing sites are, in this case, on the surface of compounds which have a metallic character.

This classification of iron and nickel compounds is supported by epidemiological data. Among the iron compounds which appear as active in this oxidation test are crocidolite, amosite, nemalite and Québec chrysotile and various dusts from industrial and mine sites where epidemiological data show an excess of lung cancers. Examples of materials active in the production of the A* species and which are probably carcinogenic in the lungs (Costa *et al.*, 1989a, 1989b, 1989c, 1990; Zalma *et al.*, 1989) are ores from the Lorraine iron mines (which contain a phyllosilicate rich in Fe(II), berthierine), granite containing biotite, the ores from gold mines containing an iron rich chlorite and rockwool rich in Fe(II).

CONCLUSION

The mechanisms for the production of the tumours which are related to exposure to fibres, mesotheliomas and lung cancers, are probably due to two different oxidative stress processes. For mesothelioma, the first step is probably the formation of very simple oxidising species, capable, in the mesothelium, of triggering lipid peroxidation and hence the formation of oxidised organic species able to act as carcinogens. In the mechanism of production of lung cancer, taking into account the higher level of protection, we must imagine the formation of more reactive and more complex oxidising species such as the ferryl species. These oxidising entities are produced from reducing surface sites (Fe(II) sites) and molecular oxygen. This process does not imply a catalytic mechanism but rather the consumption of non-renewable sites. Oxidising species, which after phagocytosis appear in intracellular medium, are capable of attacking the DNA directly.

While fibre morphology may play a role in the migration of dusts towards the mesothelial tissues and above all in their retention inside these tissues, this morphology does not appear to be a necessary condition for the production of lung cancer induced by the mineral dusts.

REFERENCES

Ames, B.N. (1988) Measuring oxidative damage in humans: relation to cancer and ageing. "IARC Scient. Publ. No. 89," Lyon, pp. 407-416.

Aust, S.D., Morehouse, L.A. and Thomas, C.E. (1985) Role of metals in oxygen radical reactions, *J. Free Rad. Biol. Med.* 1:3-25.

Costa, D., Guignard, J. and Pezerat, H. (1989a) Production of free radicals by non-fibrous materials in a cell-free buffer medium. In, "Effects of mineral dusts on cells", (eds. Mossman,B.T. & Bégin,R.) NATO-ASI Series Vol. **H30**, , Springer-Verlag, pp. 189-196.

Costa, D., Guignard, J., Zalma, R. and Pezerat, H. (1989b) Production of free radicals arising from the surface activity of minerals and oxygen. Part I. Iron mines ores. *Toxic. Indust. Hlth*,5 **5**:1061-1078.

Costa, D., Guignard, J. and Pezerat, H. (1989c) Part II. Arsenides, sulfides and sulfoarsenides of iron, nickel and copper. *Toxic. Indust. Hlth*,5 **6**:1079-1097.

Costa, D., Guignard, J. and Pezerat, H. (1990) Oxidizing surface properties of divalent iron-rich phyllosilicates in relation to their toxicity by oxidative stress mechanism. In, "Health related effects of phyllosilicates" (ed. Bignon,J.). NATO-ASI, Vol. **G21**, Springer-Verlag, pp. 129-134.

Davis, J.M.G. and Jones, A.D. (1988) Comparisons of the pathogenicity of long and short fibres of chrysotile asbestos in rats. *Br. J. Exp. Path.* **69**:717-737.

Fontecave, M., Mansuy, D., Jaouen, M. and Pezerat, H. (1987) The stimulatory effects of asbestos on NADPH-dependent lipid peroxidation in rat liver microsomes. *Biochem.J.* **241**:561-565.

Gulumian, M., Kilroe-Smith, T.A. (1987) Crocidolite-induced lipid peroxidation in rat lung microsomes. Role of different ions. *Environ. Res.* **43**:247-253.

Hesterberg, T.W. and Barrett, J.C. (1984) Dependence of asbestos and mineral dust-induced transformation of mammalian cells in culture on fibre dimension. *Cancer Res.* **44**:2170-2180.

Hesterberg, T.W., Butterick, C.J., Oshimura, M., Brody, A.R. and Barrett, J.C. (1986) Role of phagocytosis in syrian hamster cell transformation and cytogenetic effect induced by asbestos and short and long glass fibres. *Cancer Res.* **46**:5795-5802.

Kolev, K. (1982) Experimentally induced mesothelioma in white rats in response to i.p. administration of amorphous crocidolite asbestos. Preliminary report. *Environ. Res.* **29**:123-133.

Koppenol, W.H. and Liebman, J.F. (1984) The oxidizing nature of the hydroxyl radical. A comparison with the ferryl ion (FeO^{2+}). *J. Phys. Chem.* **88**:99-101.

Minotti, G. and Aust, S.D. (1987) The requirement for iron(III) in the initiation of lipid peroxidation by iron(II) and H_2O_2. *J. Biol. Chem.* **262**:1098-1109.

Okada, S., Hamazaki, S., Toyokuni, S. and Midorikawa, O. Induction of mesothelioma by intraperitoneal injections of ferric saccharate in male Wistar rats. *Br. J. Cancer* **60**:708-711.

Puntarulo, S. and Cederbaum, A.I. (1988) Comparison of the ability of ferric complexes to catalize microsomal chemiluminescence, lipid peroxidation and hydroxyl radical generation. *Arch. Biochem. Biophys.* **264**:482-491.

Yamaguchi, K., Takahara, Y. and Fueno, T. (1988) *Ab initio* MO studies on structure and reactivity of superoxo transition-metal complexes. In, "The role of oxygen in chemistry and biochemistry" (eds. Ando & Moro-Oka,Y.), Elsevier, vol. **33**, pp. 263-268.

Zalma, R. (1988) Thesis: *Contribution à l'étude de la réactivité de surface des fibres minérales. Relations possibles avec leurs propriétés cancérogénes.* Université P. et M. Curie, Paris, France.

Zalma, R., Bonneau, L., Guignard, J., Pezerat, H. and Jaurand, M.C. (1987) Formation of oxy radicals by oxygen reduction arising from the surface activity of asbestos. *Can. J. Chem.* **65**:2338-2341.

Zalma, R., Guignard, J., Pezerat, H., Jaurand, M.C. (1989) Production of radicals arising from surface activity of fibrous minerals. In, "Effects of mineral dusts on cells" (eds. Mossman,B.T. & Bégin,R.), Springer-Verlag, vol. **H30** NATO-ASI series, pp. 257-264.

IRON MOBILIZATION FROM CROCIDOLITE RESULTS IN ENHANCED IRON-CATALYZED OXYGEN CONSUMPTION AND HYDROXYL RADICAL GENERATION IN THE PRESENCE OF CYSTEINE

Ann E. Aust and Loren G. Lund

Department of Chemistry and Biochemistry
Utah State University
Logan
UT 84322-0300, U.S.A.

ABSTRACT

The reactivity of iron on crocidolite with O_2 in the presence of cysteine was determined and compared with that of iron mobilized from crocidolite by citrate, EDTA, or nitrilotriacetate (NTA). Suspension of crocidolite in 50 mM NaCl, pH 7.5, in the absence of a reducing agent, did not result in a measurable amount of O_2 consumption or OH$^\bullet$ generation. Addition of cysteine in the absence of other iron chelators increased crocidolite-dependent O_2 consumption. Therefore, iron on crocidolite had limited redox activity in the presence of cysteine. However, when iron was mobilized from crocidolite it had increased redox activity. Citrate, EDTA, or NTA (1 mM) mobilized 75, 44, or 31 μM iron, respectively, in preincubations up to 76 h, and increased O_2 consumption upon addition of cysteine to 3.1, 2.1 or 1.1 nmol O_2 consumed/mg/min, respectively. Using ESR spectroscopy with the spin trap DMPO, the amount of OH$^\bullet$ formed in the presence of crocidolite was determined and found to be directly related to the amount of iron mobilized from crocidolite by EDTA. In the presence of a chelator, both O_2 consumption and OH$^\bullet$ generation depended only upon the presence of a component(s) mobilized from crocidolite by the chelator. Pretreatment of the crocidolite with desferrioxamine B (1 mM), an iron chelator, inhibited O_2 consumption in the presence or absence of citrate. The results of this study suggest that iron is responsible for the crocidolite-dependent reactions with O_2 and that mobilization of iron by citrate, NTA, or EDTA greatly enhances its reactivity with O_2. Furthermore, mobilization of iron from crocidolite by EDTA greatly enhanced the generation of OH$^\bullet$. Therefore, intracellular mobilization of

---

Abbreviations: NTA, nitrilotriacetate ; DMPO, 5,5-dimethyl-1-pyridine-N-oxide

iron from asbestos may increase reactions with O_2 leading to $OH^{\bullet}$ formation, DNA damage, and cancer.

INTRODUCTION

After many years of research, the molecular mechanism by which asbestos causes cancer remains unknown. The most carcinogenic forms of asbestos, the amphiboles, contain as much as 36% iron by weight (Berry, 1990; Churg, 1988). Asbestos reacts with O_2 (Lund & Aust, 1991) and catalyzes many of the same biochemical reactions as iron (Aust *et al.*, 1985), e.g., generation of oxygen radicals (Weitzman & Graceffa, 1984; Zalma *et al.*, 1987; Zalma, *et al.*, 1989), lipid peroxidation (Turver & Brown, 1987; Fontecave *et al.*, 1987) and DNA single strand break formation (Aust & Lund, 1990; Kasai & Nishimura, 1984). It has been proposed that oxygen radicals may be causally involved in some of the biological effects of asbestos, such as cellular cytotoxicity (Mossman *et al.*, 1986; Mossman *et al.*, 1989) and cellular DNA damage (Turver & Brown, 1987; Libbus *et al.*, 1989; Renier *et al.*, 1990). The highly reactive $OH^{\bullet}$ can be formed by the following reactions:-

$$\text{Red. Ag.} + \text{Fe(III)} \longrightarrow \text{Ox. Ag.} + \text{Fe(II)} \tag{1}$$

$$\text{Fe(II)} + O_2 \longrightarrow \text{Fe(III)} + O_2^{-} \tag{2}$$

$$2O_2^{-} + 2H^+ \longrightarrow O_2 + H_2O_2 \tag{3}$$

$$\text{Fe(II)} + H_2O_2 \longrightarrow \text{Fe(III)} + OH^- + OH^{\bullet} \tag{4}$$

In the cell the reducing agent (Red. Ag.) in reaction 1 might be ascorbate, cysteine, or glutathione, or under unusual circumstances may even be O_2^{-}. The rate at which these reactions occur is greatly affected by the chelator of iron (Miller *et al.*, 1990). Some chelators, such as desferrioxamine B or ferrozine, completely inhibit these reactions, while others, such as citrate, EDTA, or NTA, greatly stimulate the reactions of iron with O_2 and H_2O_2 resulting in the production of higher concentrations of $OH^{\bullet}$ (Miller *et al.*, 1990). Uncontrolled reactions of iron with O_2 are thought to occur rarely in living systems because iron is under strict control by transport (transferrin) and storage (ferritin) proteins (Aust *et al.*, 1985). Under unusual circumstances, such as iron overload diseases, citrate-iron chelates have been observed in the blood of patients (Grootveld *et al.*, 1989). It is thought that the unusual occurrence of low molecular weight chelates of iron may result in the toxicities and cancer that are observed in the iron overload patients (Grootveld *et al.*, 1989).

We are proposing that the phagocytosis of asbestos represents an "uncontrolled" entry of iron into the cell. Holmes and Morgan (1967) have previously reported that iron dissociates from asbestos fibers in treated animals. We have shown that mobilization of iron from asbestos *in vitro* requires a chelator (Lund & Aust, 1990). If mobilization of iron from asbestos *in vivo* is the result of chelation by low-molecular-weight compounds, such as citrate, then it may be possible for the iron to generate oxygen radicals through the process of redox cycling. Mobilization of iron by chelators may also allow iron to access many areas of the cell, such as the nucleus, thus increasing the risk of DNA damage.

The experiments reported here were designed to determine the reactivity of the iron on crocidolite with O_2 to produce OH· and to compare this with the reactivity of iron mobilized from crocidolite by low-molecular-weight chelators. Iron on crocidolite reacted with cysteine and O_2 and produced low levels of OH· when H_2O_2 was present. However, mobilization of iron from crocidolite by a chelator greatly increased cysteine-dependent O_2 consumption and OH· formation. These results suggest that mobilization of iron from asbestos *in vivo* by low-molecular-weight chelators may contribute in a significant way to oxidation of biomolecules, such as DNA.

METHODS AND MATERIALS

Asbestos and Reagents

Crocidolite asbestos was obtained from Dr. Richard Griesemer, NIEHS/NTP, Research Triangle Park, NC and contained 36% iron by weight (Campbell *et al.*, 1980).

Monohydrated L-cysteine hydrochloride and disodium nitrilotriacetate (NTA) were obtained from Sigma Chemical Co. (St. Louis, MO). Sodium citrate and the disodium salt of EDTA were obtained from Mallinckrodt, Inc. (Paris, KY). Deferoxamine mesylate USP (desferrioxamine B) was obtained from CIBA (Summit, NJ). Chelex 100 was obtained from Bio-Rad Laboratories (Richmond, CA). The ESR spin trap, 5,5-dimethyl-1-pyridine-N-oxide (DMPO) was obtained from Aldrich Chemical Co. (Milwaukee, WI) and was further purified by treatment with charcoal (Buettner & Oberley, 1978) and Chelex 100. Hydrogen peroxide was obtained from VWR Scientific (San Francisco, CA).

Preincubation of Crocidolite with Chelators

Crocidolite (1 mg/ml) was incubated in 50 mM NaCl, pH 7.5, with or without citrate, NTA, or EDTA (1 mM) on a wrist-action shaker in the dark. Duplicate samples were examined in order to compare the O_2 consumption or OH· generating capability of the iron on crocidolite with the iron mobilized by chelators. This was done by comparing the complete sample containing crocidolite with the supernatant which contained only the mobilized iron. The supernatant samples were prepared by removing the crocidolite by centrifugation. The amount of total iron mobilized into the supernatant was determined as previously described (Lund & Aust, 1990).

Crocidolite-Dependent O_2 Consumption

Preincubated samples prepared as described above were examined for O_2 consumption for up to 15 min upon addition of cysteine (1 mM).

All O_2 consumption experiments were conducted at 28°C using a Gilson oxygraph 5/6 equipped with a Clark-type electrode and a water-jacketed reaction chamber. In 50 mM NaCl (pH 7.5) at 28°C and at an elevation of 4500 feet, the dissolved O_2 concentration was 250 μM. The rates reported represent the results for a single experiment, however, each experiment was repeated at least twice. Initial rates are expressed as nmol O_2 consumed/mg crocidolite/min.

Crocidolite-Dependent Generation of OH·

Crocidolite (1 mg/ml) was suspended in 50 mM NaCl, pH 7.5, containing DMPO (50 mM) in the presence or absence of EDTA (1 mM), H_2O_2 (25μM), and/or cysteine

(0.1 mM). After a 5 min incubation, spectra were then recorded at 25°C using a Varian Century E-109 ESR spectrometer equipped with a TM_{110} cavity. Spectrometer settings were 3360 G magnetic field, 20 milliwatts microwave power, 9.5 GHz, 100 kHz modulation frequency, 1.0 G modulation amplitude, 0.5 sec time constant and 4 min scan time. The gain setting ranged from 3.2×10^3 to 8×10^3. The height of the second of the four peaks was corrected for gain and was recorded as signal intensity, expressed in cm.

For preincubation experiments, samples were prepared as described above. DMPO (50 mM), H_2O_2 (25 μM), and cysteine (0.1 mM) were added, in that order. After a 30 min incubation, the spectra were recorded.

RESULTS

Effect of Chelators and Total Iron Mobilization on Crocidolite-Dependent O_2 Consumption

No O_2 consumption was detected in the absence of cysteine for crocidolite alone or crocidolite preincubated with citrate, EDTA, or NTA (data not shown). The results in Table I show that the rate of O_2 consumption after addition of 1 mM cysteine in the absence of a chelator was very low for the complete sample containing crocidolite and not detectable for the supernatant with crocidolite removed. However, preincubation of crocidolite with citrate, EDTA, or NTA resulted in significant iron mobilization and increased the rates of O_2 consumption by a factor of 16, 11, or 6, respectively. The rates of O_2 consumption appeared to be directly proportional to the amount of iron that had been mobilized by the chelators with citrate > EDTA > NTA.

To determine whether O_2 consumption was due to a component exposed on the crocidolite or iron mobilized from crocidolite by chelators, the rate of O_2 consumption catalyzed by the complete sample was compared with the supernatant from which the

TABLE I   Effect of Chelators and Total Iron Mobilization on Crocidolite-Dependent O_2 Consumption

| Chelator (1 mM) Time with Crocidolite | Rate of O_2 Consumption (nmole/mg/min)[a] | | |
	Total Iron Mobilized (μM)	Sample with Crocidolite	Supernatant with Mobilized Component(s)
None 76 h	ND[b]	0.2[a]	ND
Citrate 76 h	75	3.1	3.5
EDTA 3 h	44	2.1	1.8
NTA 3 h	31	1.1	1.2

[a] All incubations contained 1 mM cysteine. Each experiment was repeated at least twice. The reported rates represent results from a single experiment. [b] ND, not detectable.

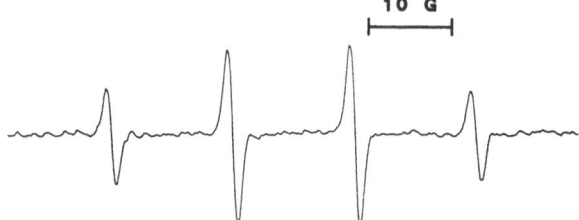

10 G

Figure 1. ESR Spectra of Crocidolite in the Presence of EDTA, H_2O_2, Cysteine and DMPO. Crocidolite (1 mg/ml) was suspended in 50 mM NaCl, pH 7.5, in the presence of DMPO (50 mM), EDTA (1 mM), and H_2O_2 (25 μM). Cysteine (0.1 mM) was added to initiate the reaction, and the spectra was recorded 20 min later, as described in **Materials and Methods**.

crocidolite had been removed, leaving only the mobilized component. The results in Table I show that the O_2 consumption was due to the mobilized component.

To determine more conclusively whether iron was responsible for the O_2 consumption being observed, iron chelators (1 mM desferrioxamine B or ferrozine) were added to the samples in the presence or absence of citrate. No O_2 consumption was observed in the presence of either of these iron chelators (data not shown).

TABLE II   Effect of Chelators on Crocidolite-Dependent Hydroxyl Radical Generation

Chelator (1 mM) Additions	Signal Intensity (cm)
None	ND[a]
+ H_2O_2	ND
+ H_2O_2 + cysteine	1.9
Citrate	ND
+ H_2O_2	ND
+ H_2O_2 + cysteine	1.0
EDTA	ND
+ H_2O_2	1.5
+ H_2O_2 + cysteine	3.0

[a] ND, not detectable.

Crocidolite (1 mg/ml) was suspended in 50 mM NaCl, pH 7.5, containing 50 mM DMPO. Reactions were initiated by addition of 25 μM H_2O_2 or 0.1 mM cysteine (when both H_2O_2 and cysteine were to be added). ESR spectra were taken 5 min after the addition of the final component.

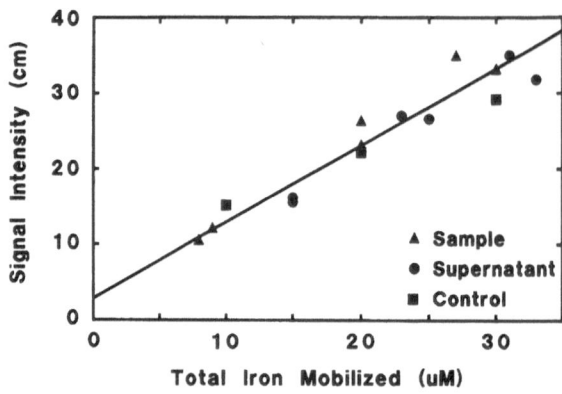

Figure 2. Effect of Iron Mobilization from Crocidolite by EDTA on Crocidolite-Dependent OH$^\cdot$ Formation. Crocidolite (1 mg/ml) was incubated with EDTA (1 mM) in 50 mM NaCl, pH 7.5. At regular time intervals two samples were removed. The first sample (▲) contained all of the incubation components. DMPO (50 mM) and H_2O_2 (25 μM) were added, followed by cysteine (0.1 mM) and the sample was incubated for 30 min before recording the ESR spectra. Crocidolite was removed from the second sample (●) by centrifugation. DMPO and H_2O_2 were added to the supernatant followed by cysteine, as described above. After a 30 min incubation, the ESR spectra was recorded. Controls (■) of freshly prepared solutions of $FeCl_3$ and EDTA (1 mM) were also examined at the indicated iron concentrations for the formation of OH$^\cdot$ after a 30 min incubation with DMPO, H_2O_2 and cysteine, as described above. The ESR spectra were recorded under the conditions defined in **Materials and Methods.**

Effect of Chelators on Crocidolite-Dependent OH$^\cdot$ Generation

It has been reported that asbestos can catalyze the formation of OH$^\cdot$ (Weitzman & Graceffa, 1984; Zalma *et al.*, 1987; Zalma *et al.*, 1989). To determine whether iron chelators enhance the generation of OH$^\cdot$, conditions were optimized to detect crocidolite-dependent formation of OH$^\cdot$ in the presence of the spin trap DMPO and the chelators, citrate or EDTA. Hydroxyl radical is a product of the reaction of Fe(II) with H_2O_2 (Equation 4, Fenton reaction). The effect of H_2O_2 and/or cysteine on crocidolite-dependent OH$^\cdot$ generation was investigated to determine whether a reducing agent, i.e. cysteine, was required for generation of Fe(II) and whether addition of H_2O_2 was necessary for detection of OH$^\cdot$ formation. Figure 1 shows the typical OH$^\cdot$-DMPO spin adduct formed when crocidolite, EDTA, H_2O_2, cysteine and DMPO were incubated together. Crocidolite in the presence of cysteine and DMPO, with or without a chelator, did not generate detectable levels of OH$^\cdot$ (data not shown). However, when both cysteine and H_2O_2 were added to crocidolite in the presence or absence of chelators, a low level of OH$^\cdot$ production was observed (Table II). Generation of OH$^\cdot$ was also observed with crocidolite, EDTA, and H_2O_2 in the absence of a reducing agent (Table II).

To determine whether the apparent enhancement of crocidolite-dependent OH$^\cdot$ generation in the presence of EDTA was due to iron mobilization, the generation of OH$^\cdot$

in samples containing crocidolite was compared with supernatants from which the crocidolite had been removed, leaving only mobilized iron. The results in Figure 2 show that the amount of $OH^{\cdot}$ produced was directly proportional to the amount of iron mobilized by EDTA. Since there was no difference between the amount of $OH^{\cdot}$ generated in the complete sample compared with the supernatant, it appeared that $OH^{\cdot}$ generation was totally due to iron mobilized from crocidolite. In addition, controls run with prepared solutions of Fe(III) chelated by EDTA in the absence of crocidolite generated almost identical amounts of $OH^{\cdot}$ as when equal amounts of iron were mobilized from crocidolite (see Figure 2).

DISCUSSION

Results with desferrioxamine B and ferrozine strongly suggest that iron was required for crocidolite-dependent O_2 consumption in the presence of cysteine. This is consistent with what we reported previously for crocidolite-dependent O_2 consumption in the presence of ascorbate (Lund & Aust, 1991). This iron-catalyzed O_2 consumption was increased by mobilization from crocidolite by citrate, EDTA, or NTA. Mobilization of iron from crocidolite by EDTA also greatly increased the generation of $OH^{\cdot}$. This may explain why we previously observed an increase in crocidolite-dependent single strand break formation in ϕX174 DNA after preincubation of crocidolite with 1 mM EDTA (Aust & Lund, 1990). Iron is mobilized from asbestos *in vivo* (Holmes & Morgan, 1967). If this is the result of chelation by low-molecular-weight compounds, this may lead to increased $OH^{\cdot}$ generation and DNA damage.

The enhanced reactivity of iron mobilized from crocidolite with O_2 in the presence of cysteine appeared to depend only upon the amount of iron mobilized and not upon the chelator being used. This is in contrast with what we observed when similar experiments were done using ascorbate instead of cysteine as the reducing agent (Lund & Aust, 1991). When ascorbate was used, the O_2 consumption appeared to be dependent upon the redox capability of the iron chelate that was formed, as well as, the amount of iron mobilized. These differences probably reflect an inherent difference in the way in which the reducing agents interact with the iron chelates.

Although EDTA is unlikely to be present within living cells, citrate is present and may serve to mobilize iron from asbestos *in vivo*. If this occurs, this may lead to reactions of iron with O_2, resulting in $OH^{\cdot}$ formation. NTA may be present in cells after environmental exposures. Humans are exposed to NTA as a result of its being used to chelate metals in a variety of applications (Anderson *et al.*, 1985). The NTA:iron chelate has been shown to be cytotoxic to cells in culture (Yamada *et al.*, 1987) and carcinogenic to laboratory animals (Ebina *et al.*, 1986). The ability of this chelate to catalyze the production of reactive oxygen species (Kawabata *et al.*, 1986) which can damage DNA (Umemura *et al.*, 1990) is thought to be the way in which it manifests its biological effects. Therefore, the presence of NTA or any xenobiotic which can enter cells and chelate iron to make it more redox active may increase the health risks associated with asbestos exposure.

Other laboratories have shown that asbestos can catalyze the formation of $OH^{\cdot}$ radicals using ESR spectroscopy with DMPO spin trapping. Using the iron chelator desferrioxamine B, Weitzman and Graceffa (1984) concluded that iron was responsible for catalyzing $OH^{\cdot}$ formation by crocidolite, chrysotile and amosite in the presence of H_2O_2. Zalma and coworkers (1987) showed that exposure of new asbestos surfaces after grinding of the fibers could increase the asbestos-catalyzed generation of OH$^•$ at high concentrations of asbestos (22 mg/ml) in the absence of H_2O_2. They concluded that grinding of the fibers exposed Fe(II) which could react with O_2 to produce OH$^•$. The studies that we have reported here demonstrate that iron can be mobilized from the asbestos and that this may increase its ability to catalyze the formation of OH$^•$.

In conclusion, the results presented here strongly suggest that iron was responsible for crocidolite-dependent O_2 consumption in the presence of cysteine. Mobilization of iron from crocidolite by citrate, EDTA, or NTA was shown to increase its reactivity with O_2. Moreover, mobilization of iron by EDTA greatly enhanced the formation of OH$^•$. If iron is mobilized from asbestos by low-molecular-weight chelators *in vivo*, then this may lead to increased reactivity of iron with O_2, producing reactive oxygen species and DNA damage.

ACKNOWLEDGEMENTS

The authors wish to thank Dr. Thomas Grover for his help in conducting the ESR studies. This work was supported by a Willard Eccles Foundation Grant and a Faculty Research Grant from Utah State University.

REFERENCES

Anderson,R.L., Bishop,W.E. and Campbell,R.L. (1985) A review of the environmental and mammalian toxicology of nitrilotriacetic acid. *CRC Critical Reviews in Toxicology* **15**:1-102.

Aust,A.E. and Lund,L.G. (1990) The role of iron in asbestos- induced damage to lipids and DNA. In, "Biological Oxidation Systems, Vol. II" eds. Reddy,C.C., Hamilton,G.A. and Madastha,K.M., Academic Press, San Diego,CA, pp. 597-605.

Aust,S.D., Morehouse,L.A., Thomas,C.E. (1985) Role of metals in oxygen radical reactions. *J. Free Rad. Biol. Med.* **1**:3-25.

Berry,G. (1990) Crocidolite and mesothelioma. *Med. J. Austral.* **152**:330-331.

Buettner,G.R. and Oberley,L.W. (1978) Considerations in the spin trapping of superoxide and hydroxyl radical in aqueous systems using 5,5-dimethyl-1-pyrroline-1-oxide. *Biochem. Biophys. Res. Comm.* **83**:69-74.

Campbell,W.J., Huggins,C.W. and Wylie,A.G. (1980) Chemical and physical characterization of amosite, chrysotile, crocidolite and nonfibrous tremolite. For "Oral Ingestion Studies", NIEHS, Bureau of Mines Report of Investigations 8452.

Churg,A. (1988) Chrysotile, tremolite and malignant mesothelioma in man. *Chest* **93**:621-628.

Ebina,Y., Okada,S., Hamazaki,S., Ogino,F., Li,J.-L., Mldorikowa,O. (1986) Nephrotoxicity and renal cell carcinoma after use of iron- and aluminum-nitrilotriacetate complexes in rats. *J. Natl. Cancer Inst.* **76**:107-113.

Fontecave,M., Mansuy,D., Jaouen,M. and Pezerat,H. (1987) The stimulatory effects of asbestos on NADPH-dependent lipid peroxidation in rat liver microsomes. *Biochem. J.* **241**:561-565.

Grootveld,M., Bell,J.D., Halliwell,B., Aruoma,O.I., Bomford,A. and Sadler,P.J. (1989) Non-transferrin bound iron in plasma or serum from patients with idiopathic hemochromatosis. *J. Biol. Chem.* **264**:4417-4422.

Holmes,A. and Morgan,A. (1967) Leaching of constituents of chrysotile asbestos *in vivo*. *Nature* **215**:441-442.

Kasai,H. and Nishimura,S. (1984) DNA damage induced by asbestos in the presence of hydrogen peroxide. *Gann* **75**:841-844.

Kawabata,T., Awai,M. and Kohno,M. (1986) Generation of active oxygen species by iron nitrilotriacetate (Fe-NTA). *Acta Med. Okayama* **40**:163-173.

Libbus,B.L., Illenye,S.A. and Craighead,J.E. (1989) Induction of DNA strand breaks in cultured rat embryo cells by crocidolite asbestos as assessed by nick translation. *Cancer Res.* **49**:5713-5718.

Lund,L.G. and Aust,A.E. (1990) Iron mobilization from asbestos by chelators and ascorbic acid. *Arch. Biochem. Biophys.* **278**:60-64.

Lund,L.G. and Aust,A.E. (1991) Mobilization of iron from crocidolite asbestos by certain chelators results in enhanced crocidolite-dependent oxygen consumption. *Arch. Biochem. Biophys.*, in press.

Miller,D.M., Buettner,G.R. and Aust,S.D. (1990) Transition metals as catalysts of "autoxidation" reactions. *Free Rad. Biol. Med.* **8**:95-108.

Mossman,B.T. and Marsh,J.P. (1989) Evidence supporting a role for active oxygen species in asbestos-induced toxicity and lung disease. *Environ. Hlth Perspect.* **81**:91-94.

Mossman,B.T., Marsh,J.P. and Shatos,M.A. (1986) Alteration of superoxide activity in tracheal epithelial cells by asbestos and inhibition of cytotoxicity by antioxidants. *Lab. Invest.* **54**:204-212.

Renier,A., Levy,F., Pilliere,F. and Jaurand,M.C. (1990) Unscheduled DNA synthesis in rat pleural mesothelial cells treated with mineral fibres. *Mutat. Res.* **241**:361-367.

Turver,C.J. and Brown,R.C. (1987) The role of catalytic iron in asbestos induced lipid peroxidation and DNA-strand breakage in C3H10T1/2 cells. *Br. J. Cancer* **56**:133-136.

Umemura,T., Sai,K., Takagi,A., Hasegawa,R. and Kurokawa,Y. (1990) Formation of 8-hydroxydeoxyguanosine (8-OH-dG) in rat kidney DNA after intraperitoneal administration of ferric nitrilotriacetate. *Carcinogenesis* **11**, 345-347.

Weitzman,S.A. and Graceffa,P. (1984) Asbestos catalyzes hydroxyl and superoxide radical generation from hydrogen peroxide. *Arch. Biochem. Biophys.* **228**:373-376.

Yamada,M., Okigaki,T. and Awai,M. (1987) Role of superoxide radicals in cytotoxic effects of Fe-NTA on cultured normal liver epithelial cells. *Cell Struct. Func.* **12**:407-420.

Zalma,R., Bonneau,L., Guignard,J., Pezerat,H. and Jaurand,M.C. (1987) Formation of oxy radicals by oxygen reduction arising from the surface activity of asbestos. *Can. J. Chem.* **65**:2338-2341.

Zalma,R., Guignard,J., Pezerat,H. and Jaurand,M.C. (1989) Production of radicals arising from surface activity of fibrous minerals. In, "Effects of Mineral Dusts on Cells", eds. Mossman,B.T. and Begin,R.O. NATO ASI Series, Vol. **H30**, Springer-Verlag, Berlin. pp. 257-264.

THE ROLE OF IRON IN THE REDOX SURFACE ACTIVITY OF FIBERS. RELATION TO CARCINOGENICITY

J. Fournier, J. Guignard, A. Nejjari, R. Zalma and H. Pezerat

Laboratoire de Réactivité de Surface et Structure
Université P.et M. Curie, CNRS
URA 1106, 4 place Jussieu
75252 Paris, Cedex 05
France

INTRODUCTION

A mechanistic study of the carcinogenicity induced by inorganic fibrous dusts should include research on the possible intervention of redox sites on the surface of these materials when immersed in an aerated aqueous medium (see the general paper by H. Pezerat, this volume). It can be shown that electrophilic species may be produced, by reaction of molecular oxygen with reducing surface sites (S_R), on inorganic particles according to the general scheme below:

$$S_R \xrightarrow[\text{H}_2\text{O}]{\text{O}_2} O^*$$

S_R being, for the purposes of the present discussion, divalent iron at the particle-medium interface and O^* defining the set of all the activated species of oxygen.

Among the numerous oxidizing species belonging to the O^* set, we can distinguish:

Ones capable of abstracting an H atom from a target molecule, such as formate, with a high rate constant. These represent the subset A^* and include for example $OH^\cdot$, iron-oxo ($Fe(V)=O$), nickel-oxo, etc. (Mimoun, 1986; Yamaguchi et al., 1988).

Others, not capable of abstracting an H atom from this type of molecule, but capable of triggering lipid peroxidation. These represent the subset P^* and include for example such species as $M(III)-O_2^{\tau}$ (Yamaguchi et al., 1988).

Since the species of the A* subset are also able to trigger lipid peroxidation, the group A* and P* represent in the O* set all the oxidizing species responsible for lipid peroxidation.

Pezerat, in his companion paper (this volume), hypothesises that the members of the subset A* are capable of inducing a lung cancer by direct attack on the genome. Members of the second subset P*, with a weaker oxidising ability, could be responsible for the carcinogenicity of fibers in the mesothelial tissue. These species could operate through an indirect mechanism involving the degradation products of lipids after peroxidation (hydroperoxides, aldehydes, etc.) (Halliwell and Gutteridge, 1985; Kappus, 1985; Ames, 1988).

In this paper we present, for the set of iron containing materials, the various parameters which play a role in the appearance of the A* species on the particle surface. We include examples of oxidising activity, particularly with reference to some samples of man-made mineral fibers (MMMF). We present also, the preliminary results of a general screening of inorganic materials in a test of lipid peroxidation, sensitive to the A* and P* species.

METHODS

Formate oxidation (Test A*):

The quantitation of the oxidizing power necessary for the appearance of an A* species is obtained through a reaction of H atom abstraction between A* and a target molecule, the formate anion (HCO_2^-), in a buffered medium (pH=7.4):

$$A^* + HCO_2^- \longrightarrow AH + CO_2^{\overline{\cdot}}$$

Production of the carboxylate radical anion ($CO_2^{\overline{\cdot}}$) is quantitatively evaluated by the intensity of the EPR signal corresponding to the radical adduct of $CO_2^{\overline{\cdot}}$ with a spin trap agent, DMPO (5-5'-dimethyl-1-pyrroline-N-oxide). The experimental procedure for the detection of these radical species has been described elsewhere, with different inorganic materials (Zalma et al., 1987; Zalma, 1988; Costa et al., 1989a).

The same scale of intensities for the EPR signals is used in all our studies. By reference to the epidemiological data concerning the excesses of lung cancer we consider samples giving a $(DMPO, CO_2^-)^{\cdot}$ signal with an intensity lower than 300 as inactive. Those samples giving a signal with an intensity higher than 500 are considered to be "active" and the samples giving a signal intensity between 300 and 500 are "slightly active".

Lipid peroxidation:

This test allows the identification of materials giving the A* species as well as the materials giving the P* species. Knowing the results of the first test, we can deduce - by difference - the materials able to generate P* oxidising species. We measure linolenic acid peroxidation after a literature method (Gavino et al., 1981). In this the formation of malonaldehyde (MDA) is detected by UV spectroscopy after reaction with thiobarbituric acid (TBA) to form a chromogen. To allow an efficient recovery of the MDA adsorbed on the solid phase the duration of the reaction with the TBA is increased to 30 min. and the final organic extraction of the chromogen is performed with butanol.

MATERIALS

Asbestos samples

These were several Canadian samples coming from different mines in Québec; UICC Canadian chrysotile B; UICC Rhodesian chrysotile A; Calidria chrysotile (USA) and three commercial chrysotile samples from Greece, South Africa and the USSR and UICC crocidolite.

Asbestos substitutes

Xonotlite (Promaxon), a synthetic calcium silicate. Several natural wollastonite samples: ($CaSiO_3$).

Man-Made Mineral Fibres (MMMF)

Various industrial samples of glass wool, slag/rock wool and ceramic fibers have been studied. They were gifts from J. Cherrie (IOM, Edinburgh) with several coming from plants included in the European epidemiological study (Simonato et al., 1987).

Other fibrous materials

Erionite (Oregon), a fibrous iron containing zeolite. Nemalite (Asbestos, Québec), a fibrous brucite, $Mg(OH)_2$, containing Fe(II).

Non fibrous materials

Siderite ($FeCO_3$) (La Mure, France). Biotite, a Fe(II) phyllosilicate (Razès, France).

PARAMETERS PLAYING A ROLE IN TEST A*

a) The absence of Fe(II)

The materials without Fe(II) are not active. This is the case with many compounds, for example xonotlite did not show any oxidizing activity in this test.

b) Traces of Fe(II)

Easily leached materials:

Some materials containing only traces of Fe(II) can be active. For example the wollastonite samples in which the Fe(II) content is low (about 0.5% expressed as % FeO) have an oxidising activity close to that obtained with the Canadian chrysotile samples (see below) after handgrinding the samples for 2 min immediately before the test.

This surprising result can be explained by the great fragility of this material in an aqueous medium rich in potassium phosphate. After some days the material is completely transformed into an amorphous phase by a process of leaching. After one hour, the duration of the test, the progression of the leaching front in the particles is sufficiently advanced so as to allow dissolved oxygen access to a great quantity of Fe(II).

It is uncertain whether the intracellular medium can allow so rapid a leaching phenomenon after phagocytosis of such particles.

TABLE I. Oxidizing surface activity of some chrysotile samples as a function of the presence of nemalite

Origin	Nemalite	Oxidizing Activity Intensity of the $(DMPO,CO_2)^{\mp}$ signal		%FeO
Quebec (several samples)	+	1600 - 1000	active	1.3 to 2.7
Canadian UICC(B)	+	1200	active	1.44
USSR (commercial)	traces	510	active	1.44
South Africa (commercial)	ND	90		0.48
Calidria	ND	80	inactive	1.15
Rhodesian UICC(A)	ND	60		0.98
Greece (commercial)	ND	30		1.14
Control	-	0-10		-

+ :Presence($>1\%$)
ND:not detected

Materials not easily leached:

In this case the materials are inactive because there is no access of molecular oxygen to Fe(II) sites.

Materials with Fe(II) containing contaminants:

Some inorganic phases which contain Fe(II) and are easily leached in the phosphate buffer can occur as contaminants of industrial materials. This is the case with nemalite, one of the principal contaminants in the chrysotile samples. Nemalite, in which up to 8% Mg^{2+} may be substituted by Fe^{2+}, is very active in our test.

In Table 1 we have summarized the results obtained with some of the chrysotile samples. For each sample the values of the Fe(II) content (expressed as %FeO) and the presence or absence of nemalite (detected by X-Ray Diffraction) are given.

We have observed that all the Québec chrysotile samples studied in our test which are active contain nemalite as an impurity. The others are either slightly active or not active and in these cases nemalite is present only in traces (less than 1%) or is not detected. In all these samples the presence of fibrous amphiboles has not been detected.

Part of the Fe(II) content of the chrysotiles is contained in another contaminant, magnetite. Since this material is inactive in this test (Costa *et al.*, 1989a) the oxidising activity of the chrysotile samples cannot be correlated with their total Fe(II) content. However this activity appears to be related to the content of "ferrous nemalite".

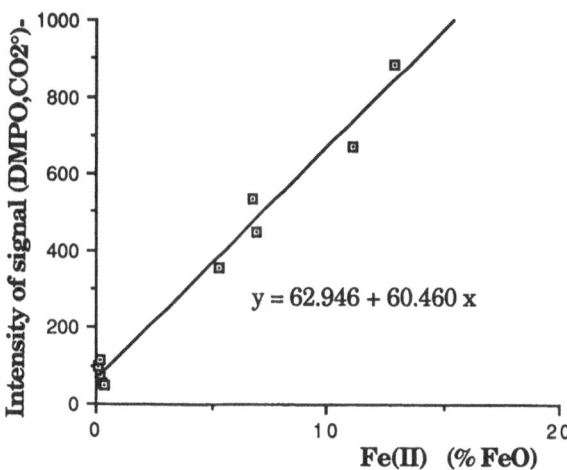

$$y = 62.946 + 60.460 \, x$$

Figure 1. Correlation between the oxidising surface activity and the total Fe(II) content

c) Materials containing a moderate quantitity of Fe(II): the case of the MMMF

We have tested - without any prior information about their origin - twelve samples collected by J. Cherrie. These materials, all manufactured several years earlier, were partially oxidized. To increase the probability that oxygen would be accessible to the Fe(II) in the solids we handground the samples for 5 min., in a mortar, immediately before the A* test.

These materials, which have the same vitreous structure and a closely related chemical composition, present the same kinetics of leaching. The access of oxygen to the Fe(II) depends on the leaching process and we obtained a good correlation (R = 0.99) between the total Fe(II) content and the oxidising activity as measured in the A* test (Fig. 1). This result is a good confirmation of the role of Fe(II) in the surface oxidising activity of the samples in a buffered aqueous medium.

The aim of this part of our studies was to examine if it was possible to distinguish, by our test, the samples coming from the factories over the periods when an excess of lung cancers had been observed in the European epidemiological study (Simonato *et al.*, 1987).

Thus, only after our test, did we learn the nature of the samples (rock-wool, glass fibers, etc.), their production year and the epidemiological results (Table 2). Six samples were too recent to have yielded epidemiological data, but two other groups are very interesting. The first were the glass fibers and glass wool, which had not produced an excess of lung cancer and were without significant activity in our test. The second group, three samples of rock/slag wool, were active in our test and they could be classified in the same order as the excess of lung cancer found in the epidemiological data.

d) Materials rich in Fe(II)

The characteristic feature of the samples of this group is a readily formed oxidised coating on the surface of the particles after ageing in air. This coating has a high Fe(III)

TABLE II. Oxidizing surface activity of some MMMF in relation to the epidemiological data on the exposed populations

Samples	Production Year	Fibre Material	Oxidizing Activity[*]	Fe(II) (%FeO)	Epidemiological Results
2	1949	rockwool	885	12	Distinct excess of lung cancers
9	1960	rockwool	670 Active	11.1	Lower excess of lung cancers
6	1974	rockwool	535	6.75	
1	1984	rockwool	450 Little	6.9	No possible data
10	1984	rockwool	350 Active	5.2	
3,5,11,12	1987	ceramic			
8	1968	glass	Inactive		No excess of lung cancers
4	1944	glass		<0.4%	
7	1949	glasswool			

[*] as TABLE I

content and a thickness which varies according to the material. In general it is not easy to leach this coating or to solubilise it. Therefore, the oxidizing activity of these samples will be related to: (i) the time elapsed between the production of dust particles with fresh surfaces, in general through grinding, and the test; (ii) the kinetics of the formation of this oxidized coating; and (iii) the kinetics of destruction of this coating in an aqueous medium.

To illustrate these factors, we have considered three examples: siderite, crocidolite and biotite with 62, 23 and 19% respectively of Fe(II) (expressed as FeO). All these materials are inactive without prior grinding, and one - the siderite, $(FeCO_3)$ - has no activity even after grinding because the rate of formation of the oxidised insoluble, unleachable coating is very rapid. On the other hand, the two silicates (crocidolite and biotite) are active after a short period of grinding (e.g. 2 to 5 min.), but since their structures are very different, the kinetics of formation of an insoluble coating is more rapid with biotite than with crocidolite. We observed the passage from the active category to the inactive one, after an ageing in air of less than one hour for the biotite and less than twenty-four hours for the crocidolite.

These results confirm that the surface oxidizing activity due to the A^* species is related to the accessibility of molecular oxygen to the Fe(II) on the surface and in the bulk of the particles and not related to the total Fe(II) content. In terms of toxicity by an oxidative stress mechanism these results emphasise, for some compounds, the great difference between the effects induced by inhaling freshly generated particles and those of particles after a long air ageing.

RESULTS OF THE TEST OF LIPID PEROXIDATION

This test of the capacity of a dust to trigger lipid peroxidation in a buffer medium, rich in potassium phosphate, seems more difficult to use than the A^* test because of possible artefacts. These may result from interactions between the MDA (or its reaction products with TBA) and the solid phase. Therefore, as a complement to this test it will be necessary, in the future, to develop a further test (e.g. of oxygen consumption) to check the validity of the classification of the materials by their lipid peroxidation activity.

With the exception of the compounds active in the formation of the A^* species it is difficult to know exactly the nature of the original sites and of the oxidizing P^* species. These may be of the type Fe(III)-O_2^- (Yamaguchi *et al.*, 1988) with a particular iron coordination. It is also possible that the activity measured in this test is related to certain catalytic surface sites and not only to sites destroyed by reaction.

In this initial study we have only looked to see if some materials, which are very active as carcinogens in the mesothelial tissue but not active in the A^* test, were active in the test for lipid peroxidation. We have observed that the erionite fibers from Oregon and the UICC crocidolite (without grinding) are not active in the A^* test but are active in the linolenic acid peroxidation test.

CONCLUSION

From our studies of iron containing minerals we have shown that materials able to generate an oxidizing surface activity of the A^* type are, apparently, present as pollutants at various occupational sites where an excess of lung cancer has been observed (Costa *et al.*, 1989a,b). But, we are conscious that we lack the animal data from which to assume this finding as a causal relationship. If we consider the range of iron containing materials which are capable of generating the P^* oxidizing species (a species complementary to the A^* species and able to trigger lipid peroxidation) it can be seen to include some fibers which are very carcinogenic in mesothelial tissues. Therefore, we feel that it is possible to advance an hypothesis of mesothelioma induction founded on a lipid peroxidation process.

ACKNOWLEDGEMENT:

Thanks are due to E.Copin who participated in the experiments.

REFERENCES

Ames,B.N. (1988) Measuring oxidative damage in humans: Relation to cancer and ageing. In,"IARC Scient. Publ. No. 89", Lyon, pp. 407-416.

Costa,D., Guignard,J., Zalma,R. and Pezerat,H. (1989a) Production of free radicals arising from the surface activity of minerals and oxygen. Part I. Iron mine ores. *Toxic. & Ind. Health, 5* **5**:1061-1078.

Costa,D., Guignard,J. and Pezerat,H. (1989b) Production of free radicals arising from the surface activity of minerals and oxygen. Part II. Arsenides, sulfides and sulfo-arsenides of iron, nickel and copper. *Toxic. & Ind. Health, 5* **6**:1079-1097.

Gavino,V.C., Miller,J.S., Ikharebha,S.O., Milo,G.E. and Cornwell,D.G. (1981) Effect of polyunsaturated fatty acids and antioxidants on lipid peroxidation in tissue cultures. *J. Lipid Res.* **22**:763-769.

Halliwell,B. and Gutteridge,J.M.C. (1985) "Free radicals in biology and Medicine", Clarendon Press, Oxford.

Kappus,H. (1985) Lipid peroxidation: Mechanisms, analysis, enzymology and biological relevance. In, "Oxidative Stress", (ed. Sies), Academic Press, New York, pp. 273-310.

Mimoun,H. (1986) Metal complexes in oxidation. In "Comprehensive coordination chemistry", (ed. Wilkinson), Pergamon Press, pp. 317-410.

Simonato,L., Fletcher,A.C., Cherrie,J., Andersen,A., Bertazzi,P., Charnay,N., Claude,J., Dodgson,J., Esteve,J., Frentzel-Beyme,R., Gardner,M.J., Jensen,O., Olsen,J., Teppo,L., Winkelmann,R., Westerholm,P., Winter,P.D., Zocchetti,C. and Saracci,R. (1987) The International Agency for research on cancer historical cohort study of MMMF production workers in seven European countries: extension of the follow up. *Ann. Occup. Hyg.* **31**:603-623.

Yamaguchi,J., Takahara,Y. and Fueno,Y. (1988) *Ab initio* MO studies on structure and reactivity of superoxo transition-metal complexes. In, "The role of oxygen in chemistry and biochemistry, Studies in organic chemistry", (eds. Ando W., Moro-Oka Y.), Elsevier, 33, pp. 263-268.

Zalma,R., Bonneau,L., Guignard,J., Pezerat,H. and Jaurand,M.C. (1987) Formation of oxy radicals by oxygen reduction arising from the surface activity of asbestos. *Can. J. Chem.* **65**:2338-2341.

Zalma,R. (1988) Thesis: *Contribution à l'étude de la réactivité de surface des fibres minérales. Relations possibles avec leurs propriétés cancérogènes*, Université P. et M. Curie, Paris, France.

CHEMICAL FUNCTIONALITIES AT THE BROKEN FIBRE SURFACE RELATABLE TO FREE RADICALS PRODUCTION

Bice Fubini, Vera Bolis, Elio Giamello and Marco Volante

Dipartimento di Chimica Inorganica, Chimica Fisica e
Chimica dei Materiali
Università di Torino
Via P.Giuria 9 - 10125, Torino
Italy

ABSTRACT

The surface chemistry of freshly created surfaces is discussed from the stand point of their potential toxicity. The most reactive sites are surface radicals (dangling bonds) which originate from homolytic cleavage of covalent bonds and transition metal ions in low oxidation states. Several types of mineral fibres will be discussed: SiC whiskers, asbestos, nemalite, glass and rockwool fibres.

INTRODUCTION

It is now widely accepted that the toxicity of mineral fibres is related to both morphological and chemical factors. The aspect ratio criteria proposed by Stanton and others (1977) govern carcinogenic activity with fibres of the same or a similar crystallo-chemical nature. With fibres of different chemical compositions, different active sites will be present at the surface and hence there will be different carcinogenic potentials for fibres of the same size (Bonneau *et al.*, 1986a,b).

In the present paper we will focus on a particular physico-chemical aspect of the surface chemistry of fibrous materials. We will consider the possible role played by the "broken fibre" and the chemical functionalities which exist at the fibre ends and edges, or on a freshly created surface. A fresh surface may be produced by mechanical rupture of the mineral fibre, by separation of a fibre bundle in liquid media, through leaching of some components and through chemical reactions involving transformation of the active sites and surface functionalities.

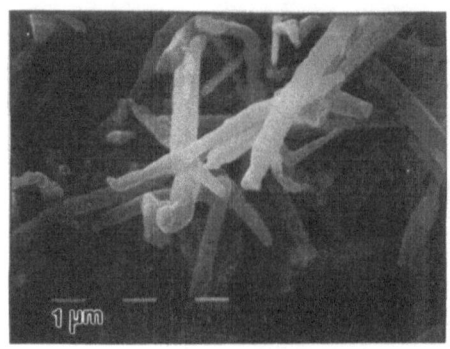
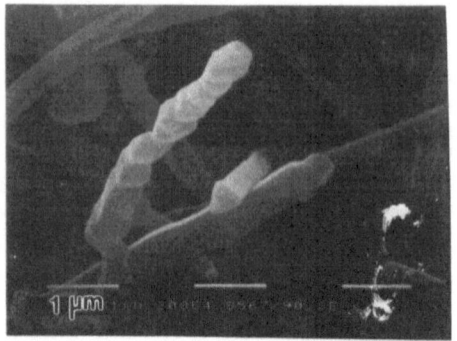
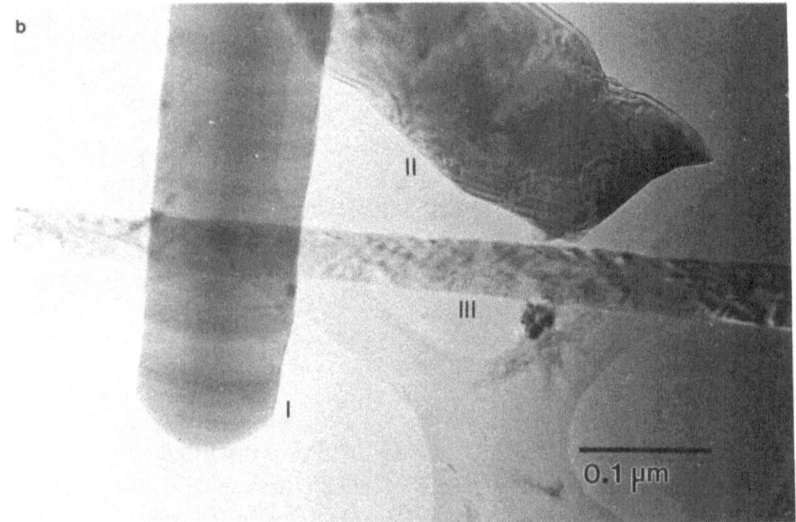
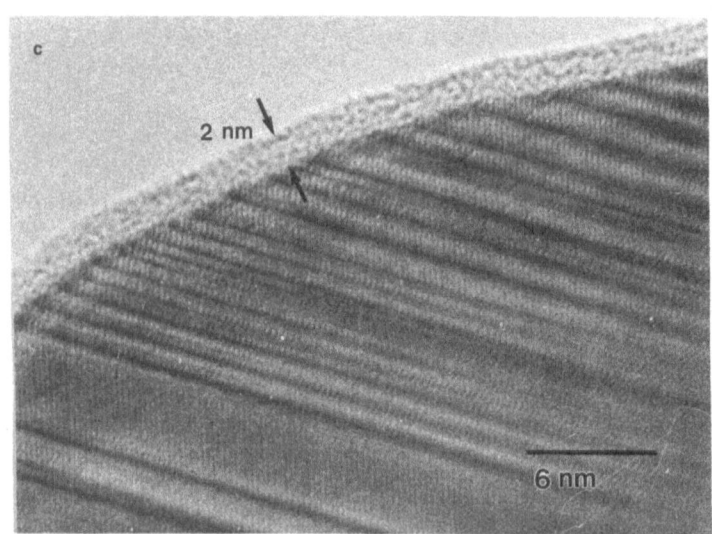

Figure 1. Electron micrographs of SiC whiskers (Tokamax).

a) Scanning electron micrograph: left side 10,400x; right side 20,800x.

b) Transmission electron micrograph at low magnification (148,000x).

c) The same sample as b) at high magnification (HRTEM image) (4,950,000x).

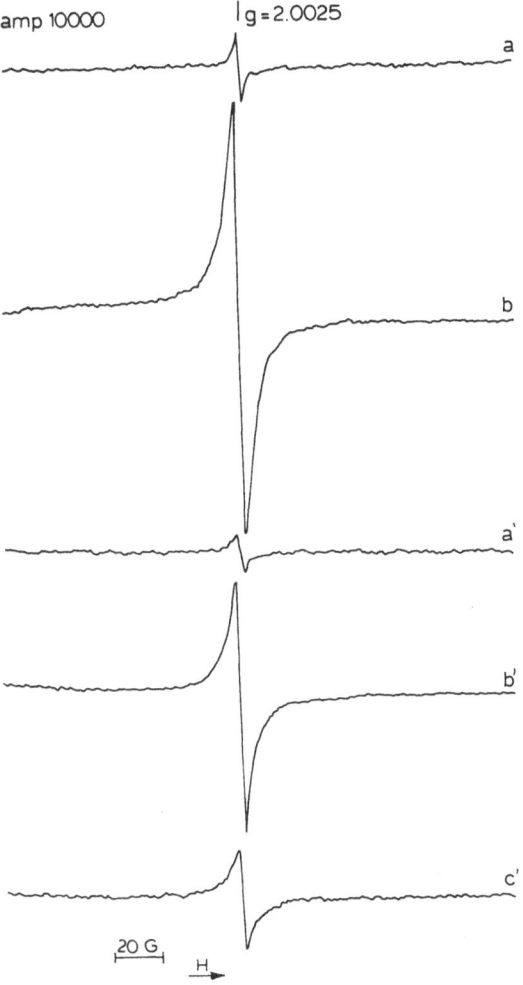

amp 10000 | g = 2.0025

a

b

a'

b'

c'

20 G

H →

Figure 2. EPR spectra at room temperature of two
commercial SiC whiskers Tokamax (a,b) and
Tateho (a',b',c'), before (a,a') and after grinding in
an inert atmosphere (b,b') and in air (c').

It is well known that the chemical composition of the surface of a given solid may,
for various reasons (e.g. electroneutrality, reaction with atmospheric components, etc.) be
different from that in the bulk. When a fibre is broken and a fresh surface is exposed,
unsaturated valencies are produced and the surface has to readjust itself (surface reconstruction) in order to attain a new equilibrium. How can this be related to the potential
toxicity of the solid?

It is difficult to assess whether inhaled particles are already in chemical equilibrium with the atmosphere. This depends on the route of inhalation and whether we are
discussing exposure in humans or administration to experimental animals. Also we must
know the kinetics of surface reconstruction which depend on the crystallochemical
composition of the particle and on the surrounding media. In many cases fibres may be
inhaled which have particularly reactive functionalities at their surface. Moreover the

fibre may fragment within the lung: often fibres found in the lung are much shorter than those inhaled.

The shape of a particle may in some cases impart particular chemical properties. It has been shown (Bolis *et al.*, 1989) that zinc oxide particles, with similar dimensions and exposed crystal faces, have different reactivities. The most reactive particles being those with sharp and well defined corners and edges. The critical aspect ratio for fibre carcinogenicity may also therefore be due to physico-chemical factors. For example: i) long thin fibres are more easily broken; ii) fibres with a high length to diameter ratio, may exhibit particular reactivity at the edges. To confirm these speculative hypotheses we have started a systematic study of the surface properties of solids of established or suspected toxicity, comparing, when possible, ground to unground fibres, isometric to fibrous morphologies.

The kind of active sites present at the broken surface largely depends on the ionic or covalent structure of the solid. Covalent solids, such as silica, can have reactive radicals at the surface (dangling bonds), which have been proposed as a possible cause of quartz pathogenicity (Fubini *et al.* 1989a,b). Ionic solids will have steps, kinks, edges and ion vacancies, created to balance surface charge (electroneutrality) and which are the most reactive sites. Moreover, particularly in the case of minerals, ions of similar dimension and charge may be substituted in the crystal lattice. For example Fe^{2+} may substitute for Mg^{2+} in chrysotile and act, when at the surface, as a particularly reactive site.

The present paper reports results obtained with a variety of fibrous materials whose structures vary from ionic to covalent:

a) Silicon carbide (SiC)

This is an essentially covalent solid which is used both in the form of whiskers and isometric particles. The toxicity to humans of the fibrous material has already been reported (Peters *et al.*, 1984; Funahashi *et al.*, 1984; Bégin *et al.*, 1989) and the whiskers have been found to be cytotoxic in *in vitro* experiments (Birchall *et al.*, 1988).

b) Chrysotile and crocidolite

Minerals composed of a covalent silicon-oxygen framework alternating with ionic layers of different cations and anions. The danger of these minerals to humans is well known.

c) Nemalite, $(Mg(OH)_2)$

An ionic hydroxide consisting of fibrous brucite which is often found associated with Canadian chrysotile. It is one of the most active solids known for the production of OH^- radicals in aqueous buffer (Zalma *et al.*, 1987; Zalma, 1988).

d) Man made mineral fibres (MMMF)

These are non-crystalline, partially ionic materials used as asbestos substitutes. Their toxicity is still under debate (IARC, 1988).

We have focussed our attention on the one hand on dangling bonds and their possible reactivity and on the other on the presence of transition metal ions in a low oxidation state. These may be active in the production of free radicals in solution and hence possibly involved in the formation of active oxygen species (AOS) *in vivo*. The role of iron in asbestos toxicity, is not limited to its ability to take part in the "Fenton reaction" in the presence of H_2O_2 as reported by Weitzmann and Graceffa (1984). It has now been established by various authors (Mossman et al., 1988; Pezerat et al., 1989; Zalma, 1988; Zalma et al., 1987) that the production of radicals occurs in the presence of Fe^{2+} at the solid-liquid interphase by direct reduction of molecular oxygen. Desferroxamine, a potent iron chelator, inhibits the toxicity of compounds that are believed to act via free radical production.

For Fe(II) to be active, either in a liquid medium or in vivo, it needs to be accessible to O_2. This means that it needs to be at or near the surface of a solid, but not readily oxidised by atmospheric agents before inhalation. We have tried to follow, using EPR, the appearance at the surface and the oxidation of Fe^{2+} when a fibre is ground. As only Fe^{3+} can be measured under our experimental conditions, the presence of Fe^{2+} is inferred by its oxidation to Fe^{3+}. We have paid particular attention to the detection of Fe^{2+} and Fe^{3+} ions in close proximity to each other. These are hard to detect by simple EPR analysis of minerals, where the spectra are made up of the signals of all paramagnetic ions in the solid. The appearance of a broad band with particular spectral features has been

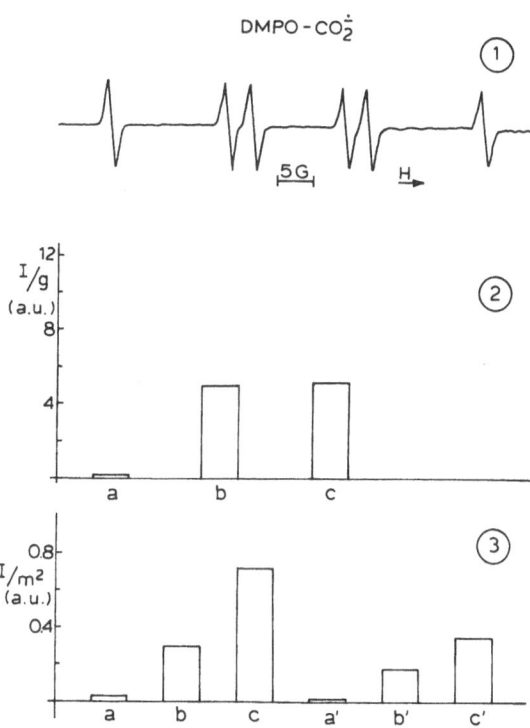

Figure 3. 1) Typical EPR spectrum of the DMPO-CO$_2$⁻ adduct in buffer.
2) Relative intensity of DMPO-CO$_2$⁻ spectra: a) blank, b) ground UICC crocidolite, c) ground isometric SiC
3) Relative intensity of DMPO-CO$_2$⁻ spectra per unit surface area: whiskers (Tokamax, a,b,c) and isometric particles (a',b',c'). Before grinding a,a', after 1 min grinding b,b' and 2 min grinding c,c'.

419

assigned in some cases (Friebele *et al.*, 1971) to ferrous and ferric ions interacting magnetically and physically close to each other in the solid. We can imagine that a solid containing iron in these two oxidation states, when broken, may have, at least for a while, the two ions at the surface. This is of particular interest in view of the role played by these two ions acting together in the initiation of lipid peroxidation (Minotti & Aust, 1987).

MATERIALS AND METHODS

Materials.

Isometric ultrafine SiC powders (α-SiC) were from Lonza. SiC whiskers were Tokawhiskers from Tokai Carbon limited and Tateho from Tateho Chemical.
Asbestos: UICC Canadian Chrysotile and crocidolite; Nemalite was from Asbestos (Quebec) kindly provided by Dr. Pezerat.
Commercial glasswool and some rockwool samples from the European historical cohort (Simonato *et al.* 1987) kindly provided by Dr. Pezerat.
Johns-Manville code 104 borosilicate glasswool was kindly provided by Dr. Pott.
DMPO (5,5'-dimethyl-1-pyrroline-N-oxide) was from Sigma.

Techniques.

The crystallinity of the whiskers was checked by X-ray diffractometry using a Guinier camera (Cu K_α radiation). For morphological studies scanning electron microscopy (SEM) was performed with a Philips instrument. High resolution transmission electron microscopy (HRTEM) with performed with a Jeol 200CX instrument equipped with a top entry stage (Bolis *et al.*, 1989a).

EPR spectra were recorded at room temperature on a Varian E 109 using appropriate cells. All spectra in the figures are shown with magnetic fields increasing from left to right: a corresponding scale for the g values is also given, together with the amplification used in the recording. In the case of iron-rich samples a remarkable temperature dependence of the spectral features was found and spectra recorded at 77 K are also reported.

The production of CO_2^- in phosphate buffer was monitored by the EPR spectrum of the DMPO-CO_2 adduct (Zalma *et al.*, 1987; Zalma, 1988).

The reactivity towards water vapour was measured by adsorption calorimetry following a technique previously described (Fubini, 1988).

RESULTS AND DISCUSSION

a) Silicon carbide

Silicon carbide is manufactured as whiskers and as isometric particles. Pneumoconiosis has been found in workers exposed to SiC (Furnahashi *et al.*, 1984; Peters *et al.*, 1984) and the toxicity of SiC whiskers has recently been suspected following *in vitro* experiments with cells (Birchall *et al.* 1988). Any danger from isometric particles is believed to be confined to fibrous by-products which have been often found during industrial production (Bye *et al.*, 1984).

The surface chemistry of Silicon Carbide has been investigated in only a few cases, the investigations being mainly concerned with the presence of silica at the surface

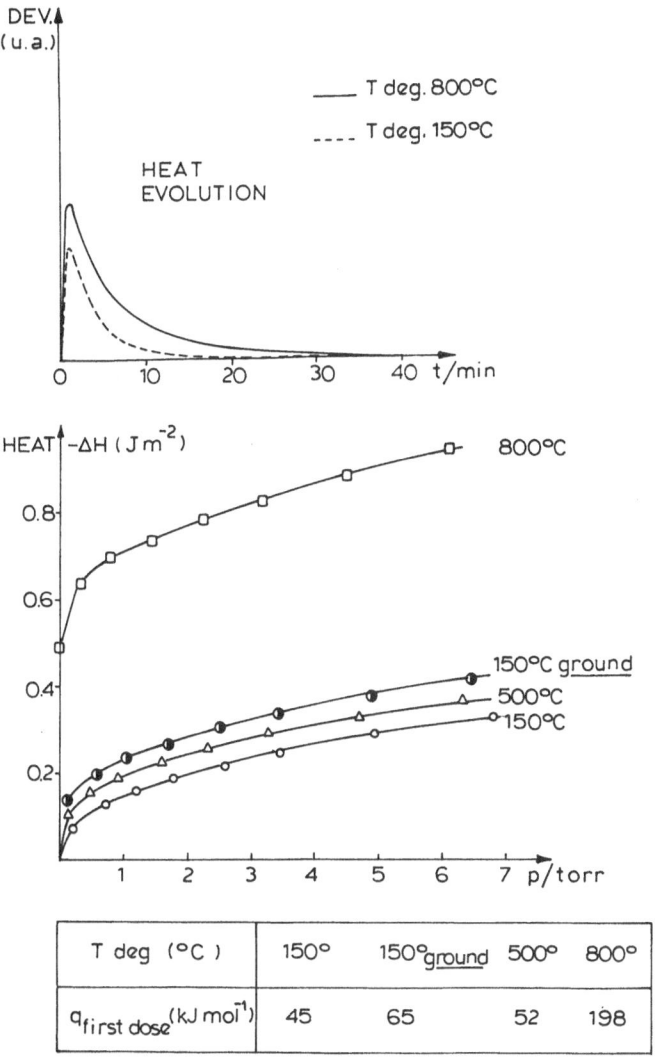

T deg (°C)	150°	150° ground	500°	800°
$q_{first\ dose}$ (kJ mol^{-1})	45	65	52	198

Figure 4. The heat released during progressive reaction of water with SiC whiskers (Tateho). Total heat is given as a function of the equilibrium pressure of water vapour, for samples degassed at different temperatures and, in one case (T_{deg} 150°C), before and after grinding. Curves indicate the typical heat evolution (kinetics) at two different degassing temperatures (150° and 800°C).

or in the outermost layers of the solid (Perrault *et al.*, 1987). Some silica is always found by ESCA analysis, but the extent of it is limited and SiC is also present on the surface. We have started a comparative study of both isometric particles and whiskers using different techniques to find which reactive functionalities are present at the surface and what is the effect of grinding or heating. We report here preliminary results from commercial whiskers (Tateho and Tokamax) which have a similar method of preparation and size.

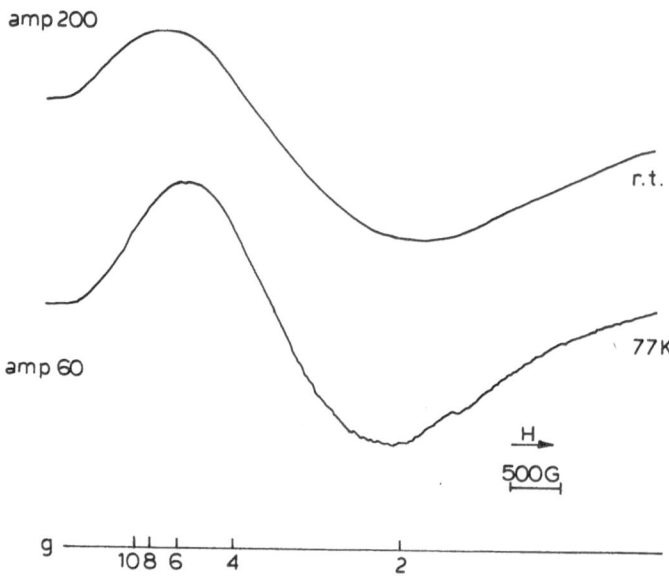

Figure 5. EPR spectrum of UICC crocidolite recorded at (a) room temperature, (b) 77 K.

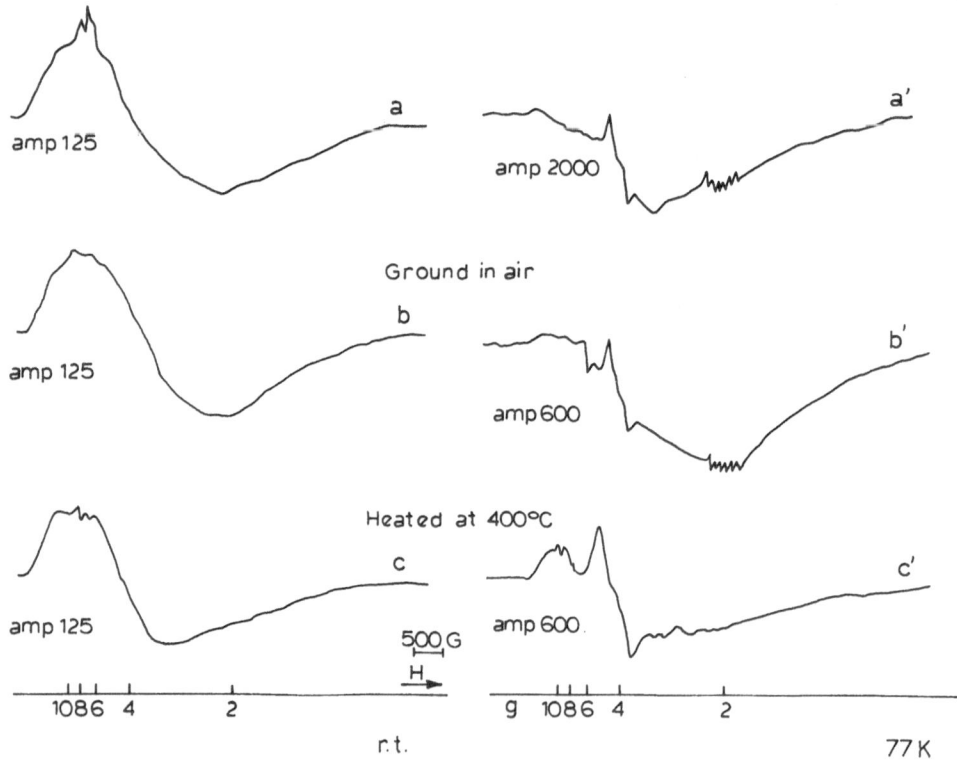

Figure 6. EPR spectra of UICC Canadian chrysotile at room temperature and 77 K. (a,a') without treatment; (b,b') after grinding in air; (c,c') as b,b' after subsequently heating at 400 °C.

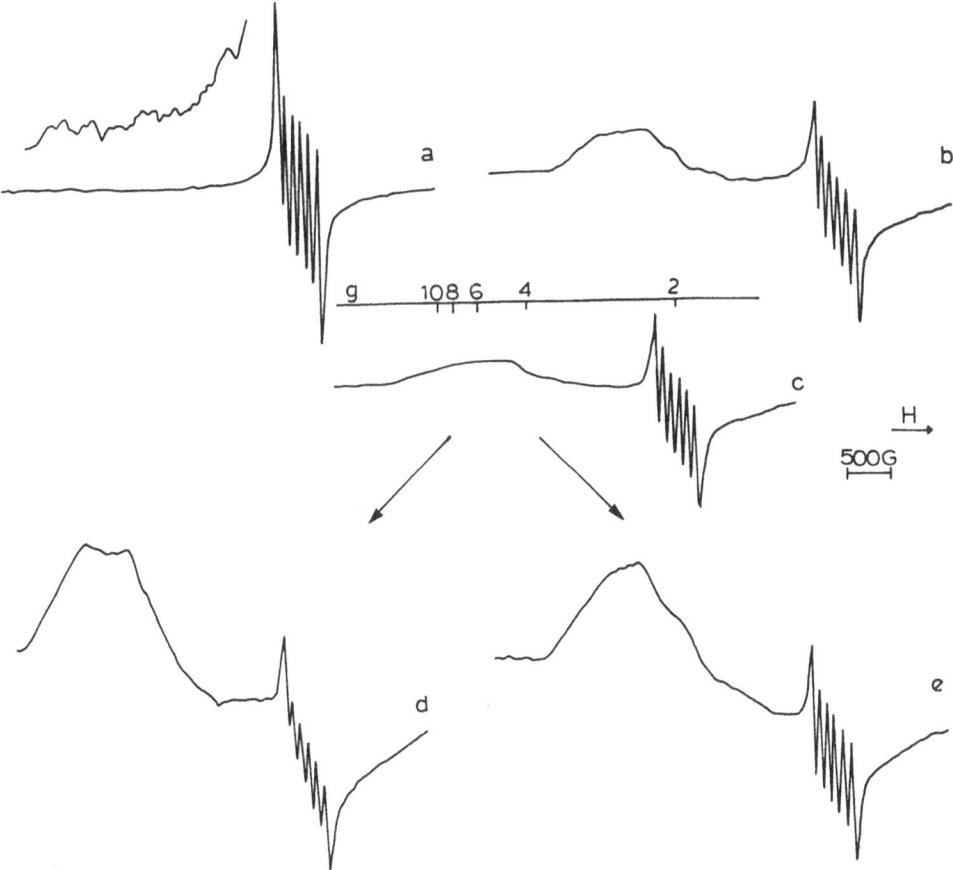

Figure 7. EPR spectra of Nemalite: a) original fibres; b) fibres ground in air; c) fibres ground in an inert atmosphere and exposed to pure oxygen; d) fibres as in c) plus water; e) fibres as in c) after standing in air for six months.

Morphological aspects

The micromorphology of SiC whiskers (Tokamax) obtained by trough scanning electron microscopy (SEM) is shown in Figure 1a. The material is made up of whiskers whose length is $> 5 \mu$m. Their shape is irregular: some are well developed along one preferential axis, while others have irregular edges and seem to be made from small irregular fragments. Both circular and polygonal cross-sections are observed with diameters in the range 0.04-0.15 μm (Figure 1b). Whisker I is more regularly grown than whisker II and III: it seems to have originated through the progressive (stepwise) growth of thin platelets normal to the main axis. At least two kinds of whiskers exist, differing in size, shape and number of structural defects which are probably due to different growth patterns. More detailed information is revealed in the high resolution electron micrograph (HRTEM) (Figure 1c), which clearly shows that the microcrystal is entirely covered by a layer ($\approx$ 2 nm thick) of an unknown amorphous material (possibly SiC or SiO_2). This cannot be seen by SEM. Several stacking faults are evident in the TEM and HRTEM images.

Production of radicals at the surface and in the bulk by grinding

Figure 2 shows the EPR spectra of the two types of whiskers before (a and a') and after grinding in an inert atmosphere (b and b') or air (c'). The peak centred at g = 2.0025, i.e. very close to the free electron value, which is small in spectra a) and a'), increases remarkably after grinding in an inert atmosphere in both samples. This signal is assigned to dangling bonds which originate from homolytic rupture of silicon-carbon bonds. The mechanical origin of these sites is confirmed by the increase of the signal intensity with grinding time. If grinding is performed in air, atmospheric components assist surface reconstruction, and the signal intensity is less.

The spectra are very different from that found with quartz (Fubini *et al.*, 1989a) but closely resemble signals observed from materials such as silicon nitride and partially hydrogenated carbon and silicon employed in optoelectronic devices. Radicals which arise from the mechanical disruption of the fibres and are possibly related to a partial removal of the outermost amorphous layer, may impart a specific reactivity to the solid, playing some role in its toxicity.

Production of free radicals in solution

The reactivity of SiC has been studied in a buffered solution with the DMPO spin trap, following the method used for mineral fibres (Zalma *et al.*, 1987; Zalma, 1988). This is based on the formation of DMPO-CO_2^- adducts following abstraction of hydrogen from the formate ion either by OH$\cdot$ or by surface sites.

Section 1 of Figure 3 shows the spectrum of the DMPO-CO_2^- adduct; section 2 compares the signal intensity of freshly ground isometric SiC particles with that of a blank experiment without any solid and that found with ground UICC crocidolite, a very active solid in this kind of test (Pezerat *et al.*, 1989). SiC is clearly also active. In section 3 of Figure 3, the activity of whiskers is compared with that of isometric particles by reporting the intensity of the signal per square meter of exposed surface. In both cases no activity is found before grinding (a, a') while activity appears after grinding (b, b') and increases with grinding time (c, c'). The activity per unit area of whiskers is higher than that of isometric particles. No DMPO-OH$\cdot$ adducts have been found in the absence of formate. This could mean that a direct reaction between surface sites on the SiC and formate ions takes place:

$$\equiv Si\cdot \; + \; HCOO^- \longrightarrow \; \equiv Si\text{-}H \; + \; COO^-$$
$$\text{or}$$
$$\equiv C\cdot \; + \; HCOO^- \longrightarrow \; \equiv C\text{-}H \; + \; COO^-$$

More work is needed to confirm that dangling bonds are involved in the reaction. It is possible that activity towards DMPO is due to impurities such as iron ions. This cannot be totally disregarded but, since fewer impurities are present in whiskers than in isometric samples, it is unlikely that the whole activity should be ascribed to them.

Reactivity towards water:

The reactivity of SiC towards water has been tested by adsorption calorimetry. Water may either be hydrogen-bonded or dissociated at the surface: these processes are characterised by their different kinetics and reaction enthalpy (Fubini, 1988). Our results show that both grinding and heating increase the uptake and energy of reaction with water vapour (Figure 4). The heat of reaction measured for subsequent doses of adsorbed water

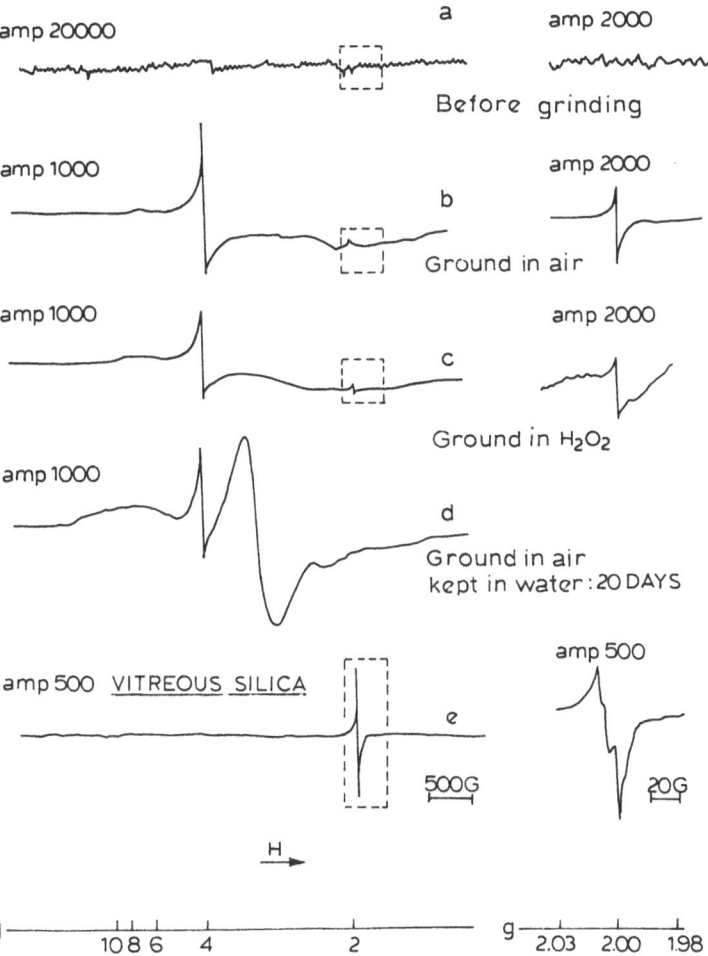

Figure 8. EPR spectra with the part of the spectrum around g = 2 is shown at higher amplification on the right.

Upper part: commercial glass fibre, (a) before and (b) after grinding, (c) after grinding in H_2O_2, (d) as (b) after standing in water for 20 days.
Lower part: pure silica glass.

as a function of equilibrium pressure, and the energy per mole of water adsorbed by the first addition of water indicate that upon grinding the mechanically activated surface reacts more readily with water, i.e. new more reactive sites have been formed. Heating at 800°C considerably modifies the surface reactivity. The interaction energy and the kinetics are consistent with the dissociation of water.

Far from being inert, SiC exhibits a remarkable surface reactivity particularly after mechanical or thermal activation. The presence of dangling bonds, the formation of free radicals in solution and the ability to dissociate water may well be related to radical reactions *in vivo* and possible strong interactions with cell membranes. We would stress that the surface behaviour of SiC is quite different from that of amorphous silica (Bolis *et al.*, 1989). Amorphous silica becomes progressively less reactive towards water upon heating and no surface radicals nor any reactivity that could be ascribed to radicals has ever been found with it (Fubini *et al.*, in press; Costa, 1989). The presence of areas of silica on the SiC surface is therefore not related to its peculiar reactivity.

b) **Asbestos**

Typical EPR spectra of fibres whose toxicity is well known are shown in Figures 5 & 6, recorded both at room temperature and 77 K. UICC crocidolite gives an extremely intense spectrum made up of one large asymmetric band only. This type of spectrum is typical of paramagnetic ions in high concentration in a solid with magnetic interactions occurring between neighbouring ions. The large band is due mainly to iron ions both in the bulk and at the surface. Because of the extreme intensity of the signal, any changes through reactions at the surface are not visible with this technique. The spectrum of UICC Canadian crysotile is quite different. At low temperature the signal can be resolved into two components superimposed on a broad band, one centred at $g = 4.3$ is due to isolated Fe^{3+} in a low symmetry crystal field while a sextet of lines near $g = 2$ is due to Mn^{2+}. The features of the spectrum are similar to those reported by Sharrock (1982).

Both Fe^{2+} and Mn^{2+} can substitute for Mg^{2+} in the brucite layer of chrysotile. The Fe^{3+} present may originate either by oxidation of Fe^{2+}, or be present in the crystal lattice as Fe^{3+} substituting for Si in the silicate tetrahedra (Stroink *et al.*, 1980). If the spectrum of the same sample is recorded at room temperature, its intensity is remarkably higher and is seen as a broad band superimposed on the various lines which are hardly visible. The reason for this is the interactions between neighbouring Fe^{3+} ions, whose broad spectral line increases in intensity with temperature (Friebele *et al.*, 1971). Some magnetite impurities, often found with crysotile, may also account for this signal. In agreement with Loveridge and Parke (1971) we assign the lines at $g = 4.3$ to Fe^{3+} isolated in a tetrahedral or octahedral distorted position (C_{2v} symmetry) and the broad signal to oxidised iron at the surface. Grinding (Fig. 6 b,b') causes a marked increase in the intensity of the broad signal. This fact confirms the idea that Fe^{2+} is brought to the surface when crysotile fibres are broken where it undergoes oxidation by the surrounding medium. In these conditions, in aqueous solution, Fe^{2+} at the interface activates molecular oxygen with the consequent release of radicals into the solution (Pezerat *et al.*, 1989).

It can be hypothesised that radical production reactions involving Fe^{2+} may occur *in vivo*. Prolonged grinding of a sample, in contrast to mild comminution, drastically reduces radical production in solution (Zalma, 1988) and the toxicity found in animal experimentation (Stanton & Wrench, 1972). In such a case we may infer that all the Fe(II) has been oxidised to Fe(III). Minerals containing only ferric ions are not carcinogenic nor able to reduce oxygen to form active radicals (Costa *et al.*, 1989).

It has been reported (Fisher *et al.*, 1988) that upon heating at 400°C crysotile changed its surface properties and was less active in several biological tests. Thermal analysis, not reported here, shows that at this temperature pure crysotile does not undergo substantial weight loss or phase modifications, whereas pure nemalite ($Mg(OH)_2$) is decomposed to MgO (Fisher *et al.*, 1988). This does not mean that surface reactions may not occur on crysotile which are not detectable by XRD or microscopical techniques. A progressive slight weight loss between 150° and 400°C suggests that the external brucite layer, which constitutes most of the exposed surface of crysotile, is converted to an oxide like state, following the reaction:

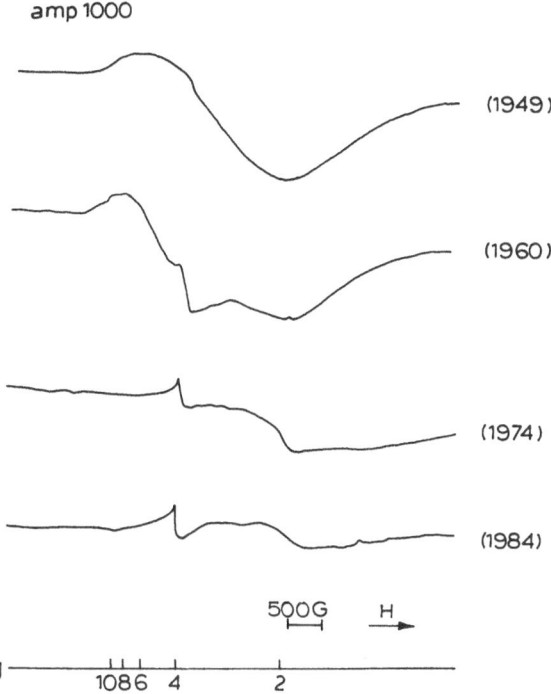

amp 1000

(1949)

(1960)

(1974)

(1984)

500G

H

g

1086 4   2

Figure 9. EPR spectra of "Historical rockwool fibres".

$$2OH^- \xrightarrow{\text{heating}} H_2O + O^{2-}$$

This behaviour is typical of oxides, covered by an external hydroxide layer. The influence of such a process on the location, oxidation state and co-ordination of iron and manganese ions originally at the surface is not straightforward and work is in progress to understand this. The spectra in Fig. 6 c,c', recorded with crysotile heated in the same way as Fisher and others (1988), show changes with respect to the original spectrum (b,b') which indicate variations in the state of the iron and manganese ions.

c) Nemalite

Nemalite is a fibrous brucite ($Mg(OH)_2$) which often occurs in association with chrysotile, where it has been thought to grow by epitaxial replacement of parachrysotile (Whittaker & Middleton, 1979). It is found particularly diffused in Canadian crysotile. It is a typical ionic solid made up from Mg^{2+} and OH^- ions, in octahedral layers where the main unit is $Mg(OH)_6^{4-}$ in which Mg occupies the centre of the octahedra. Because of the similar dimensions and charge the Mg^{2+} are often replaced by Mn^{2+} and Fe^{2+}, the replacement being easier in this ionic solid than in the complex structure of crysotile.

The EPR spectrum of a bundle of fibres (Figure. 7a) examined over a large scan range shows only a sextet of lines centred at $g = 2$ which is typical of Mn^{2+} in an octahedral field. No other paramagnetic centre is detectable under these conditions, which rules out the presence of any Fe^{3+} in the original mineral. Upon grinding in air (Figure

427

7b) a new resonance appears as a broad band with a maximum around g = 6 which is assigned to ferric ions originating by oxidation of ferrous ions exposed at the surface during the grinding process. In order to follow the kinetics and mechanism of this oxidation, a sample was ground in an inert atmosphere and subsequently exposed to oxygen (fig. 7c). Oxidation of ferrous ions occurred to a lesser extent than in a moist atmosphere. The spectrum of this sample developed to resemble that from the sample ground in air, although more intense, either after contact with liquid water (1 day, Fig. 7 d) or after standing for several months in air (Fig. 7e). The sample kept in water, developed an even more intense spectrum (not shown here) after standing in air, as a consequence of further oxidation of ferrous ions.

Three main facts can be drawn from the above experiments:

i) grinding brings to the surface ferrous ions that are not all immediately oxidised to ferric;

ii) extensive oxidation requires the presence of both water and oxygen;

iii) a slow oxidation of ferrous ions takes place over a period of months. The ionic character of the solid, which allows substitution and mobility of ions and charges, may account for this behaviour.

From the standpoint of potential toxicity it is clear that inhaled nemalite fibres will provide a **long lasting source of ferrous ions**, which at the solid-liquid interface may reduce oxygen with the subsequent formation of reactive radicals (Pezerat *et al.*, 1989). *In vivo* the Fe^{2+} becomes progressively bioavailable and is not readily oxidised to Fe^{3+}. This may be one of the reasons for the higher toxicity of Canadian Crysotile by comparison with the mineral from other sources. It is noteworthy that Nemalite was classified by Zalma (1988) as very active in the production of OH· radicals in aqueous solutions.

d) MMMF - Man made mineral fibres

This kind of fibre differs from the other fibrous minerals discussed so far in that they are glassy and the fibrous shape originates not through a crystalline structure but from the way of preparation (e.g. extrusion, etc.). These fibres fracture in a different way and, in contrast to asbestos, the product of grinding progressively creates isometric particles, not very different from quartz or vitreous silica dusts. Because of the glassy structure they have a relatively high solubility so that impurities in subsurface layers, or even in the bulk, may appear at the surface after standing in a biological medium.

We have examined the EPR spectra (Fig. 8 upper part) of a commercial glass fibre which was reported to produce a certain amount of hydroxy radical in a buffered solution (Costa, 1989) and compared it with the spectrum of pure vitreous silica (Fig. 8, lower part). The integral fibre does not show any paramagnetic signal even at a high amplification. Upon grinding a typical EPR spectrum develops, made up of two well defined sharp signals one at g = 4.3 and the other at g = 2.0 (enlarged on the right of the figure). The component at g = 4.3 is similar to the one reported for chrysotile asbestos (Fig. 6) which may be assigned to isolated Fe^{3+} in distorted tetrahedral or octahedral configurations (C_{2v} symmetry, Loveridge & Parke, 1971). The sharp signal at g = 2.0 is assigned to surface radicals produced by the fracture of the silica framework. While the Fe^{3+} signal is virtually un-modified by chemical and thermal treatments, that at g = 2.0 is sensitive to the presence of O_2, changed by heating and modified in intensity by contact with various gases. It is noteworthy that the spectrum of glass fibre is very different from the typical spectrum of silica glass (lower part of Fig. 8). This is composed of several signals assigned to SiO_2 and O_2^- similar to that found from quartz (Fubini *et al.*, in press). It should be noted that the intensity of the signal at g = 2, arising from fracture of the silica framework, is much lower in intensity than the spectrum of silica glass.

The appearance of Fe^{3+} indicates that Fe^{2+} was present in the interior of the fibre and exposed and oxidised by grinding in air. This process occurs rapidly and no variation in the intensity of the g = 4.3 signal occurs on standing in air or in aqueous solutions. Moreover the signal also appears when grinding is carried out in a glove box which is virtually free of air. A possible interpretation of these two facts is that the contemporary presence of ions in low oxidation states and dangling bonds may give rise through a complex pathway to an overall reaction between these species. E.g.:

$$SiO^{\cdot} + Fe^{2+} \longrightarrow SiO^- + Fe^{3+}$$

The short borosilicate glass fibres, JM 104, showed a very similar behaviour, with consequently few dangling bonds and gave the typical spectrum of isolated Fe^{3+}. When a glass fibre sample was ground in a solution of hydrogen peroxide, or left standing in water at 37°C Fig. 8 c,d) a broad band developed in the spectrum which was similar to, though less intense than, the spectrum of crocidolite (Fig. 5). This band is produced by physically close ions in magnetic interaction which have been brought to the surface and slowly oxidised. We regard such surface ions as possible contributors to fibre toxicity. The question then arises whether similar mechanisms operating *in vivo* can be envisaged to explain the excess of cancer found in workers exposed to artificial fibres containing substantial amounts of iron.

We have examined four rockwool samples produced in succeeding years from the "historical cohort" of the European epidemiological study (Simonato *et al.*, 1987). Their EPR spectra after grinding are shown in Figure 9. (We recall that the SMR [of workers in the rockwool industry: CHECK] decreased with the year of production.) The spectra are clearly due to the superposition of two separate signals: a symmetrical broad one centred around g = 3 due to physically close iron ions magnetically interacting, as in crocidolite, and a sharp component at g = 4.3, typical of isolated poorly reactive Fe^{3+}. It is noteworthy that the decrease in the intensity of the broad signal parallels the decrease in SMR. Similarly, the ability to generate free radicals in aqueous buffered solution decreases (J. Guignard, H. Pezerat, personal communications). During the early technological phase of the industrial production not only was the ambient pollution greater but so was the iron content of the product at or near the fibre surface. Both factors may account for the excesses of cancer.

CONCLUSIONS

The fracture of a solid, particularly a fibre, produces very active surface sites which may potentially be implicated in fibre toxicity and carcinogenicity.

Two main kinds of reactive sites have been detected: i) "dangling bonds" originating from cleavage of covalent bonds; ii) transition metal ions in low oxidation states, mainly iron, which, when brought to the surface during fragmentation, or dissolution, reduce molecular oxygen with the possible formation of active oxygen species.

For each solid these sites have different properties: dangling bonds are different in SiC, SiO$_2$ or glass fibres. Iron is present in different crystal configurations in the various solidsand this determines its bioavailability in the long term. Combined surface chemistry investigations and biological tests would be required to identify the particular chemical functionalities which are related to toxicity.

ACKNOWLEDGMENTS

A part of this research was supported by the Italian CNR (Consiglio Nazionale delle Ricerche) Progetto Finalizzato "Materiali speciali per tecnologie avanzate".

The authors are indebted to Dr. H. Pezerat and his group for encouragement, gifts of samples and fruitful discussions. Particular thanks are due to Dr. R. Zalma and Dr. D. Costa for information on the DMPO test and to Dr. J. Guignard for discussions on MMMF. The Anorganische-Chemisches Institut of Zürich University (CH) is acknowledged for giving access to the electron microscopy equipment.

REFERENCES

Bégin,R., Dufresne,A., Cantin,A., Massé,S., Sebastien,P., Durand,P. and Perrault,G. (1989) Carborundum Pneumoconiosis. In, "Effects of Mineral Dusts on Cells" Mossman,B.T. and Begin,R. eds. NATO ASI H-30 Springer Verlag pp. 81-84.

Birchall,J.D., Stanley,D.R., Mockford,M.J., Pigott,G.H. and Pinto,P.J. (1988) Toxicity of silicon carbide whiskers, *J. Mat. Sci. Lett.* **7**:350-352.

Bolis,V., Fubini,B., Giamello,E. and Reller,A. (1989a) Effect of form on the surface reactivity of differently prepared zinc oxides, *J. Chem. Soc. Faraday Trans. I* **85**:855-867.

Bolis,V., Marchese,L., Coluccia,S. and Fubini,B. (1989b) Surface properties of a pyrogenic low surface area silica: a microcalorimetric and IR spectroscopic investigation, *Adsorption Sci. Technol.* **5**:239-256.

Bonneau,L., Suquet,H., Malard,C. and Pezerat,H. (1986a) Studies on surface properties of asbestos. I Active sites on the surface of Chrysotile and Amphiboles., *Environ. Res.* **41**:251-267.

Bonneau,L., Suguet,H., Malard,C. and Pezerat,H. (1986a) Studies on the surface properties of asbestos, II. Role of dimensional characteristics and surface properties of mineral fibres in the induction of pleural tumors. *Environ. Res.*, **41**:268-275.

Bye,E., Gjionnes,J., Edward,W. and Sorbnoden,E. (1984) Occurrence of airborne silicon carbide fibres during industrial production of silicon carbide, *Scand. J. Work. Environ. Health* **II**:111-155.

Costa,D., Guignard,J., Zalma,R. and Pezerat,H. (1989) Production of free radicals arising from the surface activity of minerals and oxygen. Part I, iron mine ores., *Toxicol. Ind. Health* **5**:1061-1078.

Costa,D. (1989) Propriétés rédox de surface des poussières inorganiques en milieu aqueux. Relation avec leur propriétés cancérogènes et/ou fibrosantes. Thesis. Paris: Université P. et M. Curie.

Fisher,G.L., Mossman,B.T., Mc Farland,A.R. and Hart,R.W. (1988) A possible mechanism of crysotile asbestos toxicity. In, "Asbestos Toxicity", eds. Fisher,G.L. and Gallo,M.A. Dekker, N.Y. pp. 109-131.

Friebele,E.J., Wilson,L.K., Dozier,A.W. and Kinser,D.L. (1971) Antiferromagnetism in an oxide semiconducting glass, *Phys. Stat. Sol.* **45**:323-331 and references therein.

Fubini,B. (1988) Adsorption calorimetry in surface chemistry *Thermochimica Acta*, **135**:19-29.

Fubini,B., Giamello,E., Pugliese,L. and Volante,M. (1989a) Mechanically induced defects in quartz and their impact on pathogenicity, *Solid State Ionics*, **32-33**:334-343.

Fubini,B., Bolis,V., Giamello,E., Pugliese,L. and Volante,M. (1989b) The formation of oxygen reactive radicals at the surface of the crushed quartz dusts as a possible cause of silica pathogenicity. In, "Effects of mineral dusts on cells", eds. Mossman,B.T. and Bégin,R.O. NATO ASI Series Vol. **H30** Springer-Verlag, Berlin-Heidelberg, pp. 205-214.

Fubini,B., Giamello,E., Volante,M. and Bolis,V. (1991) Chemical functionalities at the silica surface determining its reactivity when inhaled. Formation and reactivity of surface radicals, *Toxicol. Ind. Hlth*(in the press).

Funahashi,A., Schlneter,O.P., Pintar,K., Siegesmund,K.A., Mandel,G.S. and Mandel,N.S. (1984) Pneumoconiosis in workers exposed to Silicon Carbide., *Ann. rev. Respir. Dis.* **129**: 635-640.

IARC monographs (1988) Evaluation of carcinogenic risks to humans. Man-made Mineral Fibres and Radon. Vol. **43**, Lyon.

Kennedy,T.P., Dodson,R., Rao,N.V., Ky,H., Hopkins,C., Baser,M., Tolley,E. and Hoidal,J.R. (1989) Dusts causing pneumoconiosis generate OH˙ and produce hemolysis by acting as Fenton Catalysts, *Arch. Biochem. Biophys.* **269**:359-364.

Loveridge,D. and Parke,S. (1971) Electron spin resonance of Fe^{3+}, Mn^{2+} and Cr^{3+} in glasses, *Physics Chem. Glasses* **12**:19-27.

Minotti,G. and Aust,S.D. (1987) The requirement for iron(III) in the initiation of lipid peroxidation by iron(II) and hydrogen peroxide, *J. Biol. Chem.* **262**:1098-1104.

Mossman,B.T., Marsh,J.P., Shatos,M.A., Doherty,J., Gilbert,R. and Hill,S. (1988) Implication of active oxygen species as second messengers of asbestos toxicity. In "Asbestos toxicity", eds. Fisher,G.L. and Gallo,M.A. Dekker, N.Y. pp. 157-181 and references therein.

Perrault,G., Dufresne,A., Sebastien,P., Adnot,A. and Baril,M. (1987) Caractérisation des poussieres des usines de carbure de silicium. In, "Silicosis and mixed-dusts pneumoconiosis", ed. Le Bouffant,L. INSERM **155**:301-308.

Peters,J.M., Smith,T.J., Bernstein,L., Wright,W.E. and Hammond,S.K. (1984) Pulmonary effects of exposures in silicon carbide manufacturing, *Br. J. Ind. Med.* **41**:109-115.

Pezerat,H., Zalma,R., Guignard,J. and Jaurand,M.C. (1989) Production of oxygen radicals by the reduction of oxygen arising from the surface activity of minerals fibres. IARC Scient. Publications No. **90**:100-111.

Sharrock,P. (1982) Chrysotile asbestos fibres from Quebec: electron magnetic resonance identifications, *Geochim. Cosmochim. Acta* **46**:1311-1315.

Simonato,L., Fletcher,A.C., Cherrie,J., Andersen,A., Bertazzi,P., Charnay,N., Claude,J., Dodgson,J., Estève,J., Frentzel-Beyme,R., Gardner,M.J., Jensen,O., Olsen,J., Teppo,L., Winkelmann,R., Westerholm,P., Winter,P.D., Zocchetti,C., Saracci,R. (1987), The International Agency for Research on Cancer historical cohort study of MMMF production workers in seven European countries: extension of the follow-up, *Ann. occup. Hyg.* **31**:603-623.

Stanton,M.F. and Wrench,C. (1972) Mechanism of mesothelioma induction with asbestos and fibrous glass, *J. Nat. Cancer. Inst.* **48**:797-821.

Stanton,M.F., Layard,M., Tegeris,A., Miller,E., May,H. and Kent,E. (1977) Carcinogenicity of fibrous glass: pleural response in the rat in relation to fibre dimension, *J. Nat. Cancer Inst.* **58**:587-603.

431

Stroink,G., Blaau,W.C., White,C.G. and Leiper,W. (1980) Mossbauer characteristics of UICC standard reference asbestos sample, *Can. Mineral* **18**:285-290.

Weitzmann,S.A. and Graceffa,P. (1984) Asbestos catalyzes hydroxyl and superoxide radical generation from hydrogen peroxide, *Arch. Biochem. Biophys.* **228**:373-376.

Whittaker,E.J.W. and Middleton,A.P. (1979) The intergrowth of fibrous brucite and fibrous magnesite with chrysotile, *Can. Mineral* **17**:699-702.

Zalma,R., Bonneau,L., Jaurand,M.C., Guignard,J. and Pezerat,H. (1987) Production of hydroxyl radicals by iron solid compounds, *Toxicol. Environ. Chem.* **13**:171-188; Formation of oxy-radicals by oxygen reduction arising from the surface activity of asbestos, *Can. J. Chem.* **65**:2938-2341.

Zalma,R. (1988) *Contribution à l'étude de la reactivité de surface des fibres minerales. Relations possibles avec leurs proprietés cancerogénes.* Thesis, Paris: Université P. et M. Curie.

REACTIVE OXYGEN METABOLITE PRODUCTION INDUCED BY MINERAL FIBRES

E. Yano, N. Urano and *P.H. Evans

Department of Public Health
Teikyo University School of Medicine
Tokyo, Japan

*MRC Dunn Nutrition Unit
Downhams Lane
Cambridge CB4 1XJ, U.K.

INTRODUCTION

Recently, increasing attention has been devoted to the role of reactive oxygen metabolites (ROM) in the pathogenic mechanism of various carcinogens. For example, chemical carcinogens such as benzo-(a)-pyrene, 4-nitroquinoline-N-oxide and naphthylamines have been shown to produce free radicals in biologic systems, and the carcinogenicity of these chemicals have been related to such interactions. Carcinogenicity of ionising radiation which had been believed to cause direct modification of DNA strands, is now attributed to the interactions with hydroxyl radical ($OH^{\cdot}$) produced from cytosolic water.

While ROM have also been implicated in the carcinogenic mechanism of asbestos and other mineral fibres the tumourigenic potency of such fibres has been recognised to be dependent on physical characteristics such as fibre diameter, length, and morphology. This idea, known as the Stanton hypothesis, was first formulated after epidemiological observation (Timbrell *et al.*, 1969; Timbrell, 1973) and later, established by *in vivo* studies involving the intrapleural injection or implantation of fibres (Stanton *et al.*, 1972, 1977, 1979, 1981). The aerodynamic properties of fibres needed to ensure that they reach target sites in the lung and the role of size in determining clearance of fibres from the lung are also important (Harington, 1981). However, in several *in vitro* studies, in which such mechanisms can not be important, biological effect has been shown to be size dependent (Brown *et al.*, 1978; Hesterberg & Barrett, 1984).

In the present study, which examines the relevance of ROM production to the tumourigenicity of mineral fibres, the dependence of ROM production on fibre size was examined in an *in vitro* system. As an index of ROM production by mineral fibres, luminol dependent chemiluminescence (CL) from polymorphonuclear cells was employed.

METHODS

Sample preparation

UICC Standard samples of crocidolite and micro glass fibre (JM 100) were ball milled and their fibre sizes measured using transmission electron microscopy.

Cell preparations

The stability of the results was maintained by one of the investigators providing his blood throughout the complete series of experiments. Approximately 10 ml of blood was used for each set of experiments (one day). PMN were isolated from the heparinised blood by sedimentation in dextran (3%) and subsequent centrifugation with Ficoll-Paque (Pharmacia, Sweden) to remove erythrocytes and mononuclear cells. Residual erythrocytes were lysed with an isotonic NH_4Cl solution. The resultant purified PMN were washed twice with Ca^{2+}-free buffer before being finally resuspended in Krebs-Ringer Hepes buffer pH 7.35.

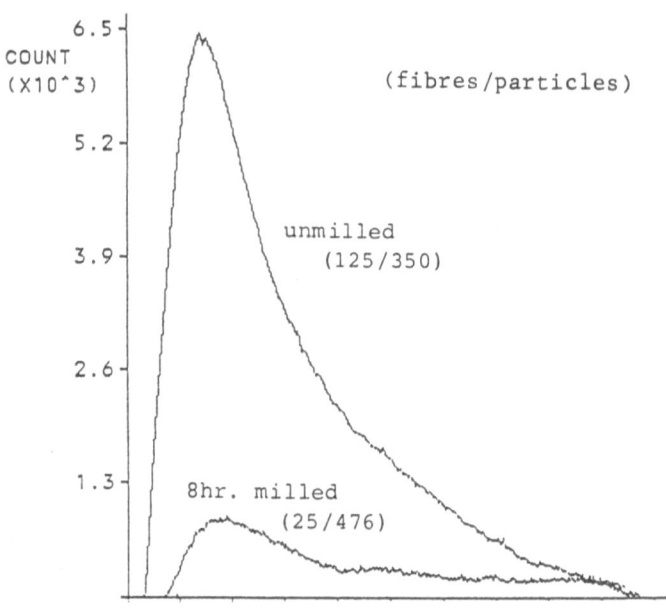

Figure 1. Luminol-dependent chemiluminescence production by milled (8 hrs) and unmilled crocidolite from PMN. Number of fibres and non-fibrous particulate is shown in parenthesis.

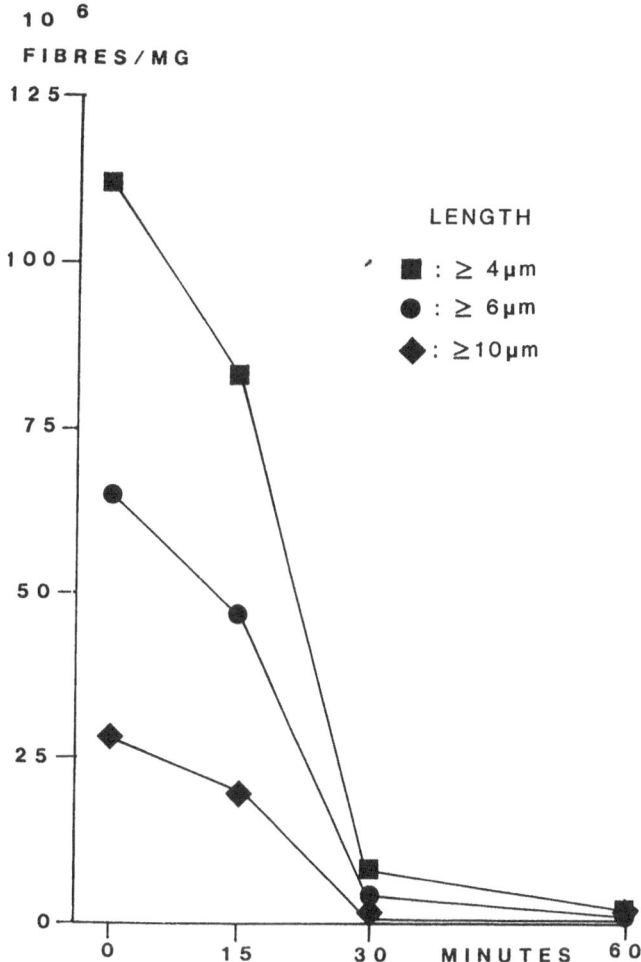

Figure 2. Effect of milling time on the length distribution of
JM 100.

Chemiluminescent measurement

The production of ROM was monitored by an automated microcomputer controlled luminometer (BIOLUMAT LB 9505, Berthold, Germany) as luminol-dependent chemiluminescence (CL). The final reaction mixture (1 ml) comprised 1×10^6 PMN with luminol (10 μM) and the reaction was initiated by the addition of 250 μl of test sample suspension in buffer. CL was followed for 20 min after the addition of the test sample to the cells.

Due to the decreasing viability of PMN after cell preparation, the absolute CL value varied depending on the time after cell isolation. However, the rank order of CL values among the samples was stable and reproducible over different sets of experiment. Therefore, all reactions were tested in parallel within the same experiment, and each experiment was replicated a minimum of nine times. All figures shown are typical examples of each such experiment.

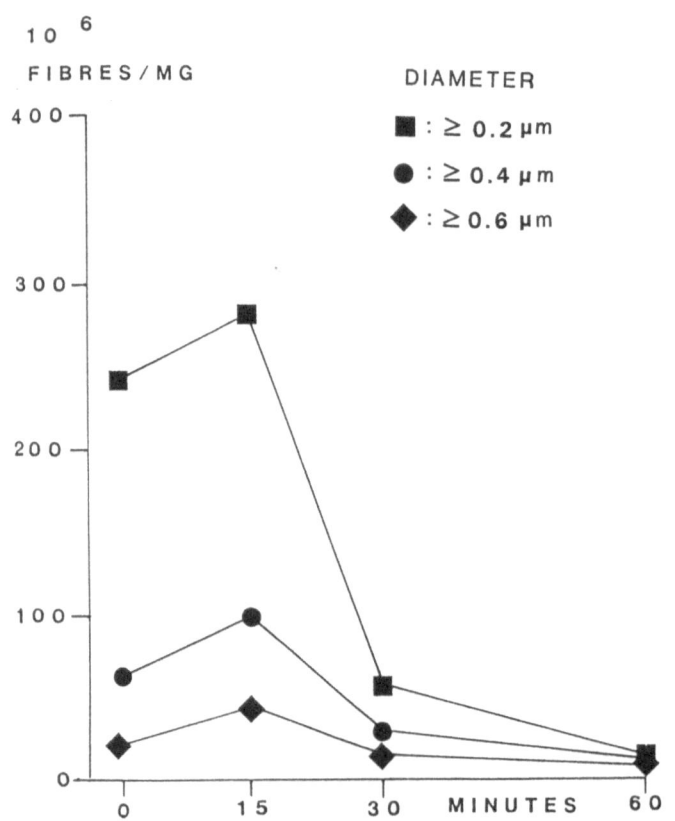

Figure 3. Effect of milling time on the diameter distribution of JM 100.

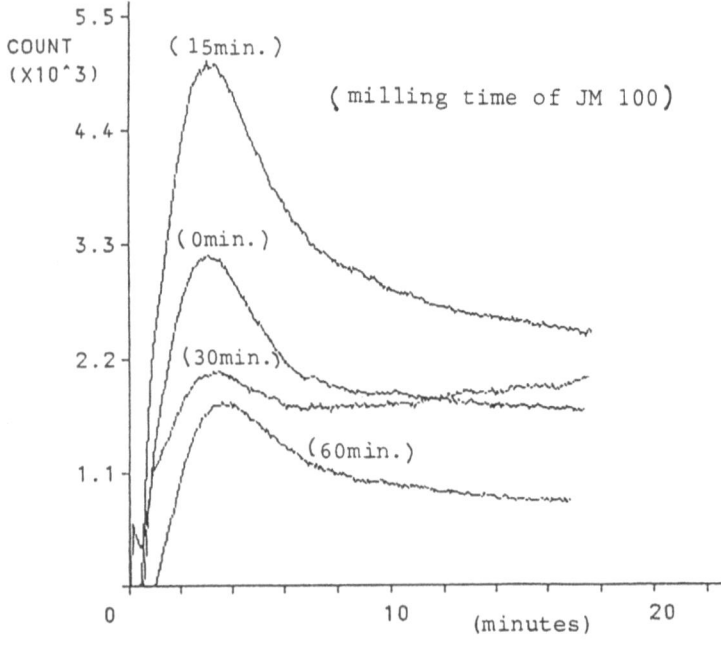

Figure 4. CL production from PMN by JM 100 milled for 0, 15, 30, and 60 min.

RESULTS and DISCUSSION

Figure 1 shows the time course of PMN CL production by milled and unmilled sample of crocidolite from PMN. In parenthesis, the number of particles and fibrous shape are shown. The final concentration of both samples in the reaction mixture was 200 μg/ml. As can be seen from the figure, the amount of CL production depends not on the number of total particulate or mass but on the number of fibres.

The detailed relationship between the number of fibres and CL production was examined by measuring the size distribution of fibres and the CL production, using JM 100 micro glass fibre milled for various periods of time.

Figure 2 demonstrates time course of milling and the number of fibres in each threshold length range. It is apparent that the number of long fibres decreased with the milling time. Similarly, figure 3 shows time course of milling and the number of fibres in each threshold diameter range. In this figure the number of thin fibre increased once until the milling time of 15 min and then decreased with the further milling. This biphasic phenomenon is explained by the vitreous property of the sample. Glass fibres fracture transversely but seldom longitudinally. Therefore at the initial phase of milling, thin fibres increase in number by reduction in length.

Figure 4 shows time courses of CL production from PMN by the glass fibre samples with various milling times. The order of the maximum CL production corresponds with the number of thin fibres demonstrated in Figure 3. Harington (1981) and Stanton and co-workers (1977) demonstrated a good correlation between the induction of pleural mesotheliomas in rats and the number of fibres less than 0.25 μm in diameter and longer than 8 μm. It is noteworthy that in our experiment, the maximum CL generation was obtained at 15 min of milling when the number of fibres around this diameter range was maximal.

The phagocytosis of asbestos has been shown to produce cell transformation and cytogenetic effects (Hesterberg *et al.*, 1986) and stimulated phagocytes produce ROM which directly modify DNA (Weitberg *et al.*, 1983). Modification of DNA can lead to mutation and tumour cell formation. The ability of the fibre to be phagocytosed is important for this explanation of the carcinogenic mechanism. Considering the size of the phagocytes, there should be an ideal size range to be phagocytosed. However, as Stanton suggested (Stanton *et al.*, 1981), fibres that are readily phagocytosed may be easily cleared away. Thus another consideration is required in the *in vivo* situation.

With regard to *in vitro* experiments, we need to consider sample weight, the number of fibres in each sample, and the surface area of the samples, in addition to the chemical properties of the sample. In the present study, the same weights of glass fibre samples were used for comparison and non-fibrous particulate was not excluded. Therefore the samples with the longer milling time contains fewer fibres. Although the number of fibres was found to change in a manner of biphasic relationship with the milling time, the total surface area appears to increase with the milling time. Thus the number of fibres rather than the surface area seems to be related to the production of CL.

This is in agreement with the Stanton's hypothesis and supports the relevance of the method of CL measurement as the assay system for the pathogenicity of mineral fibres.

REFERENCES

Brown,R.C., Chamberlain,M., Griffiths,D.M. and Timbrell,V. (1978) The effect of fibre size on the *in vitro* biological activity of three types of amphibole asbestos. *Int. J. Cancer* **22**:721-727.

Harington,J.S. (1981) Fiber carcinogenesis: epidemiologic observations and the Stanton hypothesis. *J. Natl. Cancer Inst.* **67**:977-987.

Hesterberg,T.W. and Barrett,J.C. (1984) Dependence of asbestos- and mineral dust-induced transformation of mammalian cells in culture on fiber dimension. *Cancer Res.* **44**:2170-2180.

Hesterberg,T.W., Butterick,C.J., Oshimura,M., Brody,A.R. and Barrett,J.C. (1986) Role of Phagocytosis in Syrian hamster cell transformation and cytogenetic effects induced by asbestos and short and long glass fibers. *Cancer Res.* **46**:5795-5802.

Stanton,M.F., Wrench,C. (1972) Mechanisms of mesothelioma induction with asbestos and fibrous glass. *J. Natl. Cancer Inst.* **48**:797-821.

Stanton,M.F., Layard,M., Tegeris,A., Miller,E., May,M. and Kent,E. (1977) Carcinogenicity of fibrous glass: Pleural responses in the rat in relation to fiber dimension. *J. Natl. Cancer Inst.* **58**:587-603.

Stanton,M.F. and Layard,M.W. (1979) Carcinogenicity of natural and man-made fibers. In, "Carcinogenesis: Advances in Medical Oncology Research and Education, Vol. 1", ed. Margison,G.P. Pergamon Press, New York. pp. 181-187.

Stanton,M.F., Layard,M., Tegaris,A., Miller,E., May,M., Morgan,E. and Smith,A. (1981) Relation of particle dimension to carcinogenicity in amphibole asbestos and other fibrous minerals. *J. Natl. Cancer Inst.* **67**:165-175.

Timbrell,V., Pooley,F. and Wagner,J.C. (1969) Characteristics of respirable asbestos fibers. In, "Pneumoconiosis," Proc. Ind. Co., Johannesburg. pp. 120-125.

Timbrell,V. (1973) Physical factors as etiological mechanisms. In, "Biological Effects of Asbestos", eds.Bogovski,P., Gilson,J.C., Timbrell,V. and Wagner,J.C. IARC Scientific Publications. WHO, Lyon, France. pp. 295-303.

Weitberg,A.B., Weitzman,S.A., Destrempes,M., Latt,S.A. and Stossel,T.P. (1983) Stimulated human phagocytes produce cytogenetic changes in cultured mammalian cells. *NE J. Med.* **308**:26-30.

Yano,E. (1988) Mineral fiber-induced malondialdehyde formation and effects of oxidant scavengers in phagocytic cells. *Int. Arch. Occup. Environ. Hlth* **61**:19-23.

OXYGEN CONSUMPTION, LIPID PEROXIDATION AND MINERAL FIBRES

Mairam Gulumian and Jan A. van Wyk[*]

National Centre for Occupational Health
P.O.Box 4788, Johannesburg 2000
South Africa

[*] Department of Physics
P. O. Wits
Johannesburg 2050
South Africa

ABSTRACT

Asbestos-induced lipid peroxidation in different systems is well documented in the literature. Many reports note the detection of malondialdehyde as a measure of peroxidation of unsaturated fatty acids by the thiobarbituric acid (TBA) method. Although the TBA method is quite simple and easy to use, it has many limitations and therefore has to be cross-checked by reference to one or more other methods.

Electron spin resonance was employed to monitor the peroxidation of lipids in multilamellar liposomal suspensions in the presence of hydrogen peroxide and different asbestos fibres. A correlation could be found between the ability of these minerals to generate hydroxyl radical and the ability to support peroxidation of lipids in multilamellar liposomes. This correlation agreed well with the fibrogenicity and/or carcinogenicity of the fibres tested.

INTRODUCTION

Asbestos-induced lipid peroxidation in different systems is well documented in the literature (Gabor & Anca, 1975; Gulumian et al., 1983; Weitzman & Weitberg, 1985;

Jajte *et al.*, 1987; Wydler *et al.*, 1988). These reports describe the detection of malondialdehyde, by the thiobarbituric acid (TBA) method, as a measure of peroxidation of unsaturated fatty acids. Although the TBA method is quite simple and easy to use, it has many limitations (Esterbauer *et al.*, 1984) and therefore has to be cross-checked by reference to one or more of the other methods. These include the estimation of lipid hydroperoxides, assay of conjugated dienes, chemiluminescence and oxygen uptake.

The conventional method used to monitor oxygen consumption is the Clark electrode. When employed in this laboratory to study the consumption of oxygen during mineral fibre-induced peroxidation, large quantities of samples and constant stirring were required. Hydrogen peroxide, as reported by Petersen and coworkers (1977), and mineral fibres were found to interfere with the Clark electrode (results are not shown).

It is well established that the interaction of dissolved oxygen and nitroxide spin probe through Heisenberg spin-exchange causes broadening of the electron spin resonance (ESR) lines (Backer *et al.*, 1977). In the present study 3-carbamoyl-2,2,5,5-tetramethyl-3-pyrroline-1-yloxy (CTPO) nitroxide spin probe was employed and the oxygen uptake in mineral fibre-induced oxidation of soya bean phosphatidylcholine (PC) was investigated by the ESR closed-chamber method (Sarna *et al.*, 1988; Lai *et al.*, 1982; Subczynski & Kusumi, 1985). With this method, we were able to detect increased oxygen uptake during peroxidation of PC in the presence of mineral fibres alone or mineral fibres with hydrogen peroxide to generate hydroxyl radicals which initiate lipid peroxidation.

MATERIALS AND METHODS

Crocidolite, amosite, anthophyllite and chrysotile A were UICC Reference Samples (Timbrell & Rendall, 1971/72). Tremolite fibres came from northern Italy and had lengths of 10 μm - 20 μm and diameters of 1 μm - 1.5 μm. Titanium dioxide, 1.9 μm-5.9 μm in diameter, was used. Soya bean phosphatidyl choline was obtained from Calbiochem Hoechst Laboratories, U.S.A. The spin trap 5,5-dimethyl-1-pyrroline-N-oxide (DMPO) was purchased from Sigma, St. Louis, MO. The CTPO came from Aldrich, Milwaukee, U.S.A. The rest of the reagents used were of analytical grade.

Multilamellar liposome suspensions were prepared as described by Subczynski and Kusumi (1985). Samples were prepared by adding 0.2 ml of PC containing 15 mg of the lipid, 0.6×10^{-4} M CTPO, and 3 mM NaN$_3$ to a total volume of 0.5 ml. Experimental samples contained either 0.1 ml of 0.2 mg/ml mineral fibre or/and 0.1 ml of 17.6 mM hydrogen peroxide. The mixtures were then incubated at 37°C with constant aeration for 20 hrs. At the end of the incubation, 0.05 ml of the reaction mixture were transferred into glass capillary tubes with a constant volume mark. The tubes were sealed off at both ends with teflon caps and covered with a thin layer of parafilm to stop the diffusion of oxygen to or from the mixture inside the capillary tube. The tubes were then put inside a quartz tube and placed into the ESR cavity. The signal for the CTPO spin probe was recorded after certain intervals as shown in the figures.

All ESR measurements were carried out at room temperature using a Varian E-line spectrometer. The spin probe signals were recorded at a microwave power of 1 mW, modulation amplitude of 0.04 G, and a scan range of 10 G. The oxygen consumption was measured by the degree of broadening of the superhyperfine components of the ESR signal obtained from CTPO (Fig. 1) (Hyde & Subczynski, 1984). Hydroxyl radicals were detected with DMPO as reported earlier (Gulumian & van Wyk, 1987).

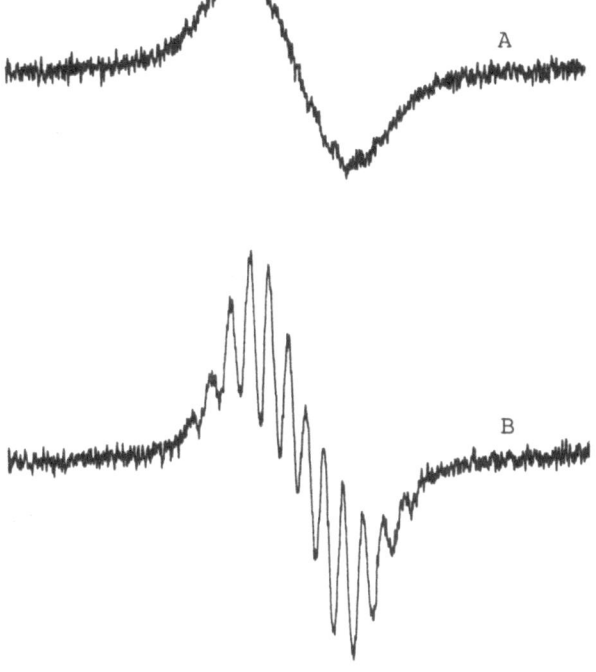

Figure 1. The ESR spectra of the spin probe in control reaction mixtures saturated with air (A) and with 100% nitrogen (B) at room temperature.

RESULTS AND DISCUSSION

Loss of molecular oxygen is an indication of the peroxidation of lipids. By monitoring this process by the spin probe CTPO, in a closed-chamber, we were able to observe the autoxidation of lipids in multilamellar liposomes (Fig.2). This was a slow process and in our closed-chamber technique, it required 25 days to complete the consumption of oxygen present in the chamber. Lipid peroxidation was however accelerated by the presence of asbestos fibres due to the formation of small amounts of oxygen-centred radicals on the surface of these minerals which have the ability to reduce oxygen (Zalma et al., 1987). This ability was different for each type of asbestos fibre used and could be ranked as follows:

UICC Amosite > UICC Crocidolite > Italian Tremolite > Transvaal Crocidolite > UICC Anthophyllite > UICC Chrysotile A (> TiO$_2$ > PC)

Hydroxyl radicals are generated by mineral fibres in the presence of hydrogen peroxide (Weitzman & Graceffa, 1984; Gulumian & van Wyk, 1987). These radicals are well known initiators of lipid peroxidation:

$$OH\cdot + RH \longrightarrow R\cdot + H_2O$$

When 1.76×10^{-2} M hydrogen peroxide was added to the reaction mixture containing the spin trap DMPO, equal concentrations of different fibres generated different levels of hydroxyl radical (Fig.3). A concomitant acceleration of oxygen consumption in the

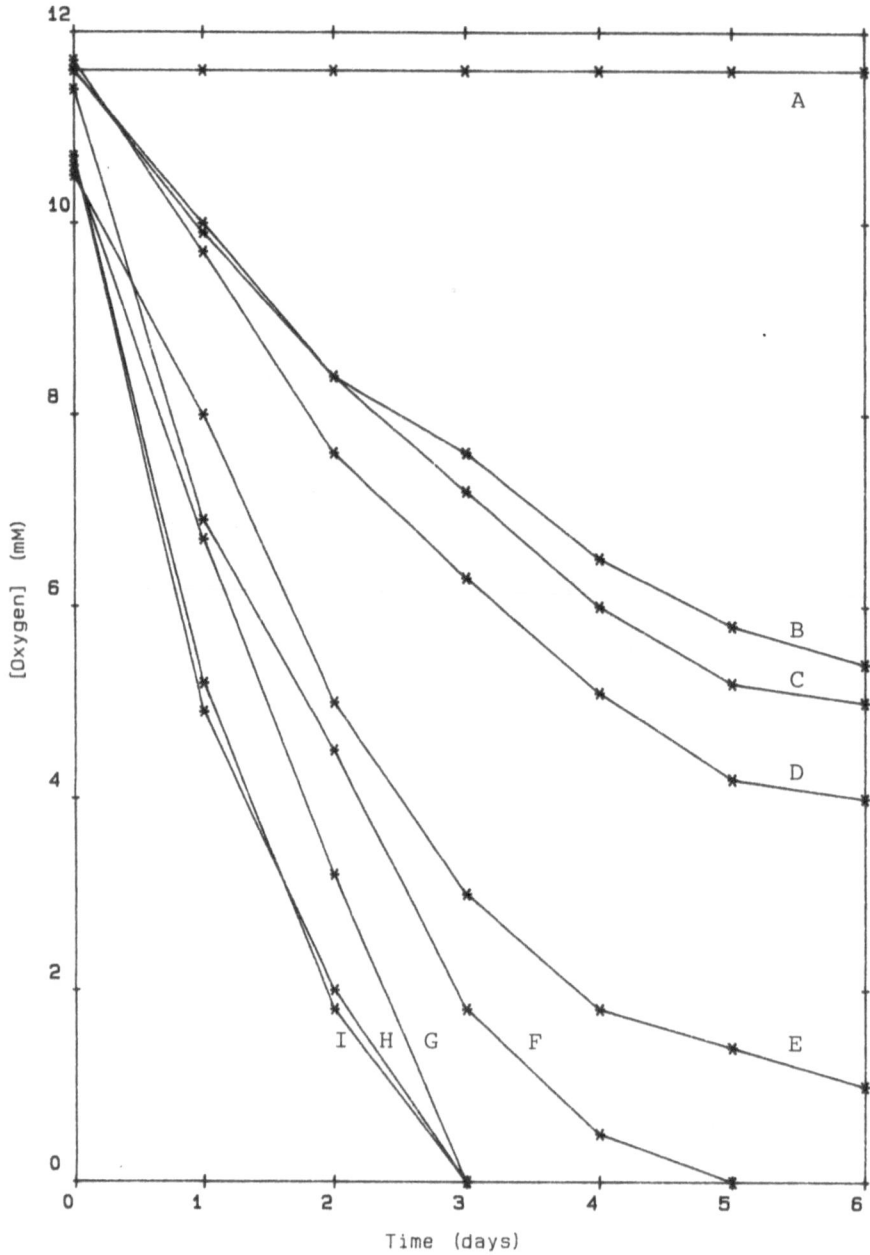

Figure 2. Oxygen consumption measured for up to six days in the presence of :-
    A. The spin probe CTPO alone,
    B. The multilamellar liposomes and CTPO.

The following dusts or asbestos fibres were added to the reaction mixture B:
    C. Titanium dioxide   G. Italian tremolite
    D. UICC chrysotile A   H. UICC crocidolite
    E. UICC anthophyllite   I. UICC amosite
    F. Transvaal crocidolite

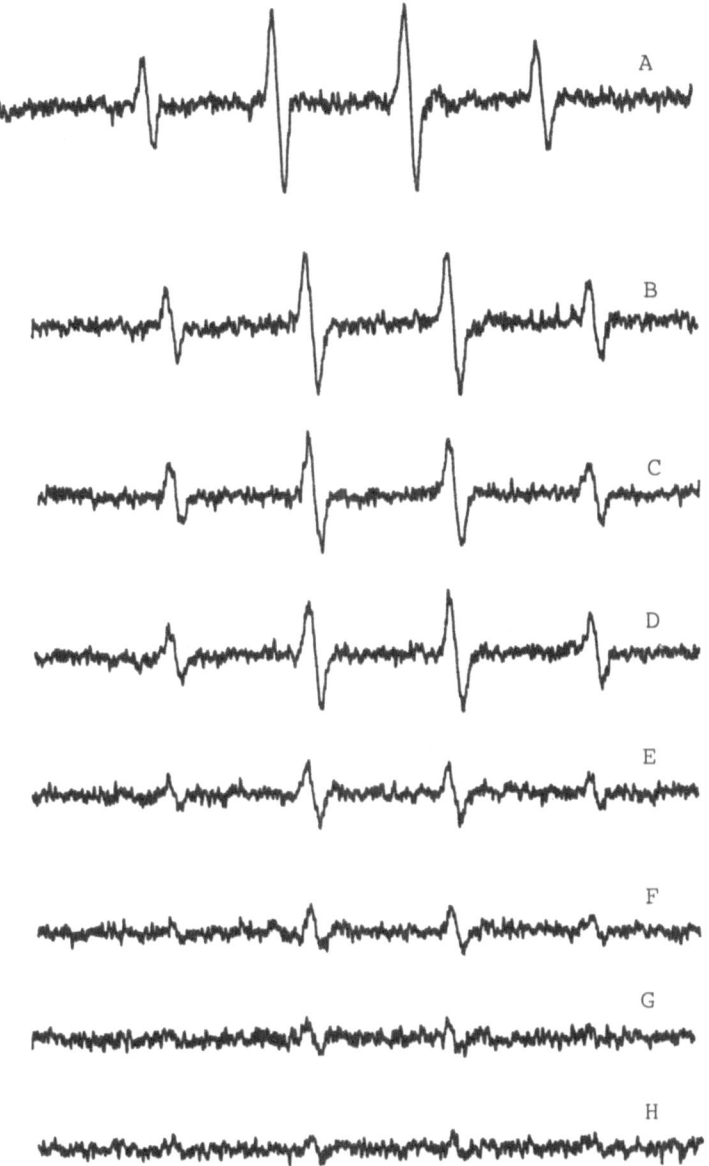

Figure 3. ESR spectra of the hydroxyl radical generated in the presence of 17.6 mM hydrogen peroxide and 0.2 mg/ml of:

A. UICC crocidolite

B. UICC amosite

C. Transvaal crocidolite

D. Italian tremolite

E. UICC anthophyllite

F. UICC chrysotile A

G. TiO_2

H. Water

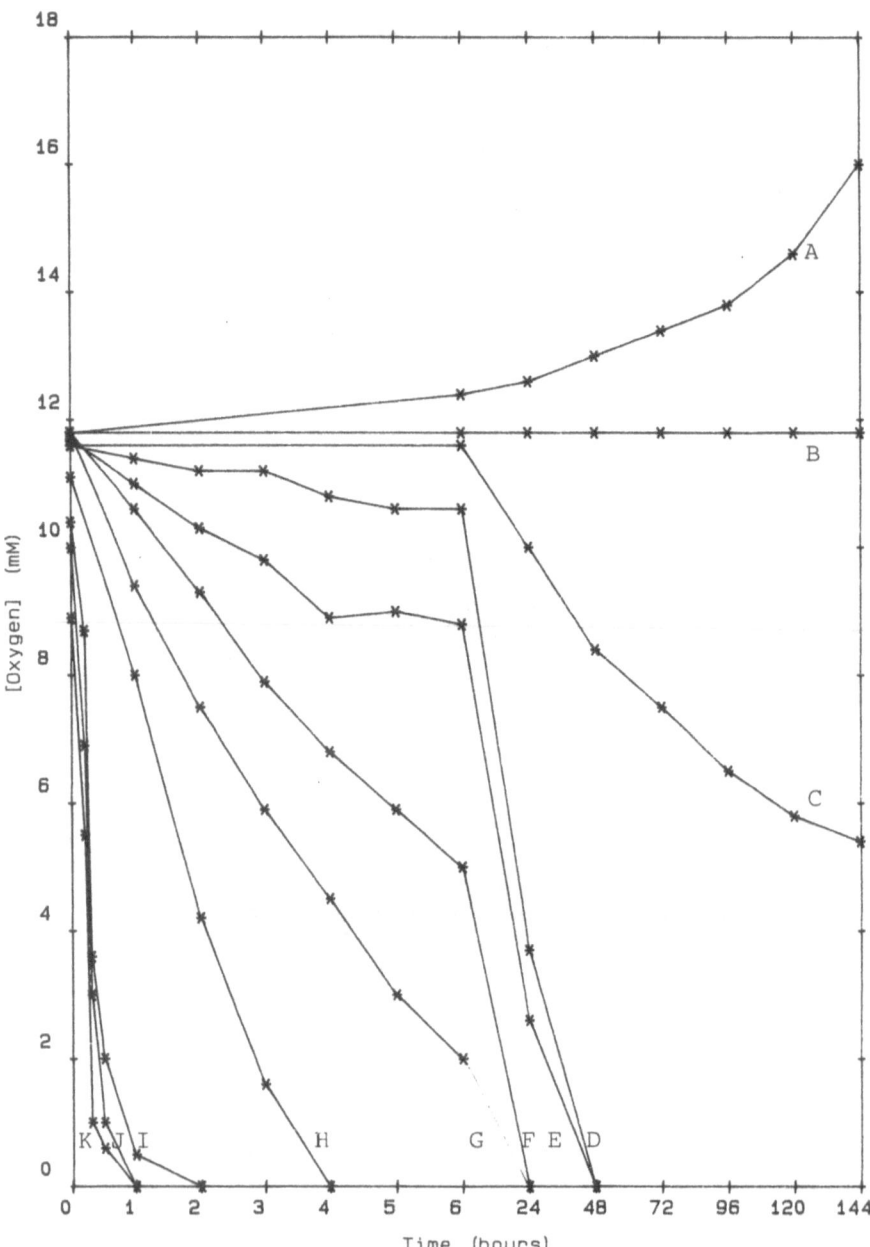

Figure 4. Oxygen consumption measured for up to six days in the presence of:

 A. 17.6 mM hydrogen peroxide and CTPO

 B. CTPO alone

 C. The multilamellar liposome and CTPO

 D. 17.6 mM hydrogen peroxide, PC and CTPO

The following dusts or asbestos fibres were added to the reaction mixture D:

 E. TiO_2 I. Transvaal crocidolite

 F. UICC chrysotile A   J. UICC amosite

 G. UICC anthophyllite   K. UICC crocidolite

 H. Italian tremolite

closed chamber was produced indicating an increase in lipid peroxidation (Fig.4). This consumption was complete within hours as compared to the same process which took days to be completed in the absence of hydrogen peroxide (Fig.2). This reagent alone with PC at the concentration used, consumed oxygen at a slower rate compared to the reaction mixtures which contained different fibres but at a faster rate when compared to the reaction mixture which contained PC alone (Fig.4). The same concentration of hydrogen peroxide (1.76×10^{-2} M), when added to CTPO, generated oxygen in the closed chamber due to its decomposition:

$$2H_2O_2 \longrightarrow 2H_2O + O_2$$

Moreover, this reagent had no effect on the CTPO level in the reaction mixture. In the presence of very active fibres however, part of the CTPO was destroyed due to the generation of high levels of hydroxyl radical. This should be kept in mind when hydroxyl radicals are involved.

Attempts have been made, in the literature, to correlate the ability of different dusts and fibres to generate OH· radicals to their ability to cause pneumoconiosis (Kennedy et al., 1989), or their ability to support lipid peroxidation to their carcinogenic potency (Wydler et al., 1988). By monitoring the consumption of oxygen as an indicator, we were able to confirm these results. Crocidolite and amosite are two mineral fibres which are very well known inducers of mesotheliomas and other malignancies (Webster, 1973; Kolev, 1982; Churg & Wiggs, 1989). Exposure to anthophyllite on the other hand, failed to produce mesotheliomas although pleural plaques occurred commonly in populations exposed to this mineral (Meurman et al., 1974). Figures 2 and 4 show the ability of anthophyllite to support peroxidation of liposomal lipids at a slower rate compared to more active fibres such as amosite, crocidolite, tremolite and Transvaal crocidolite. The latter is reported to be relatively inactive as far as inducing mesotheliomas are concerned (Webster, 1964). Tremolite, a contaminant of chrysotile and a suspected cause of cancer (Wagner et al., 1982), is found on the other hand to be as active as the very carcinogenic fibres, amosite and crocidolite. Rhodesian chrysotile was the least active asbestos fibre tested which is in good agreement with the observations made by Gelfand and Morton (1970). Finally, in the presence of titanium dioxide, a well known non-fibrogenic and non-carcinogenic dust, the rate of lipid peroxidation was almost the same as when PC was used alone. It is therefore possible, with this method of monitoring lipid peroxidation, to classify mineral dusts and fibres into those of high and low activity which can be correlated to the fibrogenic and/or carcinogenic abilities of these minerals. The closed-chamber technique thus gives a new approach to the study of lipid peroxidation induced by mineral fibres.

REFERENCES

Backer,J.M., Budker,V.G., Eremenko,S.I. and Molin. Yu,N. (1977) Detection of biochemical reactions with oxygen using exchange broadening in the ESR spectra of nitroxide radicals. *Biochem. Biophys. Acta* **460**:152-156.

Churg,A. and Wiggs,B. (1989) The distribution of amosite asbestos fibres in the lungs of workers with mesothelioma or carcinoma. *Exp. Lung Res.* **15**:771-783.

Esterbauer,H., Lang,J., Zadravec,S. and Slater,T.F. (1984) Detection of malonaldehyde by high performance chromatography. *Methods Enzymol.* **105**:319-328.

Gabor,S. and Anca,Z. (1975) Effect of asbestos on lipid peroxidation in the red cells. *Br. J. Indust. Med.* **32**:39-41.

Gelfand,M. and Morton,S.A. (1970) In, "Proceedings of the International Conference on pneumoconiosis", Johannesburg, 1969, Oxford University Press, Cape Town.

Gulumian,M.G., Sardianos,F., Kilroe-Smith,T.A. and Ockerse,G. (1983) Lipid peroxidation in microsomes induced by crocidolite fibres. *Chem.- Biol. Interact.* **44**:111-118.

Gulumian,M.G. and van Wyk,J.A. (1987) Hydroxyl radical production in the presence of fibres by a Fenton type reaction. *Chem.- Biol. Interact.* **62**:89-97.

Hyde,J.S. and Subczynski,W.K. (1984) Stimulation of ESR spectra of the oxygen-sensitive spin-label probe CTPO. *J. Magn. Res.* **56**:125-130.

Jajte,J., Lao,I. and Wilniewska-Knypl,J. M. (1987) Enhanced lipid peroxidation and lysosomal enzyme activity in the lungs of rats with prolonged pulmonary deposition crocidolite asbestos. *Br. J. Indust. Med.* **44**:180-186.

Kennedy,T.P., Dodson,R., Rao,N.V., Ky,H., Hopkins,C., Baser,M. Tolley,E. and Heidal,J.R. (1989) Dusts causing pneumoconiosis generate OH· and produce hemolysis by acting as Fenton catalysts. *Archiv. Biochem. Biophys.* **269**:359-364.

Kolev,K. (1982) Experimentally induced mesothelioma in white rats in response to intraperitoneal administration of amorphous crocidolite asbestos: preliminary report. *Environ. Res.* **29**:123-133.

Lai, C.-S., Hopwood, L.E., Hyde, J.S., Lukiewicz, S. (1982) ESR studies of O_2 uptake by chinese hamster ovary cells during the cell cycle. Cell Biol. 79:1166-1170.

Meurman,L.O., Kiviluoto,R. and Hakama,M. (1974) Mortality and morbidity among the working population of anthophyllite asbestos miners in Finland. *Br. J. Indust. Med.* **31**:105-112.

Petersen,L.C., Degu,H. and Nicholls,P. (1977) Kinetics of the cytochrome oxidase and reductase reactions in energized and de-energized mitochondria. *Can. J. Biochem.* **55**:706-713.

Sarna,T., Duleba,A., Kerytowski,W. and Swartz,H. (1988) Interaction of melanin with oxygen. *Archiv. Biochem. Biophys.* **200**:146-148.

Subczynski,W.K. and Kusumi,A. (1985) Detection of oxygen consumption during very early stages of lipid peroxidation by ESR nitroxide spin probe method. *Biochem. Biophys. Acta* **821**:259-263.

Timbrell,V. and Rendall,R.E.G. (1971/72) Preparation of UICC standard reference samples of asbestos. *Powder Technol.* **5**:279-287.

Wagner,J.C., Chamberlain,M., Brown,R.C., Berry,G., Pooley,F.D., Davies,R. and Griffiths,D.M. (1982) Biological effects of Tremolite. *Br. J. Cancer* **45**:352-360.

Webster,I. (1964) Asbestosis. *S.A. J. Science* **38**:870-872.

Webster,I. (1973) Malignancy in relation to crocidolite and amosite. In, "Biological Effects of Asbestos", International agency for Research on Cancer, Lyon. pp. 195-198.

Weitzman,S.A. and Graceffa,P. (1984) Asbestos catalyzes hydroxyl and superoxide radical generation from hydrogen peroxide. *Archiv. Biochem. Biophys.* **228**:373-376.

Weitzman,S.A. and Weitberg,A.B. (1985) Asbestos-catalysed lipid peroxidation and its inhibition by desferrioxamine. *Biochem. J.* **225**:259-262.

Wydler,M., Maier,P. and Zbinden,G. (1988) Differential cytotoxic, growth-inhibiting and lipid peroxidation activities of four different asbestos fibres *in vitro*. *Toxic. in vitro* **2**:297-302.

Zalma,R., Bonneau,J. Guignard,J. and Pezerat,H. (1987) Production of hydroxyl radicals by iron solid compounds. *Toxicol. Environ. Chem.* **13**:171-187.

EFFECTS OF CIGARETTE SMOKE ON UPTAKE OF ASBESTOS FIBRES BY TRACHEAL ORGAN CULTURES: THE ROLE OF ACTIVE OXYGEN SPECIES

Andrew Churg, Jane Hobson, Joanne L Wright

Department of Pathology and University Hospital
University of British Columbia
Vancouver, BC, Canada

ABSTRACT

In order to investigate how cigarette smoke enhances asbestos fibre retention and tissue penetration, we employed a tracheal organ culture system in which the tracheal segments were initially exposed to smoke or air, subsequently exposed to a suspension of amosite asbestos, and then maintained in organ culture for varying time periods. Exposure to smoke consistently increased both the uptake of asbestos fibres by the tracheal epithelial cells and the proliferative response to asbestos by tracheal epithelial cells compared to exposure to air. This effect was seen when asbestos was added to the tracheas immediately after smoke exposure, or when asbestos exposure was delayed for up to 48 hours. Smoke enhanced uptake could be totally abolished by inclusion of superoxide dismutase or catalase with the asbestos or pretreatment of the asbestos with deferoxamine. These scavengers of active oxygen species also decreased uptake of fibres by tracheal epithelium in the absence of smoke, although, in contrast to the effects seen with smoke, protection was never complete. We conclude that: (1) Smoke enhancement of asbestos fibre uptake is a direct effect of cigarette smoke on the epithelium and does not require the participation of inflammatory cells; (2) Brief exposure to smoke primes the cell for uptake even when there is a considerable delay in asbestos exposure; (3) Active oxygen species play a role in fibre uptake and the proximate species is probably $OH^{\cdot}$ derived from H_2O_2. We speculate that H_2O_2 is generated for long periods from smoke tar deposited on the epithelial surface.

Supported by grant MA8051 from the National Cancer Institute of Canada.

INTRODUCTION

Cigarette smoke increases the incidence of lung cancer and asbestosis in heavily exposed asbestos workers (Hammond *et al.*, 1979; Becklake, 1976; Weiss, 1984)). Although the mechanism(s) of this process is unclear, smoke in general increases the retention of mineral particles of all types in man and animals (Cohen *et al.*, 1979; Bohning *et al.*, 1982; Mauderly *et al.*, 1989). We have shown that both total fibre burden and the number of fibres penetrating tissue are increased in guinea pigs exposed to smoke and asbestos compared to animals exposed to air and asbestos (McFadden *et al.*, 1986a, 1986b). Since asbestos-induced diseases are dose-related, the effect of smoke may be to increase the functional dose to tissues.

Little information is available on the important question of how smoke enhances fibre retention and penetration. Three broad possibilities exist: (1) Smoke impairs the normal clearance of fibres by inflammatory cells and, since any mineral particle not removed by macrophages is likely to cross epithelial barriers (Brain, 1980; Adamson & Bowden, 1981, 1982), fibre uptake by epithelium is a totally passive event; (2) Smoke and asbestos-evoked inflammatory cells release mediators which increase fibre uptake by tissues; (3) Smoke directly affects epithelium to cause enhanced fibre uptake.

These hypotheses are not mutually exclusive, but, because of the intense inflammatory reaction evoked in the lung by both smoke and asbestos, are difficult to separate *in vivo*. We report here a series of experiments designed to investigate these issues using tracheal organ cultures, a system free from inflammatory cells.

MATERIALS AND METHODS

Details of the procedures may be found in the papers by Hobson and coworkers (1988) and Churg and coworkers (1989). Tracheal segments were prepared from rats (Mossman *et al.*, 1979) and exposed to smoke or air, mucosal side up, in a humidified chamber. The smoke dose was varied but total exposure time was maintained at 10 minutes. The segments were then submerged in a suspension of UICC amosite asbestos for 1 hour; the time delay between the end of smoke exposure and the asbestos exposure varied from none (immediate asbestos exposure) up to 48 hours. In some experiments catalase or superoxide dismutase, direct scavengers of active oxygen species, were included with the asbestos suspension, or the asbestos was first incubated overnight with the iron chelator deferoxamine.

After asbestos exposure the segments were maintained in organ culture using basal feeding and constant exposure of the mucosa to 95% air/5% CO_2 for periods of 1 to 7 days. This incubation period is required for uptake of asbestos fibres to occur. At the end of each incubation period segments were processed for histology. Fibre uptake was evaluated with a light microscope by counting fibres longer than 4 μm and measuring the length of underlying basement membrane to provide a value for fibres/mm of epithelium. In some experiments hyperplastic lesions (masses of heaped up epithelial cells) were separately counted as lengths of hyperplastic lesions/total length of epithelial basement membrane.

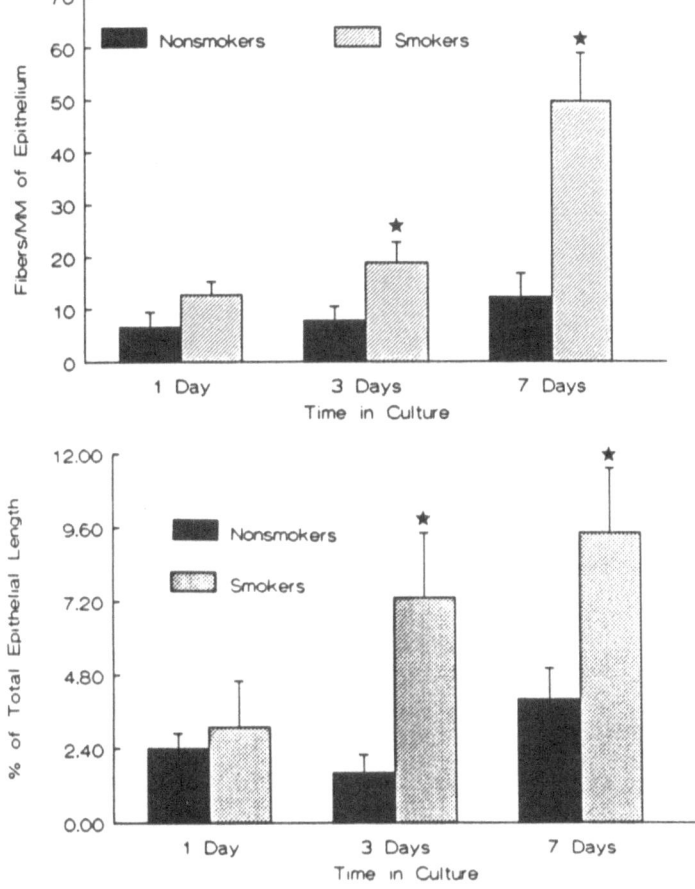

Figure 1 and Figure 2: Enhancement by smoke of asbestos fibre uptake and extent of asbestos-induced hyperplastic lesions after exposure to 10 puffs of smoke. Time in culture refers to number of days of organ culture after asbestos exposure. * = significantly greater than nonsmoker. Values are mean + SE. (Data from Hobson *et al.*, 1988.)

RESULTS

Effects of smoke on fibre uptake

Figures 1 and 2 show a typical experiment documenting the effects of smoke on both fibre uptake and percentage of epithelium occupied by hyperplastic lesions when asbestos exposure immediately follows smoke exposure. In the absence of smoke both fibre uptake and hyperplastic lesions slowly increase over time between 1 and 7 days of culture. When the tracheal segments are first exposed to smoke there is a marked increase in both total fibre uptake and total extent of hyperplastic lesions and the rate of increase is much steeper than with air exposure (Hobson *et al.*, 1988). Figure 3 shows a similar experiment in which the dose of smoke was varied from 0 (air) to 6 puffs: there is a distinct dose response effect (Churg *et al.*, 1990).

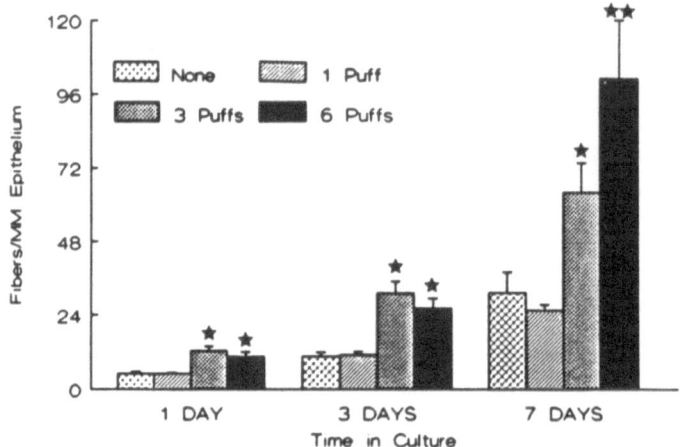

Figure 3: Effects of varying numbers of puffs of cigarettes on fibre uptake. In general the more puffs the greater the fibre uptake. * = significantly greater than nonsmoker. ** = significantly greater than 3 puffs of smoke. Values are mean + SE. (Data from Churg *et al.*, 1990.)

Effects of delaying exposure to asbestos

In these experiments the tracheal segments were exposed to smoke in the usual fashion and exposure to asbestos was delayed for 0, 3, 18, or 48 hours. Figure 4 shows the effects of a 48 hour delay: despite the delay, smoke enhancement of asbestos fibre uptake is still present at all culture periods. Similar results were seen for the other time periods (Churg *et al.*, 1990).

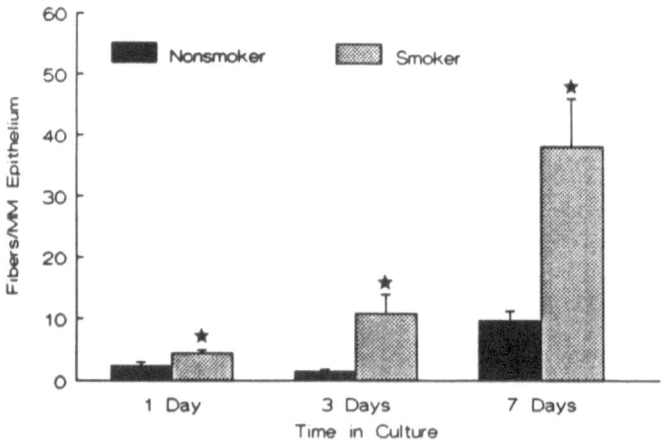

Figure 4: Effect of a 48 hour delay between smoke and asbestos exposure; even with this delay smoke still enhances fibre uptake. * = significantly greater than nonsmoker. Values are mean + SE. (Data from Churg *et al.*, 1990.)

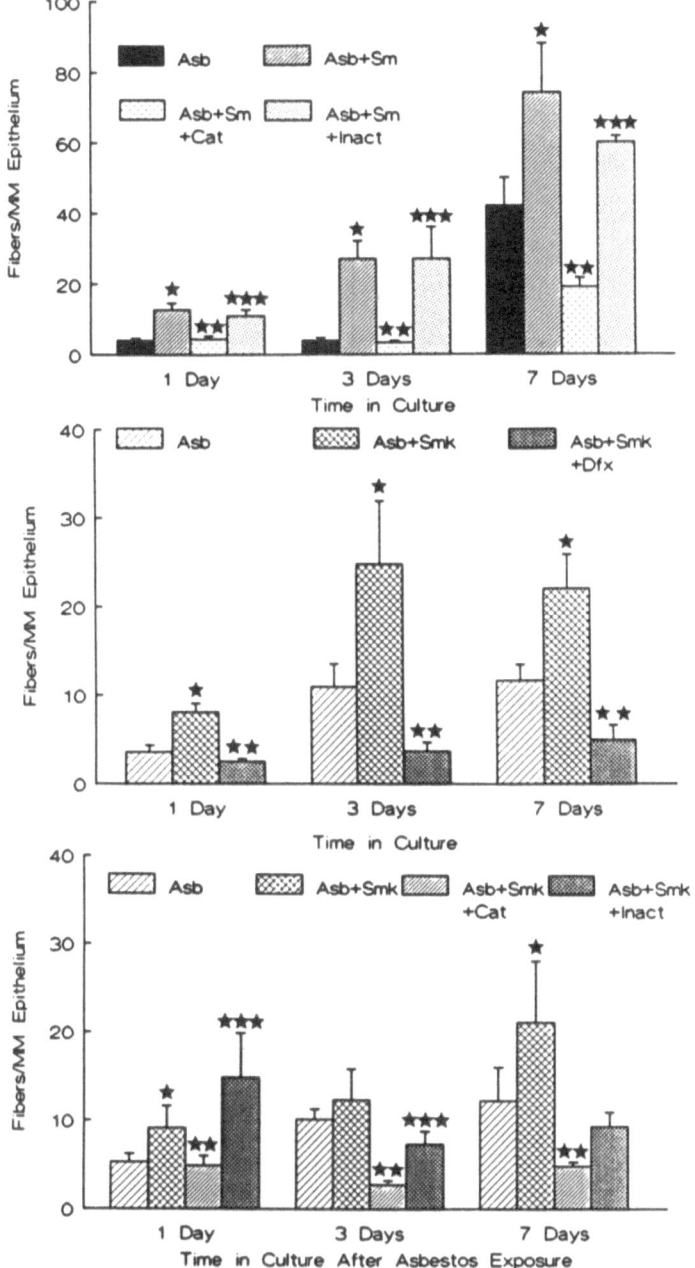

Figures 5-7: Effects of adding catalase to the asbestos suspension or preincubating the asbestos with deferoxamine. Both these agents completely abolish the smoke effects. In Figures 5 and 6 the asbestos + scavenger is added immediately after smoke exposure; in figure 7 the asbestos + catalase is added 48 hours after smoke exposures. * = significantly greater than nonsmoker. ** = significantly less than asbestos plus smoke. *** = significantly greater than asbestos plus smoke plus catalase. Asb = asbestos only. Sm = smoke. Cat = catalase. Dfx = deferoxamine. Inact = inactivated catalase. Values are mean + SE. (Data from Churg *et al.*, 1989, 1990.)

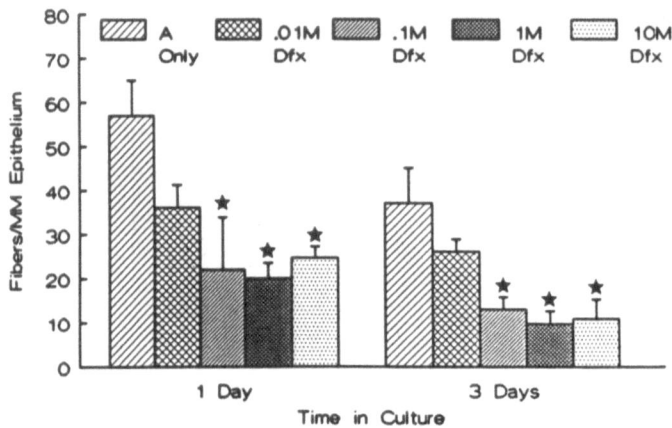

Figure 8: Effects of deferoxamine on uptake of asbestos fibres by tracheal epithelium in the absence of smoke. A definite protective effect is seen and there is a dose response, but protection is never complete no matter how high the dose. * = significantly less than asbestos alone. Dfx = deferoxamine in concentration shown. Values are mean + SE. (Data from Hobson *et al.*, 1990.)

Effects of scavengers of active oxygen species on smoke enhancement

Figures 5 and 6 show the effects of catalase and deferoxamine on fibre uptake when exposure to asbestos occurred immediately after smoke and Figure 7 shows the effect of catalase when asbestos exposure was delayed for 48 hours. In all instances the scavengers totally abolished the smoke enhancement of fibre uptake. Similar results were obtained with superoxide dismutase. Inactivating the superoxide dismutase or catalase abolished the protective effect (Churg *et al.*, 1989, 1990).

Effects of scavengers of active oxygen species on fibre uptake in the absence of smoke

Figure 8 shows the effects of deferoxamine on uptake of asbestos fibres by tracheal epithelial cells. There is a dose related protective effect but never complete protection. Identical results were obtained with catalase and superoxide dismutase and protection could be abolished by inactivating the enzymes (Hobson *et al.*, 1990).

DISCUSSION

Although the mechanism of fibre carcinogenesis is unknown, there is ample evidence that penetration of asbestos fibres into epithelial cells has deleterious effects on both cell division and chromosomal structure. Fibre uptake in itself appears to cause cell proliferation (Haugen *et al.*, 1982; Mossman *et al.*, 1977). Intracellular asbestos fibres produce DNA strand breaks (Turver & Brown, 1987) and in cell culture systems fibres accumulate around the cell nucleus and apparently interact directly with the mitotic apparatus to cause aneuploidy (Hesterberg *et al.*, 1986; Hesterberg & Barrett, 1985). Asbestos fibres which penetrate cells also increase the uptake of adsorbed carcinogens and enhance the metabolism of such substances such that greater quantities of active carcinogen are produced (Mossman *et al.*, 1985). Asbestos fibre uptake also leads to squamous metaplasia in some culture systems, and this is probably a preneoplastic event (Cameron *et al.*, 1989).

Oddly enough, even though fibre uptake is clearly an important event for asbestos fibre toxicity to epithelia, little is known about the mechanisms governing fibre penetrations and even less about how smoke augments this process. The experiments reported here were designed to start investigating this process and they lead to several basic conclusions:

(1) Enhancement of fibre uptake is a direct effect of smoke on the tracheal epithelial cell and does not require inflammatory cells or inflammatory cell mediators. However, it is possible that, *in vivo*, inflammatory cells also play a role; for example, we have shown that, after exposure to smoke, there is decreased clearance of fibre bearing macrophages from the lung (Churg *et al.*, 1991). Such macrophages may die and release fibres and a greater number of fibres might well be available to penetrate epithelia.

(2) The effect of smoke on fibre uptake is very long lasting. We found that even with delays of up to 48 hours (the longest period tested) between asbestos and smoke exposure there was still enhancement of fibre uptake. It is possible that the same effect would be seen with longer delays. This finding implies that smoke affects the structure of the epithelial cell in such a way that the cell is "primed" for fibre uptake. This observation has obvious implications for cigarette smoking humans and emphasises the dangers of even relatively remote cigarette exposure.

(3) Active oxygen species mediate smoke-enhanced fibre uptake. Because deferoxamine abolishes the smoke effect, it is likely that hydroxyl radical ($OH^{\cdot}$) is the active species in this regard (Mossman & Marsh, 1989). This radical is probably generated by the reaction: $O_2^{\cdot} + H_2O_2 \longrightarrow OH^{\cdot}$ with the iron in the asbestos fibre serving as a Fenton type catalyst. The major source of active oxygen species is probably cigarette smoke, since it has been shown that smoke bubbled into solution generates both the superoxide anion and hydrogen peroxide (Nakayama *et al.*, 1984). It is particularly interesting that hydrogen peroxide production continues for many hours in aqueous solutions of smoke tar (Nakayama *et al.*, 1989) and this is most likely the reason that long delays between smoke and asbestos exposure are still effective in increasing fibre uptake. This idea is supported by the ability of catalase to block uptake in such delayed exposure systems.

(4) Active oxygen species also mediate fibre uptake in the absence of smoke. Again, the important reaction is probably the ability of iron containing asbestos fibres to generate hydroxyl radical, probably from hydrogen peroxide produced as a product of cellular metabolism. It is interesting, however, that while scavenging of active oxygen species completely protects against smoke enhancement of fibre uptake, it never completely abolishes the much lower level of uptake seen in the absence of smoke. This observation implies that other mechanisms of fibre penetration must exist.

(5) The fact that scavenging $OH^{\cdot}$ is effective in preventing fibre uptake suggests that lipid peroxidation is playing a role (Farber *et al.*, 1990). Asbestos fibres have been shown to cause lipid peroxidation in both pure lipid membrane and pure cell culture systems (Weitzman & Graceffa, 1984; Weitzman & Weitberg, 1985); however, this supposition must be tempered by the observations of Goodglick and coworkers (1989) that lipid peroxidation and cell toxicity to cultured macrophages appeared to be separate processes.

It is also interesting to speculate on whether the effects shown here with asbestos also are true of other types of minerals, since smoke appears to inhibit the clearance of all types of mineral particles. We were unable to show a direct smoke enhancement effect for very finely divided iron oxide using this type of tracheal organ culture system (Churg &

Wright, 1989), even though there is an increase in iron oxide uptake *in vivo* by the bronchial epithelium of smoke exposed guinea pigs (Gilks *et al.*, 1988). This observation may imply that impairment of particle clearance by smoke has a number of different mechanisms. In this context it is of interest to note the report of Kennedy and coworkers (1989) who found that some, but not all, mineral dusts could generate hydroxyl radical from hydrogen peroxide in solution. It is possible that direct smoke enhancement of particle uptake only occurs with minerals that can catalyse this reaction.

REFERENCES

Adamson,I.Y.R. and Bowden,D.H. (1981) Dose response of the pulmonary macrophagic system to various particulates and its relationship to transepithelial passage of free particles. *Exper. Lung Res.* **2**:165-171.

Adamson,I.Y. and Bowden,D.H. (1982) Effects of irradiation on macrophagic response and transport of particles across the alveolar epithelium. *Am. J. Pathol.* **106**:16-22.

Becklake,M.R. (1976) Asbestos-related diseases of the lung and other organs: Their epidemiology and implications for clinical practice. *Am. Rev. Respir. Dis.* **114**:187-227.

Bohning,D.E., Atkins,H.L. and Cohn,S.H. (1982) Long-term particle clearance in man: normal and impaired. *Ann. Occup. Hyg.* **26**:259-271.

Brain,J.D. (1980) Macrophage damage in relation to the pathogenesis of lung diseases. *Environ. Hlth Perspect.* **35**:21-28.

Cameron,G., Woodworth,C.D., Edmonson,S. and Mossman,B.T. (1989) Mechanisms of asbestos-induced squamous metaplasia in tracheobronchial epithelial cells. *Environ. Hlth Perspect.* **80**:101-108.

Churg,A. and Wright,J.L. (1989) Effects of cigarette smoke on uptake of asbestos and iron oxide in rat tracheal explants. In, "Effects of Mineral Dusts on Cells", eds. Mossman,B.T. and Begin,R. Springer-Verlag, Berlin. pp. 1-6.

Churg,A., Hobson,J., Berean,K. and Wright,J. (1989) Scavengers of active oxygen species prevent cigarette smoke-induced asbestos fiber penetration in rat tracheal explants. *Am. J. Pathol.* **135**:599-603.

Churg,A., Wright,J.L., Hobson,J. and Stevens,B. (1991) Effects of cigarette smoke on the clearance of long and short asbestos fibers by macrophages. in preparation.

Churg,A., Hobson,J. and Wright,J.L. (1990) Effects of cigarette smoke dose and time after smoke exposure on uptake of asbestos fibers by rat tracheal epithelial cells. *Am. J. Resp. Cell Molec. Biol.* **3**:265-269.

Cohen,D., Arai,S.F. and Brain,J.D. (1979) Smoking impairs long-term clearance from lung. *Science* **204**:514-516.

Farber,J.L., Kyle,M.E. and Coleman,J.B. (1990) Mechanisms of cell injury by activated oxygen species. *Lab. Invest.* **62**:670-679.

Gilks,B., Wright,J.L. and Churg,A. (1988) Effects of cigarette smoke on tissue uptake and retention of iron oxide in the guinea pig. *Am. Rev. Respir. Dis.* **137**:1382-1384.

Goodglick,L.A., Pietras,L.A. and Kane,A.B. (1989) Evaluation of the causal relationship between crocidolite asbestos induced lipid peroxidation and toxicity to macrophages. *Am. Rev. Respir. Dis.* **139**:1265-1273.

Hammond,E.C., Selikoff,I.J. and Seidman,H. (1979) Asbestos exposure, smoking, and death rates. *Ann. NY Acad. Sci.* **330**:473-492.

Haugen,A., Schafer,P.W., Lechner,J.F., Stoner,G.F., Trump,B.F., and Harris,C.C. (1982) Cellular ingestion, toxic effects, and lesions observed in human bronchial epithelial tissue and cells cultured with asbestos and glass fibers. *Int. J. Cancer* 30:265-272.

Hesterberg,T.W., Butterick,C.J., Oshimura,M., Brody,A.R. and Barrett,J.C. (1986) Role of phagocytosis in Syrian hamster cell transformation and cytogenetic effects induced by asbestos and short and long glass fibers. *Cancer Res.* 46:5795-5802.

Hesterberg,T.W. and Barrett,J.C. (1985) Induction by asbestos fibers of anaphase abnormalities: Mechanism for aneuploidy induction and possibly carcinogenesis. *Carcinogenesis* 6:473-475.

Hobson,J., Gilks,B., Wright,J. and Churg,A. (1988) Cigarette smoke directly enhances asbestos fiber penetration and asbestos induced epithelial proliferation in rat tracheal explants. *JCNI* 80:518-521.

Hobson,J., Wright,J.L. and Churg,A. (1990) Active oxygen species mediate asbestos fibre uptake by tracheal epithelial cells. *FASEB J.* 4:3135-3139.

Kennedy,T.P., Dodson,R., Rao,N.V., Ky,H., Hopkins,C., Baser,M., Tolley,E. and Hoidal,J.R. (1989) Dust causing pneumoconiosis generate OH˙ and produce hemolysis by acting as Fenton catalysts. *Arch. Biochem. Biophys.* 269j:359-364.

McFadden,D., Wright,J.L., Wiggs,B. and Churg,A. (1986a) Smoking inhibits asbestos clearance. *Am. Rev. Respir. Dis.* 133:372-374.

McFadden,D., Wright,J., Wiggs,B. and Churg,A. (1986b) Cigarette smoke increases the penetration of asbestos fibers into airway walls. *Am. J. Pathol.* 123:95-99.

Mauderly,J.M., Chen,B.T., Hahn,F.F., Lundgren,D.L., Cuddihy,R.G., Namenyi,J. and Rebar,A.H. The effect of chronic cigarette smoke inhalation on the long-term pulmonary clearance of inhaled particles in the rat. In, "Biological Interactions of Inhaled Mineral Fibres and Cigarette Smoke", ed. Wehner,A.P. Battelle Press, Columbus, Ohio. pp. 223-240.

Mossman,B.T., Kessler,J.B., Ley,B.W. and Craighead,J.E. (1977) Interaction of crocidolite asbestos with hamster respiratory mucosa in organ culture. *Lab. Invest.* 36:131-138.

Mossman,B.T. and Craighead,J.E. (1979) Use of hamster tracheal organ cultures for assessing the cocarcinogenic effects of inorganic particulates on the respiratory epithelium. *Prog. Exptl. Tumor Res.* 24:37-47.

Mossman,B.T., Camerson,G.S., Yotti,L.P. (1985) Co-carcinogenic and tumor promoting properties of asbestos and other minerals in tracheobronchial epithelium. *Carcinogenesis* 8:217-238.

Mossman,B.T. and Marsh,J.P. (1989) Evidence supporting a role for active oxygen species in asbestos-induced toxicity and lung disease. *Environ. Hlth Perspect.* 81:91-94.

Nakayama,T., Kodama,M. and Nagata,C. (1984) Generation of hydrogen peroxide and superoxide anion radical from cigarette smoke. *Gann* 75:95-98.

Nakayama,T., Church,D. and Pryor,W.A. (1989) Quantitative analysis of the hydrogen peroxide formed in aqueous cigarette tar extracts. *Free Radical Biol. Med.* 7:9-15.

Turver,C.J. and Brown,R.C. (1987) The role of catalytic iron in asbestos induced lipid peroxidation and DNA strand breakage in C3H10T 1/2 cells. *Br. J. Cancer* 56:133-136.

Weiss,W. (1984) Cigarette smoke, asbestos and small irregular opacities. *Am. Rev. Respir. Dis.* 130:293-301.

Weitzman,S.A. and Graceffa,P. (1984) Asbestos catalyzes hydroxyl and superoxide radical release from hydrogen peroxide. *Arch. Biochem. Biophys.* 228:373-376.

Weitzman,S.A. and Weitberg,A.B. (1985) Asbestos-catalysed lipid peroxidation and its inhibition by desferrioxamine. *Biochem. J.* 225:259-262.

MECHANISMS OF PATHOGENESIS

4. Other effects and promotion

PROMOTING EFFECTS OF FIBRES. FIBRES AND THE SECOND MESSENGER PATHWAYS

J.A. Hoskins, R.C. Brown and C.E. Evans

MRC Toxicology Unit
Woodmansterne Road
Carshalton SM5 4EF, U.K.

INTRODUCTION

Promoters of carcinogenesis, are themselves neither mutagenic nor carcinogenic,one possible mechanism of their action is through enhanced cellular division and proliferation, processes necessary for neoplastic transformation. There is considerable evidence that asbestos, and other mineral fibres, have tumour promoting properties. Several *in vivo* experimental studies have demonstrated synergy between asbestos and a chemical carcinogen (e.g. Harrison & Heath, 1988) and this might indicate that asbestos may function as a tumour promoter. There is also epidemiological evidence for its tumour-promoting activities including the synergism between asbestos exposure and smoking in the aetiology of lung cancer in humans (NRC, 1984).

Asbestos is not usually classified as an initiator for bronchogenic tumours it is known that *in vitro* fibres can enhance the effect of chemical or physical carcinogens causing cell transformation (Brown *et al.*, 1983; Hei *et al.*, 1985). In bronchogenic cancer asbestos appears to act solely as a promoter but in the case of mesothelioma fibres seem to have both initiating and promoting activities and act as a complete carcinogen (Brown *et al.*, 1990a). Unlike most carcinogens asbestos does not cause base substitution and frameshift mutations in bacterial mutation assays (Chamberlain & Tarmy, 1977) or in mammalian cell systems (Shelby, 1988). It has been reported to be weakly mutagenic in Chinese hamster lung cells (Huang, 1979) but not in liver epithelial cells (Reiss *et al.*, 1983) or in Syrian hamster embryo fibroblasts (Oshimura *et al.*, 1984).

In spite of the evidence for tumour promoting activity the mechanism of promotion is unknown. It appears, however, that indirect activation of protein kinase C (PKC), a membrane-associated calcium and phospholipid-dependent enzyme activated by diacylglycerols (DAGs), may be involved. This same enzyme is also activated directly by the classic tumour promoter 12-*O*-tetradecanoylphorbol-13-acetate (TPA) which appears to be a DAG analogue. This is not a simple analogy since under some circumstances TPA stimulation of cells can induce release of DAGs. PKC is a part of the phosphatidylinositol second messenger system (Berridge & Irvine, 1989) and stimulation of receptors associated with this pathway by agonists induces release of DAGs. Asbestos minerals seem to act as if they were agonists. For example, addition of chrysotile, or amosite, to P388D1 cells causes diacylglycerol release in a manner which is quantitatively similar to that produced by agonists but occurring much more slowly possibly due to the asynchrony of cell-fibre interaction.(Brown *et al.*, 1989).

Another effect of tumour promoters including TPA is to interfere with the formation or integrity of gap-junctions between cells (Yotti *et al.*, 1979). Fibres do not duplicate this short term effect of tumour promoters (Chamberlain, 1982) but longer term treatment of cells with TPA, or fibres, produces a down regulation which reduces the short term effect of subsequent treatment with TPA (Blackshear *et al.*, 1985).

MATERIALS AND METHODS

UICC samples of asbestos were used.

Coated and modified fibres

Amosite was surface-modified by reacting in boiling toluene with octyldimethyl-chlorosilane or octadecyldimethylchlorosilane as described previously (Brown *et al.*, 1990b). The two materials produced were designated C_8-amosite and C_{18}-amosite.

Fibres were coated with fibronectin by adding a 1 mg/ml solution of fibronectin to a 2 mg/ml suspension of fibres, the mixture shaken vigorously and allowed to stand at room temperature for 10-30 min. The mixture was added to cells to give a final concentration of fibronectin of 30 μg/ml.

Fibres were coated with poly-lysine by suspension in a 0.1 mg/ml aqueous solution of the polypeptide and then centrifuging at 15,000 g for 15 min. Coated fibres were washed in water and again centrifuged. All fibres were resuspended in medium and sonicated immediately before use.

Cells and their culture

V79-4 cells were grown in MEM supplemented with glutamine and 15% heat inactivated foetal calf serum (HIFCS), and incubated at 37°C in 5% CO_2.

C3H10T1/2 cells were grown in LGA (low glucose Dulbecco's modified Eagle's) medium supplemented with glutamine and 10% HIFCS, and incubated at 37°C in 8% CO_2.

P388D1 cells were grown in DMEM supplemented with glutamine and 10% HIFCS and incubated at 37°C in 8% CO_2.

BL9L cells (Skilleter *et al.*, 1983) were grown in William's Medium E supplemented with 5% heat inactivated foetal calf serum in an atmosphere of 5% CO_2.

Radio-labelling of cells

Either: Cells were grown in 1 ml/well medium and incubated overnight in a 24-well plate with [^{3}H]-glycerol to a final concentration of 4 μCi/well. They were washed twice in complete medium then left 5 h before adding TPA or fibres.

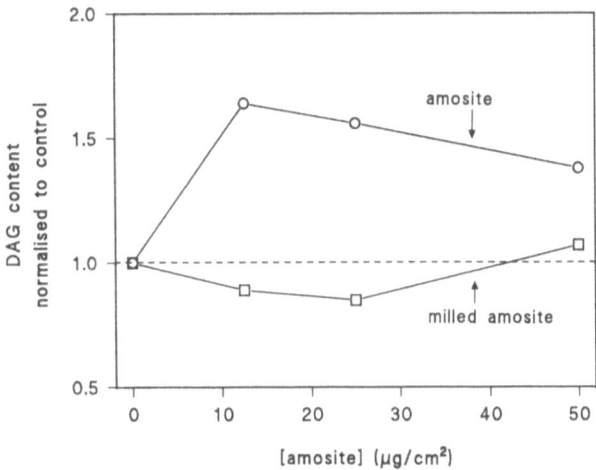

Figure 1. Dose-response relationship of DAG content of
C3H10T1/2 cells treated with amosite or milled amosite for 18h.

Or: Cells in three 8-well plates containing 1.5 ml medium/ well were incubated overnight with [^{14}C]-glycerol at a final concentration of 5 μCi/well. The cells were washed twice with complete medium then fibres added and incubated with the cells overnight.

Exposure of cells to fibres

Autoclaved fibres were made up in medium to give a suspension of 2 mg/ml, sonicated for 5 min and added to cells to give the required concentration.

Either: The fibre suspension was added to near confluent cell cultures and after incubation for 18 h the cells were washed twice with complete medium. The medium was decanted and the cells killed by adding 150 μl of a 10% TCA solution followed by suspension in 1 ml of PBS.

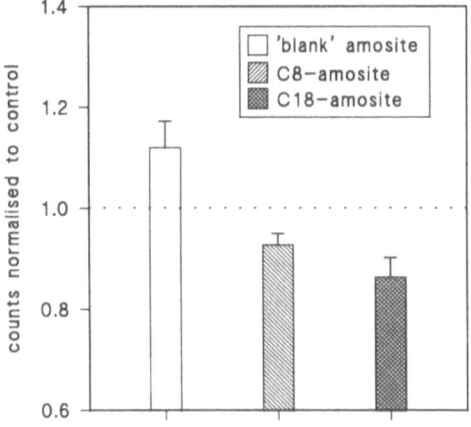

Figure 2. DAG content of C3H10T1/2 cells after treatment for
18 h with amosite, C8 or C18 coated amosite.

Or: With cells in suspension 1 ml of the fibre suspension was added to a 9 ml volume of cells and medium in a shaking waterbath at 37°C to give a final concentration of fibres of 200 μg/ml. Two ml aliquots were taken at desired time points, spun at 2000g for 5 min, the medium removed, 300 μl of 10% perchloric acid added and the mixture resuspended in 2 ml PBS.

The mixtures were spun at 2000g for 5 min, the supernatent discarded and the pellet extracted with 2 ml of $CHCl_3$/MeOH (2:1) (Bligh & Dyer, 1959) for 1 h. 300 μl of 5 mM EDTA were added and the bottom layer removed and evaporated to dryness.

Derivatisation

The diacyl glycerols in the lipid extract were acetylated with [^{14}C]-acetic anhydride in the presence of pyridine (Hoskins & Evans, 1989). Mono-, di- and tri-glycerides were separated by tlc on silica gel in chloroform/acetone (96:4) and the final analysis and quantitation of the DAGs was on a Vanguard autoscanner. Standards and controls were run on the plates to monitor the analytical procedure.

Integrity of cell communication

A method based on that of Oh and co-workers (1988) was used: essentially confluent cultures of BL9L cells were treated with fibres or TPA in medium containing 1% serum. After exposure the cultures were treated with TPA for 10 minutes, washed several times with PBS and scrape lines made with a specially constructed comb through a layer of 0.05% Lucifer yellow CH. After 3 minutes at room temperature, the monolayers were again washed to remove excess dye. The scratches and surrounding cell layers were examined immediately with a fluorescence inverted microscope. The width of the fluorescent band (in number of cells) was estimated at at least 10 randomly selected sites along the scratches.

RESULTS AND DISCUSSION

Integrity of cell communication:

When a scratch (or scrape) is made through a cell layer immersed in a solution of the dye Lucifer yellow CH the membranes of cells at the edge of the scratch are damaged, allowing dye into the cell, but quickly reseal. These cells continue to communicate with adjacent cells through gap-junctions and the dyestuff enables this process to be visualised as dye diffuses from cell to cell. Under the fluorescence microscope it is possible to see dye in cells up to 10 cells away from the loaded cell. Addition of TPA immediately prior to scrape-loading reduces the functional integrity of the gap-junctions between cells and thus limits the spread of the dyestuff. Longterm treatment with TPA reduces, or down-regulates, the short-term effect. Cells that have been treated overnight with TPA, washed with medium and then scrape-loaded no longer show the reduction in gap-junction integrity following further treatment with TPA. Asbestos does not interfere in the short-term with the formation of gap-junctions between cells. However long-term, 18 h, treatment with fibres does down-regulate any subsequent short-term effect produced by the addition of TPA.

Intracellular DAG levels

The reaction of a soluble agonist with cells in culture produces a transient increase in intracellular DAGs (Huang *et al.*, 1988). Mineral fibres behave similarly but the reac-

TABLE 1. Intracellular DAG levels in C3H10T1/2 cells after treatment for 18h with amosite

Expt. no.	[amosite] (μg/cm^2)	DAG levels $\pm$ SEM (normalised to controls)
1	3.9	1.04 $\pm$ 0.11
	11.7	1.03 $\pm$ 0.05
	23.4	1.55 $\pm$ 0.11*
2	12.5	1.64 $\pm$ 0.11**
	25.0	1.56 $\pm$ 0.14*
	50.0	1.38 $\pm$ 0.08*

* $P < 0.05$, ** $P < 0.01$

tion is less consistent probably because of the heterogeneity of the fibre samples at the concentrations used though it is likely that there are other contributory factors responsible for this variability.

In contrast to the effect of a soluble agonist which produces a DAG response within seconds, the response to fibres is slow. We have treated a variety of cell types in culture with asbestos and shown a sustained increase in intracellular DAGs which occurs over a period of hours. Initially these experiments were carried out using cell monolayers but more recently we have used cell suspensions to eliminate any part of the signal due to attachment to a plastic surface. When a near confluent layer of P388D1 cells were treated for 18h with amosite there was a dose-dependent increase in DAG levels. This was significant ($P < 0.05$) at a fibre load of 23.4 μg/cm^2. Similar results were obtained with C3H10T1/2 cells (table 1) although dose dependence was not seen at the test concentrations.

When P388D1 cells were treated with chrysotile the increase in DAG release only became significant after 8 hours. Measurements at 2 and 8 h were little different to controls. After 18 h both P388D1 and V79-4 cells treated with chrysotile showed significant increases in DAG levels although not in a dose-dependent fashion (table 2). The lack of dose dependence in most of our experiments is not surprising. A suspension of fibres settling onto a monolayer of cells which only present at most half of their surface for interaction will, in time, cover the exposed area completely. The duration of our experiments is not long enough for settlement to be complete. UICC samples are only homogeneous down to 100 mg. Therefore when cells are treated with fibres there is asynchrony in any response because of the time taken for cells and fibres to meet and the heterogeneity at that level of the fibre sample. These factors can obscure any dose relationship especially at low concentrations. Also in the case of amphibole fibres there is a charge-barrier to be overcome which further complicates cell/fibre interaction.

In early experiments cells were labelled with either ^{14}C- or tritium labelled glycerol in the cell culture medium. An excess of glycerol carrier was added to the medium to ensure maximum labelling of cell lipids. This sort of approach was not satisfactory because of expense and the problem of the disposal of quantities of radioactive waste.

TABLE 2. Intracellular DAG levels in cells after treatment for 18h with chrysotile

cell type	[amosite] (μg/cm^2)	DAG levels $\pm$ SEM (normalise to controls)
P388D1	3.9	1.40 ± 0.06
	7.8	1.76 ± 0.19 **
	15.6	1.66 ± 0.15 *
V79-4	3.9	1.40 ± 0.10
	7.8	1.39 ± 0.11 *
	15.6	1.35 ± 0.13 *

* $P < 0.05$, ** $P < 0.01$

Additionally growing cells in medium containing radioactive glycerol results in the labelling of all the cell lipids which is unnecessary and may contribute to a high background during radiochemical analysis. Therefore we used an analytical strategy in which only the final analyte containing the non-polar glycerides was made radioactive by acetylation with [^{14}C]-acetic anhydride (Hoskins & Evans, 1989). The equivalence of the two methods was demonstrated by first growing cells in a medium containing tritiated glycerol and then

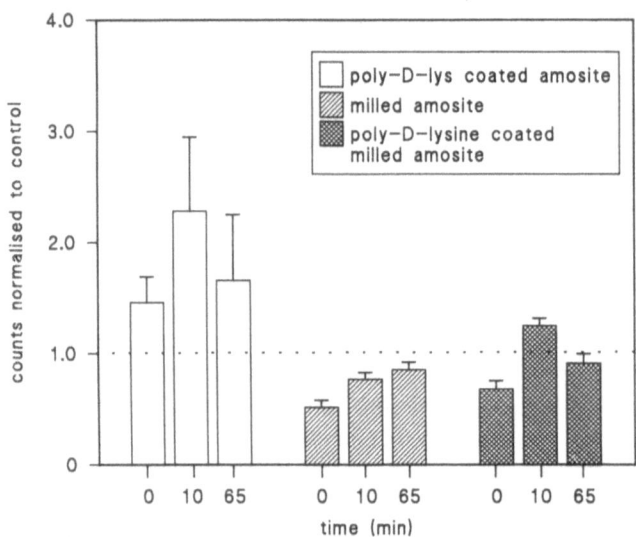

Figure 3. DAG content of C3H10T1/2 cells after treatment with poly-D-lysine coated amosite, milled amosite or milled amosite coated with poly-D-lysine.

acetylating the lipid fraction with $[^{14}C]$-acetic anhydride. The doubly labelled acetyldiacylglycerols were analysed after chromatography on acetate plates, cutting out the areas of interest and counting in a liquid scintillation counter. The R_f values measured for each of the two isotopes were the same.

To examine the importance of fibre morphology on DAG release cells were treated with amosite milled, using an agate ball mill to avoid metal contamination. After milling the particles have a similar diameter to the original fibres but few have an aspect ratio greater than 3:1, which is usually taken to define a fibre. The particles, which have identical chemistry to the amosite fibres, do not cause DAG release. Treatment of C3H10T1/2 cells for 18h with milled amosite at doses of 12.5 to 50 $\mu g/cm^2$ produced no increase in intracellular DAG levels (figure 1).

Shielding the anionic polysilicate of the amosite surface by effectively bonding hydrocarbon chains to it (C_8- and C_{18}-amosite) produces a product which does not stick to cells (Brown et al., 1990b). Reaction of these surface modified materials with C3H10T1/2 cells for 18 h did not produce an increase in DAG levels (figure 2).

Since an increase in intracellular DAG levels appears to signal an interaction with a surface, part of the measured DAGs in a cell monolayer must result from the attachment of the cell to the plastic surface of the dish. This provides a constant background in any measurements of DAG release. It can be removed by using cells in suspension. All of the subsequent work described here has been carried out using cells in this way. The attachment of cells to fibres, and also the attachment of cells to plastic culture dishes, involves the serum protein fibronectin (Brown et al., 1991a). When a cell culture medium is depleted of fibronectin it supports less cell/fibre interaction. It is believed that the first step in cell fibre interaction is the coating of the fibre with protein from the medium. This will change the surface charge of the fibre which may assist in its interaction with a cell. Fibronectin binds to a cell through receptors on the cell surface which recognise the tripeptide sequence RGD. This binding can be blocked by the peptide GRGDS (Brown et al., 1991b).

When amosite was added to a suspension of C3H10T1/2 cells in medium and fibronectin added to a final concentration of 50 $\mu g/ml$ immediately before the fibres there was generally a slight increase in DAG levels after 75 mins but this was not statistically significant in all experiments. Other end-points of cell/fibre interaction had given a positive indication by this time (see Brown et al., this volume). In some experiments fibronectin alone produced the same result. Pre-coating the fibre with fibronectin before adding it to the cells also gave the same result. The degree of cell/fibre reaction is very dependent upon the concentration of fibronectin which may explain these negative results.

To reduce the asynchrony mediated by charge fibres were coated with poly-lysine which renders them strongly positively charged. Fibres so treated reacted almost immediately with cells which have a negatively charged surface. When amosite is coated with poly-lysine and added to C3H10T1/2 cells the protein-fibre complex sticks more quickly to cells than native amosite and produces a faster response and a greater release of DAGs (figure 3). The treatment of C3H10T1/2 cells with poly-D-lysine (equivalent to poly-L-lysine but used because of its use in concomitant in vivo experiments not reported here) coated amosite produced about a 50% increase in DAG levels. Interaction was immediate and levels were sustained for at least 75 mins. When milled amosite coated with poly-lysine was used there was no increase in DAG levels.

CONCLUSIONS

Overall our results show that the phosphatidylinositol second messenger pathway, which when activated by a soluble agonist is a fast response and short duration system, is switched on by fibres for hours rather than seconds. This is probably because cell/fibre interaction is irreversible while soluble agonists have a finite residence time on a receptor. Amosite and chrysotile fibres cause DAG release from cells in culture. Changing the surface chemistry of the fibre is important in determining the degree of response that occurs. When the surface of the fibre is effectively coated with hydrocarbons cells and fibres do not interact and there is no increase in intracellular DAGs. Coating the fibre with a cell attachment protein such as fibronectin should increase the speed of cell/fibre interaction and the amount of intracellular DAGs. However, we were unable to demonstrate an increase in cell DAG levels probably because the concentration of fibronectin in the medium is critical. Fibres coated with a synthetic protein, poly-D-lysine, did cause an immediate rise in DAG levels. Destroying the fibre morphology by milling produced a material which was inactive in all our experiments.

REFERENCES

Berridge,M.J. and Irvine,R.F. (1989) Inositol phosphates and cell signalling. *Nature* **341**:197-205.

Blackshear,P.J., Witters,L.A., Girard,P.R., Kuo,J.F. and Quamo,S.N. (1985) Growth factor-stimulated protein phosphorylation in 3T3-L1 cells. *J. Biol. Chem.* **260**:13304-13315.

Bligh,E.G. and Dyer,W.J. (1959) A rapid method of total lipid extraction and purification. *Canad. J. Biochem. Physiol.* **37**:911-917.

Brown,R.C., Poole,A. and Fleming,G.T.A. (1983) The influence of asbestos dust on the oncogenic transformation of C3H10T1/2 cells. *Cancer Letters* **18**:221-227.

Brown,R.C., Hoskins,J.A., Cole,K.J., Evans,C.E. and Sara,E. (1989) Effects of Asbestos on some aspects of the Cell Second Messenger System. In "Effects of Mineral Dusts on Cells", Mossman,B.T. and Bégin,R.O., eds. NATO ASI Series Vol **H30**, pp. 375-381.

Brown,R.C., Hoskins,J.A., Miller,K. and Mossman,B.T. (1990a) Pathogenetic Mechanisms of Asbestos and other Mineral Fibres. *Molec. Aspects Med.* **11**:325-349.

Brown,R.C., Carthew,P., Hoskins,J.A., Sara,E. and Simpson,C.F. (1990b) Surface modification can affect the carcinogenicity of asbestos. *Carcinogenesis* **11**:1883-1885.

Brown,R.C., Hoskins,J.A., Sara,E.A., Cole,K.J. and Evans,C.E. (1991a) *In vitro* methods to determine the biological activities of particulate mineral pollutants. *In Vitro Toxicol.* (in press).

Brown,R.C., Hoskins,J.A. and Evans,C.E. (1991b) Factors affecting the interrelationship of asbestos fibres with mammalian cells: a study using cells in suspension. *Ann. Occup. Hyg.* **35**:25-34.

Chamberlain,M. and Tarmy,E.M. (1977) Asbestos and glass fibres in bacterial mutation tests. *Mutation Res.* **43**:159-164.

Chamberlain,M. (1982). The influence of mineral dusts on metabolic co-operation between mammalian cells in tissue culture. *Carcinogenesis* **3**:337-339.

Harrison,P.T.C. and Heath,J.C. (1988) Apparent synergy between chrysotile asbestos and *N*-nitrosoheptamethyleneimine in the induction of pulmonary tumours in rats. *Carcinogenesis* **9**: 2165-2171.

Hei,T.K., Geard,D.E., Osmak,R.S. and Travisano,M. (1985) Correlation of *in vitro* genotoxicity and oncogenicity induced by radiation and asbestos fibres. *Br. J. Cancer* **52**:591-597.

Hoskins,J.A. and Evans,C.E. (1989) Measurement of intracellular diacylglycerols by acetylation and thin-layer chromatography. *J. Chromatog. Biomed. Appl.* **490**:439-443.

Huang,S.J., Monk,P.N., Downes,C.P. and Whetton,A.D. (1988) Platelet-activating factor-induced hydrolysis of phosphatidylinositol-4,5-bisphosphate stimulates the production of reactive oxygen intermediates in macrophages, *Biochem. J.* **249**:839-845.

Huang,S.L. (1979) Amosite, chrysotile and crocidolite asbestos are mutagenic in Chinese hamster lung cells. *Mutation Res.* **68**:265-274.

National Research Council (1984) "Asbestiform Fibers: Non-occupational Health Risks" National Academy Press, Washington DC.

Oh,S.Y., Madhukar,B.V. and Trosko,J.E. (1988) Inhibition of gap junctional blockage by palmitoyl carnitine and TMB-8 in a rat liver epithelial cell line. *Carcinogenesis* **9**:135-139.

Oshimura,M., Hesterberg,T.W., Tsutsui,T. and Barrett,J.C. (1984) Correlation of asbestos-induced cytogenetic effects with cell transformation of Syrian hamster embryo cells in culture, *Cancer Res.* **44**:5017-5022.

Reiss,B., Tong,C., Telany,S. and Williams,G.M. (1983) Enhancement of Benzo[a]pyrene mutagenicity in rat liver epithelial cells. *Environ. Res.* **31**:100-104.

Shelby,M.D. (1988) The genetic toxicity of human carcinogens and its implications. *Mutation Res.* **204**:3-15.

Skilleter, D.N., Price, R.J. and Legg, R.F. (1983) Specific G1-S phase cell cycle block by beryllium as demonstrated by cytofluorometric analysis. *Biochem. J.* **216**:773-776.

Yotti,L.P., Chang,C.C. and Trosko,J.E. (1979) Elimination of metabolic co-operation in Chinese hamster cells by a tumor promotor. *Science* **206**:1089-1091.

APPARENT PROMOTION BY CHRYSOTILE ASBESTOS OF NHMI-INITIATED LUNG TUMOURS IN THE RAT

P.T.C.Harrison and *J.C.Heath

BP Group Occupational Health Centre
Surrey Research Park, Guildford
Surrey, GU2 5YQ, U.K.

*The Mill
Elsworth
Cambridge, CB3 8LJ, U.K.

ABSTRACT

The carcinogenic potency of chrysotile asbestos in man has long been the subject of debate. It is now generally accepted that asbestos can play a significant role as a promoter of lung carcinogenesis, as demonstrated for example by the smoking-asbestos interaction in humans, but there is relatively little experimental evidence corroborating this well-documented epidemiological observation.

In a study designed to investigate the carcinogenic effects of chrysotile and N-nitrosoheptamethyleneimine (NHMI; a proven lung carcinogen in rats), alone and in combination, it was found that the overall lung tumour incidence rate in the group of animals receiving chrysotile and NHMI together (16.0%) was substantially higher than in the groups receiving chrysotile alone (1.2%) or NHMI alone (4.2%). The incidence of significant pulmonary hyperplastic lesions paralleled the trend in tumour incidence (32.0%, 1.1% and 6.1% respectively). Further analysis of the data revealed an apparent synergy between chrysotile and NHMI in the causation of lung tumours, but because in this experiment the number of tumours induced by chrysotile alone did not exceed a likely background rate, it is considered more appropriate to regard the results of this experiment as confirming a promoting effect of chrysotile on NHMI-initiated lung tumours.

Additional studies are now in progress to further investigate these findings, using the inhalation route for chrysotile exposure (instead of intratracheal instillation) and incorporating multiple NHMI dose levels. It is hoped that these experiments will lead to the establishment of a protocol for investigating the lung tumour promoting potential of inhaled particulates.

INTRODUCTION

The role of asbestos as a significant risk factor in human lung carcinogenesis has been conclusively demonstrated (Lee, 1985), and although the specific role played by asbestos is still the subject of debate, it is generally believed to act as a powerful promoter rather than as a complete carcinogen for tumours of the respiratory tract (Topping & Nettesheim, 1980; Mossman et al.,1985). A possible exception is crocidolite, which is known to act as a complete carcinogen for tumours of the mesothelium (Mossman et al., 1983). A synergistic relation between smoking and asbestos exposure has been clearly established (Saracci, 1977; Hammond et al., 1979), and it is likely that in addition to the known effects of smoking, concomitant exposure to low levels of any number of other air-borne or systemic putative lung carcinogens could result in an elevated risk of lung cancer. Indeed, some epidemiological studies have identified possible synergistic effects between asbestos exposure and carcinogenic agents present in urban air in the causation of lung cancer (Selikoff, 1977).

The aim of this study was to investigate interactive effects between small doses of chrysotile asbestos, administered intra-tracheally, and N-nitrosoheptamethyleneimine (NHMI) administered by subcutaneous injection. N-nitrosamines and their precursors are widespread in the environment and experimental work in animals has shown some N-nitrosamines to be very potent carcinogens, often affecting particular target organs (Odashima, 1980). NHMI is a specific lung carcinogen in rats (Lijinsky, et al., 1969, Taylor & Nettesheim, 1975).

MATERIALS AND METHODS

Female Lister hooded rats of the Strangeways strain were used; the animals were maintained at an initial density of five per cage, had free access to drinking (tap) water and were fed *ad libitum* with Pillsbury's modified rat and mouse breeding diet (Pillsbury Ltd., Edgbaston, UK). The animals were 10-12 weeks old at the time of first treatment.

UICC Canadian chrysotile (Timbrell et al.,1968), provided by Dr.J.C.Wagner, was used after brief milling in a 'Spex' agate ball mill (Glen Creston, Stanmore, UK). NHMI was kindly provided by Dr.W.Lijinsky of the Frederick Cancer Research Centre, Maryland, USA, who first synthesised it and showed it to be a potent carcinogen practically specific for lung cancer in rats (Lijinsky et al., 1969). A 5% (w/v) solution of NHMI in ethanol was prepared and aliquots of this were added to 0.85% aqueous NaCl solution to produce a 0.4% dilution.

Experimental groups were set up as shown in Table 1. Sterile 2 mg samples of the milled asbestos were suspended in 0.4 ml sterile Tyrode's solution (Difco Laboratories, Detroit, USA) which also contained 60 μg/ml penicillin and 100 μg/ml streptomycin, and single aliquots were delivered by intratracheal instillation (i/t) according to the method of Saffiotti and co-workers (1968). Animals were first anaesthetised by intraperi-

TABLE 1. Treatment Groups

A)	Control	0.4 ml physiological saline (i/t) only
B)	Chrysotile	2 mg chrysotile in 0.4 ml physiological saline (i/t)
C)	NHMI	10 x 1 mg *N*-nitrosoheptamethyleneimine (s/c)
D)	Chrysotile + NHMI	2 mg chrysotile in 0.4 ml physiological saline (i/t) plus 10 x 1 mg *N*-nitrosoheptamethyleneimine (s/c)

toneal injection of Brietal Sodium (Lilly, Basingstoke, UK). Preliminary experiments using carbon black suspensions confirmed the i/t technique to be successful in distributing particulate material widely throughout the lungs. NHMI was administered by weekly subcutaneous injection of 0.25 ml of 0.4% NHMI solution for 10 weeks, beginning 6 weeks after i/t instillation of chrysotile. Six weeks had previously been determined, by sequential histological examination, to be sufficient time for the induction of tissue damage and attempts at repair (unpublished data).

The number of animals in each treatment group is shown in Table 2. In the statistical analysis of tumour incidence, the number of animals alive 250 days after initial treatment was used as the effective number of animals at risk; previous studies of pulmonary carcinogenesis in laboratory animals, in this laboratory and elsewhere (Laskin *et al.*, 1970), have demonstrated latent periods invariably in excess of 250 days for the development of experimental lung tumours, especially when associated with asbestos exposure. In all other calculations the effective number of animals is taken to be the total number of animals treated but not including those dying within 7 days of treatment or those for which lung tissue could not be adequately assessed.

Except for animals found dead or killed earlier for humane reasons, the rats were killed by ether inhalation 104 weeks after first treatment. A full post-mortem examination of the thoracic and abdominal cavities was performed on each animal. Lungs (and other abnormal tissues) were excised and fixed in Carnoy's fluid or EAFS (ethanol-acetic acid-formol-saline) (Harrison, 1984). Representative samples (including specimens from each lung lobe) were processed and embedded in paraffin wax according to standard methods. Five micron sections were stained with Heidenhain's azan (substitute) stain and/or haemotoxylin and eosin (H&E). Detailed histopathological observations were recorded, with particular emphasis on the identification and description of epithelial metaplastic, hyperplastic and neoplastic changes. Epithelial hyperplastic changes were classified as either alveolar type II cell hyperplasia, focal bronchiolar-alveolar cell hyperplasia, or adenomatosis. Focal bronchiolar-alveolar cell hyperplasia and adenomatosis are both regarded as significant hyperplastic lesions. Tumours (excluding hamartomas or metastases from distant primaries) were ultimately classified as benign or malignant, of epithelial (including mixed cell type) or stromal/mesodermal type. Full details of the adopted classification of pulmonary lesions are given elsewhere (Harrison & Heath, 1988).

Statistical analysis of the incidence of tumours and hyperplastic lesions was performed using the *G* statistic, incorporating Williams' correction (Sokal & Rohlf, 1981).

TABLE 2. Treatment Group Sizes and Animal Longevity

Group	Initial Group size	Effective no. of animals (1)[*]	Effective no. of animals (2)[#]	Survivors to termination (104 Weeks)
A) Control	50	48	47	22 (44%)
B) Chrysotile	100	90	86	42 (42%)
C) NHMI	50	49	48	17 (34%)
E) Chrysotile + NHMI	50	50	50	18 (36%)

[*] Excludes all animals which died within 7 days of initial treatment or tissues were too autolytic for reasonable assessment of pulmonary changes.

[#] Excludes in addition all animals which died within 250 days of initial treatment. These figures are used as effective numbers of animals at risk for development of tumours.

The method of demonstrating synergistic interactions is based on the work of Reif (1984, 1985) who considers apparent synergy to occur if:

$$(p_{AB} - p_0) > (p_A - p_0) + (p_B - p_0)$$

$$\text{i.e. } p_{AB} > p_A + p_B - p_0$$

where p_A, p_B, p_{AB} and p_0 are the incidence rates for sub-groups exposed, respectively, to A alone, B alone, A and B together, or neither A nor B. Evaluation of the statistical significance of data demonstrating apparent synergy was performed using an adapted z-test method.

RESULTS

Animal Longevity

The number and percentage of animals in each group surviving to 104 weeks are shown in Table 2. The longevity of animals treated with NHMI (with or without chrysotile) was less than that of control animals or those treated with chrysotile alone. There was little or no difference between control and chrysotile-treated animals.

Pulmonary Lesions

Incidences of hyperplastic and neoplastic lesions and other changes held to be of significance, namely alveolar epithelialisation (cuboidal metaplasia of the alveolar epithelium) and focal fibrosis, are shown in Table 3. Alveolar epithelialisation showed no apparent relation to treatment; the incidences of alveolar type II cell hyperplasia and focal bronchiolar-alveolar cell hyperplasia were increased in animals treated with NHMI;

TABLE 3. Pulmonary Lesions

Lesion incidences (percentage incidences shown in parenthesis)

Lesion	A Control	B Chrysotile	C NHMI	D Chrysotile +NHMI
Type II cell hyperplasia	12(25.0)	15(16.7)	22(44.9)$^{\#}$	17(34.0)*
Alveolar epithelialization	20(41.7)	32(35.6)	23(46.7)	25(50.0)
Focal fibrosis	5(10.4)	18(20.0)	10(20.4)	36(72.0)$^{**,++}$
Focal bronchiolo-alveolar cell hyperplasia	0(0.0)	0(0.0)	3(6.1)$^{(\#)}$	13(26.0)$^{**,++}$
Adenomatosis	1(2.1)	1(1.1)	0(0.0)	3(6.0)$^{(+)}$
Total:	**1(2.1)**	**1(1.1)**	**3(6.1)**	**16(32.0)$^{**,++}$**
Tumours - Benign	0(0.0)	0(0.0)	1(2.1)	1(2.0)
Malignant§	0(0.0)	1(1.2)	1(2.1)	7(14.0)$^{**,+}$
Total:	**0(0.0)**	**1(1.2)**	**2(4.2)**	**8(16.0)$^{**,(+)}$**

Statistical Probabilities:

(x)   0.05<P<0.1;   # = with respect to control group (A)

x   0.01<P<0.05;   * = with respect to chrysotile group (B)

xx   P<0.01;   + = with respect to NHMI group (C)

§ Malignant tumours were squamous cell carcinomas except for the single tumour in Group B which was an adenocarcinoma.

the incidence of focal bronchiolar-alveolar cell hyperplasia in animals treated with both NHMI and chrysotile (compared to those receiving either treatment alone) was highly significant (P<0.01). Incidences of focal fibrosis and malignant tumours (and, to a lesser extent, adenomatosis) showed a similar pattern to that of focal bronchiolar-alveolar cell hyperplasia, with significant increases in animals treated with both chrysotile and NHMI. Animals treated with chrysotile or NHMI alone had a higher incidence of focal fibrosis compared to the control group, but this was not statistically significant. Animals treated with NHMI alone also showed slight, statistically insignificant, increased incidences of benign and malignant tumours. Apparent synergy between chrysotile and NHMI (confirmed statistically) is indicated by the incidence values for tumours (benign plus malignant) and hyperplastic lesions (focal bronchiolar-alveolar cell hyperplasia plus adenomatosis).

Thus, in summary, chrysotile treatment alone and NHMI alone resulted in a small increase in focal fibrosis; NHMI alone also caused elevated incidences of alveolar type II cell hyperplasia, focal bronchiolar-alveolar cell hyperplasia and both benign and malignant lung tumours. Marked increases in the incidence of focal fibrosis, focal bronchiolar-alveolar cell hyperplasia and benign, malignant and total lung tumours occurred in the group of animals treated with both chrysotile and NHMI. Apparent synergy occurred between chrysotile and NHMI in the causation of pulmonary tumours and hyperplastic lesions. Also (not shown in Table 3), there was one lung fibroma in the group of animals treated with chrysotile and several malignant tumours of the upper respiratory tract in animals treated with NHMI (with or without chrysotile). The propensity of NHMI (and other nitrosamines) to produce tumours of the upper respiratory tract has been reported previously (Pour et al., 1976).

DISCUSSION

Although the pulmonary effects of asbestos exposure in man and in experimental animals are well documented, the effect of combined exposures to asbestos and other putative lung carcinogens has received relatively scant attention and few animal experiments have been performed to investigate this phenomenon.

In the present experiment, combined exposure to chrysotile asbestos and NHMI resulted in increased incidences of pulmonary fibrosis, focal epithelial hyperplasias and tumours compared to animals treated with either agent alone. With regard to tumour incidence, the results are in broad agreement with those of a similar study performed by Küng-Vösamäe & Vinkmann (1980), who found a significantly higher incidence of lung tumours in hamsters treated with both N-nitrosodiethylamine (NDEA) and chrysotile than those treated with NDEA alone, and indicate an apparent synergy between chrysotile and NHMI. However, because in the present experiment the tumour incidence in the group receiving chrysotile alone (1.2%) was no higher than a possible background rate (a background rate of 2.2% has been quoted for F344 rats (Sacksteder, 1976), and the figure obtained in this study is compatible with a broadly comparable study by Pylev and Shabad (1973) and with previous studies at the Strangeways Laboratory - unpublished data), it is considered more appropriate to regard the results of this study as indicating promotion by chrysotile of NHMI-initiated lung tumours rather than implying a specific synergistic relationship. NHMI is presumed to possess initiating potential since it is a known carcinogen practically specific for lung tumours (Lijinsky, et al., 1969, Taylor & Nettesheim, 1975).

The concept of asbestos as a promoting agent and results of experiments investigating synergistic relationships between asbestos and known carcinogens were recently reviewed by Jaurand (1989). Key among these studies are investigations of interactions between asbestos and radon gas, an ionising agent formed by natural decay of radium in soil, and benzo(a)pyrene (B(a)P), a known genotoxic carcinogen which is formed predominantly from the combustion of fossil fuels.

In experiments by Bignon and co-workers (1983) and Moncheaux and co-workers (1989), intrapleural or subcutaneous injection of chrysotile after inhalation exposure to radon gas (^{222}Ra) resulted in increased numbers of pulmonary malignancies compared to either treatment alone. The latter study demonstrated translocation of asbestos to the lung from a distant site; the chronic inflammatory reaction to the translocated fibres was considered to act as a possible promoter of pulmonary carcinogenesis (see below). Miller and co-workers (1965) administered chrysotile or amosite suspensions to hamsters by intratracheal instillation, with or without B(a)P, and recorded the time of appearance and yield of tumours of the respiratory tract. The authors concluded that the results obtained were consistent with the hypothesis that chrysotile (but not amosite) promoted benzo(a)pyrene carcinogenesis in the respiratory tract; similar conclusions were presented by Smith and co-workers (1970). Churg and co-workers (this volume), using a rat tracheal explant system, have demonstrated various interactive effects between asbestos and cigarette smoke but combined exposure of hamsters to asbestos and cigarette smoke by inhalation (Wehner *et al.*, 1975) have failed to provide any evidence for co-carcinogenic effects. Apparent synergy has been demonstrated between NHMI and cadmium metal in the presence of crocidolite in the induction of lung tumours in rats, but not directly between crocidolite and NHMI (Harrison & Heath, 1986), possibly because of the ability of crocidolite to act as a complete carcinogen.

Several *in vitro* experiments have attempted to investigate combination effects between asbestos and B(a)P. *In vitro* systems have also been used to determine whether asbestos acts as a promoter or initiator, but have generally failed to give definitive answers (Michiels, 1989). DiPaolo and co-workers (1983) demonstrated apparent enhancement of asbestos-induced morphologic transformation in Syrian hamster cells by subsequent exposure to benzo(a)pyrene, and Brown and co-workers (1983) found a putative synergistic effect demonstrated by an augmentation by asbestos dusts of the cell transforming ability of B(a)P in cultured murine fibroblasts. Hei and co-workers (this volume) have described a synergistic interaction between radon and asbestos in the oncogenic transformation of C3H10T1/2 cells. In contrast to these results, Mikalsen and co-workers (1988) found no effect of B(a)P adsorbed onto asbestos fibres on the transformation frequency of Syrian hamster embryo cells, and crocidolite did not promote the transformation of cells pre-exposed to B(a)P. Similarly Paterour and co-workers (1985) found no synergistic effect between chrysotile and B(a)P in the transformation of rat pleural mesothelial cells in culture and no clear interaction was demonstrated in a UDS/hepatocyte assay between chrysotile or xontolite (a fibrous calcium silicate) and dimethylnitrosamine, a genotoxic component of tobacco smoke (Denizeau *et al.*, 1985). Michiels and co-workers (1989) studied the effect of chrysotile and crocidolite in an initiation-promotion model on the Fischer rat embryo lung and demonstrated promoting activity but not initiating activity by the fibres. It thus appears that some *in vitro* experiments have demonstrated promoting activity by asbestos fibres, but results are dependent on the materials and models used.

While the results of the present study are considered to demonstrate apparent synergy between chrysotile and NHMI and to provide evidence that chrysotile shows promoting properties, it is acknowledged that for the clear demonstration of synergy dose-response relationships are required for each of the interacting materials, and that the experimental design does not allow for the proper use of the terms promotion and initiation as generally defined in relation to tumour induction in skin-painting experiments (Shubik, 1984). Nevertheless the results of this study indicate an enhancement by chrysotile of the carcinogenic response to the administration of a known chemical lung carcinogen, this is consistent with the action of asbestos as a promoter. The mechanism by which asbestos may act as a promoter is not known, but its ability to cause chronic lung irritation is perhaps significant; any agent which causes cellular proliferation is a potential enhancer of the carcinogenic process. The finding of increased fibrosis in animals treated with both chrysotile and NHMI is an interesting result possibly supporting the hypothesis that the processes of fibrosis and carcinogenesis are associated (Kotin, 1984; Kuschner, 1984). Perhaps chronic inflammation is indeed the link, resulting in fibrosis and at the same time creating a bed of proliferating cells on which initiating agents may act; it was

Browne (1986) who underlined the fact that it may not be asbestos *per se* but asbestosis (even sub-clinical asbestosis) which renders the lung susceptible to subsequent malignancy. Also of potential importance are the observations that certain asbestos fibres can interfere with cell division (Hesterberg & Barrett, 1985) and that asbestos can affect immune function (Hartmann, 1985), thereby perhaps acting as an epigenetic agent in tumour promotion.

In addition to tumour incidence and fibrosis, the incidence of bronchiolar-alveolar cell hyperplasia also was clearly related to combined treatment with NHMI and chrysotile; no animals receiving chrysotile alone had this lesion. This is a significant observation because it implies that bronchiolar-alveolar cell hyperplasia is not a direct result of non-specific "irritation" by chrysotile but is very much a consequence of NHMI and, especially, of NHMI and chrysotile co-treatment. It is proposed that this lesion is a true pre-neoplastic change which may be used as a marker for pulmonary tumourigenesis.

In conclusion, results of this study appear to confirm that a single small i/t dose of chrysotile asbestos can significantly enhance the carcinogenic potential of small doses of NHMI, perhaps by promoting the initiating activity of the nitrosamine. Induction of bronchiolar-alveolar cell hyperplasia and fibrosis appear to be associated with this process.

To follow up the results of this study, a further experiment investigating interactions between chrysotile asbestos and NHMI is now underway. In this experiment, animals are exposed to the asbestos by inhalation instead of by intratracheal instillation, and several different dose levels of nitrosamine are used. The ultimate aim of this work is to devise a protocol by which the potential synergistic or promoting activities of any particulate to which man may be exposed can be determined. This is considered important because although asbestos use is in decline it is possible that exposure to other particulate materials, especially other fibres, may result in promotion of tumours initiated by smoking or low level environmental exposure to other air-borne or systemic lung carcinogens.

ACKNOWLEDGEMENTS

The experimental work on which this paper is based was performed at Strangeways Research Laboratory, Cambridge, UK and formed part of a programme of work which the UK Health and Safety Executive commissioned J.C.H. to perform. The authors wish to thank Mr.W.G.Stebbings and Mr.D.C.Rogers for their invaluable assistance throughout the project.

REFERENCES

Bignon,J., Monchaux,G., Chamaud,J., Jaurand,M.C., Lafuma,J. and Masse,R. (1983) Evidence of various types of thoracic malignancies induced in rats by intrapleural injection of 2mg of various mineral dusts after inhalation of ^{222}Ra. *Carcinogenesis*, **4**:621-628.

Brown, R.C., Poole,A. and Fleming,G.T.A. (1983) The influence of asbestos dust on the oncogenic transformation of C3H10T½ cells. *Cancer Lett.* **18**:221-227.

Browne,K. (1986) Editorial: Is asbestos or asbestosis the cause of increased lung cancer in asbestos workers? *Br. J. Indust. Med.*, **43**:145-149.

Denizeau,F., Marion,M., Chevalier,G. and Cote,M.G. (1985) Genotoxicity of dimethylnitrosamine in the presence of chrysotile asbestos UICC B and xonotlite. *Carcinogenesis* **6**:1815-1817.

DiPaolo,J.A., DeMarinis,A.J. and Doniger,J. (1983) Asbestos and benzo(a)pyrene synergism in the transformation of Syrian golden hamster cells. *Pharmacology* **27**:67-73.

Hammond,E.C., Selikoff,I.J. and Seidman,H. (1979) Asbestos exposure, cigarette smoking and death rates. *Ann N Y Acad. Sci.* **330**:473-490.

Harrison,P.T.C. (1984) An ethanol-acetic acid-formol saline fixative for routine use with special application to the fixation of non-perfused rat lung. *Lab. Animals* **18**:325-331.

Harrison,P.T.C. and Heath,J.C. (1986) Apparent synergy in lung carcinogenesis: interactions between *N*-nitrosoheptamethyleneimine, particulate cadmium and crocidolite asbestos fibres in rats. *Carcinogenesis* **7**:1903-1908.

Harrison,P.T.C. and Heath,J.C. (1988) Apparent synergy between chrysotile asbestos and *N*-nitrosoheptamethyleneimine in the induction of pulmonary tumours in rats. *Carcinogenesis* **9**:2165-2171.

Hartmann,D.-P. (1985) Immunological consequences of asbestos exposure. *Surv. Immunol. Res.* **4**:65-68.

Hesterberg,H.W. and Barrett,J.C. (1985) Induction by asbestos fibres of anaphase abnormalities: mechanism for aneuploidy induction and possible carcinogenesis. *Carcinogenesis* **6**:32-47.

Jaurand,M.C. (1989) Particulate-state carcinogenesis: a survey of recent studies on the mechanisms of action of fibres. In, "Non-Occupational Exposure to Mineral Fibres", Bignon,J., Peto,J. and Saracci,R. (eds) IARC Scientific Publication No.90. IARC, Lyon. pp 54-73.

Kotin,P. (1984) Historical review of fibrogenicity and carcinogenicity of MMMF in experimental animals. In, "Biological Effects of Man-Made Mineral Fibres." WHO, Copenhagen. pp. 199-208.

Küng-Vösamäe,A. and Vinkmann,F. (1980) Combined carcinogenic action of chrysotile asbestos dust and *N*-nitrosodiethylamine on the respiratory tract of Syrian golden hamsters. In, "Biological Effects of Mineral Fibres." Wagner, J.C., ed., IARC, Lyon, Vol. 1. IARC Scientific Publications No.30/INSERM Symposia Series, Vol. **92**, pp 305-310.

Kuschner,M. (1984) Peer Review: Pathogenicity of MMMF in contrast to natural fibres. In, "Biological Effects of Man-Made mineral fibres." WHO, Copenhagen. pp 367-269.

Laskin,S., Kuschner,M., and Drew,R.T. (1970) Studies in pulmonary carcinogenesis. In, "Hanna,M.G., Nettesheim,P. and Gilbert,J.R., eds. *Inhalation Carcinogenesis.* US Atomic Energy Commission, Symposium Series, No.18. pp 321-352.

Lee,K.P.(1985) Lung response to particulates with particular emphasis on asbestos and other fibrous dusts. *CRC Crit. Rev. Toxicol.* **14**:33-86.

Lijinsky,W., Tomatis,L. and Weynon,C.E.M. (1969) Lung tumours in rats treated with *N*-nitrosoheptamethyleneimine and *N*-nitrosooctamethyleneimine. *Proc. Soc. Exp. Biol. Med. (NY)* **130**:945-949.

Michiels,F.M., Moëns,G., Montage,J.J. and Chouroulinkov,I. (1989) Biological effects of asbestos fibres on rat lung maintained *in vitro*. In, "Non-Occupational Exposure to Mineral Fibres", Bignon,J., Peto,J. and Saracci,R., eds. IARC Scientific Publications, No.90. IARC, Lyon. pp 156-160.

Mikalsen,S.-O., Rivedal,E. and Sanner,T. (1988) Morphological transformation of Syrian hamster embryo cells induced by mineral fibres and the alleged enhancement of benzo[a]pyrene. *Carcinogenesis* **9**:891-899.

Miller,L., Smith,W.E. and Berliner,S.W. (1965) Tests for effects of asbestos on benzo(a)pyrene carcinogenesis in the respiratory tract. *Ann. NY Acad. Sci.* **132**:489-500.

Monchaux,G., Chamaud,J., Morlier,J.P., Janson,X., Morin,M. and Bignon,J. (1989) Translocation of subcutaneously injected chrysotile fibres: potential cocarcinogenic effect on lung-cancer induced in rats by inhalation of radon and its daughters. In, "Non-Occupational Exposure to Mineral Fibres", Bignon,J., Peto,J. and Saracci,R., eds. Scientific Publications, No.90. IARC, Lyon. pp. 161-184.

Mossman,B.T., Cameron,G.S. and Yotti,L.P. (1985) Co-carcinogenic and tumour promoting properties of asbestos and other minerals in tracheobronchial epithelium. In, "Carcinogenesis - A Comprehensive survey, Vol. 8, Cancer of the Respiratory Tract, Predisposing Factors", Mass,M.G., Kaufmann,D.G., Siegfried,J.M., Steele,V.E. and Nesnow,S. (eds.) . Raven Press, New York.

Mossman,B.T., Light,W. and Wei,E. (1983) Asbestos: mechanisms of toxicity and carcinogenicity in the respiratory tract. *Ann. Rev. Pharmacol. Toxicol.* **23**:595-615.

Odashima,S. (1980) Overview: *N*-nitroso compounds as carcinogens for experimental animals and man. *Oncology* **37**:282-286.

Paterour,M.J., Bignon,J. and Jaurand,M.C. (1985) *In vitro* transformation of rat pleural mesothelial cells by chrysotile fibres and/or benzo[a]pyrene. *Carcinogenesis* **6**:523-529.

Pour,P., Stanton,M.F., Kuschner,M., Laskin,S. and Shabad,L.M. (1976) Tumours of the respiratory tract. In, "Tumours in Laboratory Animals", Turusov,V.S. ed.   IARC, Lyon. Vol.**1**., *Tumours of the Rat,* Part 2, pp 1-40.

Pylev,L.N. and Shabad,L.M. (1973) Some results of experimental studies in asbestos carcinogenesis. In, "Biological Effects of Asbestos", Bogovski,P. Gilson,J.C., Timbrell,V. and Wagner,J.C., eds. IARC, Lyon. IARC Scientific Publication No.8, pp. 99-107.

Reif,A.E. (1984) Synergism in carcinogenesis. *J. Natl. Cancer Inst.* **73**:325-39.

Reif,A.E. (1985) Letters to the editor: Synergism in carcinogenesis. *J. Natl. Cancer Inst.* **74**:729-731.

Sacksteder,M.R. (1976) Occurrence of spontaneous tumours in the germfree F344 rat. *J. Natl. Cancer Inst.* **57**:1371-1377.

Saffiotti,U., Cefis,F. and Kolb,L.H. (1968) A method for the experimental induction of bronchogenic carcinoma. *Cancer Res.* **28**:104-124.

Saracci,R. (1977) Asbestos and lung cancer: an analysis of the epidemiological evidence on the asbestos-smoking interaction. *Int. J. Cancer* **20**:323-331.

Selikoff,I.J. (1977) Air pollution and asbestos carcinogenesis: investigation of possible synergism. In, "Air Pollution and Cancer in Man", Mohr,U., Schmahl,D. and Tomatis,L., eds.   IARC Scientific Publications, No.16. IARC, Lyon. pp. 247-251.

Shubik,P. (1984) Progression and promotion. *J.Natl. Cancer Inst.* **73**:1005-1011.

Smith,W.E., Miller,L. and Churg,J. (1970) An experimental model for the study of cocarcinogenesis in the respiratory tract. In, "Morphology of Experimental Respiratory Carcinogenesis", Nettesheim,P., Hanna,M.G., Jr., and Deatherage,J.W., Jr., eds.   US Atomic Energy Commission, Oak Ridge. pp. 299-316.

Sokal,R.R. and Rohlf,F.J. (1981) *Biometry* (2nd Edition). W.H.Freeman and Co., New York. pp. 731-747.

Taylor,H.W. and Nettesheim,P. (1975) Influence of administration route and dosage schedule on tumour response to *N*-nitrosoheptamethyleneimine in rats. *Int. J. Cancer* **15**:301-307.

Timbrell,V., Gilson,J.C. and Webster,I. (1968) UICC standard reference samples of asbestos. *Int. J. Cancer* **3**:406-408.

Topping,D.C. and Nettesheim,P. (1980) Two-stage carcinogenesis studies with asbestos in Fischer 344 rats. *J. Natl. Cancer Inst.* **65**:627-630.

Wehner,A.P., Busch,R.H., Olson,R.J. and Craig,D.K. (1975) Chronic inhalation os asbestos and cigarette smoke by hamsters. *Environ.Res.* **10**:368-383.

EFFECTS OF AMOSITE ASBESTOS FIBERS ON THE FILAMENTS PRESENT IN THE CYTOSKELETON OF PRIMARY HUMAN MESOTHELIAL CELLS

Angela N.A. Somers[1], Elizabeth A. Mason, Brenda I. Gerwin,
Curtis C. Harris and John F. Lechner[2]

Laboratory of Human Carcinogenesis
Division of Cancer Etiology
National Cancer Institute
National Institutes of Health
Bethesda, MD 20892, USA

[1] Department of Epithelial Biology
Paterson Institute for Cancer Research
Manchester, U.K.

[2]Inhalation Toxicology Research Institute
P.O. Box 5890, Albuquerque
NM 87158, USA

ABSTRACT

In order to examine the possible relationship between asbestiform fiber caused cytotoxicity/carcinogenicity of human mesothelial cells and the plastic nature of their cytoskeleton, we have observed the positioning of intermediate filaments and microfilaments in cultured normal human mesothelial cells before and after ingestion of amosite. In addition, we have examined the relative fidelity of the mitotic spindle of these cells. We have found that asbestiform fibers had no significant affect upon the distribution of cytoskeletal filaments. However, the filaments and/or their subunits appear to adsorb to asbestos fibers. Further, the physical presence of the fibers within the cell appears to cause mitotic spindle dysfunction. Therefore, in human mesothelial cells, the mechanism of asbestos damage may include interaction of cytoskeleton filaments with the fibers which, in turn, potentiates disruption of normal spindle function.

INTRODUCTION

Compared to human airway epithelial and lung stromal fibroblastic cells, pleural mesothelial cells are exceptionally sensitive to the cytotoxic effects of asbestos fibers (Lechner et al., 1983, 1985). This cytotoxicity directly parallels epidemiological data that have established that, for non-smoking individuals exposed to asbestos, the risk of the mesothelioma is significantly greater than it is for airway epithelial carcinomas or lung sarcomas (Mossman & Gee, 1989). However, and in contrast to other types of cells (Setoyama et al., 1986), free-radical formation does not occur when normal human mesothelial cells phagocytose asbestiform fibers (Gabrielson et al., 1988). Thus, it is difficult to include free-radical formation in the mechanism of asbestos-caused cytotoxicity or carcinogenicity of human mesothelial cells. On the other hand, both the cytotoxicity and carcinogenicity data are directly reflected in the unique cytoskeletal nature of mesothelial cells.

The cytoskeleton consists of actin and tubulin microfilaments, both of which are present in virtually all cells and one type of at least five immunological and biochemically distinct classes of intermediate filament (Franke et al., 1978; Osborn et al., 1982; Steinert et al., 1985). The intermediate filament composition in a particular cell type appears to be regulated by the specific structural needs of the cell (Moll et al., 1982; Osborn et al., 1985) and, in general, only one of these filaments is present in any particular type of cell. Thus, keratins are found in epithelial cells, vimentin in mesenchymal cells, desmin in muscle cells, neurofilaments in neural cells and glial filaments in glial cells. In the lung, as expected, the epithelial and stromal fibroblastic cells have keratin and vimentin tonofilaments, respectively. However, the mesothelial cells are unique, i.e. they have a plastic tonofilament composition. Thus, under conditions of rapid proliferation vimentin fibers are prominent and keratin tonofilaments are rare. In contrast, when the cells reach confluence or when the cells are starved for growth factors and hydrocortisone, vimentin is difficult to detect and cytokeratin filaments are prominent (Wu et al., 1982; Connell & Rheinwald, 1983; Rheinwald et al., 1983; LaVeck et al., 1988).

There is a growing body of evidence to suggest that mitotic aneuploidy may play an important role in cellular transformation and tumor development (Tsutsui et al., 1983, 1984; Oshimura & Barrett, 1986). In agreement with the importance of aneuploidy in carcinogenesis are epidemiological and experimental studies that have indicated that agents that interfere with intermediate filament polymerization can result in karyotypic instability (Heston & White, 1978; Puck, 1979; Tsutsui et al., 1984), with the induction of aneuploidy in the resulting daughter cells (Somers et al., 1986; Parry et al., 1984; Parry, 1985). We have suggested that the unique and remarkably plastic cytoskeleton of human mesothelial cells may contribute both to the cytotoxicity of asbestiform fibers and contribute to the rapid evolution of aneuploidy and chromosome aberrations (Lechner et al., 1985). This study was performed to determine the effects of asbestos fibers on the distribution of the filaments present in the cytoskeleton of mesothelial cells in an effort to gain further insight into the mechanisms of asbestos cytotoxicity and carcinogenicity.

MATERIALS AND METHODS

Cell Culture

Primary mesothelial cell cultures were initiated from pleural effusions of non-cancerous adult patients with a medical history of thoracentesis. The fluid was centrifuged and the resulting pellet resuspended in growth media. This process was repeated a further two times before

482

the cells were inoculated into fibronectin coated 10 cm dishes at a ratio of 1 dish per 50 ml original fluid (Lechner *et al.*, 1985). The mesothelial origin of these primary cell cultures was certified by the presence of keratin, a variable cell morphology which was dependent upon the factors present in the growth media (Connell & Rheinwald, 1983), the production of hyaluronic acid (Wagner *et al.*, 1962) and the presence of long branched microvilli (Andrews & Porter, 1973), prior to use in experimental protocols.

Growth Media

The culture media used in the following studies has been described in detail elsewhere (LaVeck *et al.*, 1988). Briefly, nutrient medium was supplemented with 0.87 μM zinc-free insulin, 5 ng/ml epidermal growth factor, 10 μg/ml transferrin, 0.2 μM hydrocortisone, and 3% fetal calf serum.

Immunofluorescent Antibodies

(i) Keratin. Anti-keratin anti-sera raised in guinea pig (ICN) was used at a dilution of 1:15 in 0.01 M PBS at pH 7.6; fluorescein-conjugated anti-guinea pig IgG raised in rabbit (ICN) was used at a dilution of 1:64 in 0.01 M PBS at pH 7.6; rabbit serum was used as a blocking reagent to inhibit non-specific staining.

(ii) Vimentin. Anti-vimentin anti-sera raised in goat (ICN) was used at a concentration of 1:15 in 0.01 M PBS at pH 7.6; fluorescein-conjugated anti-goat IgG raised in rabbit (ICN) was used at a dilution of 1:32 in 0.01 M PBS at pH 7.6; and rabbit serum was used as a blocking reagent to prevent non-specific staining.

(iii) Tubulin. Anti-tubulin anti-sera raised in rabbit (ICN) was used at a dilution of 1:10 in 0.01 M PBS at pH 7.6; fluorescein-conjugated anti-rabbit IgG raised in goat (ICN) was used at a dilution of 1:32 in 0.01 M PBS at pH 7.6 and goat serum was used as a blocking reagent to prevent non-specific staining.

Fibers

UICC standard reference amosite asbestos was provided by Dr. V. Timbell at the Medical Research Council, U.K.. Stock concentrations of these materials were prepared by the procedure of (Haugen *et al.*, 1982) and were sterilized by autoclaving prior to use.

Exposure of Cells to Amosite

Double chamber glass microscope slides were inoculated with primary normal human mesothelial cells at a concentration of 1 x 10^5 cells per chamber (5 x 10^4 cells per cm^2). After a 24 hr incubation period to allow attachment of the cells and re-establishment of growth, non-confluent cultures were exposed to varying cytotoxic concentrations of UICC standard reference amosite asbestos. Untreated control cultures were given fresh media. The cells were incubated until 85-90% confluent, then fixed and stained by indirect immunofluorescent techniques for the presence of keratin, vimentin or tubulin.

Indirect Immunofluorescent Staining

Cells were fixed in cold methanol (-20°C) for 10 min, then in cold acetone (-20°C) for 10 sec. After air drying the slides were immediately placed in a solution of 0.5% Triton X-100 in Hanks balanced salt solution for 30 min. They were rinsed three times with 0.01 M PBS at pH 7.6, each wash lasting 5 min, placed in a moist chamber and incubated with the primary antibody solution for 60 min at 37°C. After rinsing with 0.01 M PBS at pH 7.6 three times, each wash lasting 10 min, they were incubated with

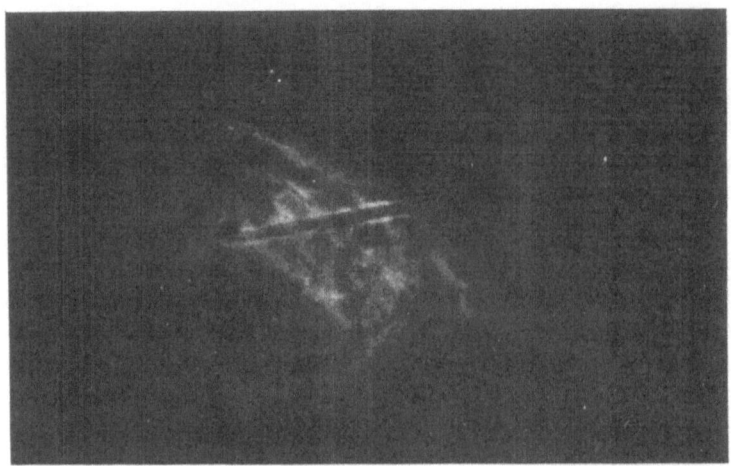

Figure 1. Immunofluorescence staining of cells, exposed to amosite, for keratin.

the appropriate blocking serum for 60 min to inhibit non-specific staining of antibodies present in the polyclonal secondary antibody solution. Following this they were rinsed again in 0.01 M PBS at pH 7.6: three washes each lasting 10 min. The slides were then incubated with the appropriate fluorescein-conjugated secondary antibody solution for 60 min and finally washed three times in 0.01 M PBS at pH 7.6: each wash lasting 10 min. The slides were finally rinsed in double distilled water three times: each wash lasting 5 min. After mounting with fluoromount, the distribution of the filaments in the cytoskeleton of untreated and amosite-treated cultures were compared.

Spindle Assay

Glass chamber microscope slides were inoculated with immortalized human mesothelial cells (MeT-5A) (Ke *et al.*, 1989) at a concentration of 1×10^5 cells per slide. After a 24 hr incubation period to allow attachment of cells and re-establishment of growth, non-confluent cultures were exposed for a 36-38 hr period or a 60-64 hr period to enable the observation of the cells at their first or second division. There were three replicate slides at each concentration of the test agents investigated. Following treatment the cells were harvested, fixed and stained by differential staining (Somers *et al.*, 1986). Briefly the cells were fixed at room temperature in a 3:1 methanol/acetic acid solution supplemented with 4 mM Ca^{++} and 1.5 mM Mg^{++} to preserve spindle proteins. There were three fixations each lasting 14 min. After air drying the cells were immersed in a pre-chilled 5% perchloric acid solution for 24 h at 4°C to remove RNA from the cells. After perchloric acid treatment the cells were rinsed in several changes of distilled water for 10 min and air dried once more. The cells were then stained for 24 hr in a solution of 0.5% brilliant blue R and 0.5% safranin O dissolved in 15% acetic acid. Brilliant blue R stains the spindle proteins blue and safranin O stains the mitotic chromosomes red.

The following criteria were used to assess the affects of the test agents on the fidelity of mesothelial cell division:

(i) an increase in abnormally dividing cells with one of the following phenotypes; chromosome clustering, chromosome scattering, chromatin mass or scattered chromatin;

(ii)   an increase in chromosome clustering alone - previous studies suggest that this is a very sensitive indicator of spindle damage (Parry, 1985);

(iii)   an increase in the mitotic index of the cells which indicates mitotic arrest;

(iv)   an alteration in the AT/M ratio of the cells which allows the determination of the stage at which mitotic arrest (if any) has occurred;

(v)   an increase in multipolar spindle formation which indicates irregular centriole function;

(vi)   an increase in chromosome dislocation from the spindle at metaphase, which would indicate irregular spindle function and/or centromere/kinetochore damage;

(vii)   an increase in lagging and bridging chromosomes at anaphase or telophase, which are also indicators of irregular spindle function and/or centromere/ kinetochore damage.

RESULTS AND DISCUSSION

The effect of UICC standard reference amosite asbestos (at concentrations ranging from 0.07 to 0.7 μg/cm^2) on the pattern of keratin and vimentin intermediate filaments present in the cytoskeleton of mesothelial cells was investigated first. For the cytoplasmic microtubules the concentration of amosite was increased, the range being from 0.07 to 1.4 μg/cm^2. Similar results were obtained for all three filament types. There were no obvious differences in the overall appearances of the cytoskeletons of untreated and amosite-treated mesothelial cells. Regularly, however, fibers were detected in the cytoskeletons of treated cells that were surrounded by an aura of fluorescence (Fig. 1).

It is possible that this observation is an artifact of the fixation and staining process. On the other hand, the fluorescence may represent the binding of filament subunits or the filaments themselves to the asbestos fiber. This latter interpretation is supported both by the absence of a fluorescence aura on fibers comparably fixed in buffer and preliminary *in vitro* studies (W. Brinkley, unpublished observations) that show that cytoskeletal proteins absorb to asbestos fibers. If the fluorescent aura is due to the association between cytoskeletal proteins and asbestos fibers, then tonofilament and mitotic tubulin polymerization could be affected. Since the tonofilament composition of mesothelial cells is highly plastic, with significant rates of subunit polymerization and depolymerization, absorption of the subunits to asbestos may, in a manner analogous to colchicine, result in a shift towards depolymerization which, in turn, has been suggested to result in karyotypic instability (Heston & White, 1978; Puck, 1979; Tsutsui *et al.*, 1984). Also, since the mitotic spindle forms via repolymerization of depolymerized cytoplasmic tubulin subunits, absorption of the tubulin subunits to asbestos may disrupt the kinetics of spindle assembly. As a possible consequence, the mitotic spindle may be defective.

Ingested asbestos fibers accumulate in the perinuclear area of the cell (Lechner *et al.*, 1985). Thus, it is possible that asbestos also directly affects the formation and/or function of the mitotic spindle. This possibility has not been investigated using normal mesothelial cells because, in culture, their mitotic index is too low for statistical quantification of mitotic spindle activites. However, we have recently initiated studies with near diploid, SV40 T-antigen gene immortalized human mesothelial cells (MeT-5A) (Ke *et al.*, 1989). Like normal human mesothelial cells, MeT-5A cells exhibit tonofilament plasticity (LaVeck *et al.*, 1988). A second common feature that both MeT-5A normal human mesothelial cell cultures share is a relatively high percentage of binucleate cells (Lechner

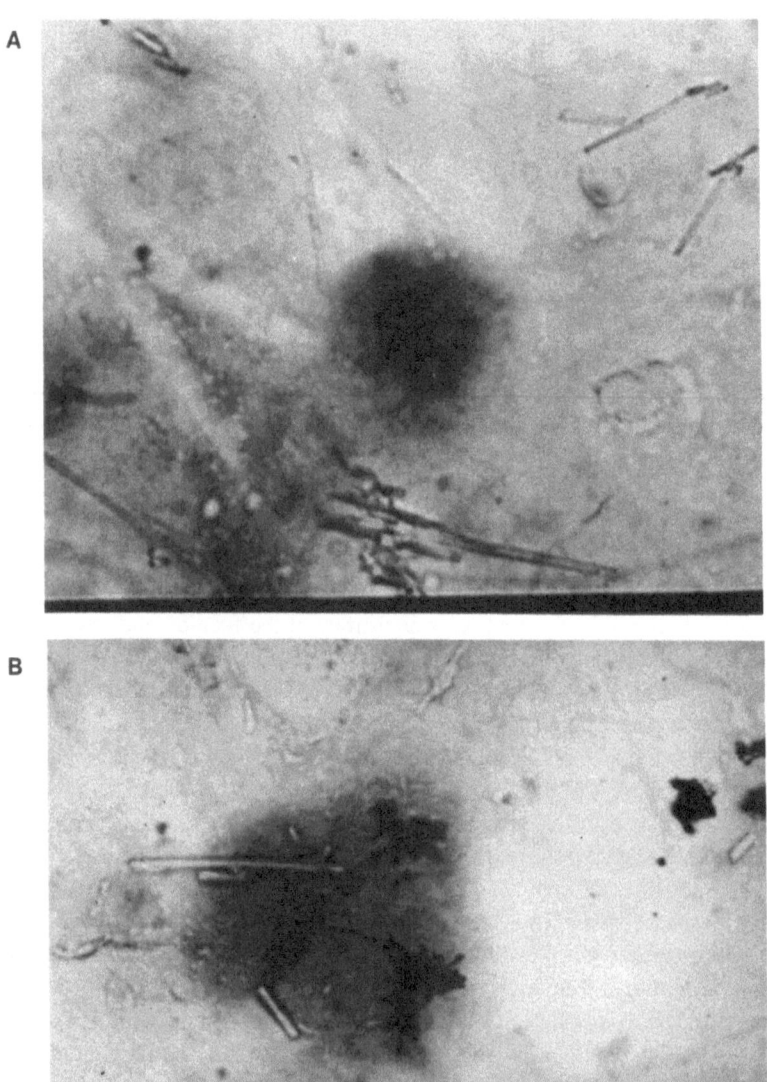

Figure 2. SV40 T-antigen gene immortalized human mesothelial cells (MeT-5a) (Ke *et al.*, 1989) exposed to amosite. A: Chromosome clustering; B: Multipolar anaphase.

TABLE 1. Fidelity of mitosis of human mesothelial cells exposed to asbestos

Condition	MI[1]	Aberrations			Abnormalities		
		D_M	$L_{A/T}$	MP	CM	SC	CC
Control	4.9	3.1	1.2	2.3		0.8	2.7
+ Amosite (1st mitosis)	5.8	4.2	2.0	3.0	0.5	2.6	6.5
+ Amosite (2nd mitosis)	4.9	4.0	0.4	6.9	0.9	0.9	5.0

[1] MI = Mitosis index; D_M = chromosome dislocated from mitotic plate; $L_{A/T}$ = lagging chromosomes at anaphase and metaphase; MP = multipolar metaphase; CM = chromatin mass; SC = scattered chromosomes; CC = chromosome clusters (Lechner *et al.*, 1990).

et al., 1990). Finally, and in marked contrast to human lung fibroblastic cells, both MeT-5A cells and normal human mesothelial cells have a high innate level of chromosomal aberrations, even when cultured in the absence of fibers. Specifically, the background level of chromosomal aberrations in normal human lung fibroblastic cells is >3%. In contrast, aberrations in the metaphases of normal human mesothelial cells ranges from 4-12% and the level for the MeT-5A cells is comparable (Table 1) (Lechner *et al.*, 1990).

The effects of amosite on MeT-5A mitotic fidelity are shown in Table 1. At the first mitosis, in cells containing asbestos, centriole separation is occasionally inhibited. This results in the formation of a single aster, which produces a monopolar spindle. When the chromosomes attach to the monopolar spindle they form the abnormal structure termed the chromosome cluster (Fig. 2A), which is considered to be an indicator of spindle damage (Somers *et al.*, 1986; Parry *et al.*, 1984; Parry, 1985). As division proceeds, although centromere separation is normal, anaphase and telophase chromosome movements do not take place. The net result is, that at the end of division, two discrete chromosome sets are present within one binucleate cell. When these binucleate cells enter a second division a multipolar spindle (Fig. 2B) would be formed and, as expected, a marked increase in multipolar metaphases was seen (Table 1). In such cells the disjunction and segregation of the chromosomes at anaphase and telophase would be irregular, resulting in the formation of aneuploid daughter cells.

Our results differ in specifics from those of Hesterberg and Barrett (1985) who demonstrated an increase in lagging chromosomes and other anaphase abnormalities in mammalian cells after exposure to asbestos. These types of dysfunction were not observed in human mesothelial cell cultures. In contrast, our results, especially the increase in chromosome clustering, permit us to hypothesize that amosite asbestos fibers disturb the fidelity of mesothelial cell division by inhibiting mitotic tubulin polymerization and/or centriole separation.

The hypothesis that a chromosomal imbalance may be important in tumor development stems from the consistent finding of non-random numerical chromosome changes in many forms of cancer (Barrett *et al.*, 1990). When aneuploidy arises in a cell it may alter cellular regulation by a number of mechanisms. For example, the number of genes as well as chromosomes will have been altered and therefore gene dosage effects may cause dysfunction of cellular regulation (Knudson, 1985; Oshimura *et al.*, 1988). Alternatively, the expression of a recessive gene may be allowed and this may alter cellular regulation (Knudson, 1985; Lee *et al.*, 1987; Vogelstein *et al.*, 1988). Finally there is the possibility of an indirect effect upon the cellular genetic complement, perhaps by an alteration of the rate of DNA synthesis or repair. Both of these processes precede cell division and dysregulation of these functions may, therefore, induce further genetic instability which in turn could alter cellular regulation (Oshimura *et al.*, 1988).

In summary, our observations suggest a dual effect of asbestos in human mesothelial cells. One is a shift in tonofilament and microtubulin depolymerization/polymerization kinetics due to the absorption of the subunits to asbestiform fibers. This, in turn, results in deformed and dysfunctioning mitotic spindles. Secondly, asbestos appears to physically interrupt centriole separation leading to aberrant mitotic mechanics, binucleate cells and eventually aneuploidy. This dual mechanism is in agreement with the increase in random hypodiploidy (Linnainma *et al.*, 1986) after exposure to asbestos and also correlates with the relatively heterogeneous nature of human mesothelioma (Corson, 1987). Finally, both the absorption of cytoskeletal elements to asbestos and asbestos-caused impairment of mitotic spindle mechanics may partially account for the enhanced cytotoxicity of asbestiform fibers for human mesothelial cells.

REFERENCES

Andrews,P.M. and Porter,K.R. (1973). The ultrastructural morphology and possible functional significance of mesothelial microvilli. *Anat. Rec.* **177**:409-426.

Barrett,J.C., Tsutsui,T., Tlsty,T. and Oshimura, M. (1990). Role of genetic instability in carcinogenesis. In, "Genetic Mechanisms in Carcinogenesis and Tumor Progression", eds. Harris,C.C. and Liotta,L.A. Wiley-Liss, New York, pp. 97-114.

Connell,N.D. and Rheinwald,J.G. (1983). Regulation of the cytoskeleton of mesothelial cells: reversible loss of keratin and increase of vimentin during rapid growth in culture. *Cell* **34**:245-253.

Corson, J.M. (1987). Pathology of malignant mesothelioma. In, "Asbestos-Related Malignancy", eds. Antman,K. and Aisner,J. Grune and Stratton, Orlando, pp. 179-199.

Franke,W.W., Schmid,E., Osborn,M. and Weber,K. (1978). Different intermediate-sized filaments distinguished by immunofluorescence microscopy. *Proc. Natl Acad. Sci. USA* **75**:5034-5038.

Gabrielson,E.W., Rosen,G.M., Grafstrom,R.C., Strauss,K.E., Miyashita,M. and Harris,C.C. (1988). Role of oxygen radicals in 12-0-tetradecanoylphorbol-13-acetate-induced squamous differentiation of cultured normal human bronchial epithelial cells. *Cancer Res.* **48**:822-825.

Haugen,A., Schafer,P.W., Lechner,J.F., Stoner,G.D., Trump,B.F. and Harris,C.C. (1982). Cellular ingestion, toxic effects, and lesions observed in human bronchial epithelial tissue and cells cultured with asbestos and glass fibers. *Int. J. Cancer* **30**:265-272.

Hesterberg,T.W. and Barrett,J.C. (1985). Induction by asbestos fibers of anaphase abnormalities: mechanism for aneuploidy induction and possibly carcinogenesis. *Carcinogenesis* **6**:473-475.

Heston,L.L. and White,J. (1978). Pedigrees of 30 families with Alzheimer disease: associations with defective organization of microfilaments and microtubules. *Behav. Genet.* **8**:315-331.

Ke,Y., Reddel,R.R., Gerwin,B.I., Reddel,H.K., Somers,A.N.A., McMenamin,M.G., LaVeck,M.A., Stahel,R., Lechner,J.F. and Harris,C.C. (1989). Establishment of a human *in vitro* mesothelial cell model system for investigating mechanisms of asbestos-induced mesothelioma. *Am. J. Pathol.* **134**:979-991.

Knudson,A.G. Jr. (1985). Hereditary cancer, oncogenes, and antioncogenes. *Cancer Res.* **45**:1437-1443.

LaVeck,M.A., Somers,A.N.A., Moore,L.L., Gerwin,B.I. and Lechner,J.F. (1988). Dissimilar peptide growth factors can induce normal human mesothelial cell multiplication. *In Vitro* **24**:1077-1084.

Lechner,J.F., Haugen,A., Tokiwa,T., Trump,B.F. and Harris,C.C. (1983). Effects of asbestos and carcinogenic metals on cultured human bronchial epithelium. In, "Human Carcinogenesis", eds. Harris,C.C. and Autrup,H. Academic Press, New York, pp. 561-585.

Lechner,J.F., Tokiwa,T., LaVeck,M.A., Benedict,W.F., Banks-Schlegel,S.P., Yeager,H. Jr., Banerjee,A. and Harris,C.C. (1985). Asbestos-associated chromosomal changes in human mesothelial cells. *Proc. Natl. Acad. Sci. USA* **82**:3884-3888.

Lechner,J.F., Gerwin,B.I., Reddel,R.R., Gabrielson,E.W., Van der Meeren,A., Linnainmaa K., Somers,A.N.A. and Harris,C.C. (1990) Studies on human mesothelial cells: Effects of growth factors and asbestiform fibers. In, "Cellular and Molecular Fiber Carcinogenesis", eds. Brinkley, B., Lechner,J.F. and Harris,C.C., Cold Spring Harbor Laboratory Press, New York, in press.

Lee,W.H., Bookstein,R., Hong,F., Young,L.J, Shew,J.Y. and Lee,E.Y. (1987) Human retinoblastoma susceptibility gene: cloning, identification and sequence. *Science* **235**:1394-1399.

Linnainmaa,K., Gerwin,B.I, Pelin,K., Jantuene,K., LaVeck,M.A., Lechner,J.F. and Harris,C.C. (1986) Asbestos-induced mesothelioma and chromosomal abnormalities In human mesothelial cells *in vitro*. In, "The Changing nature of Work and the Work place", NIOSH, Cincinnatti, pp. 119-122.

Moll,R., Franke,W.W., Schiller,D.L, Giegger,B.and Krepler,R. (1982) The catalog of human cytokeratins: patterns of expression in normal epithelia, tumors and cultured cells. *Cell* **31**:11-24.

Mossman,B.T. and Gee,J.B. (1989) Asbestos-related diseases. *N. Engl. J. Med.* **320**:1721-1730.

Osborn,M., Geisler,N., Shaw,G., Sharp,G. and Weber,K. (1982) Intermediate filaments. Cold Spring Harbor Symp. Quant. Biol. **46**:413-429.

Osborn,M., Altmannsberger,M., Debus,E. and Weber,K. (1982) Differentiation of the major human tumor groups using conventional and monoclonal antibodies specific for individual intermediate filament proteins. *Ann. N.Y. Acad. Sci.* **455**:649-668.

Oshimura,M. and Barrett,J.C. (1986) Chemically induced aneuploidy in mammalian cells: mechanisms and biological significance in cancer. *Environ. Mutagen.* **8**:129-159.

Oshimura,M., Koi,M., Ozawa,N., Sugawara,O., Lamb,P.W. and Barrett,J.C. (1988) Role of chromosome lose in ras/myc-induced Syrian hamster tumors. *Cancer Res.* **48**:1629-1692.

Parry,E.M. (1985) Tests for effects on miosis and mitotic spindle in Chinese hamster primary liver cells (CH 1-L) in culture. In, "Progress in Mutation Research 5", eds. Ashbey,J. and de Serres,F.J., Elsevier Science Publishing, Amsterdam, pp. 479-485.

Parry,J.M., Danford,N. and Parry,E.M. (1984) *In vitro* techniques for the detection of chemicals capable of inducing mitotic chromosome aneuploidy. *ATLA* **11**:117-128.

Puck,T.T. (1979) Studies on cell transformation. *Somatic Cell. Genet.* **5**:973-990.

Rheinwald,J.G., Germain,E. and Beckett,M.A. (1983) Expression of keratins and envelope proteins in normal and malignant human keratinocytes and mesothelial cells. In, "Human Carcinogenesis", eds. Harris,C.C. and Autrup,H., Academic Press, New York, pp. 85-96.

Setoyama,C., Frunzio,R., Liau,G., Mudryj,M. and de Crombrugghe,B. (1986) Transcriptional activation encoded by the v-*fos* gene. *Proc. Natl. Acad. Sci. USA* **83**:3213-3217.

Steinert,P.M., Steven,A.C. and Roop,D.R. (1985) The molecular biology of intermediate filaments. *Cell* **42**:411-420.

Somers,A., Parry,J.M., Parry,E.M., Stafford,A. and Kelly,S.L. (1986) Detection of natural products that induce aberrations of the mitotic spindle. *J.Biotech.* **4**:219-233.

Tsutsui,T, Maizumi,H., McLachlan,J,A., Barrett,J.C. (1983) Aneuploidy induction and cell transformation by diethylstilbestrol: a possible chromosomal mechanism in carcinogenesis. *Cancer Res.* **43**:3814-3821.

Tsutsui,T., Matzumi,H. and Barrett,J.C. (1984) Colcemid-induced neoplastic transformation and aneuploidy in Syrian hamster embryo cells. *Carcinogenesis* **5**:89-93.

Vogelstein,B., Fearon,E.R., Hamilton,S.R., Kern,S.E., Preisinger,A.C., Leppert,M., Nakamur,Y., White,R., Smits,A.M. and Bos,J.L. (1988) Genetic alterations during colorectal-tumor development. *N. Engl. J. Med.* **319**:525- 532.

Wagner,J.C., Munday,D.E. and Harington,J.S. (1962) Histochemical demonstration of hyaluronic acid in pleural mesotheliomas. *J. Pathol. Bacteriol.* **84**:73-78.

Wu,Y.J., Parker,L.M., Binder,N.E., Beckett,M.A., Sinard,J.H., Griffiths,C.T. and Rhein wald,J.G. (1982) The mesothelial keratins: a new famlly of cytoskeletal proteins identified in cultured mesothelial cells and nonkeratinizing epithelia. *Cell* **31**:693-703.

A COMPARISON OF THE OCCURRENCE AND DISTRIBUTION OF IRON DEPOSITS CAUSED BY THE PRESENCE OF INTRAPLEURAL ERIONITE AND CROCIDOLITE FIBRES IN RATS

P. Carthew, R.E. Edwards and R.J. Hill

MRC Toxicology Unit
MRC Laboratories, Woodmansterne Road
Carshalton, Surrey, SM5 4EF, U.K.

ABSTRACT

Changes in the pleura of rat lungs after intrapleural inoculation of the fibrous zeolite, erionite, have been compared to those occurring after the intrapleural inoculation of crocidolite asbestos. While the acute response of the pleura to both dusts is similar in terms of damage and inflammatory response, there was more marked proliferation of the pleura in response to erionite, with the formation of new blood vessels in the areolar space. Both dusts cause haemorrhaging of the pleura, although the effect of erionite was appreciably greater. There was also a distinct difference between the amount of iron deposited as haemosiderin in the granulomas which formed on the pleura and diaphragm in response to inoculated dusts.

While heavy haemosiderin deposits were common in crocidolite induced granulomas these were absent from the erionite induced granulomas. The greater oncogenicity of erionite may in part be due to the larger amounts of non-storage iron released by erionite. This, iron could participate in free radical reactions, both increasing the initiation of cellular transformation and also promoting tumour formation, as has been found with iron overload in the liver.

INTRODUCTION

The early changes in the pleural mesothelium in response to the injection of erionite are characterised by pleural inflammation and proliferation with localised destruction of the elastic membrane under the visceral pleura (Hill *et al.*,1990). The chronic stimulation of the pleura eventually produces tumours which are invasive or compressing. There

TABLE 1. Size distribution of the fibres used

	No. of fibres/μg ($\times 10^{-4}$)	
	Crocidolite	Erionite
Total fibres	160.0	24.0
>6 μm long and <1.6 μm diameter	7.5	2.0
>6 μm long and <0.5 μm diameter	5.1	1.2
>6 μm long and <0.2 μm diameter	0.8	0.6

Measurements by D.M. Griffiths, University of Wales, College of Cardiff, personal communication

is a notable haemorrhaging component to the reaction of the pleura to erionite which is more evident at post mortem than with the asbestos fibre crocidolite. However, both fibres produce granulomas on the pleura and the diaphragm after intrapleural injection.

The haemorrhaging and granuloma formation could be expected to lead to the deposition in the granulomas of iron from released red cell haemoglobin. This iron could be either in the form of free iron (Gutteridge, 1986), ferritin or haemosiderin and could be present in macrophages, mesothelial cells or subserosal cells. In the case of non storage iron this could lead to the generation of free radicals which damage DNA (Halliwell & Gutteridge, 1984) and cause lipid peroxidation which leads to cell death (Gutteridge, 1986; Britton et al., 1987). Subsequent cell proliferation would tend to "promote" tumour formation.

Recently it has been found that peritoneal mesotheliomas are induced after repeated injections of ferric saccharate and nitrilotriacetic acid in rats (Okada et al., 1989). To examine the forms of iron present during early changes in the mesothelium of the rat pleura, we have compared the deposition of iron in the rat pleura in response to fibres, for up to a month after the intrapleural inoculation of erionite and crocidolite dusts. By staining for haemosiderin iron by the histochemical Perls' reaction, ferrous iron by Turnbull's reaction and immunohistochemically for ferritin using an anti-rat ferritin antibody, we have determined the relative amounts of the safer storage forms of iron in both cases and correlated this to differences in the pleural response to the two dusts.

MATERIALS AND METHODS

Oregon erionite (sterile, 20 mg) or UICC crocidolite (sterile, 20 mg) were suspended by sonication in saline and inoculated intrapleurally (0.4 ml/rat) into 200g, male Porton rats under ether anaesthetic. The numbers of fibre in certain size ranges for Oregon erionite and crocidolite are given in Table 1.

Control groups of saline inoculated and non-inoculated rats were also included. Groups of four rats were sacrificed at days 1, 3, 7, 14 and 28 after inoculation. The lungs (inflated), hearts, livers, spleens, kidneys and adrenals were fixed in 10% neutral buff-ered formalin prior to paraffin embedding and the cutting of 5 μm sections. Sections were stained with haematoxylin and eosin, orcein Van Giesen for elastin, MSB for collagen and Perls' Prussian Blue reaction for haemosiderin. Ferrous iron was demonstrated by the Turnbull's Blue reaction. Ferritin was demonstrated by immunocytochemical staining with a guinea-pig antibody to rat ferritin (Sigma) by the indirect immunoperoxidase method after trypsinization of formalin fixed tissues as previously described (Johnson *et al.*, 1984).

RESULTS

Acute response of the pleura to erionite and crocidolite inoculated intrapleurally

Twenty-four hours after the intrapleural injection of both dusts, aggregates of hyaline material with fibrin, inflammatory cells and fibres were evident attached to the visceral pleura. Haemorrhaging and fibrin deposition was more pronounced in response to erionite. In crocidolite injected animals Perls' positive macrophages were evident extra-pleurally and subpleurally. However in erionite treated animals such macrophages were only seen in subpleural alveolae. The elastic membrane under the pleura was damaged in erionite treated rats, allowing fibres to penetrate beneath.

On day three there was mesothelial hyperplasia and heavy Perls' positive iron deposits in the pleural inflammatory cell aggregates induced by crocidolite. Mesothelial hyperplasia was more pronounced with erionite and although inflammatory cell aggre-gates were induced to a greater extent they were completely Perls'negative. The subpleu-ral alveolar macrophages were still positive by the Perls' reaction in erionite inoculated rats.

By day seven after the injection of crocidolite the pleura had prominent iron deposits (Fig. 1), especially in the fibre filled giant cells of the granulomas (Fig. 2). However, again the granulomas in the erionite treated rats were negative for Perls' iron staining, although the subpleural alveolar macrophages were still positive. The mesotheli-al hyperplasia in erionite treated rats was more pronounced than the crocidolite inoculated animals.

After fourteen days mesothelial hyperplasia was reduced, with repair to the elastic membrane under the pleura evident by the demonstration of increased collagen deposits. Again Perls' positive iron was abundant in the crocidolite induced granuloma giant cells, but not in the erionite induced pleural granulomas.

Twenty-eight days after the intrapleural inoculation of crocidolite, Perls' positive stain was also evident in the subserosal cells adjacent to the elastic membrane (Fig. 3). This staining was found to be ferritin, by immunohistochemical demonstration (Fig. 4). In high concentrations ferritin stains homogeneously blue in the cytoplasm of cells rather than the granular stain seen with haemosiderin deposits (Carthew *et al.*, 1991). No ferri-tin stained cells in the pleura were found at any earlier time points in the crocidolite treat-ed animals. At twenty-eight days, for the first time, some small deposits of homogeneous Perls' stain were seen in spindle cell proliferations adjacent to the granulomas induced by erionite. There was no Perls' staining of the giant or other inflammatory cells containing fibres in the erionite treated animals although the subpleural alveolar macrophages were still seen to have demonstrable Perls' reactive iron present. The Turnbull's Blue reaction for ferrous iron was negative on all sections throughout this study.

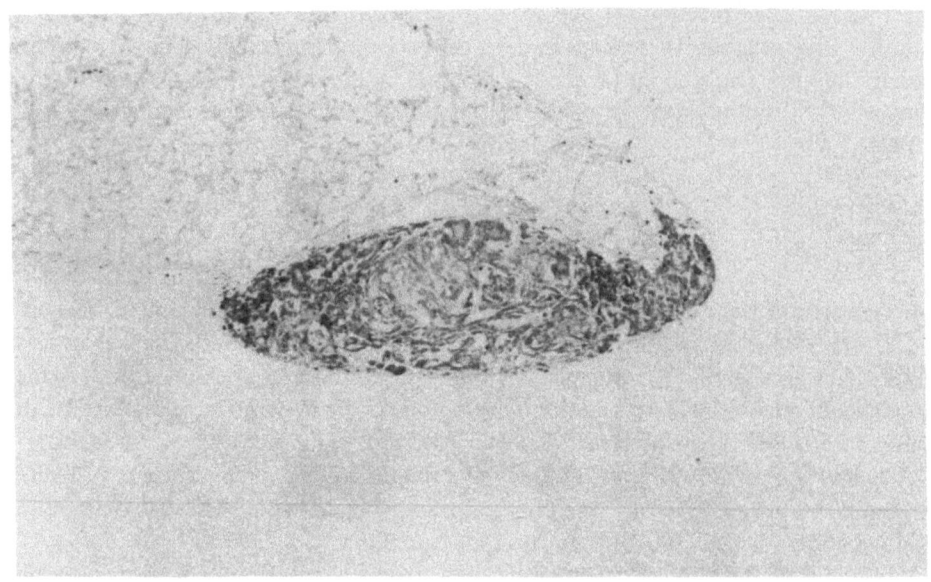

Figure 1. A haemosiderin containing granuloma formed on the pleura seven days after the intrapleural injection of crocidolite asbestos. Perls' Prussian Blue reaction for iron x35.

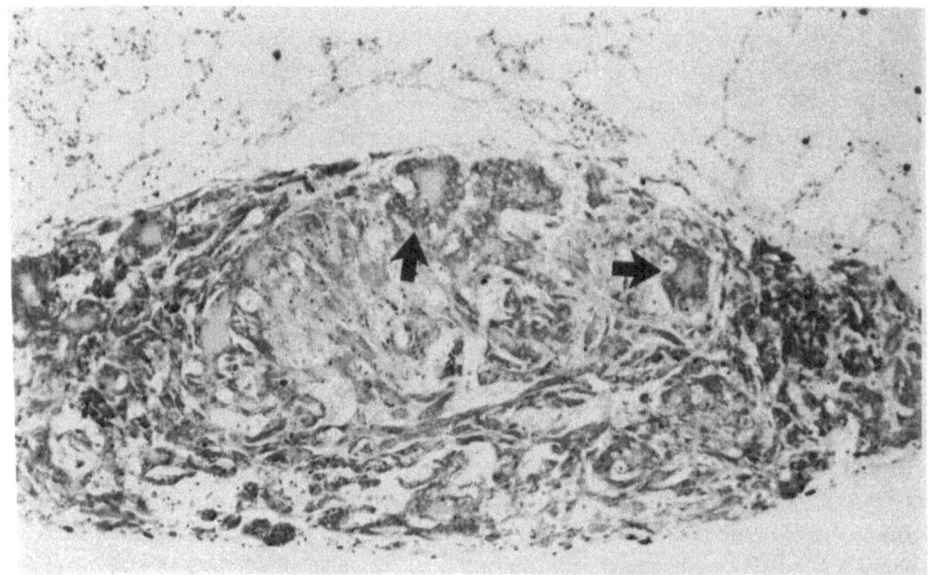

Figure 2. Higher power of Fig. 1 showing the Perls' positive multinucleate giant cells (arrows) present in the pleural granuloma. Perls' Prussian Blue reaction for iron x360.

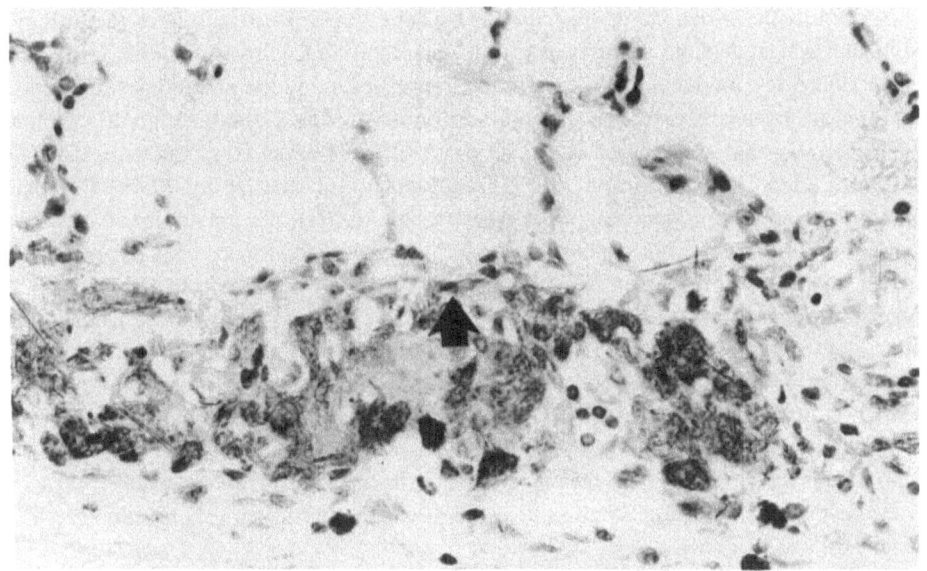

Figure 3. Pleural inflammatory response to crocidolite asbestos twenty-eight days after intrapleural inoculation. Note the iron containing subserosal cell adjacent to the elastic membrane (arrow). Perls' Prussian Blue reaction for iron x360.

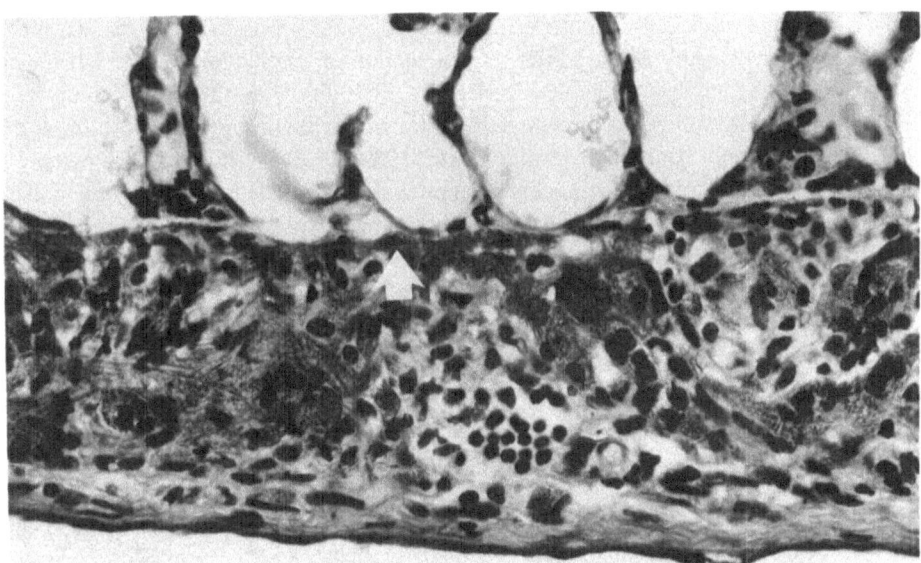

Figure 4. A serial section of the same inflammatory areas as in Fig. 3 immunostained for ferritin. Again subserosal cells adjacent to the elastic membrane are stained positive for ferritin (arrow). Immunoperoxidase stain x360.

DISCUSSION

Whilst the inflammatory reaction and pleural hyperplasia seen after the intrapleural inoculation of erionite in rats was greater than that induced by crocidolite (Hill *et al.*, 1990), there was a notable difference in the iron deposited in the granulomas induced on the pleura by these two fibres. Although erionite induces a greater amount of pleural haemorrhaging than crocidolite, and this extravasation of blood is undoubtedly the source of the iron deposited as storage iron in granulomas (induced also by other pathological conditions involving haemorrhaging (Carthew *et al.*, 1991)), the granulomas we observed induced by erionite did not contain any iron deposited as ferritin or haemosiderin. These are the forms of storage iron considered to be relatively less toxic to tissues than low molecular weight iron (Okada *et al.*, 1989; Bacon & Britton, 1990). If the haemoglobin-iron released from red cells due to haemorrhaging is not incorporated into ferritin or haemosiderin, it could be bound to transferrin, which is known to be important in the regulation of cell growth (Sussman, 1989), and could be involved in the enhanced proliferative response of the pleura to erionite. It was observed that the proliferation of the pleural cells to erionite was always much greater than with crocidolite, even to the extent of the formation of notable size blood vessels in the areolar space two months after treatment (Hill *et al.*, 1990).

Another possible effect of iron is the generation of hydroxyl free radicals via the Fenton reaction (Aust & Svingen, 1982) with subsequent damage to DNA (Shires, 1982), resulting in the transformation of cells. It has been shown *in vitro* that haemoglobin will release iron to participate in the Fenton reaction and cause lipid peroxidation in the presence of peroxides (Gutteridge, 1986). The possible susceptible cell type responsible for some spindle form mesotheliomas would be the subserosal cell (Bolen *et al.*, 1987). The ferritin storage form of iron, which is associated with greater potential free radical production than haemosiderin iron (O'Connell *et al.*, 1986), was found in the subserosal cells of the pleura of rats 28 days after the inoculation of crocidolite. Where ferritin has been found in the cytoplasm of mouse liver cells, it has also been found to accumulate in the nucleus, even forming ferritin containing intranuclear inclusion bodies (Smith *et al.*, 1990) prior to tumour induction (Smith *et al.*, 1989, 1990b). This may also occur in subserosal cells leading to free radical generation within the nucleus. Finally, low molecular weight iron may initiate free radical induced lipid peroxidation leading to cell damage and death (Halliwell & Gutteridge, 1984). This would further stimulate a hyperplastic response of the pleural cells and assist in the promotion and progression of any transformed cells through clonal expansion. In this respect the iron would be acting as a promoter of mesothelial tumour formation.

REFERENCES

Aust,S.D. and Svingen,B.A. (1982) The role of iron in enzymatic lipid peroxidation, Chapter 1. In, "Free radicals in biology, Vol. V." Academic Press, pp.1-28.

Bacon,B.R. and Britton,R.S. (1990) The pathology of hepatic iron overload: A free radical-mediated process. *Hepatology* **11**:127-137.

Britton,R.S., Bacon,B.R. and Recknagel,R.O. (1987) Lipid peroxidation and associated hepatic organelle dysfunction in iron overload. *Chem. Phys. Lipids* **45**:207-239.

Bolen,J.W., Hammer,S.P. and McNutt,M.A. (1987) Serosal tissue: reactive tissue as a model for understanding mesothelioma. *Ultrastruct. Pathol.* **11**:251-262.

Carthew,P., Edwards,R.E., Smith,A.G., Dorman,B. and Francis,J.E. (1991) Rapid induction of hepatic fibrosis in the gerbil after parenteral administration of iron dextran complex. *Hepatology*, (in press).

Gutteridge,J.M.C. (1986) Iron promoters of the Fenton reaction and lipid peroxidation can be released from haemoglobin by peroxidase. *FEBS* 201:291-295.

Halliwell,B. and Gutteridge,J.M.C. (1984) Oxygen toxicity, oxygen radicals, transition metals and disease. *Biochem. J.* **219**:1-14.

Hill,R.J., Edwards,R.E. and Carthew,P. (1990) Early changes in the pleural mesothelioma following intrapleural inoculation of the mineral fibre erionite and the subsequent development of mesotheliomas. *J. Exp. Path.* **71**:65-118.

Johnson,N.F., Edwards,R.E., Munday,D.E., Rowe,N. and Wagner,J.C. (1984) Pluripotential nature of mesotheliomata induced by inhalation of erionite in rats. *Br. J. Exp. Path.* **65**:377-388.

O'Connell,M, Halliwell,B., Moorhouse,C.P., Aruoma,O.I., Baum,H. and Peters,T.J. (1986) Formation of hydroxyl radicals in the presence of ferritin and haemosiderin. *Biochem. J.* **234**:727-731.

Okada,S., Hamazaki,S., Toyokuni,S. and Midorikawa,O. (1989) Induction of mesothelioma by intraperitoneal injections of ferric saccharate in male Wistar rats. *Br. J. Cancer* **60**:708-711.

Shires,T.K. (1982) Iron-induced DNA damage and synthesis in isolated rat liver nuclei. *Biochem. J.* **205**:321-329.

Smith,A.G., Cabral,J.R.P., Carthew,P., Francis,J.E. and Manson,M.M. (1989) Carcinogenicity of iron in conjunction with a chlorinated environmental chemical, hexachlorobenzene, in C57BL/10ScSn mice. *Int. J. Cancer* **43**:492-496.

Smith,A.G., Carthew,P., Francis,J.E., Edwards,R.E. and Dinsdale,D. (1990) Accumulation of ferritin by hepatic nuclei in mice with iron overload. *Hepatology* **12**:1399-1406.

Smith,A.G., Francis,J.E. and Carthew,P. (1990b) Iron as a synergist for hepatocellular carcinoma induced by polychlorinated biphenyls in *Ah*-responsive C57BL/10ScSn mice. *Carcinogenesis* **11**:437-444.

Sussman,H.H. (1989) Iron and tumour cell growth. Chapter 12. in, "Iron in immunity cancer and inflammation", De Sousa,M. & Brock,J.H., eds. Wiley, N.Y., pp.261-282.

SECRETION OF INTERLEUKIN 1 AND TUMOUR NECROSIS FACTOR BY ALVEOLAR MACROPHAGES FOLLOWING EXPOSURE TO PARTICULATE AND FIBROUS DUSTS

G.M. Brown, X.Y. Li, and K. Donaldson

Institute of Occupational Medicine
Roxburgh Place
Edinburgh, Scotland

ABSTRACT

Inflammatory leukocytes have been implicated in the pathogenesis of lung disease in persons exposed to asbestos. Inflammatory leukocyte-derived cytokines may play an important part in the disease process through their ability to modulate the function of other cell types. We have previously shown that long and short fibre amosite asbestos samples have different pathological potential. In this study, we have therefore examined cytokine secretion by bronchoalveolar leukocytes exposed to long and short fibre amosite asbestos *in vivo* and *in vitro*. We have shown that both *in vivo* and *in vitro* exposure to amosite results in increased secretion of IL-1 and TNF which was greater with the long fibre sample. The relationship between long term pathology and increased secretion of TNF and IL-1 suggests that cytokines may play a part in the pathogenesis of lung disease in workers exposed to asbestos.

INTRODUCTION

A common finding in individuals occupationally exposed to asbestos is increased numbers of macrophages in the bronchoalveolar space (Begin *et al.*, 1986). These macrophages are likely to be activated due to phagocytosis of dust particles (Schnyder & Baggiolini, 1978) and perhaps also to interaction of the intracellular fibres with the cell membrane (Brown *et al.*, 1991). The cytokines interleukin 1 (IL-1) and tumour necrosis factor (TNF) are secreted by inflammatory macrophages (Billingham, 1987) and may play a role in the development of lung fibrosis and neoplasia in persons exposed to asbestos (Bignon & Jaurand, 1984) through their ability to cause proliferation and oxidant production in certain cell types. A major factor governing the harmfulness of fibrous dusts is thought to be their geometry although surface chemistry may also be important. Increasing fibre length has been shown to correlate with increasing carcinogenic potential *in vivo* (Stanton & Wrench, 1972) and with increasing toxicity *in vitro* (Brown *et al.*, 1986). We have also demonstrated experimentally that long asbestos fibres have significantly greater inflammatory potential than short fibres which will result in the recruitment of large numbers of inflammatory macrophages to the bronchoalveolar space (Donaldson *et al.*, 1989).

In the present study, we have investigated production of IL-1 and TNF by rat alveolar macrophages at various time points after intratracheal instillation of long and short amosite asbestos. In order to investigate the direct effects of asbestos on macrophage cytokine production, alveolar macrophages from untreated rats were exposed *in vitro* to amosite asbestos and then IL-1 and TNF secretion were measured.

Recent work in the Institute of Occupational Medicine has indicated that particulate dusts may potentiate the carcinogenic effects of fibres (Davis *et al.*, manuscript in preparation). We therefore also assessed production of TNF by alveolar macrophages exposed *in vitro* to mixed dusts of long and short amosite in combination with quartz or titanium dioxide.

MATERIALS AND METHODS

Dusts

Long and short fibre amosite were prepared from a commercially available South African amosite as previously described (Donaldson *et al.*, 1989). Titanium dioxide (TiO_2) was the rutile form (Tioxide Ltd., Stockton on Tees) and quartz was the DQ12 standard.

Exposure regimen and preparation of bronchoalveolar macrophage supernatants

Male HAN rats were exposed to dust by intratracheal instillation of 1 mg as a single bolus in 0.5 ml phosphate buffered saline (PBS); control rats were injected with PBS alone. At selected time points thereafter, the rats were killed and the bronchoalveolar (BAL) leukocytes retrieved by lavage as previously described (Brown & Donaldson, 1988). Total cell and differential counts were performed on the BAL leukocyte populations using Trypan blue and May Grunwald-Geimsa stains respectively.

For *in vitro* exposure, BAL leukocytes were obtained from untreated rats and these were set up in 24-well culture plates as 1 ml of Ham's F10 medium (Gibco, Paisley) plus 2% bovine serum albumin (BSA) containing 1×10^6 cells/ml and 50, 100 or 500 μg dust/ml. Cultures exposed to mixed dusts received 50 μg/ml asbestos and 50 μg/ml TiO_2 or quartz. Supernatant medium from these cultures was obtained 24 h later and analysed for content of IL-1 and TNF. Supernatants from BAL leukocytes from rats exposed to dust *in vivo* were obtained under the same conditions but in the absence of dust.

Cytokine measurement

IL-1 in leukocyte supernatants was measured using the standard mouse thymocyte bioassay (Kusaka *et al.*, 1990). The presence of TNF in the supernatants was assessed by measuring their ability to cause lysis of the TNF-sensitive L929 cell line (Manson *et al.*, 1989).

RESULTS

Effect of *in vivo* exposure to long and short amosite asbestos on bronchoalveolar leukocyte numbers

Short-fibre amosite elicited an alveolitis which was evident as recruitment of PMN to the bronchoalveolar region by 3 days after intratracheal injection (Table 1). The greatest influx of inflammatory leukocytes was apparent at 7 days and was comprised largely of macrophages; inflammation in the short-fibre amosite-exposed rats persisted up to 14 days. Long fibre amosite did not elicit a measurable increase in numbers of leukocytes in the BAL at any time point and at 3 days there was a reduction in the number of lavageable leukocytes in these animals. However, the presence of PMN in the BAL at each time point with long fibre amosite was indicative of an inflammatory response

IL-1 and TNF secretion

Although there was evidence of an inflammatory response in both long and short amosite-exposed rats at 3 days post-injection, there was no evidence of increased secretion of TNF by BAL leukocytes at this time-point with either amosite sample (control-46.91($\pm$4.98); long-41.53($\pm$0); short-44.58 ($\pm$0) (mean (sd) units of TNF/ml in three rats per group). By 14 days, however, BAL leukocytes elicited by both amosite samples secreted more TNF than did control BAL macrophages; this was greater with long fibre than with short (control - 25.94 ($\pm$31.0); long - 54.48 ($\pm$18.8); short - 46.22 ($\pm$2.31).

TABLE 1. Total number and differential count of cells in the bronchoalveolar region of rats exposed by intratracheal injection to 1 mg long and short fibre amosite asbestos

Days exposed		Total(x10^6)	[1]Mϕ	Differential count (%) PMN	Lymph[a]
	Control[2]	5.29(1.0)	99(0.6)	0	1(0.6)
3	Short	7.34(2.5)	96(4.7)	3(4.2)	1(0.6)
	Long	2.03(1.13)	92(6.1)	4(2.0)	1(0.6)
	Control	6.5(1.6)	98(5.5)	0	2(0.6)
7	Short	24.7(18.8)	92(2.7)	7(2.9)	1(0.6)
	Long	6.2(1.8)	91(4.6)	5(5.7)	1(1)
	Control	7.3(1.0)	99(0.6)	0	1(0.6)
14	Short	12.1(6.9)	97(4.7)	2(4.3)	1(0.5)
	Long	5.8(2.5)	94(4.1)	5(3.5)	0

[1] Macrophages, [2] 3 rats/group at each time point.

IL-1 secretion by long fibre amosite-elicited BAL leukocytes was greater than that of control BAL macrophages 7 days post-injection (control - 4892 (±1291); long - 6819 (±1093) (mean (sd) cpm for 3 rats per group) but returned to near control levels by 14 days - 5731 (±1714). Short fibre was no different from the control at 7 days (4348 (±408)), or 14 days (4732 (±607)).

Bronchoalveolar leukocytes from untreated control rats comprised > 95% macrophages. When these cells were exposed to dust *in vitro* at 500 µg/ml, only the long fibre amosite induced increased IL-1 secretion compared with control BAL leukocytes (control -2260; long - 4038; short - 2513 cpm). TNF secretion by macrophages exposed to 500 µg/ml amosite *in vitro* was greater with long (31.48 (±22.4) units/ml) than with short fibre amosite (16.16 (±21.99) units/ml).

Exposing control bronchoalveolar leukocytes to mixed dusts composed of long and short fibre amosite and toxic or non-toxic dust *in vitro*, did not result in potentiation of TNF secretion with either asbestos sample (Table 2).

DISCUSSION

In this study we have investigated the ability of long and short fibre amosite preparations to elicit secretion of cytokines by bronchoalveolar leukocytes. We used *in vivo* and *in vitro* exposure to examine the effects of long and short fibre amosite asbestos on secretion of IL-1 and TNF by rat alveolar macrophages. We have previously demonstrated that long fibre amosite causes greater recruitment of inflammatory leukocytes into the mouse peritoneal cavity (Donaldson *et al.*, 1989) than does short fibre. In rats exposed by inhalation (Davis *et al.*, 1986), we have also shown that there is greater tumour production with long fibre amosite. In the present study, the bronchoalveolar lavage data showed more inflammation with short amosite than with long and although we did observe an inflammatory response in terms of PMN recruitment into the BAL with long fibre amosite, there was no measurable increase

TABLE 2. TNF secretion by control bronchoalveolar macrophages following *in vitro* exposure to long and short fibre amosite asbestos, either alone or in combination with quartz and TiO$_2$

| | Amosite alone | Mixed dusts | |
		+TiO$_2$	+quartz
Short	23.58	29.16	21.97[*]
Long	44.58	29.31	36.05

[*] Units of TNF/ml.

All dusts are at a concentration of 50 µg/ml which gives a final concentration of 100 µg/ml with the mixed dusts (mean of 3 wells/sample in a single experiment).

in total numbers of leukocytes. This apparently disagrees with the hypothesis that lung injury is related to the recruitment of inflammatory leukocytes into the bronchoalveolar region. The apparent lack of a macrophage response in the long fibre-exposed rats, however, is probably due to the problems of instilling a long fibre sample which may block the airways, so making lavage difficult. We have previously shown that there is an inflammatory response in rats to inhaled asbestos (Donaldson *et al.*, 1988) that leads to long term pathology.

Those leukocytes which could be lavaged from the long fibre amosite-exposed animals were different from both control BAL leukocytes and the short-fibre amosite-elicited BAL leukocytes in terms of cytokine secretion. Both TNF and IL-1 secretion by long amosite-elicited leukocytes was greater than secretion by short fibre-elicited cells. The most likely source of these cytokines is the macrophage since the percentage of PMN was small and was similar in both short and long fibre- elicited BAL leukocytes. The finding of increased cytokine production with long fibre is in agreement with the study of (Bissonette & Rola-Pleszczynski, 1989) who demonstrated increased secretion of TNF in mice exposed to chrysotile asbestos but not silica.

In order to investigate the direct dust particles/leukocyte interaction in cytokine production, we exposed control bronchoalveolar leukocytes *in vitro* to amosite asbestos. Secretion of both TNF and IL-1 was greater with exposure to long fibre amosite than to short fibre. Thus, increased secretion of cytokines by BAL leukocytes exposed to dust *in vivo* may be due to direct interaction between fibres and leukocytes in the lung. Both TNF and IL-1 can act as chemotaxins and so the direct interaction of dust and alveolar macrophages to induce secretion of these cytokines, may be involved in initiation of the inflammatory response following deposition of dust in the lung.

In long term pathology studies, we have recently shown that exposure to non-fibrous dust in combination with asbestos potentiates the carcinogenic potential of the asbestos. In the present study therefore, we assessed the ability of non-toxic (TiO_2) and toxic (quartz) particulate dusts to potentiate secretion of TNF by leukocytes exposed to amosite *in vitro*. There was no evidence in the present study that the additional particulate burden could stimulate increased TNF secretion.

This study has shown that amosite asbestos can cause increased release of IL-1 and TNF from BAL leukocytes following *in vivo* and *in vitro* exposure. We have shown that long fibre amosite, which causes long term pathology in our rat model, induces greater secretion of both cytokines. The inflammogenic and immunostimulatory roles of TNF and IL-1 are central to normal host defence in the lung but they may also contribute to the pathogenesis of lung disease in situations where there is excessive secretion of these cytokines (Beissert *et al.*, 1989; Updyke *et al.*, 1989). In workers occupationally exposed to asbestos, increased secretion of cytokines in the bronchoalveolar region of the lung may occur in response to fibre deposition and thus may ultimately contribute to the pathogenesis of asbestosis and neoplasia.

REFERENCES

Begin,R., Bisson,G., Boileau,R. and Masse,S. (1986) Assessment of disease activity by Gallium-67 scan and lung lavage in the pneumoconioses. *Sem. Respir. Med.* 7:271-280.

Beissert,S., Berghotz,M., Waase,I., Lepsein,G., Schauer,A., Pfizenmaier,K. and Kronke,M. (1989) Regulation of tumor necrosis factor gene expression in colorectal adenocarcinoma: *in vivo* analysis by *in situ* hybridisation. *Proc. Natl. Acad. Sci. USA* 86:5064-5068.

Bignon,J. and Jaurand,M.C. (1984) Asbestos fiber toxicity and lung disease. In, "Occupational Lung Disease", ed. Gee,J.B.L., pp. 51-73.

Billingham,M.E.J. (1987) Cytokines as inflammatory mediators. *Br. Med. Bull.* **43**:350-370.

Bissonette,E. and Rola-Pleszynski,M. (1989) Pulmonary inflammation and fibrosis in a murine model of asbestosis and silicosis. *Inflammation* **13**:329-339.

Brown,G.M., Cowie,H., Davis,J.M.G. and Donaldson,K. (1986) *In vitro* assays for detecting carcinogenic mineral fibres: a comparison of two assays and the role of fibre size. *Carcinogenesis* **7**:1971-1974.

Brown,G.M. and Donaldson,K. (1988) Degradation of connective tissue components by lung-derived leukocytes *in vitro*: role of proteases and oxidants. *Thorax* **43**:132-139.

Brown,G.M., Brown,D.M., Slight,J. and Donaldson,K. (1991) Persistent biological reactivity of quartz in the lung: raised protease burden compared with a non-pathogenic mineral dust and microbial particles. *Br. J. Indust. Med.* **48**:61-69.

Davis,J.M.G., Addison,J., Bolton,R.E., Jones,A. and Smith,T. (1986) The pathogenicity of long versus short fibre samples of amosite asbestos administered to rats by inhalation and intraperitoneal injection. *Br. J. Exp. Path.* **67**:415-430.

Donaldson,K., Bolton,R.E., Jones,A., Brown,G.M., Robertson,M.D., Slight,J., Cowie,H. and Davis,J.M.G. (1988) Kinetics of the bronchoalveolar leukocyte response in rats during exposure to equal airborne mass concentrations of quartz, chrysotile asbestos or titanium dioxide. *Thorax* **43**:525-533.

Donaldson,K., Brown,G.M., Brown,D.M., Bolton,R.E. and Davis,J.M.G. (1989) Inflammation generating potential of long and short fibre amosite samples. *Br. J. Indust. Med.* **46**:271-276.

Kusaka,Y., Cullen,R.T.C. and Donaldson,K. (1990) Immunomodulation in mineral dust-exposed lungs: stimulatory effect and interleukin-1 release by neutrophils from quartz-elicited alveolitis. *Clin. Exp. Immunol.* **80**:293-298.

Manson,J.C., Symons,J.A., Di Giovine,F.S., Poole,S. and Duff,G.W. (1989) Autoregulation of interleukin-1 production. *Eur. J Immunol.* **19**:261-265.

Schnyder,J. and Baggiolini,M. (1978) Role of phagocytosis in the activation of macrophages. *J. Exp. Med.* **148**:1449-1457.

Stanton,M.F. and Wrench,C. (1972) Mechanisms of mesothelioma induction with asbestos and fibrous glass. *J. Natl. Cancer Inst.* **48**:797-821.

Updyke,L.W., Yoon,H.L., Chuthaputti,A., Pfeifer,R.W. and Yim,G.K. (1989) Induction of interleukin-1 and tumor necrosis factor by 12-O-tetradecanoylphorbol-13-acetate in phorbol ester-sensitive (SENCAR) and resistant (B6C3F1) mice. *Carcinogenesis* **10**:1107-1111.

CULTURED HUMAN MESOTHELIAL CELLS ARE SELECTIVELY SENSITIVE TO CELL KILLING BY ASBESTOS AND RELATED FIBERS: A POTENTIAL *IN VITRO* ASSAY FOR CARCINOGENICITY

*E.W. Gabrielson, J.F. Lechner, B.I. Gerwin, and C.C. Harris

*Department of Pathology
Johns Hopkins University School of Medicine, Baltimore, MD 21224
U.S.A.

Laboratory of Human Carcinogenesis
National Cancer Institute
Bethesda, MD 20892, U.S.A.

ABSTRACT

The use of man-made fibers as substitutes for asbestos is increasing, and it is important to determine the potential health hazards, including potential carcinogenicity, of these fibers. It is of particular interest to develop short-term *in vitro* assays that could significantly decrease the need for animal testing as well as provide rapid and reliable information to industry during the time that applications for new fibers are being developed. One potential rapid screening assay for mesothelioma-causing fibers is selective cytotoxicity for cultured human mesothelial cells. Cell killing of cultured normal human mesothelial cells by amosite, chrysotile, or crocidolite fibers occurs at levels of exposure approximately 50 fold less those required to produce cell killing cultured human lung fibroblasts. This differential cytotoxic effect is also observed for erionite, a fibrous zeolite linked to mesothelioma in Turkey, and Code 100 glass fiber found to cause mesothelioma in laboratory animals. In contrast, both mesothelial cells and fibroblasts require similar high levels of exposure to fibers such as Wollastonite, glass wool, and refractory ceramic fibers for cell killing. Non-fibrous particulates such as aluminum oxide, and a variety of chemicals are also toxic for both cell types at similar concentrations. Thus, human mesothelial cells, a selective *in vivo* target to the carcinogenic effects of asbestiform fibers, are also a selective *in vitro* target of the cytotoxic effects of these fibers. Differential cytotoxicity for cultured mesothelial cells and fibroblasts may be useful for predicting the potential of new man-made fibers to cause mesothelioma.

INTRODUCTION

It is well recognized that asbestos and related fibrous particulates have important human health effects when inhaled, including carcinogenic effects for the bronchus and mesothelium. These deleterious health effects have resulted in a ban on the use of asbestos in the United States; however, a number of other fiber materials are increasingly used for a variety of applications in industry. Little is known about the potential health effects of these fibers.

Long-term bioassays using laboratory animals, which are now being used to test selected new fibers, are costly and take 2-4 years for test completion. A number of short term screening tests including bacterial, yeast, and *in vitro* mammalian test systems have been developed for the testing of the carcinogenic potential of chemicals. While these are cost effective and relatively rapid to perform (Lave & Omenn, 1986) they are not well-suited for testing of fibers.

With a goal of establishing *in vitro* short term tests for identifying fibers with the potential for causing mesothelioma, we have compared the *in vitro* toxicity of a number of different types of fibers using cultured human mesothelial cells and bronchial fibroblasts. In addition to the mesothelium being a selective target for the *in vivo* carcinogenic effects of asbestos, our previous studies have found cultured human mesothelial cells to be selectively sensitive to the cytotoxic and clastogenic effects of asbestos fibers (Lechner *et al.*, 1985). It was the purpose of this study to determine if the sensitivity of mesothelial cells to toxic effects is specific for asbestos related compounds or a general sensitivity of mesothelial cells to all toxic agents.

HUMAN MESOTHELIAL CELL CULTURE

Human mesothelial cells were obtained by centrifuging the pleural fluids of hospital patients with non-malignant disease (e.g., congestive heart failure), and cultured in an enriched media previously described (Gabrielson *et al.*, 1988). In culture, these cells express both vimentin and keratin, replicate with a doubling time of 24-36 hours, and can be passaged up to 25 population doublings. In some experiments, asbestos effects on cultured mesothelial cells have been compared to the effects on cultured human lung fibroblasts, a number of human mesothelioma cell lines (Gerwin *et al.*, 1987) and human mesothelial cells that have been immortalized by SV40 T-antigen (Ke *et al.*, 1989).

ASBESTOS-INDUCED CYTOTOXICITY OF MESOTHELIAL CELLS AND FIBROBLASTS

Mesothelial cells and fibroblasts plated at clonal density (4×10^6 cells/ 60 mm. dish) were exposed to U.I.C.C. samples of amosite, crocidolite, chrysotile A (Rhodesian) and chrysotile B (Canadian). All of these fibers resulted in dose-dependent decreased colony forming efficiency of both cultured mesothelial cells and cultured fibroblasts. The cytotoxicity of amosite for mesothelial cells and fibroblasts is demonstrated in figure 1A; data for the four types of asbestos fibers are summarized in figure 1B. All types of asbestos tested had LD_{50}'s for mesothelial cells of less than 1 $\mu g/cm^2$, and were toxic at 20-100 fold lower levels of exposure for mesothelial cells than for fibroblasts.

TABLE 1. Transmission electron microscopy quantitation of amosite fiber phagocytosis by cultured human lung cells

cell type	number of cells counted	cross sectional area examined per cell (μm^2)	number of fibers observed per cell	number of fibers observed per μm^2
mesothelial	50	896±350	3.1±2.9	.0034
bronchial fibroblasts	20	1000±314	3.8±2.7	.0027

The basis for the sensitivity of the mesothelial cells is not known; in this study, phagocytosis of asbestos fibers was studied in mesothelial cell and fibroblast cultures exposed to 0.3 $\mu g/cm^2$ amosite by transmission electron microscopy. As noted in table 1, similar numbers of amosite fibers were quantified within the cytoplasm of exposed mesothelial cells and fibroblasts.

The high degree of sensitivity of normal mesothelial cells to the cytotoxic effects of amosite was also observed in the immortalized Met 5A human mesothelial cell line; in contrast, all human mesothelioma cell lines examined were markedly less sensitive to cell killing by amosite, with 10 to 20 fold higher level of amosite exposure required for cytotoxicity to these cells (data not shown).

CYTOTOXICITY INDUCED BY OTHER FIBERS AND CHEMICALS

The observation that cultured human mesothelial cells are much more sensitive to the cytotoxic effects of asbestos fibers than cultured human lung fibroblasts correlates with the clinical observation that the mesothelium is a selective *in vivo* target for the carcinogenic effects of asbestos fibers. Additional experiments were performed to determine if the high degree of sensitivity of mesothelial cells to asbestos is simply a result of these cells being fragile, or if the sensitivity is specific for fibers with properties similar to asbestos.

In these experiments, mesothelial cells and fibroblasts were exposed to a series of other fibers, non-fibrous particulates and chemicals. Non-fibrous particulates tested were nickel subsulfide and aluminum oxide, and the soluble chemicals used in these comparative cytotoxicity assays were sodium azide, hydrogen peroxide and nickel sulfate. The fibers used in these experiments were Wollastonite 400 and 1250, erionite, 0.45 μm mean diameter (MD) Manville Code 100 glass fibers, 1.8 μm MD refractory ceramic fibers, 3.1 μm MD glass wool, and 2.7 μm MD mineral wool provided by N.F. Johnson (Smith *et al.*, 1985). Erionite is a fibrous mineral that has been associated with mesothelioma in humans and experimental animals (Baris *et al.*, 1979). Manville code 100 fibers have been reported to cause peritoneal mesothelioma in rats and hamsters after implantation, but mesothelioma was not observed among animals that inhaled these fibers (Smith *et al.*, 1985). Refractory ceramic fibers have similarly been reported by Smith and co-workers (1985) to induce animal mesothelioma after implantation into the peritoneum, but not after inhalation. In other recent inhalation experiments, however, high incidences of mesothelioma were observed in rats after inhalation of refractory ceramic fibers

(Hesterberg, 1990); no human exposure data is available for these fibers. The glass wool and mineral wool implantations have not induced experimental animal mesothelioma, and limited human exposure data has not implicated Wollastonite as a cause of mesothelioma.

The results of cytotoxicity assays are presented in figure 2. The non-fibrous particulates aluminum oxide and nickel subsulfide, and the soluble chemicals hydrogen peroxide and nickel sulfate, were equally toxic to mesothelial cells and fibroblasts. Two of the fibers, erionite and Manville code 100, were highly toxic to mesothelial cells (LD_{50} less than 1 $\mu g/cm^2$), with a large differential toxicity for mesothelial cells compared to fibroblasts. The other fibers examined had LD_{50}'s of greater than 10 $\mu g/cm^2$ for mesothelial cells, and were no more toxic for mesothelial cells than for fibroblasts. The sensitivity of the mesothelial cells therefore appears to be selective for asbestiform fibers, and is not a result of a general delicate nature of the mesothelial cells.

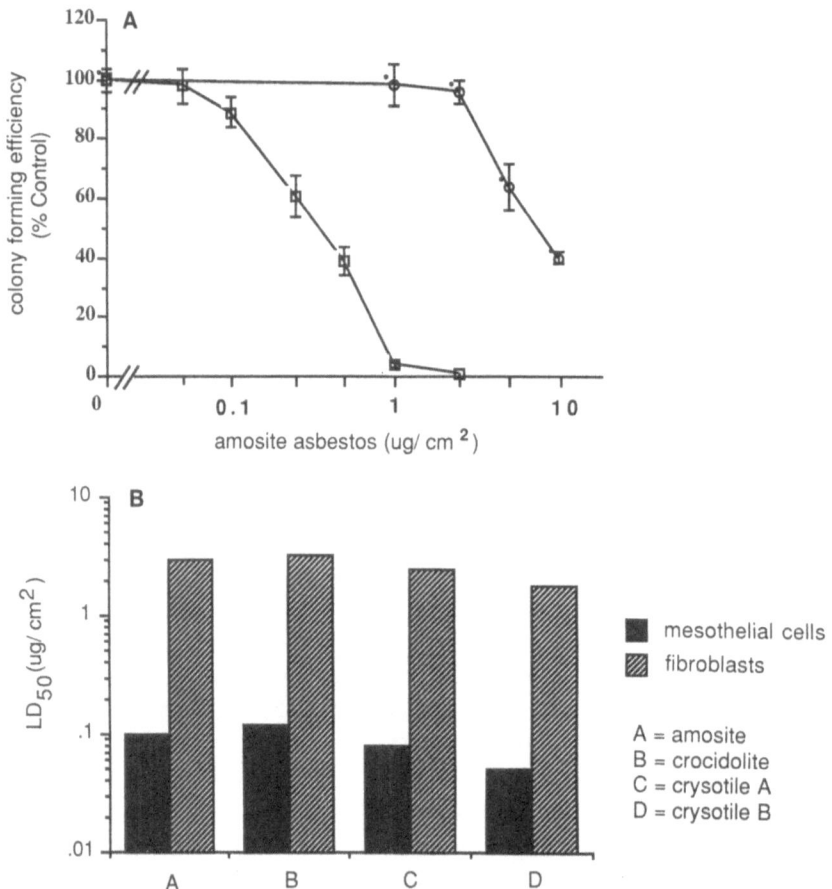

Figure 1. A) Cell killing of cultured human mesothelial cells and fibroblasts by amosite fibers. ■ mesothelial cells, O bronchial fibroblasts.
B) Cytotoxicity of U.I.C.C. asbestos samples for cultured human mesothelial cells and human lung fibroblasts.

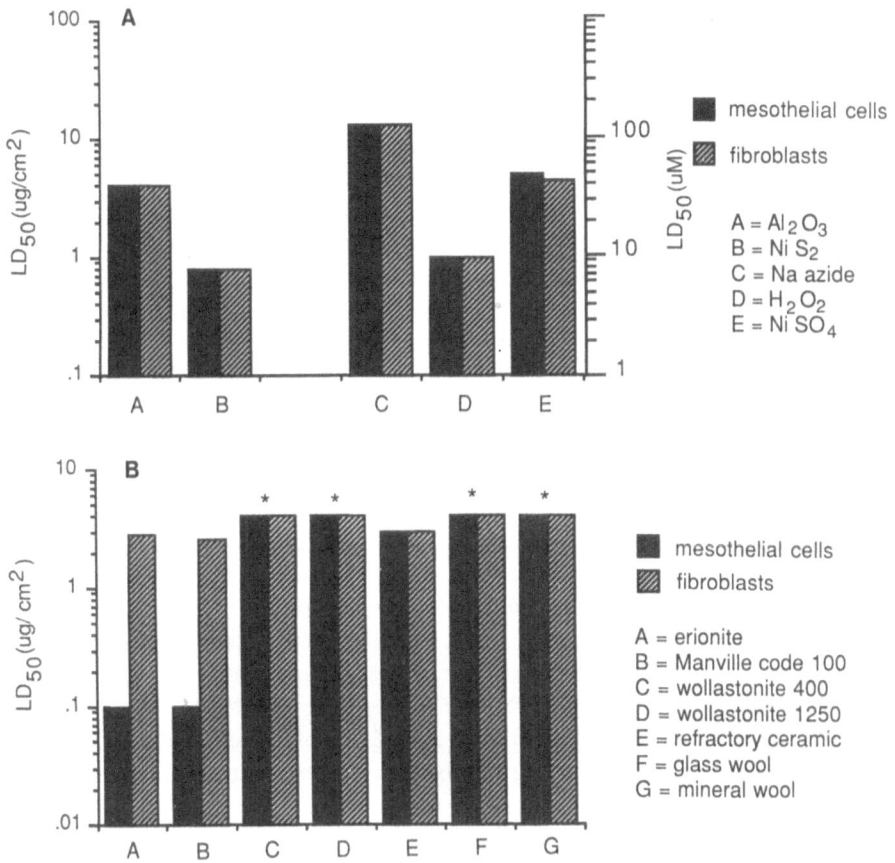

Figure 2. A) Cytotoxicity of non-fibrous particulates (aluminium oxide and nickel subsulphide) and water soluble chemicals (sodium azide, hydrogen peroxide and nickel sulphate) for cultured human mesothelial cells and human lung fibroblasts.

DISCUSSION

The properties of fibers that are thought to be important in causing human mesothelioma include the "respirability" of the fibers, translocation of inhaled fibers to the target cells, durability of the fibers in tissue, and some intrinsic transforming effect of the fibers on the mesothelial cells. Intrapleural and intraperitoneal injections of fibers into animals ignore the first two of these factors, and to some extent ignore the durability issue, since the latency period for animal mesothelioma is much less than for human mesothelioma; thus in these assays all fibers with an intrinsic transforming effect on the mesothelium are likely to produce tumors.

It is attractive to speculate that the intrinsic transforming potential could also be determined by examining *in vitro* effects of fibers on mesothelial cells. Unfortunately, for *in vitro* exposures of human mesothelial cells, malignant transformation is not a realistic endpoint, since to date we have been unable to fully transform human mesothelial cells by asbestos exposure. Furthermore, transformation of cultured human mesothelial cells by asbestos to phenotypically and chromosomally abnormal populations is an unusual event that would be difficult to adapt to a quantititative assay.

Our laboratories are investigating the possibility that *in vitro* cytotoxicity, particularly differential cytotoxicity to human mesothelial cells and fibroblasts, will be useful for predicting intrinsic transforming potential of fibers. Asbestos cytotoxicity to mesothelial cells is likely to be important in the carcinogenesis process, because: 1) cell killing is typically followed by replication of the injured cell population, and replicating populations are targets for DNA damaging events; 2) selective cytotoxicity of a normal cell population compared to an initiated cell population, as we have observed for cultured human mesothelial cells (manuscript in preparation), may result in an expansion of an initiated cell population, *i.e.* tumor promotion; and 3) DNA/ chromosomal damage is likely to be the mechanism for killing of cultured human mesothelial cells by asbestos, and thus a measure of cytotoxicity is likely to reflect DNA and chromosomal damage.

A selective cytotoxicity for mesothelial cells (when compared with fibroblasts) appears to be positively associated with the carcinogenic potential of a fiber with the possible exception of Manville code 100, if this fiber is interpreted as being non-carcinogenic based on animal inhalation data. Similarly, a fiber not being selectively cytotoxic to mesothelial cells correlates with the fiber not causing mesothelioma, with the possible exception of refractory ceramic fibers, which have been found to cause mesothelioma in some animal inhalation experiments. However, as noted above, the animal data is not unequivocal for either of these types of fibers. It should also be noted when comparing *in vitro* experiments to animal experiments that cultured cells were exposed to samples of fibers similar to those that mesothelial surfaces were exposed to during implantation experiments, but substantially different than fiber samples inhaled by animals and transported to the pleura during inhalation experiments. Exposing cell cultures to "respirable fractions" of fiber samples may result in a better correlation between *in vitro* assays and animal inhalation assays.

In summary, we have demonstrated that cultured human mesothelial cells have a marked selective hypersensitivity to the cytotoxic effects of asbestos and some related fibers. Additional comparisons of *in vitro* cytotoxicity data to animal studies and human epidemiological data will help determine the usefulness of the cytotoxicity assay for predicting carcinogenic potential of specific fibers. It is cautioned that any *in vitro* assay will be limited in usefulness by not accounting for the respirable nature of a fiber sample or the durability of the fiber within human tissue.

REFERENCES

Baris,Y.I., Artvinli,M. and Sahin,A.A. (1979) Environmental mesothelioma in Turkey. *Ann. N.Y. Acad. Sci.* **330**:423-432.

Gabrielson,E.W., Gerwin,B.I., Harris,C.C., Roberts,A.B., Sporn,M.B. and Lechner,J.F. (1988) Stimulation of DNA synthesis in cultured primary human mesothelial cells by specific growth factors. *F.A.S.E.B. Journal* **2**:2717-2721.

Gerwin,B.I., Lechner,J.F., Reddel,R.R., Roberts,A.B., Robbins,K.C., Gabrielson,E.W. and Harris,C.C. (1987) Comparison of production of transforming growth factor-B and platelet-derived growth factor by normal mesothelial cells and mesothelioma cell lines. *Cancer Res.* **47**:6180-6184.

Hesterberg,T.(1990) Fiber carcinogenesis in animal models. In, "Cellular and Molecular Aspects of Fiber Carcinogenesis", Brinkley,B.R., Lechner,J.F. and Harris,C.C., eds), Cold Spring Harbor Laboratory Press, (in press).

Ke,Y., Reddel,R.R., Gerwin,B.I., Reddel,H.K., Somers,A.N., McMenamin,M.G., LaVeck,M.A., Stahel,R.A., Lechner,J.F. and Harris,C.C. (1989) Establishment of a human *in vitro* mesothelial cell model system for investigating mechanisms of asbestos-induced mesothelioma. *Am. J. Pathol.* **134**:979-991.

Lave,L.B. and Omenn,G.S. (1986) Cost-effectiveness of short term tests for carcinogenicity. *Nature* **324**:29-34.

Lechner,J.F., Tokiwa,T., LaVeck,M., Benedict,W.F., Banks-Schlegel,S., Yeager,H., Banerjee,A. and Harris,C.C. (1985) Asbestos-associated changes in human mesothelial cells. *Proc. Natl Acad. Sci. USA* **82**:3884-3888.

Smith,D.M., Ortiz,L.W., Archuletta,R.F. and Johnson,N.F. (1987) Long-term health effects in hamsters and rats exposed chronically to man-made vitreous fibers. *Ann. Occup. Hyg.* **31**: 731-754.

HUMAN RISK ASSESSMENT

ANIMAL AND CELL MODELS FOR UNDERSTANDING AND PREDICTING FIBRE-RELATED MESOTHELIOMA IN MAN

J. Bignon and P. Brochard

Unité INSERM 139
Hôpital Henri Mondor
94010 Créteil, Cédex
France

SUMMARY

The epidemiologic, animal and *in vitro* data concerning mesothelioma associated with exposure to fibres will be briefly addressed followed by a discussion of the various animal and cell models currently used to assay the carcinogenic potential of fibrous materials for the mesothelium. Among the animal models there are those which use a non-physiological route of fibre administration (intrapleural and intraperitoneal inoculations) and those which are based on physiological penetration, i.e. inhalation. In addition, the strategies proposed by experimenters vary according to whether the fibrous materials are commercially available or in the development stage. For materials already on the market, the results of animal experiments using non-physiological routes of administration have to be confirmed by inhalation studies. However, for several mentioned reasons, the intrapleural model (IPl) seems more appropriate than the intraperitoneal one (IPe). Moreover, the results obtained would be useful for validating *in vitro* assays that represent alternative methods for the assessment of carcinogenic potential. *In vitro* assays would be extremely useful for materials under development giving rapid and cheap initial screening for cytotoxicity, genotoxicity, carcinogenicity and durability.

INTRODUCTION

Given both the economic importance of fibrous materials in the modern world and the known hazard associated with some fibres, the evaluation of the potential health risk is of major concern and is currently in a critical transition period. During the past three decades, extensive epidemiological data concerning asbestos-associated diseases have accumulated, mostly in relation to occupational exposure at relatively high doses. In coming years, we shall have to deal with two other situations. Firstly, the risk of cancers of the respiratory tract following past or present exposure to relatively low doses (around the present exposure limit values) of commercialised fibres such as asbestos, asbestiform or man-made mineral fibres (MMMF) and secondly, the potential hazards of natural or synthetic fibres before their introduction into the environment.

Among the diseases related to inhalation of mineral fibres, malignant mesothelioma is becoming a major concern because of its strong association with past asbestos exposure and its increasing incidence in Western countries, the reasons for which are presently unknown (McDonald & McDonald 1986, Bignon *et al.*, 1990). This review will consider the human, animal and *in vitro* data concerning mesothelioma and the usefulness of animal and cell studies. Several points will be taken into account, including the relevance of the different routes of exposure for studying pleural carcinogenesis, the physico-chemical characteristics of fibrous particles relevant to the induction of mesothelioma (geometry, size, surface properties, chemistry, mineralogical types) and the biological mechanisms involved in mesothelial carcinogenesis.

SCIENTIFIC BACKGROUND

1 - The main issues related to fibre-associated mesothelioma in man

Since the discovery by Wagner *et al.*, (1960) of the relationship between mesothelioma and occupational and environmental exposure to crocidolite asbestos in South Africa, many human studies have confirmed this association with other varieties of asbestos. In mesothelioma registers, asbestos exposure has been found in 40 to 80 % of cases (McDonald & Mc Donald, 1986; Bignon *et al.*, 1979). In some cohort studies of asbestos workers exposed to amphiboles, the proportional mortality due to mesothelioma has reached 15 to 18 percent (McDonald & McDonald 1978; Jones *et al.*, 1980; Finkelstein, 1984). Nevertheless, data demonstrating a dose-response relationship between pleural mesothelioma and occupational asbestos exposure are rare (McDonald & McDonald, 1986). Attempts have been made to model the dose-specific risk for this tumour. The time since the beginning of exposure seems to be more important in the determination of incidence than dust levels themselves (Doll & Peto, 1985).

The potential risk of developing mesothelioma after non-occupational exposure to low doses of mineral fibres, particularly inside buildings contaminated by friable asbestos (Mossman *et al.*, 1990) and in various occupational situations where MMMF have been used (WHO/IARC, 1989), has not been assessed. We do not know the significance of mesothelioma in the absence of specific asbestos exposure and the question is whether these cases could be due to environmental (background) contamination or to other unrecognised agents and/or genetic factors (Hirsch *et al.*, 1982, Peterson *et al.*, 1984).

Convergent epidemiologic data of the past ten years have raised an important issue concerning the influence of fibre type on the incidence of mesothelioma. The carcinogenic potency of representative fibres with regard to mesothelioma would appear to be, in decreasing order, as follows: erionite > tremolite-actinolite > crocidolite-amosite > chrysotile. Such pathogenic gradients cannot be ascertained with total certitude, because several questions remain to be answered. There is general agreement, supported by animal experiments, that erionite is the strongest carcinogen at the level of the mesothelium. Natural tremolite-actinolite is very common in the geological environment

(WHO/IARC and CEC 1990). Tremolite fibres are considered by several authors to be the cause of the excess of mesotheliomas observed among workers of the vermiculite mines and mills in Montana (USA) (McDonald et al., 1986) and among Quebec chrysotile miners and millers (Moatamed et al., 1986; McDonald, 1988). Although only a small proportion of tremolite is present in the ore of these mines, it is the fibre type found predominantly in the lungs of deceased workers (McDonald et al., 1982; Churg et al., 1984). The "amphibole hypothesis" is also supported by data concerning cohorts of workers exposed to pure chrysotile in the asbestos cement industry where no cases of mesothelioma have been observed (Thomas et al., 1982, Gardner and Powell 1986). Mesothelioma has occurred in other cohorts where chrysotile was admixed with amphiboles (crocidolite or amosite) (Churg & Wright, 1989; WHO, 1989; Albin et al., 1990). This strong association of mesothelioma with amphiboles has been supported by residual lung fibre analysis where tremolite, crocidolite, amosite and anthophyllite were the main fibre types found and at higher concentrations than chrysotile (Gibbs, 1990). However, there is no consensus on the pathogenic significance of residual fibres many years after exposure and we do not really see how this issue can be clarified.

The two large cohorts of workers exposed to MMMF studied in Europe and in the USA contained no cases of mesothelioma unless there was associated asbestos exposure (WHO/IARC, 1988). However, the group of workers exposed to fibres for long enough for these to represent a potential hazard to mesothelial cells is small. A longer follow-up time is required.

The mechanisms involved in the induction of pleural tumours in man remain unclear. In particular, we do not know if fibres directly affect mesothelial cells, or if they act through indirect mechanisms. There is some evidence that asbestos fibres can reach the pleura in man, including parietal tissue and fibres have been found in pleural plaques. However, only short microfibres of chrysotile (< 3 microns in length) are found in these locations (Bignon et al., 1978; Boutin et al., 1981; Dodson et al., 1990). Such fibres are not considered to be biologically relevant to pleural carcinogenesis (Stanton et al., 1981). A second hypothesis implies that mesothelial cells are stimulated to grow and transform under the action of clastogenic factors and mediators. These can be oxidants and growth factors released by other cells, particularly alveolar (possibly also pleural) macrophages which are stimulated by the durable long fibres remaining in lung tissue (Sebastien et al., 1979; Churg & Wright, 1989). Both hypotheses are in accordance with two features of human mesothelioma. The long latency period, that could be due to the time required for fibres to migrate to the pleura and the strong association of this tumour with amphiboles, fibre types known to be much more durable than chrysotile and MMMF.

2 - Experimental data

2.1 - Rodents

Since the first animal study using asbestos (Wagner, 1962) many animal experiments have been carried out, mainly in rodents, in order to reproduce the human diseases induced by asbestos and other fibres: lung fibrosis, lung cancer and mesothelioma. The choice of animal species is a crucial one and may be relevant to the carcinogenic potential for a given organ. The rat has been the most widely used species for research on the carcinogenic potential of asbestos in the lung as well as in the pleura (Reeves et al., 1974; Wagner et al., 1974; Davis et al., 1978; Monchaux et al., 1981, 1985). The hamster has been rarely used, but this species might be particularly sensitive to pleural carcinogenesis, at least after inhalation exposure to ceramic fibres (Smith *et al.*, 1987; Hesterberg *et al.*, this volume). Mice can also be used, particularly for the induction of mesothelioma after intraperitoneal inoculation of fibres (Kane *et al.*, 1991). According to Davis (1991), the guinea pig is rarely used and appears to be resistant to experimental carcinogenesis.

The intrapleural and intraperitoneal routes have been used successfully to examine the carcinogenic potential of various fibres in the pleural or peritoneal cavity (Stanton & Wrench 1972; Pott & Friedrichs, 1972; Stanton *et al.*, 1977; Pott *et al.*, 1989; Wagner *et al.*, 1973; Monchaux *et al.*, 1981; Davis, 1991). These intracoelomic models do not allow one to derive a scale of mesothelial carcinogenicity according to fibre type. Indeed, within the same study and from study to study, different chrysotile fibre types gave different percentages, depending on their origin and size (Davis *et al.*, 1986; Jaurand *et al.*, 1987; Davis, 1991). Carcinogenic potency has also been confirmed in various studies with MMMF, but with variations depending on type, size, chemistry and durability (Stanton *et al.*, 1977, 1981; Pott *et al.*, 1989; Wagner *et al.*, 1984; Davis *et al.*, 1984).

The enormous amount of experimental work by Stanton and co-workers (1981), involving the intrapleural implantation of various fibre types in rats, led to clear cut conclusions concerning the predominant role played by size parameters in pleural carcinogenesis. With fibres less than 1.5 μm in diameter, the best correlation with the incidence of cancer was obtained with fibres of a diameter equal to or less than 0.25 μm and a length greater than 8 μm. However, results obtained after intrapleural injection of acid-leached chrysotile in the rat have shown that other parameters than fibre type and size are probably involved in mesothelial carcinogenesis (Morgan *et al.*, 1977; Monchaux *et al.*, 1981; Jaurand *et al.*, 1987).

Long-term asbestos inhalation by rodents can produce pleural mesothelioma, as well as lung cancer and asbestosis, but at a low rate (Wagner *et al.*, 1974; Davis *et al.*, 1978). Fibre penetration by this physiological route produces fewer peritoneal mesotheliomas (one case in the study of Davis *et al.*, 1978).

In the published animal studies on the induction of mesothelioma after inhalation of a fibrous aerosol it has not been possible to reproduce in rodents the differences between lung and pleural fibre-related carcinogenesis observed in man. These are, a longer latency period for mesothelioma than for lung cancer and the synergistic effect of smoking and asbestos in lung cancer but not in mesothelioma (Doll & Peto, 1985). The difference in the latency period between animals and man might be related to differences either in the kinetics of fibre translocation to the pleura or to the life span. The synergy between asbestos and smoking for lung cancer (and the absence of synergy for mesothelioma) has not been demonstrated in rodents, probably because of problems with long-term experiments involving tobacco smoke in these species.

2.2 - In vitro transformation assays with rodent cells

Several cell systems have recently been used to assess the transforming potency of asbestos and other mineral fibres and to explore their mechanisms of action (Jaurand *et al.*, 1991). The transformation of rodent cells by fibres has been assessed using three standard cell systems: Syrian hamster embryo cells (SHE) (DiPaolo *et al.*, 1983; Hesterberg & Barrett, 1984; Mikalsen *et al.*, 1988), two types of mouse fibroblasts: C3H10T1/2 (Brown *et al.*, 1983; Hei *et al.*, 1985) and Balb 3T3 (Lu *et al.*, 1988). In addition, a new model using rat pleural mesothelial cells (RPMC) has been developed in our laboratory to model the relationship between asbestos exposure and mesothelioma occurrence in populations exposed to asbestos (Jaurand *et al.*, 1983).

Data obtained with these models clearly demonstrate that asbestos fibres can produce morphologic and/or neoplastic transformation of rodent cells, in agreement with experimental findings *in vivo*. In addition, these models are ideal for studying mechanisms of genotoxicity and cell transformation. It appears that mineral fibres may play a role in inducing chromosomal mutations or abnormalities (Barrett, this volume) and might act at several stages of cell transformation. It has been shown that asbestos fibres can transform RPMC in the same way as chemicals like benzo-[*a*]-pyrene (Paterour *et al.*, 1985).

CRITICAL APPRAISAL OF TOOLS FOR ASSAYING THE MESOTHELIAL CARCINOGENIC POTENTIAL OF FIBROUS MATERIALS

Characterisation of dust samples

Before discussing the issues raised by biological tests, the problem of the correct characterisation of dust samples has to be emphasised. Fibre concentrations must be accurately assessed in terms of both mass and number. A careful examination of the published *in vitro* and *in vivo* experiments on fibres (including asbestos, MMMF and organic fibres) shows that the greatest inconsistency is in the assessment of the dose. Most often the only parameter given is the weight. The percentage, both mass and number, of associated non-fibrous particles is seldom given in the early experiments. At best, a detailed analysis of the fibres morphology is provided. Even then it is difficult to make a comparison between experiments because of the variation of the particle size distribution from one type of fibre to another.

It is well known that a modification of fibre size distribution can induce significant changes in biological response (Stanton *et al.*, 1977, 1981; Jaurand *et al.*, 1987; Davis, 1991). Moreover, the methods used to generate the fibre samples for *in vivo* and *in vitro* tests, can produce side-products such as non-fibrous material derived from the fibres, or organic or metallic contaminants from the generation system. These contaminants are seldom taken into account and it has been shown that they can interfere with biological responses.

To gain a more accurate assessment of the dose administered, in order to compare the different tests, the physico-chemical parameters of the fibres must be assessed. These include length and diameter, which gives the equivalent aerodynamic diameter (EAD), chemical composition, surface charge, surface active redox sites, durability and associated products such as oils added to the surface of synthetic fibres during their manufacture.

Animal experiments

This section is a detailed discussion of the advantages and disadvantages of the various routes which can be used to expose rodents to fibres. These routes fall into non-physiological means such as intracoelomic inoculation and intratracheal injection and physiological approaches such as inhalation.

Intracoelomic inoculation

Intrapleural inoculation

The intrapleural (IPl) injection of fibrous particles (Wagner, 1962) is a valuable method for comparative studies. It provides results reproducible within the same laboratory and comparable from one laboratory to another (Jaurand, personal communication). The low background level of pleural mesothelioma in the rat (0.3 to 0.8 %) (Ilgren 1989) means that generally no cases of mesothelioma are found in sham controls (Wagner *et al.*, 1973, Monchaux *et al.*, 1981, Maltoni *et al.*, 1982, Jaurand *et al.*, 1987) and the IPl model can be used as a screening test for assessing intrinsic carcinogenicity of mineral fibres, with apparently a much higher sensitivity than inhalation studies.

There are a number of limitations associated with the IPI test. It does not seem to allow the relationship between dose and tumorigenic response to be assessed and extrapolation to a no effect level for carcinogenicity (threshold dose) made. Data from quantitative studies which attempt this remain rare, often inconsistent and questionable, particularly with regard to soluble and splittable fibres. Fibres are often concentrated in certain areas within the pleural cavity, implying that high concentrations are in direct contact with mesothelial cells. This makes it difficult to interpret the carcinogenic effect at the cellular level. The associated granulomatous inflammatory reaction, which is dependent on fibre length, may increase the carcinogenic effect of long fibres (Stanton & Layard, 1978). For all these reasons, extrapolation to man is questionable.

The intrapleural route appears effective for studying the physico-chemical parameters of fibres and their relationship with carcinogenic potency (Wagner *et al.*, 1973; Stanton *et al.*, 1981; Davis *et al.*, 1986). Also, it has been used to study the modification of the carcinogenic potential of asbestos after surface changes produced by acid treatment (Jaurand *et al.*, 1987) or phosphatation (Jaurand, 1991).

Intraperitoneal inoculation (IPe)
The induction of peritoneal tumours by inoculating fibres into the peritoneal cavity of rodents, particularly female Sprague Dawley rats, has been used routinely by Pott and co-workers during the past 20 years for studying peritoneal carcinogenesis of various dust samples (Pott & Friedrichs, 1972). Recently, this test has been recommended as reliable and sensitive for predicting the intrinsic carcinogenic potency of mineral fibres and other minerals (Pott *et al.*, 1990). However, this method does not seem to be ideal, for the following reasons:

1) The IPe route is not the most appropriate, since peritoneal mesothelioma is rare compared to pleural mesothelioma following asbestos exposure.

2) The incidence of spontaneous peritoneal mesothelioma is variable but not negligible. According to Ilgren (1989), the rate of spontaneous peritoneal tumours in the rat is 0.9 - 1.7%. In Fischer rats, tumour rates of about 3 - 4% (mainly abdominal and/or scrotal) have been reported (Tanigawa *et al.*, 1987).

3) The histological patterns of peritoneal malignancies in animals may be confusing since some tumours may derive from the tunica vaginalis of the testis or mimic interstitial cell tumours which possibly involve hormonal imbalance. Sex-related differences in the histogenesis and incidence of omentum tumours in rats are not well documented, raising questions as to the use of only female rats (Friemann *et al.*, 1990). Proliferative lesions of the serous membranes have been observed in ovariectomised female and stilboestrol-treated male dogs (O'Shea & Jabara, 1971).

4) The IPe route, although highly sensitive, appears to be a non-specific model for studying fibre-induced mesothelial cancer. Indeed, in Fischer rats, peritoneal mesotheliomas have been induced at variable rates, sometimes exceeding 50%, following oral administration of chemical carcinogens (Kurokawa et al., 1983).

5) The fate and durability of particles have only been studied in the pleural cavity and not the peritoneal cavity (Monchaux et al., 1982).

6) Last, but not least, the background level of tumours in sham controls appears to be much higher in the IPe model than in the IPl model; indeed, Pott states that the tumour yield should be at least 15% of animals for a response to be considered positive (Pott, 1989).

In addition to these disadvantages, and although the IPl model has been used extensively in many different laboratories (Wagner et al., 1973, 1984; Morgan et al., 1977; Davis, 1979; Monchaux et al., 1981; Pigott & Ishmael, 1982; Coffin et al., 1982; Jaurand et al., 1987), the IPe model has only been used in a few laboratories (Pott, et al., 1972, 1978, 1982, 1989, 1990; Smith et al., 1987; Davis 1991). As a result, the reproducibility of the results obtained using intraperitoneal inoculation is not well known, even when standardised samples of mineral are used.

Conclusions concerning intracoelomic inoculation

The advantages of experiments involving intracoelomic inoculation (IPl or IPe) can be summarised as follows :

1) Although it involves a non-physiological route, this model can predict the intrinsic carcinogenic potency of dust samples. However, the results can only be extrapolated to humans with caution and positive results must be confirmed by inhalation experiments.

2) If toxicological assays in animals are required for the evaluation of solid particles, there is no scientific basis for choosing IPe rather than IPl. Indeed, it seems more appropriate to use IPl, particularly because it gives a much lower background frequency of tumours in sham controls.

3) Intercomparison studies between independent laboratories, using the same mineral samples at the same doses in the same animal strains (both sexes) and by the same route of inoculation, are warranted. However, it must be emphasised that, while data obtained after intraserosal injection of minerals may be relevant to their mesothelial carcinogenic potency, the extrapolation to lung carcinogenesis has not been validated and any attempt must be viewed with extreme caution.

4) IPl and to some extent IPe, appear to be useful tools for defining the fibre parameters involved in fibre-related carcinogenesis. They may also be useful for exploring mechanisms, particularly lymphatic clearance (Brinkmann & Muller, 1989; Kane et al., 1991).

Intratracheal injection

This technique is not useful for long-term experiments exploring the carcinogenic potency of fibrous materials in rodents. Heterogeneous deposition and local overload effects are the usual pitfalls and prevent this route from being used to assess carcinogenicity in animals.

By contrast, intratracheal injection can be used for short-term experiments which assess *in vivo* toxicity and/or inflammatory responses. In *ex vivo* studies, cells exposed *in vivo* are recovered by bronchoalveolar lavage (Begin *et al.*, 1983). This model permits the investigation of mechanisms of lung fibrosis and possibly also lung cancer, particularly those related to cooperation between phagocytic cells and epithelial cells. It can also be used for testing the durability of mineral dusts after various residence times within the lung.

Inhalation studies

These studies are theoretically the most valuable, because they attempt to reproduce the human situation, where the pathogenicity of fibres, as far as mesothelioma is concerned, is dependent on several factors, e.g. the respirability, clearance, relocation and dissolution of fibres.

Before starting an inhalation experiment, a detailed characterisation of the aerosols is required, with particular attention paid to adapting the EAD to the geometry and size of the respiratory tract of the experimental animal.

The procedure used for inhalation is also important: whether exposure is in a chamber or nose-only exposure; if there should be size-selection of fibres, e.g., excluding those with an EAD above 1 μm, which theoretically do not penetrate into the distal airways of rodents (Bernstein *et al.*, this volume). Size selection for experimental conditions, however, may introduce a bias when extrapolating results to man.

One crucial problem is that we do not know exactly by which pathway and how long it takes fibres to reach their remote target cells at the level of the parietal pleura to cause the induction of mesothelioma by inhalation. The fibres must maintain their carcinogenic potency over the long latency period necessary for the induction. This period is much longer in humans, about 40 years, than in small rodents with a life-span of about 2 years. This difference in life-span and perhaps the uncertainties concerning the inhaled doses, might explain the discrepancies between human and rodent data on the incidence of mesothelioma after inhalation of asbestos. This is much lower in rats than in some cohorts of asbestos workers exposed to crocidolite or amosite (see above).

Among the factors which limit the reliability of inhalation studies in rodents, there may be an overload effect when excessive doses are used. In addition, there will be a low yield of tumours at low doses or with particles with low carcinogenic potency. This then requires an increase in the number of animals (and also the cost) needed in order to reach statistical significance. Moreover, the need for special equipment for nose-only or chamber exposure and for selecting fibre dimensions makes inhalation experiments very expensive.

In spite of these disadvantages, inhalation studies are the only ones which can be extrapolated to man, because they reproduce the physiological route. Mesothelioma incidence following inhalation, experiments in rats or hamsters is in accordance with human data. The exception at the present is studies with refractory ceramic fibres where there has been insufficient follow-up in man. However, because of their cost, inhalation studies are performed only in a second phase after the results of screening tests for intrinsic carcinogenic potency (intracoelemic inoculations and *in vitro* assays).

Inhalation-based experiments are also ideal for exploring *in vivo* the physico-chemical characteristics related to the pathogenetic potential of mineral dusts. These are the involvement of shape and size in the deposition, clearance and relocation of the dusts, their surface chemistry and their stability in lung tissue. As the mineral chemistry of fibres seems to be a major determinant for their potential dissolution *in vitro*, a significant residence time in lung tissue after physiological penetration should be the best method for investigating their durability *in vivo*.

In vitro tests

Transformation assays with rodent cells

The most suitable cell systems for *in vitro* tests for assessing the carcinogenic potency of fibres and exploring their mechanisms of action seem to be Syrian hamster embryo cells (SHE) and rat pleural mesothelial cells (RPMC). This last model is particularly appropriate as it involves cells specific for mesothelioma carcinogenesis.

These tests are still in the development stage; they need to be validated by inter-comparison with animal studies designed to explore intrinsic carcinogenic potency. They are promising alternative methods to animal experiments with several advantages: lower cost and the results obtained in a shorter time. Moreover, the possibility of using human mesothelial cells *in vitro* should reproduce ideal conditions for studying the mechanisms of human mesothelioma induction. However, as for tests based on intracoelemic inoculation, these assays explore only intrinsic carcinogenic potency. As cells in culture are usually exposed to high doses, with a large number of fibres per cell, the relevance of

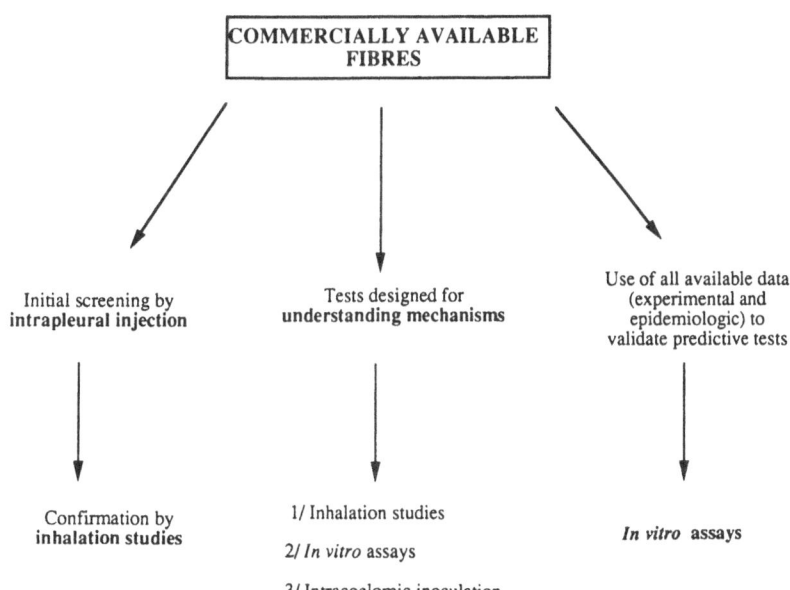

Figure 1: Strategy proposals for assessing the carcinogenic risk of commercially available fibres (asbestos, asbestiforms, man-made mineral fibres).

these methods and their extrapolation to the *in vivo* situation needs to be investigated. Positive results require confirmation by animal experiments, particularly inhalation.

Acellular assays for studying the durability of minerals

We have already suggested that the residence time of mineral dusts in the lung may have a strong influence on the ability of inhaled fibrous materials to cause mesothelioma. This raises the "durability" hypothesis, which depends on the clearance rate and on the dissolution rate of the fibrous material. The dissolution rate can be assessed in *in vitro* systems, but there is a need to design the most appropriate assays after evaluating which lung compartment is most involved in mineral leaching. The interstitial or pleural fluid has a pH around 7.6 (Sahn, 1985) while the pH of the intralysosomal compartment is acid (Gordon, 1973; Jaurand *et al.*, 1984).

CONCLUSION : A STRATEGY FOR THE EVALUATION AND DESIGN OF "SAFE" FIBRES

Figures 1 and 2 show schematic representations of possible strategies for two different situations in fibre assessment: (i.) with commercially available fibres (asbestos, natural asbestiforms, MMMF), where the toxicological objectives are risk-assessment and determination of carcinogenic mechanisms; (i.i.) new natural or synthetic fibres, prior to commercial release.

Commercial fibres.(Figure 1)

Only animal experiments based on inhalation are predictive or can confirm health hazards in man. They allow the study of dose-response curves for lung fibrosis, lung

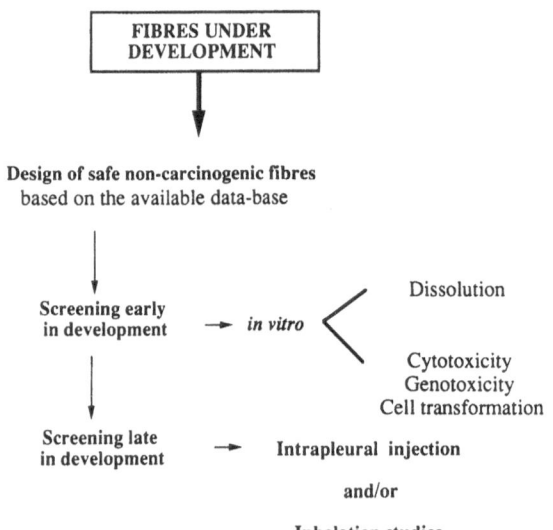

Figure 2: Strategy proposals for new fibers under development (natural or synthetic).

cancer and mesothelioma and thus could help to determine threshold doses. However, when possible, priority should be given to epidemiology, because any extrapolation to man is hazardous even after experiments by inhalation. When positive results are obtained in inhalation tests regulation of the fibrous material should be considered, taking into account the existing regulations for asbestos in the different countries.

Animal studies can also help in understanding the role played by physico-chemical fibre parameters in relation to pleural carcinogenesis. The results obtained in inhalation experiments must be compared with results from other experimental models, e.g. intraserosal injections and *in vitro* tests using various cells in culture, including mesothelial cells. More animal and *in vitro* studies are warranted for studying the durability of fibres and the role of this factor in the induction of mesothelioma. *In vitro* assays should be validated by using fibrous materials that have already been assessed in animal experiments and epidemiological studies and by intercomparison studies.

New fibres. (Figure 2)

The industrial development of new, synthetic or natural fibres must now take into account two different and possibly opposed requirements: (i.) that they should obtain the optimal technical and economic performance ; (i.i.) that they avoid potential biological hazards, particularly regarding carcinogenesis.

In order to achieve the second aim, decision-makers in industry must proceed according to the following steps. They must use the available data-base concerning fibre toxicity in order to design a theoretically "safe" fibre, taking into account the various physico-chemical parameters mentioned above. Durability in particular is important and must be investigated using assays mimicking the biological environment encountered in the lung, especially pH. If *in vitro* test data is strongly positive this can help in the decision to stop the development of a new product at an early stage. If there is any doubt about the safety of a product before *in vitro* tests are completed, intraserosal tests (mainly intrapleural inoculation) should be used as a second screening tool before commercialisation. If the results of intraserosal tests are inconclusive, inhalation studies should be performed before releasing new fibres into the market.

ACKNOWLEDGEMENTS

The authors are grateful to Dr Marie-Claude Jaurand for reviewing the manuscript and to Dr Jeff Everitt for his invaluable advice and bibliographic contributions. Thanks are also due to Mrs. Joan Beaurain for typing the manuscript and Mr D. Young for help with the English translation.

REFERENCES

Albin,M., Jakobsson,K., Hattewel,R., Johansson,L. and Welinder,H. (1990) Mortality and cancer morbidity in cohorts of asbestos cement workers and referents. *Br. J. Indust. Med.* **47**:602-610.

Barrett J.C. (1991) Mechanisms of asbestos induced neoplastic transformation. *This volume.*

Begin,R., Rola-Pleszczynski,M., Masse,S., Lemaire,I., Sirois,P., Boctor,M., Nadeau,D., Drapeau,G. and Bureau,M.A. (1983) Asbestos-induced lung injury in the sheep model: the initial alveolitis. *Environ. Res.* **30**:195-210.

Bernstein,D., Fleissner,H., Bouvier,C., Hesterberg,T., and Mast,R. (1991) An inhalation model for evaluation of fiber oncogenicity in rodents. *This volume.*

Bignon,J., Brochard,P., De Cremoux,H., Nebut,M. and Jaurand,M.C. (1990) Contribution of epidemiology and biology to the comprehension of causes and mechanisms of mesothelioma. In Deslauriers J., Lacquets L.K., eds. "International trends in general thoracic surgery. Thoracic surgery : surgical management of pleural diseases." The CV Mosby Company. Toronto.Vol. **6**:327-335.

Bignon,J., Sébastien,P. and Gaudichet,A. (1978) Measurement of asbestos retention in the human respiratory system related to health effects. In "Proceedings of the workshop on asbestos : definition and measurement methods". National Bureau of Standard, Gaithersburg, Maryland, Special Publication n° **506**:95-119.

Bignon,J., Sébastien,P., Di menza,L., Nebut,M. and Payan,H. (1979) French mesothelioma register. *Ann. NY Acad. Sci.* **330**:455-466.

Boutin,C., Sébastien,P., Janson,X. and Viallat,J.R. (1981) Métrologie des fibres minérales dans des biopsies thoracoscopiques pulmonaires et pleurales. Résultats préliminaires. *Poumon Coeur* **37**:253-257.

Brinkmann,O.A. and Muller,K.M. (1989) What's new in intraperitoneal test on kevlar. *Path. Res. Pract.* **185**:412-417.

Brown,R.C., Poole,A. and Fleming,G.T.A. (1983) The influence of asbestos dust on the oncogenic transformation of C3H 10T1/2 cells. *Cancer Lett.* **18**:221- 227.

Churg,A., Wiggs,B., De Paoli,L., Kampe,B. and Stevens,B. (1984) Lung asbestos content in chrysotile workers. *Am. Rev. Respir. Dis.* **130**:1042-1045.

Churg,A. and Wright,J.L. (1989) Fibre content of lung in amphibole and chrysotile induced mesothelioma : implications for environmental exposure. In Bignon J.,Peto J. and Saracci R., eds. "Non occupational exposure to mineral fibres." IARC Scientific Publications, n°90:314-318.

Coffin,D.L., Palekar,L.D. and Cook,P.M. (1982) Tumorigenesis by a ferroactinolite mineral. *Toxicol. Lett.* **13**:143-150.

Davis,J.M.G. (1979) The histopathology and ultrastructure of pleural mesotheliomas produced in the rat by injections of crocidolite asbestos. *Br. J. Exp. Path.* **60**:642-652.

Davis,J.M.G. (1991) Information obtained from fibre-induced lesions in animals. In,"Mineral fibers and health", eds. Miller K., Liddell F.D.K. : CRC Press. (in press)

Davis,J.M.G., Addison,J., Bolton,R.E., Donaldson,K., Jones,A.D. and Wright,A. (1984) The pathogenic effects of fibrous ceramic aluminium silicate glass administered to rats by inhalation or peritoneal injection. In, "Biological effects of man-made mineral fibres". WHO/IARC conference, WHO publication. Vol. **2**, pp. 303-322.

Davis,J.M.G., Addison,J., Bolton,R.E., Donaldson,K., Jones,A.D, and Smith,T. (1986) The pathogenicity of long versus short fibre samples of amosite asbestos administered to rats by inhalation and intraperitoneal injection. *Br.J.Exp.Path.* **67**:415-430.

Davis,J.M.G., Beckett,S., Bolton,R.E., Collings,P. and Middleton,A.P. (1978) Mass and number of fibres in the pathogenesis of asbestos-related lung disease in rats. *Br.J.Cancer* **37**:673-688.

Di Paolo,J.A., De Marinis,A.J. and Doniger,J. (1983) Asbestos and benzo(a)pyrene synergism in the transformation of Syrian hamster embryo cells. *Pharmacol.* **27**:65-73.

Dodson,R.F., Williams,M.G., Corn,C.J., Brollo,A. and Bianchi,C. (1990) Asbestos content of lung tissue, lymph nodes, and pleural plaques from former shipyard workers. *Am. Rev. Respir. Dis.* **142**:843-847.

Doll,R. and Peto,J. (1985) Effects on health of exposure to asbestos. London: Health and Safety Commission. Her Majesty's Stationery Office.

526

Finkelstein,M.M. (1984) Mortality among employees of an Ontario factory manufacturing insulation materials from amosite asbestos. *Am. J. Indust. Med.* **15**:477-481.

Friemann,J. and Muller,K.M. (1990) Mesothelial proliferation due to asbestos and man-made fibres. Experimental studies on rat omentum. *Path. Res. Pract.* **186**:117-123.

Gardner,M.J. and Powell,C.A. (1986) Mortality of asbestos cement workers using almost exclusively chrysotile fibre. *J. Soc. Occup. Med.* **36**:124-126.

Gibbs,A.R. (1990) Role of asbestos and other fibres in the development of diffuse malignant mesothelioma. *Thorax* **45**:649-654.

Gordon,A.H. (1973) The role of lysosomes in protein catabolism. In, "Lysosomes in biology and pathology" ed. Dingle,J.T. North Holland publ. Cy, Amsterdam, London. **3**:89-137.

Hei,T.K., Geard,C.R., Osmak,R.S. and Travisano,M. (1985) Correlation of *in vitro* genotoxicity and oncogenicity induced by radiation and asbestos fibres. *Br. J. Cancer* **52**:591-597.

Hesterberg,T.W. and Barrett,J.C. (1984) Dependence of asbestos and mineral dust-induced transformation of mammalian cells in culture on fibre dimension. *Cancer Res.* **44**:2170-2180.

Hesterberg,T.W., Mast,R., McConnell,E.E., Chevalier,J., Bernstein,D.M., Bunn,W.B. and Anderson,R. (1991) Chronic inhalation toxicity of refractory ceramic fibers in Syrian hamsters. This workshop. *This volume.*

Hirsch,A., Brochard,P., De Cremoux,H., Erkan,L., Sebastien,P., Di Menza,L. and Bignon,J. (1982) Features of asbestos-exposed and unexposed mesothelioma. *Am. J. Indust. Med.* **3**:413-422.

Ilgren,E.B. (1989) Mesothelioma threshold. In, "Effects of mineral dusts on cells", eds. Mossman,B.T. and Begin,R.O. NATO ASI. Springer Verlag - Heidelberg. Vol. **H30**:455-464.

Jaurand,M.C. (1991) Observations on the carcinogenicity of asbestos fibers. *Ann. N.Y. Acad. Sci.*, in press.

Jaurand,M.C., Gaudichet,A., Halpern,S. and Bignon,J. (1984) *In vitro* biodegradation of chrysotile fibres by alveolar macrophages and mesothelial cells in culture : comparison with a pH effect. *Br. J. Indust. Med.* **41**:389-395.

Jaurand,M.C., Bastie-Sigeac,I., Paterour,M.J., Renier,A. and Bignon,J. (1983) Possibility of using rat mesothelial cells in culture to test cytotoxicity, clastogenicity and cancerogenicity of asbestos fibers. *Ann. N.Y. Acad. Sci.* **407**:409-411.

Jaurand,M.C., Fleury,J., Monchaux,G., Nebut,M. and Bignon J. (1987) Pleural carcinogenic potency of mineral fibers (asbestos, attapulgite) and their cytotoxicity on cultured cells. *J. Natl. Cancer Inst.* **79**:797-804.

Jaurand,M.C., Saint-Etienne,L., Van Der Meeren,A., Endo-Capron,S., Renier,A. and Bignon,J. (1991) Neoplastic transformation of rodent cells. In, " Cellular and molecular aspects of fiber carcinogenesis". Current communications in molecular biology . Cold Spring Harbor. n° 2. (in press).

Jones,J.S.P., Pooley,F.D., Sawle,S.W., Madely,R.J., Smith,P.G., Berry,G. and Wignal,B.K. (1980) The consequences of exposure to asbestos dust in a wartime gas-mask factory. In,"Biological effects of mineral fibres", ed. Wagner,J.C. IARC. Scientific Publications n° **30**:637-653.

Kane,A.B., Gleva,G.F. and Goodlick,L.A. (1991) Transforming growth factor-beta: a potential mediator of asbestos-induced disease. Proceedings of the VIIth International Pneumoconioses Conference. Pittsburgh, August 23-26 1982. (in press)

Kurokawa,Y., Hayashi,Y., Maekawa,A., Yakahashi,M., Kokubo,T. and Odashima,S. (1983) Carcinogenicity of potassium bromate administered orally to F344 rats. *J. Natl. Cancer Inst.* **71**:965-972.

Lu,Y.P., Lasne,C., Lowy,R. and Chouroulinkov,I. (1988) Use of the orthogonal tetradecanoyl-phorbol-13-acetate (TPA) in the BALB/3T3 cell transformation system. *Mutagenesis* **3**:355-362.

Maltoni,C., Minardi,F., Morisi,L. (1982) Pleural mesotheliomas in Sprague-Dawley rats by erionite: first experimental evidence. *Environ. Res.* **29**:238-244.

McDonald,J.C. (1988) Tremolite, other amphiboles and mesothelioma. *Am. J. Indust. Med.* **14** 247-249.

McDonald,A.D. and McDonald,J.C. (1978) Mesothelioma in persons exposed to crocidolite in gas mask manufacture. *Environ. Res.* **17**:340-346.

McDonald,A.D. and McDonald,J.C. (1986) Epidemiology of malignant mesothelioma. In, "Asbestos related malignancy", eds. Antman,K. and Aisner,M.J. Grune and Stratton, New York, pp.57-79.

McDonald,A.D., McDonald,J.C. and Pooley,F.D. (1982) Mineral fibre content of lung in mesothelial tumours in North America. *Ann. Occup. Hyg.* **26**: 417-422.

McDonald,J.C., McDonald,A.D., Amstrong,B. and Sebastien,P. (1986) Cohort study of mortality of vermiculite miners exposed to tremolite. *Br. J. Indust. Med.* **43**:436-444.

Mikalsen,S.O., Rivedal,E. and Sanner,T. (1988) Morphological transformation of Syrian hamster embryo cells induced by mineral fibres and the alleged enhancement of benzo(a)pyrene. *Carcinogenesis* **9**:891-899.

Moatamed,F., Lockey,J.E. and Parry,W.T. (1986) Fibre contamination of vermiculites : a potential occupational and environmental health hazard. *Environ. Res.* **41**:207-218.

Monchaux,G., Bignon,J., Jaurand,M.C., Lafuma,J., Sebastien,P., Masse,R. and Goni,J. (1981) Mesotheliomas in rats following inoculation with acid-leached chrysotile asbestos and other mineral fibres. *Carcinogenesis* **2**:229-236.

Monchaux,G., Bignon,J., Hirsch,A., Lafuma,J. and Sebastien,P. (1982) Translocation of mineral fibres through the respiratory system after injection into the pleural cavity of rats. *Ann. Occup. Hyg.* **26**:309-318.

Monchaux,G., Lafuma,J., Bignon,J. and Hirsch,A. (1985) Experimental pleural carcinogenesis induced by mineral fibers. In, "The pleura in health and disease", eds. Chrétien,J., Bignon,J. and Hirsch,A.. Marcel Dekker, inc., New York. Basel. Vol. **30**:551-570.

Morgan,A., Davies,P., Wagner,J.C., Berry,G. and Holmes,A. (1977) The biological effects of magnesium-leached chrysotile asbestos. *Br. J. Exp. Pathol.* **58**:465-473.

Mossman,B.T., Bignon,J., Corn,M., Seaton,A. and Gee,J.B.L. (1990) Asbestos : scientific developments and implications for public policy. *Science* **247**:294-301.

O'Shea,J.D. and Jabara,A.G (1971) Proliferative lesions of serous membranes in ovariectomised female and entire male dogs after stilbestrol administration. *Vet. Pathol.* **8**:81-90.

Paterour,M.J., Renier,A., Bignon,J. and Jaurand,M.C. (1985) Induction of transformation in cultured rat pleural mesothelial cells by chrysotile fibres. In, "*In Vitro* effects of mineral dusts", eds. Beck,E.G. and Bignon,J. NATO ASI series, Springer Verlag - Heidelberg. Vol.**G3**:203-207.

Peterson,J.T., Greenberg,S.D. and Buffler,P.A. (1984) Non asbestos related malignant mesothelioma. A Review. *Cancer* **54**:951-960.

Pigott,G.H. and Ishmael,J. (1982) A strategy for the design and evaluation of a "safe" inorganic fibre. *Ann. Occup. Hyg.* **26**:371-380.

Pott F. (1978) Some aspects on the dosimetry of the carcinogenic potency of asbestos and other fibrous dusts. *Staub-Reinholt Luft* **38**:486.

Pott,F. and Friedrichs,K.H. (1972) Tumours in rats after intraperitoneal injection of asbestos dusts. *Naturwissenschaften* **59**:S 318.

Pott,F., Schlipkoter,H.W., Ziem,U., Spurny,K. and Huth,F. (1984) New results from implantation experiments with mineral fibres. In, "Biological effects of man-made mineral fibres", WHO/IARC conference, WHO publication. Vol. 2, pp.286-302.

Pott,F., Roller,M., Ziem,U., Reiffer,F.J., Bellmann,B., Rosenbruch,M. and Huth,F. (1989) Carcinogenicity studies on natural and man-made fibres with the intraperitoneal test in rats. In, "Non occupational exposure to mineral fibres", Bignon,J., Peto,J. and Saracci,R., eds. IARC Scientific Publications, n° **90**:173-179.

Pott,F., Bellmann,G., Muhle,H., Rödelsperger,K., Rippe,R.M, Roller,M. and Rosenbruch,M. (1990) Intraperitoneal injection studies for the evaluation of the carcinogenicity of fibrous phyllosilicates. In, "Health related effects of phyllosilicates", ed. Bignon,J. NATO ASI Series. Springer Verlag - Heidelberg. Vol. **G21**:319-329.

Reeves,A.K., Puro,H.E. and Smith,R.G. (1974) Inhalation carcinogenesis from various forms of asbestos. *Environ. Res.* **8**:178-202.

Sahn,S.A. (1985) Pleural fluid pH in the normal state and in diseases affecting the pleural space. In, "The pleura in health and disease", eds. Chrétien,J., Bignon,J. and Hirsch,A. Marcel Dekker, inc., New York. Basel. Vol. 30:253- 266.

Sebastien,P., Janson,X., Bonnaud,G., Riba,G., Masse,R. and Bignon,J. (1979) Translocation of asbestos fibers through respiratory tract and gastrointestinal tract according to fiber and size. In, "Dusts and disease", eds. Lemen,R. and Dement,J., Park Forest South, Pathotox, Publishers Inc. pp. 65-85.

Smith,D., Ortiz,L., Archuleta,R. and Johnson,N. (1987) Long term health effects on hamsters and rats exposed chronically to MMVF. *Ann. Occup. Hyg.* 32: 731-754.

Stanton,M.F. and Layard,M. (1978) The carcinogenicity of fibrous minerals. In, "Proceedings of the workshop on asbestos: definition and measurement methods". National Bureau of Standard, Gaithersburg, Maryland, Special Publication n° **506**:143-151.

Stanton,M.F., Layard,M., Tegaris,A., Miller,E., May,M. and Kent,E. (1977) Carcinogenicity of fibrous glass: Pleural response in the rat in relation to fiber dimension. *J. Natl. Cancer Inst.* **58**:587-603.

Stanton,M.F., Layard,M., Tegeris,A., Miller,E., May,M., Morgan,E. and Smith A. (1981) Relation of particle dimension to carcinogenicity in amphibole asbestos and other fibrous minerals. *J. Natl. Cancer Inst.* **67**:165-175.

Stanton,M.F. and Wrench,C. (1972) Mechanisms of mesothelioma induction with asbestos and fibrous glass. *J. Natl. Cancer Inst.* **48**:797-821.

Tanigawa,H., Onodera,H. and Maekawa A. (1987) Spontaneous mesotheliomas in Fischer rats - a histological and electron microscopic study. *Toxicol. Pathol.* **15**:157-163.

Thomas,H.F., Benjamin,I.T., Elwood,P.C. and Sweetnam,P.M. (1982) Further follow up study of workers from an asbestos-cement factory. *Br. J. Indust. Med.* **39**:273-276.

Wagner,J.C. (1962) Experimental production of mesothelial tumours of the pleura by implantation of dusts in laboratory animals. *Nature* **196**:180-181.

Wagner,J.C., Berry,G.B., Hill,R.J., Munday,E.E. and Skidmore J.W. (1984) Animal experiments with MMM (V) F - Effects of inhalation and intrapleural inoculation in rats. In, "Biological effects of man-made mineral fibres", WHO/IARC conference, WHO publication. Vol. 2:209-233.

Wagner,J.C., Berry,G., Skidmore,J.W. and Timbrell,V. (1974) The effects of the inhalation of asbestos in rats. *Br. J. Cancer* **29**:252-269.

Wagner,J.C., Berry,G. and Timbrell,V. (1973) Mesotheliomas in rats after inoculation with asbestos and other materials. *Br.J.Cancer* **28**:173-185.

Wagner,J.C., Sleggs,C.A. and Marchand,P. (1960) Diffuse pleural mesothelioma and asbestos exposure in the North Western Cape Province. *Br. J. Indust. Med.* **17**:260-271.

WHO/IARC and CEC (1990) Non occupational exposure to mineral fibres. Eds. Bignon,J., Peto,J. and Saracci,R., Lyon: IARC Scientific Publications n° 90.

WHO/IARC (1988) IARC monographs on the evaluation of carcinogenic risks to humans : Man-made mineral fibres and radon. Vol. **43**.

WHO (1989) Occupational exposure limit for asbestos. WHO Meeting, Oxford, United Kingdom, WHO/OCH/89. 1 vol.: 56 pages.

CHRONIC INHALATION TOXICITY OF REFRACTORY CERAMIC FIBERS IN SYRIAN HAMSTERS

T.W. Hesterberg, R. Mast[1], E.E. McConnell, J. Chevalier[2], D.M. Bernstein[2], W.B. Bunn and R. Anderson

Manville Technical Center
Littleton
CO, U.S.A.

[1]Carborundum Corporation
Niagra Falls
NY, U.S.A.

[2]Research and Consulting Co.
Geneva
Switzerland

ABSTRACT

This study was initiated to assess the potential for kaolin based refractory ceramic fibers (RCF) to induce toxicity and tumors in Syrian golden hamsters after chronic exposure. To simulate the fiber dimensions found in workplace air and to increase fiber respirability in the rodent, RCF fibers were size-selected before aerosolization to be approximately 1 μm in diameter and ~20 μm in length. Three groups of animals were exposed in nose-only inhalation chambers, 6 hrs/day, 5 days/week, for 18 months to 30 mg/m^3 of RCF, 10 mg/m^3 of chrysotile asbestos (positive control; average diameter = 0.09 μm, average length = 2.2 μm), or to filtered air (negative control). Animals from each group were sacrificed at three month intervals for evaluation of the sequential changes in lung pathology over the course of the study. Microscopic examination of the RCF exposed lungs revealed pulmonary interstitial fibrosis (grade 4 on the Wagner Pathology Grading Scale) starting at 6 months and showed little progression of the lesion through the remainder of the exposure. Examination of the asbestos exposed hamster lungs revealed an average pathology grade of 4.3 at 3 months, which progressed to an average score of 5.0

at 6 months and plateaued at that level. The latter change was characterized by interlobular linking of the lesions observed in the grade 4 lungs. In addition, a time-dependent increase in collagen deposition in the pleura was observed in the RCF exposed animals and to a lesser extent in the positive control animals. RCF exposure resulted in no pulmonary neoplasms. However, pleural mesotheliomas were found in 43 out of 102 RCF exposed hamsters (42%). No lung neoplasms or mesotheliomas were observed in either the chrysotile exposed or negative control hamsters.

INTRODUCTION

There is a natural concern about the safety of any material which has the potential to give off particles or vapors which can be inhaled and/or ingested. There is a heightened concern about all such fiber materials, and, for that reason, there have been and continue to be many studies dealing with man-made vitreous fibers (MMVF) (Lippmann, et al., 1990; Hesterberg et al., 1991b). There have also been reviews of those studies and other pertinent information, such as exposure levels, by various national and international agencies and organizations (IARC, 1988; IPCS, 1988).

MMVF are a class of insulating materials which have found widespread applications in both residential and industrial settings. MMVF is a generic expression for fibrous inorganic substances made primarily from rock, clay, slag, or glass (Pundsack, 1976). Since the initial development of some of these fibers in the late 1800s, large numbers of individuals in various occupational settings have been exposed to these materials (Bunn et al., 1990).

The three general types of MMVF are fibrous glass, mineral wool, and refractory ceramic fibers (RCFs). A variety of RCF types are produced depending on the intended application. While applications for RCFs vary they are all used in high temperature environments (Boyd & Thompson, 1980). All types are blends of alumina and silica, with other refractory oxides added. A kaolin clay based RCF was used in the present study.

There have been two previous chronic inhalation studies of RCF, one using rats (Davis, et al., 1984) and the other using rats and hamsters (Smith, et al., 1987). One mesothelioma was observed in a rat in the Davis study and in a hamster in the Smith study, while there were sharp differences in the induction of rat pulmonary tumors in the two studies. Therefore, the present RCF inhalation study in hamsters was initiated to determine whether the mesothelioma induction in the previous studies could be reproduced and to address the issue concerning pulmonary tumor induction. This study was designed to use more advanced techniques of fiber preparation and sizing, non-destructive aerosolization, and fiber measurement (Hesterberg et al., 1991b). In addition, larger numbers of animals were used and a nose-only exposure method was utilized.

The chronic oncogenicity of the maximum tolerated dose (30 mg/m^3) of RCF is also being assessed using the rat; this study is still in progress and results will be reported elsewhere (Hesterberg et al., 1991a). Furthermore, a chronic inhalation study of multiple concentrations of RCF in the rat has been initiated. Finally, several other chronic inhalation studies of multiple concentrations of other MMVFs (ie., fibrous glass and mineral wool) are in progress. All of these studies are being sponsored by the Thermal Insulation Manufacturer's Association (TIMA).

MATERIALS AND METHODS

Fibers

The kaolin based RCF used in this study was obtained from Carborundum Corp. and came in a single lot. It was made by melting kaolin clay and spinning into fibers. The fibers were then size separated having an average diameter of about 0.95 ± 0.1 μm and and average length of about 22.1 ± 1.7 μm. Intermediate range fiber length chrysotile asbestos (Jeffrey Mine, Asbestos, Quebec, Canada) served as a putative positive control (described previously in McConnell *et al.*, 1984).

Fiber Aerosol Exposure

Hamsters were exposed by nose-only inhalation to 30 mg/m^3 RCF (average of 210 ± 59 fibers/cc, using phase contrast optical microscopy), which was determined to be the maximum tolerated dose in a preliminary 28-day range finding study. The chrysotile asbestos exposure was chosen to be 10 mg/m^3 based on the previous studies with this fiber in rats, where pulmonary fibrosis and lung tumors resulted (McConnell *et al.*, 1984). Exposure was for 6 hours/day, 5 days/week for 18 months. Aerosol concentrations were monitored at the level of the hamster's nose for both mass (milligrams/cubic meter [mg/m^3]) and fiber levels (fibers/cubic centimeter [f/cc]). Fiber size distributions of RCF and Chrysotile aerosols were determined on a quarterly basis using scanning electron microscopy.

Animals

Weanling Syrian Golden male hamsters, obtained from Charles River, were held for a two week quarantine period, and then were randomly distributed into exposure groups. Following the 18 month exposure, the animals were held for lifetime observation (until ~20% survived). Sick or moribund animals were killed via intraperitoneal injection of pentobarbitol sodium as were those remaining at the end of the study.

Pathology

A complete necropsy was performed on each animal. The lungs were removed *in toto* and perfused with Karnovsky's fixative via the trachea at a pressure of 20 cm water for 2 hours. After fixation a consistently uniform transverse section, 2 mm diameter, of the left lung was obtained from each animal for histopathology. Replicate lung sections were routinely stained with hematoxylin and eosin (H&E) and Masson's trichrome stain for collagen deposition.

Groups of 3-6 randomly selected hamsters from each exposure group were killed at 3, 6, 9, 12, and 18 months to follow the progression of the pulmonary lesions. Histopathology of the lungs was examined and given a Wagner Pathology Grading Score in accordance with the guidelines presented at the WHO conference on "Biologic Effects of Man-made Mineral Fibres" in 1982 (McConnell *et al.*, 1984). In this system a grade of 1.0 is considered normal, grades 2 to 3 are evidence of focal cellular change, and grades 4 to 8 represent increasing degrees of non-reversible fibrosis. In accordance with these guidelines, proliferative lesions of the pulmonary parenchyma are designated as "bronchoalveolar hyperplasia, pulmonary adenoma or adenocarcinoma". Other types of neoplasm, including those in the pleura, were noted where appropriate.

TABLE 1. Aerosol Fiber Characteristics[a]

Fiber	mg/m³	F/cc	Diameter (μm)	Length (μm)
RCF1	29±8	210±59	0.95±0.13	22.1±1.7
Chrysotile	11±6	n/d	0.09±0.01	2.2±0.9

[a]from Hesterberg *et al.* (1991)

RESULTS

The average RCF and Chrysotile fiber concentrations and dimensions during the 18 month exposure are given in Table 1 and the RCF aerosol concentrations (mass and fiber levels) over this period are shown in Figure 1.

Histopathologic review of the lungs of the hamsters sacrificed at 3, 6, 9, 12, 18, and 21 months showed that RCF at 30 mg/m³ produced progressive, prominent pulmonary pathology. At 3 months the lesions were restricted to the terminal bronchioles and proximal alveoli and were characterized by the influx of large numbers of pulmonary macrophages, which, in this region, were present both on the epithelial lining and in the lumina. By light microscopy numerous distinct fibers were observed in the macrophages, microgranulomas and in the walls of the alveolar ducts and alveoli. Occasional microgranulomas were noted in the walls of the proximal alveoli. The peripheral alveoli showed minimal pathology at this time. The hamsters exposed to chrysotile asbestos for 3 months showed qualitative and quantitative differences to those exposed to RCF. While the macrophage response was less intense in the asbestos exposed lungs, early fibrosis was already evident in the walls of the terminal bronchioles and proximal alveoli.

All lung sections were given a "Pathology Score" according to the following Wagner pathology grading scale (Figure 2).

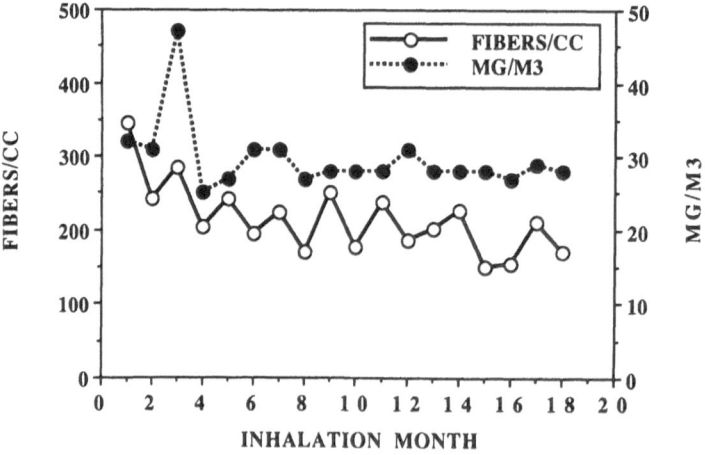

Figure 1. RCF Fiber Aerosol Data.

Cellular change:

	Normal	1	No lesion
	Minimal	2	Macrophage Response
	Mild	3	Bronchiolization, inflammation

Fibrosis:

	Minimal	4	Minimal fibrosis
	Mild	5	Linking of Fibrosis
	Moderate	6	Consolidation
	Severe	7	Marked fibrosis and consolidation.
		8	Complete obstruction of most airways.

Figure 2. Wagner Pathology Grading Scale.

Microscopic examination of the RCF exposed lungs revealed interstitial pulmonary fibrosis (grade 4 on the Pathology Grading Scale) starting at 6 months and showed little progression of the lesion through the remainder of the exposure (Figure 3). Examination of the asbestos exposed hamster lungs revealed an average pathology grade of 4.3 at 3 months, which progressed to an average score of 5.0 at 6 months and plateaued thereafter (Figure 3). The latter change was characterized by interlobular linking of the lesions observed in the grade 4 lungs.

After 12 months of exposure, the influx of macrophages was more prominent in the region of the terminal bronchioles and proximal alveoli and had spread to peripheral alveoli, especially in the RCF exposed hamsters. Microgranulomas, interstitial fibrosis and bronchiolization were more prominent, and were slightly more evident in the asbestos exposed rats.

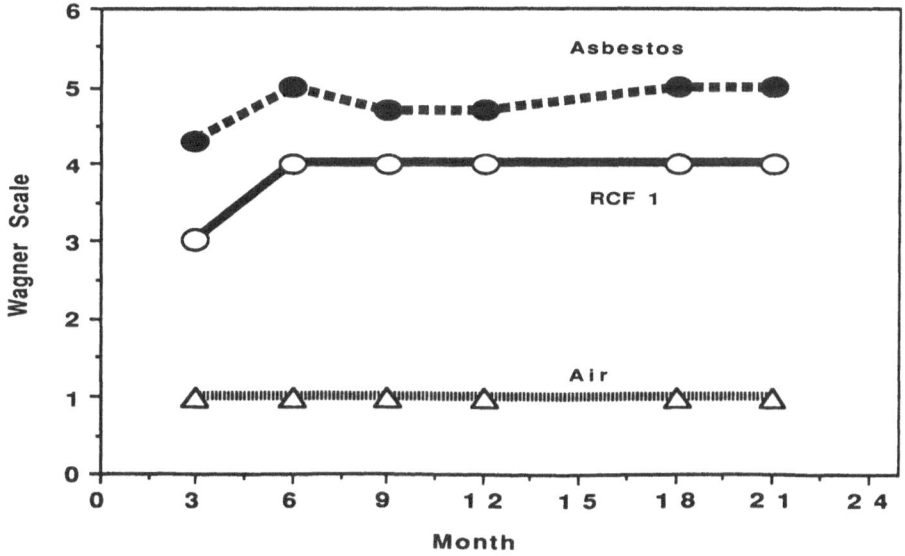

Figure 3. Hamster Lung Pathology Scores.

TABLE 2. Lung and Pleural Tumors[a]

Exposure Group	Adenomas	Carcinomas	Mesotheliomas
Air Control	0	0	0
Asbestos	0	0	0
RCF	1	0	43 (42%)

[a]from Hesterberg *et al.* (1991)

Pleural fibrosis was prominently visible in the RCF exposed hamsters. It varied from diffuse thickening to focal raised nodules. The mesothelium covering the collagen was often cuboidal and appeared hyperplastic.

The most significant findings from the hamster study were the occurrence of a total of 43 pleural mesotheliomas (42% incidence) in the lungs of kaolin RCF exposed animals. No significant increase in the incidence of pulmonary neoplasms was observed in the RCF exposed animals. In the chrysotile exposed and negative control animals, no lung tumors or mesotheliomas were observed (Table 2).

DISCUSSION

There have been two previous chronic inhalation studies of RCF (Davis, *et al.*, 1984; Smith *et al.*, 1987). The results from both of these studies differ sharply from the studies presented here. In the Davis study, 48 rats were exposed to RCF by inhalation for seven hours a day, five days a week, over a period of 224 days. The airborne dose of fibers longer than 5 μm was reported to be 95 f/cc. Animals sacrificed at the end of the study were reported to have an average of 5% pulmonary fibrosis, and eight of the rats were found to have pulmonary tumors with three animals demonstrating lung carcinomas. There was also one peritoneal mesothelioma reported.

The lower fibrosis and tumor response in the Davis study could be due to the following: 1) the test species in the Davis study, rats, may not be as responsive in terms of pleural reactions and mesotheliomas as hamsters (this does not explain the minimal findings of pulmonary fibrosis, however); 2) the Davis study was a whole-body exposure with fewer fibers to the target tissue and the generation technique (Timbrell Generator) created shorter fibers; 3) the RCF exposure in the Davis study was only 12 months, and the observation period was only 24 months, compared to 18 months exposure and lifetime observation in the present study; and 4) the exposure concentration in the Davis study was 8.4 mg/m^3 compared to 30 mg/m^3 in the present study.

Smith and co-workers (1987) exposed hamsters and rats to RCF at 200 f/cc, 6 hours a day, 5 days a week, for 24 months. The experiments involving rats showed no significant increase in neoplasms and minimal pulmonary fibrosis (22% had Wagner Scores > 4.0). Inhalation experiments with hamsters showed one cancer (mesothelioma) in fifty animals, but no fibrosis was observed. One of 157 control animals developed a spontaneous lung neoplasm without exposure to fibers. This study is more comparable to the present study because: 1) hamsters as well as rats were studied; 2) nose-only exposure techniques were employed; and 3) animals were exposed for 24 months and observed for lifetime. It is difficult to explain why more fibrosis and mesothelioma induction were observed in the present study, but it may be related to the higher mass exposure (30 mg/m^3 compared to 12 mg/m^3 in the Smith *et al.* study), the smaller average RCF fiber diameter and resulting increase in respirability in the present study, and the lower ratio of non-fibrous to fibrous particulates in the present study. Thus, in the present study, one would predict that more fibers and longer (more biologically active) fibers reached the target tissues of the lung and pleura. Lung digestions, and fiber recovery and analysis (using SEM) are being conducted on tissues from the present study, which will provide information to address this issue.

A number of chronic inhalation studies have been reported using man-made fibers (Lippmann, 1990, Hesterberg *et al.*, 1991b). The results vary, depending on protocol and fiber type. In addition to the study in hamsters reported here, the Thermal Insulation Manufacturer's Association (TIMA) is sponsoring several additional chronic inhalation studies using multiple concentrations of different types and compositions of MMVF; these studies are currently in progress. One of these studies was designed to assess the toxic and tumorigenic potential of the maximum tolerated dose of RCF in rats. The protocol is similar to the hamster study except that the exposure is for 24 months and several different compositions of RCF are being used. Pathology results from interim sacrifices from the rat study show similar levels of inflammation and fibrosis as was seen in the hamster study, but since the study is not complete, tumor results are not yet reported (Hesterberg, *et al.*, 1991a). The rat will be the likely model of choice for future chronic fiber inhalation studies because there is a much larger database on the effects of asbestos and other fibers in the rat, and because the rat model is capable of producing not only mesotheliomas, but also pulmonary neoplasms in response to inhalation exposure to fibers. Results of these and other chronic MMVF inhalation studies will further the understanding of the determinants of fiber carcinogenesis.

ACKNOWLEDGEMENTS

We should like to acknowledge sponsorship by the Thermal Insulation Manufacturers Association (TIMA).

REFERENCES

Boyd,D.C. and Thompson,D.A. (1980) Glass. In, " Kirk-Othmer Encyclopedia of Chemical Technology, 3rd ed.", Grayson,M., Mark,H.F., Othmer,D.F., Overberger,C.G. and Seaborg,G.T., eds. Vol. **11**, New York, John Wiley & Sons, pp. 807-880.

Bunn,W.B., Hesterberg,T.W., Chase,G. and Anderson,R. (1990) Man-made Mineral Fibers. In, "Medical Toxicology of Hazardous Materials". Williams and Wilkins, Baltimore, (in press).

Davis,J.M.G., Addison,J., Bolton,R.E., Donaldson,K., Jones,A.D. and Wright,A. (1984) The pathogenic effects of fibrous ceramic aluminum silicate glass administered to rats by inhalation or peritoneal injection. In, "Biological Effects of Man-made Mineral Fibres" (Proceedings of a WHO/IARC Conference). Vol. 2, Copenhagen, World Health Organization, pp. 303-322.

Hesterberg,T.W., Mast,R., McConnell,E.E., Vogel,O., Chevalier,J. Bernstein,D.M. and Anderson,R. (1991a) Chronic inhalation toxicity and oncogenicity study of refractory ceramic fibers in Fisher 344 rats. *The Toxicologist* **11**:85.

Hesterberg,T.W., Vu,V., McConnell,E.E., Chase,G.R., Bunn,W.B. and Anderson,R. (1991b) Use of Animal Models to Study Man-made Fiber Carcinogenesis. In, "Current Communications in Molecular Biology: Molecular Mechanisms of Fiber Carcinogenesis", Brinkley,B., Lechner,J. and Harris,C., eds. Cold Spring Harbor Laboratory Press (in press).

International Agency for Research on Cancer (IARC). (1988) IARC Monographs on the Evaluation of Carcinogenic Risks to Humans. Man-made Mineral Fibres and Radon. Lyon, France.

International Programme on Chemical Safety (IPCS). (1988) Man-made Mineral Fibres. Environmental Health Criteria 77. World Health Organization, Geneva, Switzerland. [copy provided]

Lippmann,M. (1990) Man-Made Mineral Fibers (MMMF): Human exposures and health risk assessment. *Toxicol. Indust. Hlth* **6**:225-.

McConnell,E.E., Wagner,J.C., Skidmore,J.W. and Moore,J.A. (1984) A comparative study of the fibrogenic and carcinogenic effects of UICC Canadian chrysotile B asbestos and glass microfibre (JM 100). In, "Biological Effects of Man-made Mineral Fibres", (Proceedings of a WHO/IARC Conference), Vol. **2**, Copenhagen, World Health Organization, pp. 234-252.

Pundsack,F.L. (1976) Fibrous glass - manufacture, use, and physical properties. In, "Occupational Exposure to Fibrous Glass", LeVee,W.N. and Schulte,P.A., eds. (DHEW [NIOSH] Publ. NO. 76-151; NTIS Publ. No. PB-258869), Cincinnati, OH, National Institute for Occupational Safety and Health, pp. 11-18.

Smith,D.M., Ortiz,L.W., Archuleta,R.F. and Johnson,N.F. (1987) Long-term health effects in hamsters and rats exposed chronically to man-made vitreous fibers. *Ann. Occup. Hyg.* **31**:731-754.

A CLASSIFICATION SYSTEM FOR NON ASBESTIFORM FIBRES

G. H. Pigott

ICI Central Toxicology Laboratory
Alderly Park, Macclesfield
Cheshire SK10 4TJ, U.K.

Regulation of substances may be divided into three phases; hazard assessment, risk assessment and risk management. Hazard assessment and risk assessment are both processes where scientific input should predominate, risk management must take into account a multiplicity of factors and cannot be regarded as solely a province for scientific debate.

Nevertheless it is incumbent on the scientific community to present both hazard and risk assessment data in a format which can be readily assimilated and applied in the risk management process. The characteristics of a successful risk management process are that it is:

  a) simple to understand and operate,
  b) easy to monitor for compliance,
  c) fail-safe,
  d) cost-effective and
  e) credible to the work force which must operate the process.

Any process which aims to contain practical risks associated with mineral fibres would initially seek a single standard embracing all materials. A standard based on the most toxic fibre conceivable would fulfil objectives a-c above. However, the process would not be cost-effective and over-regulating materials which are in fact relatively safe to handle, risks both devaluing the safety precautions essential for the more dangerous materials and generating a climate of public apprehension which is unwarranted by the risks involved. This paper aims to present a classification system for non-asbestiform mineral fibres which may aid the development of a rational risk management process taking into account the known characteristics of fibre which determine biological activity. Consideration will be limited to fibres which are non-asbestiform and in this context non-asbestiform is defined as fibres not exhibiting a fibrous cleavage; *i.e.* comminution of bulk material does not result in a reduction in fibre diameter.

Mechanisms in Fibre Carcinogenisis
Edited by R.C. Brown *et al.*, Plenum Press, New York, 1991

Man made mineral fibres have been produced for about a century. The major production has always centred on amorphous silicate materials produced from glass, rock or other minerals. These are more correctly termed man-made vitreous fibres (MMVF). These materials were originally used exclusively for thermal insulation at relatively low temperatures and this remains one of the dominant uses. The methods of production all give rise to a spectrum of diameters. However, it was evident from the early days of the industry that the "coarse end" of the fibre distribution in the product counted for much of the weight but little of the insulation efficiency. Thus, process engineering development concentrated on the reduction of the amount of fibre at this coarse end. This has been achieved by tightening the spectrum of diameters rather than increasing the percentage of low diameter fibres (Hartung, 1982). Though, theoretically, the latter give better insulation efficiency, with the materials used low mechanical strength and high energy inputs required to produce very fine fibres have restricted these to a very small specialised market sector. The bulk production has settled to a median diameter in the 6-8 μm region as representing a reasonable compromise between the various properties desired.

During the 1950's similar technology was applied to a different range of chemistry to give materials which were capable of insulating at very much higher temperatures. These refractory fibres (usually termed ceramics) are extensively used in industrial applications at high temperatures. The drive for energy efficiency associated with increased energy costs over the recent 2-3 decades has resulted in a substantial increase in their use. Nevertheless, they remain a minority product in tonnage terms.

Continuous filament glass fibres which are of relatively coarse diameter find use in reinforcement, particularly of resins and cement, electrical insulation and also in some textile production. The spectrum of diameter is narrower than that for insulation wools and the coarse diameter precludes significant atmospheric concentration during production and use.

Thus, it is not surprising that investigations into the possible health effect of these fibres and discussions on control measures and standards to be adopted have centred on the insulation wools as these represent: i) the predominant tonnage production over the entire history of the industry and ii) the major fibres produced in sufficient quantities for long enough to allow meaningful epidemiological surveys to be undertaken at this time.

Given this background it is understandable that for classification purposes vitreous fibres have assumed a dominant position. Thus Head and Wagg (1980) surveying exposure to man-made mineral fibres, proposed a classification scheme for fibres in general. This divided fibres into organic or inorganic composition with each category subdivided into natural and man-made fibres. The man-made mineral fibre category was further subdivided as shown in figure 1. This scheme recognises different categories of fibre on the basis of a mix of material, manufacturing process and application. Subsequent reviews of health effects of man-made mineral fibres have built on this scheme (IPCS, 1988; IARC, 1988) although there has been a move towards a dominance of material in the classification used. Thus IARC categorised man-made mineral fibres into three classes; glass fibre, rock/slag wool and ceramic fibres. The latter categorisation includes "a wide range of amorphous or crystalline synthetic mineral fibres characterised by their refractory properties". This categorisation is so wide as to have little practical utility.

Given the problems associated with the industrial and commercial use of asbestos there has been, over the last 20 years or so, a search for suitable replacement materials for the multitude of applications of these fibres. While this has not always been successful the process has resulted in a new generation of products which have found novel applica-

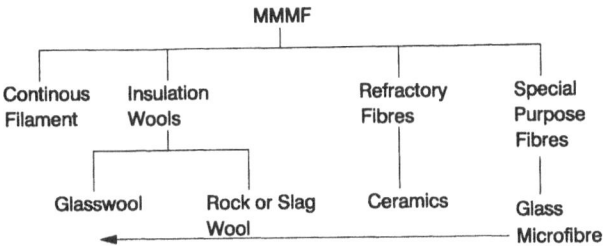

Figure 1. Classification of mineral fibres (after Head and Wagg, 1980).

tions especially in the field of materials reinforcement. These consist of both mineral fibres and more durable organic materials. In the absence of any generally agreed method of assessment of the toxicological properties of these fibres, some have been introduced into the market place with little or no assessment of their possible health effects. Given the diversity of these materials and the increasing pressure to regulate the use of synthetic fibres it is appropriate to examine the continuing utility of materials-based classifications as an aid to regulation.

There has emerged over several years a consensus within the scientific community that for insoluble mineral fibres the dimension of the fibre is more important than its chemical composition in determining biological reaction. Both length and diameter of fibres are important (Pott *et al.*, 1987; Stanton *et al.*, 1977). The inter-relationship of these parameters with chemical durability in tissue is complex and results in a spectrum of potential effects, particularly for carcinogenicity (Pott, 1987). However in the context of fibre specification length is not generally relevant. Most fibres have a initial length measured in centimetres and the spectrum of length encountered in normal exposure depends on the degree and type of attrition encountered in use. By contrast the spectrum of diameter is amenable to engineering control during manufacture. For manufactured fibre which does not exhibit fibrous cleavage (to yield particles of decreased diameter) this parameter may form the basis of a specification.

Those diameters of potential importance in fibre specification are shown in figure 2 (derived from Pigott & Ishmael, 1982). This is a synthesis of data from a variety of sources in the literature. The values were first suggested by experimentation in the 1970's and although subsequent research including data presented at this conference, has refined the values somewhat and demonstrated the complexity of some of the inter-relationships they remain valid as a first approximation. Inspection of these values leads to the following considerations for a fibre in which biological effects are minimised:

    i) **No fibres < 3 μm diameter to avoid respirable fraction.**
    ii) **Eliminate fibres < 1 μm and therefore mesothelioma risk.**
    iii) **No fibres > 5 μm in diameter to reduce irritant potential.**
    iv) **Avoid materials which are fibrogenic in particulate form.**

In practice these requirements conflict to a degree because of the difficulties of close control of fibre diameter. Clearly the need is for a material with a majority of fibres in the range of 3-5 μm diameter. There is a question of the relative importance of the various effects. Thus, though skin irritation may be an inconvenience it is generally considered not to represent a serious health risk even over a long period. Therefore a bias towards or beyond the upper limit of the specified diameter range is sometimes advocated

Inhalable Fibres	<7.5 m diameter
Respirable Fibres	<3 m diameter (3:1)
	(about 2 m 20:1)
Mesothelioma Induction	<1 m diameter (?1:5)
Skin Irritation	<5 m non-irritant
	5-10 m irritant
	>10 m highly irritant
Fibrogenic Potential	Non respirable/short
Lung Cancer	? As for fibrogenic potential

Figure 2. Diameters of importance in fibre specification.

to increase the certainty that problems associated with deposition and retention in the deep lung are minimised. This philosophy is not without its problems. Respirable fractions of fibres have been most extensively studied and therefore their problems are best understood. This reflects the common view that the carcinogenic and fibrogenic potential of fibres resides solely with material which is deposited in the lung for a long period. This may well be a valid assumption. There is no epidemiological evidence that irritant fibres produce any specific health risk and in particular no consistent evidence for excess cancers in the naso-pharyngeal region, which may be expected if irritant fibres acted as promoters of carcinogenic activity (IARC, 1988). Nevertheless, there may be a size at which this becomes important. Bhatt and coworkers (1984) demonstrated that a biogenic silica fibre of mean diameter 15 μm but with tapering ends is a carcinogenic promoter, at least in mice. It should also be remembered that skin irritation is an important feature in acceptability of materials to those who must handle them.

This property alone is sufficient to warrant the use of protective clothing, with its attendant inconvenience, in many applications involving mineral fibres. Also coarser fibres represent less efficient use of material particularly for insulation and in these terms are less cost effective. Thus the pressures for the reduction of these properties and the drive to finer diameters, providing these do not provoke an unacceptable health risk, should not be underestimated. This balance of requirements has lead to a number of fibres with median diameters at or around the 3 μm range.

Given that there is some degree of consensus for the parameters associated with biological reaction to fibres there remains the question of the means by which exposure can be best assessed in practice. Measurement of exposure is an integral part of both risk assessment for retrospective studies and risk management prospectively. Practical considerations reduce the means of monitoring to a direct choice between gravimetric assessment of the atmosphere and some form of fibre counting. These are not necessarily mutually exclusive. Gravimetric assessment suffers from non-specificity and may therefore be insensitive to quite wide variations in biologically important components of the atmosphere. It is often discounted for this reason. However the merits of simplicity and immediacy should not be overlooked. Modern equipment giving continuous digital readouts can allow corrective action to be taken rapidly should a problem arise which causes a hygiene standard to be exceeded.

Fibre counting is frequently cited as being "more relevant" in the control of the important component within the atmosphere. This is manifestly true in that it is fibres which we seek to control. However the situation is not so clear cut as may at first appear.

In the first place all fibres are counted irrespective of composition or origin. For practical reasons counting is limited to the light microscope except in rare cases. This results in an effective fibre cut-off of about 0.2 μm diameter, as fibres below this level are not visible. In practice there may also be some under estimation of fibres slightly above the limit of resolution. Since fibres in this area may be of importance in mesothelioma induction this omission is significant. The convention whereby all fibres of less than 3 μm diameter are counted as potentially respirable results in an overestimate of the true respirable fraction for coarser fibres. Fibre deposition in the lung is a complex process which may be affected by many factors including breathing patterns, activity levels, species, sex, age etc and attempts to devise a universal model to predict deposition have been largely unsuccessful. Nevertheless it is clear that deposition may be modelled on the Monte Carlo principle (Koblinger & Hoffman, 1988) and that the probability for a particle of a given diameter penetrating to the alveolar space is effectively zero for particles greater than 10 μm aerodynamic diameter (Miller *et al.*, 1975). Below this level penetration and deposition increases to a peak which in humans is quoted to be approximately 2-3 μm aerodynamic diameter.

For a mineral fibre of density 2.5 g/cm^3 this corresponds to an actual diameter of approximately 1-1.5 μm (Bernstein, this volume). The respirable limit for such fibres corresponding to a 10 μm aerodynamic equivalent diameter particle is approximately 3 μm. It is evident that only a small proportion of 3 μm fibres will penetrate to the alveolar space, whereas a much greater proportion of 1 μm fibres will do so.

Thus the fibre counting method overestimates those fibres at the coarser end of the range which make a substantial contribution to the mass of airborne fibre. However it must be recognised that, for the fibres in the finer diameter ranges, this is probably the only valid method of assessment of the biologically important component of the atmosphere. Nevertheless given the additional resource requirements and the current retrospective nature of fibre counting it is clear that this form of monitoring should be confined to situations where its use is considered especially appropriate. The availability of automated fibre aerosol monitors which also incorporate a degree of size selection would improve the situation, though realistically a cheap reliable method is some way off and for practical purposes can be discounted at present.

A useful classification system would differentiate those fibres with significant potential health risk which warrant the extra effort involved in control by fibre count from those where general dust control by a gravimetric limit would provide adequate protection. From the foregoing it is evident that relevant indices could be either inhalable fibres (i) respirable fibres (r) or submicron fibres (s). It is obvious that for any given fibre $i \geq r \geq s$. Thus it is proposed that a single parameter, the percentage submicron fibre present in the bulk material, should suffice as a classification criterion.

It is accepted that this is a relatively crude measurement. The relative concentration of the fibre in air may be somewhat different to that seen in the bulk material as there is no reason to suppose that fragility is independent of diameter. Nevertheless fibre sizing consumes considerable resource and is unlikely to be acceptable on a routine basis and increased fragility of fine fibres would result in a larger proportion of short fragments of lesser toxicological importance. Thus categorisation of the bulk by a technique such as a line scan under the scanning electron microscope should provide sufficient information about the potential for concentration of relatively fine fibres in any dust cloud generated. The concentration on submicron fibres is warranted by the relatively slight contribution which these make to the total airborne mass and their biological potential; larger fibres are to a degree self-regulating within a given gravimetric limit. Three categories are proposed:

Category 1 - fibres with < 1% sub-micron fibre in bulk
Category 2 - fibres with 1-10% sub-micron fibre in bulk
Category 3 - fibres with >10% sub-micron fibre in bulk
(All percentages are by number.)

These categories could then be associated with different control measures as follows:

Category 1: Fibres where health concerns are generally low, normally regulated on a gravimetric basis.

Category 2: Fibres where there is the possibility for some concentration of fine fibres in air. These would be subject to a dual standard (as currently proposed in the UK) where either a gravimetric or a fibre count limit is imposed, following a once-and-for-all assessment to determine which is the most stringent in any particular set of conditions. (Head & Wagg, 1980).

Category 3: Identifies those fibres of potentially high hazard where a fibre-based limit would be the most appropriate control measure.

It is recognised that this system is not fully comprehensive. No account is taken of fibre durability which is clearly an important feature in the overall biological response. However until a clearer pattern emerges of the most appropriate measure of dissolution and the minimum residence time to provoke a longer term reaction, it would be prudent to ignore this property for a classification scheme. There is also the question of fibrosis. A few materials can induce a progressive fibrosis which may be different in character to the lesser fibrous reaction which is provoked by all deposited material. It may be reasonable to suppose that this property could be detected by relatively simple model systems and that the principles of containment would follow those conventionally applied to such fibrogenic minerals as crystalline silica. This might result in a more stringent (gravimetric?) standard than that applied to the more inert materials.

A major benefit of the proposed classification system would be to focus attention on those fibres which are believed to be of maximum biological activity. This would result not only in appropriate regulation of ultra-fine fibres, which are generally produced in small tonnage for specialised applications (and hence are amenable to containment), but also encourage process engineering development in bulk fibre industries to minimise the production of submicron fibre. This is in contrast to the existing situation of regulation by product type which results in a spectrum of materials from different manufacturers which at the extreme may represent different degrees of hazard for nominally similar materials.

REFERENCES

Bhatt,T., Coombs,M. and O'Neill,C. (1984) Biogenic Silica Fibre Promotes Carcinogenesis in Mouse Skin. *Int. J. Cancer* **34**:519-528.

Hartung,W.J.A. (1984) Technical History of MMMF with particular reference to fibre diameter. In, "Biological Effects of Man-Made mineral Fibres", ed. Guthe,T. Proceeding of WHO IARC Conference Vol 1 World Health Organisation Copenhagen. pp. 12-19.

Head,I.W.H. and Wagg,R.M. (1980) A Survey of Occupational Exposure to Man-Made Mineral Fibre Dust, *Ann. Occup. Hyg.* **27**:235-258.

IARC (1988) The Evaluation of Carcinogenic Risks to Humans, Vol **43** "Man-Made Mineral Fibres and Radon", IARC Lyon.

IPCS (International Programme on Chemical Safety) (1988) Environmental Health Criteria 77, Man-Made Mineral Fibres. World Health Organisation, Geneva.

Koblinger,L. and Hoffman,W. (1988) Monte Carlo model for Aerosol Deposition in Human Lungs, *Ann. Occup. Hyg.* **32**:Suppl 1:65-70.

Miller,F.J., Gardner,D.E., Graham,J.A., Lee,R.E.,Jr, Wilson,W.E. and Bachman,J.D. (1979) Size considerations of Establishing a Standard for Inhalable Particles, *APCA Journal* **29**:610-615.

Pigott,G.H. and Ishmael,J. (1982) A Strategy for the Design and Evaluation of a 'Safe' Inorganic Fibre, *Ann. Occup. Hyg.* **26**:371-380.

Pott,F., Ziem,U., Reiffer,F.J., Huth,F., Ernst,H. and Mohr,U. (1987) Carcinogenicity studies on fibres, metal compounds and some other dusts in rats, *Exp. Pathol.* **32**:129-152.

Stanton,M.F., Layard,M., Tegaris,A., Miller,E., May,M. and Kent,E. (1977) Carcinogenicity of fibrous glass: pleural response in the rat in relation to fibre dimension, *J. Natl Cancer Inst.* **58**:587-603.

TUMOURS BY THE INTRAPERITONEAL AND INTRAPLEURAL ROUTES AND THEIR SIGNIFICANCE FOR THE CLASSIFICATION OF MINERAL FIBRES

F. Pott[1], M. Roller[1], R.M. Rippe[1], P.-G. Germann[2] and B. Bellmann[2]

[1]Medical Institute for Environmental Hygiene
Heinrich Heine University of Düsseldorf
Auf'm Hennekamp 50, D-4000 Düsseldorf, Germany

[2]Fraunhofer Institute of Toxicology and Aerosol Research
Nikolai-Fuchs-Straße 1
D-3000 Hannover 61, Germany

1. On the Low Sensitivity of Inhalative Carcinogenicity Studies with Mineral Fibres in Rats

In 1967, the first author of this paper started intraperitoneal studies on the carcinogenicity of asbestos. Later Wagner and coworkers (1974) published their large inhalation experiment with five asbestos types and this provided the incentive to perform similar inhalation studies with man-made mineral fibres. This aim was achieved in 1983 after some years of planning and the construction of the facilities for inhalation toxicology at the Fraunhofer Institute in Hannover. It was a great disappointment that Muhle and coworkers (1987) at the new facilities could not even find a clear carcinogenic effect in the groups exposed to chrysotile or crocidolite. Even in a parallel injection study with about 10^9 chrysotile fibres there was no clear carcinogenic effect. In these experiments Californian chrysotile was used as a positive control following the recommendation of NIOSH. The crocidolite sample used gave an unequivocally positive result after intraperitoneal injection of a relatively low number of fibres but no significant carcinogenic effect could be detected in the inhalation study. This compares with the work of Wagner and others (1985, 1987) who also failed to reproduce the high tumour rates observed with crocidolite in their previous inhalation study.

Figure 1 shows the dose response relationship which resulted from some inhalation studies carried out by Davis and coworkers (1978, 1986, 1988). It can be concluded that, under Davis' experimental conditions, non-smoking rats should be exposed to a concentration of about 1000 fibres/ml to detect their carcinogenic potential, under the pre-condition that their carcinogenic potency is in the same range as the fibre samples used. This concentration is 1000 times higher than the earlier "Technical Guiding Concentration" for chrysotile in Germany, which has recently been reduced. Such high concentrations of man-made fibres cannot be generated as these are mostly thicker than asbestos fibres. Therefore there is a probability of obtaining false negative results in inhalation studies with man-made fibres which is large if they are not more carcinogenic than the asbestos fibres used by Davis. The risk for workers exposed to fibres not showing a carcinogenic effect in inhalation studies may be higher than the risk accepted by our society. Therefore, we have to look out for criteria other than positive inhalation studies to evaluate the carcinogenic potential of mineral fibres.

Investigations on the carcinogenicity of mineral fibres by the intrapleural (i.pl.) or intraperitoneal (i.p.) routes in rats have shown that the i.p. test is much more sensitive than inhalation exposure. Results with these routes demonstrate remarkable differences between the effects of long and short or durable and non-durable fibres.

In this paper, new results with different durable glass fibres and several other mineral fibre types are reported and also, an experiment on the possible role of surface properties. Moreover, older data, including some of Stanton and coworkers are reviewed. On the basis of all these results, a classification of natural and man-made mineral fibres according to their carcinogenicity is proposed and has already been published (Pott, 1989a, 1990). Finally, some objections to this proposal are quoted and discussed.

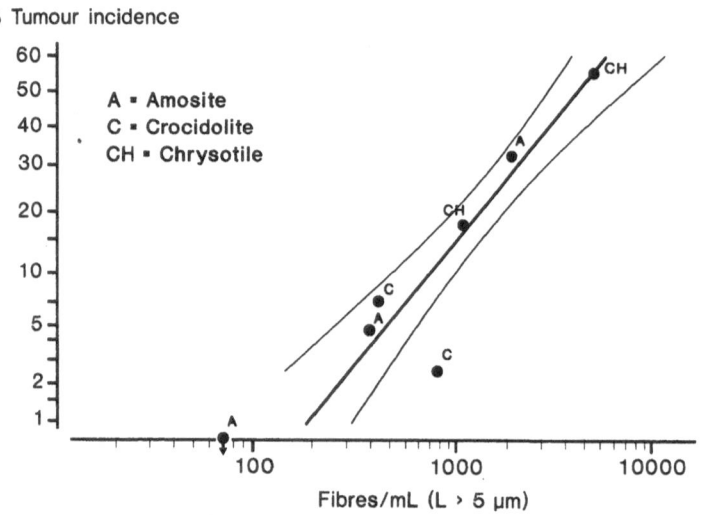

Figure 1. **Relationship between fibre concentration and tumour incidence** (benign or malignant lung tumours and pleural mesotheliomas in rats) after exposure to an aerosol of asbestos (5 x 7 hours/week, 12 months; different preparations of amosite and chrysotile, one preparation of crocidolite for the two concentrations; according to data of Davis *et al.*, 1978, 1986; Davis & Jones, 1988). Under these conditions, fibre concentrations of about 1000/ml are necessary for detecting the carcinogenicity of fibres having a deposition pattern and a carcinogenic potency similar to the fibre samples applied by Davis and coworkers.

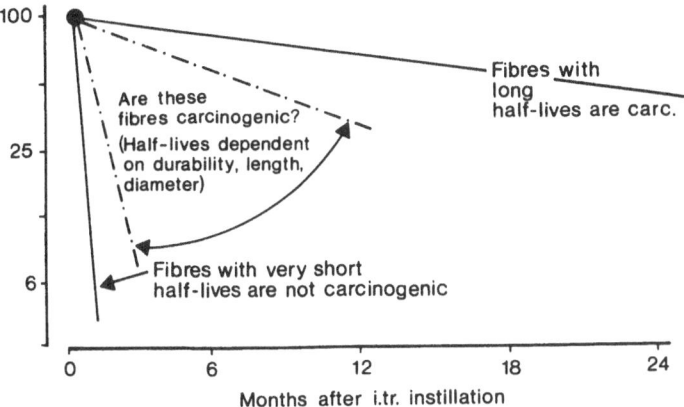

Figure 2. Illustration of the wide span of possible half-lives of different fibre types in the lung. The half-life of a fibre sample depends on durability, diameter, and length of the fibres. The bio-durability of a fibre depends on its chemical composition, on the location of its persistence in the body, the contact with cells, etc. The disappearance of a fibre from the lung can occur by bronchial clearance, migration to the lymph nodes or other locations in the body, by disintegration into non-fibrous particles or by dissolution. A carcinogenicity test is hardly necessary for fibres with a very short or with a long half-life.

2. The Significance of Durability of Mineral Fibres for Their Carcinogenic Activity

In 1972, Stanton and Wrench concluded from their results, after intrapleural implantation of several fibrous dusts, that "the simplest incriminating feature for both carcinogenesis and fibrogenesis seems to be a durable fibrous shape". At the same time, other authors also concluded from their experiments with several fibre types and non-fibrous dusts, which have a similar chemical composition to the fibrous dusts, that the elongated particle shape is the primary carcinogenic agent and that the negative effect of gypsum fibres could be caused by their solubility in the abdominal cavity in rats (Pott & Friedrichs, 1972). At the first NIOSH symposium on occupational exposure to fibrous glass held in 1974, experts from the glass fibre industry could not answer the question whether fibrous glass can change or break down to smaller particles in the lung.

It seemed, even then, that it should be possible to find natural and synthetic fibres with a short survival time in the body which would therefore not be carcinogenic. But it was an open question whether such fibres would be suitable for practical use. Certainly, it is possible to produce wool even from sugar but such fibres, which dissolve readily are not suitable for insulation wool. We found that chrysotile and glass microfibres treated *in vitro* with HCl lost cations but not their fibrous shape. Their half-life in the lung was shortened substantially and a carcinogenic effect could not be detected (Bellmann *et al.*, 1987; Pott *et al.*, 1987).

In 1976 Klingholz from the German glass fibre industry argued at the WHO-meeting on man-made mineral fibres in Copenhagen that insulation glass wool dissolves

TABLE 1. Chemical glass composition [%] of fibres examined for carcinogenicity

Glass type	Median diam. (μm)	Abbr.[1]	SiO_2	B_2O_3	Na_2O	Al_2O_3	$FeO + Fe_2O_3$	BaO	ZnO	K_2O	CaO	MgO
3101	~1.5	B-1	60.7	3.3	15.4	-	0.2	-	-	0.7	16.5	3.2
3101	~0.5	B-2	60.7	3.3	15.4	-	0.2	-	-	0.7	16.5	3.2
3102	~0.35	B-3	58.5	11.0	9.8	5.8	0.1	5.0	3.9	2.9	3.0	-
475	~0.14	M-475	57.9	10.7	10.1	5.8	0.1	5.0	3.9	2.9	——— 3.0 ———	

B = Bayer AG

M = Manville. Tested fibre: Code 104

[1] Abbreviations used in Tables 2 and 3

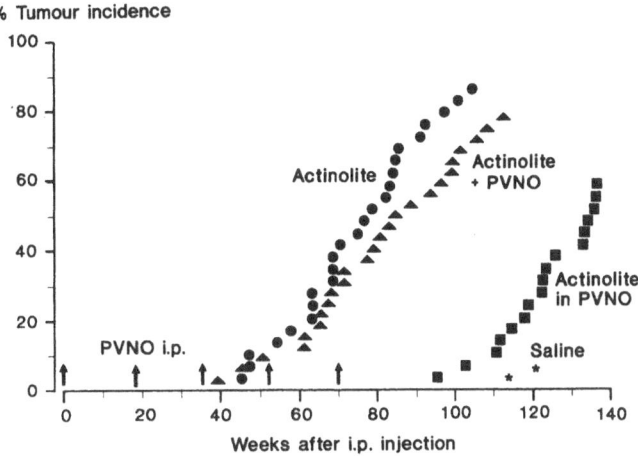

Figure 3. Tumour rates and latency periods after one intraperitoneal injection of 0.3 mg actinolite in 1 ml 0.9% saline (groups "actinolite" and "actinolite + PVNO") or in 1 ml saline together with 20 mg PVNO (group "actinolite in PVNO"). Additionally, the latter group was injected with 1 ml 2% PVNO solution after 18, 35, 52 and 70 weeks. The group "actinolite + PVNO" received actinolite in saline i. p., additionally 1 ml 2% PVNO i.p. 1 day before the treatment with actinolite as well as 18, 35, 52 and 70 weeks later (Pott *et al.*, 1987 and additional data for the latency periods).

in the body but there is no published data to support this suggestion. We tried for some years, without success, to cooperate with the MMMF industry in investigations of *in vivo* fibre durability. The aim was to find fibres which are suitable for some technical purposes and which additionally have only a very low or no carcinogenic potency.

In 1985 an outsider of the MMMF industry consulted us about the production of non-carcinogenic glass fibres and one year later the company offered us the new fibre type expected to have a relatively low durability in the body (Table 1). Moreover, we obtained a glass fibre with the same chemical composition as the glass type 475 from Manville. The durability of these materials has been studied by Bellmann *et al.* (1990) and Muhle *et al.* (this volume). One year after intratracheal instillation less than one percent of the new fibre type (related to the durable one) could be found in the ashed lung. The results of the carcinogenicity study were recently published in part in a short communication (Pott *et al.*, 1990b). Table 2 shows the essentials of the experimental design as well as the results. A carcinogenic effect could be detected only in the groups injected with durable glass fibres. We do not have any data on surface properties, but even if surface properties (rather than particle morphology) were responsible for tumour induction then persistence would still be important. A fibre which disintegrates in a short time can no longer act as a carcinogen. This means that fibres with a low durability induce a lower cancer risk. This might be true in rats as in humans and these fibres can be recommended for use instead of durable ones.

The durability of a fibre sample in the lung can be studied by estimating its half-life after intratracheal instillation (Bellmann *et al.*, 1987; Muhle *et al.*, this workshop).

TABLE 2. Results after intraperitoneal injection of durable and slightly durable glass fibres (see Table 1) into female Wistar rats[1]

Fibre type	Dose injected[2] (mg)	fibres[3] x 10^9	Fibre size[4] median L [μm]	D	Rats with tumour[5] no./exam.	[%]	Lifespan [weeks][6] 20% <	50% <	80% <	Rats with tum., mean
B-1K	60	0.24	7.4	1.06	3/46	6.5	74	107	127	89
B-1M	20	0.05	10.7	1.68	1/48	2.1	104	119	131	107
B-1M	60	0.16			1/46	2.2	102	127	131	75
B-1L	20	0.04	17.8	1.40	1/48	2.1	100	112	131	107
B-1L	60	0.11			5/46	10.9	92	120	131	106
B-2K	6.7	0.29	4.2	0.49	0/48	0.0	98	120	131	-
B-2K	20	0.86			0/46	0.0	91	113	130	-
B-2L	6.7	0.39	6.0	0.51	0/45	0.0	89	126	132	-
B-2L	20	1.16			2/44	4.5	81	121	131	109
B-3K	6.7	0.38	3.3	0.37	10/48	20.8	89	108	131	104
B-3K	20	1.14			30/47	63.8	76	94	119	97
B-3L	6.7	0.15	5.6	0.34	19/48	39.6	83	99	125	94
B-3L	20	0.46			31/47	66.0	69	87	117	83
M-475	2	0.32	2.3	0.14	8/48	16.7	85	115	132	109
TiO2	20	-	non-fibrous		0/47	0.0	97	122	132	-

[1] Mean body weights of the groups at start 158 - 166 g, standard deviations 8 - 10 g.
[2] Injection of up to 20 mg dust in 2 ml saline; 60 mg divided into 3 weekly injections.
[3] Definition of fibres: aspect ratio > 5/1, length (L) > 5 μm, diameter (D) < 2 μm.
[4] More details on fibre sizes and measurement method see Bellmann et al., (1990).
[5] Rats with mesothelioma or sarcoma in the abdominal cavity amongst rats examined; tumours of the uterus are excluded but rats with mesothelioma or sarcoma and a simultaneous tumour of the uterus are included. Rats dead within 40 weeks after first injection are excluded; tumours of the uterus are excluded but rats with mesothelioma or examined (this means: only "rats at risk" were evaluated for tumour rate); the first tumour-bearing animal died in the 45th week of the experiment.
[6] Weeks after first injection including rats which died spontaneously within 40 weeks (48 rats per group at start; 13 of 720 rats died spontaneously within 40 weeks, 4 were killed for examination of early lesions).

552

TABLE 3. Results after intraperitoneal injection of mineral fibres and non-fibrous materials into female Wistar rats

Fibre type (material)	Dose injected[1] [mg]	fibres[2] x 10^9	Fibre size[3] median L [μm]	D	Rats with tumour[4] No./exam.	[%]	Lifespan [weeks][5] 50% ≤	80% ≤	Rats with tum., mean
Glass B-1K	3 x 50	0.60	7.4	1.06	1/32 (48)[6]	3.1	106	124	106
Glass B-1ML	2 x 50	0.51	11.0	1.19	1/39 (48)[6]	2.6	106	123	124
Glass B-2L	2 x 50	5.80	6.0	0.51	1/35 (48)[6]	2.9	104	119	88
Ca-Na-metaphosphate	1 x 50	0.26	2.8	0.30	3/17 (36)	17.6	100	117	108
Ca-Na-metaphosphate	5 x 50	1.29	2.8	0.30	4/16 (36)	25.0	89	102	81
Gypsum A 30	5 x 50	0.19	11.2	1.34	1/24 (36)	4.2	109	128	121
Gypsum H 30	5 x 50	0.16	9.7	0.98	0/12 (36)	0.0	118	130	-
Mg-oxide-sulphate	1 x 50	5.98	2.2	0.19	1/21 (44)	4.8	129	130	19
Mg-oxide-sulphate	10 x 15	17.9	2.2	0.19	0/10 (36)	0.0	112	130	-
Sepiolite, Uicaluaro	1 x 50	7.56	1.0	0.06	0/23 (36)	0.0	105	131	-
Sepiolite, Uicaluaro	5 x 50	37.8	1.0	0.06	2/21 (36)	9.5	126	130	99
Basalt	1 x 25	0.005	13.8	1.08	1/38 (48)[6]	2.6	110	127	128
Basalt	5 x 30	0.030	13.8	1.08	15/21 (36)	71.4	84	101	89
Slag	5 x 30	0.25	9.0	1.21	2/28 (36)	7.1	106	131	77
Silicon carbide	1 x 0.05	0.005	3.1	0.31	2/16 (36)	12.5	115	130	115
Silicon carbide	1 x 0.25	0.027	3.1	0.31	5/23 (36)	21.7	109	120	111
Silicon carbide	1 x 1.25	0.13	3.1	0.31	13/21 (36)	61.9	66	109	61
Silicon carbide	1 x 6.25	0.67	3.1	0.31	23/30 (36)	76.7	54	86	54
Silicon carbide	1 x 25	2.68	3.1	0.31	36/37 (48)[6]	97.3	36	48	39
Carbon	1 x 50	0	193	17.7	0/25 (36)	0.0	102	124	-
Carbon	5 x 50	0	193	17.7	0/20 (36)	0.0	120	130	-
Chlorite	2 x 25	non-fibrous			1/42 (48)[6]	2.4	97	124	124
Silicon carbide	5 x 50	non-fibrous			1/22 (36)	4.5	130	131	121
Carbon activated	1 x 50	non-fibrous			0/22 (36)	0.0	105	130	-
Carbon activated	5 x 50	non-fibrous			1/25 (36)	4.0	122	131	92
NaCl-solution	5 x 2 ml				2/50 (72)	4.0	106	130	78

Notes to table overleaf.

Footnotes to TABLE 3.

[1] Repeated injections were carried out in weekly intervals.

[2] Definition: aspect ratio > 5/1, length (L) > 5 μm, diameter (D) < 2 μm.

[3] Method of measurement technique see Bellmann et al., 1987. All measurements were made by scanning electron microscopy, except for sepiolite which was analysed by transmission electron microscopy.

[4] Rats with mesothelioma or sarcoma in the abdominal cavity related to rats examined which either survived at least 56 weeks of the experiment or died earlier and were diagnosed as tumour-bearing. Figures in parentheses give the numbers of rats at start. Mean body weights of the groups at start were about 160 g (standard deviations 9 - 17 g) except for the groups described in footnote 6. The mean lifespan per group was strongly reduced by an infectious disease of the lung in the 12th and 13th month (the cause could not be diagnosed unequivocally by bacteriological and virological examinations of many animals). Tumours of the uterus are excluded but rats with mesothelioma or sarcoma and a simultaneous tumour of the uterus are included.

[5] Weeks after first injection. Percentages refer to the rats examined as described in footnote 3.

[6] Mean body weights at start about 190 g, standard deviations 11 - 15 g.

We have demonstrated that fibre samples with a long half-life induced tumours and samples with a short half-life were not carcinogenic after intraperitoneal injection (Fig. 2). However half-life does not depend only on the durability but also on the fibre dimensions. It would be of great interest to know the difference of the cancer risk which results from a certain difference of the half-lives of two fibre types having the same particle size distribution. It is also important to extrapolate these differences from rats to humans but even if the carcinogenicity studies could be performed by inhalation, this kind of extrapolation remains very difficult.

The principal question arising from these studies is, how long do fibres have to stay in the bronchial wall or serosal tissue in order to cause an alteration that can lead to the development of a tumour without further presence of fibres? This question and some others have been put over the last few years (Pott, 1987b; 1989b) and while several colleagues promised to answer them only a few did. It may be that the persistence of fibres for just one cell division phase is, in principle, sufficient for a transforming effect. Some years ago, it was discussed whether fibres can be exonerated from the suspicion of being carcinogenic when the fibres do not survive more than two or three years in the body (Pott, 1987a, 1988). In view of the negative results obtained with slightly durable glass fibres and with wollastonite, both of the answers mentioned are compatible. It may that the persistence of fibres in the tissue for just one cell division phase is sufficient for tumour induction but fibre induced cell transformation seems to be a very rare event. Therefore, it is not surprising that the injection of 10^9 or more of the slightly durable glass fibres did not prove to be carcinogenic in a 2½ year animal experiment. Even two years after intratracheal instillation some of these fibres were found in the ashed lung (Muhle et al., this volume).

TABLE 4. Results after intraperitoneal (or subcutaneous) injection of mineral fibres and non-fibrous materials1 into female Wistar rats

Fibre type (material)	Dose injected[1] [mg]	fibres[2] x 10^9	Fibre size[3] median L [μm]	D	Rats with tumour No./exam.	[4] [%]	Lifespan [weeks] [5] 20% ≤	50% ≤	80% ≤	rats with tum., mean
Al-silicate "Fiberfrax" I	1 x 12	0.029	5.5	0.47	15/35	42.9	77	93	112	102
Al-silicate "Fiberfrax" II	1 x 12	0.021	13.1	0.84	17/36	47.2	80	99	119	104
Al-silicate "Fiberfrax" II	2 x 20	0.069	13.1	0.84	29/36	80.6	64	77	97	79
Al-silicate, Manville5	2 x 20	0.009	16.4	1.35	6/36	16.7	87	108	131	103
Potassium titanate	1 x 0.5	0.045	3.2	0.22	1/34	2.9	92	118	131	105
Potassium titanate	1 x 2	0.18	3.2	0.22	11/36	30.6	86			
DEE6, particulate matter	1 x 20	-	non-fibrous		3/31	9.7	98	112	132	106
DEE6, particulate matter	4 x 20	-	non-fibrous		0/34	0.0	100	128	132	-
Carbon black "Corax L"	4 x 20	-	non-fibrous		1/35	2.9	101	119	132	124
Magnetite	1 x 40	-	non-fibrous		0/36	0.0	96	124	130	-
Magnetite	4 x 40	-	non-fibrous		2/34	5.9	73	102	130	116
Iron(II)oxide	1 x 40	-	non-fibrous		1/36	2.8	95	118	130	130
Iron(III)oxide	4 x 40	-	non-fibrous		0/33	0.0	98	122	130	-
Ni-alloy (29% Ni)	2 x 156	-	non-fibrous		1/36	2.8	84	114	124	53
Ni-subsulfide	1 x 8.2	-	non-fibrous		20/36	55.6	27	45	105	40
Ni-carbonate	50 x 2.1	-	non-fibrous		3/33	9.1	84	113	130	90
NaCl-solution	50 x 1 mL	-	-		0/34	0.0	87	105	130	-
Crocidolite s.c.[7]	1 x 20	1.97	1.8	0.19	1/34 [7]	2.97	101	115	130	130
Chrysotile B (UICC) s.c.[7]	1 x 20	16.2	0.7	0.06	0/31 [7]	0.07	94	123	130	-

Notes to table overleaf.

Footnotes to TABLE 4.

This raises the question: does a "true" threshold exist or can we only try to define an "acceptable" risk? This question has a scientific and a political viewpoint. Scientifically, each single fibre is the elementary unit of the physical carcinogen "fibre", like one molecule of a chemical carcinogen. But it is also important to avoid the misunderstanding that the inhalation of a very few fibres can lead to an unacceptable cancer risk. This can be illustrated by reference to sexual reproduction: there is no doubt that only one single sperm is necessary to fertilize an ovum, but an ejaculation containing less than 20 million sperm/ml is called oligospermia, which can be a cause of infertility.

3. Further New Results of the Intraperitoneal Test of Several Fibre Types

Tables 3 and 4 show further new results after intraperitoneal injection of dusts in rats. Unfortunately, in part of the experiments listed in Table 3, many animals were killed or died of an infectious lung disease but we have tried to recover as much information as possible. In any case, of the fibre types shown in Table 3, only silicon carbide and basalt fibres gave unequivocally positive results; Ca,Na-metaphosphate is very probably carcinogenic. Therefore, a mode of evaluation was chosen which should avoid false negative interpretations of the results found in the other groups (explained in a footnote of Table 3).

The main results can be summarized as follows:

(1) Slightly durable glass fibres: the findings do not indicate a carcinogenic effect with 0.5×10^9 fibres longer than 5 μm and a median diameter of 1.2 μm nor with 5.8×10^9 fibres with a median diameter of 0.5 μm.

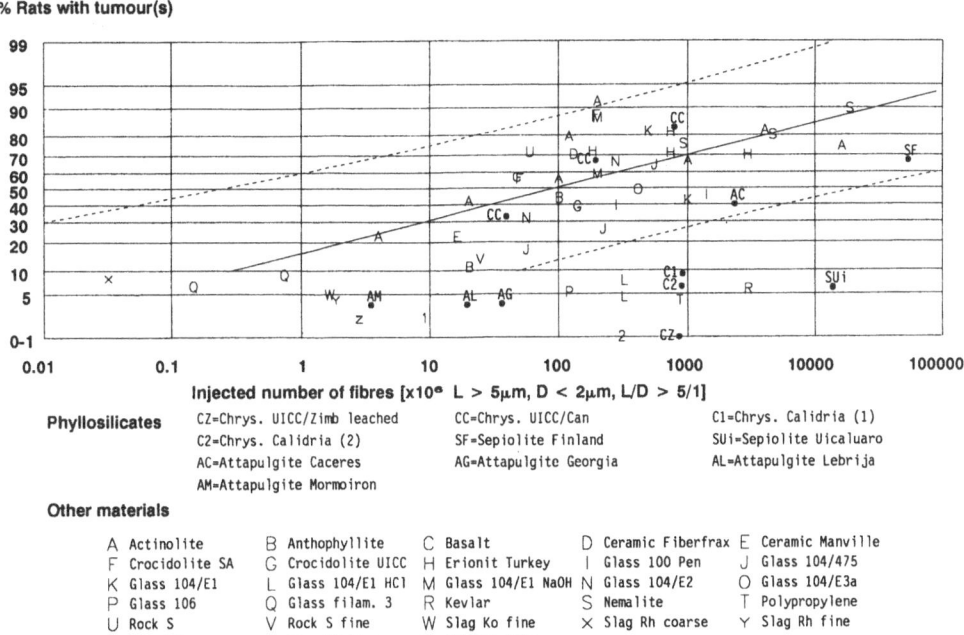

% Rats with tumour(s)

Injected number of fibres [x10⁶ L > 5μm, D < 2μm, L/D > 5/1]

Phyllosilicates

CZ=Chrys. UICC/Zimb leached	CC=Chrys. UICC/Can	C1=Chrys. Calidria (1)
C2=Chrys. Calidria (2)	SF=Sepiolite Finland	SUi=Sepiolite Uicaluaro
AC=Attapulgite Caceres	AG=Attapulgite Georgia	AL=Attapulgite Lebrija
AM=Attapulgite Mormoiron		

Other materials

A Actinolite	B Anthophyllite	C Basalt	D Ceramic Fiberfrax	E Ceramic Manville
F Crocidolite SA	G Crocidolite UICC	H Erionit Turkey	I Glass 100 Pen	J Glass 104/475
K Glass 104/E1	L Glass 104/E1 HCl	M Glass 104/E1 NaOH	N Glass 104/E2	O Glass 104/E3a
P Glass 106	Q Glass filam. 3	R Kevlar	S Nemalite	T Polypropylene
U Rock S	V Rock S fine	W Slag Ko fine	X Slag Rh coarse	Y Slag Rh fine
Z Slag Zi coarse	1 Slag Zi fine	2 Wollastonite		

Figure 4. Dose response relationship between fibre numbers injected intraperitoneally and tumour incidences. Summary of results after Pott *et al.*, (1990a). The regression line and the 95% confidence limit were calculated for all measurement points with a tumour incidence > 10%. Tumour rates < 10% are evaluated as non-specific. After i. p. injection of 0.5×10^9 or more relatively durable mineral fibres, a tumour incidence of at least 20% can be expected.

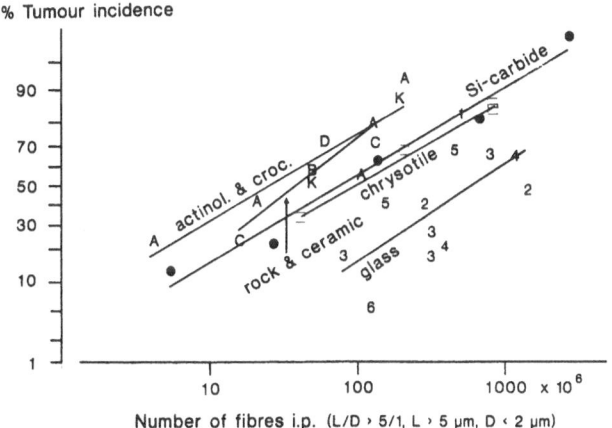

% Tumour incidence

Number of fibres i.p. (L/D > 5/1, L > 5 μm, D < 2 μm)

Figure 5. Dose response relationship of fibre types or groups after intraperitoneal injection (from Pott *et al.*, 1987, 1989 and Tables 2 and 3). The true dose response relationship of silicon carbide may take its course shifted a little more towards higher doses (see text). Symbols:A = actinolite, K =crocidolite, B = basalt, D = diabase, C = ceramic (2 types), (squares) = chrysotile, (closed circles) = silicon carbide, 1-6 = glass microfibres, 1 = (Manville) M-104/E, 2 = M-100/475, 3 = M-104/475, 4 = (Bayer) B-3K, 5 = B-3L, 6 = M-106

(2) Silicon carbide fibres: a clear dose response relationship was found. However, the type of evaluation applied, chosen because of the infectious disease, overestimated tumour incidence; e.g., the true percentage of tumour bearing animals of the highest dose group is less than 97% but higher than 75%. Stanton et al. (1981) previously reported the carcinogenic potential of silicon carbide fibres.

(3) Basalt fibres: the carcinogenic potential detected in earlier i.p. experiments was confirmed (Nikitina et al., 1989; Pott et al., 1989).

(4) Ceramic fibres: the carcinogenic potential detected in earlier i.p. experiments was confirmed (Smith et al., 1987; Pott et al., 1989).

(5) Sepiolite fibres: injection of a high number of fibres (approx 4×10^{10}) did not produce tumours and this confirmed an earlier result (Pott et al., 1990a).

(6) Non-fibrous dusts, relatively thick fibres and repeated injections: neither the administration of a high mass (up to 5 injections of 50 mg) nor up to 50 injections per rat produced a non-specific carcinogenic effect.

4. An Experiment on the Role of the Surface Properties for the Mechanism in Fibre Carcinogenesis

Many authors concluded from the experimental results of the last two decades that the elongated particle shape is the decisive factor for tumour induction by fibrous materials. Consequently, much work has been done on determining the aspect ratios, minimal lengths and maximum diameters necessary to define the "carcinogenic fibre sizes". The experiments described above show that chemical composition is also important as it determines the potential of a fibre to persist in the tissue for a certain length of time.

In addition to fibre dimensions and fibre durability, surface properties are often discussed as the third essential factor determining fibre carcinogenesis. Surface properties have proved to be important with regard to the cytotoxicity. For example, the polymer polyvinylpyridine-N-oxide (PVNO) showed a clear inhibition of the cytotoxic effect of quartz and other dusts in vitro and of the fibrogenic effect of quartz in rats. Therefore, we examined the influence of PVNO on the carcinogenicity of actinolite and observed a significant retardation of tumour induction by the combined administration of actinolite asbestos with PVNO. The result is shown in Table 3. In a second experiment, the effect, while not as high as in the first experiment, was in principle reproduced (Pott et al., 1989). However, the delayed appearance of tumours in the group treated with a mixture of actinolite and PVNO is not necessarily caused by an inactivation of the surface of the fibres. The fibres tend to agglomerate more when they are suspended in a solution of PVNO. Possibly, the influence of the polymer on the kind of fibre deposition on the peritoneum modifies the cell reactions.

It should be asked in the present state of knowledge on the mechanisms in fibre carcinogenesis: does anyone know a plausible hypothesis which could support the idea that long, thin and durable fibres do not have a carcinogenic potential? If so which additional properties should such fibres have?

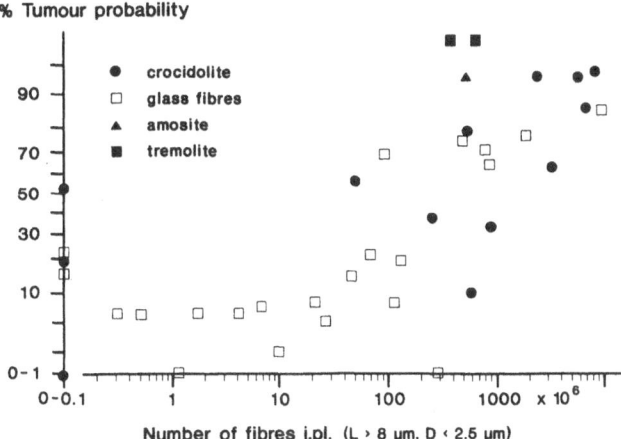

% Tumour probability

Number of fibres i.pl. (L > 8 µm, D < 2.5 µm)

Figure 6. Dose response relationship between different 40 mg samples of crocidolite, glass fibres, amosite, and tremolite and the tumour probability after intrapleural administration of these dusts. The tumour probability was given by Stanton *et al.* (1981) on the basis of the observed number of tumour-bearing animals per treated group, which was corrected in consideration of the life-span of the animals. The publication of Stanton and coworkers also contains the common log of the number of particles per microgramme for each of 34 dimensional categories. The data were used for the calculation of the numbers of fibres > 8 µm in length and < 2.5 µm in diameter which were administered intrapleurally per rat and which were applied for the dose scale in the diagram.

5. Dose Response Relationships of Mineral Fibres in the Serosal Tests as a Basis for Their Classification According to Their Carcinogenicity

Ten years ago, the German Commission for the Investigation of Health Hazards of Chemical Compounds in the Work Area classified man-made mineral fibres with diameters less than 1 µm into the group of compounds which are suspected of having carcinogenic potential. The bases of this classification were the results of the serosal tests. For several years physicians, toxicologists, hygienists and other persons responsible for health protection, who were independent from the fibre industry, asked us to develop a proposal for the classification of non-asbestos fibres. The proposal should include not only man-made mineral fibres but also inorganic and organic man-made and natural fibres if they were inhalable.

No earlier than last year, the results of fibre measurements of Bellmann (published in Pott *et al.*, 1990b) and Rödelsperger (1990) led to dose response relationships on the basis of fibre numbers instead of fibre masses. Figure 4 shows the state of knowledge from April 1989. The regression line and the 95 % confidence interval were calculated and drawn by a computer; they consider all tumour rates higher than 10 %. Tumour incidences below 10 % may be spontaneous or induced non-specifically. The average of these tumour incidences in several hundred rats was lower than 5 %, however, in a particular experimental group of about 40 animals it may be about 10%. In a few cases, it is not possible to be sure of the diagnoses. This problem is, in principle, the same in rats as in humans. In any case, after intraperitoneal injection of at least 0.5 to 1 x 10^9 durable mineral fibres in rats, an incidence of sarcomas or mesotheliomas of more than 20 to 30% can be expected. Fibres having a low durability did not show a carcinogenic effect at this dose.

Until now the maximum tolerable mass of mineral fibres has not been examined. Five weekly intraperitoneal injections of 50 mg of durable dusts did not result in a loss of weight or an increase in mortality (see Table 3). The danger of false positive results by high doses seems to be lower than the danger of false negative results. From our experience with organic fibres it can be concluded that a high fibre mass leads to an agglomeration of fibres which decreases the effect.

Many of the data points shown in Figure 5 are also contained in Figure 4. But in Figure 5, the regression lines for five fibre types or groups of fibre types are drawn in. In the cases of the three kinds of asbestos and of silicon carbide only one dust preparation has been used; the measurement points for ceramic, rock and glass fibres contain two to six different types. The dose response curve for silicon carbide is shifted to the left side because of the mode of evaluation of the groups with the reduced survival time due to the infectious disease.

In principle, the measured data resulted in clear dose response curves. For most of the materials used, tumour rates of more or much more than 50% have been observed in the highest dose groups. However, there is a dose difference of about twenty between the regression line for the two amphiboles and the regression line for the group of six glass fibre types. The question arises: What could be the cause for this remarkable factor? There are some ideas for its explanation, but this question is undoubtedly one of the points which needs further research.

Figure 6 shows the dose response relationship from many preparations of amphiboles and glass fibres calculated from the data of Stanton *et al.* (1981). These intrapleural studies do not show a clear difference between the carcinogenicity of crocidolite and glass fibres. However, better knowledge of the characteristics of the particle size distributions and other properties of all the different preparations could give more detailed information.

At present, the main question is not, does the comparative cancer risk for glass fibres relate to that for asbestos, in animals and humans, but what is the decision as to which fibre types have to be classified as possibly or probably carcinogenic to humans and which fibre types can be exonerated from the suspicion of being carcinogenic.

Table 5 contains many fibre types which showed unequivocally a carcinogenic potential in rats after intrapleural or intraperitoneal administration. It is proposed to classify fibres of these positive fibre types into the group "probably carcinogenic to humans" if they have an aspect ratio of more than 5 to 1 and if they are longer than 5 μm and thinner than 2 μm, or if they can split up into such thin fibres. This classification of fibre types according to their carcinogenic potential should not be confused with carcinogenic potency or risk assessment. Table 5 also contains evaluations of a few fibre types which lead, together with additional knowledge, to the hope that these fibre types can be classified as probably not carcinogenic. The durability studies on wollastonite are not yet completed.

There are many more fibre types and variations of fibre types than listed in Table 5. It appears that it is not necessary to test all these fibres because there are a great many similarities. With the growing rate of knowledge it should be possible to classify more

TABLE 5. Carcinogenic potential of natural and man-made inorganic fibres after intrapleural or intraperitoneal administration in rats (for proposed classification see remarks)

Natural fibres (mineralog. name)	Result	Refer.	Man-made fibres (material)	Result	Refer.
asbestos			aluminium oxide	+ +	14
all 6 types	+ +		aluminium silicate		
attapulgite			("ceramic")	+ +	8,10,11,15
Caceres	+	7	dawsonite	+ +	14
Leicester	+	17	glass		
Torrejon	+	17	relatively durable	+ +	3,5,6,7,10,
dawsonite	+	14			11,12,13,16
erionite	+ +	1,2,7,18	slightly durable	−	10
nemalite	+ +	5,7	gypsum	−	10
sepiolite			potassium		
Finland	+	9	octatitanate	+ +	10,14
Uicaluaro	−	9,10	rock		
wollastonite			basalt	+ +	4,8,10
Canada	(+)	14	diabase	+	7
India	−	8	silicon carbide	+ +	10,14

+ + $\geq$ 2 experiments unequivocally positive

+   1 experiment unequivocally positive

(+)   experiment weakly positive

−   experiment(s) and additional knowledge allow the conclusion that the fibres are "probably not carcinogenic" to humans. (Regarding the "probably not carcinogenic" natural mineral fibres, the problem of impurities with more durable or longer fibre types has to be considered in any case. Moreover, the durability studies on wollastonite are not yet completed.)

Remarks: The positive studies can already be explained by the portion of fibres $> 5 \mu m$ in length, $< 2 \mu m$ in diameter, with an aspect ratio of $> 5/1$. If a fibrous material evaluated as "positive" in this Table does not contain fibres within the mentioned dimensions, then the material is "probably not carcinogenic to humans" according to this proposal. Even if the material contains such fibres, it has not necessarily to be labelled as carcinogenic. The percentage of such fibres which has to be exceeded for labelling has to be set by regulators.

References to table 5 (without the numerous references of asbestos studies)

1   Coffin *et al.*, 1989
2   Maltoni and Minardi, 1989
3   Monchaux *et al.*, 1981
4   Nikitina *et al.*, 1989
5   Pott and Friedrichs, 1972
6   Pott *et al.*, 1976
7   Pott *et al.*, 1987
8   Pott *et al.*, 1989
9   Pott *et al.*, 1990a

10   Pott *et al.*, (this paper)
11   Smith *et al.*, 1987
12   Stanton and Wrench, 1972
13   Stanton *et al.*, 1977
14   Stanton *et al.*, 1981
15   Wagner *et al.*, 1973
16   Wagner *et al.*, 1976
17   Wagner *et al.*, 1987
18   Wagner, 1988

and more fibre types by analogy to well known fibres. For example: if silicon nitride fibres are as durable and in the same size range as the tested silicon carbide fibres, preventive measures should be introduced for both fibre types. On the other hand, if a new glass fibre has a lower durability than the tested glass fibre which did not induce tumours, both glass fibre types could be exonerated from the suspicion of being carcinogenic.

6. Discussion of the Proposal for the Classification of Fibres

Professor Pott: My proposal for the classification of fibres (Pott, 1989a) induced the Chairman of the Joint European Medical Research Board, Prof. Rossiter, to request statements from several scientists. I received a copy of these statements, bound as a book, from the German Federal Institute for Occupational Safety and Health. I feel that there are obviously misunderstandings, mistakes and strong emotions in the arguments. In the following, I reply only to some objections of Dr. McClellan which are representive of the serious arguments.

Dr. McClellan: In the absence of adequate human experience, it is necessary to utilize exclusively data from animal bio-assays and *in vitro* studies to assess carcinogenicity. For such material, the predictive value of the animal bioassays will be strongest when the tests have been conducted with relevant routes of exposure. In the case of airborne materials, that is inhalation.

Reply: Certainly, the statement of Dr. McClellan describes a good general rule. But this rule is no law of nature like "all humans have to die". It is only a rule for toxicologic work which has remarkable exceptions. The exceptions can be explained by the differences between humans and animals and the requirements and limitations of the methods of toxicology. One of the exceptions is cigarette smoke, another regards mineral fibres. Therefore, it would be prudent to examine in any case whether the rule is adequate or not.

Dr. McClellan: If positive results are obtained in an inhalation exposure bioassay using very high fiber exposure concentrations, I suggest the material be identified as "possibly carcinogenic to humans".

Reply: The experience with asbestos fibres obviously conflicts with this statement. In relation to human exposure, a concentration of a few hundred fibres/ml has to be evaluated as very high: the exposure of humans to this would induce a very high risk. Nevertheless, the carcinogenic activity cannot be detected with certainty in inhalation experiments. Therefore, I think it is inadequate to classify a fibre type as "possibly carcinogenic to humans", if positive results are obtained under these exposure concentrations. In my opinion, such a result should lead to immediate preventive measures.

Dr. McClellan: I further hypothesize that the inflammatory response with associated changes in cell proliferation may play a critical role in the carcinogenic response of the peritoneal cavity to bolus "doses" of fibers irrespective of the true potential of the fibers producing a carcinogenic response in the lung when inhaled.

Reply: The experimental results obtained after intraperitoneal injection of mineral dusts up to 250 mg in our experiments do not support the "bolus hypothesis". The "bolus hypothesis" is a misinterpretation of the so-called Oppenheimer effect. Oppenheimer and many others observed tumours after implantation of materials within the dimensions of a coin. But the administration of the same material (about 500 to 1500 mg) induced tumours only in a low percentage when it was given in small fragments or as a powder.

Dr. McClellan: I have serious reservations as to the appropriateness of using intraperitoneal injections to assay fibers for their carcinogenic potency.

Reply: We do not have to decide on potency but only on potential. However, Dr. McClellan's statement may also refer to the carcinogenic potential of fibres for the lung. Other authors also hold this view (e.g. WHO, 1988: "... these routes are primarily relevant to the mesothelium and do not necessarily predict what happens in the pulmonary parenchyma or airways"). However, the carcinogenic reaction of the lung tissue as well as the serosal tissue to asbestos fibres has been well-known for decades. Therefore, a carcinogenic potential of other fibres for the lung can be expected, if these fibres induce tumours in the serosal test and if they can be inhaled by humans. This conclusion by analogy is plausible. I think it has to be drawn from the point of view of preventive medicine.

It should be kept in mind that the extrapolation from results received in an experimental model to a very different organ in humans is not unique. Here are two examples: the reaction of an asthmatic patient to allergens is normally not tested by inhalation but by intracutaneous injection, although the skin and the lung obviously differ in many properties. Moreover, the development of the so-called low risk cigarette is based essentially on results with cigarette smoke condensate on the skin of mice. Of course, the skin of mice and the lung of humans are quite different organs. Nevertheless, both tissues react similarly to some substances, for example to polycyclic aromatic hydrocarbons.

In the case of the extrapolation from the carcinogenic effect of fibres in the serosal tissue to the lung, it should be argued scientifically why an analogous activity probably does not exist.

Dr. McClellan: Pott proposes a third step in which fibers may be classified by analogy. I disagree with this approach except as an approach to prioritizing fibers for evaluation of biological assays.

Reply: Classification by analogy is no new idea, it is a fact. Actinolite asbestos is regulated only by analogy with other asbestos types because of the similarity of the fibrous particle shape and durability. There are no epidemiologic results of workers exposed to actinolite and no inhalation tests but only some positive injection studies. According to the statements of Dr. McClellan, actinolite should not be classified as a carcinogen. Should we use it as a substitute for crocidolite and amosite?

ACKNOWLEDGEMENTS

We are grateful to all our colleagues who have contributed with their effort. We especially acknowledge with gratitude the skilled technical work of Mrs. Y. Steinfartz and Mrs. H. Struppeck and the great secretarial assistance of Mrs. B. Piotrowski and Mrs. E. Saygin.

REFERENCES:

Bellmann,B., Muhle,H., Pott,F., König,H., Klöppel,H. and Spurny,K. (1987) Persistence of man-made mineral fibres (MMMF) and asbestos in rat lungs. *Ann. Occup. Hyg.* **31**:693-709.

Bellmann,B., Muhle,H. and Pott,F. (1990) Study on the durability of chemically different glass fibres in lungs on rats. *Zbl.Hyg.* **190**:310-314. (German)

Coffin, D.L., Palekar,L.D., Cook,P.M. and Creason,J.P. (1989) Comparison of mesothelioma induction in rats by asbestos and nonasbestos mineral fibers: Possible correlation with human exposure data. In, "Biological interaction of Inhaled Mineral Fibers and Cigarette Smoke", eds. Wehner,A.P., Felton,D.-L. Proceedings of an International Symposium/Workshop held at the Battelle Seattle Conference Center April 10-14, 1988, Seattle, Washington. Battelle Press, Columbus, Ohio. pp. 313-323.

Davis,J.M.G. and Jones,A.D. (1988) Comparisons of the pathogenicity of long and short fibres of chrysotile asbestos in rats. *Br. J. Exp. Path.* **69**:717-737.

Davis, J.M.G., Beckett,S.T., Bolton,R.E., Collings,P. and Middleton,A.P. (1978) Mass and number of fibres in the pathogenesis of asbestos-related lung disease in rats. *Br. J. Cancer* **37**:673-687.

Davis, J.M.G., Addison,J., Bolton,R.E., Donaldson,K., Jones,A.D. and Smith,T. (1986) The pathogenicity of long versus short fibre sample of amosite asbestos administered to rats by inhalation and intraperitoneal injection. *Br. J. Exp. Path.* **67**:415-430.

IARC (International Agency for Research on Cancer, ed.) (1987) IARC Monographs on the Evaluation of the Carcinogenic Risk of Chemicals to Humans. Vol. **42**, "Silica and Some Silicates". World Health Organization International Agency for Research on Cancer, Lyon.

IARC (International Agency for Research on Cancer, ed.) (1988) IARC Monographs on the Evaluation of the Carcinogenic Risk of Chemicals to Humans. Vol. **43**, "Man-Made Mineral Fibres and Radon". World Health Organization International Agency for Research on Cancer, Lyon.

Maltoni,C. and Minardi,F. (1989) Recent results of carcinogenicity bioassays of fibres and other particulate materials. In, "Non-Occupational Exposure to Mineral Fibres", eds. Bignon,J., Peto,J. and Saracci,R. (IARC Scientific Publ. No. 90), International Agency for Research on Cancer, Lyon. pp. 46-53.

McClellan,R.O. (1990) Comments on proposal of Professor F. Pott for classifying the carcinogenicity of fibers. (Manuscript).

Monchaux,G., Bignon,J., Jaurand,M.C., Lafuma,J., Sebastien,P., Masse,R., Hirsch,A. and Goni,J. (1981) Mesotheliomas in rats following inoculation with acid-leached chrysotile asbestos and other mineral fibres. *Carcinogenesis* 2:229-236.

Muhle,H., Pott,F., Bellmann,B., Takenaka,S. and Ziem,U. (1987) Inhalation and injection experi- ments in rats for testing MMMF on carcinogenicity. *Ann. Occup. Hyg.* **31** No.4B:755-764.

Muhle,H., Bellmann,B. and Pott,F. (1990) Durability of various mineral fibers in rat lungs. (This volume).

Nikitina,O.V., Kogan,F.M., Vanchugova,N.N. and Frash,V.N. (1989) Comparative oncogenicity of basalt fibres and chrysotile asbestos. *Gig. Tr. Prof. Zabol.* **4**:7-11. (Russian)

Pott,F. (1987a) The fibre as a carcinogenic agent. *Zbl. Bakt. Hyg. B* 184:1-23. (German)

Pott,F. (1987b) Problems in defining carcinogenic fibres. *Ann. Occup. Hyg.* **31** No.4B:799-802.

Pott,F. (1988) Umweltkanzerogene. In, "Ges.z.Förderung d. Lufthygiene u. Silikoseforschung e.V.", ed. Düsseldorf, Umwelthygiene. Bd 20. Medizinisches Institut für Umwelthygiene. Jahresbericht 1987/88. - Stefan W. Albers, Düsseldorf. pp. 22-26.

Pott,F. (1989a) Carcinogenicity of fibers in experimental animals - Data and evaluation. In, "Assessment of Inhalation Hazards", eds. Mohr,U. *et al..* Integration and Extrapolation Using Diverse Data (ILSI Monographs), Springer, Berlin a.o. pp. 243-253.

Pott,F. (1989b) Questions to the cell biologists and proposals for a definition and classification of carcinogenic fibers derived from animal studies. In, "Effects of Mineral Dusts on Cells", eds. Mossmann,B.T. and Begin,R.O. NATO ASI Series, Vol. **H30**, Springer, Berlin, Heidelberg. pp. 439-444.

Pott,F. (1990) Beurteilung der Kanzerogenität von Fasern aufgrund von Tierversuchen. Tagung "Faserförmige Stäube - Vorschriften, Wirkungen, Messung, Minderung" der Kommission Reinhaltung der Luft (KRdL) im VDI und DIN, Heidelberg, 11-13. Sept. 1990.

Pott,F. and Friedrichs,K.H. (1972) Tumoren der Ratte nach i.p.-Injektion faserförmiger Stäube. *Naturwissenschaften* **59**:318.

Pott,F., Friedrichs,K.H. and Huth,F. (1976) Ergebnisse aus Tierversuchen zur kanzerogenen Wirkung faserförmiger Stäube und ihre Deutung im Hinblick auf die Tumorentstehung beim Menschen. *Zbl. Bakt. Hyg., I. Abt. Orig. B* **162**:467-505.

Pott,F., Ziem,U., Reiffer,F.-J., Huth,F., Ernst,H. and Mohr,U. (1987) Carcinogenicity studies on fibres, metal compounds, and some other dusts in rats. *Exp. Pathol. (Jena)* **32**:129-152

Pott,F., Roller,M., Ziem,U., Reiffer,F.-J., Bellmann,B., Rosenbruch,M. and Huth,F. (1989) Carcinogenicity studies on natural and man-made fibres with the intraperitoneal test in rats. In, "Non- Occupational Exposure to Mineral Fibres", eds. Bignon,J., Peto,J. and Saracci,R. IARC Sci. Publ. No. 90. Intern. Agency for Research on Cancer, Lyon. pp. 173-179.

Pott,F., Bellmann,B., Muhle,H., Rödelsperger,K., Rippe,R.M., Roller,M. and Rosenbruch,M. (1990a) Intraperitoneal injection studies for the evaluation of the carcinogenicity of fibrous phyllosilicates. In, "Health Related Effects of Phyllosilicates", ed. Bignon,J. NATO ASI Series G: Ecological Sciences. Vol. 21, Springer, Berlin a.o. pp. 319-329.

Pott,F., Schlipköter,H.-W., Roller,M., Rippe,R.M., Germann,P.G. Mohr,U. and Bellmann,B. (1990b) Carcinogenicity of glass fibres with different durability. *Zbl. Hyg.* **189**:563-566. (German)

Pott,F., Rippe,R.M., Roller,M., Rosenbruch,M. and Huth,F. (1990c) Vergleichende Untersuchungen über die Kanzerogenität verschiedener Nickellegierungen (Schriftenreihe der Bundesanstalt für Arbeitsschutz. Forschung), Wirtschaftsverl. NW, Verl. f. neue Wissenschaft, Bremerhaven, Dortmund. (in press).

Rödelsperger,K., Brückel,B., Patrzich,R., Pott,F. and Woitowitz,H.-J. (1990) Concentrations of fibres in phyllosilicates. In, "Health Related Effects of Phyllosilicates", ed. Bignon,J., NATO ASI Series G: Ecological Sciences Vol. 21, Springer, Berlin a.o. pp. 75-83.

Smith,D.M., Ortiz,L.W., Archuleta,R.F. and Johnson,N.F. (1987) Health effects in hamsters and rats exposed chronically to man-made vitreous fibres. *Ann. Occup. Hyg.* **31**:731-754.

Stanton,M.F. and Wrench,C. (1972) Mechanisms of mesothelioma induction with asbestos and fibrous glass. *J. Natl. Cancer Inst.* **48**:797-821.

Stanton,M.F., Layard,M., Tegeris,A., Miller,E., May,M. and Kent,E. (1977) Carcinogenicity of fibrous glass: pleural response in the rat in relation to fiber dimension. *J. Natl. Cancer Inst.* **58**:587-603.

Stanton,M.F., Layard,M., Tegeris,A., Miller,E., May,M., Morgan,E. and Smith,A. (1981) Relation of particle dimension to carcinogenicity in amphibole asbestoses and other fibrous minerals. *J. Natl. Cancer Inst.* **67**:965-975.

Wagner,J.C. (1988) Animal models of pneumoconiosis I. Biological effects of short fibres. VIIth International Conference on Pneumoconioses, Pittsburgh, U.S.A., August, 23-26, 1988.

Wagner,J.C., Berrry,G. and Timbrell,V. (1973) Mesotheliomata in rats after inoculation with asbestos and other materials. *Br. J. Cancer* **28**:173-185.

Wagner,J.C., Berry,G. and Skidmore,J.W. (1976) Studies of the carcinogenic effects of fiber glass of different diameters following intrapleural inoculation in experimental animals. In, "Occupational Exposure to Fibrous Glass", ed. US Department of Health, Education and Welfare, National Institute for Occupational Safety and Health. Proceedings of a Symposium, College Park, Maryland, June 26-27, 1974 (HEW Publ. No. NIOSH 76-151), U.S. Government, Printing Office, Washington, D.C. pp. 193-204.

Wagner,J.C., Griffiths,D.M. and Munday,D.E. (1987) Experimental studies with palygorskite dusts. *Br. J. Indust. Med.* **44**:749-753.

Wagner,J.C., Berry,G., Skidmore,J.W. and Timbrell,V. (1974) The effects of the inhalation of asbestos in rats. *Br. J. Cancer* **29**:252-269.

Wagner,J.C., Skidmore,J.W., Hill,R.J. and Griffiths,D.M. (1985) Erionite exposure and mesotheliomas in rats. *Br. J. Cancer* **51**:727-730.

WHO (World Health Organization, ed.) (1988) Man-Made Mineral Fibres (Environmental Health Criteria 77), World Health Organization, Geneva.

FIBRE CARCINOGENESIS: INTRA-CAVITARY STUDIES CANNOT ASSESS RISK TO MAN

Charles E. Rossiter

Joint European Medical Research Board
10 Mynchen Road, Knotty Green
BEACONSFIELD, Bucks, HP9 2AS, UK

ABSTRACT

Intra-pleural and intra-peritoneal studies in rats have demonstrated convincingly that long/fine fibres have the greatest potential to cause mesothelioma. Recent studies have also shown that low fibre durability in the peritoneum reduces the mesothelioma rate. The results of these studies have been used to infer that fibres with the same critical dimensions and durability must be considered as lung carcinogens. The extension of that argument is that all fibres with these properties must be formally classified as probable human carcinogens, with legislative consequences. The inference, and the extension, are both untenable.

Intra-cavitary studies can only assess mesothelioma hazard of fibres which reach the pleura or peritoneum of the rat. The non-physiological route of administration bypasses natural defence mechanisms of the experimental animal. In particular, it excludes consideration of nasal filtration and of biopersistence in the lung. Intra-cavitary studies cannot assess the risk of lung cancer, because the target organ of the experiments is the mesothelium.

Marked discrepancies exist between the epidemiological evidence for the risk of mesothelioma and the intra-cavitary study results. Chrysotile very rarely causes mesothelioma in man, yet the intra-cavitary studies have been used to point to a very high risk. For glass microfibres, no human mesothelioma has been reported and none of the animal inhalation studies has been positive for lung cancer or mesothelioma, yet several intra-cavitary studies have produced significant mesothelioma rates.

The role of intra-pleural and intra-peritoneal studies in establishing critical fibre dimensions for mesothelioma induction was very valuable. These studies can also be used for assaying the effect of durability. However, they cannot be used for the study of mechanisms of action of fibres in the lung and they cannot replace epidemiological and animal inhalation studies in the assessment of risk to man.

BACKGROUND

Since the recognition of the human health effects of asbestos exposure and the association of these effects with the fibrous nature of asbestos, questions have been raised as to the safety of the use of non-asbestos fibres. Throughout the past 30 years, research in experimental animals has sought to reproduce the chronic pulmonary diseases associated with human asbestos exposure including interstitial fibrosis, lung cancer, and for some asbestos varieties, mesothelioma. Additional research has been directed at elucidating the mechanisms responsible for the biological activity of asbestos and other fibres.

The chronic diseases associated with airborne asbestos are all associated with the respiratory system. It is therefore logical that inhalation studies in laboratory animals have been the preferred experimental model in almost all laboratories for the assessment of airborne fibre activity. These models allow estimation of deposition, clearance, and translocation of inhaled fibres. Inhalation is also the only route of administration which closely models the route of human exposure.

The most significant feature of inhalation studies is that chronic inhalation studies with a variety of asbestos fibres have produced all the signs of asbestos related pulmonary disease: fibrosis, lung cancer and mesothelioma. These inhalation studies have used airborne fibre levels not greatly different from those associated with heavy exposures in the early days of the asbestos industry (Wagner *et al.*, 1974; Davis *et al.*, 1986a, 1986b).

A number of man made mineral fibres have also been evaluated by inhalation. For glass, rock and slag wools the results of these studies have been consistently negative. Chronically exposed animals have shown no significant malignant or nonmalignant pulmonary disease following exposure to fibre concentrations much higher (about 10,000 fold) than in the work environment (IARC, 1988; Gross *et al.*, 1970).

In addition to the inhalation studies a series of studies has been performed in which fibres have been administered by injection or implantation into either the pleural or peritoneal cavity. This research has produced positive results in a number of studies in which the same fibre was found not to produce tumours by inhalation (Muhle *et al.*, 1987; Wagner *et al.*, 1984; Smith *et al.*, 1987). It has been argued that the rodent inhalation model is not adequately sensitive to detect possible carcinogenic potential of fibres (Pott, 1989, 1990). In particular, Pott stresses that because some inhalation studies with crocidolite asbestos have not produced significant tumour formation, the inhalation model is not appropriate for the evaluation of biological activity of airborne fibres. The evidence is that the UICC crocidolite used in most experiments was too short to produce tumours by inhalation (Wagner (1990) in discussion at this workshop). This does not apply to the experiment of Wagner and coworkers (1974), which Pott described as unrepeatable (Pott, 1990). Pott also ignored the validation of the inhalation model by the studies of size-selected samples of refractory ceramic fibres (Hesterberg *et al.*, 1991).

REGULATORY ISSUE

The proposal by Pott

Because of his belief that the inhalation model is not adequately sensitive, Pott (1989) proposed formally that the classification of fibres, for regulatory purposes in West Germany, should be based solely on the results of intra-peritoneal tests, ignoring all evidence from human studies and from other animal experimental studies. Pott expected this proposal to be considered in 1990 by the Deutsche Forschungsgemeinschaft Kommission zur Prüfung gesundheitsschädlicher Arbeitsstoffe (Senatskommission) and he advised many people accordingly. However, the Senatskommission did not consider this proposal then. Nevertheless, many criticisms were made of that proposal, both on concept and on content.

A new regulatory proposal has now been prepared by Pott and others (1990a) to be considered by the Senatskommission. Pott (1990) has also recently presented a conference paper in Heidelberg, which provides the arguments on which he bases his new proposal.

Despite strong attacks against his critics, Pott has clearly accepted many of the criticisms, incorporating corresponding amendments. As examples. he has accepted that the intraperitoneal test can produce false negative results and he has accepted that his published summary dose-response relation for the intra-peritoneal test is invalid because different dose-response relations apply for different fibre types.

However, he still argues that non-physiological exposure leads to relevant results which are suitable for the classification of fibres according to their carcinogenicity by inhalation to man. He still argues that human evidence is irrelevant.

In brief, the regulatory proposal now recommends the following:

(1)   There are only two types of fibre for regulatory purposes: those probably carcinogenic to man and those probably not carcinogenic to man. Untested fibres must be placed on a "waiting list of possibles" until tested, or until carcinogenic status can be determined by analogy with other tested fibres. Analogy is undefined, although an example is given in the Heidelberg paper and proposals for classifying certain glass fibres and basalt rock wool are made based on analogy.

(2)   The classification of fibres should be based on the results of the intra-peritoneal test administered in one dose of 10^9 fibres, with length greater than 5 μm, diameter less than 2 μm and aspect ratio greater than 5:1.

Only if this test shows no evidence of carcinogenicity can a fibre be cleared of being carcinogenic to man.

(3)   Formation of tumours of any type in 12% or more of the animals in an intra-peritoneal test is considered positive evidence of carcinogenicity to man. Any positive results from any other animal experiment (inhalation, intra-tracheal instillation, intra-pleural implantation) would also show that the fibre is carcinogenic to man.

The figure of 12% in (3) above is the upper limit of a "grey-zone". This concept was first introduced in 1990, when Pott and others (1990b) reported the results of an intraperitoneal test comparison between a new glass fibre of low durability and JM104/475 glass fibres. In that study, the highest tumour rate for the low durability fibre was 10.9%. The dividing line has been reduced to just above that figure from the previous value of up to 15% (Pott, 1987).

International opinion

No support has been found outside Germany for the regulatory proposal by Pott, except perhaps in the classification by the International Agency for Research on Cancer of glass fibres and rock wool as "possibly carcinogenic to man" (IARC, 1988). IARC's classification of glass fibres included glass wool, although the only evidence of possible carcinogenic potential was provided by serosal tests of special purpose durable fine glass fibres. The classification of rock wool was based solely on two intra-peritoneal implantation studies by Pott of experimental non-commercial fibres (Brown *et al.*, 1991). The IARC classification for MMMF appears not to take into account its own criteria, requiring physicochemical characterisation of fibrous particulates "for two reasons: (i) to ensure that the sample to be tested is representative of the materials to which humans are exposed; and (ii) to permit evaluation of specific physical and chemical characteristics of the fibres which may be important in the induction of cancer" (IARC Monograph (1988), page 33).

Following the IARC evaluation, the International Programme on Chemical Safety (IPCS, 1988) also reviewed the animal study literature and warned against the overinterpretation of the results of injection and implantation studies:
"The need for caution in the extrapolation of the results of studies involving injection or implantation in body cavities to predict the potency of various fibre samples, even with respect to the induction of mesotheliomas, cannot be overemphasised. The relevance of these types of studies to other types of cancer, such as lung cancer, has not been established".

As NIOSH (1977) has acknowledged:
"The routes of exposure used in many of the intrapleural and intraperitoneal experiments have been considered to be inappropriate to indicate the effects of fibrous glass after inhalation. It is not valid to extrapolate from the results from these intracavitary exposures to humans in the workplace."

More recently NIOSH (1987) has reconsidered the evidence on the carcinogenicity of fibrous glass:
"Carcinogenic responses in animals, particularly rodents, following intrapleural or intraperitoneal fibrous glass administration is similar to the responses found after implantation of any foreign material such as polyethylene, asbestos, nylon, cellophane or teflon [5 references given]. Tumor development in laboratory animals following pleural or intraperitoneal administration of fibrous glass material probably represents a non-specific foreign body response. The response depends on the physical characteristics of the fibrous glass, the most important being size and shape; certain characteristics of the animal; and the length of time the fibrous glass is present in the animal (a critical factor). On the basis of present information, fibrous glass cannot be considered a carcinogenic agent."

This recognition by NIOSH that the intracavitary studies are not predictive of inhalation risks was recently affirmed by the EPA, which stated in its recent review of "Health Aspects of Asbestos Substitutes," Vu (1988):
"Positive results from studies using intrapleural or intraperitoneal injection/implantation method in the absence of positive findings from inhalation experiments do not indicate that these fibres will cause tumours in man upon inhalation".

This position is still held by the US Environmental Protection Agency: viz. in a letter from L.J. Fisher to Frank J. Rauscher, Jr., Executive Director of TIMA, Inc., June 1990 and another (May 1990) from C. Elkins to Sheldon W. Samuels, Director, Health, Safety & Environment, Industrial Union Department, AFL-CIO. For example, Elkins, the Director of EPA's Office of Toxic Substances, wrote:

"EPA believes that animal inhalation studies *per se* are universally adequate in assessing the potential of a test fibre to induce lung cancer. However [they] may not be adequately sensitive for detecting fibres which induce mesothelioma unless the test fibre is ... very potent. For this reason, more sensitive methods utilizing several non-physiological routes of exposure ...have been used by many investigators to screen for the overall potential carcinogenicity of these substances.

In spite of the major limitation discussed above, EPA believes that the results of well designed and well-conducted chronic inhalation bioassays provide the most meaningful and definitive information for the assessment of inhalational cancer hazards of the test fibre in humans. Data from injection studies cannot be used to predict inhalation hazard in humans but do provide useful information regarding the intrinsic biological activity of the test fibre."

ASSESSMENT OF RISK TO MAN

Epidemiological studies, animal bioassays, and *in vitro* investigations may all be used to evaluate the carcinogenic risk of exposure to airborne fibres. The concern is with potential human health risk, in this case, carcinogenicity. Thus, human data from epidemiological studies are of primary importance if such data exist. For erionite and the major varieties of asbestos, the human data proved the carcinogenicity of these fibres; the animal evidence confirmed the carcinogenicity, but did also provide insight into the mechanisms of carcinogenicity. The epidemiological evidence continues to provide the proof of the fibre size hypothesis, the so-called Stanton hypothesis. It also provides the specific evidence on the critical size parameters for human carcinogenic potential.

It has been argued that epidemiological studies which show no excess of lung cancer or mesothelioma cannot be used as evidence against carcinogenic potency. This is illogical (Miettinen & Rossiter, 1990). For example, if a large group of exposed workers has an SMR well under the expected value for similar, but unexposed, workers, then it is possible to conclude that there is no evidence of excess mortality at their exposure levels. This is a valid assessment of risk to man at these exposure levels. This requires no extrapolation from high dose animal experiments to lower dose human experience.

One example of the overwhelming importance of the negative human evidence is the mesothelioma risk associated with exposure to chrysotile asbestos. Despite the positive animal studies, there is widespread acceptance that the human mesothelioma risk is very low or non-existent (Mossman *et al.*, 1990).

For the insulation wool production industry, the epidemiological studies have been extensive. They show no evidence of excess lung cancer risk for special purpose fine durable glass fibres, glass wool or rock wool. There is considerable doubt that the lung cancer excess among slag wool production workers can be attributed to their fibre exposure (Brown *et al.*, 1991).

If adequate human data are not available, then only animal bioassays and *in vitro* studies are available to assess carcinogenicity. To predict human risk, the animal bioassays will be most valuable when the tests have been conducted with relevant routes

TABLE 1. Evidence of carcinogenic potential by inhalation and by serosal test

	By inhalation				By serosal test
	Lung		Mesothelium		Mesothelium
	Human	Animal	Human	Animal	Animal
Erionite	+ +	(–)	+ + +	+ + +	+ +
Crocidolite	+ +	+ +	+ +	+	+ +
Amosite	+ +	+ +	+	+	+ +
Chrysotile	+ +	+ +	(+)	+	+ +
Anthophyllite	+ +	–	– –	+ +	+ +
Tremolite	+ +	+ +	(+)	+	+ +
Ceramic Fibres	nd	+	nd	+ +	+ +
Special purpose fine fibres	–	–	–	–	+ +
Glass wool	–	– –	– –	– –	–
Rock wool	–	–	– –	–	*
Slag wool	☼	–	– –	–	–
Glass filaments	–	–	– –	–	–
Very fine Aramid fibres	nd	+ +	nd	–	–

Notes:

+ + Definite and strong evidence of effect.

+ Definite evidence of effect.

(+) The incidence of mesothelioma following chrysotile exposure is very low. Most of the few cases also involved exposure to tremolite.

☼ There is an excess of lung cancer among workers who have produced slag wool, but this cannot be attributed with any certainty to exposure to slag fibres.

* No experiment with commercial insulation rock wool has suggested any carcinogenic effects. The positive results only occurred when experimental fibres were used.

(–) Inhalation of erionite in animal studies produces so many mesotheliomas that there is almost no opportunity for lung cancer to occur.

– Data available and do not indicate an effect.

– – Data available, with strong evidence against the existence of an effect.

nd No data available.

of exposure. In the case of airborne materials, that route is inhalation. The animal inhalation studies correlate well with the human data, except for anthophyllite. The comparison in Table 1 is taken from Brown *et al.*, 1991. Pott (1990) criticised an unpublished version of this table, because it did not include actinolite; he clearly missed the point that this table provides comparisons and no comparison is available for actinolite.

Although there is also a correlation of human evidence with the results of the serosal tests, it is less good. The positive serosal tests with special purpose fine durable glass fibres and with non-commercial rock wool are not in accord with the epidemiological data or the animal inhalation studies. Matched against the "gold standard" of human risk, these serosal tests are false positives.

Pott (1990) also argued that the differentiation made in Table 1 between the special purpose fine durable glass fibres and glass wool was unjustified. If he were correct in his argument, then this would be yet another false-positive. However, he ignored the fundamental difference in biopersistence of these two classes of fibres.

The intra-cavitary studies can only yield data of a screening nature. Positive results using artificial routes of administration cannot be used as positive evidence for human risk.

CRITICAL FIBRE SIZES FOR CARCINOGENICITY

Wagner and coworkers (1960) reported a significant human mesothelioma risk following exposure to amphibole asbestos, but with differing risks for different mining areas. This early evidence led to suggestions that amphibole fibre size may be a critical parameter. Intra-cavitary studies provided the confirmation of this tentative epidemiological conclusion (Stanton & Wrench, 1972; Pott & Friedrichs, 1972). However, as the human evidence accumulated, epidemiological analyses involving lung burden data and environmental fibre size distributions also clearly demonstrate the critical importance of amphibole fibre diameter and length for both mesothelioma and lung cancer risk (Lippmann, 1988; Timbrell, 1989). They also have the major advantage of permitting specific conclusions on critical fibre sizes for human carcinogenic risk.

The human evidence from the Transvaal, North West Cape (South Africa) and Paakkila (Finland) puts an upper limit on fibre diameter for mesothelioma of 0.1 μm and a length range of 5-10 μm (Timbrell, 1983; Timbrell et al., 1987; Harington, 1981). For lung cancer, the analyses indicate the critical dimensions to be between 0.3 and 0.8 μm in diameter and between 10 and 100 μm in length (Lippmann, 1988).

The lack of mesothelioma among insulation wool production workers is thus not surprising, as virtually no airborne insulation wool fibres are under 0.2 μm in diameter.

Subsequent intra-cavitary experiments, particularly by Stanton and his colleagues (Stanton et al., 1977, 1981; Stanton & Layard, 1978), with fibres of a wide range of dimensions confirmed the importance of fibre size, but also showed that non-asbestos fibres of similar dimensions elicited similar responses. Stanton and Layard (1978) concluded that the critical fibres were less than 0.25 μm in diameter and longer than 8 μm in length. Thus the human and animal evidence on critical fibre dimensions are in reasonable agreement, although the animal-based critical fibre diameter for the development of mesothelioma appears to be too high. This is presumably because the informative intra-cavitary studies could not take into account the size selectivity by inhalation in man.

SOME LIMITATIONS OF INTRA-CAVITARY STUDIES FOR REGULATORY PURPOSES

The following discussion concerns some of the major limitations of the intra-pleural and intra-peritoneal tests insofar as they concern evidence of risk to man. The full arguments will be presented in the formal response to the proposal by Pott and others (1990a) described above. I must emphasise that these comments are not a criticism of the scientific practice, but of the conclusions drawn, for regulatory purposes, on carcinogenic risk.

Obviously, the first major concern is about the non-physiological method of administration of fibres. Inhalation studies lead to accumulation of dust in the lungs where there is a biological interaction with lung fluids and phagocytes. The intraperitoneal method involves administering large doses of fibres to parts of the animal body to which little or no fibrous dust will reach. The number of fibres reaching the pleura or peritoneum is much higher than following inhalation. These techniques are artificial and create local fibre concentrations at the mesothelial tissue which could never occur via inhalation.

Further, the injection studies by-pass the size selectivity by the lungs to inhaled fibres. And then, if any inhaled fibres reach the serosal cavities, the size, shape, surface area, surface charge, and chemical composition may be substantially changed. Finally, by natural selection, the lung phagocytes are responsive to dust deposited in the alveoli. There is no reason to believe that pleural or peritoneal phagocytes will respond effectively to an injected bolus of fibres and granular material.

There is also no reason to believe that the intraperitoneal test provides evidence about the risk of lung cancer in man from the inhalation of fibres. The agreement for asbestos of the findings in man, by animal inhalation and by serosal tests does not prove that the serosal tests can predict the lung cancer risk. The serosal tests do not even predict the very low mesothelioma risk for chrysotile (Brown *et al.*, 1991).

The formal regulatory proposal by Pott and others (1990a) chooses to use a critical diameter of 2 μm. This has been changed from earlier suggestions by Pott (1987/8) that 1 μm was appropriate. The major reason for this change and for suggestions that an even thicker fibre diameter should be defined, relates to his experiments with relatively coarse rock wool and fine actinolite. "It was found that 0.25 mg actinolite and 75 mg basalt fibres contained a similar number of fibres longer than 5 μm and gave a similar tumour incidence. This means that the carcinogenic potency does not decrease with increasing diameter as has been supposed up to now" (Pott *et al.*, 1989). This is a very tenuous basis for Pott's change of mind. It ignores all the other possible confounding factors related just to dose of material which could have caused his observed tumour incidence. In particular, the coarse rock wool could have irritated and inflamed the serosa, leading to cell death and consequent hyperplasia. The probability of cancer is directly related to the numbers of cell divisions, so the rapid extensive cell proliferation could have led to the development of tumours (Cohen & Ellwein, 1990; Ames & Gold, 1990). There is no reason to attribute the excess tumour rate observed with the very coarse fibres to a direct effect of the fibres on the serosa. Two other studies form noteworthy comparisons. Maltoni and Minardi (1989) produced insignificant mesothelioma rates in their intraperitoneal studies of rock wool fibres obtained by water fractionation. In her intraperitoneal studies, Nikitina (1989) produced only borderline evidence of the carcinogenic potential of ultrafine basalt wool. It must be noted that in his Heidelberg paper, Pott (1990) references the Maltoni and Minardi paper for the asbestos studies, but omits the rock wool results. It is also surprising that he classifies the Nikitina results (12.5% and 14% positive) as definite evidence of carcinogenic potential, which he could not have done before dropping his dividing line to 12%.

It is an inherent limitation of the intraperitoneal test, as advocated by Pott, that coarse fibres can produce "positive" results which are unrelated to the direct effect of fibres on the serosa. This, of course, is one of the several disadvantages of a test method which bypasses the natural defence mechanisms of the experimental animal against coarse fibres.

As discussed above, the intraperitoneal test is limited by its false-positives relative to human risk. Pott (1990) argues that it is the human evidence which is false-negative without specifying explicitly that he takes his own evidence as the "gold standard". He does not justify that position, except to remark that the intraperitoneal test is the most sensitive. That is illogical. If it were logical, then any test which indicated carcinogenic potential could be taken as evidence of carcinogenic risk to man. Only about 25 known human carcinogens exist, but over 1000 substances have shown an indication of carcinogenic activity in long-term bioassays (Moolenaar, 1990). If short-term and *in vitro* tests are included, most substances could test positive. Why then choose the intraperitoneal test, a test of intermediate sensitivity? It is only logical to choose the most extreme, or to rely on tests which correlate closely with the evidence of carcinogenicity in man. I prefer the latter option. The intraperitoneal test meets neither requirement.

Finally, what is the effect of fibre diameter on the outcome of the intraperitoneal test? Recent studies (Pott *et al.*, 1990b; Bellmann *et al.*, 1990) with fibres of the same composition but different diameters yield different half-lives in a long-term fibre clearance study, with corresponding tumour rates in an intraperitoneal study. The thicker fibre (median diameter 1.68 μm) had a half-life of 107 days, almost exactly three times the half-life of 38 days for the thinner fibre (median diameter 0.51 μm). Thus, half-life is proportional to diameter, as would be expected if solubility is proportional to surface area. The corresponding intraperitoneal tumour rates were 10.9% for the thicker fibre at a dose of 0.24×10^9 fibres and up to 5% for the thinner fibre at the higher dose of 10^9 fibres. It is clear that testing the same fibre at a nominal diameter of above 2 μm would produce a positive intraperitoneal result for a fibre declared to be non-carcinogenic (*"es erweis sich nach Injektion von etwa 10^9 Fasern nicht als krebserzeugend"*) (Pott, 1990), especially using Pott's re-definition of a positive carcinogenic response. Any test which concludes that a fine fibre is less carcinogenic than the same, but coarser, fibre must be fallacious.

ETHICAL CONSIDERATIONS

Because of the finding that the same material gives different evidence of carcinogenicity for different fibre diameters and lengths, Bellmann and coworkers (1990) have proposed that *"für eine vergleichende Untersuchung der Beständigkeit mehrerer Faserproben im Tierexperiment ist es daher optimal, gleiche Fasergrößen und massen vershiedener Proben zu applizieren"* (Therefore, it is best to apply the same fibre sizes and masses of different samples when carrying out a comparative study on the persistence of several fibre samples in animal experiments).

This idea arises because of the assumption that the difference between the biological effect of fibres of the same size is primarily related to biopersistence. For regulatory purposes it may be acceptable to base decision-making on fibre size and biopersistence. Testing fibre samples of the same size in the intraperitoneal test will provide evidence on the durability in the peritoneum. If peritoneal durability did correlate perfectly with biopersistence in the lung — as yet an untested correlation — then the intraperitoneal test could be useful. But the intraperitoneal test would then only measure biopersistence and this would be a very expensive, and unethical, procedure; biopersistence in the lung could be measured directly, quicker and with fewer animals in intratracheal biopersistence studies.

Further, techniques for approximating biopersistence exist in the laboratory (Potter & Mattson, 1991) which are currently being calibrated by intratracheal studies. When these, or similar techniques, have been validated, there may not be need for *in vivo* studies of biopersistence, for regulatory purposes.

An alternative classification system could regulate fibres which are finer than 1 μm, longer than 5 μm, and with an estimated standardised *in vivo* biopersistence of more than 5 years. On the European Community model, any product which contained more than 1% by weight of such fibres would be similarly regulated. This approach could be acceptable to regulators, workers and industry, as it could be easily implemented for all fibres, applied quickly for new fibres, lead to changes to less durable fibres and avoid the unethical implications of the present regulatory proposals in Germany.

This proposal would also return the intraperitoneal test to its rightful place, as an experimental research tool.

CONCLUSION

The conclusions of this analysis of the intraperitoneal test are simple. It is inappropriate for the purpose of regulatory assessment of the carcinogenic potential of fibres. It does not assess risk to man. Its limitations are too great for the purpose proposed. Its practice for regulatory purposes could be unethical.

REFERENCES

Ames,B.N. and Gold,L.S. (1990) Too many rodent carcinogens: mitogenesis increases mutagenesis. *Science* **249**:970-971.

Bellmann,B., Muhle,H. and Pott,F. (1990) Untersuchung zur Beständigkeit chemisch unterschiedlicher Glasfasern in Rattenlungen. *Zbl. Hyg.* **190**:310-314.

Brown,R.C., Davis,J.M.G., Douglas,D., Gruber,U.F., Hoskins,J.A., Ilgren,E.B., Johnson,N.F., Rossiter,C.E. and Wagner,J.C. (1991) Carcinogenicity of the insulation wools: re-assessment of the IARC evaluation. *Reg. Tox. Pharm.*, in press.

Cohen,S.M. and Ellwein,L.B. (1990) Cell proliferation in carcinogenesis. *Science* **249**:1007-1011.

Davis,J.M.G., Addison,J., Bolton,R.E., Donaldson,K. and Jones,A.D. (1986a) Inhalation and injection studies in rats using dust samples from chrysotile asbestos prepared by a wet dispersion process. *Br. J. Exp. Path.* **67**:113-129.

Davis,J.M.G., Addison,J., Bolton,R.E., Donaldson,K., Jones,A.D. and Smith,T. (1986b) The pathogenicity of long versus short fibre samples of amosite asbestos administered to rats by inhalation and intraperitoneal injection. *Br. J. Exp. Path.* **67**:415-430.

Gross,P., Kascak,M., Tolker,E.B., Babyak,M.A. and Detreville,R.T.P. (1970) The pulmonary reaction to high concentrations of fibrous glass dusts: a preliminary report. *Arch. Environ. Hlth* **20**:696-704.

Harington,J.S. (1981) Fiber carcinogenesis: Epidemiological observations and the Stanton hypothesis. *J. Nat. Cancer Inst.* **67**:977-989.

Hesterberg,T.W., Mast,R., McConnell,E.E., Chevalier,J., Bernstein,D.M., Bunn,W.B. and Anderson,R. (1991) Chronic inhalatio toxicity of refractory ceramic fibers in Syrian Hamsters. This volume.

International Agency for Research on Cancer. (1988) Monographs on the evaluation of carcinogenic risks to humans. Vol **43**: Man-made mineral fibres and radon. IARC, Lyon.

International Programme on Chemical Safety. (1988) Environmental Health Criteria 77. Man-made Mineral Fibres. World Health Organization, Geneva.

Lippmann,M. (1988) Asbestos exposure indices. *Environ. Res.* **46**:86-106.

Maltoni,C. and Minardi,F. (1989) Recent results of carcinogenicity bioassays of fibres and other particulate matters. Non-occupational exposure to mineral fibres. In, "Non-occupational exposure to mineral fibres", eds. Bignon,J., Peto,J. and Saracci,R. IARC Scientific publications No. 90. Lyon, pp. 46-53.

Miettinen,O.S. and Rossiter,C.E. (1990) Man-made mineral fibres and lung cancer: epidemiologic evidence regarding the causal hypothesis. *Scan. J. Work Environ. Hlth* **16**:221-231.

Moolenaar, R.J. (1990) Carcinogen classification systems: a time for change. The SIRC Review April 1990:43-49.

Mossman,B.T., Bignon,J., Corn,M., Seaton,A. and Gee,J.B.L. (1990) Asbestos: scientific developments and implications for public policy. *Science* **247**:294-301.

Muhle,H., Pott,F., Bellman,B., Takenaka,S. and Ziem,V. (1987) Inhalation and injection experiments in rats to test the carcinogenicity of MMMF. *Ann. Occup. Hyg.* **31**:755-764.

National Institute for Occupational Safety and Health (1977) Criteria for a recommended standard. Occupational exposure to fibrous glass. Cincinnati, Ohio, National Institute for Occupational Safety and Health (DHEW Publication No 77-152).

National Institute for Occupational Safety and Health (1987) Occupational Respiratory Diseases. Cincinnati, Ohio, National Institute for Occupational Safety and Health (Publication No DHHS(NIOSH) 86-102).

Nikitina,O.V., Kogan,F.M., Vanchugova,N.N. and Frash,V.N. (1989) (Comparative oncogenicity of basalt fibres and chrysotile asbestos). *Gigiena Truda I Professional'Nye Zabolevaniya* **4**:7-11.

Pott,F. (1990) Beurteilung der Kanzerogenität von Fasern aufgrund von Tierversuchen. Paper presented to the Heidelberg conference on "Faserförmige Stäube - Vorschriften, Wirkungen, Messung, Minderung", September 1990.

Pott,F., Blome,H., Bruch,J., Friedberg,K.D., Rödelsperger,K. and Woitowitz,H.-J. (1990a) Einstufungsvorschlag für anorganische und organische Fasern. *Arbeitsmed, Sozialmed, Praventivmed* **25**:463-466.

Pott,F., Schlipköter,H.-W., Roller,M., Rippe,R.M., Germann,P.-G., Mohr,U. and Bellmann,B. (1990b) Kanzerogenität von Glasfasern mit unterschiedlicher Beständigkeit. *Zbl. Hyg.* **189**:563-566.

Pott,F. (1989) Carcinogenicity of Fibers in experimental animals - data and evaluation. In, "Assessment of Inhalation Hazards", ed-in-chief Mohr,U. Springer Verlag, Berlin & Heidelberg, pp. 243-253.

Pott,F., Roller,M., Ziem,U., Reiffer,F.-J., Bellmann,B., Rosenbruch,M. and Huth,F. (1989) Carcinogenicity studies on natural and man-made fibres with the intraperitoneal test in rats. In, "Non-occupational exposure to mineral fibres", eds. Bignon,J., Peto,J. and Saracci,R. IARC Scientific publications No. 90. Lyon,International Agency for Research on Cancer, pp. 173-179.

Pott,F. (1987/88) Die krebserzeugende Wirkung anorganischer Fasern im Tierexperiment: Daten und Bewertung. *Umwelthyg.* **20**:97-134.

Pott,F. (1987) Die Faser als krebserzeugendes Agens. *Zbl. Bakt. Hyg., I. Abt. Orig,* **B 184**:1-23.

Pott,F. and Friedrichs,K.H. (1972) Tumoren der Ratte nach i.p Injektion faserförmiger Stäube. *Naturwissenschaften* **59**:318.

Potter,R.M. and Mattson,S.M. (1991) Glass fiber dissolution in physiological saline solution. *Glastechnische Berichte,* in press.

Smith,D.M., Ortiz,L.W., Archuleta,R.F. and Johnson,N.F. (1987) Long-term health effects in hamsters and rats exposed chronically to man-made vitreous fibres. *Ann. Occup. Hyg.* **31**:731-750.

Stanton,M.F. and Layard,M. (1978) The carcinogenicity of fibrous materials. In, "Proceedings of the Workshop on Asbestos: Definitions and Measurement Methods." National Bureau of Standards Special Publication 506, pp. 143-151.

Stanton,M.F., Layard,M., Tegeris,A., Miller,E., May,M. and Kent,E. (1977) Carcinogenicity of fibrous glass: pleural response in the rat in relation to fibre dimension. *J. Nat. Cancer Inst.* **58**:587-603.

Stanton,M.F. and Wrench, C. (1972) Mechanisms of mesothelioma induction with asbestos and fibrous glass. *J. Natl. Cancer Inst.* **48**:797-821.

Stanton,M.F., Layard,M., Tegeris,A.S., Miller,E., May,M., Morgan,E. and Smith,A. (1981) Relation of particle dimension to carcinogenicity in amphibole asbestos and other fibrous minerals. *J. Natl. Cancer Inst.* **67**:965-975.

Timbrell,V. (1989) Review of the significance of fibre size in fibre-related lung disease: a centrifuge cell for preparing accurate microscope-evaluation specimens from slurries used in inoculation studies. *Ann. Occup. Hyg.* **33**:483-505.

Timbrell,V. (1983) Fibres and carcinogenesis. *J. Occup. Hlth Soc. Australia* **3**:3-12.

Timbrell,V., Ashcroft,T., Goldstein,B., Heyworth,F., Meurman,L.O., Rendall,R.E.G., Reynolds,R.A., Shilkin,K.B. and Whitaker,D. (1987) Relationships between retained amphibole fibres and fibrosis in human lung tissue specimens. In, "Inhaled Particles VI", ed. Walton,W.H., Pergamon, Oxford. pp. 323-340.

Vu,V.T. (1988) Health hazard assessment of nonasbestos fibers. US Environmental Protection Agency, Washington, DC.

Wagner,J.C., Berry,G., Hill,R.J., Munday,D.E. and Skidmore,J.W. (1984) Animal experiments with MMM(V)F. In, "Biological Effects of Mineral Fibres". WHO, Copenhagen. pp. 209-233.

Wagner,J.C., Berry,G., Skidmore,J.W. and Timbrell,V. (1974) The effects of the inhalation of asbestos in rats. *Br. J. Cancer* **29**:252-269.

Wagner,J.C., Sleggs,C.A. and Marchand,P. (1960) Diffuse pleural mesothelioma and asbestos exposure in North Western Cape Province. *Br. J. Indust. Med.* **17**:260-271.

AUTHOR INDEX

INDEX

(**Note**: the common fibres and associated diseases occur in the text too often to be indexed)

Tubulin, 483, 484, 486
Tumorgenesis, Tumourigenesis, 71, 74,
      77, 172
   Tumorigenic, Tumourigenic, 30, 72,
      119, 162, 172, 173, 220,
      313, 332, 434, 521
   Tumorgenicity, Tumourigenicity, 61,
      119, 173, 311, 312, 335, 435
Tumoricidal, 166
T-antigen, 487, 507
T-lymphocytes, 39, 40, 121, 125, 127

Ulcer, 22
Up-regulation, 129
Uranium, 92
Uranium-plutonium, 91, 92
Uterus, 82, 553, 555, 557
   Uterine, 319, 557

Vacuole, 294, 295
Vermiculite, 232-234, 237, 240, 242, 246,
      518
Vimentin, 379, 483, 484, 486, 507
Virions, 354
Virus, 338, 341, 343
Vitronectin, 119

Volcanic, 215, 222

Wittenoom, 17, 22
Wollastonite, 46, 82, 143-146, 150-155,
      172, 181-186, 327-329, 409,
      505, 507, 508, 554, 560, 561
Wools, 11, 17, 18, 20, 24, 25, 191, 540,
      568
Wrist-action, 399

Xenobiotic(s), 390, 391, 403
Xeroderma, 291
Xerox, 213
Xonotlite, 409, 475
X-ray, 9, 33, 109, 193, 214, 232, 244,
      259, 271, 272, 370, 410, 420
  XRD, 270-272, 280, 426
  XRF, 109

Zeolite, 45, 61, 62, 71-73, 409, 491, 505
Zeta potential, 108, 111, 112, 241, 243,
      245, 249
Zinc, 336, 418
Zirconia, 192-196
Zirconium, 46
Zymosan, 135